LABORATORY MANUAL FOR

Human Anatomy

with **CAT DISSECTIONS**

LABORATORY MANUAL FOR
Human Anatomy *with* CAT DISSECTIONS

Michael G. Wood

with

William C. Ober, M.D.
Art Coordinator and Illustrator

Claire W. Garrison, R.N.
Illustrator

Ralph T. Hutchings
Biomedical Photographer

Shawn Miller
Organ and Animal Dissector

Mark Nielsen
Organ and Animal Dissection Photographer

PEARSON
Benjamin Cummings

San Francisco Boston New York
Cape Town Hong Kong London
Madrid Mexico City Montreal Munich
Paris Singapore Sydney Tokyo Toronto

Executive Editor:
Leslie Berriman

Project Editor:
Nicole George-O'Brien

Development Manager:
Claire Alexander

Development Editor:
Irene Nunes

Assistant Editor:
Robin Pille

Managing Editor:
Deborah Cogan

Production Supervisor:
Beth Masse

Production Management
and Compositor:
S4Carlisle Publishing Services

Senior Photo Manager:
Travis Amos

Interior Designer:
Studio A

Cover Designer:
Side by Side Studios

Director, Image Resource Center:
Melinda Patelli

Image Rights and Permissions Manager:
Zina Arabia

Senior Manufacturing Buyer:
Stacey Weinberger

Marketing Manager:
Gordon Lee

Text Printer:
LSC Communications

Cover printer:
LSC Communications

Cover Photographers:
Anita Impagliazzo and William Ober

Cover Photo Models:
Amanda Davis and Reggie Jackson

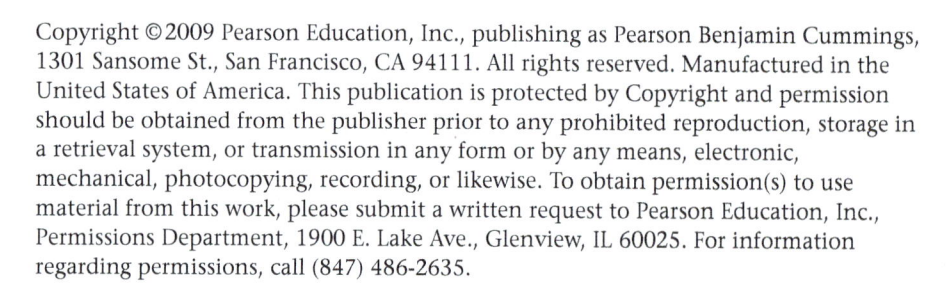

PEARSON
Benjamin Cummings

Copyright © 2009 Pearson Education, Inc., publishing as Pearson Benjamin Cummings, 1301 Sansome St., San Francisco, CA 94111. All rights reserved. Manufactured in the United States of America. This publication is protected by Copyright and permission should be obtained from the publisher prior to any prohibited reproduction, storage in a retrieval system, or transmission in any form or by any means, electronic, mechanical, photocopying, recording, or likewise. To obtain permission(s) to use material from this work, please submit a written request to Pearson Education, Inc., Permissions Department, 1900 E. Lake Ave., Glenview, IL 60025. For information regarding permissions, call (847) 486-2635.

Many of the designations used by manufacturers and sellers to distinguish their products are claimed as trademarks. Where those designations appear in this book, and the publisher was aware of a trademark claim, the designations have been printed in initial caps or all caps.

Safety Notification

The Author and Publisher believe that the laboratory experiments described in this publication, when conducted in conformity with the safety precautions described herein and according to the school's laboratory safety procedures, are reasonably safe for the student to whom this manual is directed. Nonetheless, many of the described experiments are accompanied by some degree of risk, including human error, the failure or misuses of laboratory or electrical equipment, mismeasurement, chemical spills, and exposure to sharp objects, heat, bodily fluids, blood, or other biologics. The Author and Publisher disclaim any liability arising from such risks in connection with any of the experiments contained in this manual. If students have any questions or problems with materials, procedures, or instructions on any experiment, they should always ask their instructor for help before proceeding.

Library of Congress Cataloging-in-Publication Data
Wood, Michael G.
   Laboratory manual for human anatomy : with cat dissections / Michael G. Wood ; with William C. Ober . . . [et al.].— 1st ed.
       p. cm.
   Includes index.
   ISBN-13: 978-0-8053-7375-2
 1. Human anatomy—Laboratory manuals. 2.   Cats—Dissection—Laboratory manuals. I. Title.
   QM34.W74 2009
   611.0078—dc22

2007024589

ISBN-10:   0-8053-7375-6
ISBN-13: 978-0-8053-7375-2

16 2020
www.aw-bc.com

About the Author

Michael G. Wood received his Master of Science in 1986 at Pan American University, now the University of Texas at Pan American. His graduate studies included vertebrate physiology and freshwater ecology. Today he is a Professor of Biology at Del Mar College in Corpus Christi, Texas, where he has taught anatomy and physiology and biology for more than 20 years. He has received the "Educator of the Year" and "Teacher of the Year" awards from the Del Mar College student body and from the local business community. Professor Wood is a member of the Human Anatomy and Physiology Society. He has a passion for science, reading, and playing the guitar, and he and his family enjoy camping, Scottish and Irish dancing, and a yard full of cats, dogs, and plants.

Dedication

To my father, Ernest M. Wood, Jr.

Preface

This exciting new *Laboratory Manual for Human Anatomy with Cat Dissections* is designed to serve the laboratory course that accompanies the human anatomy lecture course. It provides students with comprehensive coverage of anatomy, beautiful four-color illustrations and photographs, and a highly-intuitive pedagogical framework. The primary goal of this manual is to guide students in hands-on laboratory studies of the body, beginning with the cellular level and progressing through each organ system, to reinforce the information they learn in the lecture course.

The manual is written to correspond to all current human anatomy textbooks, though those students and instructors who use the Martini/Timmons/Tallitsch *Human Anatomy* textbook will recognize many of the illustrations by William Ober and Claire Garrison, renowned biomedical illustrators, that appear in that book.

ORGANIZATION AND APPROACH

This manual has been crafted for optimal student learning. The organization and pedagogical tools reinforce and assess understanding of the terminology, facts, and concepts that are at the core of the study of human anatomy. The laboratory manual contains 30 exercises plus 9 cat dissection exercises toward the end of the manual. Large systems, such as the skeletal, muscular, and nervous systems, appear across several exercises; the first exercise in each series serves as an overview that introduces the general anatomical organization of the system. Programs with limited laboratory time might use the overview exercises for a hands-on summary of organ systems, as they can be completed during a short laboratory period.

Each exercise is divided into a series of "Lab Activities" that cover a range of key topics. Every exercise begins with a list of the Lab Activities it contains and a set of Objectives for student learning. A general introduction to the exercise gives students a preview of what they are about to learn. Then, individual Lab Activities focus on more specific study. Each Lab Activity is self-contained, and instructors may easily assign only certain ones within an exercise.

Each Lab Activity begins with an introduction to the activity and a review of the concepts necessary for understanding it. These are followed by two or three QuickCheck Questions that students can use to gauge their comprehension of the material before proceeding. A clearly marked list of Materials indicates what is needed to carry out the activity. Finally, a numbered list of Procedures guides students through the in-the-lab work.

After the set of Lab Activities within an exercise, the exercise concludes with a Lab Report that includes a variety of question-types—Labeling, Matching, Multiple Choice, Short Answer, and Analysis and Application—to assess student learning.

TERMINOLOGY

The anatomical terminology used in the manual follows the standard terminology adopted by the International Federation of Associations of Anatomists in the reference book titled *Terminologia Anatomica*. Other accepted or familiar terms are frequently included in parentheses after the standard term.

SUPPLEMENTS

Instructor's Manual

by Michael G. Wood
ISBN 10: 0-321-55054-4; ISBN 13: 978-0-321-55054-5
Paperback; 224 pages

This comprehensive Instructor's Manual is a useful resource for instructors at every level of experience. For each of the 30 Exercises and 9 Cat Dissection Exercises, it includes the following sections:

- Instructor's Preparation and Tips
  o Master List of Materials
  o Time Requirement
  o Laboratory Preparation
  o Teaching Tips and Students' Misconceptions
- Answers to Questions
  o QuickCheck Questions
  o Laboratory Report

Practice Anatomy Lab (PAL™) 2.0

ISBN 10: 0-321-54725-X; ISBN 13: 978-0-321-54725-5
CD-ROM

Practice Anatomy Lab (PAL™) 2.0 is an indispensable virtual anatomy practice tool that gives students 24/7 access to the most widely used lab specimens, including the human cadaver, anatomical models, histology slides, cat, and fetal pig. Each of the five specimen modules includes hundreds of images as well as interactive tools for reviewing the specimens, hearing the names of anatomical structures, and taking multiple choice quizzes and fill-in-the-blank lab practical exams. Specimen images are also linked to animations.

All practice material in PAL 2.0 is organized into three sections:

- **Self-Review** allows students to review the names of structures, see structures highlighted, hear pronunciations of anatomical terms, and turn labels on and off.
- **Quizzes** let students quiz themselves on their knowledge of anatomical structures and functional anatomy. Students can receive immediate feedback to their answers, or they can wait to see a summary of results with the correct answers next to the images.
- **Lab Practicals** simulate a real lab practical exam. Fill-in-the-blank exam questions ask students to identify and spell the names of a series of structures.

The 2.0 version of PAL is a substantial upgrade over the original version and includes new content and new features.

- **The all-new Human Cadaver module** includes hundreds of specially-commissioned cadaver photos that can be found in all three activity sections in the module. This module also provides a fully-rotatable human skull and 17 other rotatable skeletal structures.
- **3D Animations of Origins, Insertions, Actions, and Innervations** depict more than 65 muscles in the Human Cadaver module. These 3D Animations are viewable as students learn and hear the name of a muscle, thereby giving them a one-stop learning experience. Quizzes for the Animations are included.
- **A greatly expanded Histology module** includes more images of different types of tissues.
- **Two new body systems, Endocrine and Lymphatic**, have been added to the Human Cadaver and Anatomical Models modules.

The **Instructor's Resource DVD for PAL 2.0** includes labeled and unlabeled images from PAL 2.0 in both JPEG and PowerPoint® formats. The PowerPoint Label Edit feature, with editable labels and leader lines, is built into every image. PowerPoint image slides include embedded links to the relevant 3D Animations of Origins, Insertions, Actions, and Innervations.

The PAL 2.0 CD-ROM is available for individual student purchase and can also be added to the *Laboratory Manual for Human Anatomy with Cat Dissections* package for a package price. The Instructor's Resource DVD for PAL 2.0 is available at no cost to instructors who adopt the *Laboratory Manual for Human Anatomy with Cat Dissections* packaged with the PAL 2.0 CD-ROM.

viii   PREFACE

ART PROGRAM

Accurate and Engaging Illustrations

Extraordinarily accurate illustrations lead students through the highly visual subject of human anatomy. Every illustration has been carefully reviewed for correctness and consistency. Color is used to delineate structures and emphasize three-dimensional relationships. Clear labels coordinate meticulously with the descriptions that appear in the narrative text.

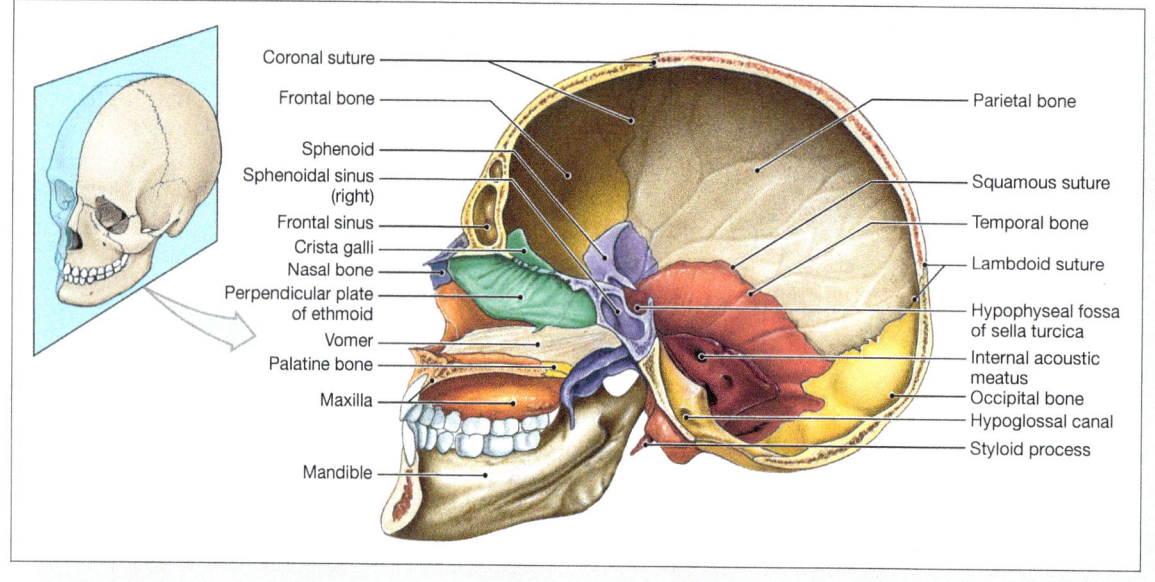

- Coronal suture
- Frontal bone
- Sphenoid
- Sphenoidal sinus (right)
- Frontal sinus
- Crista galli
- Nasal bone
- Perpendicular plate of ethmoid
- Vomer
- Palatine bone
- Maxilla
- Mandible
- Parietal bone
- Squamous suture
- Temporal bone
- Lambdoid suture
- Hypophyseal fossa of sella turcica
- Internal acoustic meatus
- Occipital bone
- Hypoglossal canal
- Styloid process

Cat Dissection Photographs with Corresponding Illustrations

The cat dissection exercises feature photographs by the expert dissector/photographer team of Shawn Miller and Mark Nielsen and illustrations by the award-winning artists William Ober and Claire Garrison. The photographs and illustrations provide superb representations of what students will see on the dissection table.

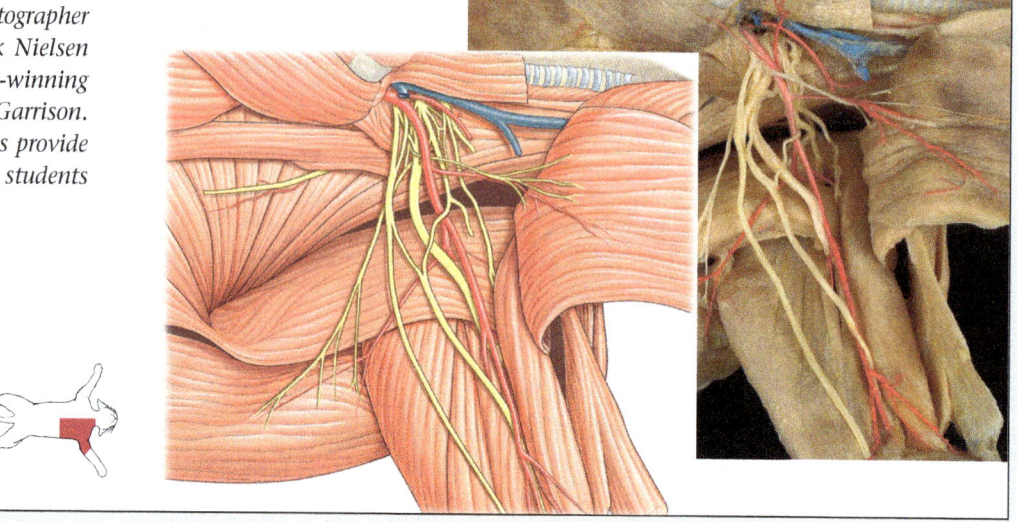

PREFACE ix

Side-by-Side Figures

Artists' drawings are paired with cadaver photographs to help students see the structures they will encounter in the laboratory and in clinical settings.

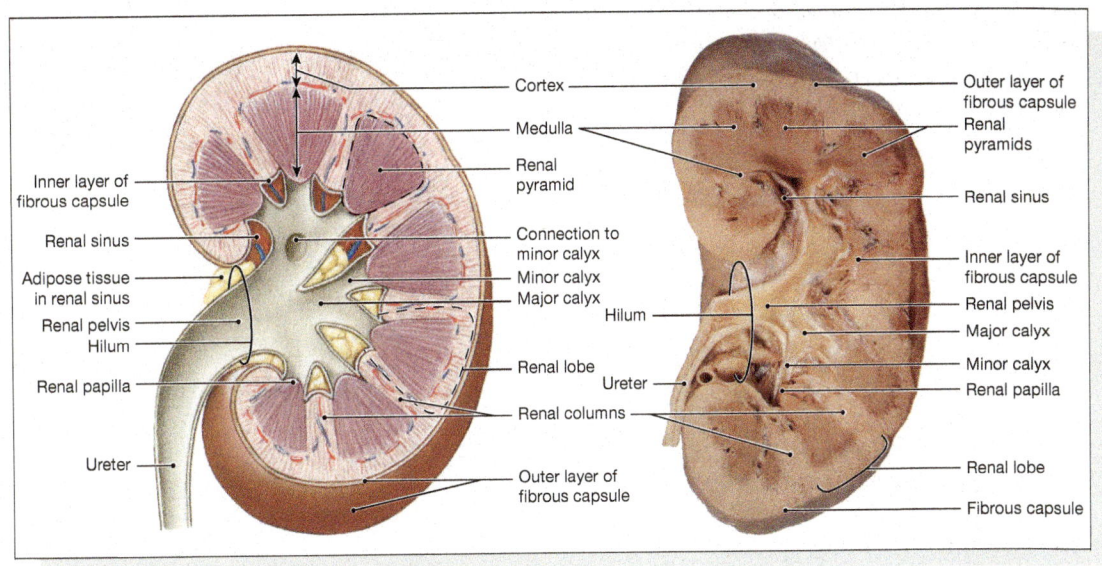

Histology Images

High-quality micrographs together with corresponding illustrations are included for activities that cover tissues. These histology images help students prepare for what they will see through the microscope and provide useful delineations of structures.

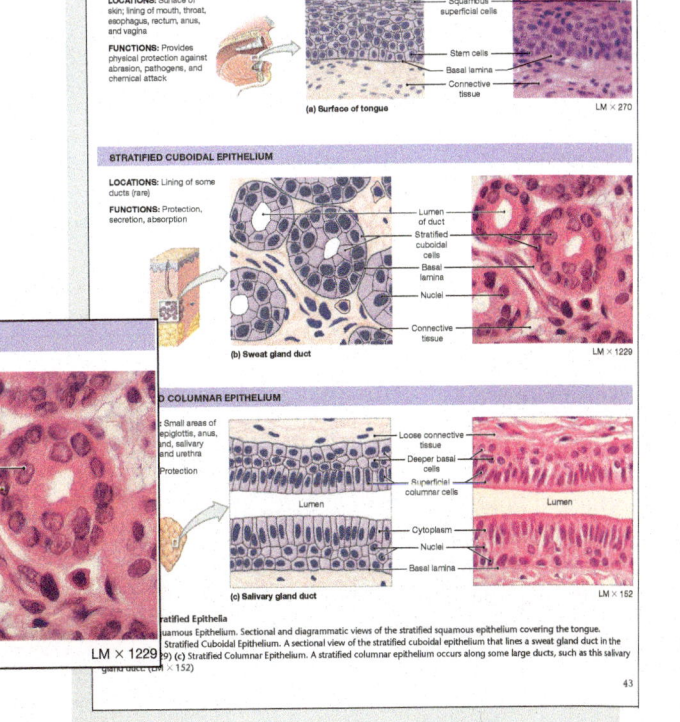

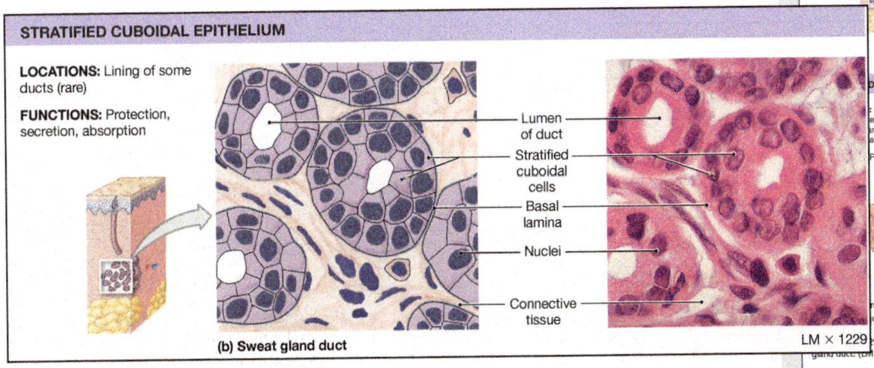

x PREFACE

FEATURES

QuickCheck Questions

The narrative introductory sections in each Lab Activity are generally followed by two or three QuickCheck Questions that check students' understanding of terminology, facts, and concepts before they proceed to the laboratory work.

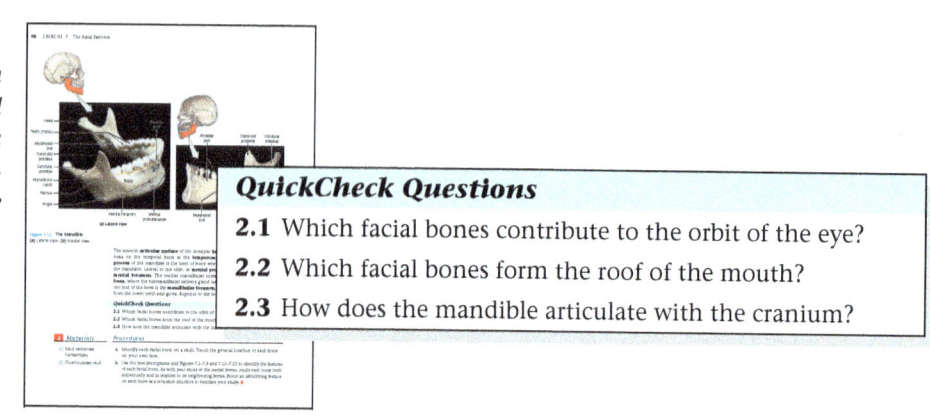

> **QuickCheck Questions**
> **2.1** Which facial bones contribute to the orbit of the eye?
> **2.2** Which facial bones form the roof of the mouth?
> **2.3** How does the mandible articulate with the cranium?

Clinical Application

To tie students' learning tightly to potential future careers in the allied health field, many exercises include relevant discussions of particular diseases and injuries in Clinical Application boxes. Thought-provoking critical thinking questions are featured in many of these boxed sections.

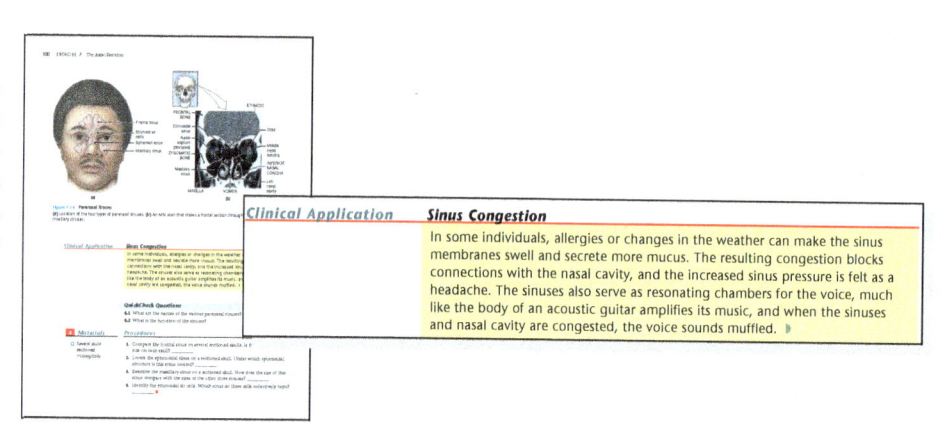

> **Clinical Application   Sinus Congestion**
> In some individuals, allergies or changes in the weather can make the sinus membranes swell and secrete more mucus. The resulting congestion blocks connections with the nasal cavity, and the increased sinus pressure is felt as a headache. The sinuses also serve as resonating chambers for the voice, much like the body of an acoustic guitar amplifies its music, and when the sinuses and nasal cavity are congested, the voice sounds muffled.

Making Connections

Making Connections boxes give students handy mnemonic devices, tricks for learning certain anatomical relationships, or general suggestions for retaining the complex concepts that underlie the study of anatomy.

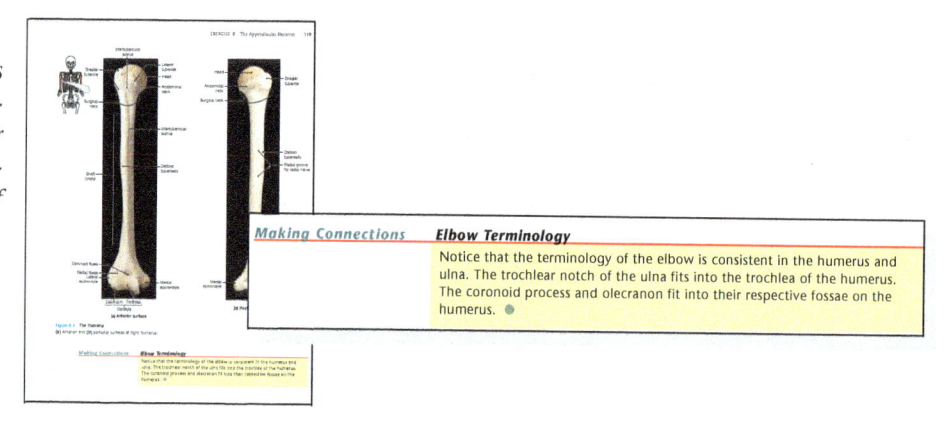

> **Making Connections   Elbow Terminology**
> Notice that the terminology of the elbow is consistent in the humerus and ulna. The trochlear notch of the ulna fits into the trochlea of the humerus. The coronoid process and olecranon fit into their respective fossae on the humerus.

Safety Alert

Safety Alerts highlight critical safety information where appropriate. These descriptions of safe procedures include proper precautions for undertaking particular activities, correct usage of laboratory instruments, and safe handling and disposal of chemical and biological wastes.

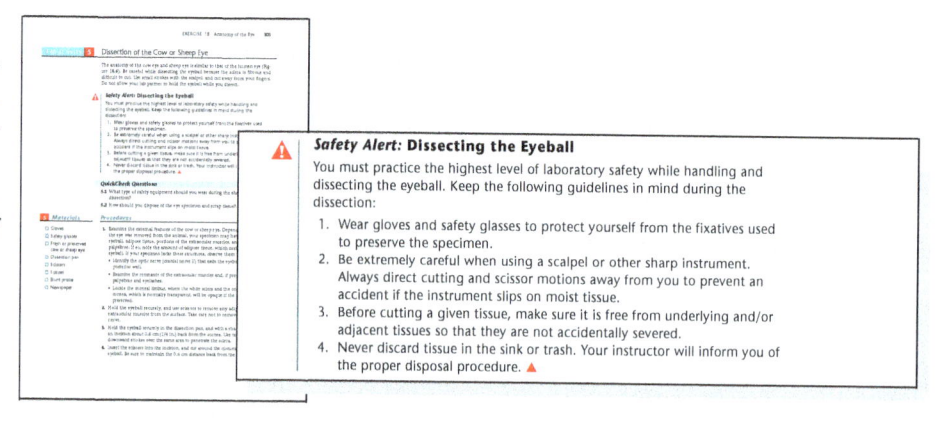

> **Safety Alert: Dissecting the Eyeball**
> You must practice the highest level of laboratory safety while handling and dissecting the eyeball. Keep the following guidelines in mind during the dissection:
> 1. Wear gloves and safety glasses to protect yourself from the fixatives used to preserve the specimen.
> 2. Be extremely careful when using a scalpel or other sharp instrument. Always direct cutting and scissor motions away from you to prevent an accident if the instrument slips on moist tissue.
> 3. Before cutting a given tissue, make sure it is free from underlying and/or adjacent tissues so that they are not accidentally severed.
> 4. Never discard tissue in the sink or trash. Your instructor will inform you of the proper disposal procedure.

A Regional Look

At the end of every body system, two-page spreads give students an introduction to a regional approach to human anatomy. Cadaver photographs, accompanied by a narrative discussion, help students see the body systems that come together in one region of the body. This feature allows students and instructors to step back from the systems orientation of the laboratory manual and see how the systems work together within regions.

face
A Regional Look

Have you ever seen a television show or movie where all the police have is the skeleton of a murder victim, but a forensic scientist is able to construct a face on the victim's skull? The process involves measuring bony landmarks to calculate the size of attached muscles and soft tissues. Reference measurements assist in determining the size of the ears, nose, and eyes, and modeling clay is then used to form all the muscles of the face and cartilage for the nose and ears. Through this painstakingly slow process, the face is re-created and used to assist in identifying the victim. Observe the cadaver photograph in Figure 5.7 and notice how even though the integument has been removed, you still can visualize the face. One feature a forensic scientist would look at is the mandible of the lower jaw. Note how the cadaver's lower jaw has soft angles, creating a smooth rather than a prominent chin. In this example, the scientist would create a clay musculature that is slight rather than heavy.

What is also evident in this cadaver dissection is the anatomical association of muscles and the integument. Identify the zygomaticus muscles that attach to the skin at the corners of the mouth. When these muscles contract, they pull on the skin and the mouth is pulled into a smile. Place your fingers on your face just inferior to your cheek bones, called the *zygomatic bones*, and smile. Did you feel your zygomaticus muscles contract?

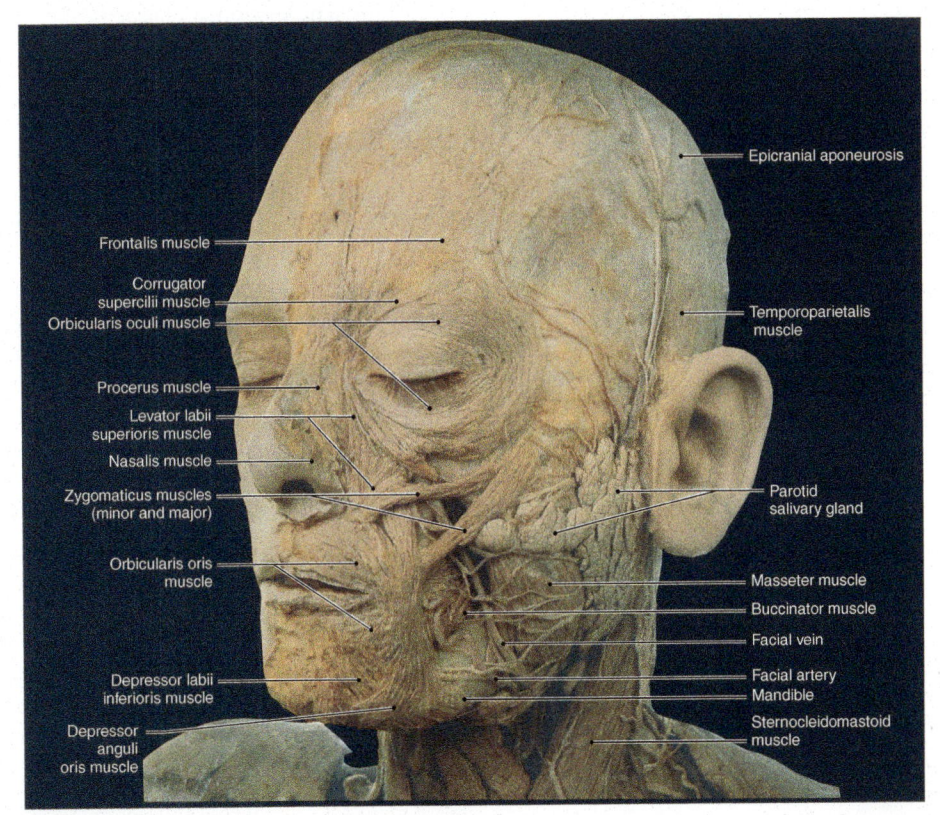

Figure 5.7 Face of a Cadaver with Integument Removed

Acknowledgments

I thank the academic reviewers whose insight and experience-based suggestions were invaluable as I prepared the manuscript for this first edition lab manual:

Debra J. Barnes, *Contra Costa College*
David M. Conley, *Washington State University*
Martha Dixon, *Diablo Valley College*
Brent Dodge, *Crowder College*
Andrew R. Ellis, *South Dakota State University*
Tim Lyerla, *Clark University*
Susan V. Monk, *Pennsylvania State University, Berks*
Karen Montgomery, *Community College of Rhode Island*
Philip Osborne, *San Diego City College*
Dean J. Scherer, *Oklahoma State University*
Robert B. Tallitsch, *Augustana College*
Christa Voss, *Tulsa Community College*
Rick Wiedenmann, *New Mexico State University*
Tony Weinhaus, *University of Minnesota*
John M. Zook, *Ohio University*

I am grateful to a number of people for this first edition's excellent illustrations and photographs. Frederic H. Martini, main author of the best-selling and award-winning *Human Anatomy,* now in its sixth edition, deserves credit for his insight and creativity in visualizing anatomical concepts with the talented biomedical illustrators William Ober and Claire Garrison. This laboratory manual benefits from their work through the inclusion of many illustrations from that book, and I was also able to work closely with Bill and Claire to create additional illustrations for the manual. Shawn Miller and Mark Nielsen of the University of Utah are a gifted dissector/photographer team whose meticulous work is coupled with the Bill Ober/Claire Garrison illustrations in the cat dissection exercises. The award-winning human photographs in the manual are by biomedical photographer Ralph Hutchings.

I was very fortunate to have Robert B. Tallitsch, one of Ric Martini's coauthors on *Human Anatomy*, critically review the manuscript and offer suggestions on terminology, pedagogy, and style. His eye for detail ensured consistency with *Human Anatomy* and throughout the exercises of this laboratory manual.

Special thanks are extended to my colleagues Albert Drumright, III; Lillian Bass; Billy Bob Long; and Joel McKinney for their suggestions, encouragement, and support of the manual.

I thank all the talented and creative individuals at Benjamin Cummings. I especially credit my project editor, Nicole George-O'Brien, for her expertise in organizing and coordinating the enormous number of details in the project and managing a challenging production schedule. I was grateful to have Irene Nunes as the development editor on the text. Her experienced eye for consistency and accuracy immensely improved the cohesiveness of the final product. Leslie Berriman, Executive Editor, provided valuable guidance and pulled together a resourceful team of editors, dissectors, photographers, and illustrators whose outstanding work shines.

Beth Masse, Production Supervisor, and Deborah Cogan, Managing Editor, masterfully coordinated all aspects of the production process.

I thank Judy Ludowitz and her fine team at S4Carlisle Publishing Services for their creative layout and attention to detail. I also thank Jana Anderson for her outstanding design, which gives this complex assemblage of text, illustrations, photographs, and procedures a user-friendly look. Mark Ong, Design Manager, oversaw the design process and provided crucial insight into our design complexities, and he also designed the cover.

My gratitude is also extended to the many students who have provided suggestions and comments to me over the years.

Most importantly, I am deeply grateful to my wife, Laurie, and children, Abi and Beth, for enduring months of my late-night writing and the pressures of never-ending deadlines.

Any errors or omissions in this first-edition publication are my responsibility and are not a reflection on the editorial and review team. Comments from faculty and students are welcomed and may be directed to me at the addresses below. I will consider each suggestion in the preparation of the subsequent edition.

Michael G. Wood
Del Mar College
101 Baldwin Blvd.
Corpus Christi, TX 78404
mwood@delmar.edu

Contents

About the Author v
Preface vi
Acknowledgments xii

EXERCISE 1 — Introduction to the Human Body — 1
1. Anatomical Position and Directional Terminology 2
2. Regional Terminology 6
3. Planes and Sections 9
4. Body Cavities and Serous Membranes 10
   Lab Report 13

EXERCISE 2 — Use of the Microscope — 17
1. Parts and Care of the Compound Microscope 17
2. Using the Microscope 20
3. Depth of Field Observation 22
4. Relationship between Magnification and Field Diameter 23
   Lab Report 25

EXERCISE 3 — Anatomy of the Cell and Cell Division — 27
1. Anatomy of the Cell 27
2. Cell Division 31
   Lab Report 35

EXERCISE 4 — Tissues — 39
1. Epithelia 39
2. Connective Tissue 46
3. Muscle Tissue 56
4. Neural Tissue 59
   Lab Report 61

EXERCISE 5 — The Integumentary System — 67
1. Epidermis and Dermis 67
2. Accessory Structures of the Integument 70
   A Regional Look: Face 74
   Lab Report 75

EXERCISE 6 — Organization of the Skeletal System — 77
1. Bone Structure 77
2. Histological Organization of Bone 79

xiv

3 The Skeleton 80
4 Bone Markings 83
Lab Report 85

EXERCISE 7 The Axial Skeleton 87

1 Cranial Bones 87
2 Facial Bones 95
3 Hyoid Bone 99
4 Sinuses of the Skull 99
5 Fetal Skull 101
6 Vertebral Column 102
7 Thoracic Cage 108
Lab Report 111

EXERCISE 8 The Appendicular Skeleton 115

1 Pectoral Girdle 115
2 Upper Limb 118
3 Pelvic Girdle 123
4 Lower Limb 126
A Regional Look: Shoulder 132
Lab Report 133

EXERCISE 9 Articulations 137

1 Joint Classification 137
2 Structure of Synovial Joints 140
3 Types of Diarthroses 141
4 Movement at Diarthrotic Joints 141
5 Selected Synovial Joints: Elbow and Knee Joints 147
Lab Report 151

EXERCISE 10 Organization of Skeletal Muscles 155

1 Skeletal Muscle Arrangement 155
2 Neuromuscular Junction 159
Lab Report 161

EXERCISE 11 Axial Muscles 163

1 Muscles of Facial Expression 164
2 Muscles of the Eye 167
3 Muscles of Mastication 169
4 Muscles of the Tongue and Pharynx 170
5 Muscles of the Anterior Neck 173
6 Muscles of the Vertebral Column 175
7 Oblique and Rectus Muscles 179
8 Muscles of the Pelvic Region 183
Lab Report 187

EXERCISE 12 Appendicular Muscles — 191

1. Muscles That Move the Pectoral Girdle 191
2. Muscles That Move the Arm 195
3. Muscles That Move the Forearm, Wrist, Hand, and Fingers 197
4. Muscles That Move the Thigh 204
5. Muscles That Move the Leg 208
6. Muscles That Move the Ankle, Foot, and Toes 213
   A Regional Look: Compartments 218
   Lab Report 221

EXERCISE 13 Organization of the Nervous System — 225

1. Histology of the Nervous System 227
2. Anatomy of a Nerve 231
3. Autonomic Nervous System 232
   Lab Report 237

EXERCISE 14 The Spinal Cord and Spinal Nerves — 239

1. Gross Anatomy of the Spinal Cord 239
2. Spinal Meninges 243
3. Spinal Nerves 243
4. Dissection of the Spinal Cord 251
   Lab Report 253

EXERCISE 15 Anatomy of the Brain — 255

1. Cranial Meninges and Ventricles of the Brain 256
2. Regions of the Brain 260
3. Cranial Nerves 267
4. Sheep Brain Dissection 271
   A Regional Look: Cranium 276
   Lab Report 277

EXERCISE 16 The General Senses — 281

1. General-Sense Receptors 281
2. Two-Point Discrimination Test 284
   Lab Report 285

EXERCISE 17 Special Senses: Olfaction and Gustation — 287

1. Olfaction 287
2. Olfactory Adaptation 289
3. Gustation 290
4. Relationship between Olfaction and Gustation 292
   Lab Report 293

EXERCISE 18 Anatomy of the Eye — 295
1. External Anatomy of the Eye   295
2. Internal Anatomy of the Eye   298
3. Cellular Organization of the Retina (Neural Tunic)   300
4. Observation of the Retina (Neural Tunic)   303
5. Dissection of the Cow or Sheep Eye   305
   Lab Report   307

EXERCISE 19 Anatomy of the Ear — 309
1. External and Middle Ear   309
2. Inner Ear   311
3. Examination of the Tympanic Membrane   317
   Lab Report   319

EXERCISE 20 The Endocrine System — 321
1. Pituitary Gland   322
2. Thyroid Gland   323
3. Parathyroid Glands   326
4. Thymus Gland   328
5. Suprarenal Glands   329
6. Pancreas   332
   A Regional Look: Suprarenal Glands   334
   Lab Report   335

EXERCISE 21 Blood — 337
1. Composition of Whole Blood   337
2. ABO and Rh Blood Groups   342
3. Hematocrit (Packed Red Cell Volume)   346
   Lab Report   349

EXERCISE 22 The Heart — 351
1. Heart Wall   351
2. External and Internal Anatomy of the Heart   354
3. Coronary Circulation   359
4. Sheep Heart Dissection   360
   Lab Report   363

EXERCISE 23 The Systemic Circuit — 365
1. Comparison of Arteries, Capillaries, and Veins   365
2. Arteries of the Head, Neck, and Upper Limb   367
3. Arteries of the Abdominopelvic Cavity and Lower Limb   370
4. Veins of the Head, Neck, and Upper Limb   374
5. Veins of the Lower Limb and Abdominopelvic Cavity   378
   A Regional Look: Limbs   381
   Lab Report   383

EXERCISE 24 — Lymphatic System — 387

1. Lymphatic Vessels 389
2. Lymphoid Tissues and Lymph Nodes 390
3. The Spleen 393
   A Regional Look: Female Breast 395
   Lab Report 397

EXERCISE 25 — The Respiratory System — 399

1. Nose and Pharynx 399
2. Larynx 401
3. Trachea and Bronchial Tree 403
4. Lungs 406
   A Regional Look: Airways 409
   Lab Report 411

EXERCISE 26 — The Digestive System — 413

1. Mouth 415
2. Pharynx and Esophagus 418
3. Stomach 419
4. Small Intestine 422
5. Large Intestine 425
6. Liver and Gallbladder 428
7. Pancreas 430
   A Regional Look: Upper Abdomen 432
   Lab Report 433

EXERCISE 27 — The Urinary System — 435

1. Kidney 435
2. Nephron 436
3. Blood Supply to the Kidney 440
4. Ureters, Urinary Bladder, and Urethra 441
5. Sheep Kidney Dissection 444
   A Regional Look: Urinary Bladder 446
   Lab Report 447

EXERCISE 28 — The Reproductive System — 449

1. Male: Testes and Spermatogenesis 449
2. Male: Epididymis and Ductus Deferens 453
3. Male: Accessory Glands 454
4. Male: Penis 456
5. Female: Ovaries and Oogenesis 456
6. Female: Uterine Tubes and Uterus 460
7. Female: Vagina and Vulva 463
8. Female: Mammary Glands 464

CONTENTS xix

A Regional Look: Pelvis   466
Lab Report   467

EXERCISE 29 Development   469

1. First Trimester: Fertilization, Cleavage, and Blastocyst Formation   469
2. First Trimester: Implantation and Gastrulation   472
3. First Trimester: Extraembryonic Membranes and the Placenta   476
4. Second and Third Trimesters and Birth   478
   Lab Report   481

EXERCISE 30 Surface Anatomy   483

1. Head, Neck, and Trunk   483
2. Shoulder and Upper Limb   485
3. Pelvis and Lower Limb   489
   Lab Report   493

DISSECTION EXERCISES

DISSECTION EXERCISE 1 Muscles of the Cat   495

1. Preparing the Cat for Dissection   496
2. Superficial Muscles of the Back and Shoulder   498
3. Deep Muscles of the Back and Shoulder   500
4. Superficial Muscles of the Neck, Abdomen, and Chest   500
5. Deep Muscles of the Chest and Abdomen   505
6. Muscles of the Forelimb   505
7. Muscles of the Thigh   509
8. Muscles of the Lower Hindlimb   512
   Lab Report   515

DISSECTION EXERCISE 2 Cat Nervous System   517

1. Preparing the Cat for Dissection   517
2. The Brachial Plexus   518
3. The Sacral Plexus   518
4. The Spinal Cord   521
   Lab Report   523

DISSECTION EXERCISE 3 Cat Endocrine System   525

1. Preparing the Cat for Dissection   525
2. Endocrine Glands of the Cat   527
   Lab Report   529

DISSECTION EXERCISE 4 Cat Circulatory System   531

1. Preparing the Cat for Dissection   532
2. Arteries That Supply the Head, Neck, and Thorax   532

3	Arteries That Supply the Shoulder and Forelimb (Medial View)	535
4	Arteries That Supply the Abdominal Cavity	536
5	Arteries That Supply the Hindlimb (Medial View)	538
6	Veins That Drain the Head, Neck, and Thorax	539
7	Veins That Drain the Forelimb (Medial View)	540
8	Veins That Drain the Abdominal Cavity	540
9	Veins That Drain the Hindlimb (Medial View)	542

Lab Report 543

DISSECTION EXERCISE 5 Cat Lymphatic System 547

1. Preparing the Cat for Dissection 547
2. The Cat Lymphatic System 549

Lab Report 551

DISSECTION EXERCISE 6 Cat Respiratory System 553

1. Preparing the Cat for Dissection 553
2. Nasal Cavity and Pharynx 554
3. Larynx and Trachea 555
4. Bronchi and Lungs 555

Lab Report 559

DISSECTION EXERCISE 7 Cat Digestive System 561

1. Preparing the Cat for Dissection 561
2. The Oral Cavity, Salivary Glands, Pharynx, and Esophagus 562
3. The Abdominal Cavity, Stomach, and Spleen 564
4. The Small and Large Intestines 565
5. The Liver, Gallbladder, and Pancreas 566

Lab Report 569

DISSECTION EXERCISE 8 Cat Urinary System 571

1. Preparing the Cat for Dissection 571
2. External Anatomy of the Kidney 573
3. Internal Anatomy of the Kidney 573

Lab Report 575

DISSECTION EXERCISE 9 Cat Reproductive System 577

1. Preparing the Cat for Dissection 577
2. The Reproductive System of the Male Cat 578
3. The Reproductive System of the Female Cat 580

Lab Report 583

APPENDIX: *Weights and Measures* 585
PHOTO CREDITS 587
INDEX 589

LABORATORY MANUAL FOR
Human Anatomy *with* CAT DISSECTIONS

Introduction to the Human Body

EXERCISE 1

OBJECTIVES

On completion of this exercise, you should be able to:

- Define *anatomy* and describe the various areas within the study of anatomy.
- Describe the anatomical position and its importance in anatomical studies.
- Use directional terminology to describe the relationships of the surface anatomy of the body.
- Identify the major planes and sections of the body.
- Locate all abdominopelvic quadrants and regions on laboratory models.
- Identify the location of the cranial, spinal, and ventral cavities.
- Describe the two main divisions of the ventral cavity.
- Describe and identify the serous membranes of the body.

LAB ACTIVITIES

1. Anatomical Position and Directional Terminology 2
2. Regional Terminology 6
3. Planes and Sections 9
4. Body Cavities and Serous Membranes 10

Knowledge about what lies beneath the skin and how the body works has been slowly amassed over a span of nearly 3,000 years. It may be obvious to us now that any practice of medicine depends on an accurate knowledge of human anatomy, yet people have not always realized this. Through most of human history, corpses were viewed with superstitious awe and dread. Dissecting corpses to study the body's anatomy was illegal, and medicine therefore remained an elusive practice that often harmed, rather than helped, the unfortunate patient. Despite these superstitions and prohibitions, however, there have always been scientists who wanted to know the human body as it really is rather than how it was imagined to be.

The founder of anatomy was the Flemish anatomist and physician Andreas Vesalius (1514–1564). Vesalius set about to describe human structure accurately. In 1543 he published his monumental work *De Humani Corporis Faberica (On the Structure of the Human Body),* the first meaningful text on human anatomy. In this work he corrected more than 200 errors from earlier anatomists and produced drawings that are still useful today. The work done by Vesalius laid the foundation for all future knowledge of the human body. At last, merely imagining the body's internal structure was unacceptable in medical literature.

Many brilliant anatomists and physiologists since the time of Vesalius have contributed significantly to the understanding of human form and function. Advances in medicine and in the understanding of the human body continue at an accelerated pace. In order to ensure accuracy and consistency, this manual follows the terminology of the publication *Terminologica Anatomica* as endorsed by the International Federation of Associations of Anatomists.

Anatomy is the study of body structures. Early anatomists described the body's **gross anatomy**, which includes the large parts such as muscles and bones. As knowledge of the body advanced and scientific tools permitted more detailed observations, the field of anatomy began to diversify into such areas as **microanatomy**, the study of microscopic structures; **cytology**, the study of cells; and **histology**, the study of **tissues**, which are groups of cells that coordinate their effort toward a common function.

Physiology is the study of how the body functions and of the work that cells must do to keep the body stable and operating efficiently. **Homeostasis** (hō-mē-ō-STĀ-sis; *homeo-*, unchanging; *stasis*, standing) is the maintenance of a relatively steady internal environment through physiological work. Stress, inadequate diet, and disease disrupt the normal physiological processes and may lead to either serious health problems or death.

The various **levels of organization** at which anatomists study the body are reflected in the fields of specialization in anatomy. Each higher level increases in structural and functional complexity, progressing from chemicals to cells, tissues, organs, and finally the organ systems that function to maintain the organism.

The human body is comprised of 11 **organ systems**, and most anatomy courses progress through the lower levels of organization first and then examine each organ system. Because organ systems work together to maintain the organism, it is important that you understand the function of each system. Examine Figure 1.1 to learn the major organs of the human body and the basic function of each organ system.

LAB ACTIVITY 1 — Anatomical Position and Directional Terminology

The human body can bend and stretch in a variety of directions. Although this flexibility allows us to move and manipulate objects in our environment, it can cause difficulty when describing and comparing structures. For example, what is the correct relationship between the wrist and the elbow? If your upper limb is raised above your head, you might reply that the wrist is above the elbow. With your upper limb at your side, you would respond that the wrist is below the elbow. Each response appears correct, but which is the proper anatomical relationship?

In order to avoid confusion, the body is always referred to as being in the **anatomical position**. In this position, the individual stands erect with the feet pointed forward, the eyes straight ahead, and the palms of the hands facing forward with the upper limbs at the sides (Figure 1.2). An individual in the anatomical position is said to be **supine** (soo-PĪN) when lying on the back, and **prone** when lying face down.

Imagine attempting to give someone directions if you could not use terms like "north" and "south" or "left" and "right." These words have a unique meaning and guide the traveler toward his or her destination. Descriptions of the body also require specific terminology. Expressions such as "near", "close to", or "on top of" are too vague for anatomical descriptions. In order to prevent misunderstandings, precise terms are used to describe the locations and spatial relationships of anatomy.

- **Superior** and **inferior** describe vertical positions. *Superior* means above, *inferior* means below. For example, on a person in the anatomical position, the head is superior to the shoulders and the knee is inferior to the hip.

- **Anterior** and **posterior** refer to front and back. *Anterior* means in front of or forward, and *posterior* means in back of or toward the back. The anterior surface of the body is comprised of all front surfaces, including the palms of the hand, and the posterior surface includes all the back surfaces. These directional terms also describe position *relationships*, which means that one body part can be described using both terms. The heart, for example, is posterior to the breastbone and anterior to the spine.

- In four-legged animals, the anatomical position is with all four limbs on the ground, and therefore, the meanings of some directional terms change. *Superior* now refers to the back, or **dorsal**, surface, and *inferior* refers to the belly, or

EXERCISE 1 Introduction to the Human Body   3

THE INTEGUMENTARY SYSTEM

Major Organs:
- Skin
- Hair
- Sweat glands
- Nails

Functions:
- Protects against environmental hazards
- Helps regulate body temperature
- Provides sensory information

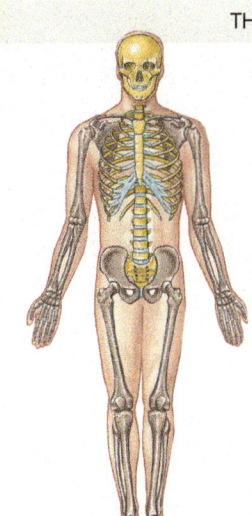

THE SKELETAL SYSTEM

Major Organs:
- Bones
- Cartilages
- Associated ligaments
- Bone marrow

Functions:
- Provides support and protection for other tissues
- Stores calcium and other minerals
- Forms blood cells

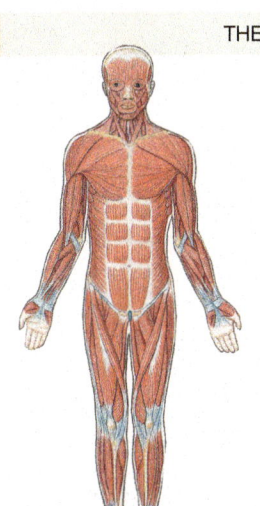

THE MUSCULAR SYSTEM

Major Organs:
- Skeletal muscles and associated tendons

Functions:
- Provides movement
- Provides protection and support for other tissues
- Generates heat that maintains body temperature

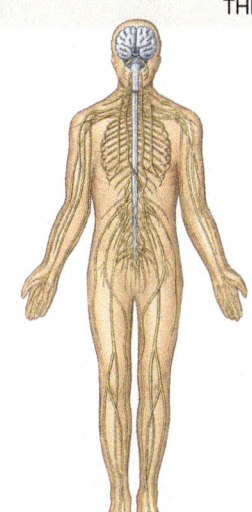

THE NERVOUS SYSTEM

Major Organs:
- Brain
- Spinal cord
- Peripheral nerves
- Sense organs

Functions:
- Directs immediate responses to stimuli
- Coordinates or moderates activities of other organ systems
- Provides and interprets sensory information about external conditions

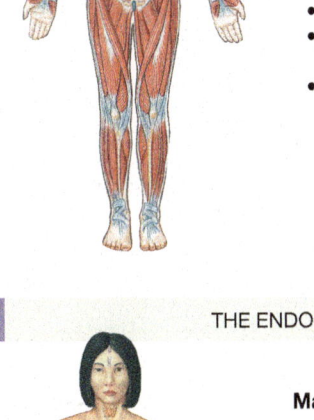

THE ENDOCRINE SYSTEM

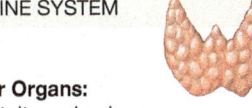

Major Organs:
- Pituitary gland
- Thyroid gland
- Pancreas
- Adrenal glands
- Gonads (testes and ovaries)
- Endocrine tissues in other systems

Functions:
- Directs long-term changes in the activities of other organ systems
- Adjusts metabolic activity and energy use by the body
- Controls many structural and functional changes during development

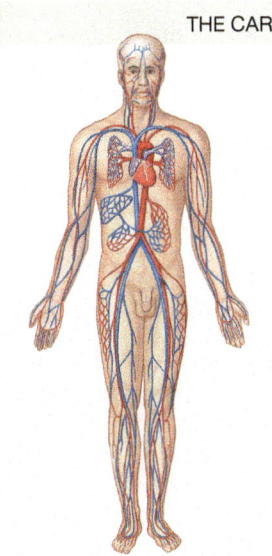

THE CARDIOVASCULAR SYSTEM

Major Organs:
- Heart
- Blood
- Blood vessels

Functions:
- Distributes blood cells, water, and dissolved materials, including nutrients, waste products, oxygen, and carbon dioxide
- Distributes heat and assists in control of body temperature

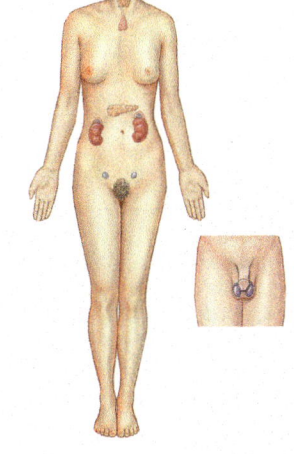

Figure 1.1 The Eleven Organ Systems of the Human Body

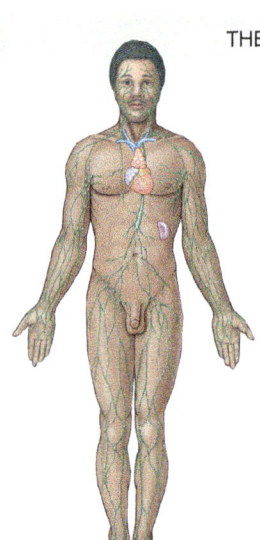

4   EXERCISE 1   Introduction to the Human Body

THE LYMPHATIC SYSTEM

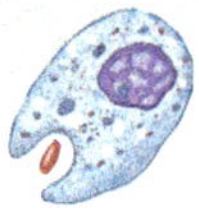

Major Organs:
- Spleen
- Thymus
- Lymphatic vessels
- Lymph nodes
- Tonsils

Functions:
- Defends against infection and disease
- Returns tissue fluids to the bloodstream

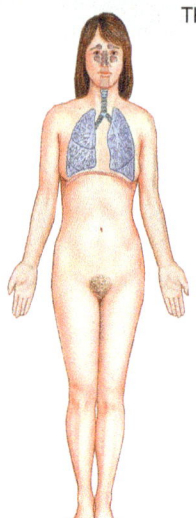

THE RESPIRATORY SYSTEM

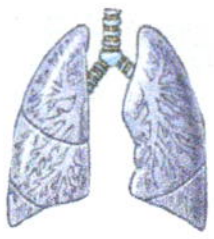

Major Organs:
- Nasal cavities
- Paranasal sinuses
- Larynx
- Trachea
- Bronchi
- Lungs
- Alveoli

Functions:
- Delivers air to alveoli (sites in lungs where gas exchange occurs)
- Provides oxygen to bloodstream
- Removes carbon dioxide from bloodstream
- Produces sounds for communication

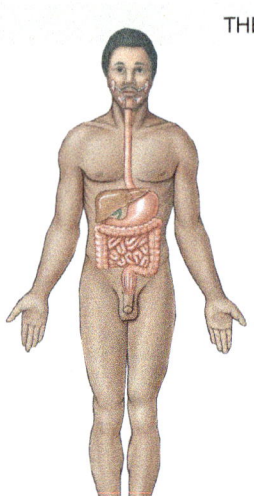

THE DIGESTIVE SYSTEM

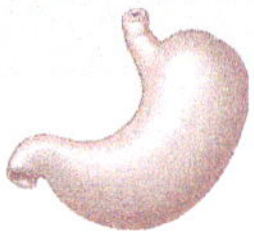

Major Organs:
- Teeth
- Tongue
- Pharynx
- Esophagus
- Stomach
- Small Intestine
- Large intestine
- Liver
- Gallbladder
- Pancreas

Functions:
- Processes and digests food
- Absorbs and conserves water
- Absorbs nutrients (ions, water, and the breakdown products of dietary sugars, proteins, and fats)
- Stores energy reserves

THE URINARY SYSTEM

Major Organs:
- Kidneys
- Ureters
- Urinary bladder
- Urethra

Functions:
- Excretes waste products from the blood
- Controls water balance by regulating volume of urine produced
- Stores urine prior to voluntary elimination
- Regulates blood ion concentrations and pH

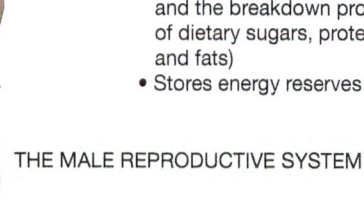

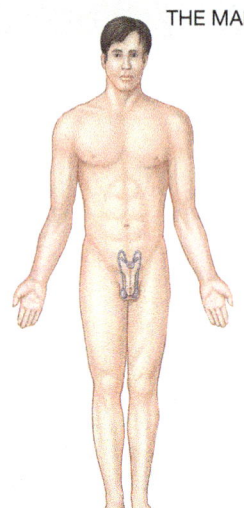

THE MALE REPRODUCTIVE SYSTEM

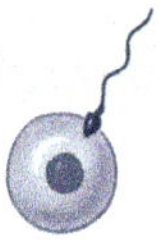

Major Organs:
- Testes
- Epididymis
- Ductus deferens
- Seminal vesicles
- Prostate gland
- Penis
- Scrotum

Functions:
- Produces male sex cells (sperm) and hormones

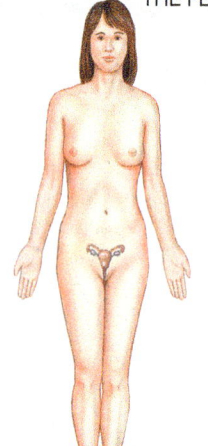

THE FEMALE REPRODUCTIVE SYSTEM

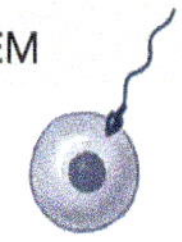

Major Organs:
- Ovaries
- Uterine tubes
- Uterus
- Vagina
- Labia
- Clitoris
- Mammary glands

Functions:
- Produces female sex cells (oocytes) and hormones
- Supports developing embryo from conception to delivery
- Provides milk to nourish newborn infant

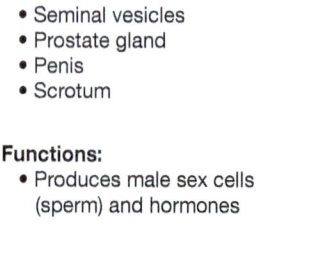

Figure 1.1   **The Eleven Organ Systems of the Human Body** *(Continued)*

EXERCISE 1 Introduction to the Human Body 5

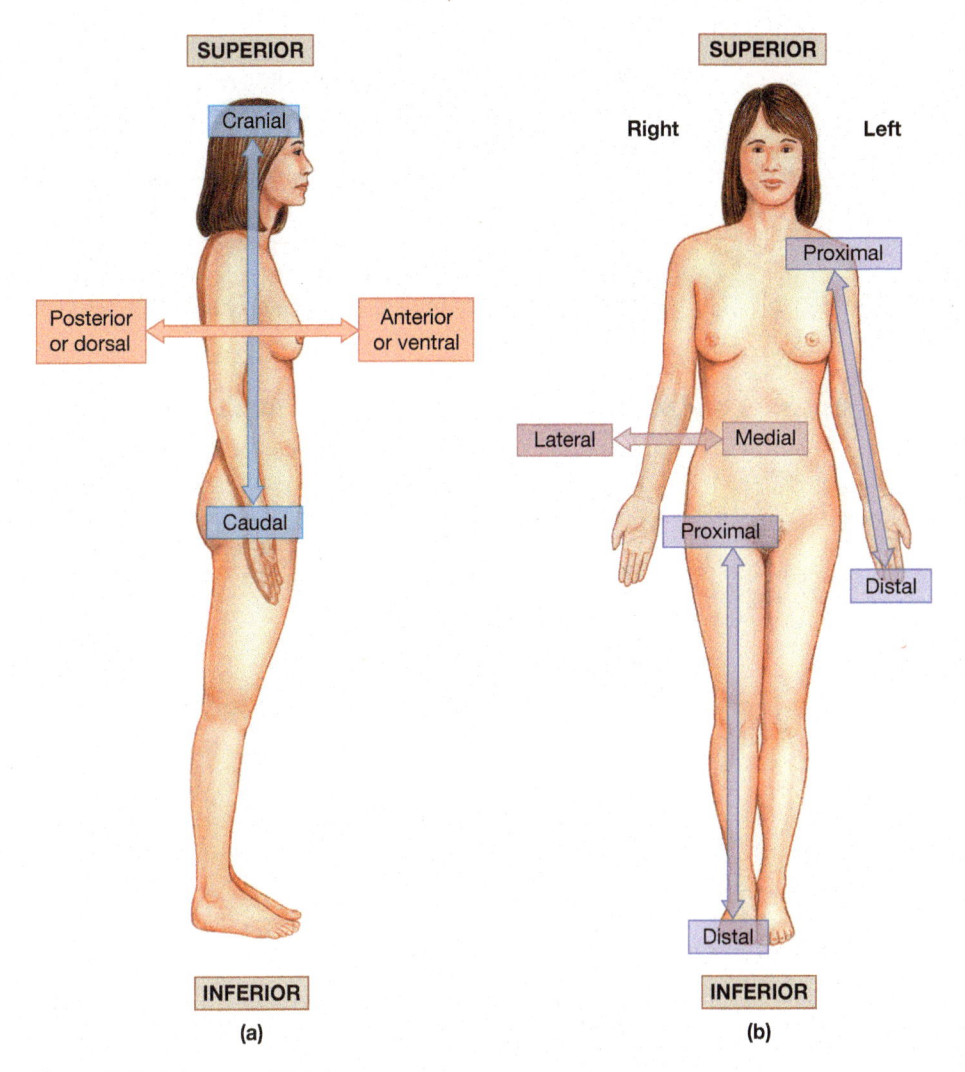

Figure 1.2 Directional References
The important directional terms used in this manual are indicated. **(a)** A lateral view. **(b)** An anterior view.

ventral, surface. **Cephalic** and **cranial** refer to the head in four-legged animals, and **caudal** refers to the tail in four-legged animals and the coccyx in humans.

- **Medial** and **lateral** describe positions relative to the body's *midline*, the vertical middle of the body or any structure in the body. *Medial* has two meanings. It describes one structure as being closer to the body's midline than some other structure; for instance, the ring finger is medial to the middle finger when the hand is held in the anatomical position. *Medial* also describes a structure that is permanently between others, as the nose is medial to the eyes. *Lateral* means either farther from the body's midline or permanently to the side of some other structure; the eyes are lateral to the nose, and, in the anatomical position, the middle finger is lateral to the ring finger.

- **Proximal** refers to parts near another structure. **Distal** describes structures that are distant from other structures. These terms are frequently used to describe the proximity of a structure to its point of attachment on the body. For example, the thigh bone (femur) has a proximal region where it attaches to the hip and a distal region toward the knee.

- **Superficial** and **deep** describe layered structures. *Superficial* refers to parts on or close to the surface. Underneath an upper layer are *deep*, or *bottom*, structures. The skin is superficial to the muscular system, and bones are usually deep to the muscles.

Some directional terms seem to be interchangeable, but there is usually a precise term for each description. For example, *superior* and *proximal* both describe the upper region of limb bones. When discussing the point of attachment of a bone, *proximal* is the more descriptive term. When describing the location of a bone relative to an inferior bone, the term *superior* is used.

Making Connections

Getting Organized for Success

You will benefit more from your laboratory studies if you prepare for each laboratory meeting. Before class, read the appropriate exercise(s) in this manual and complete the labeling of as many figures as possible. If possible, relate the laboratory exercises to the concepts in your lecture textbook. Approaching the laboratory in this manner will maximize your hands-on time with laboratory materials and improve your understanding.

QuickCheck Questions

1.1 Why is it important to have a precisely defined anatomical position in anatomical studies?

1.2 Is your natural stance while standing considered the anatomical position?

1.3 What is the relationship of the shoulder joint to the elbow joint?

1 Materials

- ☐ Yourself or a laboratory partner
- ☐ Torso models
- ☐ Anatomical charts
- ☐ Anatomical models

Procedures

1. Assume the anatomical position. Consider how this orientation differs from your normal stance.
2. Review each directional term presented in Figure 1.2.
3. Use the laboratory models and charts and your own body (or your partner's) to practice using directional terms while comparing anatomy. The Lab Report at the end of this exercise may be used as a guide for comparisons.

LAB ACTIVITY 2 — Regional Terminology

Approaching the body from a regional perspective simplifies the learning of anatomy. Body surface features are used as anatomical landmarks to locate internal structures, and as a result many internal structures are named after an overlying surface structure. For example, because the back of the knee is called the *popliteal* (pop-LIT-ē-al) region, the major artery in the knee is the popliteal artery. Figure 1.3 presents the major regions of the body.

Reference to the position of internal abdominal organs is simplified by partitioning the trunk into four equal **quadrants** (Figure 1.4). Quadrants are used to describe the positions of organs. The stomach, for example, is mostly located in the left upper quadrant. For more detailed descriptions, the lower torso is divided into nine **abdominopelvic regions**, shown in Figure 1.4b. Four imaginary planes are used to define the regions: two vertical and two horizontal planes arranged in the familiar tic-tac-toe pattern. These planes are represented by the four red lines in Figure 1.4b.

EXERCISE 1 Introduction to the Human Body 7

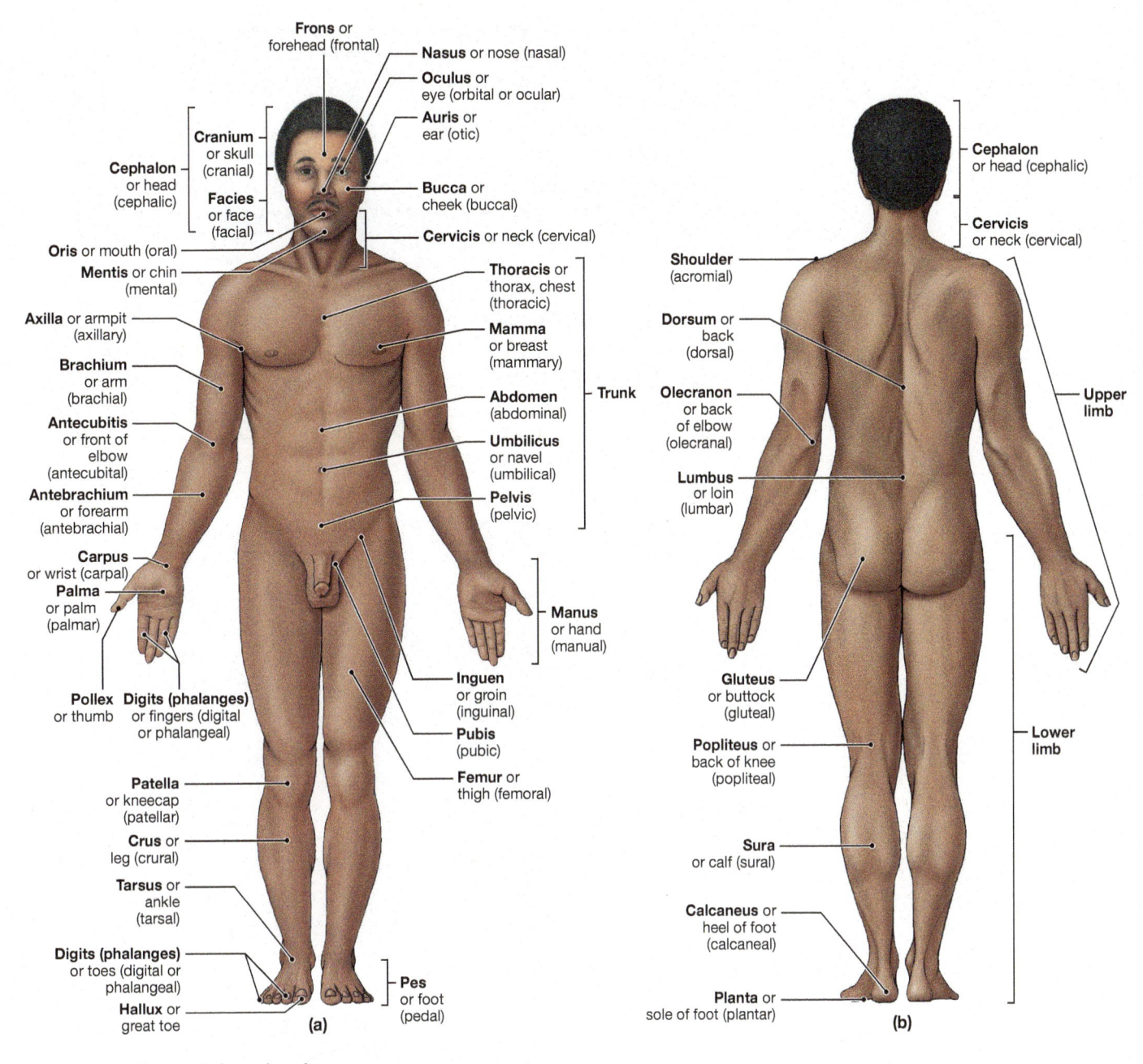

Figure 1.3 Anatomical Landmarks
Anatomical terms are shown in boldface type, common names in plain type, and anatomical adjectives in parentheses. **(a)** Anterior view. **(b)** Posterior view.

QuickCheck Questions

2.1 What are the arm, elbow, and forearm called?

2.2 What are the planes that divide the lower torso into different regions?

2 Materials

- Yourself or a laboratory partner
- Torso models
- Anatomical charts

Procedures

1. Review the regional terminology in Figure 1.3.
2. Identify on a torso model or anatomical chart and on yourself the nine abdominopelvic regions presented in Figure 1.4. On the model, observe which organs occupy each abdominopelvic region. ∎

8 EXERCISE 1 Introduction to the Human Body

Right Upper Quadrant (RUQ)
Right lobe of liver, gallbladder, right kidney, portions of stomach, small and large intestine

Left Upper Quadrant (LUQ)
Left lobe of liver, stomach, pancreas, left kidney, spleen, portions of large intestine

Right Lower Quadrant (RLQ)
Cecum, appendix, and portions of small intestine, reproductive organs (right ovary in female and right spermatic cord in male), and right ureter

Left Lower Quadrant (LLQ)
Most of small intestine, and portions of large intestine, left ureter, and reproductive organs (left ovary in female and left spermatic cord in male)

(a)

Right hypochondriac region
Right lumbar region
Right inguinal region
Epigastric region
Umbilical region
Hypogastric region
Left hypochondriac region
Left lumbar region
Left inguinal region

(b)

Liver
Gallbladder
Large intestine
Small intestine
Appendix
Stomach
Spleen
Urinary bladder

(c)

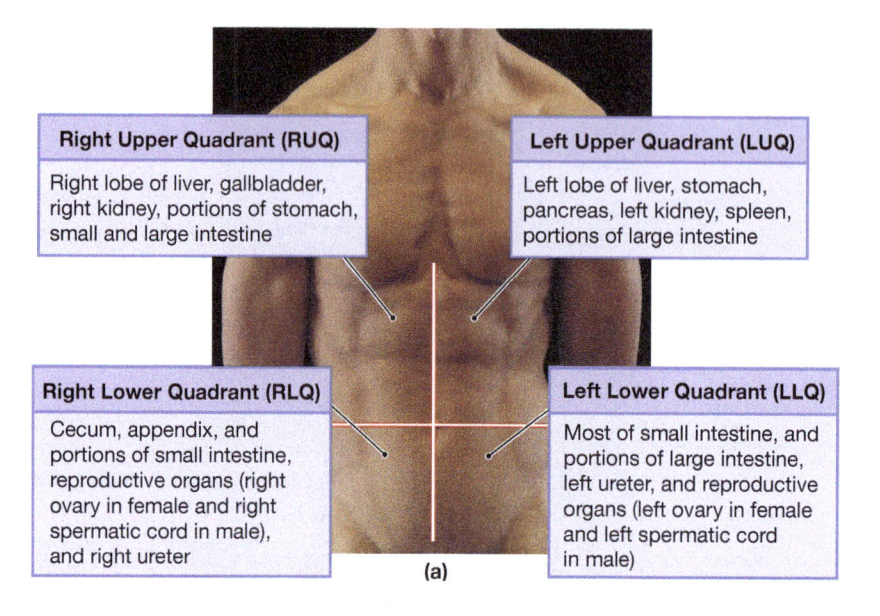

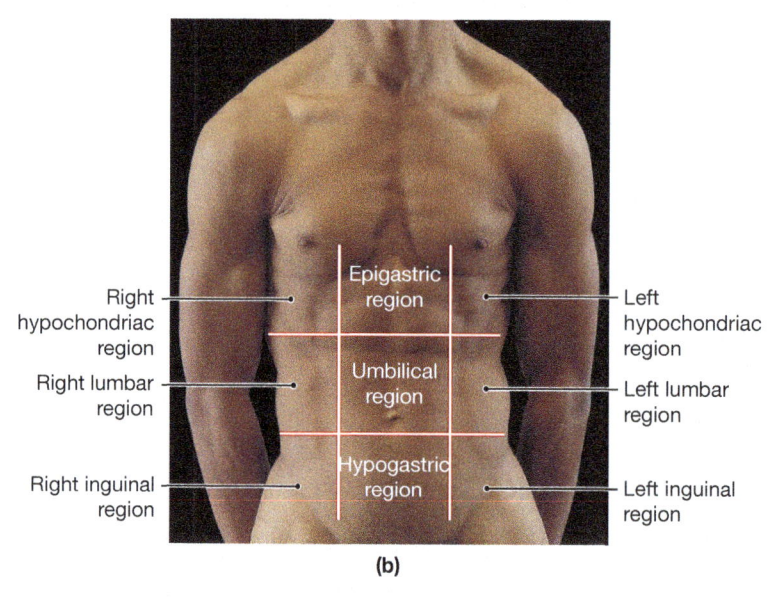

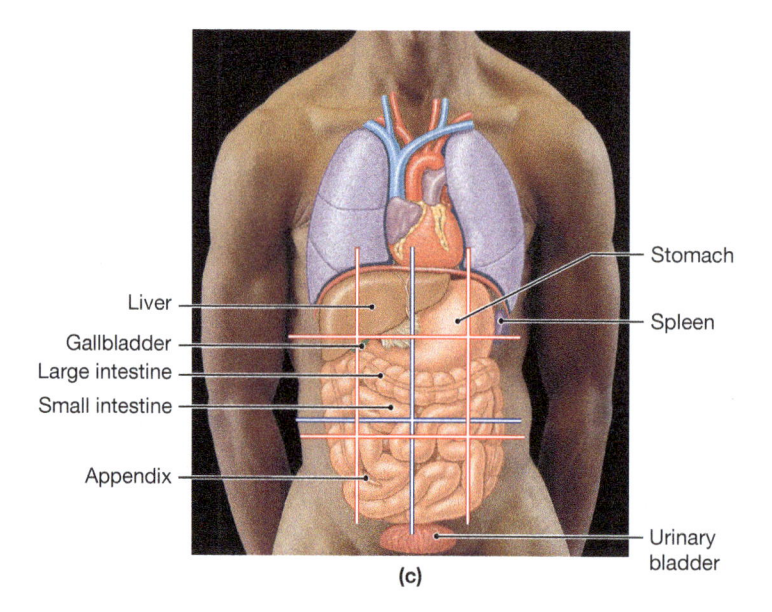

Figure 1.4 Abdominopelvic Quadrants and Regions

(a) The four abdominopelvic quadrants. These terms, or their abbreviations, are most often used in clinical discussions. (b) Further division into nine abdominopelvic regions provides more precise regional descriptions. (c) Overlapping quadrants and regions and the relationship between superficial anatomical landmarks and underlying organs.

LAB ACTIVITY 3 — Planes and Sections

The body must be cut in order to study its internal organization. The process of cutting the body is called **sectioning**. Most structures, such as the trunk, knee, arm, and eyeball, can be sectioned. The orientation of the **plane of section** (the direction in which the cut is made) determines the shape and appearance of the exposed internal region. Imagine cutting one soda straw crosswise (crosswise plane of section) and another straw lengthwise (lengthwise plane of section). The former produces a circle, and the latter produces a concave rectangular surface.

Three major types of sections are used in the study of anatomy: two vertical and one transverse (Figure 1.5). **Transverse** sections are perpendicular to the vertical orientation of the body. (The crosswise cut you made on the imaginary straw yielded a transverse section.) Transverse sections are often called **cross-sections** because they go across the body axis. A transverse section divides superior and inferior structures. **Vertical** sections are parallel to the vertical axis of the body and include sagittal and frontal sections. A **sagittal** vertical section divides a body or organ into right and left portions. A **midsagittal** vertical section equally divides structures, and a **parasagittal** vertical section produces nearly equal divisions. A **frontal**, or **coronal**, vertical section separates anterior and posterior structures.

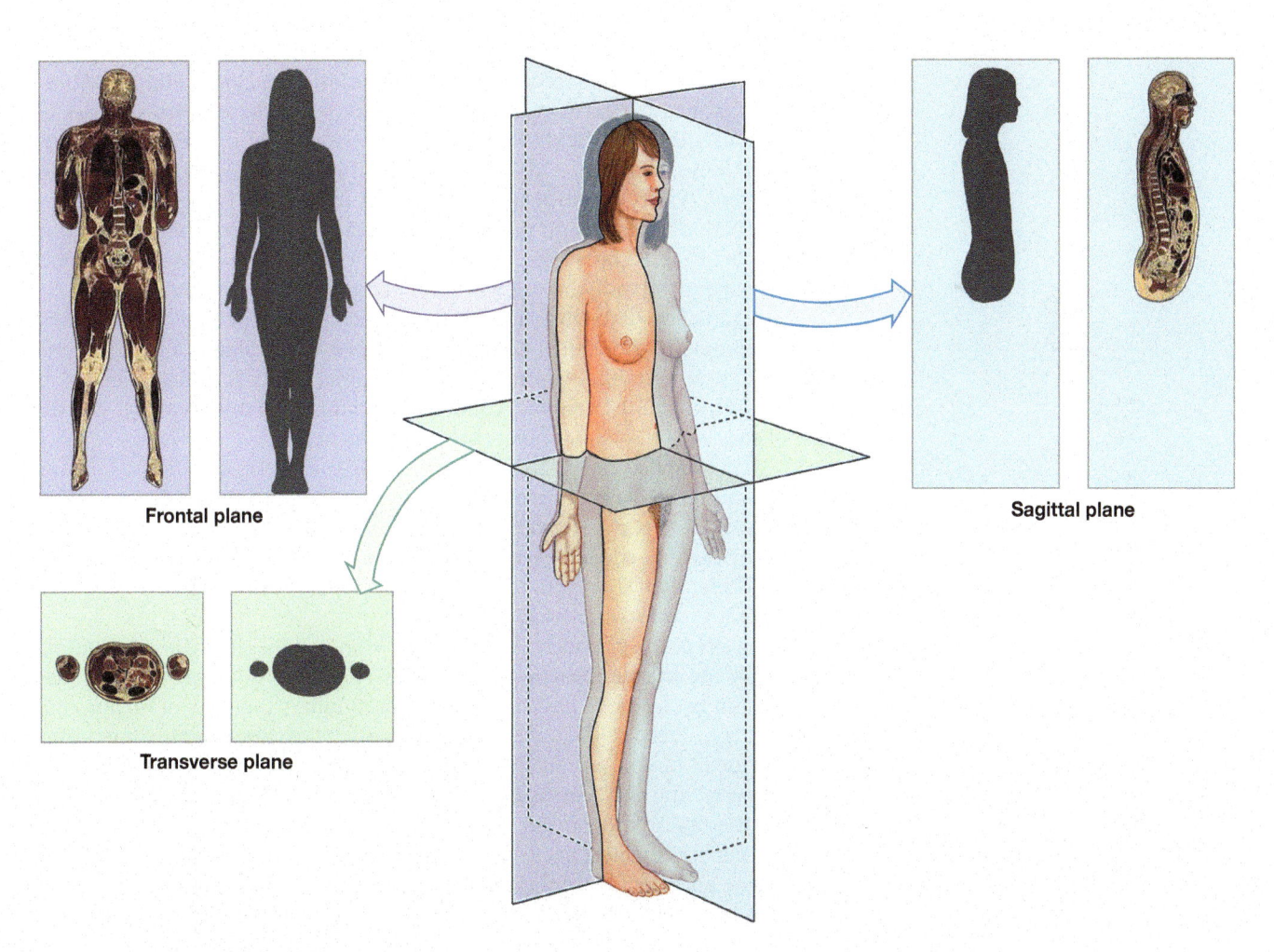

Figure 1.5 Planes of Section
The three primary planes of section.

QuickCheck Questions

3.1 Which type of section separates the kneecap from the lower limb?

3.2 Amputation of the forearm is performed by which type of section?

3 Materials

- ☐ Anatomical models with various sections
- ☐ Objects for sectioning

Procedures

1. Review each plane of section shown in Figure 1.5.
2. Identify the sections on models and other materials presented by your instructor.
3. Cut several common objects, such as an apple and a hot dog, along their sagittal and transverse planes. Compare the exposed arrangement of the interior. ■

LAB ACTIVITY 4 Body Cavities and Serous Membranes

Body cavities are internal spaces that house internal organs, such as the brain in the cranium and the digestive organs in the abdomen. The walls of a body cavity support and protect the soft organs contained in the cavity. In the trunk, large cavities are subdivided into smaller cavities that contain individual organs. The smaller cavities are enclosed by thin sacs, such as those around the heart, lungs, and intestines. Most of the space in a cavity is occupied by the enclosed organ and by a thin film of liquid.

The **cranial cavity** and **spinal cavity** (often collectively called the **dorsal body cavity**, a term not recognized by *Terminologia Anatomica* and therefore not used in this manual) contain the central nervous system, and the **ventral body cavity** houses the thoracic and abdominopelvic organs. In the ventral body cavity, the heart, lungs, stomach, and intestines are covered with a double-layered **serous** (SĒR-us; *seri-*, watery) **membrane**. Each serous membrane isolates one organ and reduces friction and abrasion on the organ surface. In the cranial and spinal cavities, the brain and spinal cord are contained within the **meninges**, a protective three-layered membrane that we will study later with the nervous system.

The cranial cavity is the space within the oval part of the skull (Figure 1.6). The brain is located in the cranial cavity, and this part of the skull, called the *cranium*, encases and protects the delicate brain. The spinal cavity is a long, slender canal that passes through the vertebral column. The cranial and spinal cavities join at the base of the skull, where the spinal cord meets the brain.

The ventral body cavity, also called the **coelom** (SĒ-lom; *koila*, cavity), is the entire space of the body trunk anterior to the vertebral column and posterior to the sternum (breastbone) and the abdominal muscle wall. This large cavity is divided into two major cavities, the **thoracic** (*thorax*, chest) **cavity** and the **abdominopelvic cavity**. These cavities, in turn, are further divided by membranes into the specific cavities that surround individual organs.

The walls of the thoracic cavity are muscle and bone. The main subdivisions of this cavity are the **mediastinum** (mē-dē-as-TĪ-num *or* mē-dē-AS-tĭ-num; *media-*, middle) and two **pleural cavities.** The mediastinum is the mass of organs and tissues that separates the pleural cavities. Each pleural cavity contains one lung. Inside the mediastinum is a smaller cavity, the **pericardial** (*peri-*, around + *kardia*, heart) **cavity**, and the heart is most often described as being contained inside this cavity rather than simply inside the mediastinum.

The abdominopelvic cavity is the space between a dome-shaped muscle, the diaphragm, and the floor of the pelvis. This cavity is subdivided into the abdominal cavity and the pelvic cavity. The **abdominal cavity** contains most of the digestive organs, such as the liver, gallbladder, stomach, pancreas, kidneys, and small and large intestines. The **pelvic cavity** is the small cavity enclosed by the pelvic girdle

EXERCISE 1 Introduction to the Human Body 11

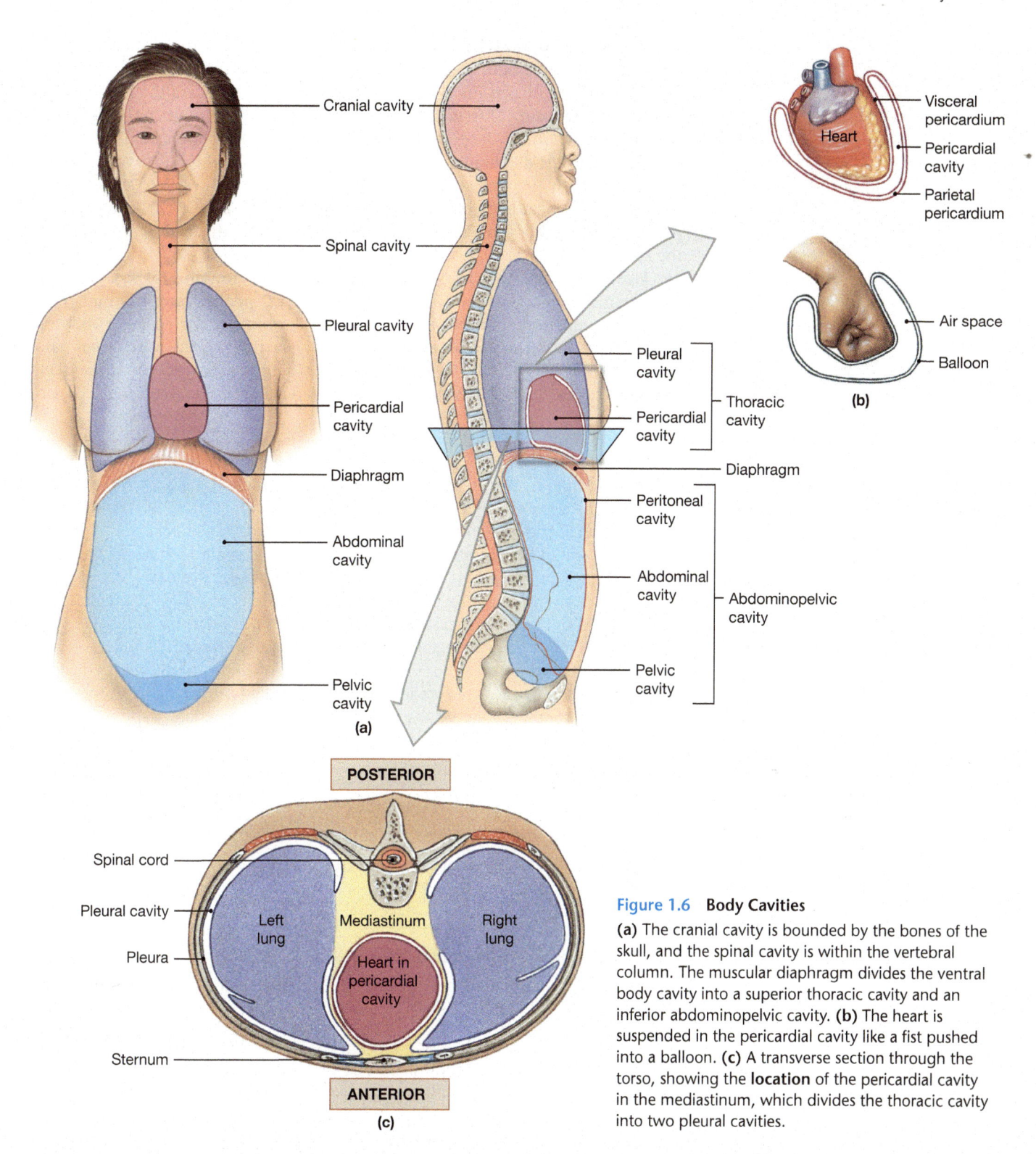

Figure 1.6 Body Cavities
(a) The cranial cavity is bounded by the bones of the skull, and the spinal cavity is within the vertebral column. The muscular diaphragm divides the ventral body cavity into a superior thoracic cavity and an inferior abdominopelvic cavity. (b) The heart is suspended in the pericardial cavity like a fist pushed into a balloon. (c) A transverse section through the torso, showing the location of the pericardial cavity in the mediastinum, which divides the thoracic cavity into two pleural cavities.

of the hips. This cavity contains the internal reproductive organs, parts of the large intestine, the rectum, and the urinary bladder.

The heart, lungs, stomach, and intestines are encased in double-layered **serous membranes** that have a minuscule fluid-filled cavity between the two layers. Directly attached to the exposed surface of an internal organ is the **visceral** (VIS-er-al; *viscera,* internal organ) **layer** of the serous membrane. The **parietal** (pah-RĪ-e-tal; *pariet-,* wall) **layer** is superficial to the visceral layer and lines the wall of the body cavity. The **serous fluid** between these layers is a lubricant that reduces friction and abrasion between the layers as the enclosed organ moves.

Figure 1.6 highlights the anatomy of the serous membrane of the heart, which is the **pericardium**. This membrane is composed of an outer **parietal pericardium** and an inner **visceral pericardium**. The parietal pericardium is a fibrous sac attached to the diaphragm and supportive tissues of the thoracic cavity. The visceral pericardium is attached to the surface of the heart.

The serous membrane of a lung is called the **pleura** (PLOO-rah). The **parietal pleura** lines the thoracic wall, and the **visceral pleura** is attached to the surface of the lung.

Most of the digestive organs are encased in the **peritoneum** (per-i-tō-NĒ-um), the serous membrane of the abdomen. The **parietal peritoneum** has numerous folds that wrap around and attach the abdominal organs to the posterior abdominal wall. The **visceral peritoneum** lines the organ surfaces. The **peritoneal cavity** is the space between the parietal and visceral peritoneal layers. The kidneys are **retroperitoneal** (*retro-*, behind) and are located outside the peritoneum.

Clinical Application **Problems with Serous Membranes**

Serous membranes may become inflamed and infected as a result of bacterial infection or damage to the underlying organ. Liquids often build up in the cavity of the serous membrane, causing additional complications. **Peritonitis** is an infection of the peritoneum that occurs when the digestive tract is damaged—often by ulceration, rupture, or a puncture wound—in a way that permits intestinal bacteria to contaminate the peritoneum. **Pleuritis**, or **pleurisy**, is an inflammation of the pleura that is often caused by tuberculosis, pneumonia, or thoracic abscess. Breathing becomes painful as the inflamed membranes move when a person inhales and exhales. **Pericarditis** is an inflammation of the pericardium that results from infection, injury, heart attack, or other causes. In advanced stages, a buildup of liquid causes the heart to compress, a condition that results in decreased cardiac function.

QuickCheck Questions

4.1 What structures form the walls of the cranial and spinal cavities?

4.2 What are the various subdivisions of the ventral body cavity?

4.3 What are the two layers of a serous membrane?

4 Materials

- ☐ Torso models
- ☐ Articulated skeleton
- ☐ Anatomical charts

Procedures

1. Review each cavity illustrated in Figure 1.6.
2. Locate each body cavity on the torso models, anatomical charts, and articulated skeleton.
3. Identify the organ(s) in the various cavities of the ventral body cavity on the torso models.
4. Identify the pericardium, pleura, and peritoneum on the torso models and charts.
5. With your hands, trace the location of the various body cavities on the surface of your body.

Name _____

Date _____

Section _____

LAB REPORT

1

Introduction to the Human Body

A. Matching

Match each directional term on the left with the correct description on the right.

_____ 1. anterior   A. to the side
_____ 2. lateral   B. away from a point of attachment
_____ 3. proximal   C. close to the body surface
_____ 4. inferior   D. front
_____ 5. posterior   E. away from the body surface
_____ 6. medial   F. above, on top of
_____ 7. distal   G. toward a point of attachment
_____ 8. superficial   H. below, a lower level
_____ 9. superior   I. back
_____ 10. deep   J. toward the middle

B. Fill in the Blanks

Use the correct term to complete each statement.

1. The heart is surrounded by a small cavity called the _____, which is inside a collection of organs and tissues called the _____.
2. The _____ cavity surrounds the digestive organs in the abdominal cavity.
3. The kidneys are _____ because they are located superficial to the _____.
4. A transverse plane at the top of the hips separates the abdominal cavity from the _____.
5. The inner membrane layer surrounding a lung is the _____.
6. The brain is contained within the _____ cavity.
7. A lubricating substance in body cavities is called _____.
8. The large medial space of the chest is called the _____.
9. The muscle that divides the ventral body cavity horizontally is the _____.
10. The outer layer of a serous membrane is the _____ layer.

C. Fill in the Blanks

Use the correct directional term to show the relationship between the structures in each statement.

1. The chin is _____ to the nose.
2. In the anatomical position, the brachium is _____ to the antecubitis.

13

EXERCISE 1 — LAB REPORT

3. The index finger is _____ to the ring finger.
4. The skin is _____ to the muscles.
5. The trunk is _____ to the pubis.
6. The middle toe is _____ to the little toe.
7. Where it attaches to the elbow, the upper humerus is _____ to the elbow.
8. The ears are _____ to the eyes.
9. The buttock is _____ to the pubis.
10. Relative to where the leg attaches to the trunk, the knee is _____ to the hip.

D. Labeling

Label the regions of the body in Figure 1.7.

E. Short-Answer Questions

1. Describe the six main levels of organization in the body.

2. List the nine abdominopelvic regions and describe the location of each.

3. Describe the mediastinum.

F. Drawing

1. Draw two pictures of a donut sectioned by a plane. In one drawing, make the sectioning plane parallel to the circular surface of the donut; in the other drawing, make the sectioning plane perpendicular to that surface.

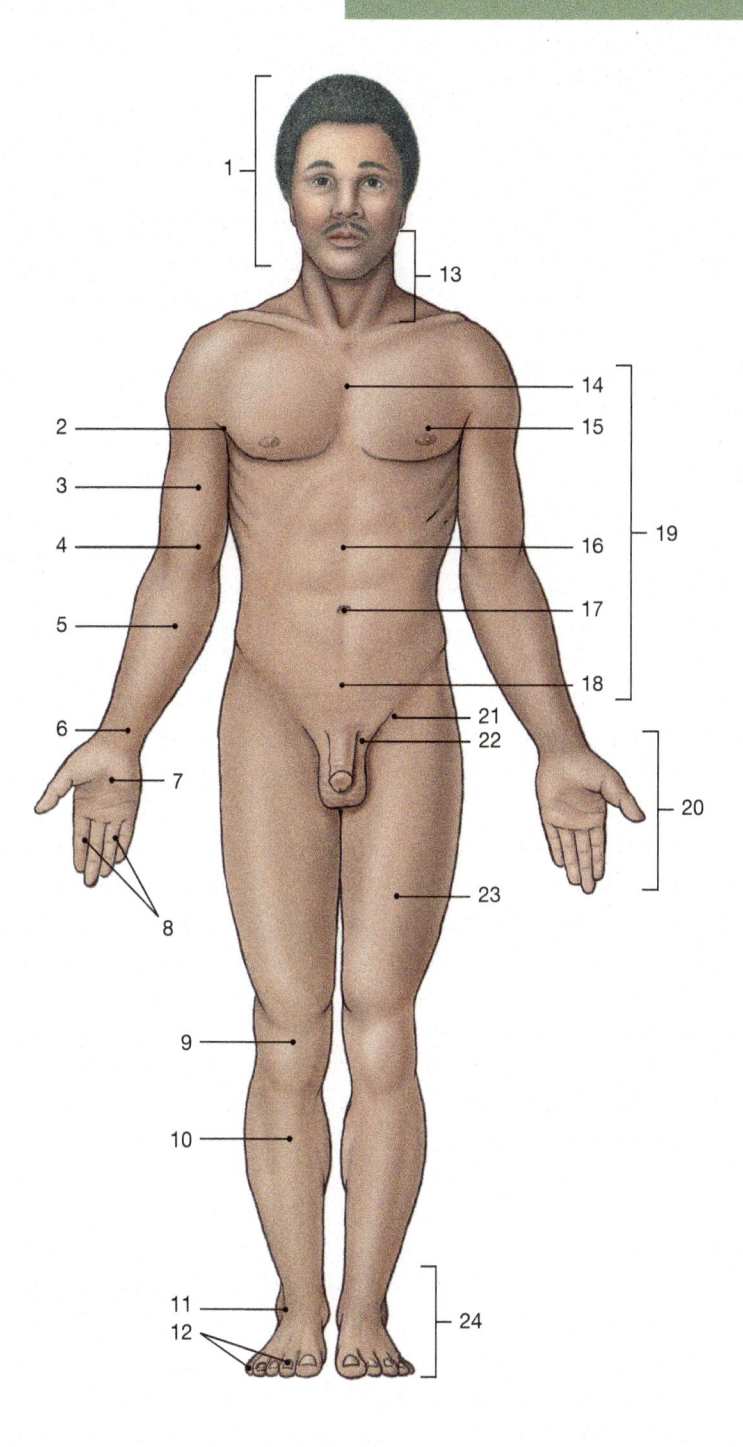

LAB REPORT   EXERCISE 1

Figure 1.7   Regional Terminology

1. _____
2. _____
3. _____
4. _____
5. _____
6. _____
7. _____
8. _____
9. _____
10. _____
11. _____
12. _____
13. _____
14. _____
15. _____
16. _____
17. _____
18. _____
19. _____
20. _____
21. _____
22. _____
23. _____
24. _____

15

EXERCISE 1

LAB REPORT

2. Sketch the body trunk and the four planes that designate the nine abdominopelvic regions.

G. Analysis and Application

1. Explain why it is important to use the anatomical position when describing body parts.

2. Which organ systems protect the body from infection?

3. Long-term coordination of body functions is regulated by which organ system?

4. Which organ system stores minerals for the body?

5. Compare the body axis of a four-legged animal to the axis of a human.

EXERCISE 2

Use of the Microscope

OBJECTIVES

On completion of this exercise, you should be able to:

- Describe how to properly carry, clean, and store a microscope.
- Identify the parts of a microscope.
- Focus a microscope on a specimen and adjust the magnification.
- Adjust the light source of a microscope.
- Calculate the total magnification for each objective lens.
- Measure the field diameter at each magnification.
- Make a wet-mount slide.

LAB ACTIVITIES

1. Parts and Care of the Compound Microscope 17
2. Using the Microscope 20
3. Depth of Field Observation 22
4. Relationship between Magnification and Field Diameter 23

As a student of anatomy, you will explore the organization and structure of cells, tissues, and organs. The basic research tool for your observations is the microscope. This instrument is easy to use once you learn its parts and how to adjust them to produce a clear image of a specimen. Therefore, it is important that you complete each activity in this exercise and that you are able to use a microscope effectively by the end of the laboratory period.

The **compound microscope** uses several lenses to direct a narrow beam of light through a thin specimen mounted on a glass slide. Focusing knobs move either the lenses or the slide to bring the specimen into focus within the round viewing area of the lenses, an area called the **field of view**. Lenses magnify objects so that the objects appear larger than they actually are. As magnification increases, the viewer can more easily see details that are close together. It is this increase in **resolution**—the ability to distinguish between two objects located close to each other—that makes the microscope a powerful observational tool.

LAB ACTIVITY 1 — Parts and Care of the Compound Microscope

Because the microscope is a precision scientific instrument with delicate optical components, you should always observe the following guidelines when using one. Your laboratory instructor will provide you with specific information regarding the use and care of microscopes in your laboratory.

1. Carry the microscope with two hands, one hand on the arm and the other hand supporting the base. (See Figure 2.1 for the parts of the microscope.) Do not swing the microscope as you carry it to your laboratory bench because such movement could cause a lens to fall out. Avoid bumping the microscope as you set it on the laboratory bench.

18   EXERCISE 2   Use of the Microscope

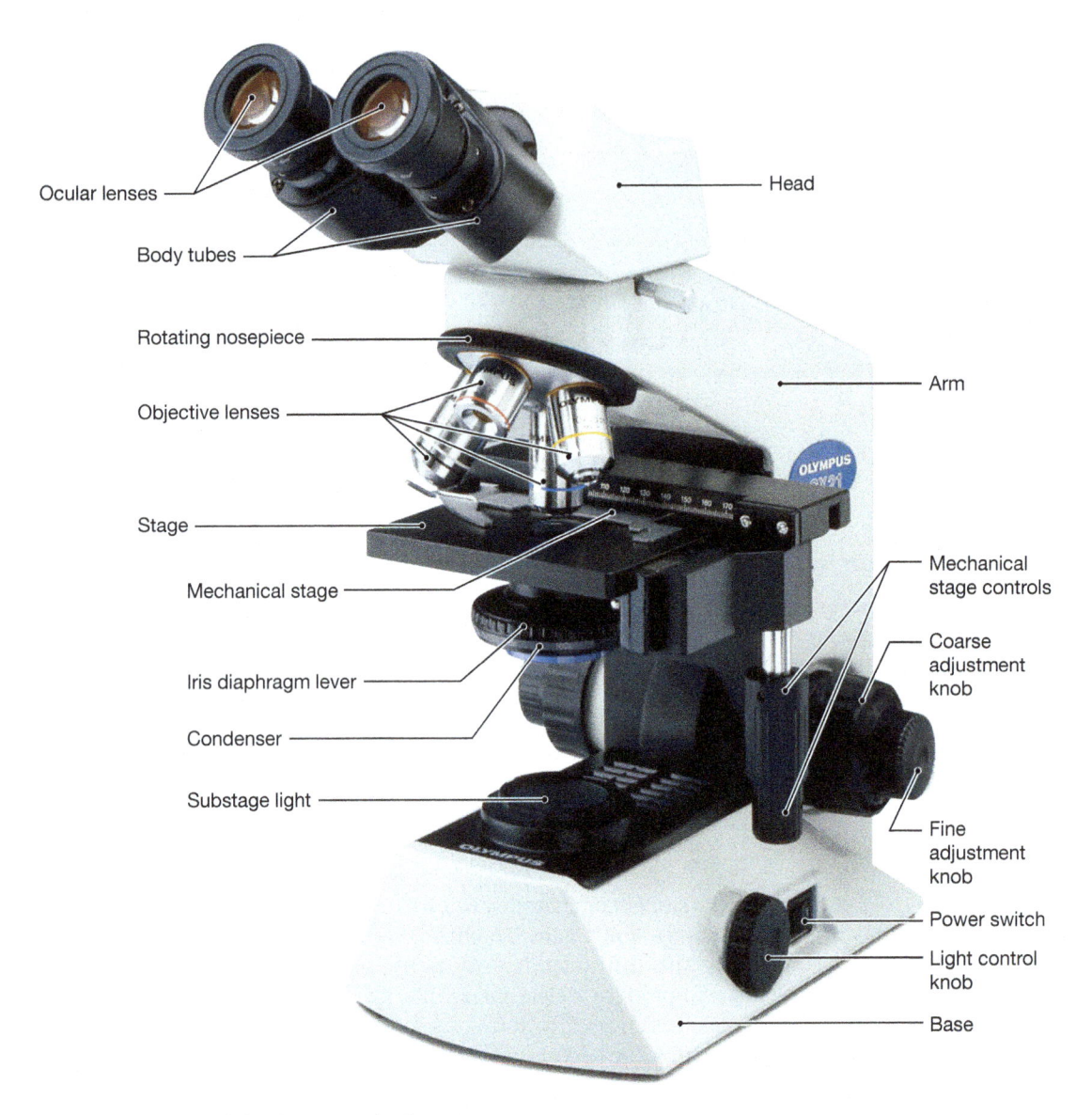

Figure 2.1 Parts of the Compound Microscope
Source: Courtesy of Olympus America, Inc.

2. If the microscope has a built-in light source, completely unwind the electric cord before plugging it in.

3. To clean the lenses, use only the lens cleaning liquid and lens paper provided by your instructor. Facial tissue is unsuitable for cleaning because it is made of small wood fibers that will damage the optical coating on the lenses.

4. When you are finished using it, store the microscope with the low-power objective lens in position and the stage in the uppermost position. Either return the microscope to the storage cabinet or cover it with a dust cover. The cord may be wrapped neatly around the base; some cords are removable for separate storage.

Figure 2.1 shows the parts of a typical compound microscope. Your laboratory may be equipped with a different type; if so, your instructor will discuss the type of microscope you will use.

- The **arm** is the supportive frame of the microscope. It joins the body tube to the base. As noted earlier, the correct way to carry the microscope is with one hand on the arm and the other on the base.

- The **base** is the broad, flat lower support of the microscope.
- The **body tube** is the cylindrical tube that supports the ocular lens and extends down to the nosepiece.
- The **ocular lens** is the eyepiece where you place your eye(s) to observe the specimen. The magnification of most ocular lenses is 10X. This results in an image ten times larger than the actual size of the specimen. **Monocular** microscopes have a single ocular lens; **binocular** microscopes have two ocular lenses, one for each eye. The ocular lenses may be moved closer together or farther apart by adjusting the body tubes.
- The **nosepiece** is a rotating disk at the base of the body tube where several objective lenses of different lengths are attached. Turning the nosepiece moves an objective lens into place over the specimen being viewed.
- **Objective lenses** are mounted on the nosepiece. Magnification of the viewed imaged is determined by the choice of objective lens. The longer the objective lens, the greater its magnifying power. The **working distance** is the distance between the tip of the lens and the top surface of the microscope slide. Your microscope may also have an **oil-immersion objective lens**, which is usually 100X. With this lens, a small drop of immersion oil is used on the slide to eliminate the air between the lens and the slide, thereby improving the resolution of the microscope. It is important to carefully clean the lens and slide to completely remove the oil.
- The **stage** is a flat, horizontal shelf under the objective lenses that supports the microscope slide. The center of the stage has an **aperture**, or hole, through which light passes to illuminate the specimen on the slide. Most microscopes have a **mechanical stage** that holds and moves the slide with more precision than is possible manually. The mechanical stage has two **controls** on the side that move the slide around on the stage in horizontal and vertical planes.
- The **coarse adjustment knob** is the large dial on the side of the microscope that is used only at low magnification to find the initial focus on a specimen. The small dial on the side of the microscope is the **fine adjustment knob**. This knob moves the objective lens for precision focusing after coarse focus has been achieved. The fine adjustment knob is the only focusing knob used at magnifications greater than low.
- The **condenser** is a small lens under the stage that narrows the beam of light and directs it through the specimen on the slide. A **condenser adjustment knob** moves the condenser vertically. For most microscope techniques, the condenser should be in the uppermost position, near the stage aperture.
- The **iris diaphragm** is a series of flat metal plates at the base of the condenser that slide together and create an aperture in the condenser to regulate the amount of light that passes through the condenser. Most microscopes have a small **diaphragm lever** that extends from the iris diaphragm; this lever is used to open or close the diaphragm to adjust the light for optimal contrast and minimal glare.
- A **lamp** provides the light that passes through the specimen, through the lenses, and finally into your eyes. Most microscopes have a built-in light source beneath the stage. The **light control knob**, a rheostat dial located on either the base or the arm, controls the brightness of the light. Microscopes without a light source use a mirror to reflect ambient light into the condenser.

Microscopes use a **compound lens system**, in which each lens consists of many pieces of optical glass. The magnification is stamped on the barrel of each objective lens, as is the magnifying power of the ocular lens. In order to calculate the **total magnification** of the microscope at a particular lens setting, multiply the ocular lens magnification by the objective lens magnification. For example, a 10X ocular lens used with a 10X objective lens produces a total magnification of 100X.

20 EXERCISE 2 Use of the Microscope

QuickCheck Questions

1.1 What is the proper way to hold a microscope while carrying it?
1.2 Why is a facial tissue not appropriate for cleaning microscope lenses?
1.3 How do you change the magnification of a microscope?
1.4 What is the function of the iris diaphragm on a microscope?

1 Materials

☐ Compound microscope

Procedures

1. Identify and describe the function of each part of the microscope.
2. Determine the magnification of the ocular lens and each objective lens on the microscope. Enter this information in the second and third columns of Table 2.1, and then fill in the fourth column by calculating the total magnification for each ocular/objective combination.
3. Use a ruler to measure the working distance between the objective lens and slide for each magnification. Record your data in Table 2.1. ■

LAB ACTIVITY 2 Using the Microscope

Four basic steps are involved in successfully viewing a specimen under the microscope: (1) setup, (2) focusing, (3) magnification control, and (4) light intensity control.

Setup

- Plug in the electrical cord and turn the microscope lamp on. If the microscope does not have a built-in light source, adjust the mirror to reflect light into the condenser.
- Check the position of the condenser; it should be in the uppermost position, near the stage aperture.
- Rotate the nosepiece to swing the low-magnification objective lens into position over the aperture.
- Place the slide on the stage and use the stage clips or mechanical slide mechanism to secure the slide. Move the slide so that the specimen is over the stage aperture.
- After you have finished your observations, reset the microscope to low magnification, remove the slide from the stage, and store the microscope.

Table 2.1 Microscope Data

	Ocular lens	Objective Lens	Total Magnification	Working Distance	Field Diameter
Low Power	_____	_____	_____	_____	_____
Medium Power	_____	_____	_____	_____	_____ *
High Power	_____	_____	_____	_____	_____ *
Oil Immersion	_____	_____	_____	_____	_____ *

*Calculated field diameter

Focusing and Ocular Lens Adjustment

- The distance between the ocular lenses is adjustable for your *interpupillary distance*, the distance between the two pupils of your eyes, so that a single image is seen in the microscope. Move the body tubes apart and look into the microscope. If two images are visible, slowly move the body tubes closer together until you see, with *both* eyes open, a single, circular field of view.
- To focus on a specimen, first move the low-power objective lens, or the stage on older microscopes, to its lowest position. Next, look into the ocular lenses and raise the low-power objective lens by slowly turning the coarse adjustment knob. The image should come into focus.
- Once the image is clear, use the fine adjustment knob to examine the detailed structure of the specimen.
- When you are ready to change magnification, do not move the adjustment knobs before changing the objective lens. Most microscopes are **parfocal**, which means they are designed to stay in focus when you change from one objective lens to another. After changing magnification, use only the fine adjustment knob to adjust the objective lens.

Magnification Control

- Always use low magnification during your initial observation of a slide. You will see more of the specimen and can quickly select areas on the slide for detailed studies at higher magnification.
- To examine part of the specimen at higher magnification, move that part of the specimen to the center over the aperture before changing to a higher-magnification objective lens. This repositioning keeps the specimen in the field of view at the higher magnification. Because a higher-magnification lens is closer to the slide, less of the slide is visible in the field of view. The image of the specimen enlarges to fill the field of view.

Light Intensity Control

- Use the light control knob to regulate the intensity of light from the bulb. Adjust the brightness so that the image has good contrast and no glare.
- Adjust the iris diaphragm by moving the diaphragm lever side to side. Notice how the field illumination is changed by different settings of the iris.
- At higher magnifications, increase the light intensity and open the iris diaphragm.

QuickCheck Questions

2.1 When is the coarse adjustment knob used on a microscope?

2.2 What is the typical view position for the condenser?

2 Materials

- ☐ Compound microscope, slide, and coverslip
- ☐ Newspaper cut into small pieces
- ☐ Dropper bottle that contains water

Procedures

Make a **wet-mount slide** of a small piece of newspaper as follows:

1. Obtain a slide, a coverslip, and a small piece of newspaper that has printing on it.
2. Place the paper on the slide and add a small drop of water to it.
3. Put the coverslip over the paper as shown in Figure 2.2. The coverslip keeps the lenses dry.

Figure 2.2 Preparing a Wet Mount

Place the object to be mounted in a drop of water or stain on the slide. Using tweezers or your fingers, touch the water or stain with the edge of the coverslip and carefully lower the coverslip until it rests flat on the slide. Use a paper towel to absorb excess liquid that has leaked out from under the coverslip.

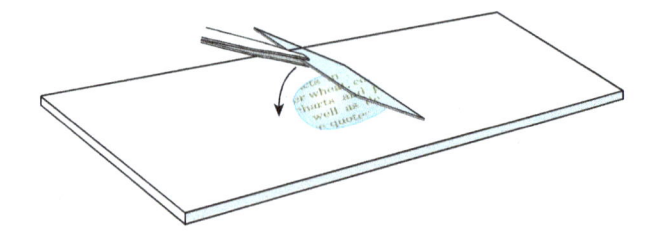

Observe the slide using the microscope:

1. Move the low-magnification objective lens into position (if it is not already there), and place the slide on the stage so that the newspaper is directly over the stage aperture.
2. Look into the ocular lens and slowly turn the coarse adjustment knob until you see the fibers of the newspaper. Once they are in focus, adjust the light source for optimal contrast and resolution. Adjust the distance between the ocular lenses for your interpupillary distance.
3. Use the fine adjustment knob to bring the image into crisp focus. Remember, the microscope you are using is a precise instrument and produces a clear image when adjusted correctly. Be patient and keep adjusting it until you get a perfectly clear image.
4. Once the image is correctly focused, do the following and record your observations in the spaces provided.
   a. Locate a letter "a" or "e". Describe the ink and the paper fibers. _____
   b. Slowly move the slide forward with the mechanical stage knob. In which direction does the image move? _____
   c. Move the slide horizontally to the left using the mechanical stage knob. In which direction does the image move? _____
   d. Is the image of the letter oriented in the same direction as the real letter on the slide? _____ ■

LAB ACTIVITY 3 Depth of Field Observation

Depth of field, or **focal depth**, is a measure of how much depth (thickness) of a specimen is in focus, and the in-focus thickness is called a **focal plane**. Depth of field is greatest at low power and decreases as magnification increases. In other words, the focal plane is thicker at low power and thinner at higher powers (Figure 2.3a). Because depth of field is reduced at higher power, you use the fine adjustment knob to move the focal plane up and down through the thickness of the specimen and scan the specimen layers. As you turn the fine adjustment knob, the objective lens moves either closer to, or farther from, the slide surface (Figure 2.3b). This lens movement causes the focal plane to move through the layers of the specimen. Most specimens

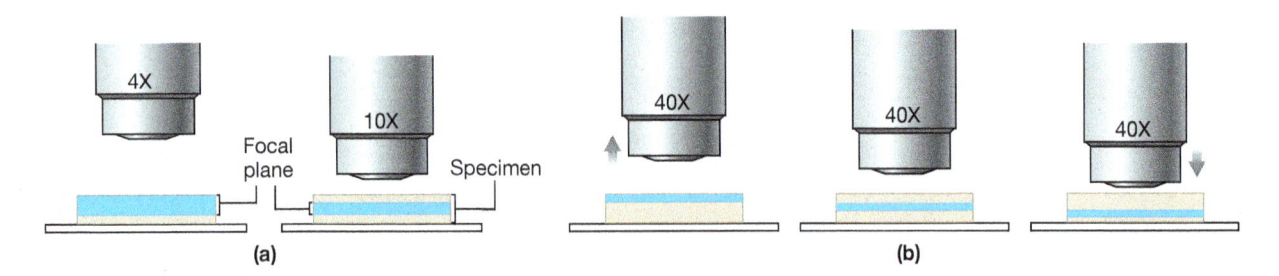

Figure 2.3 Focal Plane and Magnification

(a) Focal plane thickness, shown in blue, decreases as magnification increases. (b) The fine adjustment knob changes the distance between the objective and the stage, thereby moving the focal plane up and down through the specimen.

EXERCISE 2   Use of the Microscope   23

are many cell layers thick. By slowly rotating the fine adjustment knob back and forth, you will see different layers of the specimen come into, or go out of, focus.

QuickCheck Question

3.1 What is depth of field in a microscope?

3 Materials

- ☐ Compound microscope
- ☐ Slide of colored threads (or slide of hairs from different students if thread slides are unavailable)

Procedures

To see how depth of field works, you will examine a slide of overlapping colored threads (or hairs). While examining your slide, notice how the threads are layered and how much of each thread is in focus at each magnification.

1. Move the low-power objective lens into position, and place the slide on the stage with the threads over the aperture.
2. Use the coarse adjustment knob to bring the threads into focus. Find the area where the threads overlap.
3. Rotate the nosepiece to select the medium-power objective lens.
4. Use the fine adjustment knob to focus through the overlapping threads. After determining which thread is on top, which is in the middle, and which is on the bottom, write your observations in the space provided:

   Color of top thread: _____

   Color of middle thread: _____

   Color of bottom thread: _____ ■

LAB ACTIVITY 4   Relationship between Magnification and Field Diameter

At low magnification, the diameter of the field of view is large and most of the slide specimen is visible. As magnification increases, the field diameter decreases. This is because at higher power, the objective lens is closer to the slide and magnifies a smaller area. Figure 2.4 reviews the relationship between magnification and field diameter.

Figure 2.4
Magnification and Field Diameter

Each circle on the slide illustrates the field diameter for a particular magnification; the corresponding circle outside the slide represents that magnification.

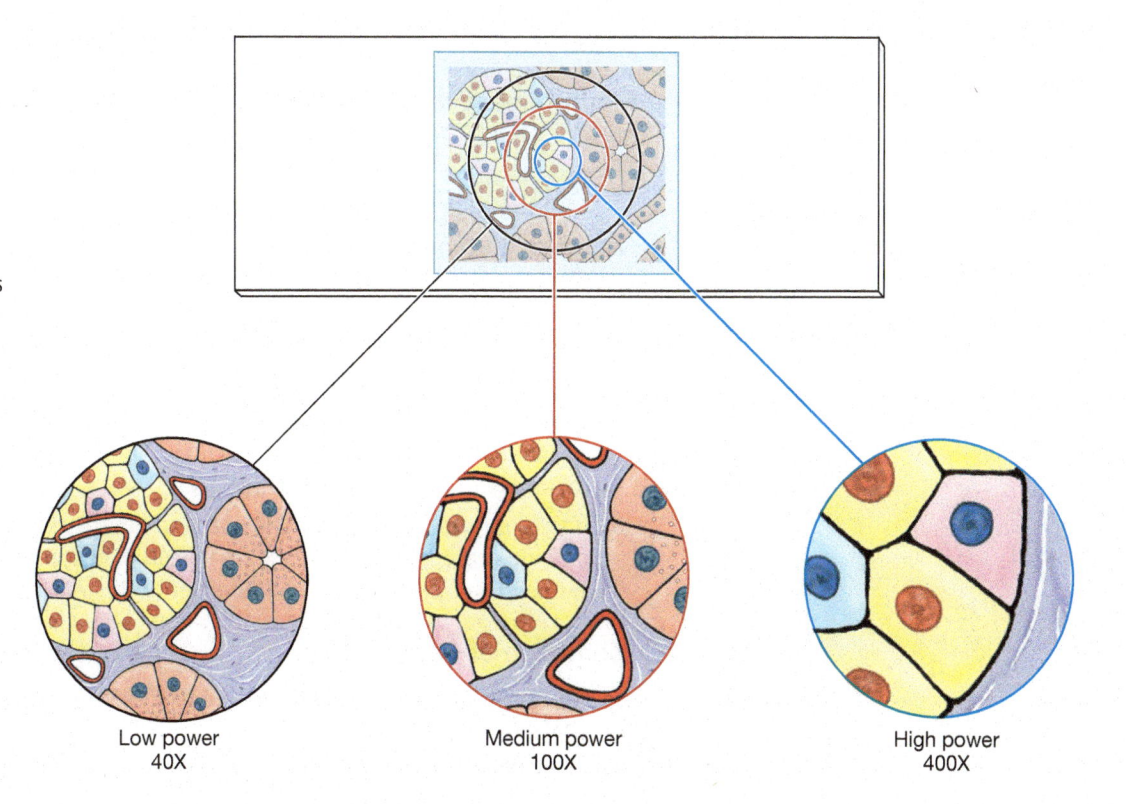

Low power
40X

Medium power
100X

High power
400X

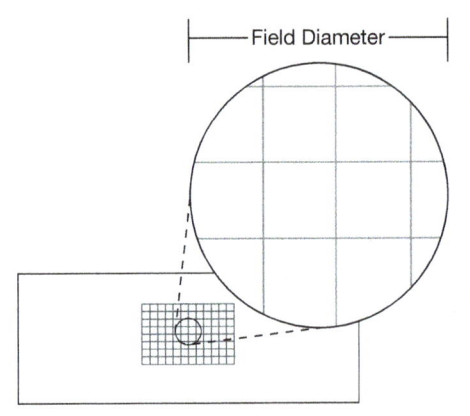

Figure 2.5 Calculation of Field Diameter Using Millimeter Graph Paper In this sample, the field is approximately 3.5 mm in diameter.

Field diameter at low and medium magnifications can be measured using millimeter graph paper glued to a microscope slide. By aligning a vertical marking on the paper with the edge of the field and then counting the number of millimeter (mm) squares across the field, you can determine the diameter (Figure 2.5). Knowing the diameter of the field of view enables you to estimate the actual size of an object. For example, if the field diameter is 4 mm and an object occupies one-half of the field, the object is approximately 2 mm wide.

Once you know the field diameter for one magnification—this is lens A in the following formula—you can calculate the field diameter for other magnifications (lens B) using the formula:

$$\text{Field diameter of lens B} = \frac{\text{Field diameter of A} \times \text{total magnification of lens A}}{\text{Total magnification of lens B}}$$

Making Connections

Field Diameter

You can demonstrate the relationship between magnification and field diameter by curling your fingers until the thumb of each hand overlaps the index and middle fingers of the same hand. The space enclosed by the curled fingers of each hand forms the barrel of a "lens". Place these two "lenses" to your eyes, and while sitting up straight in your chair, look at this page. Notice that you can see the entire page at this "low magnification". Now slowly bend forward until the "lenses" are just a few inches away from the page. In this "high magnification" view, the field of view is much smaller, and you can see only part of the page.

QuickCheck Question

4.1 What happens to the field diameter as magnification increases?

4 Materials

- ☐ Compound microscope
- ☐ Graph paper slides
- ☐ Practice slides (epithelium or cartilage recommended)

Procedures

Measurement of Field Diameter

1. Place the graph paper slide on the microscope stage and focus at low magnification. Position the slide so that a vertical line on the paper lines up with the edge of the field.
2. Count the number of millimeters across the field to measure the field diameter. Record this value in Table 2.1.
3. Use your low-power measured field diameter in the formula provided earlier to calculate the field diameter for the microscope set at medium power, and record this value in column 6 of Table 2.1.
4. Use your low-power measured field diameter in the formula provided earlier to calculate the field diameter for the microscope set at high power, and record this value in column 6 of Table 2.1.
5. If your microscope has an oil-immersion objective lens, use the formula provided earlier and any of the three field diameter values you listed in Table 2.1 to calculate the field diameter of the oil-immersion lens.

Estimation of Cell Size

1. On one of the practice slides chosen by your instructor, observe the cells at medium magnification.
2. Estimate the size of some cells using your field diameter measurements.

Name _____

Date _____

Section _____

LAB REPORT

EXERCISE 2

Use of the Microscope

A. Matching

Match each microscope part on the left with the correct description on the right.

_____	1. ocular lens	A.	used for precise focusing
_____	2. aperture	B.	lower support of microscope
_____	3. body tube	C.	narrows the beam of light
_____	4. mechanical stage	D.	hole in stage
_____	5. fine adjustment knob	E.	used only at low power
_____	6. base	F.	has knobs to move slide
_____	7. objective lens	G.	special paper for cleaning
_____	8. coarse adjustment knob	H.	eyepiece
_____	9. condenser	I.	holds ocular lens
_____	10. lens paper	J.	lens attached to nosepiece

B. Labeling

Label the parts of the microscope in Figure 2.6.

C. Short-Answer Questions

1. Which parts of a microscope are used to regulate the intensity and contrast of light? What does each of these parts do?

2. How is magnification controlled in a microscope?

3. Why should you always view a slide at low power first?

4. Briefly explain how to care for a microscope.

25

EXERCISE 2

LAB REPORT

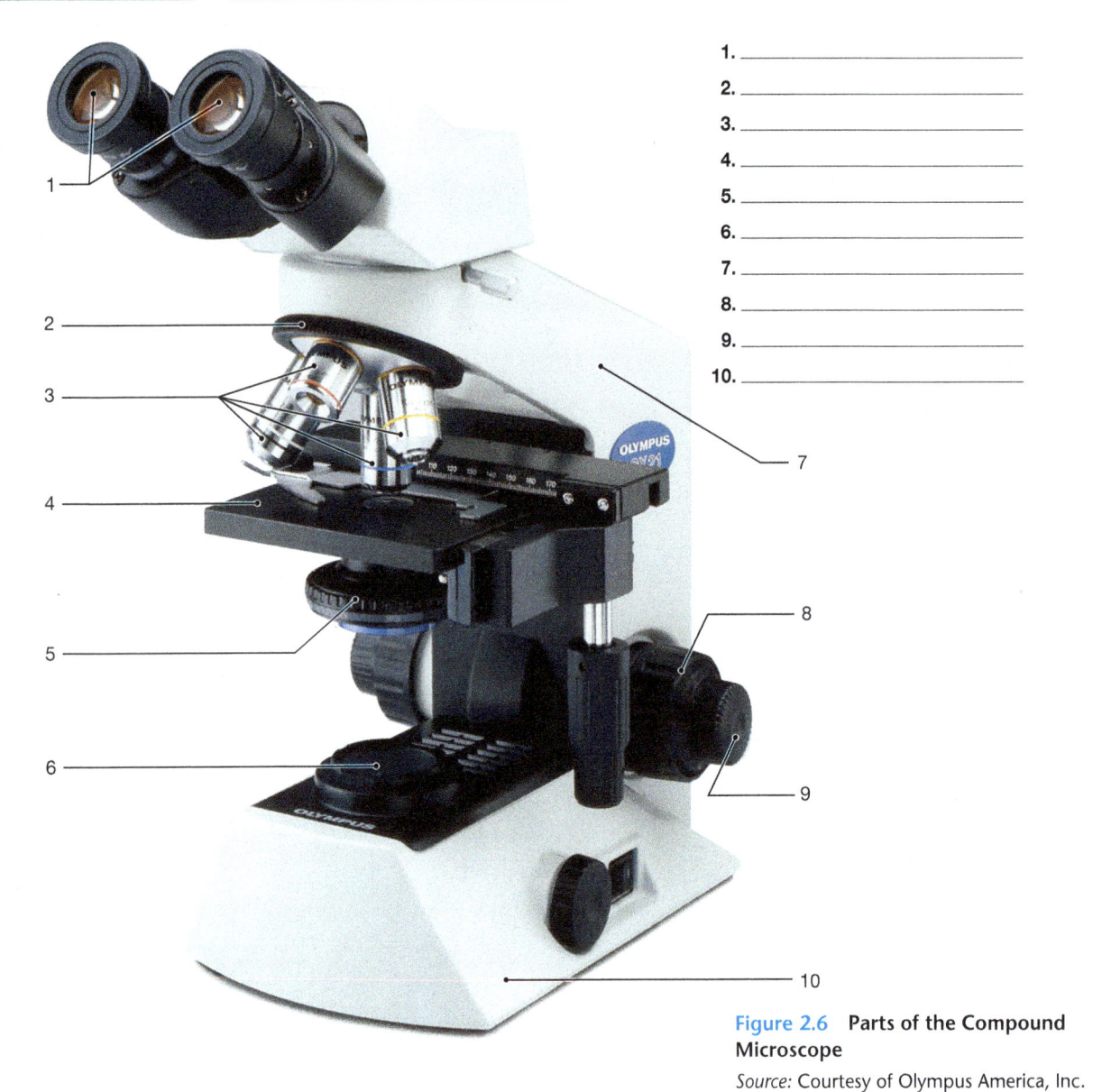

1. _____
2. _____
3. _____
4. _____
5. _____
6. _____
7. _____
8. _____
9. _____
10. _____

Figure 2.6 Parts of the Compound Microscope
Source: Courtesy of Olympus America, Inc.

D. Analysis and Application

1. You are looking at a slide under the microscope and observe a cell that occupies one-quarter of the field of view at high magnification. Use your field diameter calculation from Lab Activity 4 to estimate the size of this cell.

2. Describe how the field diameter changes when magnification is increased.

EXERCISE 3

Anatomy of the Cell and Cell Division

LAB ACTIVITIES

1 Anatomy of the Cell 27
2 Cell Division 31

OBJECTIVES

On completion of this exercise, you should be able to:

- Identify cell organelles on charts, models, and other laboratory material.
- Use the microscope to identify the nucleus and cell membrane of cells.
- State a function of each organelle.
- Discuss a cell's life cycle, including the stages of interphase and mitosis.
- Identify the stages of mitosis using the whitefish blastula slide.

Cells were first described in 1665 by a British scientist named Robert Hooke. He examined a thin slice of tree cork with a simple microscope and observed that the slice contained many small open spaces, which he called *cells*. Over the next two centuries, scientists examined cells from plants and animals and formulated the **cell theory**, which states that:

1. all plants and animals are composed of cells
2. all cells come from preexisting cells
3. cells are the smallest living units that perform physiological functions
4. each cell works to maintain itself at the cellular level
5. homeostasis is the result of the coordinated activities of all the cells in an organism.

In this exercise you will examine the structure of the cell and how cells reproduce to create the new cells the body needs for growth and repair.

LAB ACTIVITY 1 Anatomy of the Cell

Although the body is made up of a variety of cell types, a generalized composite cell is used to describe cell structure. All cells have an outer boundary, the **cell membrane**, also called the *plasma membrane*. This physical boundary separates the **extracellular fluid** that surrounds the cell from the cell interior. It regulates the movement of ions, molecules, and other substances into and out of the cell.

Inside the volume defined by the cell membrane is a central structure called the **nucleus** of the cell and other internal structures. Collectively, all of these internal structures are called **organelles** (or-gan-ELZ). The volume inside the cell membrane but outside the nucleus is referred to as the **cytoplasm**. This region is made up of solid components (all of the cell's organelles except the nucleus) suspended in a liquid called the **cytosol**.

Each organelle has a distinct anatomical organization and is specialized for a specific function (Figure 3.1). Organelles are grouped into two broad classes: nonmembranous and membranous. **Nonmembranous organelles** lack an outer membrane

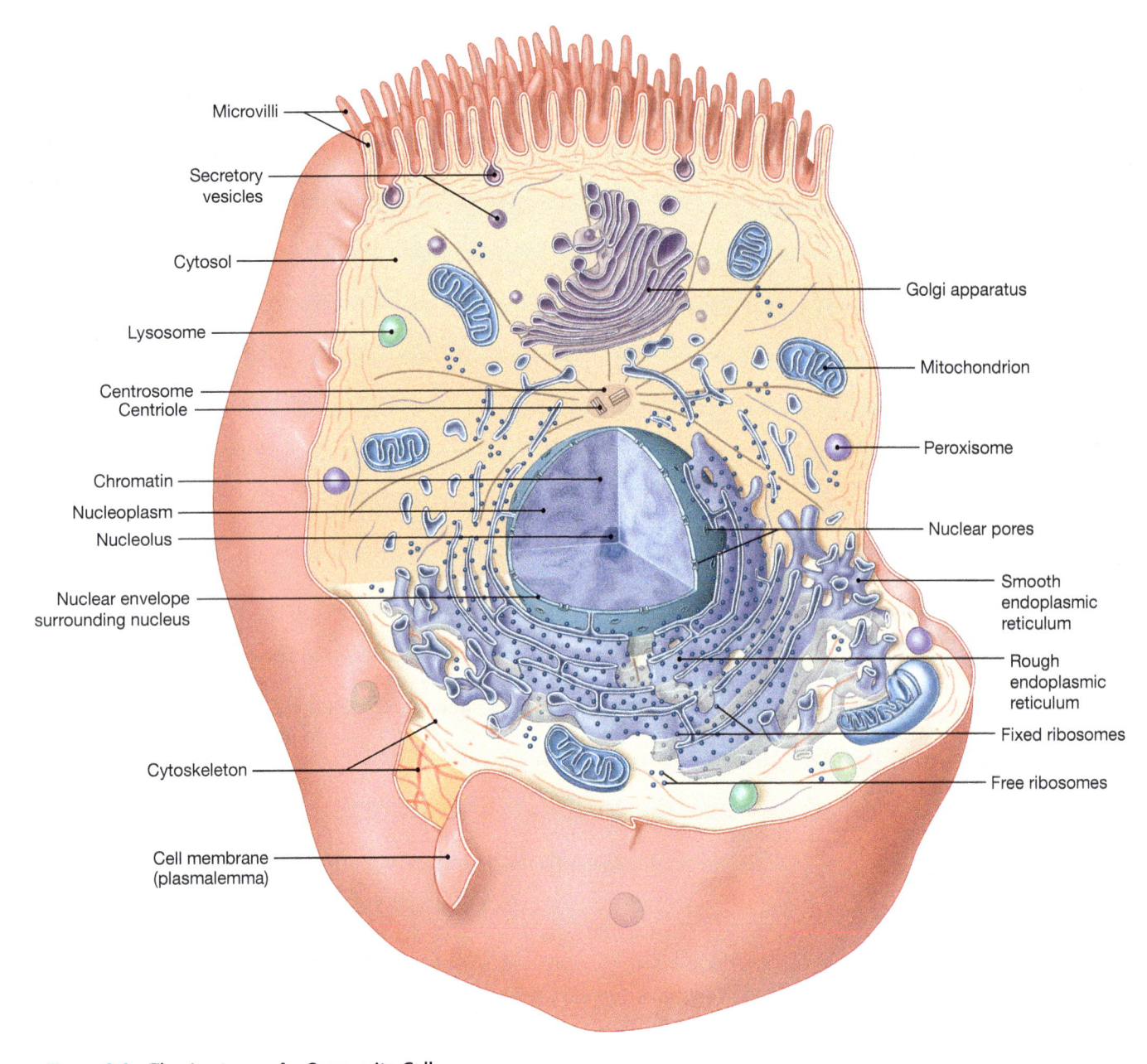

28   EXERCISE 3   Anatomy of the Cell and Cell Division

Figure 3.1 The Anatomy of a Composite Cell

and are directly exposed to the cytosol. Ribosomes, microvilli, centrioles, the cytoskeleton, cilia, and flagella are nonmembranous organelles. **Membranous organelles** are enclosed in a phospholipid membrane that isolates them from the cytosol. The nucleus, endoplasmic reticulum, Golgi apparatus, lysosomes, peroxisomes, and mitochondria are membranous organelles.

Keep in mind while studying cell models that most organelles are not visible with a light microscope. The nucleus typically is visible as a dark-stained oval. It encases and protects the **chromosomes**, which store genetic instructions for protein production by the cell.

Making Connections   **Information Linking**

Practice connecting information together rather than memorizing individual facts and terms. The cell is like a mass-production factory. Each organelle in the cell, like each station in the factory, has a specific task that integrates into the overall function of the cell. As you identify organelles on cell models, consider their functions. Once you are familiar with all the organelles, begin to associate them with one another as functional teams. For example, molecules made in the organelle known as the *endoplasmic reticulum* are transported to the organelle known as the *Golgi apparatus*, and so you should associate these two organelles with each other. Assimilating information in this way improves your ability to apply knowledge in a working context.

Nonmembranous Organelles

- **Microvilli** are projections of microfilament proteins in the cytosol that create small folds in the cell membrane. These folds increase the surface area of the cell. A larger membrane surface allows the cell to obtain extracellular materials, such as nutrients, at a greater rate.

- **Centrioles** are paired organelles that are composed of **microtubules**, which are small hollow tubes made of the protein **tubulin**. The **centrosome** is the area surrounding the pair of centrioles in a cell. When a cell is not dividing, it contains one pair of centrioles. When the cell divides, one of the first steps is replication of the centriole pair, so that the cell contains two pairs. The two centrioles in one pair migrate to one pole of the nucleus, and the two centrioles in the other pair migrate to the opposite pole of the nucleus. As the two pairs migrate, a series of **spindle fibers** radiate from them. The spindle fibers pull the chromosomes of the nucleus apart to give each of the forming daughter cells full genetic instructions.

- All cells have a **cytoskeleton** for structural support and anchorage of organelles. Like the centrioles, the cytoskeleton is made of microtubules.

- Many cells of the respiratory and reproductive system have nonmembranous organelles called **cilia**, which are short, hair-like projections that extend from the cell membrane. One type of human cell, the spermatozoon, has a single, long **flagellum** (fla-JEL-um) for locomotion.

- **Ribosomes** direct protein synthesis. Instructions for making a protein are stored in deoxyribonucleic acid (DNA) molecules in the cell nucleus. The "recipe" for a protein is called a *gene* and is copied from a segment of DNA onto a molecule of messenger RNA that then carries the instructions out of the nucleus and to the ribosome. Each ribosome consists of one large subunit and one small subunit. Both subunits clamp around the messenger RNA molecule to coordinate protein synthesis. Ribosomes occur either as **free ribosomes** in the cytoplasm or as **fixed ribosomes** attached to the endoplasmic reticulum (ER).

Membranous Organelles

- The **nucleus** controls the activities of the cell, such as protein synthesis, gene action, cell division, and metabolic rate. The material responsible for the dark appearance of the nucleus in a stained specimen is a collection of loosely-coiled chromosomes called **chromatin**. A **nuclear envelope** surrounds the nuclear material and contains pores through which instruction molecules from the nucleus pass into the cytosol. A darker-stained region inside the nucleus, the **nucleolus**, produces ribosomal RNA molecules for the creation of ribosomes.

- Surrounding the nucleus is the **endoplasmic reticulum** (en-dō-PLAZ-mik re-TIK-ū-lum). Two types of ER are found in the cell: **rough ER**, which has ribosomes attached to its surface; and **smooth ER**, which lacks ribosomes. Generally, the ER functions in the synthesis of organic molecules, transport of materials within the cell, and storage of molecules. Materials in the ER may pass into the Golgi apparatus for eventual transport out of the cell. Proteins produced by ribosomes on the rough ER surface enter the rough ER and assume its complex folded shape. Smooth ER is involved in the synthesis of cholesterol and phospholipids, and in reproductive cells it produces sex hormones. In liver cells, it synthesizes and stores glycogen, while in muscle and nerve cells it stores calcium ions. Calcium ions are stored in the smooth ER in muscle, nerve, and other types of cells.

- The **Golgi** (GŌL-jē) **apparatus** is a series of flattened disks that adjoin the ER. The ER can pass protein molecules to the Golgi apparatus for modification and secretion. Cell products such as mucus are synthesized, packaged, and secreted by the Golgi apparatus. A **vesicle** is a small membranous sac that forms either at the cell membrane or in the cytoplasm and then fuses with the cell membrane. In a process called **exocytosis**, small **secretory vesicles** pinch off from the Golgi disks, fuse with the cell membrane, and then rupture to release their contents into the extracellular fluid. The phospholipid membranes of the empty vesicles contribute to the renewal of the cell membrane.

- **Lysosomes** (LĪ-sō-sōm; *lyso-*, dissolution + *soma*, body) are vesicles produced by the Golgi apparatus. They are filled with powerful enzymes that digest worn-out cell components and destroy microbes. As certain organelles become worn out, lysosomes dissolve them, and some of the materials are used to rebuild the organelles. White blood cells trap bacteria with cell membrane extensions and pinch the membrane inward to release a vesicle inside the cell. Lysosomes fuse with the vesicle and release enzymes to digest the bacteria. Injury to a cell may result in the rupture of lysosomes, followed by destruction, or **autolysis**, of the cell. Autolysis is implicated in the aging of cells, which results from the accumulation of lysosomal enzymes in the cytosol.

- **Peroxisomes** are vesicles filled with enzymes that break down fatty acids and other organic molecules. Metabolism of organic molecules can produce toxins, such as hydrogen peroxide (H_2O_2), that damage the cell. Additional enzymes metabolize the hydrogen peroxide to oxygen and water.

- **Mitochondria** (mī-tō-KON-drē-uh) produce energy for the cell's use. Each mitochondrion is wrapped in a double-layered phospholipid membrane. The inner membrane is folded into fingerlike projections called **cristae** (the singular is *crista*). The region inside the double-layered membrane is called the **matrix** of the mitochondrion. To provide the cell with energy, molecules from nutrients are passed along a series of metabolic enzymes in the cristae to produce a molecule called *adenosine triphosphate (ATP)*, the energy currency of the cell. The abundance of mitochondria varies greatly among cell types. Muscle and nerve cells have large numbers of mitochondria that supply energy for contraction and generate nerve impulses, respectively. Mature red blood cells lack mitochondria and subsequently have a low metabolic rate.

Clinical Application   **Maternal Mitochondria**

Mitochondria are unique in that they contain mitochondrial DNA and evidently do not rely on the DNA in the cell nucleus for replication instructions. Your mitochondria were inherited from your mother's egg; the sperm from your father contributed no mitochondria and few other organelles. Thus, if a man inherited a mitochondrial disorder from his mother, he would not pass the disorder on to his children because their mitochondria are inherited from their mother.

QuickCheck Questions

1.1 What is the function of the cell membrane?

1.2 What are the two major categories of organelles?

1.3 Which organelles are involved in the production of protein molecules?

1 Materials

- ☐ Cell models and charts
- ☐ Toothpicks
- ☐ Microscope slide and coverslip
- ☐ Physiological saline in dropper bottle
- ☐ Iodine stain or methylene blue stain
- ☐ Compound microscope

Procedures

1. Review the nonmembranous and membranous organelles in Figure 3.1.
2. Identify each organelle on a cell model.
3. Prepare a wet-mount slide from cells of the inner lining of your cheek:
   a. Place a drop of saline on a microscope slide.
   b. Gently scrape the inside of your cheek with the blunt end of a toothpick.
   c. Stir the scraping into the drop of saline on the slide.
   d. Add one drop of stain, carefully stir again with the same toothpick, and add a coverslip.
   e. Dispose of your used toothpick in a biohazard bag as indicated by your instructor.
4. Examine your cheek cell slide at low power and note the many flattened epithelial cells. These cells are thin and often become folded on the slide.
5. Increase magnification to medium power and observe individual cells. Identify the nucleus, cytoplasm, and cell membrane of a cell. ■

LAB ACTIVITY 2 Cell Division

Cells must reproduce for an organism to grow and repair damaged tissue. During cell reproduction, a cell divides its genes equally and then splits into two identical cells. The division involves two major events: mitosis and cytokinesis. During **mitosis** (mī-TŌ-sis), the chromatin in the nucleus condenses into chromosomes and is equally divided between the two forming cells. Toward the end of mitosis, **cytokinesis** (sī-tō-ki-NĒ-sis; *cyto-*, cell + *kinesis*, motion) separates the cytoplasm to produce the two daughter cells. The daughter cells have the same number of chromosomes as the parent cell.

Interphase

During most of a cell's life cycle, it is not dividing and is in **interphase**. This is not a resting period for the cell, however, because during this phase the cell carries out various functions and prepares for the next cell division. Distinct phases occur during interphase, each related to cell activity (Figure 3.2). At this time the nucleus is visible, as is the darker nucleolus. Replication of DNA occurs during the **S phase**. After DNA replication, each chromosome is double-stranded and consists of two **chromatids**; one chromatid is the original strand, and the other is an identical copy. The chromatids are held together by a **centromere**.

Mitosis

The **M phase** of the cell cycle is the time of mitosis, during which the nuclear material divides. After chromosomes are duplicated in the S phase of interphase, the double-stranded chromosomes migrate to the middle of the cell, and spindle fibers attach to each chromatid. Chromosomes are divided when the spindle fibers drag sister chromatids to opposite ends of the cell. The division is complete when the cell undergoes cytokinesis and pinches inward to distribute the cytosol and chromosomes into two new daughter cells.

The four stages of mitosis are prophase, metaphase, anaphase, and telophase. Telophase and the latter part of anaphase are together referred to as cytokinesis.

- **Prophase:** Mitosis starts with prophase (PRŌ-fāz; *pro-*, before), when chromosomes become visible in the nucleus (Figure 3.3). In early prophase, the chromosomes are long and disorganized, but as prophase continues the nuclear

Figure 3.2 The Cell Life Cycle
Interphase (green chart sections, purple background) comprises the majority of the cycle, with mitosis (tan chart sections, blue background) taking place about one-fourth of the time.

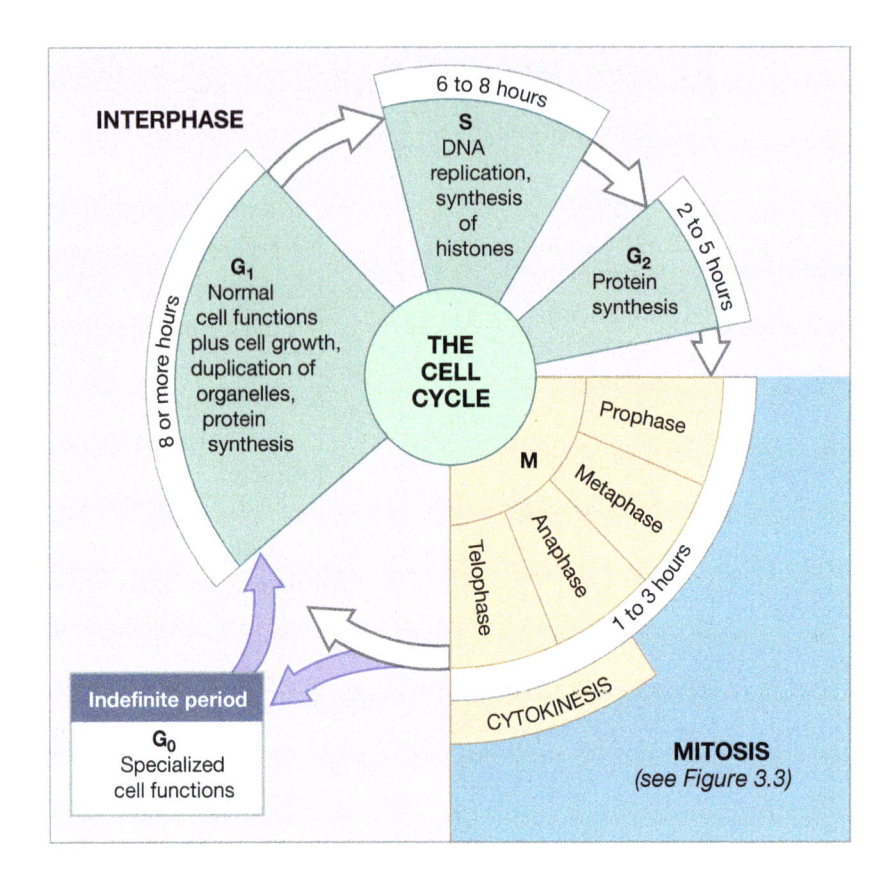

envelope breaks down, and the chromosomes shorten and move toward the middle of the cell. In the cytosol, the two centriole pairs begin moving to opposite sides of the cell. Between the centrioles, microtubules fan out as spindle fibers and extend across the cell.

- **Metaphase:** Metaphase (MET-a-fāz; *meta-*, change) occurs when the chromosomes line up in the middle of the cell at the **metaphase plate**. Spindle fibers extend across the cell from one pole to the other and attach to the centromeres of the chromosomes. The cell is now prepared to partition the genetic material and give rise to two new cells.

- **Anaphase:** The chromosome separation event defines anaphase (AN-a-fāz; *ana-*, apart). Spindle fibers pull apart the chromatids of a chromosome and drag them toward opposite poles of the cell. Once apart, individual chromatids are considered chromosomes. Toward the end of anaphase, cytokinesis begins as a **cleavage furrow** develops along the metaphase plate and the cell membrane pinches. Cytokinesis continues into the next stage of mitosis, telophase.

- **Telophase:** Mitosis nears completion during telophase (TĒL-ō-fāz; *telo-*, end), as each batch of chromosomes unwinds inside a newly-formed nuclear envelope. Each daughter cell has a set of organelles and a nucleus that contains a complete set of genes. Telophase ends as the cleavage furrow deepens along the metaphase plate and separates the cell into two identical daughter cells. These daughter cells are in interphase and, depending on their cell type, may divide again.

Clinical Application **Cell Division and Cancer**

A tumor is a mass of cells that are produced by uncontrolled cell division. The mass replaces normal cells, and cellular and tissue functions are compromised. If **metastasis** (me-TAS-ta-sis), which means spreading of abnormal cells, occurs, secondary tumors may develop. Cells that metastasize are often cancerous.

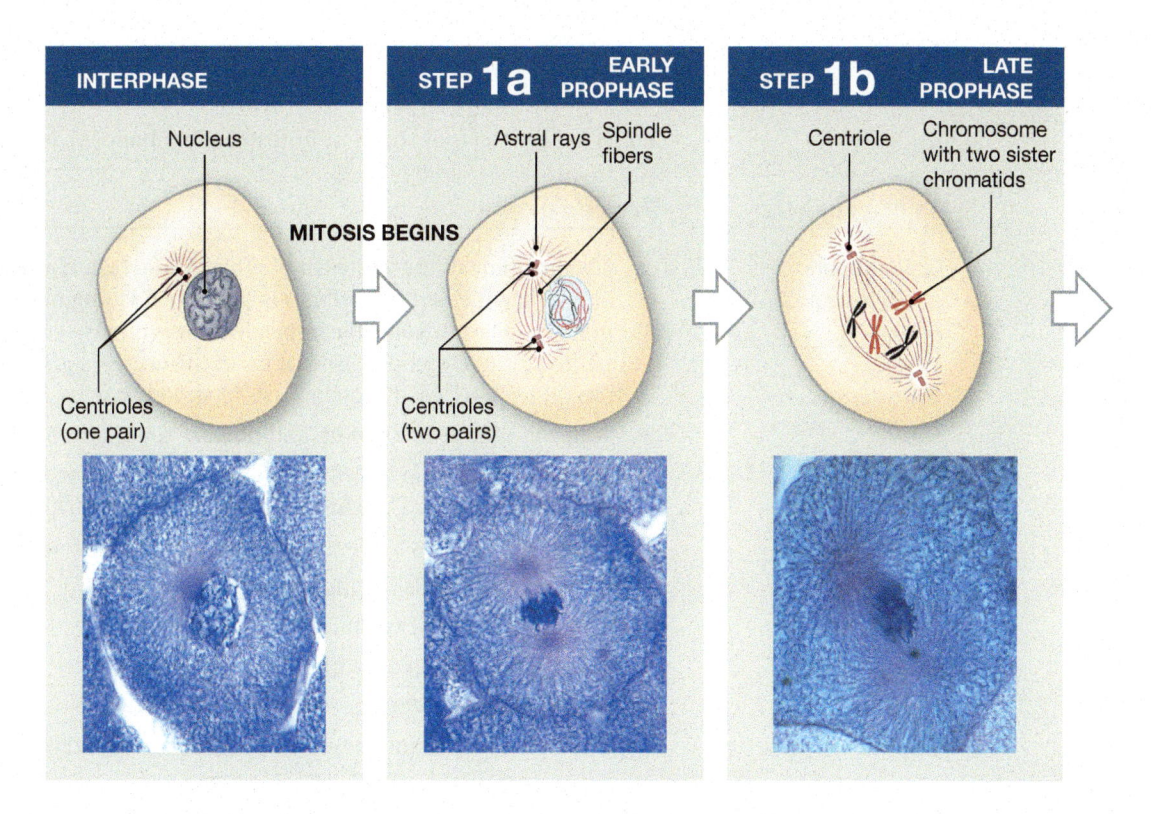

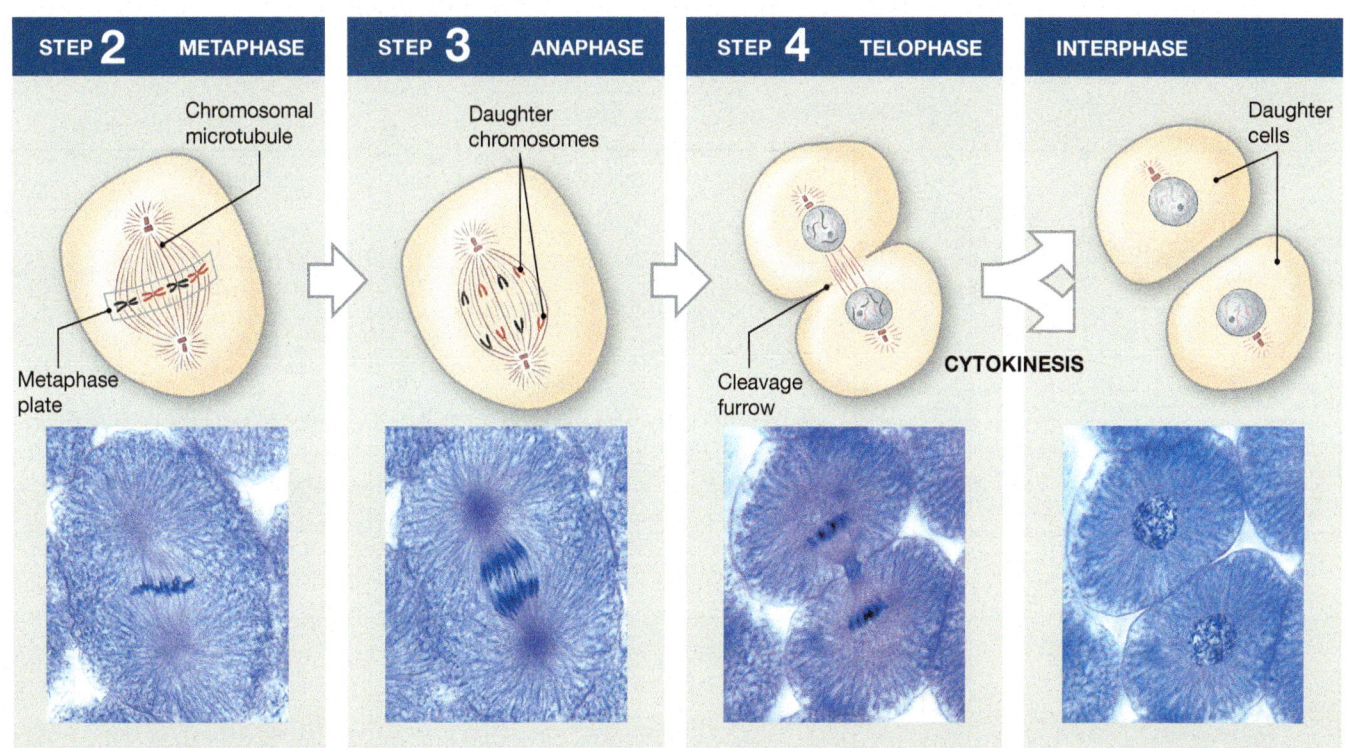

EXERCISE 3 Anatomy of the Cell and Cell Division   33

Figure 3.3 The Steps of Cell Division
Diagrammatic and microscopic views of representative cells undergoing cell division. The sequence begins with the cell ending the interphase part of its cycle and entering the mitosis part. The cytokinesis part of mitosis ends the sequence, at which point the cells return to interphase, and the cycle repeats.

34 EXERCISE 3 Anatomy of the Cell and Cell Division

QuickCheck Questions

2.1 What must the cell do with the genetic material in the nucleus prior to mitosis?

2.2 What are the four stages of mitosis? What happens in each stage?

2 Materials

- ☐ Compound microscope
- ☐ Whitefish blastula slide

Procedures

1. Obtain a slide of the whitefish blastula. A **blastula** is a stage in early development when the embryo is a rapidly dividing mass of cells that is growing in size and, eventually, in complexity. The whitefish embryo is sectioned and stained for microscopic observation of the cells. A slide preparation usually has several sections, each showing cells in various stages of mitosis.
2. Scan the slide at low power, and observe the numerous cells of the blastula.
3. Slowly scan a group of cells at medium power to locate a nucleus, centrioles, and spindle fibers. The chromosomes appear as dark, thick structures.
4. Using Figure 3.3 as a reference, locate cells in the following phases:
   - Interphase with a distinct nucleus
   - Prophase with disorganized chromosomes
   - Metaphase with equatorial chromosomes attached to spindle fibers
   - Anaphase with chromosomes separating toward opposite poles
   - Telophase with nuclear envelope forming around each set of genetic material
   - Cytokinesis in late anaphase and telophase
5. Draw and label cells in each stage of mitosis in the space provided: ■

LAB REPORT

EXERCISE 3

Anatomy of the Cell and Cell Division

Name _____

Date _____

Section _____

A. Matching

Match each cellular structure on the left with the correct description on the right.

_____ 1. cell membrane
_____ 2. centrioles
_____ 3. ribosome
_____ 4. membranous organelle
_____ 5. chromatid
_____ 6. lysosome
_____ 7. nucleus
_____ 8. cytoplasm
_____ 9. cristae
_____ 10. nonmembranous organelle
_____ 11. cytosol
_____ 12. cilia

A. copy of a chromosome
B. contains chromatin
C. short, hair-like cellular extensions
D. organelle that is exposed to cytosol
E. intracellular liquid
F. involved in mitosis
G. folds of the inner mitochondrial membrane
H. composed of a phospholipid bilayer
I. organelle isolated from cytosol
J. site for protein synthesis
K. vesicle containing powerful digestive enzymes
L. intracellular liquid plus all organelles except nucleus

B. Labeling

Label the organelles of the cell in Figure 3.4.

C. Fill in the Blanks

Use the correct term to answer each statement.

1. Replication of genetic material results in chromosomes that consist of two _____.
2. A cell in metaphase has chromosomes located in the _____ of the cell.
3. Division of the cytoplasm to produce two daughter cells is called _____.
4. Double-stranded chromosomes separate during the _____ stage of mitosis.
5. During interphase, DNA replication occurs in the _____ phase.
6. Microtubules called _____ attach to chromatids and pull them apart.
7. Chromosomes become visible during the _____ stage of mitosis.
8. The last stage of mitosis is _____.
9. Division of the nuclear material is called _____.
10. Matching chromatids are held together by a _____.

EXERCISE 3 — LAB REPORT

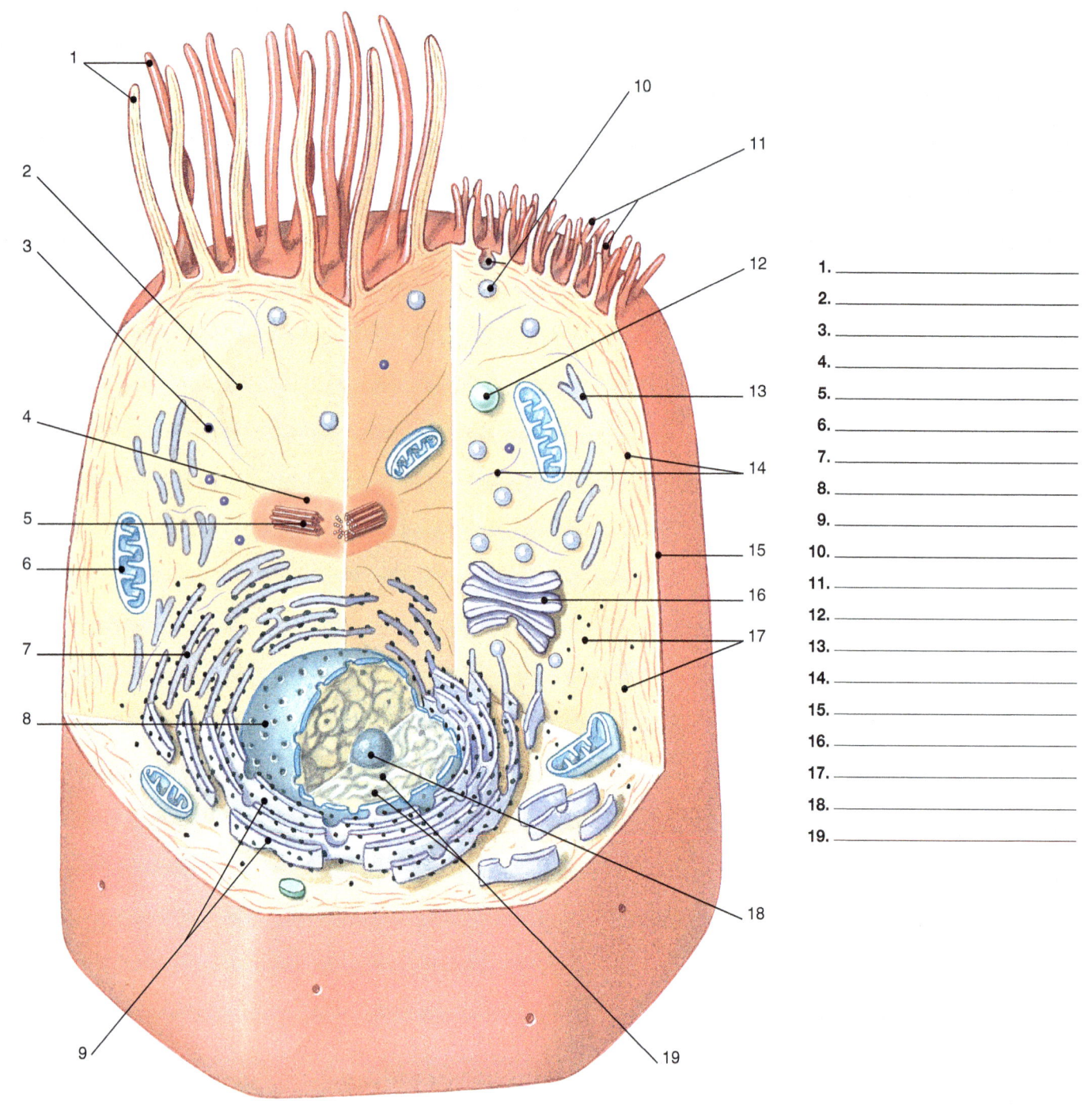

Figure 3.4 The Organelles of a Composite Cell

1. _____
2. _____
3. _____
4. _____
5. _____
6. _____
7. _____
8. _____
9. _____
10. _____
11. _____
12. _____
13. _____
14. _____
15. _____
16. _____
17. _____
18. _____
19. _____

36

LAB REPORT EXERCISE 3

D. Short-Answer Questions

1. What are the major functions of the nucleus of a cell?

2. What is the purpose of cell division?

3. What is the function of the spindle fibers during mitosis?

4. Which organelles were visible in the cells of the cheek cell slide?

E. Drawing

Draw and label a cell that contains six chromosomes and is in metaphase. How will the chromosomes appear during anaphase and telophase?

EXERCISE 3

LAB REPORT

F. **Analysis and Application**

1. Describe how the nucleus, ribosomes, rough ER, Golgi apparatus, and cell membrane interact to produce a protein molecule and then release it into the extracellular fluid.

2. Lysosomes are sometimes referred to as "suicide bags". Describe what would happen to a cell if its lysosomes ruptured.

3. What happens in a cell during the S phase of interphase?

4. Describe how chromosomes are evenly divided during mitosis.

EXERCISE 4

Tissues

LAB ACTIVITIES

1. Epithelia 39
2. Connective Tissue 46
3. Muscle Tissue 56
4. Neural Tissue 59

OBJECTIVES

On completion of this exercise, you should be able to:

- List the characteristics used to classify epithelia.
- Describe the location and function of each type of epithelium and identify each type under the microscope.
- List the major types of connective tissue and the characteristics of each.
- Describe the location and function of each type of connective tissue and identify each type under the microscope.
- Name the three types of muscle tissue, describe one function of each, and identify each type under the microscope.
- Describe the two basic types of cells in neural tissue.
- Identify a neuron under the microscope.

As noted in Exercise 1, **histology** is the study of tissues. A **tissue** is a group of similar cells that work together to accomplish a specific function. An understanding of histology is vital for the study of organ function. The stomach, for example, plays major digestive roles as it secretes digestive juices and is involved in the mixing and movement of food. Each of these functions is performed by a specialized tissue in the stomach.

Cells and their secretions compose the various tissues of the body. There are four major tissue categories: **epithelial**, **connective**, **muscle**, and **neural**. Each category includes specialized tissues that have specific locations and functions. Many tissues join with other tissues to form organs, such as the stomach, a muscle, or a bone.

As you observe tissues under the microscope in the following activities, it is important to scan each slide at low power because a single slide may contain several tissues, and you must survey the specimen to locate a particular tissue. Once you have located the tissue, increase the magnification and observe the individual cells of the tissue.

LAB ACTIVITY 1 — Epithelia

Epithelia (e-pi-THĒ-lē-a; singular **epithelium**), also called **epithelial tissues**, line and cover organs and their internal passageways. The integumentary, respiratory, digestive, reproductive, and urinary systems all are lined with an epithelium.

Epithelium is made up of sheets of cells, with the cells in a given sheet tightly joined together, like pieces in a jigsaw puzzle, by a variety of strong intercellular connections. An epithelium always has one surface where the cells are exposed either to the external environment or to an internal passageway or cavity; this surface is called the **free surface** of the epithelium. Because epithelia are surface and lining tissues, they do not contain blood vessels (they are **avascular**). The cells obtain nutrients and other necessary materials by diffusion of substances from connective

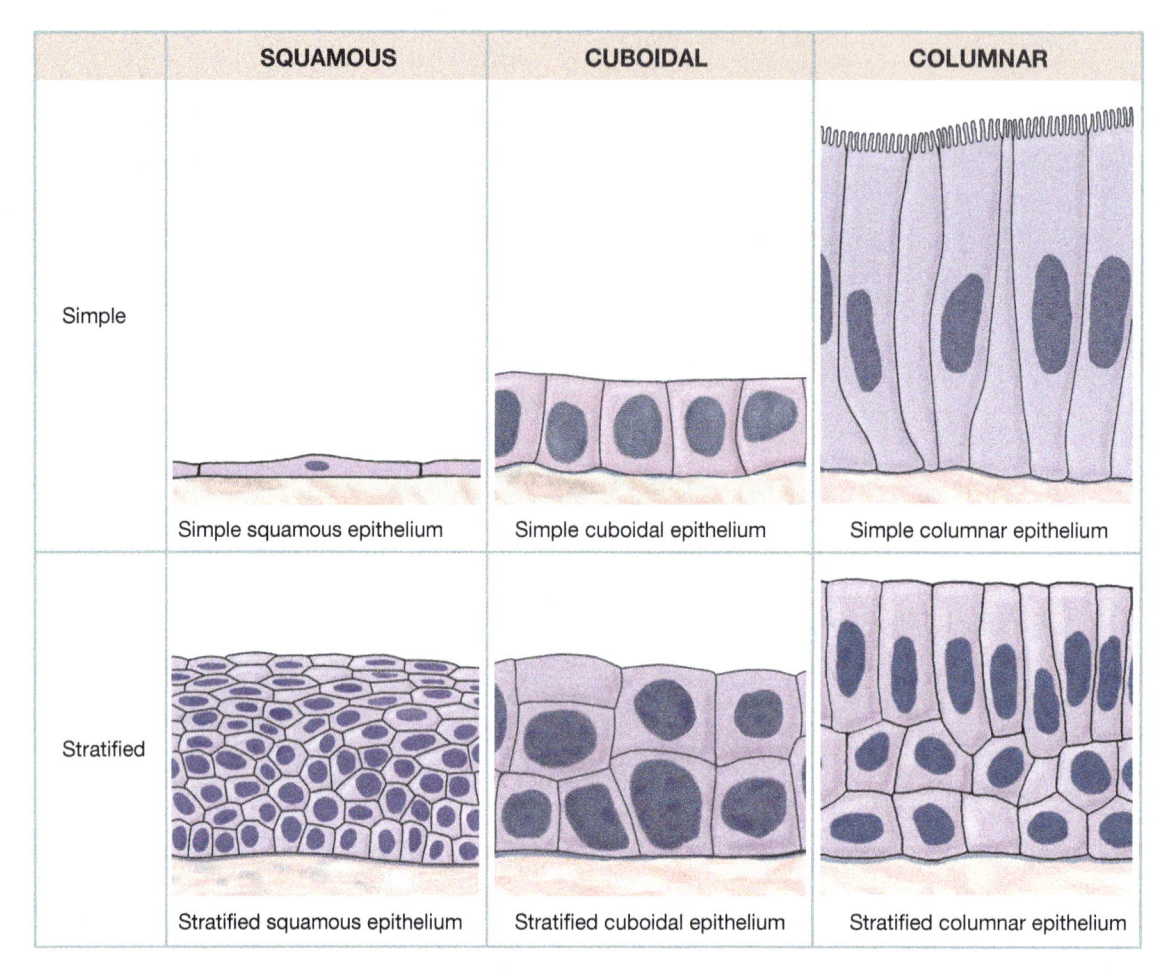

	SQUAMOUS	CUBOIDAL	COLUMNAR
Simple	Simple squamous epithelium	Simple cuboidal epithelium	Simple columnar epithelium
Stratified	Stratified squamous epithelium	Stratified cuboidal epithelium	Stratified columnar epithelium

Figure 4.1 **Classifying Epithelia**
All epithelia are classified by the number of cell layers (simple and stratified) and by the shape of the cells (squamous, cuboidal, and columnar).

tissue underlying the epithelia. Each epithelium is attached to the body by a **basal lamina** (LA-mi-nah; *lamina,* plate) located between the epithelium and its connective tissue layer. In some epithelial tissue, such as that associated with glands, the cells are short-lived, and **stem cells**, which are cells in constant mitosis, must constantly produce new epithelial cells to replenish the tissue.

Epithelia are classified and identified by the number of cell layers and by the general shape of the tissue cells (Figure 4.1). A **simple epithelium** has a single layer of cells and provides a thin surface for exchange of materials, such as the exchange of gases between the lungs and blood. At body surfaces exposed to the external environment, multiple layers of cells in **stratified epithelium** protect against friction, prevent dehydration, and keep microbes and chemicals from invading the body. A **transitional epithelium** is a special kind of stratified epithelium with cells of many shapes that permit the tissue to stretch and recoil. In a **pseudostratified epithelium,** all the cells touch the basal lamina, but the cells grow to different heights. Taller cells grow over shorter ones and cover them, thereby preventing them from reaching the free surface.

A second way of classifying epithelium is by cell shape. **Squamous** (SKWĀ-mus; *squama,* scale) epithelial cells are irregularly shaped, flat, and scale-like. These cells, depending on how they are organized, function either in protection or in secretion and diffusion. **Cuboidal** epithelial cells are cubic (that is, their cross-section is approximately square) and have a large nucleus. They are found in the tubules of the kidneys and in many glands, and can secrete and absorb materials across the tubular/glandular wall. **Columnar** epithelial cells are taller than they are wide.

Clinical Application **A Barrier Against Infection**

> Infectious organisms enter the body by penetrating both the lining epithelia and covering epithelia. AIDS, caused by HIV (human immunodeficiency virus), is transmitted when the virus enters a wound or abrasion in the epithelium. The purpose of **safe sex** is to maintain a barrier between your epithelia and other people's body fluids, which may contain **HIV**. The epithelial linings of the mouth, vagina, urethra, and rectum should all be protected to prevent exposure to a sexually transmitted disease.

Simple Epithelia

The main functions of simple epithelia are diffusion, absorption, and secretion. At the free surface, microvilli on the epithelial cell membrane increase the surface area available for absorption. To protect epithelia at the free surface, cells called **goblet cells** secrete mucus that coats the cells. Other cells in a simple epithelial layer have cilia that sweep the mucus along the free surface to remove debris.

- **Simple squamous epithelium** (Figure 4.2a) is a thin tissue that in a superficial preparation appears as a sheet of cells that look like ceramic floor tiles. In serous membranes, this tissue is called **mesothelium**. In locations where it lines blood vessels and the heart chambers, it is called **endothelium**. Simple squamous epithelium also constructs the thin walls of air sacs in the lungs, where gas exchange occurs.

- **Simple cuboidal epithelium** (Figure 4.2b) lines kidney tubules, the thyroid and other glands, and ducts. On slides of cuboidal epithelium from the kidney, the tubules sectioned longitudinally appear as two rows of square cells, in transverse sections the cuboidal cells are arranged in a ring to form the round wall of the tubule. Typically, the basal lamina is conspicuous in simple cuboidal epithelium.

- **Simple columnar epithelium** (Figure 4.2c) lines most of the digestive tract, the uterine tubes, and the renal collecting ducts. In the small intestine, the wall is folded and covered with simple columnar epithelium to increase the surface area available for digestion and absorption of nutrients. In the uterine tubes, the cilia transport released eggs to the uterus.

Making Connections **Looking at Epithelia**

> When observing epithelia microscopically, find the free surface of the tissue and then look on the opposite edge of the cells. The basal lamina is located right under this edge. It appears as a dark line between the epithelial cells and the connective tissue.

Stratified Epithelia

Stratified epithelia (Figure 4.3) are found in areas exposed to abrasion and friction, such as the body surface and upper digestive tract. When a stratified epithelium contains more than one type of epithelial cell, the type at the free surface determines the classification of the tissue.

- **Stratified squamous epithelium** (Figure 4.3a) forms the superficial region of the skin, called the **epidermis**. Stem cells produce new cells at the basal lamina and are pushed toward the free surface by the next group of new cells. The cells manufacture the protein **keratin** (KER-a-tin; *keros,* horn), which toughens the cells but also kills them. The cells then dehydrate and interlock into a broad sheet, forming a dry protective barrier against abrasion, friction, chemical exposure, and even infection. Stratified squamous epithelium of the skin is thus said to be **keratinized** and has a dry surface. Stratified squamous epithelium also lines the tongue, mouth, pharynx, esophagus, anus, and vagina. The epithelium in these regions is kept moist by lining cells on the tissue surface. This moist tissue is described as being **nonkeratinized** (mucosal type) stratified squamous epithelium.

SIMPLE SQUAMOUS EPITHELIUM

LOCATIONS: Mesothelia lining ventral body cavities; endothelia lining heart and blood vessels; portions of kidney tubules (thin sections of nephron loops); inner lining of cornea; alveoli of lungs

FUNCTIONS: Reduces friction; controls vessel permeability; performs absorption and secretion

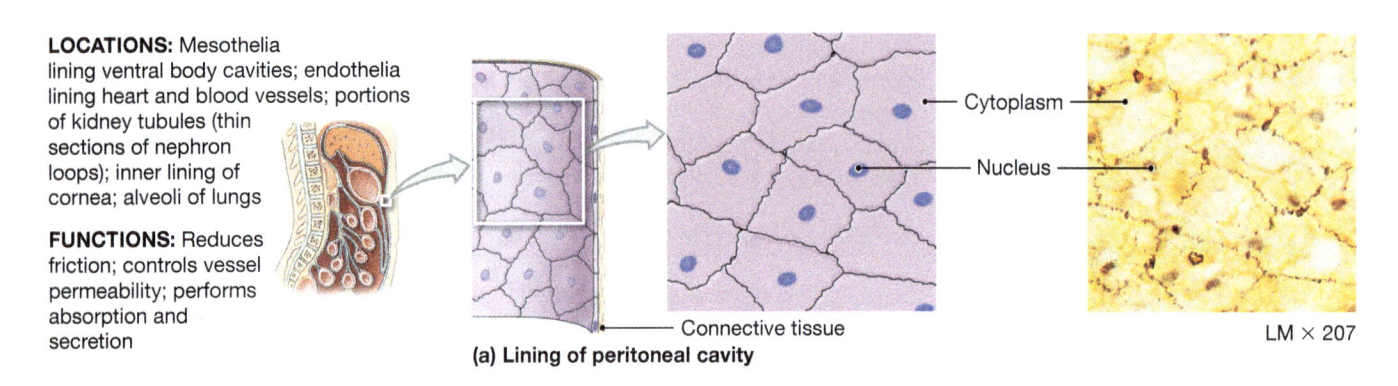

- Cytoplasm
- Nucleus
- Connective tissue

(a) Lining of peritoneal cavity

LM × 207

SIMPLE CUBOIDAL EPITHELIUM

LOCATIONS: Glands; ducts; portions of kidney tubules; thyroid gland

FUNCTIONS: Limited protection, secretion, absorption

- Connective tissue
- Nucleus
- Cuboidal cells
- Basal lamina

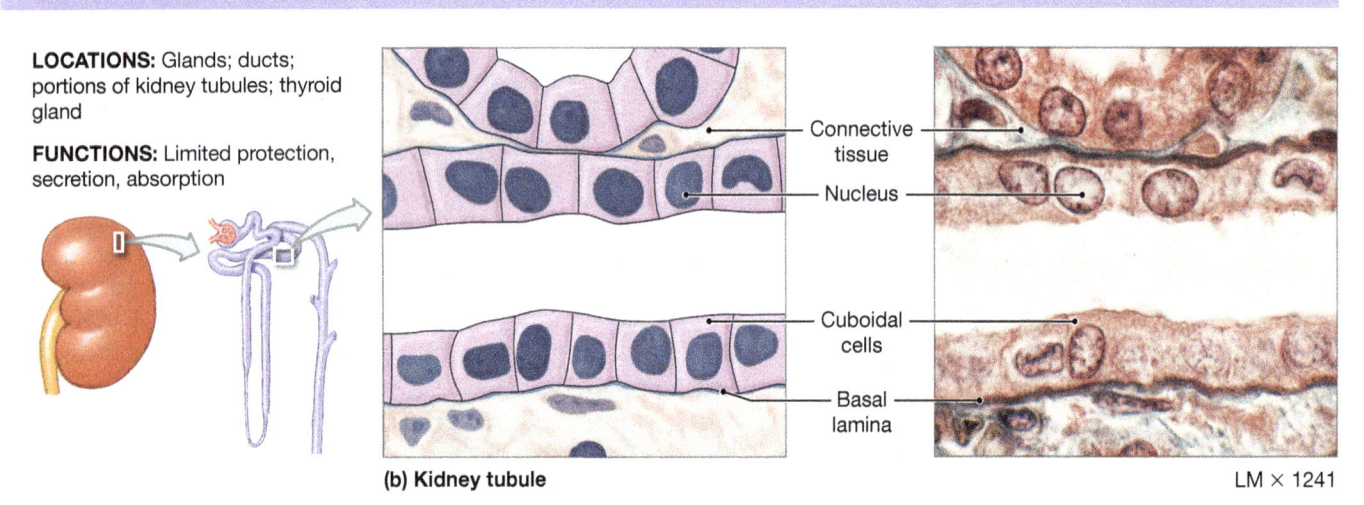

(b) Kidney tubule

LM × 1241

SIMPLE COLUMNAR EPITHELIUM

LOCATIONS: Lining of stomach, intestine, gallbladder, uterine tubes, and collecting ducts of kidneys

FUNCTIONS: Protection, secretion, absorption

- Microvilli
- Cytoplasm
- Nucleus
- Basal lamina
- Loose connective tissue

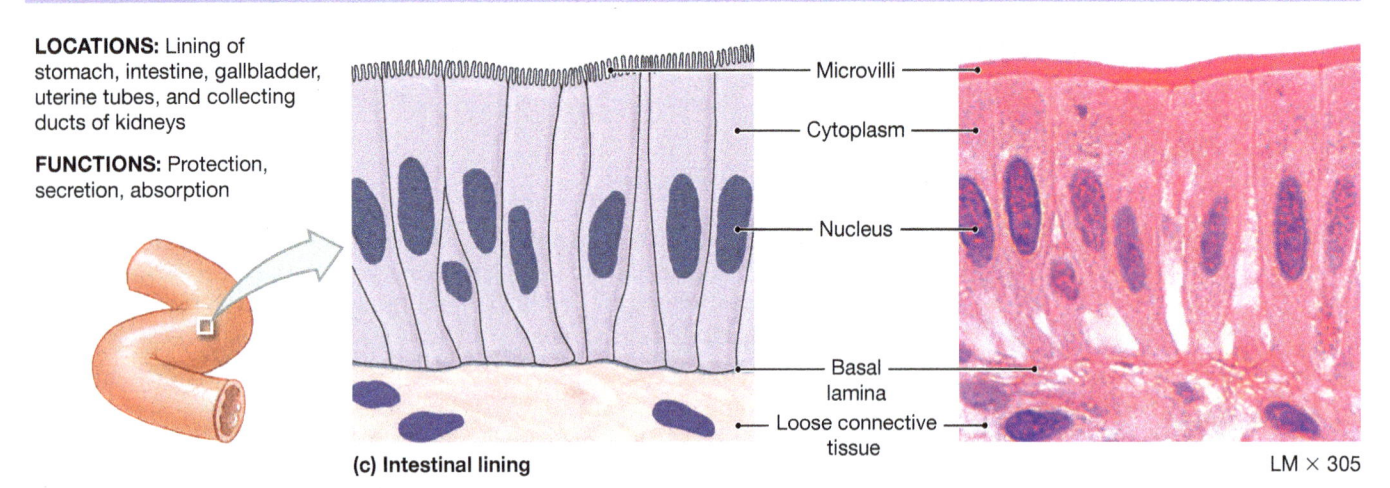

(c) Intestinal lining

LM × 305

Figure 4.2 Simple Epithelia

(a) Simple Squamous Epithelium. A superficial view of the simple squamous epithelium (mesothelium) that lines the peritoneal cavity. The three-dimensional drawing shows the epithelium in superficial and sectional views. (LM × 207) (b) Simple Cuboidal Epithelium. A section through the simple cuboidal epithelial cells of a kidney tubule. (LM × 1241) (c) Simple Columnar Epithelium. In the diagrammatic sketch, note the relationship between the height and width of each cell; the relative size, shape, and location of nuclei; and the distance between adjacent nuclei. Contrast these observations with the corresponding characteristics of simple cuboidal epithelium. (LM × 305)

STRATIFIED SQUAMOUS EPITHELIUM

LOCATIONS: Surface of skin; lining of mouth, throat, esophagus, rectum, anus, and vagina

FUNCTIONS: Provides physical protection against abrasion, pathogens, and chemical attack

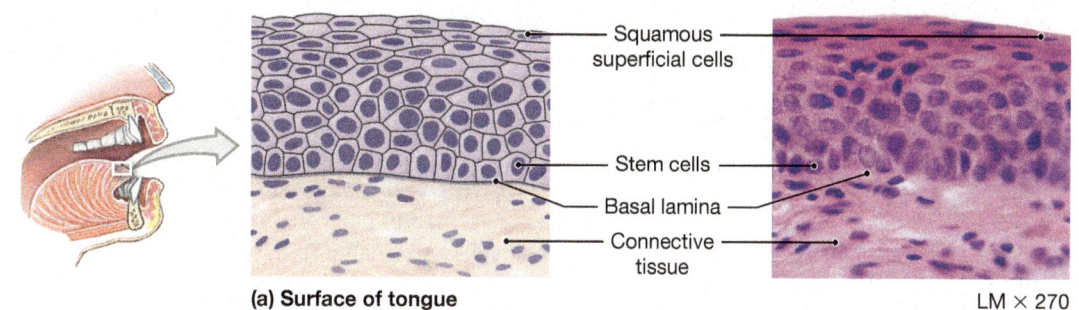

(a) Surface of tongue

- Squamous superficial cells
- Stem cells
- Basal lamina
- Connective tissue

LM × 270

STRATIFIED CUBOIDAL EPITHELIUM

LOCATIONS: Lining of some ducts (rare)

FUNCTIONS: Protection, secretion, absorption

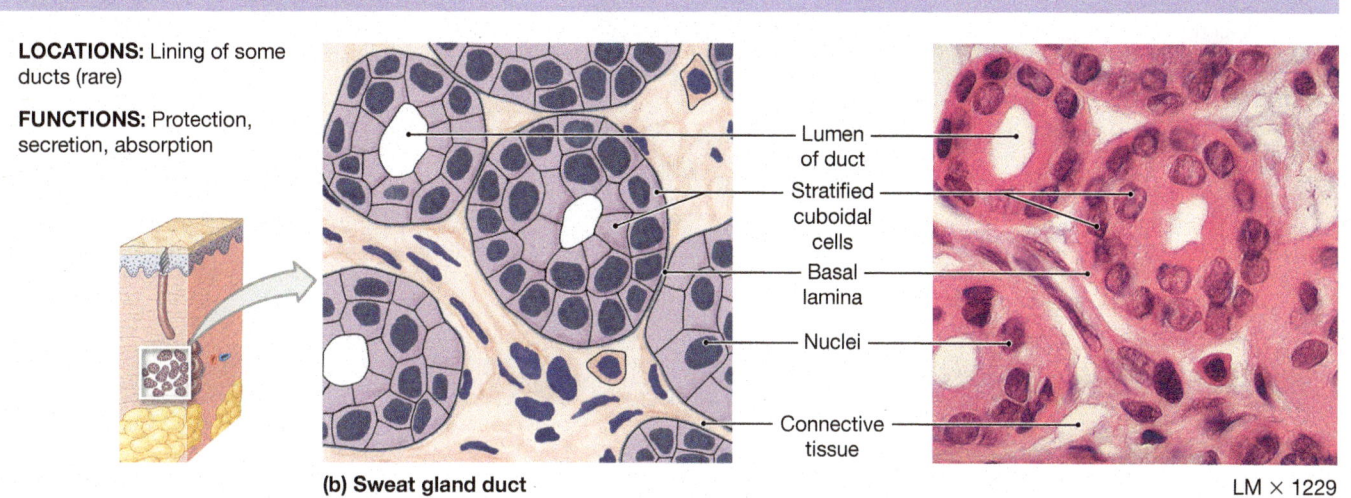

(b) Sweat gland duct

- Lumen of duct
- Stratified cuboidal cells
- Basal lamina
- Nuclei
- Connective tissue

LM × 1229

STRATIFIED COLUMNAR EPITHELIUM

LOCATIONS: Small areas of the pharynx, epiglottis, anus, mammary gland, salivary gland ducts, and urethra

FUNCTION: Protection

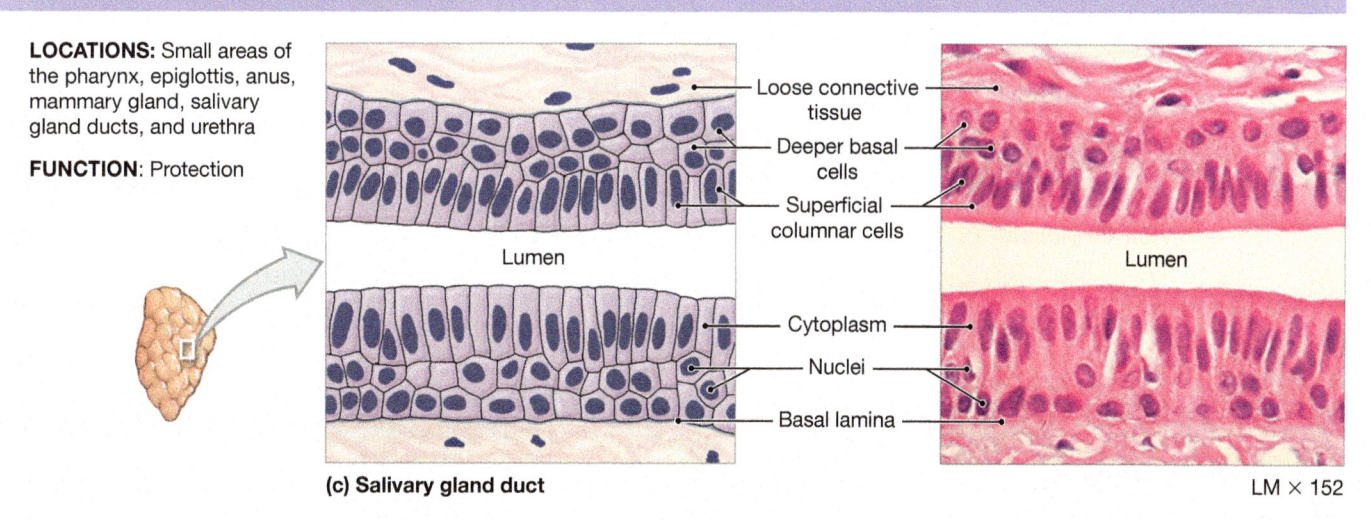

(c) Salivary gland duct

- Loose connective tissue
- Deeper basal cells
- Superficial columnar cells
- Lumen
- Cytoplasm
- Nuclei
- Basal lamina

LM × 152

Figure 4.3 Stratified Epithelia
(a) Stratified Squamous Epithelium. Sectional and diagrammatic views of the stratified squamous epithelium covering the tongue. (LM × 270) (b) Stratified Cuboidal Epithelium. A sectional view of the stratified cuboidal epithelium that lines a sweat gland duct in the skin. (LM × 1229) (c) Stratified Columnar Epithelium. A stratified columnar epithelium occurs along some large ducts, such as this salivary gland duct. (LM × 152)

44 EXERCISE 4 Tissues

- **Stratified cuboidal epithelium** (Figure 4.3b) is uncommon. It is found in the ducts of certain sweat glands.
- **Stratified columnar epithelium** (Figure 4.3c) is found in parts of the mammary glands, in salivary gland ducts, and in small regions of the pharynx, epiglottis, anus, and urethra.

Pseudostratified and Transitional Epithelia

- **Pseudostratified columnar epithelium** (Figure 4.4a) lines the nasal cavity, the trachea, bronchi, and parts of the male reproductive tract. The tissue has columnar cells and smaller stem cells, which replenish the tissue. It appears

PSEUDOSTRATIFIED CILIATED COLUMNAR EPITHELIUM

LOCATIONS: Lining of nasal cavity, trachea, and bronchi; portions of male reproductive tract

FUNCTIONS: Protection, secretion

(a) Trachea

Labels: Cilia, Cytoplasm, Nuclei, Basal lamina, Loose connective tissue

LM × 343

TRANSITIONAL EPITHELIUM

LOCATIONS: Urinary bladder; renal pelvis; ureters

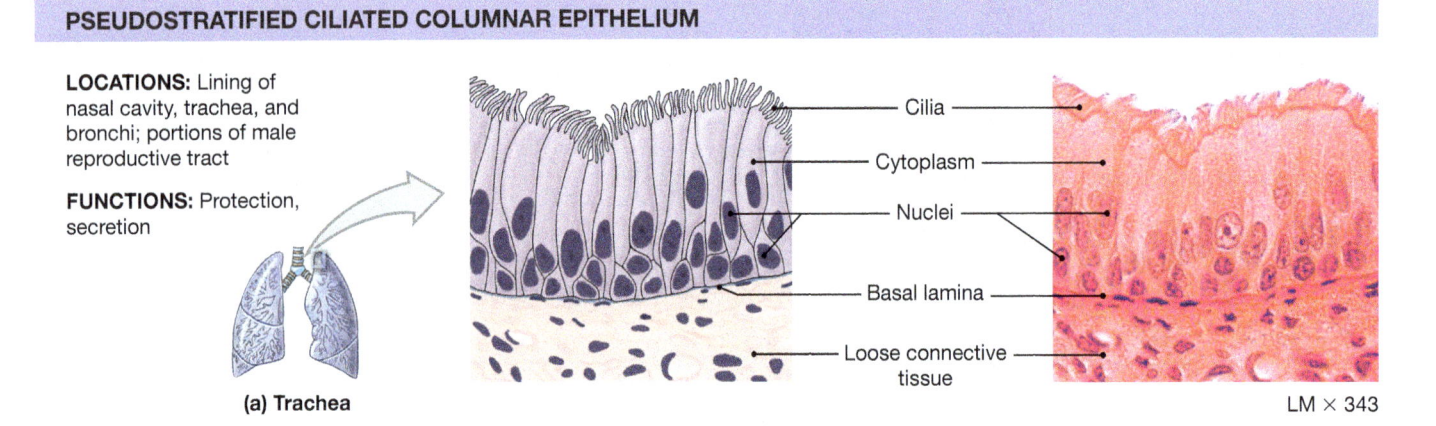

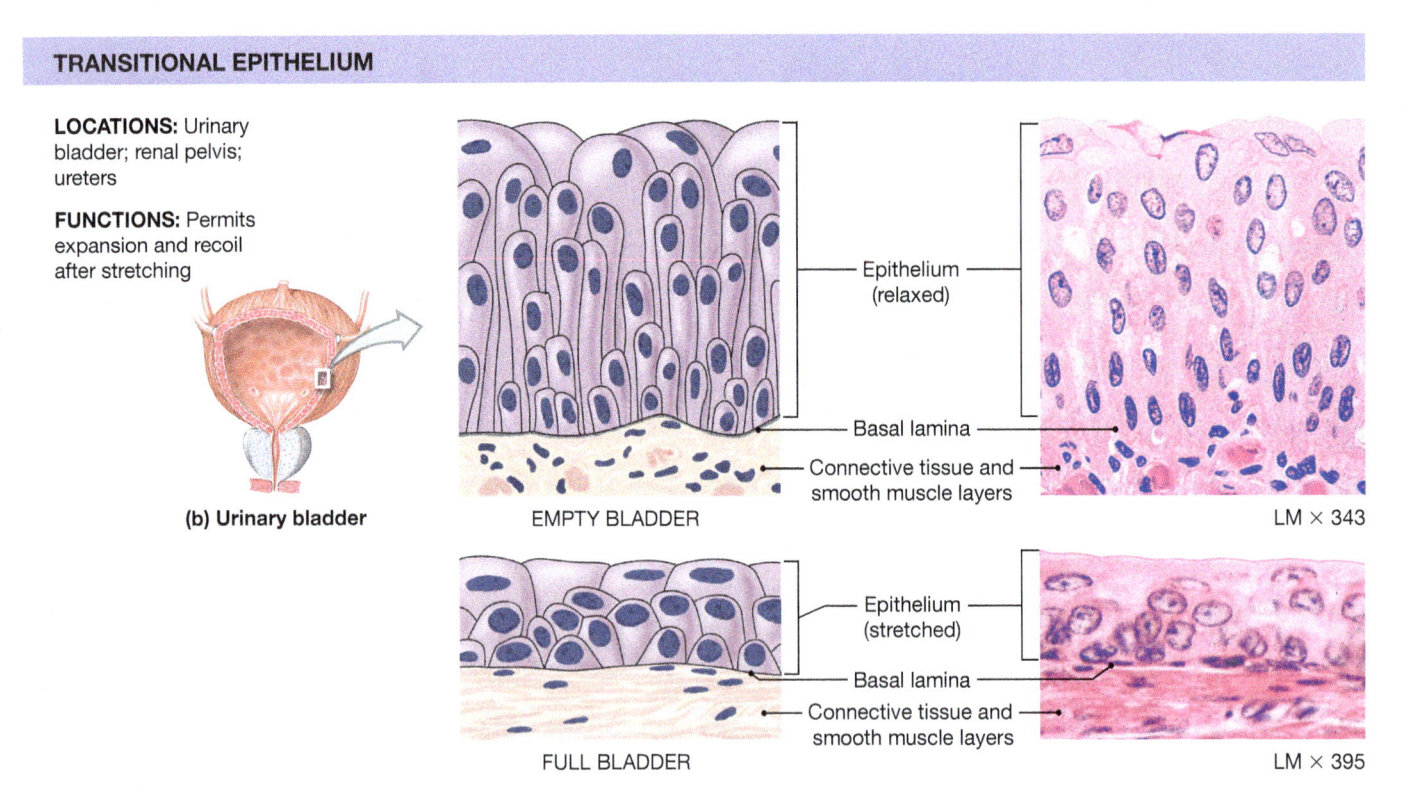

FUNCTIONS: Permits expansion and recoil after stretching

(b) Urinary bladder

EMPTY BLADDER — Epithelium (relaxed), Basal lamina, Connective tissue and smooth muscle layers

LM × 343

FULL BLADDER — Epithelium (stretched), Basal lamina, Connective tissue and smooth muscle layers

LM × 395

Figure 4.4 Pseudostratified and Transitional Epithelia
(a) Pseudostratified Columnar Epithelium. The pseudostratified columnar epithelium of the respiratory tract. Note the uneven layering of the nuclei that results from taller cells overlying shorter ones. (LM × 343) (b) Transitional Epithelium. *Left:* The lining of the empty urinary bladder, showing a transitional epithelium in the relaxed state. (LM × 343) *Right:* The lining of the full bladder, showing the effects of stretching on the appearance of cells in the transitional epithelium. (LM × 395)

stratified but is not (hence the *pseudo-* part of the name) because every cell touches the basal lamina. Typically the columnar cells are ciliated, and the tissue is called *pseudostratified ciliated columnar epithelium*. Large goblet cells interspersed among the columnar cells secrete mucus onto the epithelial free surface. The mucus traps dust and other particles in the inhaled air. Cilia at the free surface sweep the mucus to the throat, where it is swallowed and disposed of in the digestive tract.

- **Transitional epithelium** lines organs, such as the urinary bladder, that must stretch and shrink (Figure 4.4b). The cells have a variety of shapes and sizes, and not all of them touch the basal lamina. Most transitional tissue slides are prepared from relaxed transitional tissue, and the tissue appears thick, with many cells stacked one upon another. If the organ is stretched, the transitional epithelium gets thinner.

QuickCheck Questions

1.1 How are epithelia attached to the body?

1.2 How is pseudostratified epithelia different from simple epithelia?

1.3 What is the difference between keratinized and nonkeratinized epithelia?

1 Materials

- ☐ Compound microscope
- ☐ Prepared microscope slides of:
  - Simple squamous epithelium
  - Simple cuboidal epithelium
  - Simple columnar epithelium
  - Stratified squamous epithelium
  - Stratified cuboidal epithelium
  - Stratified columnar epithelium
  - Pseudostratified epithelium
  - Transitional epithelium

Procedures

1. Examine slides of each simple epithelium under the microscope at low, medium, and high magnification. Draw each tissue in the space provided.
   - Simple squamous epithelium is often observed from a superficial view on a microscope slide. Observe how the cells are closely fitted together, with little space between cells available for extracellular material.
   - Simple cuboidal epithelium is recognizable by its cube-shaped cells organized into rings to form ducts and tubules. Look for the basal lamina at the edge of the cells opposite the free surface.
   - In simple columnar epithelium, the cell nuclei are uniformly located at the base of the cells. If your slide is of the intestine, interspersed between the columnar cells are goblet cells.

2. Examine slides of each stratified epithelium under the microscope at low, medium, and high magnification. Draw each tissue in the space provided.
   - Stratified squamous epithelium is usually stained red or purple on a microscope slide. Observe that the cells at the free surface are squamous, while some of the cells in the middle layers are cuboidal and columnar cells. Remember that the cells at the free surface determine epithelium type.
   - Stratified cuboidal epithelium is normally only two cell layers thick. Locate a small duct of a sweat gland. With its thick wall, the sectioned duct will look like a donut. Increase magnification and locate the basal lamina.
   - Stratified columnar epithelium is typically only two to three cell layers thick.

3. Examine slides of pseudostratified and transitional epithelium under the microscope at low, medium, and high magnification. Draw each tissue in the space provided.
   - Notice how the nuclei in the pseudostratified columnar epithelium are unevenly distributed, creating a stratified appearance. At high magnification, slowly turn the fine focus knob back and forth approximately one-quarter turn and examine the tissue surface for cilia. Identify the goblet cells interspersed between the columnar cells. Be sure to include a few goblet cells in your sketch.
   - From the thickness of the transitional epithelium, determine which of these your specimen was made from: an empty, relaxed bladder or a full, stretched bladder. ■

46   EXERCISE 4   Tissues

Simple squamous epithelium

Simple cuboidal epithelium

Simple columnar epithelium

Stratified squamous epithelium

Stratified cuboidal epithelium

Stratified columnar epithelium

Pseudostratified epithelium

Transitional epithelium

LAB ACTIVITY 2   Connective Tissue

Connective tissue provides the body with structural support and a means of joining structural components to one another. Unlike epithelial cells, cells in connective tissue are widely scattered throughout the tissue. These cells produce and secrete protein fibers and a ground substance that together form an extracellular **matrix**. The **ground substance** is composed mainly of glycoprotein and polysaccharide

Table 4.1 Summary of Connective Tissues

Tissue	Description
Connective Tissue Proper	**Syrup matrix, various cell types**
Areolar connective tissue	Fibroblasts and mast cells, matrix with collagen and elastic fibers
Adipose tissue	Adipocytes with contents pushed to edge of cells
Recticular tissue	Network of reticular fibers supporting cells of liver, spleen, and lymph nodes
Dense connective tissue	Fibroblasts located between parallel bundles of yellow collagen fibers
Fluid Connective Tissues	**Fluid matrix, circulates in vessels**
Blood	Red blood cells transport respiratory gases, white cells provide immunity, platelets clot blood
Lymph	Fluid matrix with scattered lymphocytes
Supportive Connective Tissues	**Surrounded by outer cell-producing membrane, cells trapped in lacunae in matrix**
Hyaline cartilage	Perichondrium with chondroblasts, chondrocytes in lacunae
Elastic cartilage	Perichondrium with chondroblasts, chondrocytes in lacunae, elastic fibers visible in matrix
Fibrocartilage	Chondrocytes stacked up in columns within lacunae
Bone tissue	Concentric lamellae form osteons surrounding central canals, osteocytes trapped in lacunae, canaliculi interconnect osteocytes to central canal

molecules that surround the cells as either a thick, syrupy liquid; a gelatinous layer; or a solid, crystalline material. Suspended in the ground substance are **collagen fibers**, which give tissues strength, and **elastic fibers**, which provide flexibility. As we age, cells secrete fewer protein fibers into the matrix, which results in brittle bones and wrinkled skin.

The matrix of a connective tissue determines the physical nature of the tissue. Blood, for example, has a liquid matrix called *blood plasma* that allows the blood to flow freely through vessels. Cartilage has a thick, gelatinous matrix that allows this connective tissue to slide easily over other structures. Bone has a solid matrix and provides the structural framework for the body.

Connective tissues are divided into three groups, distinguished primarily by cellular composition and matrix characteristics (Table 4.1). **Connective tissue proper** has a thick liquid matrix and a variety of cell types. **Fluid connective tissues** are liquid tissues that flow through blood vessels and lymphatic vessels. **Supportive connective tissues** have a matrix that is either gelatinous or solid and acts as support for other tissues.

All connective tissues are produced in the embryo from an unspecialized tissue called **mesenchyme** (Figure 4.5a). Early in the third week of embryonic development, mesenchyme appears and produces the specialized cells that are needed to construct mature connective tissues. Mesenchyme is a loose meshwork of star-shaped cells. Unlike adult connective tissue, mesenchyme has no visible protein fibers in its ground substance. **Mucous connective tissue**, also called **Wharton's jelly**, is an embryonic connective tissue found only in the umbilical cord (Figure 4.5b). In the adult body, connective tissue contains only a few mesenchyme cells, which assist in tissue repair.

Connective Tissue Proper

Connective tissue proper includes two groups of tissues: loose and dense. **Loose connective tissue** has an open network of protein fibers in a thick, syrupy ground substance. *Areolar, adipose,* and *reticular* tissues are the three main types of loose connective tissue. **Dense connective tissue** is made up of two types of fibers: *protein fibers* assembled into thick bundles of collagen, and *elastic fibers* with widely

48 EXERCISE 4 Tissues

(a) Mesenchymal cells

(b) Mesenchymal cells | Blood vessel

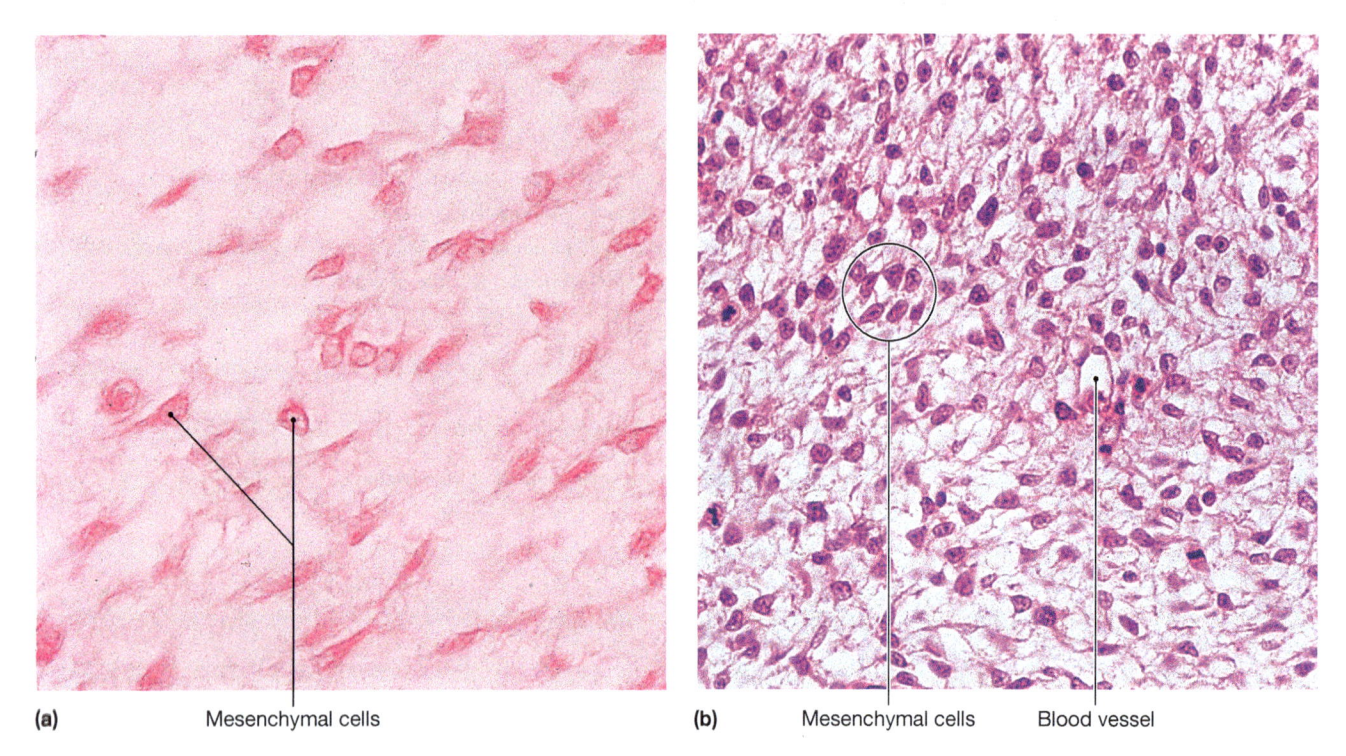

Figure 4.5 Connective Tissues in Embryos
These connective tissues give rise to all other connective tissue types. **(a)** Mesenchyme is the first connective tissue to appear in an embryo. (LM × 136) **(b)** Mucous connective tissue (*Wharton's jelly*). (LM × 136) This sample was taken from the umbilical cord of a fetus.

scattered cells. There are two types of dense connective tissue: *dense regular,* in which the protein fibers in the matrix are arranged in parallel bands; and *dense irregular,* in which the fibers are interwoven.

Connective tissue proper contains a variety of cell types in addition to the collagen fibers and elastic fibers described above (Figure 4.6). **Fibroblasts** (FĪ-brō-blasts) are fixed (stationary) cells that secrete proteins that join other molecules in the matrix to form the collagen and elastic fibers. Phagocytic **macrophages** (MAK-rō-fā-jez; *phagein,* to eat) patrol these tissues, ingesting microbes and dead cells. Macrophages are mobilized during an infection or injury, migrate to the site of disturbance, and phagocytize damaged tissue cells and microbes. **Mast cells** release histamines that cause an inflammatory response in damaged tissues. **Adipocytes** (AD-i-pō-sī-ts) are fat cells and contain vacuoles for the storage of lipids.

- **Areolar tissue** (Figure 4.6a) is distributed throughout the body. This tissue fills spaces between structures for support and protection. It is very flexible and permits muscles to move freely without pulling on the skin. Most of the cells in areolar tissue are oval-shaped fibroblasts that usually stain light. Mast cells are small and filled with dark-stained granules of histamine and heparin, both of which cause inflammation. Collagen and elastic fibers are clearly visible in the matrix.

- **Adipose tissue** (also called *fat tissue*) is distributed throughout the body and is abundant under the skin and in the buttocks, breasts, and abdomen (Figure 4.6b). Two types of adipose tissue occur in the body. Infants have **brown fat**, which is highly vascularized. Older children and adults have **white fat**, in which the adipocytes are packed more closely together than are the cells in other types of connective tissue proper. The distinguishing feature of adipose tissue is displacement of the nucleus and cytoplasm due to the storage of lipids. When an adipocyte stores fat, its vacuole expands with lipid and fills most of the cell while pushing the organelles and cytosol to the periphery.

AREOLAR TISSUE

LOCATIONS: Within and deep to the dermis of skin, and covered by the epithelial lining of the digestive, respiratory, and urinary tracts; between muscles; around blood vessels, nerves, and around joints

FUNCTIONS: Cushions organs; provides support but permits independent movement; phagocytic cells provide defense against pathogens

Areolar tissue from pleura

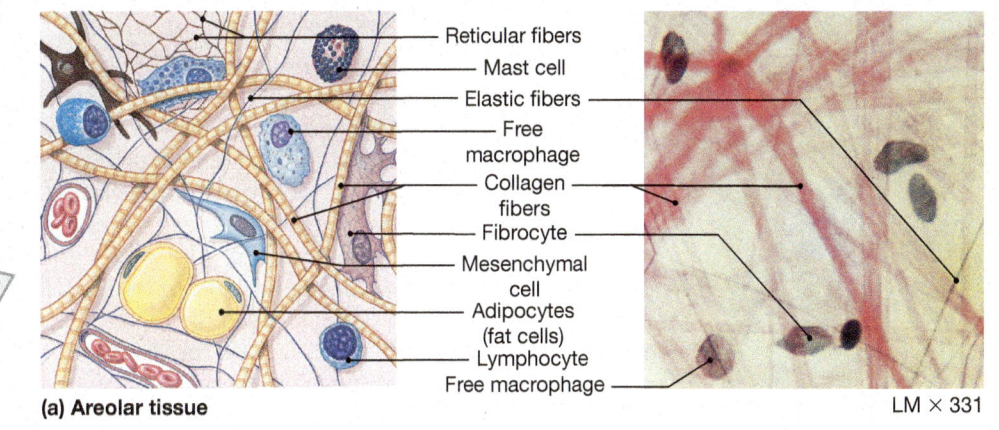

(a) Areolar tissue

- Reticular fibers
- Mast cell
- Elastic fibers
- Free macrophage
- Collagen fibers
- Fibrocyte
- Mesenchymal cell
- Adipocytes (fat cells)
- Lymphocyte
- Free macrophage

LM × 331

ADIPOSE TISSUE

LOCATIONS: Deep to the skin, especially at sides, buttocks, breasts; padding around eyes and kidneys

FUNCTIONS: Provides padding and cushions shocks; insulates (reduces heat loss); stores energy

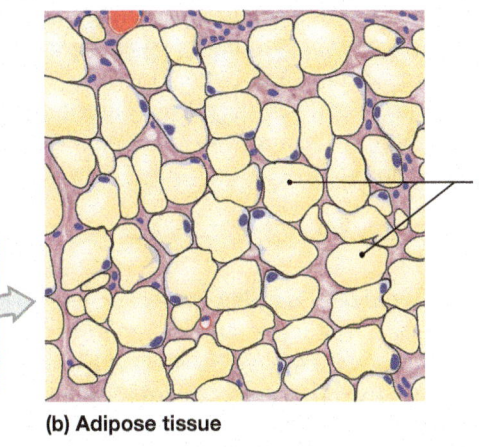

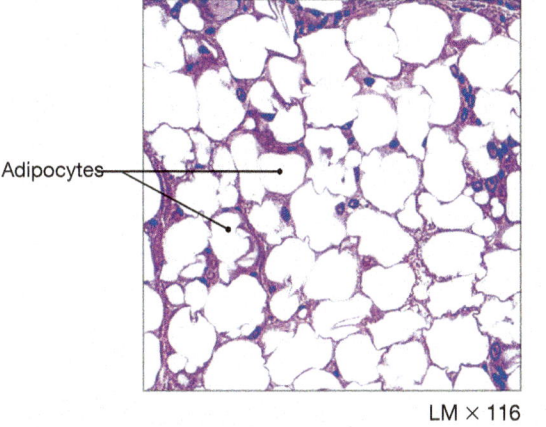

(b) Adipose tissue

Adipocytes

LM × 116

RETICULAR TISSUE

LOCATIONS: Liver, kidney, spleen, lymph nodes, and bone marrow

FUNCTIONS: Provides supporting framework

Reticular tissue from liver

(c) Reticular tissue

Reticular fibers

LM × 326

Figure 4.6 Loose Connective Tissues

This is the "packing material" of the body, filling spaces between other structures. **(a)** Areolar Tissue. Note the open framework: all the cells of connective tissue proper are found in areolar tissue. (LM × 331) **(b)** Adipose Tissue. Adipose tissue is a loose connective tissue dominated by adipocytes. In standard histological views, the tissue looks empty because the lipids in the cells dissolve during slide preparation. (LM × 116) **(c)** Reticular Tissue. Reticular tissue consists of an open framework of reticular fibers. These fibers are usually very difficult to see because of the large number of cells organized around them. (LM × 326)

- **Reticular tissue** forms the internal supporting framework for soft organs, such as the spleen, liver, and lymphatic organs. The tissue is composed of an extensive network of **reticular fibers** interspersed with small, oval **reticulocytes** (Figure 4.6c).
- **Dense regular connective tissue** consists mostly of collagen or elastic fibers organized into thick bands, with fibroblasts widely interspersed in the fibrous matrix. This strong tissue forms tendons, which join muscle to bone, and ligaments, which join bone to bone. Because tendons and ligaments conduct pulling forces mainly from one direction, the protein fibers in dense regular tissues are parallel. Tendons transfer strong pulling forces from muscle to bone and have an abundance of strong bands of collagen fibers in the matrix (Figure 4.7a). Ligaments have more elasticity than tendons and have a large quantity of elastic fibers in the matrix. Elastic ligaments support the bones of the vertebral column (Figure 4.7b). Flat layers of dense regular connective tissue called *fascia* protect and isolate muscles from surrounding structures and allow muscle movement. On slides with limited stain, the profusion of collagen fibers makes the dense regular connective tissue of tendons appear yellow under the microscope.
- **Dense irregular connective tissue** (Figure 4.7c) is a mesh of collagen fibers with interspersed fibroblasts. Dense irregular connective tissue is located in the **dermis**, which is the skin layer just above the epidermis, and in the layers surrounding cartilage and bone. The kidneys, liver, and spleen are protected inside a capsule of dense irregular connective tissue. With its meshwork of collagen fibers, this connective tissue supports areas that receive stress from many directions.

Clinical Application **Liposuction**

The surgical procedure called *liposuction* removes unwanted adipose tissue with a suction wand. The treatment is dangerous and may damage blood vessels or nerves near the site of fat removal. Overlying skin may appear pocketed and marbled after the procedure. Considering that adult connective tissues contain a few mesenchyme cells, what would most likely occur if a liposuction patient consumed excessive calories after the surgery? ▶

Fluid Connective Tissue

There are two types of fluid connective tissue: **blood** and **lymph**. These tissues have a liquid matrix and circulate in blood vessels or lymphatic vessels. Blood is composed of cells collectively called *formed elements* (Figure 4.8), which are supported in a liquid ground substance called **plasma**. Protein fibers are dissolved in the blood matrix. During blood clotting, in a process called *coagulation,* the protein fibers produce a fibrin net to trap cells as they pass through the wound. The fibers also regulate the viscosity, or thickness, of the blood. The formed elements are grouped into three general categories: red blood cells, white blood cells, and platelets. Red blood cells, called **erythrocytes**, transport blood gases. The cells are biconcave discs, with a center so thin that it often looks hollow when viewed under the microscope. White blood cells, called **leukocytes**, are the cells of the immune system and protect the body from infection. Upon injury to a blood vessel, **platelets** become sticky and form a plug to reduce bleeding.

Lymph has protein fibers dissolved in its matrix. The most numerous cells in lymph are **lymphocytes**, which are white blood cells that are produced in lymphoid tissues. A detailed study of these tissues is presented in Exercises 22 and 25.

DENSE REGULAR CONNECTIVE TISSUE: Tendon

LOCATIONS: Between skeletal muscles and skeleton (tendons and aponeuroses); between bones or stabilizing positions of internal organs (ligaments); covering skeletal muscles; deep fasciae

FUNCTIONS: Provides firm attachment; conducts pull of muscles; reduces friction between muscles; stabilizes relative positions of bones

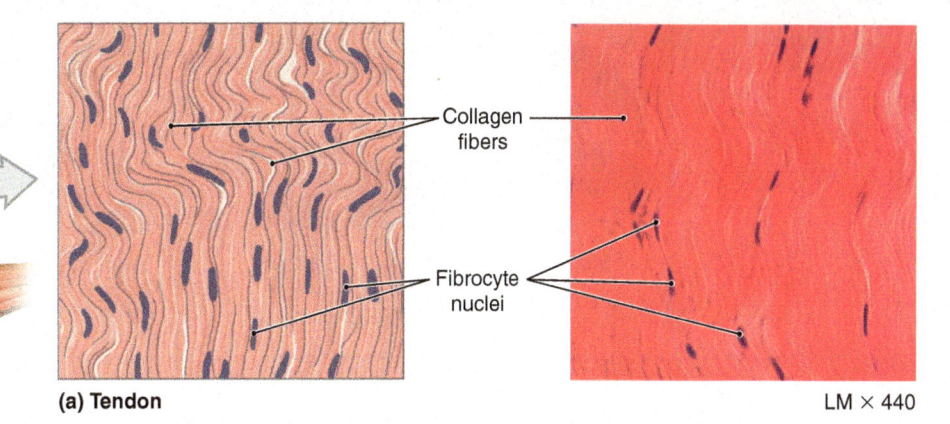

- Collagen fibers
- Fibrocyte nuclei

(a) Tendon LM × 440

DENSE REGULAR CONNECTIVE TISSUE: Elastic Ligament

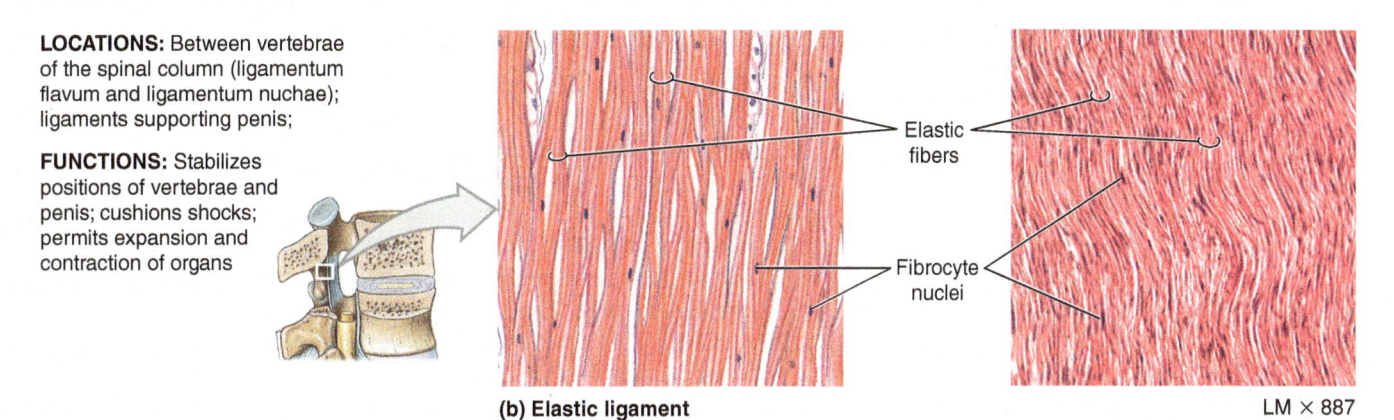

LOCATIONS: Between vertebrae of the spinal column (ligamentum flavum and ligamentum nuchae); ligaments supporting penis;

FUNCTIONS: Stabilizes positions of vertebrae and penis; cushions shocks; permits expansion and contraction of organs

- Elastic fibers
- Fibrocyte nuclei

(b) Elastic ligament LM × 887

DENSE IRREGULAR CONNECTIVE TISSUE

LOCATIONS: Dermis; capsules of visceral organs; periostea and perichondria; nerve and muscle sheaths.

FUNCTIONS: Provides strength to resist forces applied from many directions; helps prevent overexpansion of organs such as the urinary bladder

- Collagen fiber bundles

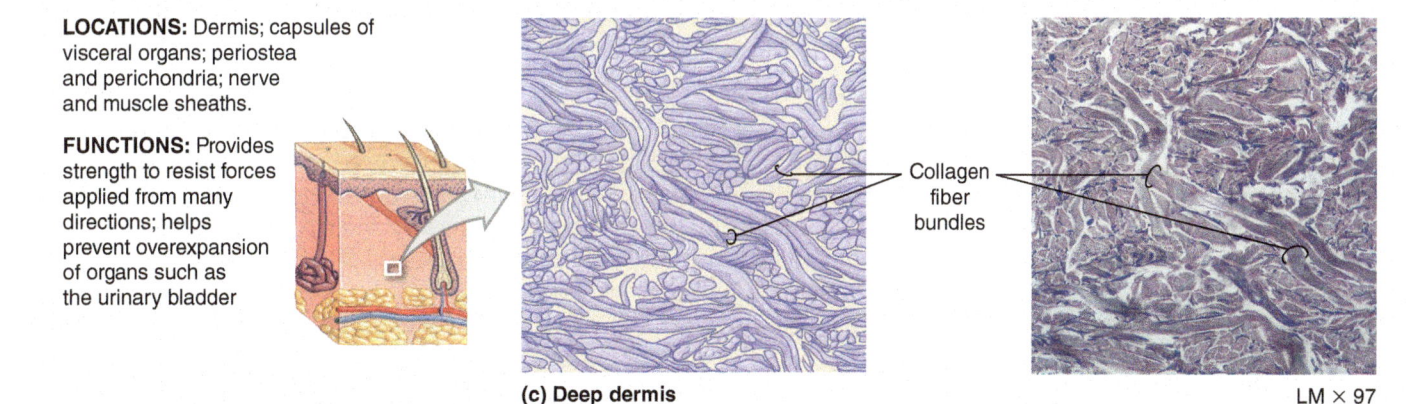

(c) Deep dermis LM × 97

Figure 4.7 Dense Connective Tissues
(a) Dense Regular Connective Tissue. The dense regular connective tissue in a tendon consists of densely packed, parallel bundles of collagen fibers. The fibroblast nuclei can be seen flattened between the bundles. Most ligaments resemble tendons in their histological organization. (LM × 440) (b) Dense Regular Connective Tissue. Elastic ligaments extend between the vertebrae of the spinal column. The bundles of elastic fibers are fatter than the collagen fiber bundles of a tendon or a typical (nonelastic) ligament. (LM × 887) (c) Dense Irregular Connective Tissue. The deep portion of the dermis contains a thick layer of interwoven collagen fibers oriented in various direction. (LM × 97)

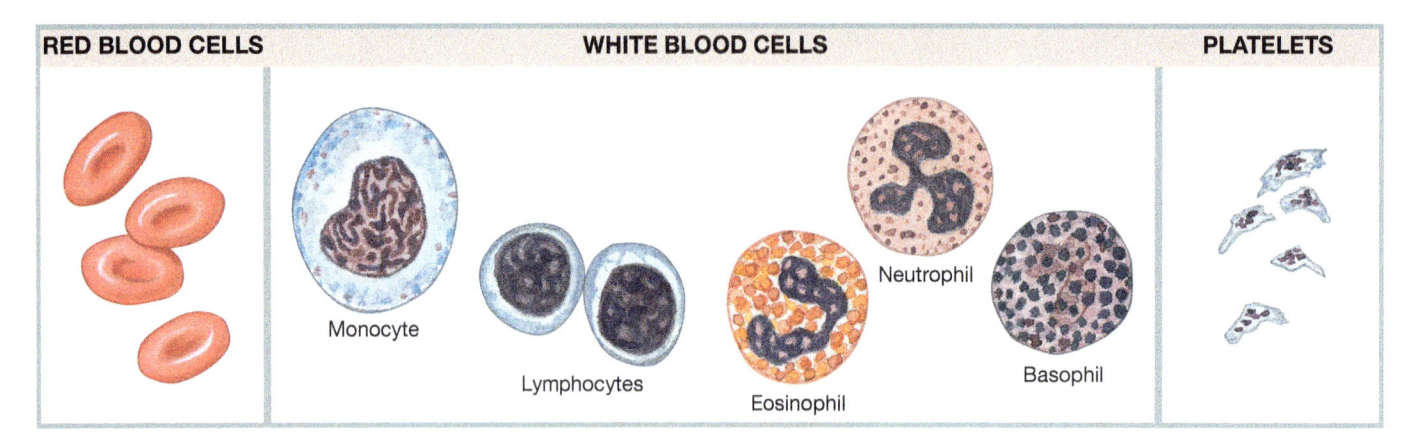

Figure 4.8 Formed Elements of the Blood
Formed elements is the collective name for all the different cell types found in blood.

Supporting Connective Tissues

Cartilage and bone, the two types of supporting connective tissues, contain a strong matrix of fibers capable of supporting body weight and stress. **Cartilage** is a rubbery, avascular tissue with a gelatinous matrix and many fibers for structural support. **Bone** has a solid matrix that is composed of calcium phosphate and calcium carbonate. These salts crystallize on collagen fibers and form a hard material called **hydroxyapatite**.

A membrane surrounds all supporting connective tissues to protect the tissue and supply new tissue-producing cells. The **perichondrium** (pe-rē-KON-drē-um) is the membrane that surrounds cartilage (Figure 4.9) and produces **chondroblasts** (KON-drō-blasts; *chondros,* cartilage), which secrete the fibers and ground substance of the cartilage matrix. Eventually, chondroblasts become trapped in the matrix in small spaces called **lacunae** (la-KOO-nē, *lacus,* pool) and lose the ability to produce additional matrix. These cells are then called **chondrocytes** and function to maintain the mature tissue.

- **Hyaline** (HĪ-uh-lin; *hyalus,* glass) **cartilage** (Figure 4.10a) is the most common cartilage in the body. The tissue is distinguishable from other cartilages by the apparent lack of fibers in the matrix. Hyaline cartilage does contain elastic and collagen fibers, but they do not stain and therefore are not visible.

- **Elastic cartilage** (Figure 4.10b) has many elastic fibers in the matrix and is therefore easily distinguished from hyaline cartilage. The elastic fibers permit considerable binding and twisting of the tissue.

- **Fibrocartilage** contains irregular collagen fibers that are visible in the matrix (Figure 4.10c). This cartilage is very strong and durable, and its function is to cushion joints and limit bone movement.

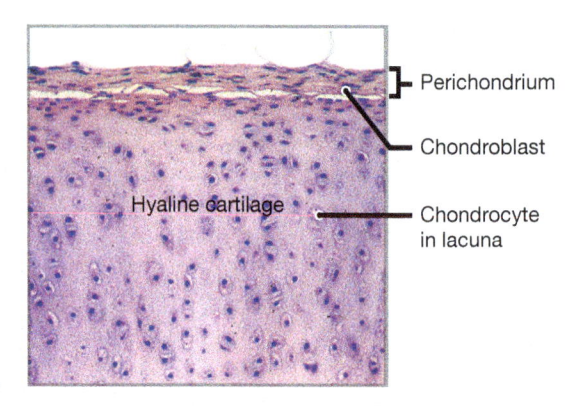

Figure 4.9 Cartilage, One Type of Supporting Connective Tissue
A perichondrium separates cartilage from other tissues.

HYALINE CARTILAGE

LOCATIONS: Between tips of ribs and bones of sternum; covering bone surfaces at synovial joints; supporting larynx (voice box), trachea, and bronchi; forming part of nasal septum

FUNCTIONS: Provides stiff but somewhat flexible support; reduces friction between bony surfaces

- Chondrocytes in lacunae
- Matrix

(a) Hyaline cartilage   LM × 435

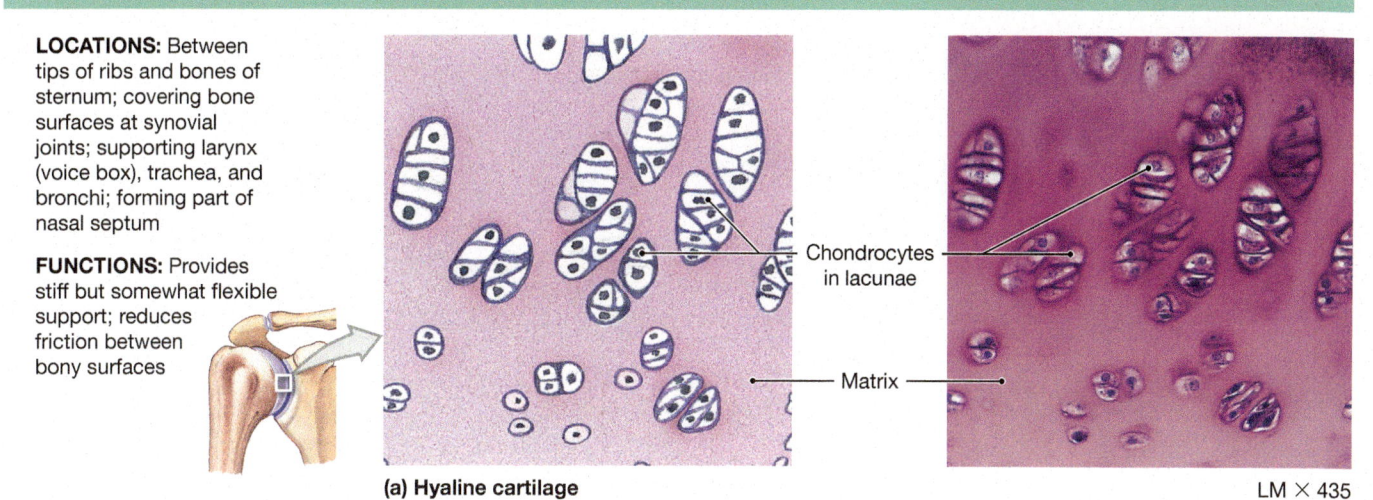

ELASTIC CARTILAGE

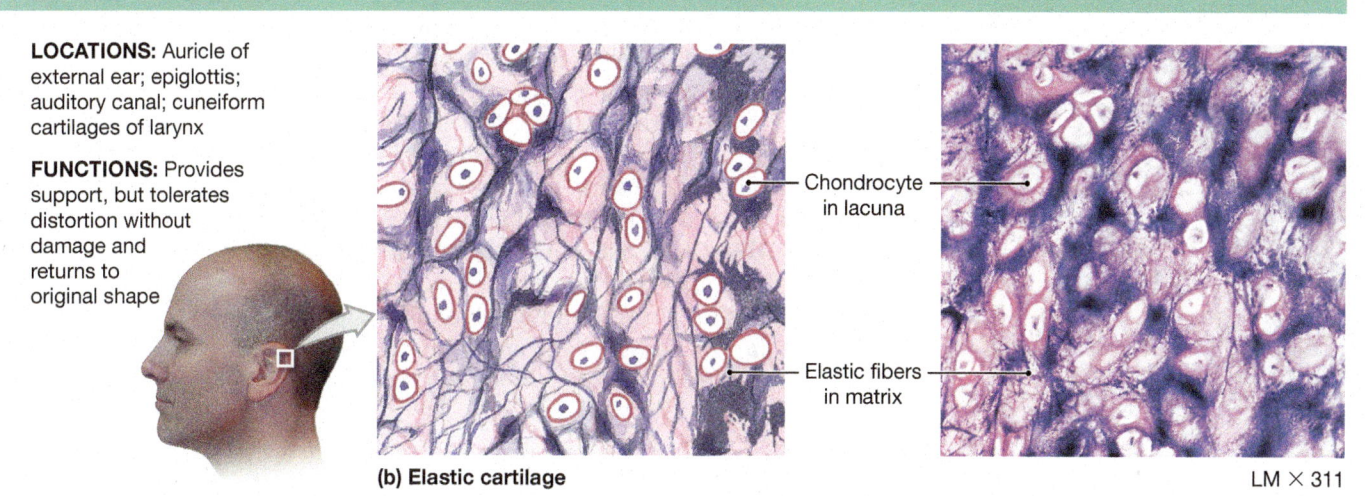

LOCATIONS: Auricle of external ear; epiglottis; auditory canal; cuneiform cartilages of larynx

FUNCTIONS: Provides support, but tolerates distortion without damage and returns to original shape

- Chondrocyte in lacuna
- Elastic fibers in matrix

(b) Elastic cartilage   LM × 311

FIBROUS CARTILAGE

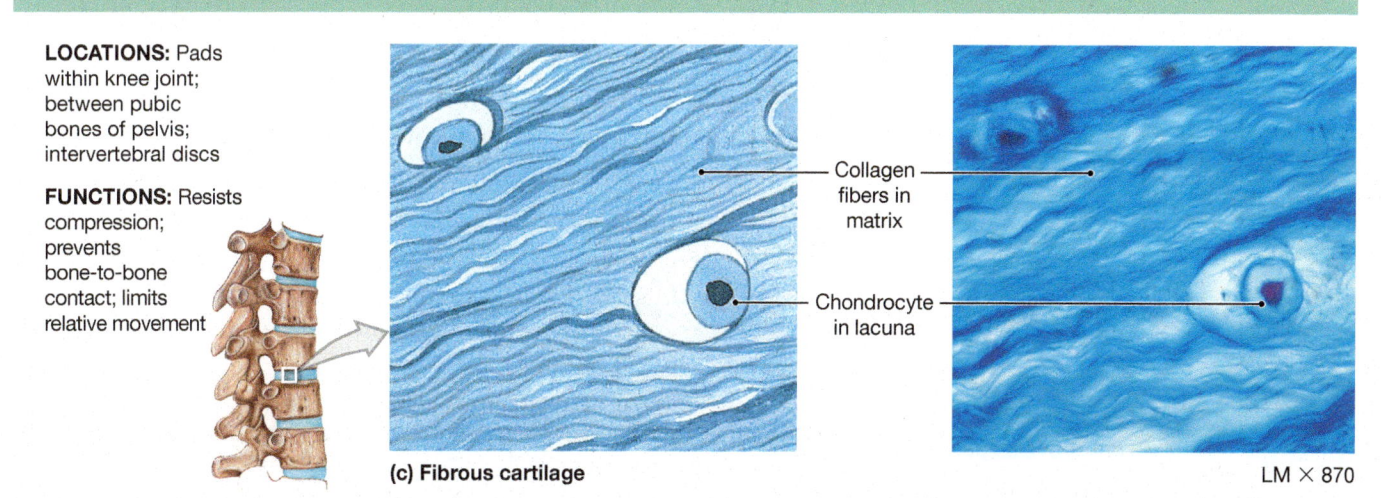

LOCATIONS: Pads within knee joint; between pubic bones of pelvis; intervertebral discs

FUNCTIONS: Resists compression; prevents bone-to-bone contact; limits relative movement

- Collagen fibers in matrix
- Chondrocyte in lacuna

(c) Fibrous cartilage   LM × 870

Figure 4.10 Types of Cartilage

(a) Hyaline Cartilage. Note the translucent matrix and the absence of prominent fibers. (LM × 435) (b) Elastic Cartilage. The closely packed elastic fibers are visible between the chondrocytes. (LM × 311) (c) Fibrocartilage. The collagen fibers are extremely dense, and the chondrocytes are relatively far apart. (LM × 870)

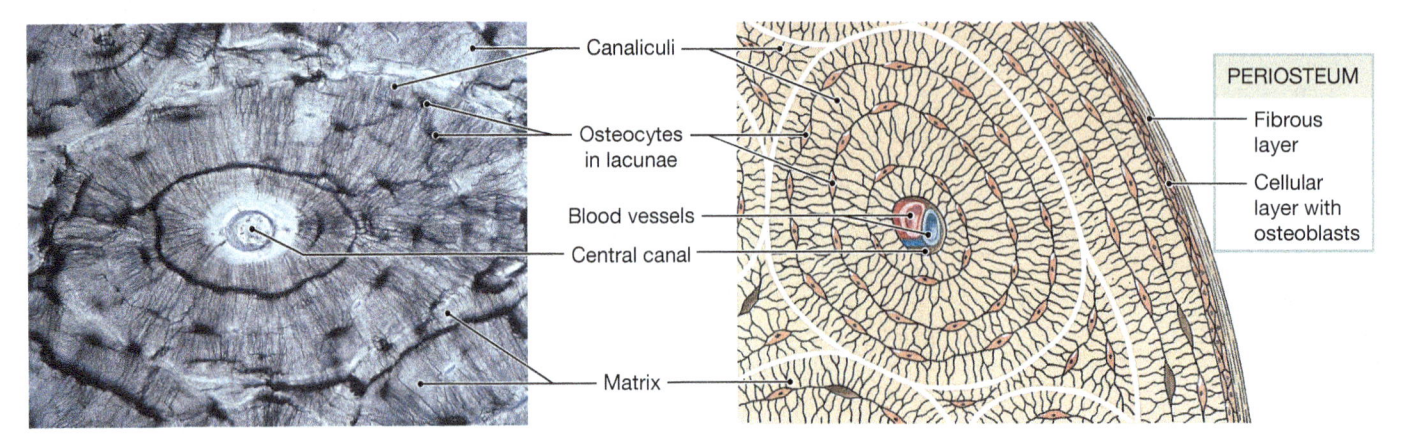

Figure 4.11 Bone
The osteocytes in bone are usually organized in groups around a central canal that contains blood vessels. For the photomicrograph, a sample of bone was ground thin enough to become transparent. Bone dust filled the lacunae and the central canal, making them appear dark. (LM × 362)

Bone supporting tissue is surrounded by a membrane called the **periosteum** (pe-rē-OS-tē-um), which contains cells called **osteoblasts** (OS-tē-ō-blasts) for bone growth and repair (Figure 4.11). Like chondroblasts, osteoblasts secrete the organic components of the matrix, become trapped in lacunae, and mature into **osteocytes**. Rings of matrix called **lamellae** (lah-MEL-lē; *lamella,* thin plate) surround a **central canal** that contains blood vessels. **Canaliculi** (kan-a-LIK-ū-lē; "little canals") are small channels in the lamellae that provide passageways through the solid matrix for diffusion of nutrients and wastes.

The main functions of bone are support of the body, attachment of skeletal muscles, and protection of internal organs. Bone tissue is studied in more detail in Exercise 6.

2 Materials

- Compound microscope
- Prepared microscope slides of:
  - Mesenchyme
  - Areolar tissue
  - Adipose tissue
  - Reticular tissue
  - Dense regular connective tissue: tendon
  - Dense regular connective tissue: elastic ligament
  - Dense irregular connective tissue
  - Blood
  - Hyaline cartilage
  - Elastic cartilage
  - Fibrocartilage
  - Bone

QuickCheck Questions

2.1 Why is mesenchyme more abundant in embryos than in adults?

2.2 Which cell in connective tissue proper manufactures the protein fibers for the matrix?

2.3 What fiber types are common in the matrix of connective tissue proper?

2.4 What are the characteristics of the matrix of cartilage?

2.5 How are cartilage and bone tissue similar to each other?

Procedures

1. Place the mesenchyme slide on the microscope stage and scan the tissue at low power. Then increase the magnification first to medium and then to high. Observe the shape and distribution of the mesenchyme cells and a matrix that lacks protein fibers.

2. Scan the slides of the different types of connective tissue proper at all magnifications and observe the variety of cells and fibers in each tissue. Draw each tissue in the space provided.

   - Areolar: Identify the prominently stained cells and fibers in the matrix.
   - Adipose: Observe individual adipocytes with their cytoplasm displaced to the cell's edge by fat vacuoles, giving the cell a "signet" appearance similar to a graduation ring.
   - Reticular: Note the reticular fibers and reticulocytes.

EXERCISE 4   Tissues   55

- Dense regular (tendon): Examine at low and medium magnifications and note the abundance of collagen fibers organized into parallel bands, with few fibroblasts scattered in between.
- Dense regular (elastic ligament): Examine at low and medium magnifications. Observe the organization of the elastic fibers and the arrangement of fibroblasts.
- Dense irregular: Examine at low and medium magnifications and note the interwoven network of collagen that distinguishes this tissue from dense regular.

3. Scan the blood slide at low power, then increase magnification to medium power and identify the three types of blood cells. Sketch several erythrocytes and leukocytes in the space provided.
   - Erythrocytes: The majority of the cells on the slide are red blood cells. How do these cells differ from most other cells in the body?
   - Leukocytes: These are the dark-stained cells. Note the variation in the morphology, or shape, of the nucleus in the different types of leukocytes.
   - Platelets: Look closely between the erythrocytes and leukocytes and observe these fragile formed elements.

4. Scan the slides of the various kinds of supporting connective tissue. Make careful comparisons between the matrix and fibers of the cartilages. Sketch each supporting connective tissue in the space provided.
   - Hyaline cartilage: Locate the perichondrium and chondroblasts on the hyaline cartilage slide. Examine the deeper middle region of the cartilage, where the chondroblasts have migrated and become chondrocytes inside lacunae.
   - Elastic cartilage: Observe the many elastic fibers in the matrix of the elastic cartilage. Identify the perichondrium with its many small chondroblasts and the chondrocytes trapped in lacunae deeper in the tissue.
   - Fibrocartilage: Observe the thick bundles of collagen and the absence of a perichondrium. Chondrocytes are in lacunae and may be either scattered or stacked in small groups.
   - Bone: Scan the bone slide at low and medium magnifications to observe the organization of lamellae around central canals. ■

Areolar tissue

Adipose tissue

Reticular tissue

56   EXERCISE 4   Tissues

Dense regular connective tissue: tendon	Dense regular connective tissue: elastic ligament	Dense irregular connective tissue
Erythrocytes and leukocytes	Hyaline cartilage	Elastic cartilage
Fibrocartilage	Bone	

LAB ACTIVITY 3   Muscle Tissue

There are three types of muscle tissue, each named for its location in the body. **Skeletal muscle tissue** is found in skeletal muscles, which are attached to bone and provide the means by which the body skeleton moves. **Cardiac muscle tissue** makes up cardiac muscle, which forms the walls of the heart and pumps blood through the vascular system. **Smooth muscle tissue** is found in the muscles that are present inside hollow organs and control such functions as the movement of material through the digestive system and uterine contraction during labor.

Muscle tissue specializes in contraction. Muscle cells shorten during contraction, and this shortening produces a force that causes movement. During the contraction phase of a heartbeat, for example, the force created by the contraction pumps blood through the vascular system to supply cells with oxygen, nutrients, and other essential materials.

- Skeletal muscle tissue is composed of long cells called **muscle fibers** (Figure 4.12a). During development, a number of embryonic cells called *myoblasts* (*myo-*, muscle + *-blast*, precursor) fuse into one large cell that is the muscle fiber. Because each fiber forms from numerous embryonic cells, it is *multinucleated*. The nuclei are clustered under the **sarcolemma** (sar-cō-LEM-uh; *sarco*, flesh), which is the muscle fiber's cell membrane. Muscle fibers are striated, with a distinct banded pattern that results from the organization of internal contractile proteins called *filaments*.

- Cardiac muscle tissue occurs only in the muscles that form the walls of the heart. This tissue is striated like skeletal muscle tissue. However, each cardiac muscle cell, called a **cardiocyte**, has a single nucleus and is branched (Figure 4.12b). Cardiocytes are connected to one another by **intercalated** (in-TER-ka-lā-ted) **discs**, which are junctions that conduct contraction stimuli from one cardiocyte to the next. Unlike skeletal muscles, cardiac muscle is under involuntary control. For example, your autonomic nervous system causes an increase in your heart rate when you exercise and a decrease in your heart rate when you sleep, all with no conscious input from you.

- Smooth muscle tissue is found in the smooth muscles of the body's soft organs, such as the stomach, intestines, eyes, and blood vessels. The cells of smooth muscle tissue are not striated (Figure 4.12c). Each cell has a single nucleus and is spindle-shaped—thick in the middle and tapered at the ends. The tissue usually occurs in double sheets of muscle, with one sheet positioned at a right angle to the other. This arrangement permits the tissue to shorten structures and decrease the diameter of vessels and passageways. Smooth muscle is under involuntary control.

Making Connections   **Muscle Terminology**

In reference to muscle control, the terms *voluntary* and *involuntary* are used more for convenience than for description. Skeletal muscle is said to be voluntary, and yet you cannot stop your skeletal muscles from shivering when you are cold. Additionally, once you voluntarily start a muscle contraction, the brain assumes control of the muscle. The heart muscle is classified as involuntary, but some individuals can control their heart rate. Generally, the term *voluntary* is associated with skeletal muscle and *involuntary* with cardiac and smooth muscle.

The terms *muscle fiber* and *muscle cell* might also seem confusing. These two terms mean essentially the same thing, but the general convention is to say muscle fiber when referring to skeletal muscles, and muscle cell when referring to cardiac and smooth muscles. We follow that convention in this manual.

QuickCheck Questions

3.1 What are the three types of muscle tissue in the body?

3.2 How are cardiocytes connected to one another?

3.3 What are the two locations where smooth muscle tissue is found in the body?

3.4 How are smooth muscle cells different from skeletal muscle fibers and cardiac muscle cells?

SKELETAL MUSCLE TISSUE

Cells are long, cylindrical, striated, and multinucleate.

LOCATIONS: Combined with connective tissues and neural tissue in skeletal muscles

FUNCTIONS: Moves or stabilizes the position of the skeleton; guards entrances and exits to the digestive, respiratory, and urinary tracts; generates heat; protects internal organs

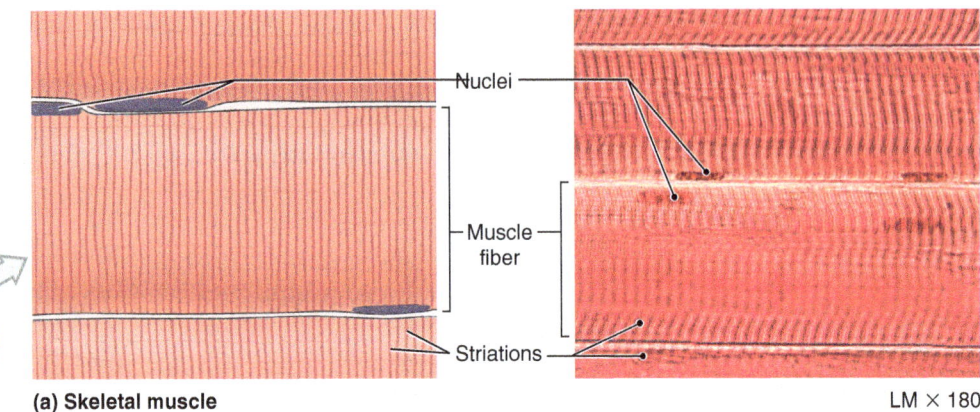

(a) Skeletal muscle

LM × 180

CARDIAC MUSCLE TISSUE

Cells are short, branched, and striated, usually with a single nucleus; cells are interconnected by intercalated discs.

LOCATION: Heart

FUNCTIONS: Circulates blood; maintains blood (hydrostatic) pressure

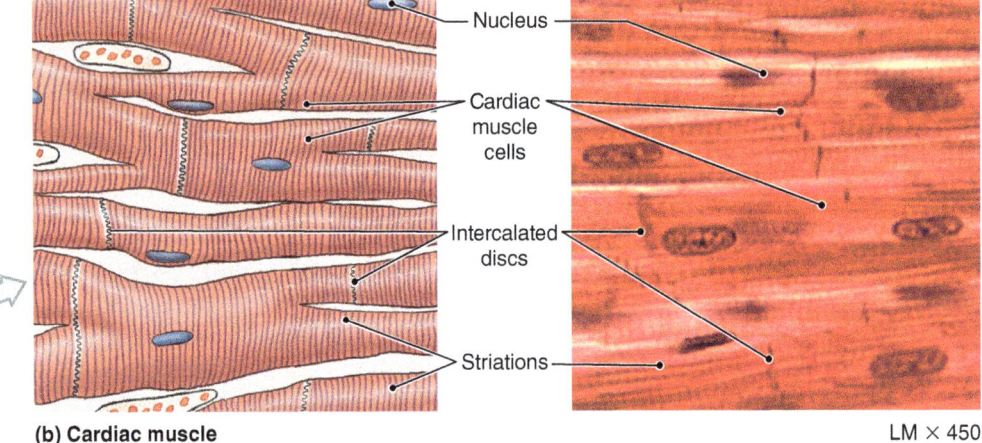

(b) Cardiac muscle

LM × 450

SMOOTH MUSCLE TISSUE

Cells are short, spindle-shaped, and nonstriated, with a single, central nucleus

LOCATIONS: Found in the walls of blood vessels and in digestive, respiratory, urinary, and reproductive organs

FUNCTIONS: Moves food, urine, and reproductive tract secretions; controls diameter of respiratory passageways; regulates diameter of blood vessels

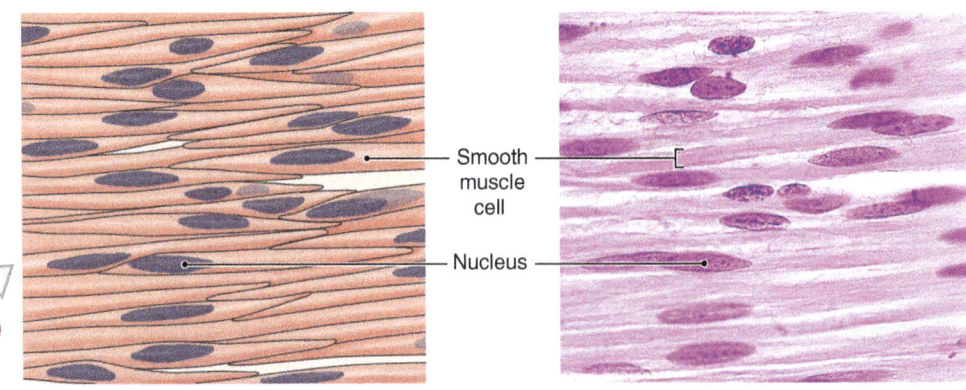

(c) Smooth muscle

LM × 204

Figure 4.12 Muscle Tissues
(a) Skeletal Muscle Fibers. Note the large fiber size, prominent banding pattern, multiple nuclei, and unbranched arrangement. (LM × 180) (b) Cardiac Muscle Cells. Cardiac muscle cells differ from skeletal muscle fibers in three major ways: size (cardiac muscle cells are smaller), organization (cardiac muscle cells branch), and number and location of nuclei (a typical cardiac muscle cell has one centrally placed nucleus). (LM × 450) (c) Smooth Muscle Cells. Smooth muscle cells are small and spindle-shaped, with a central nucleus. They do not branch, and there are no striations. (LM × 204)

EXERCISE 4 Tissues 59

3 Materials

- ☐ Compound microscope
- ☐ Skeletal muscle slide
- ☐ Cardiac muscle slide
- ☐ Smooth muscle slide

Procedures

1. Examine the skeletal muscle slide at various magnifications.
   - Identify an individual muscle fiber and note the many nuclei and the striations.
   - If both transverse and longitudinal sections are on your slide, compare the appearance of the muscle fibers in the two sections.
   - Draw and label the microscopic structure of skeletal muscle tissue in the space provided.
2. Examine the cardiac muscle slide at various magnifications. If your slide has different sections, observe the longitudinal section first.
   - Compare the nucleation and striations of cardiac muscle with what you observed in skeletal muscle.
   - Observe how cardiocytes are interconnected by intercalated discs.
   - Draw and label the microscopic structure of cardiac muscle tissue in the space provided.
3. Examine the smooth muscle slide at various magnifications.
   - Note the position of the nucleus in each cell and the shape of the cells.
   - Draw and label the microscopic structure of smooth muscle tissue in the space provided. ■

Skeletal muscle tissue   Cardiac muscle tissue   Smooth muscle tissue

LAB ACTIVITY 4 Neural Tissue

The nervous system is made up of cells called **neurons** and cells called **glial cells**. Together these two types of cells are collectively referred to as either **nerve tissue** or **neural tissue**. (The two terms are synonyms, and you will see both terms in textbooks and in scientific literature.) Glial cells are presented in more detail in Exercise 13, and here we focus on neurons.

To maintain homeostasis, the body must constantly evaluate internal and external conditions and respond quickly and appropriately to environmental changes. The nervous system processes information from sensory organs and responds with motor instructions to muscles and glands, which are collectively called the body's **effectors**. Cells responsible for receiving, interpreting, and sending the electrical signals of the nervous system are called **neurons**.

Neurons are *excitable*, which means they can respond to environmental changes by converting sensory stimuli to electrical impulses called **action potentials**. Sensory neurons detect changes in the environment and communicate these changes to the central nervous system (CNS), which consists of the brain and spinal cord. The CNS responds to the sensory input with motor commands to glands and muscles. This constant monitoring and adjustment play a vital role in homeostasis.

60 EXERCISE 4 Tissues

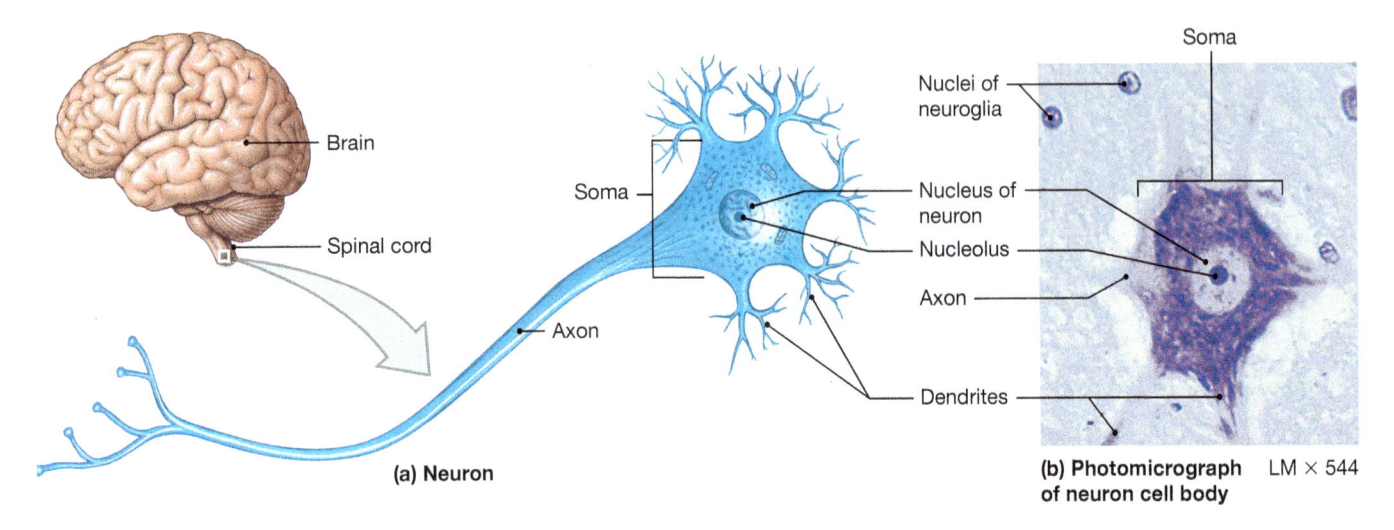

Figure 4.13 Neural Tissue
(a) Diagrammatic and (b) histological views of a representitive neuron. (LM × 544) Neurons are specialized for conduction of electrical impulses over relatively long distances in the body.

A typical neuron has several distinct regions (Figure 4.13). A central nucleus is surrounded by a region called either the **cell body** or the **soma**, which contains most of the neuron's organelles. Radiating out from the soma, many fine extensions called **dendrites** (DEN-drīts; *dendron*, a tree) receive signals from other cells and send this information to the soma. The impulse is then conducted into a single **axon** that carries information away from the soma, either to other neurons or to effector cells.

On most slides of neural tissue, it is difficult to distinguish axons from dendrites. Locate one neuron that is isolated from the others and look for the longest extension off the soma. This is most likely the axon.

QuickCheck Questions

4.1 Which part of a neuron sends information to another neuron?

4.2 In general, how do neurons communicate with other cells?

4 Materials

☐ Compound microscope
☐ Neural tissue slide

Procedures

1. Scan the slide at low magnification to locate the neurons. Center one neuron in the field of view and increase the magnification. Adjust the light setting of the microscope if necessary.
2. On the neuron you have chosen, identify the soma, nucleus, dendrites (thin extensions), and axon (thicker extension).
3. Draw and label several neurons in the space provided. ■

Neural tissue

Name _____

Date _____

Section _____

LAB REPORT

EXERCISE 4

Tissues

A. Matching

Match each term in the left column with its correct description from the right column.

_____ 1. collagen fiber
_____ 2. perichondrium
_____ 3. osteon
_____ 4. lacuna
_____ 5. fibroblast
_____ 6. matrix
_____ 7. elastic fiber
_____ 8. ground substance
_____ 9. mast cell
_____ 10. canaliculi
_____ 11. muscle fiber cell membrane
_____ 12. cardiocyte connectors

A. extracellular material
B. column of bone tissue
C. produces matrix fibers
D. nutrient channels in bone matrix
E. protein fiber that provides flexibility
F. small space that surrounds a cell
G. outer membrane of cartilage
H. causes inflammation
I. sarcolemma
J. syrupy fluid of matrix
K. protein fiber that provides strength
L. intercalated disc

B. Identification

Identify the tissue or structure described in each statement.

1. Cells in which nucleus and cytoplasm are pushed against cell membrane: _____
2. Cell types in this tissue include mast cells, fibroblasts, and macrophages: _____
3. Solid matrix lamellae that surround central canals: _____
4. Epithelium that occurs in a single layer: _____
5. Found in intervertebral discs, chondrocytes stacked inside lacunae: _____
6. Parallel bundles of collagen fibers with fibroblasts between fibers: _____
7. Gelatinous matrix, chondrocytes in lacunae, elastic fibers in matrix: _____
8. Many cells among network of reticular fibers: _____
9. Tissue deep to epithelium: _____
10. Epithelium that stretches and relaxes: _____
11. Cells that secrete mucus: _____
12. Tissue made of multinucleated striated cells: _____

61

EXERCISE 4 — LAB REPORT

C. **Completion**

List the cell type(s) found in the following connective tissues.

1. Hyaline cartilage: _____
2. Bone tissue: _____
3. Adipose tissue: _____
4. Areolar tissue: _____
5. Elastic cartilage: _____
6. Dense regular connective tissue: _____
7. Reticular tissue: _____

D. **Labeling**

In the six numbered areas of Figure 4.14, label the following:

1. tissue
2. white line
3. tissue
4. black lines
5. ring system
6. tissue

E. **Labeling**

In the twelve numbered areas of Figure 4.15, label the following:

1. thin line
2. thick line
3. cell
4. cell
5. cell
6. layer
7. extensions
8. main branch
9. elongated white structures
10. vertical stripes
11. cell
12. tissue

LAB REPORT

EXERCISE 4

(a)

(b)

(c)

(d)

(e)

(f)

1. _____

2. _____

3. _____

4. _____

5. _____

6. _____

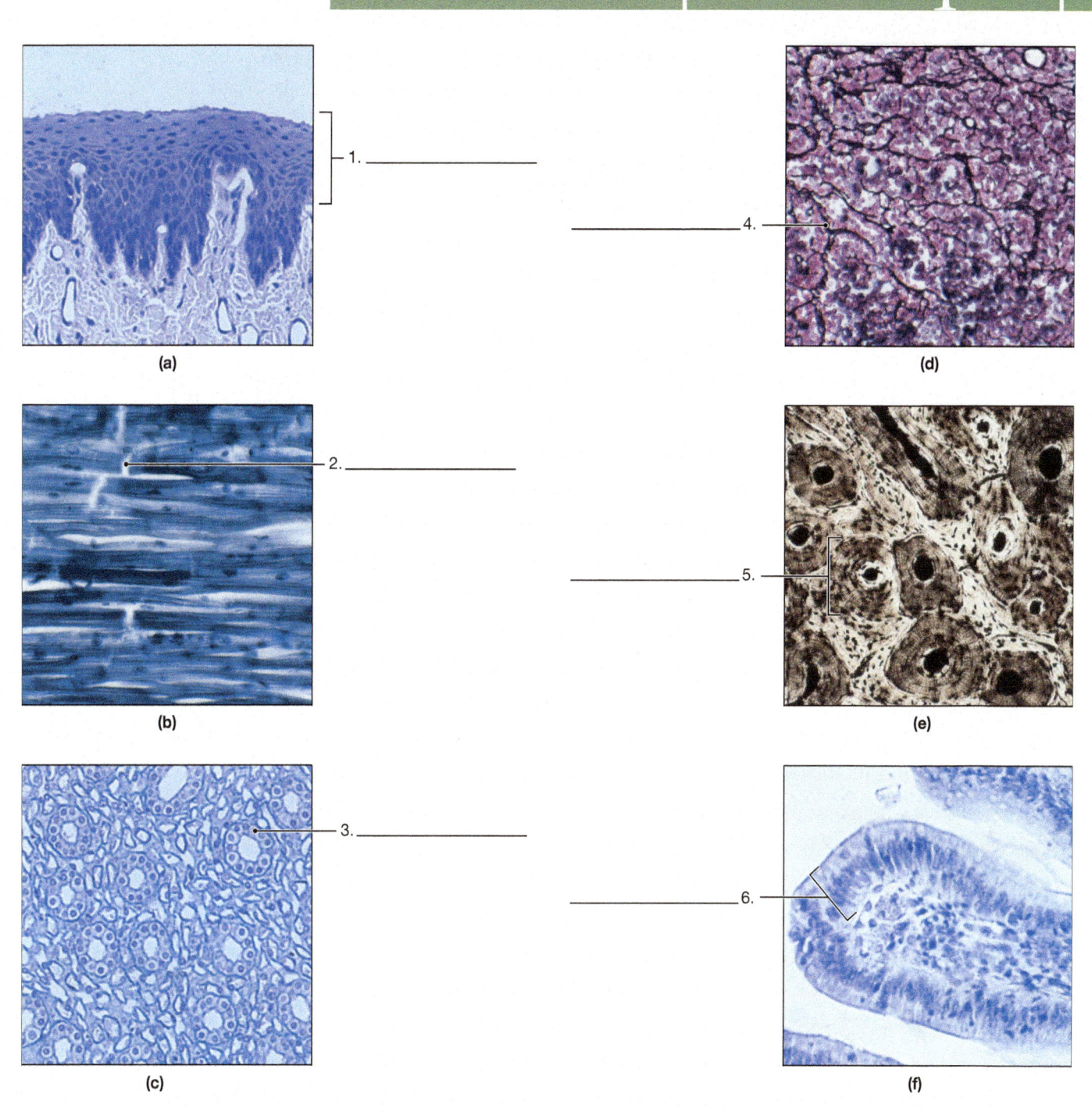

Figure 4.14 Histology Review 1

63

EXERCISE 4

LAB REPORT

(a)

1. _____
2. _____
3. _____

(b)

4. _____
5. _____
6. _____

(c)

7. _____
8. _____

(d)

9. _____
10. _____

(e)

11. _____

(f)

12. _____

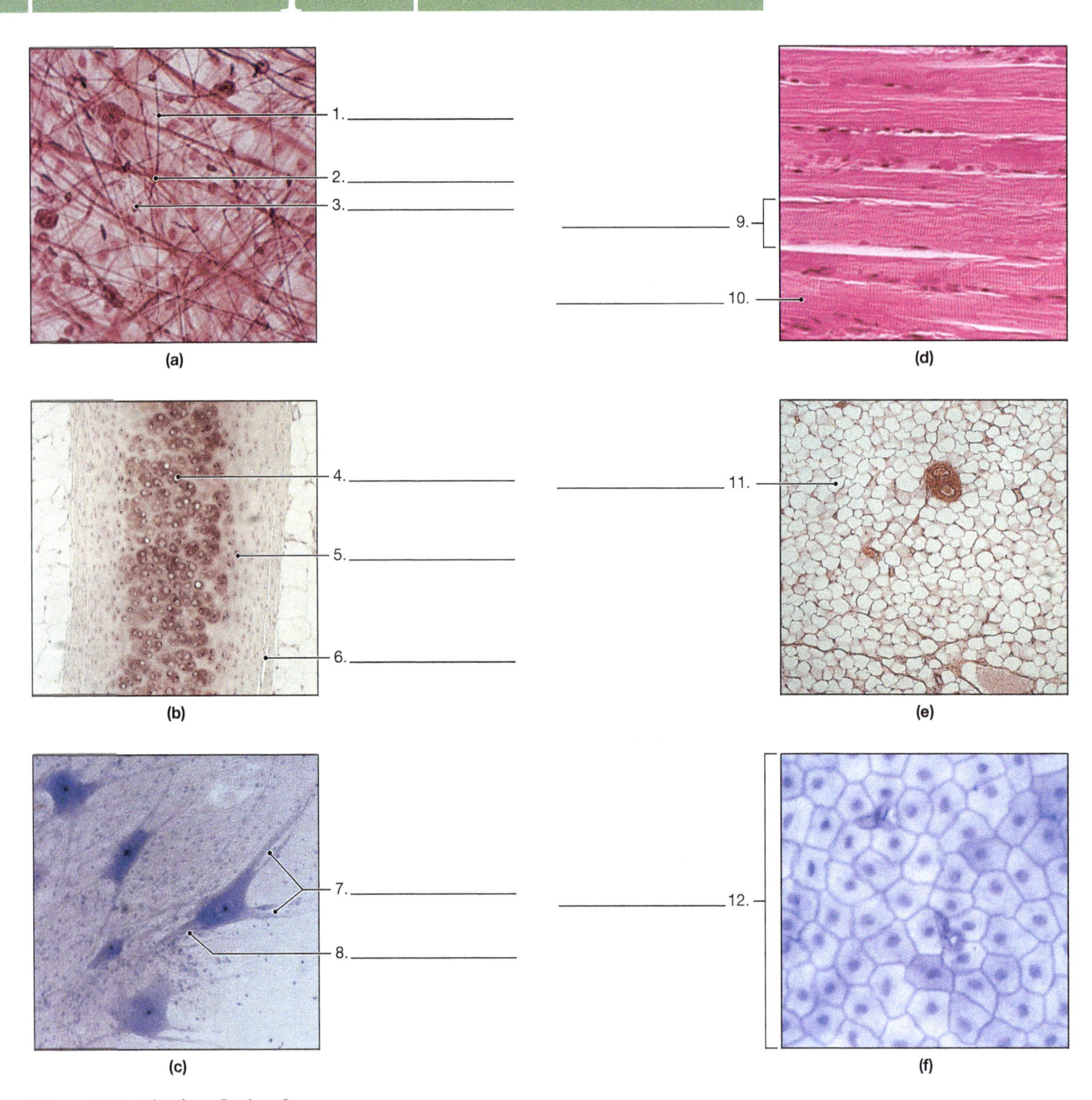

Figure 4.15 Histology Review 2

64

LAB REPORT — EXERCISE 4

F. **Short-Answer Questions**

1. Describe the different kinds of cells and layering of epithelia.

2. How is epithelium attached to underlying tissue?

3. List the three major groups of connective tissue and give an example of each.

4. What type of fibers are embedded in loose connective tissue?

5. What comprises the matrix in elastic cartilage?

6. Which types of muscle tissue are striated?

7. What is the function of intercalated discs in cardiac muscle?

8. Which muscle tissues are controlled involuntarily?

9. What are the basic functions of neural tissue?

EXERCISE 4

LAB REPORT

G. Analysis and Application

1. Suppose you are examining the lining of the stomach. Describe the tissue you are observing and relate its structure to its function in that location.

2. Compare and contrast the epithelium that covers the skin with the epithelium that lines the inner surface of the cheeks.

3. List the parts of a neuron through which an action potential passes, starting at a dendrite and ending where neurotransmitter molecules are released.

4. How are smooth muscle cells similar to skeletal muscle fibers? How are they different from skeletal muscle fibers?

5. How are skeletal and cardiac muscle tissues similar to each other? How do they differ from each other?

6. Tendons are made of dense regular connective tissue, and dermis is made of dense irregular connective tissue. Explain how this difference is related to the functions of tendons and dermis.

7. Describe the ground substance and fibers found in each type of cartilage.

EXERCISE 5

The Integumentary System

LAB ACTIVITIES

1. Epidermis and Dermis 67
2. Accessory Structures of the Integument 70

A Regional Look: Face 74

OBJECTIVES

On completion of this exercise, you should be able to:

- Identify the two major layers of the integument and their sublayers.
- Identify the accessory structures of the integument.
- Identify a hair follicle, the parts of a hair, and an arrector pili muscle.
- Distinguish between sebaceous glands and sweat glands.
- Describe three sensory organs of the integument.

The **integumentary** (in-TEG-ū-MEN-ta-ree) **system** is the most visible organ system of the human body. The **integument** (in-TEG-ū-ment), or skin, is classified as an organ system because it is composed of many types of tissues and organs. Organs of the integument include oil-, wax-, and sweat-producing glands; sensory organs for touch; muscles attached to hair follicles; and blood and lymphatic vessels.

The integument seals the body in a protective barrier that is flexible yet resistant to abrasion and evaporative water loss. People interact with the external environment via the integument. Caressing a baby's head, feeling the texture of granite, and testing the temperature of bath water all involve sensory organs of the integumentary system. Sweat glands cool the body to regulate body temperature. When exposed to sunlight, the integument manufactures vitamin D_3, a vitamin essential for calcium and phosphorus balance.

LAB ACTIVITY 1 — Epidermis and Dermis

There are two principal tissue layers in the integument: a superficial layer of epithelium called the *epidermis* and a deeper layer of connective tissue, the *dermis* (Figure 5.1). The **epidermis** consists of a stratified squamous epithelium that is organized into many distinct layers, or *strata*, of cells, as shown in Figure 5.2. Thick-skinned areas, such as the palms of the hands and soles of the feet, have five layers; thin-skinned areas have only four. Cells called **keratinocytes** are produced deep in the epidermis and are pushed superficially toward the surface of the skin. It takes from 15 to 30 days for a cell to migrate from the basal region to the surface of the epidermis. During this migration, the keratinocytes synthesize and accumulate the protein keratin, which disrupts the internal organization of the cell, and the cells die. These dry, scale-like keratinized cells on the surface of the epidermis are resistant to dehydration and friction.

68 EXERCISE 5 The Integumentary System

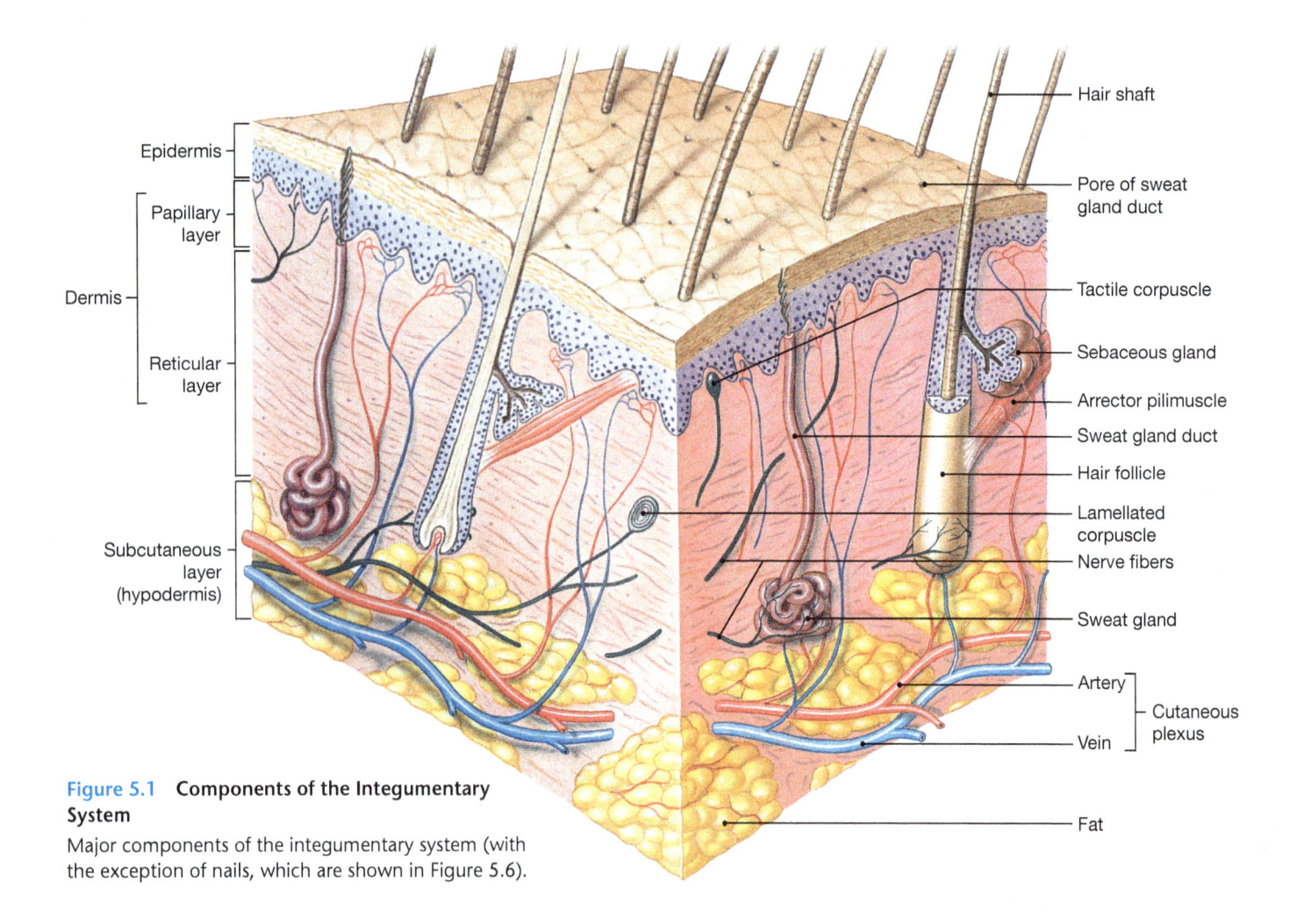

Figure 5.1 Components of the Integumentary System
Major components of the integumentary system (with the exception of nails, which are shown in Figure 5.6).

Moving superficially from the basal lamina, the five layers of the epidermis are as follows:

1. The **stratum germinativum** (STRA-tum jer-mi-na-TĒ-vum), or **stratum basale**, is a layer just one cell thick that connects the basal lamina of the epidermis to the upper surface of the dermis. The cells in this stratum are stem cells and so are in a constant state of mitosis, replacing cells that have rubbed off the epidermal surface. Other cells in this layer, called **melanocytes**, produce the pigment **melanin** (ME-la-nin), which protects deeper cells from the harmful effects of ultraviolet (UV) radiation from the sun. Prolonged exposure to UV light causes an increase in melanin synthesis, which results in a darkening, or tanning, of the integument.

2. Superficial to the stratum germinativum is the **stratum spinosum**, which consists of five to seven layers of cells interconnected by strong protein molecules between cell membranes that form cell attachments called **desmosomes**. When a slide of epidermal tissue is prepared, cells in this layer often shrink, but the desmosome bridges between cells remain intact. This results in cells with a spiny outline; hence the name "spinosum".

3. Superficial to the stratum spinosum is a layer of darker cells that make up the **stratum granulosum**. As cells from the stratum germinativum are pushed superficially, they synthesize the protein **keratohyalin** (ker-a-tō-HĪ-a-lin), which increases durability and reduces water loss from the integument surface. Keratohyalin granules stain dark and give this layer its color.

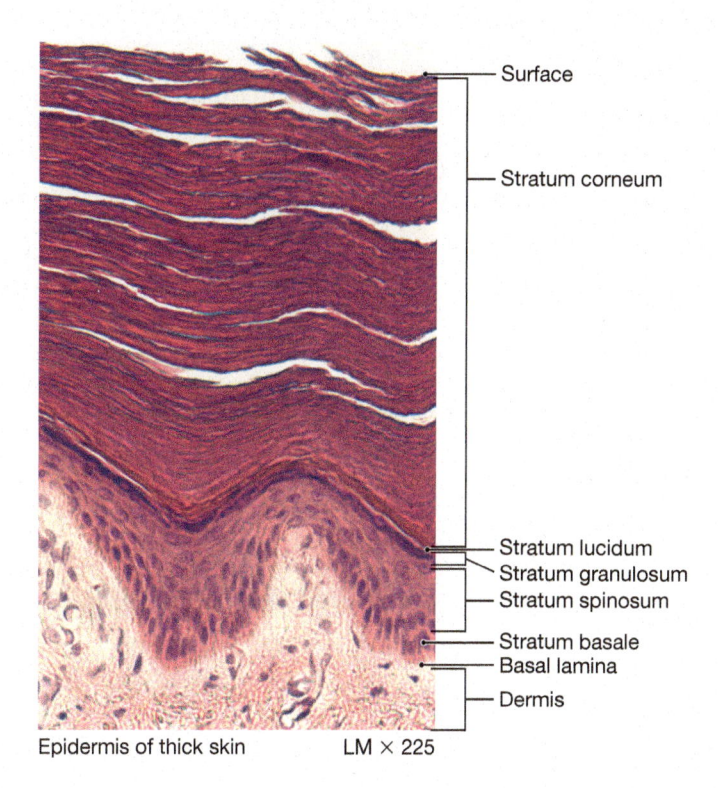

Figure 5.2 Layers of the Epidermis
A light micrograph through a portion of the epidermis, showing the major stratified layers of epidermal cells. (LM × 225)

4. In thick skin, a thin, transparent layer of cells called the **stratum lucidum** lies superficial to the stratum granulosum. Only the thick skin of the palms and the soles of the feet have the stratum lucidum; the rest of the skin is considered thin and lacks this layer.

5. The **stratum corneum** (KOR-nē-um; *cornu,* horn) is the most superficial layer of the epidermis and contains many layers of flattened, dead cells. As cells from the stratum granulosum migrate superficially, keratohyalin granules are converted to the fibrous protein keratin. Cells in the stratum corneum also accumulate the yellow-orange pigment **carotene**, which is common in light-skinned individuals.

Deep in the epidermis is the second of the two layers of the integument, the **dermis**, which is a thick layer of irregularly arranged connective tissue that supports and nourishes the epidermis and secures the integument to the underlying structures (Figures 5.1 and 5.2). The dermis is divided into two layers: *papillary* and *reticular*. Although there is no distinct boundary between these layers, the superficial portion of the dermis is designated the **papillary layer**. It consists of areolar tissue that contains numerous collagen and elastic fibers. Folds in the tissue are called **dermal papillae** (pa-PIL-lē; singular, *papilla,* a small cone) and project into the epidermis as the swirls of fingerprints. Within the dermal papillae are small sensory receptors for light touch, movement, and vibration called **tactile corpuscles** (also called *Meissner's corpuscles*).

Deep in the papillary layer is the **reticular layer** of the dermis. This layer is distinguished by a meshwork of thick bands of collagen fibers in dense, irregular connective tissue. Hair follicles and glands from the epidermis penetrate deep into the reticular layer. Sensory receptors in this layer, called **lamellated corpuscles** (*pacinian corpuscles*) detect deep pressure.

Attaching the dermis to underlying structures is the **hypodermis**, or **subcutaneous layer**, which is composed primarily of adipose tissue and areolar tissue. (Recall from Exercise 4 that *cutaneous membrane* is yet another name for the skin, thus the name *subcutaneous* for this layer.) The hypodermis is not part of the integumentary system.

Clinical Application   Burns

Burns are classified by the damage they cause to the layers of the integument. First-degree burns injure cells of the epidermis, no deeper than the stratum germinativum. Sunburns and other topical burns are first-degree burns. Second-degree burns destroy the entire epidermis and portions of the dermis but do not injure hair follicles and glands in the dermis. This destruction of portions of the dermis causes blistering, and the wound is extremely painful. Third-degree burns penetrate completely through the integument, severely damaging epidermis, dermis, and subcutaneous structures. This type of wound cannot heal because the restorative layers of the epidermis are lost. To prevent infection in cases of third-degree burns and to reestablish the barrier formed by the skin, a skin graft is used to cover the wound. Nerves are usually damaged by third-degree burns, so these more serious burns may not be as painful as first- and second-degree burns.

QuickCheck Questions

1.1 Describe the two layers of the integument.

1.2 How does the epidermis constantly replace its cells?

1 Materials

- ☐ Compound microscope
- ☐ Skin slide (cross-section)

Procedures

1. Scan the skin slide at low magnification and identify the epidermis, dermis, and hypodermis.

2. Increase the magnification to medium and examine the epidermis. Locate the epidermal layers, beginning with the deepest epidermal layer, the stratum germinativum.
   - What is the shape of the cells in the stratum spinosum?
   - What color is the stratum granulosum?
   - Does the specimen have a stratum lucidum?
   - What is the top layer of cells called? Are these cells alive?

3. Study the dermis at low, medium, and high magnifications.
   - Distinguish between the areolar tissue in the papillary layer and the dense, irregularly-arranged tissue in the reticular layer.
   - Are tactile corpuscles visible at the papillary folds?
   - What type of connective tissue is in the reticular layer?

LAB ACTIVITY 2   Accessory Structures of the Integument

During embryonic development, the epidermis produces accessory integumentary structures called **epidermal derivatives**, which include oil and sweat glands, hair, and nails. These structures are exposed on the surface of the skin and project deep into the dermis.

- **Sebaceous** (se-BĀ-shus) **glands** are associated with hair follicles and secrete the oily substance **sebum**, which coats the hair shafts and the epidermal surface to reduce brittleness and prevent excessive drying of the hair and epidermis (Figure 5.3). **Sebaceous follicles** secrete sebum onto the surface of the skin to lubricate the skin and provide limited antibacterial action. These follicles are not associated with hair and are distributed on the face, most of the trunk, and the male reproductive organs.

EXERCISE 5 The Integumentary System 71

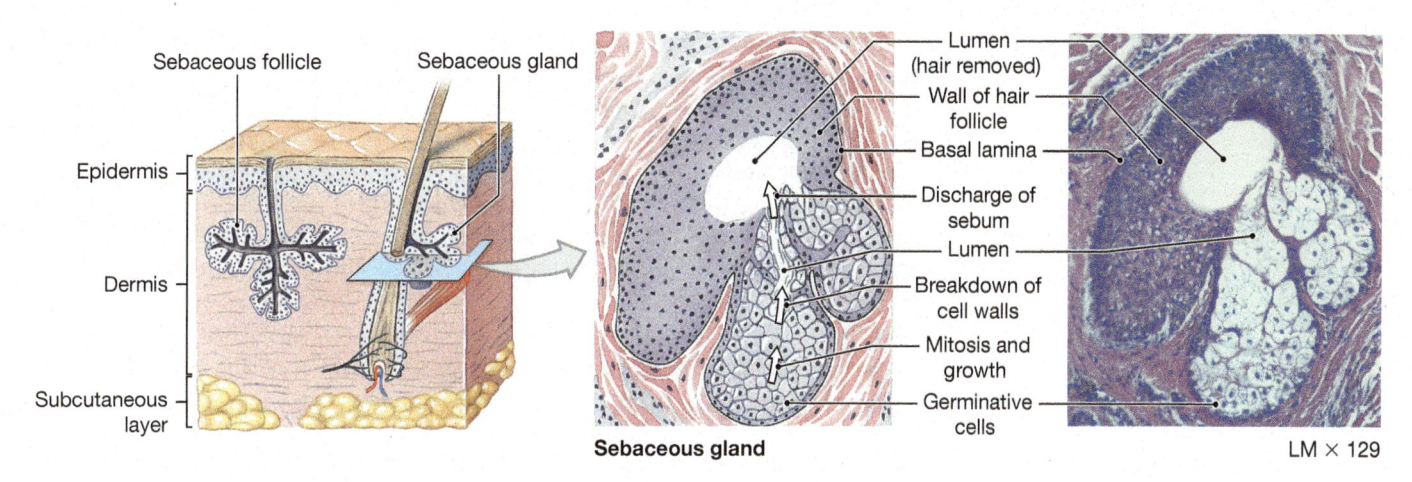

Figure 5.3 Sebaceous Glands and Follicles
The structure of sebaceous glands and sebaceous follicles in the skin. (LM × 129)

Clinical Application **Acne**

Many teenagers have dealt with skin blemishes called *acne*. During puberty, hormone levels increase, which activates sebaceous glands to produce more sebum. If a gland's duct becomes blocked, sebum accumulates and causes inflammation, which results in a pimple. A pimple with a white head indicates that a duct is blocked and full of sebum. A black head forms when an open sebaceous duct contains solid material that is infected with bacteria.

- **Sweat glands**, or **sudoriferous** (sū-dor-IF-er-us) **glands**, are scattered throughout the dermis of most of the integument. They are exocrine glands that secrete their liquid either into sweat ducts leading to the skin surface or into sweat ducts leading to hair follicles (Figure 5.4).

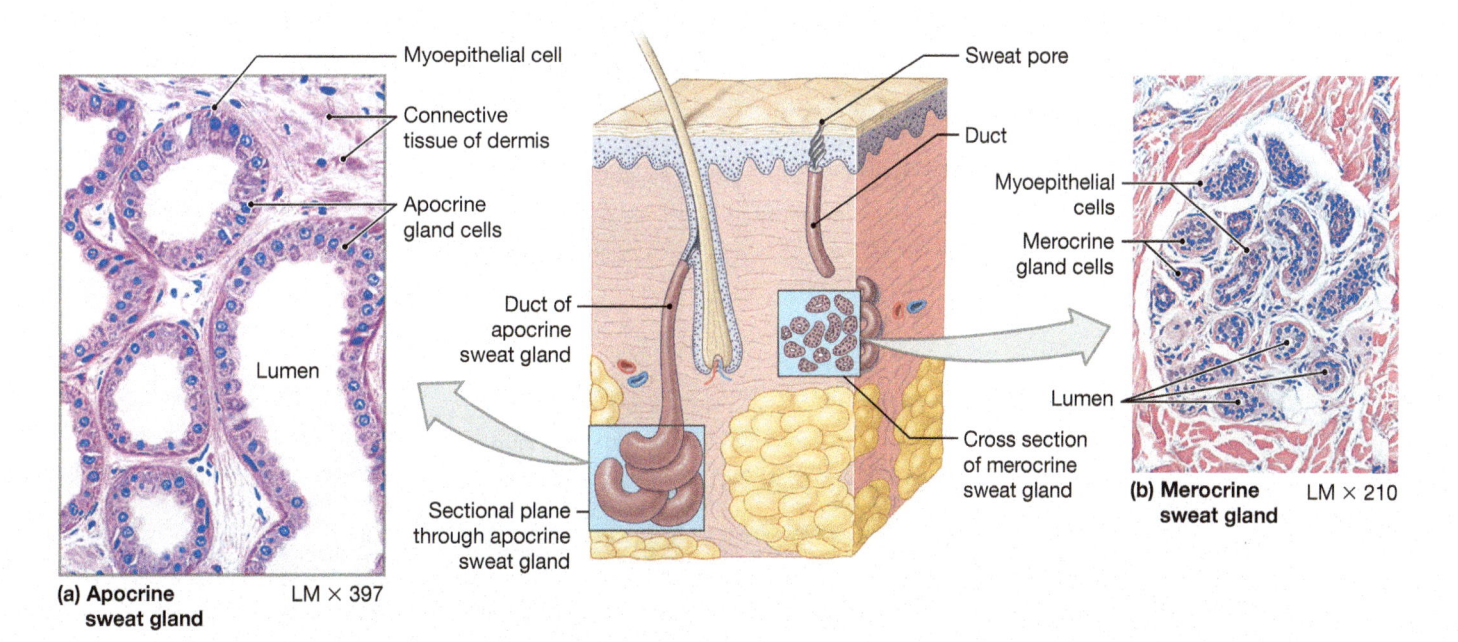

Figure 5.4 Sweat Glands
(a) Apocrine sweat glands, located in the axillae, groin, and nipples, produce a thick, odorous sweat. (LM × 397) **(b)** Merocrine sweat glands produce a thinner, watery sweat. (LM × 210)

The liquid we call **sweat** can be a thick or a thin substance. To cool the body, **merocrine** (MER-ō-krin) **sweat glands** secrete onto the skin surface a thin sweat that contains electrolytes, proteins, urea, and other compounds. The sweat absorbs body heat and evaporates from the skin, cooling the body. It also contributes to body odor because of the presence of urea and other wastes. Merocrine glands, also called *eccrine* (EK-rin) *glands*, are not associated with hair follicles and are distributed throughout most of the skin. **Apocrine sweat glands** are found in the groin, nipples, and axillae. These glands secrete a thick sweat into ducts that are associated with hair follicles. Bacteria on the hair metabolize the sweat and produce the characteristic body odor of, for example, axillary sweat.

- **Hair** covers most of the skin, with the exception of the lips, nipples, portions of the external genitalia, soles, palms, fingers, and toes. Three major types of hair are found in humans. **Terminal hairs** are the thick, heavy hairs on the scalp, eyebrows, and eyelashes. **Vellus hairs** are lightly pigmented and distributed over much of the skin as fine "peach fuzz". **Intermediate hairs** are the hairs on the arms and legs. Hair generally serves a protective function. It cushions the scalp and prevents foreign objects from entering the eyes, ears, and nose. Hair also serves as a sensory receptor. Wrapped around the base of each hair is a **root hair plexus**, a sensory neuron that is sensitive to movement of the hair.

  Each hair has a **hair root** embedded deep in a hair follicle (Figure 5.5). At the root tip is a **hair papilla** that contains nerves, blood vessels, and the hair **matrix**, which is the living, proliferative part of the hair. Cells in the matrix undergo mitotic divisions that cause the hair to elongate (i.e., it "grows"). Above the matrix, when the hair cells keratinize, they harden and die. The **hair shaft** that remains contains an outer **cortex** and an inner **medulla**.

  A smooth muscle called the **arrector pili** (a-REK-tor PI-lē) **muscle** is attached to each hair follicle. When fur-covered animals are cold, this muscle contracts, which raises the hair and traps a layer of warm air next to the skin. In

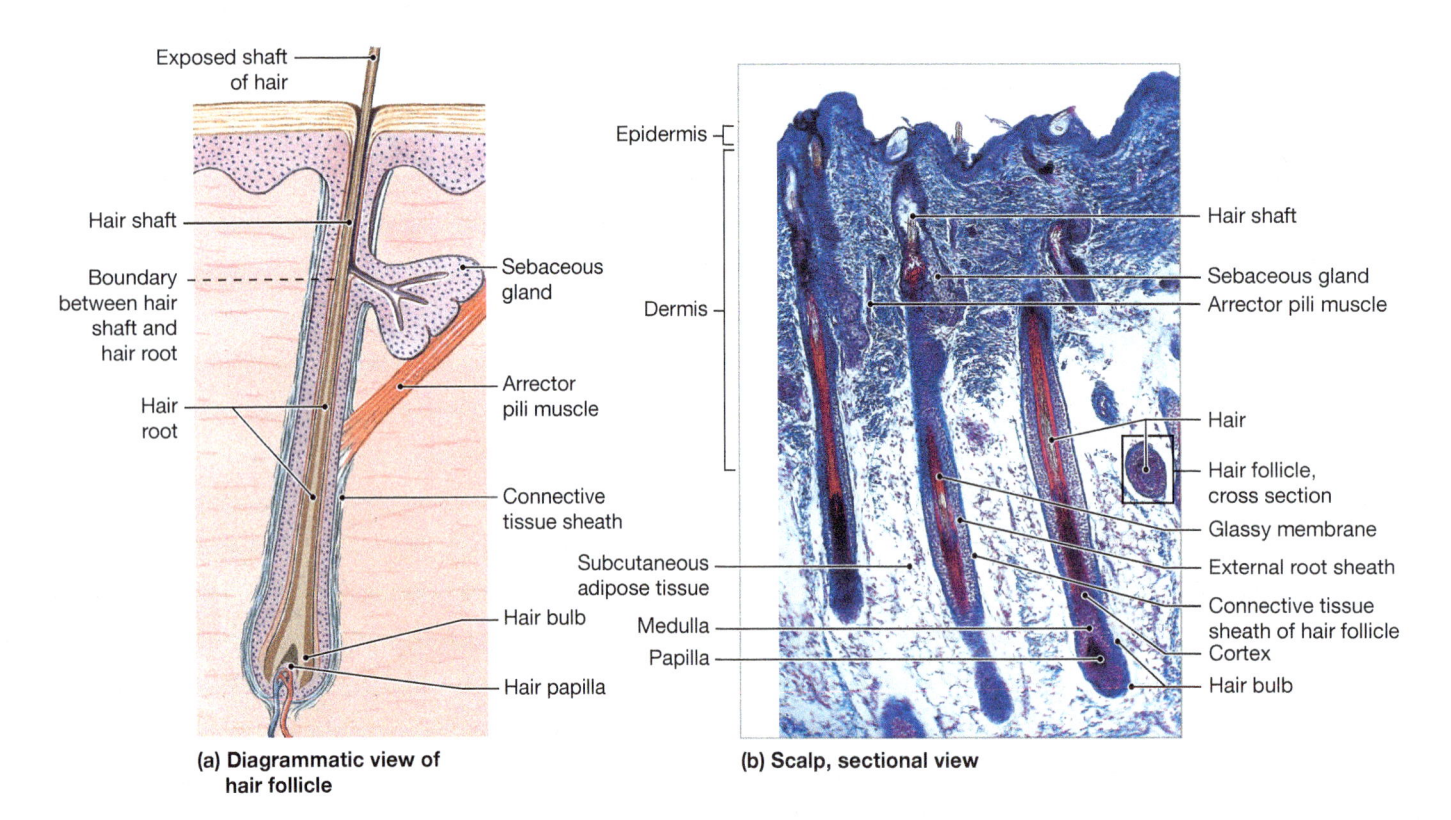

(a) Diagrammatic view of hair follicle

(b) Scalp, sectional view

Figure 5.5 Structure of a Hair
(a) The anatomy of a hair. (b) A light micrograph showing the sectional appearance of the skin of the scalp.

humans, the muscle has no known thermoregulatory use because humans do not have enough hair to gain an insulation benefit. We do have arrector pili muscles, though, and they contract when we are cold, which produces "gooseflesh".

- **Nails**, which protect the dorsal surface and tips of the fingers and toes, consist of tightly packed keratinized cells (Figure 5.6). The visible part of the nail, called the **nail body**, protects the underlying **nail bed** of the integument. Blood vessels underneath the nail body give the nail its pinkish color. The **free edge** of the nail body extends past the end of the digit. The **nail root** is at the base of the nail and is where new growth occurs. The **lunula** (LOO-nū-la; *luna*, moon) is a whitish portion of the proximal nail body where blood vessels do not show through the layer of keratinized cells. The epidermis around the nail is the **eponychium** (ep-ō-NIK-ē-um; *epi*, over + *onyx*, nail), what is commonly called the cuticle. At the cuticle the epidermis seals the nail with the nail groove and the raised nail fold. Under the free edge of the nail is the **hyponychium** (hī-pō-NIK-ē-um), a thicker region of the epidermis.

QuickCheck Questions

2.1 List the four main accessory structures of the integumentary system.

2.2 What are the two major types of glands in the integumentary system?

2 Materials

- ☐ Compound microscope
- ☐ Scalp slide (cross-section)

Procedures

1. On the scalp slide, locate a hair follicle.
   - What is the shape of the follicle? In which layer of the integument is it located?
   - Identify the hair shaft and, if possible, the cortex and medulla of a hair.
   - Identify a sebaceous gland. Where does it empty its secretions?
2. Scan the dermis of the slide for a sudoriferous gland.
   - Trace the duct from the gland to the surface of the skin. ■

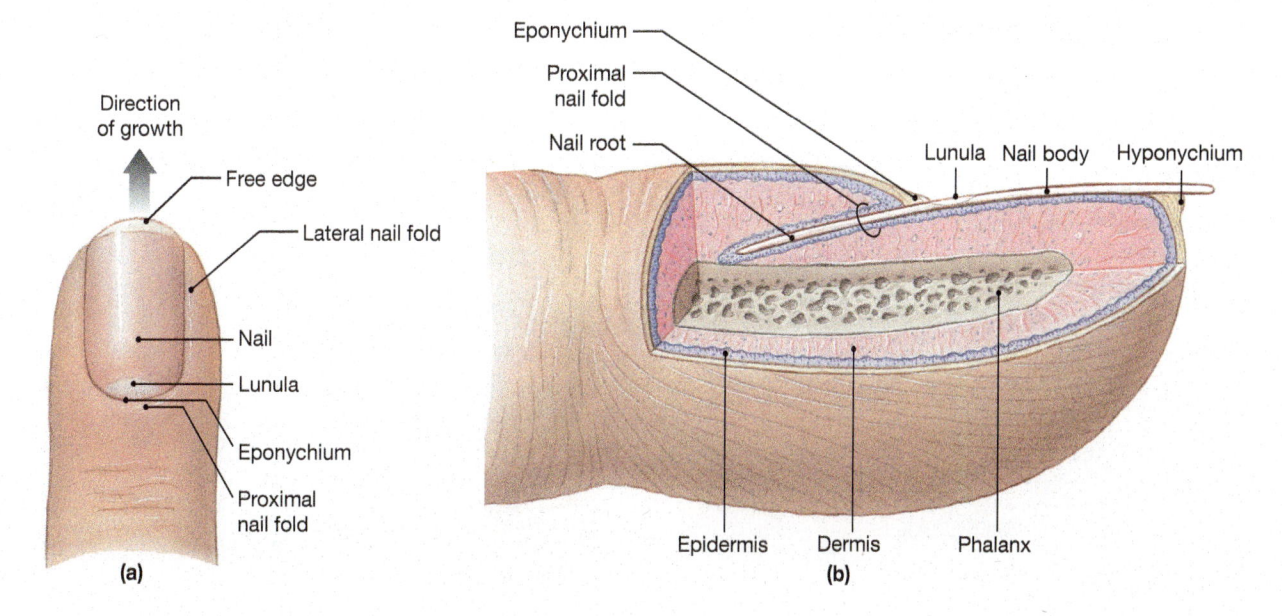

Figure 5.6 Structure of a Nail

The prominent features of a typical fingernail as viewed **(a)** from the surface and **(b)** in longitudinal section.

face   A Regional Look

Have you ever seen a television show or movie where all the police have is the skeleton of a murder victim, but a forensic scientist is able to construct a face on the victim's skull? The process involves measuring bony landmarks to calculate the size of attached muscles and soft tissues. Reference measurements assist in determining the size of the ears, nose, and eyes, and modeling clay is then used to form all the muscles of the face and cartilage for the nose and ears. Through this painstakingly slow process, the face is re-created and used to assist in identifying the victim. Observe the cadaver photograph in Figure 5.7 and notice how even though the integument has been removed, you still can visualize the face. One feature a forensic scientist would look at is the mandible of the lower jaw. Note how the cadaver's lower jaw has soft angles, creating a smooth rather than a prominent chin. In this example, the scientist would create a clay musculature that is slight rather than heavy.

What is also evident in this cadaver dissection is the anatomical association of muscles and the integument. Identify the zygomaticus muscles that attach to the skin at the corners of the mouth. When these muscles contract, they pull on the skin and the mouth is pulled into a smile. Place your fingers on your face just inferior to your cheek bones, called the *zygomatic bones*, and smile. Did you feel your zygomaticus muscles contract?

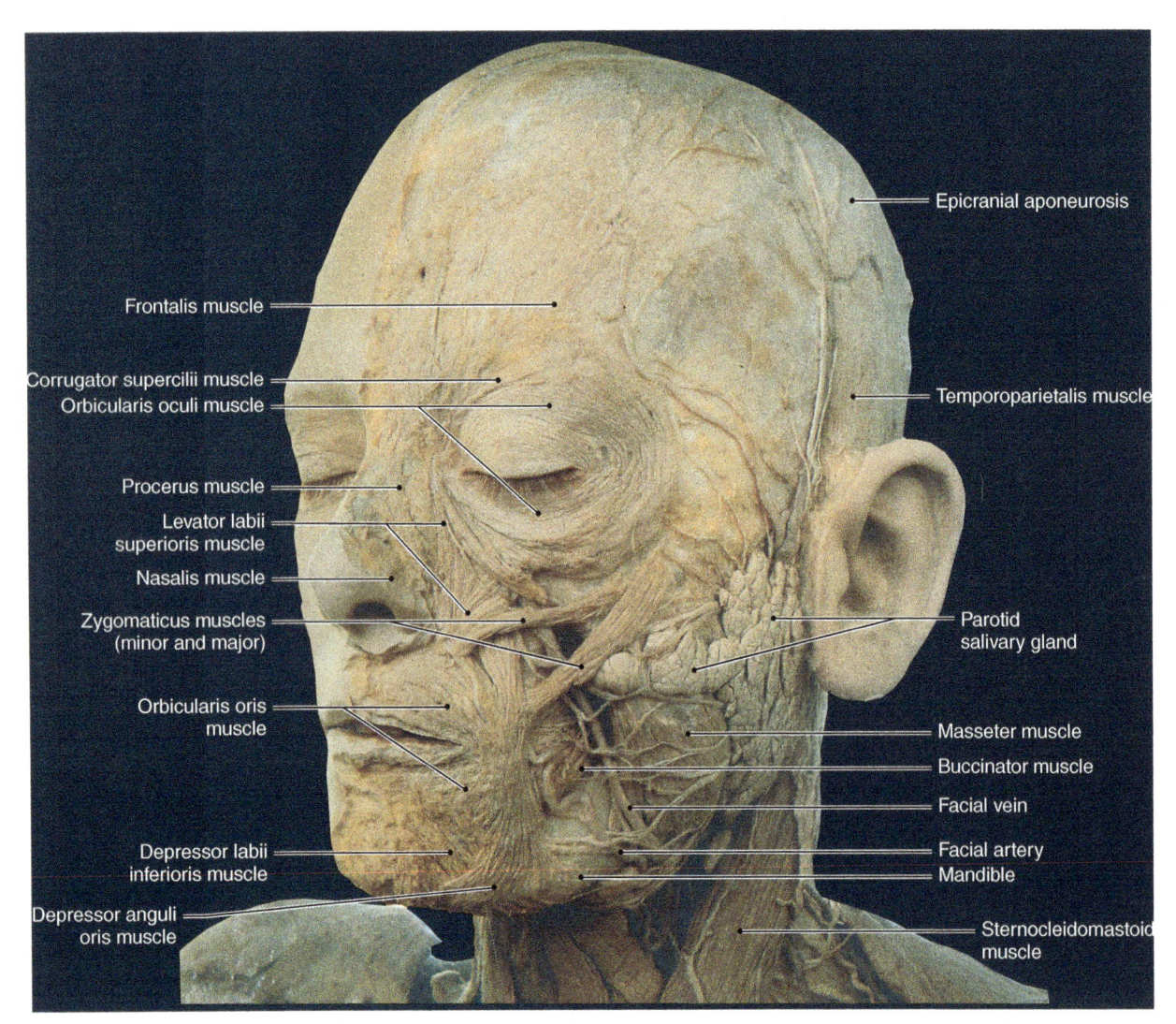

Figure 5.7 Face of a Cadaver with Integument Removed

LAB REPORT

The Integumentary System

Name _____
Date _____
Section _____

A. Matching

Match each skin structure in the left column with its correct description from the right column.

_____ 1. sebaceous gland
_____ 2. apocrine sweat gland
_____ 3. keratinocyte
_____ 4. arrector pili
_____ 5. stratum corneum
_____ 6. papillary layer
_____ 7. stratum germinativum
_____ 8. reticular layer
_____ 9. subcutaneous layer
_____ 10. stratum lucidum
_____ 11. eccrine sweat gland

A. layer in thickened areas of epidermis
B. deep to dermis, contains adipose tissue
C. sweat gland associated with hair follicle
D. produces new epidermal cells
E. deep layer of dermis
F. produces a protein that reduces water loss
G. functions in thermoregulation
H. surface layer of epidermis
I. muscle attached to hair follicle
J. folded layer of dermis next to epidermis
K. produces sebum

B. Labeling

Label the structures of the skin represented by the numbers in Figure 5.8.

C. Short-Answer Questions

1. Describe the layers of epidermis in an area where the integument is thick.

2. How does the integument tan when exposed to sunlight?

EXERCISE 5

LAB REPORT

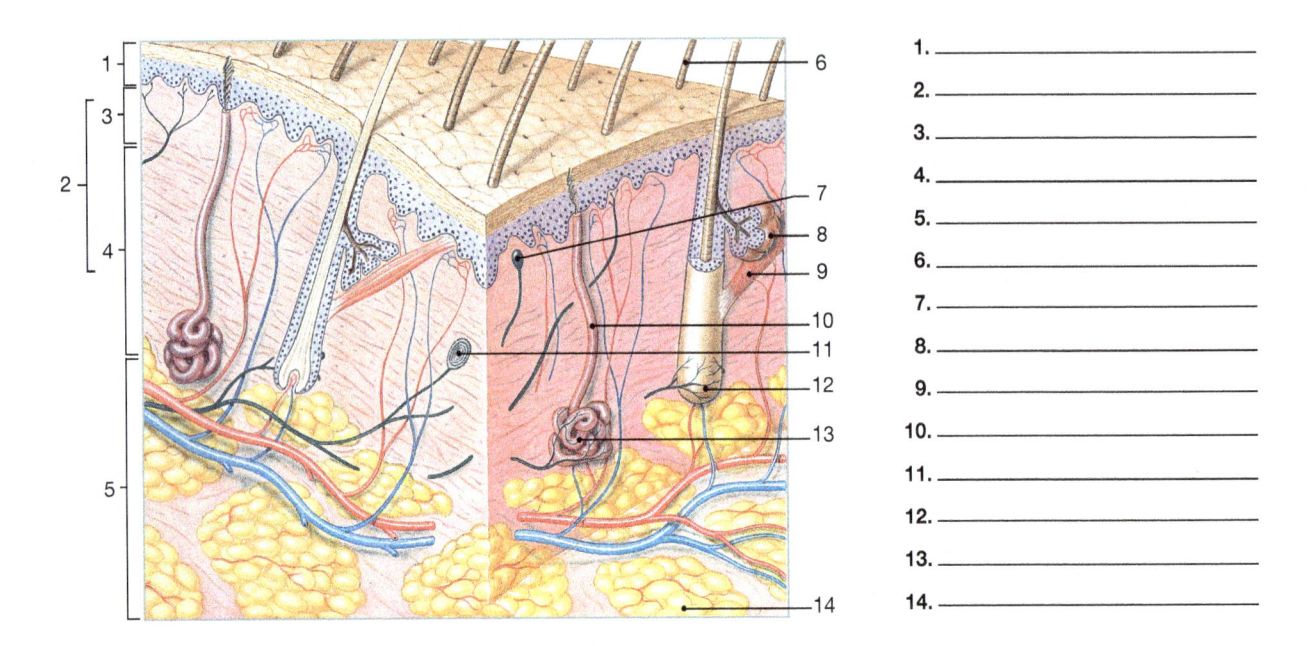

1. _____
2. _____
3. _____
4. _____
5. _____
6. _____
7. _____
8. _____
9. _____
10. _____
11. _____
12. _____
13. _____
14. _____

Figure 5.8   Components of the Integumentary System

3. List the types of sweat glands associated with the integumentary system.

4. What is the function of arrector pili muscles in animals other than humans?

D. Analysis and Application

1. What does keratinizing do to an epidermal cell?

2. What is the main cause of acne, and in which part of the skin does it occur?

3. How are cells replaced in the epidermis?

EXERCISE 6

Organization of the Skeletal System

OBJECTIVES

On completion of this exercise, you should be able to:

- Describe the gross anatomy of a long bone.
- Describe the histological organization of compact bone and of spongy bone.
- List the components of the axial skeleton and those of the appendicular skeleton.
- Describe the bone markings visible on the skeleton.

LAB ACTIVITIES

1. Bone Structure 77
2. Histological Organization of Bone 79
3. The Skeleton 80
4. Bone Markings 83

The skeletal system serves many functions. Bones support the body's soft tissues and protect internal organs. Calcium, lipids, and other materials are stored in the bones, and blood cells are manufactured in them. Bones serve as levers that allow the muscular system to produce movement or maintain posture.

Bones may be grouped according to shape. **Long bones**, which are greater in length than in width, and **flat bones**, which are thin and plate-like. **Sutural**, or **Wormian, bones** occur where the interlocking joints of the skull, called **sutures**, branch and isolate a small piece of bone. The number of sutural bones varies from one person to another, and for this reason this number is not included when counting the number of bones in the skeletal system. The vertebrae of the spine are classified as **irregular bones** because they fit into no other category. Bones of the wrist and ankle are **short bones**, almost as wide as they are long. **Sesamoid bones** form inside tendons. The sesamoid bone known as the patella (kneecap), for instance, develops inside the tendons anterior to the knee and is the largest sesamoid bone.

LAB ACTIVITY 1 Bone Structure

Two types of bone tissue are found in the skeleton: **compact bone** and **spongy bone**. Compact bone is denser than spongy bone because the latter has an open framework that resembles a lattice.

As described in Exercise 4, bones are encapsulated in a *periosteum*. This tough, fibrous membrane, which covers the compact bone surface, appears shiny and glossy and is sometimes visible on a chicken bone or on the bone in a steak. Histologically, the periosteum is composed of two layers: an outer fibrous layer where muscle tendons and bone ligaments attach, and an inner cellular layer that produces the osteoblasts needed for bone growth and repair. Osteoblasts are eventually cemented in the bone matrix they produce and become the osteocytes that maintain the bone tissue.

Long bones, such as the femur of the thigh, have a long central shaft called the **diaphysis** (dī-AF-i-sis), with an **epiphysis** (ē-PIF-i-sis) on each end (Figure 6.1a).

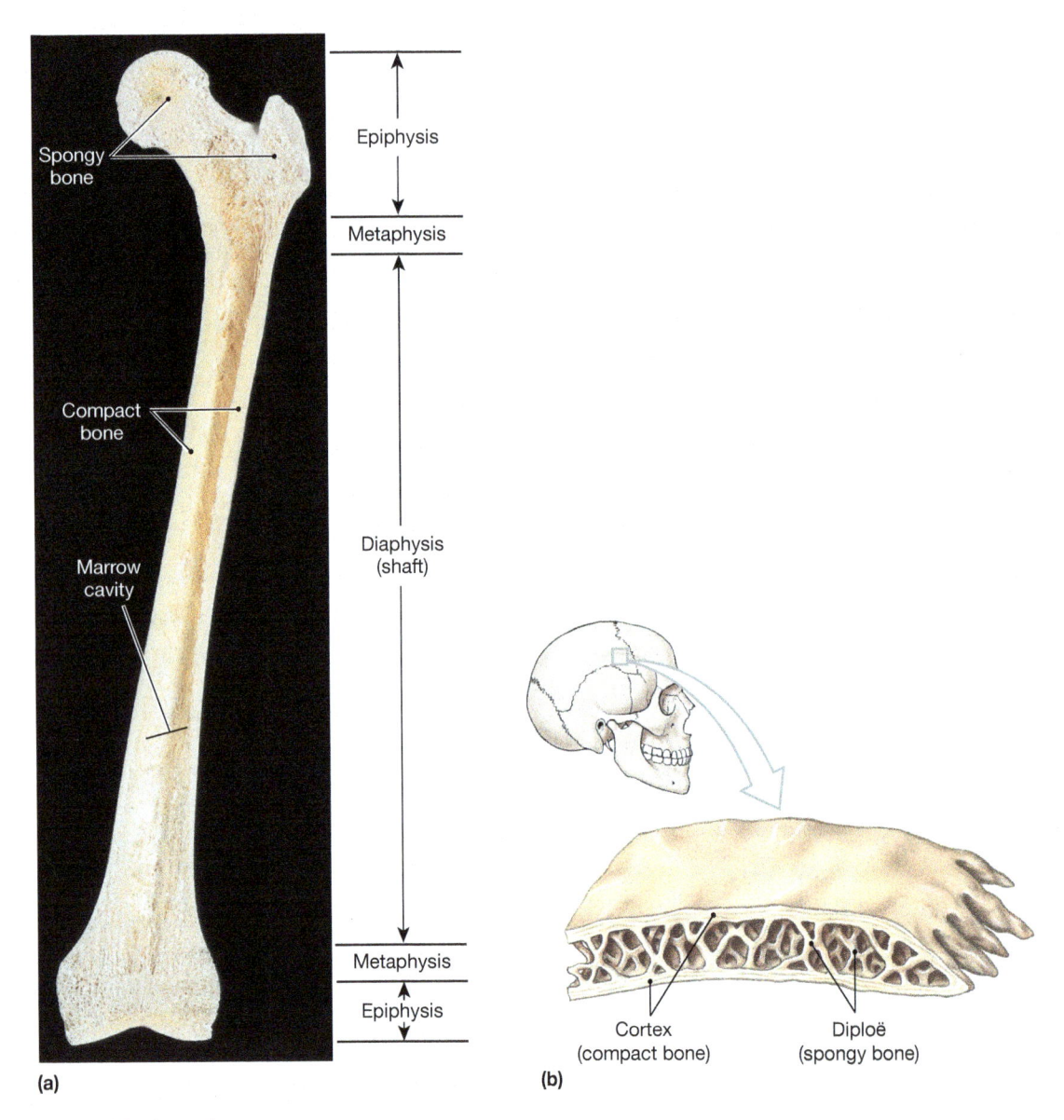

Figure 6.1 Bone Structure
(a) The structure of a representative long bone in longitudinal section. (b) The structure of a flat bone.

The proximal epiphysis is on the superior end of the diaphysis, and the distal epiphysis is on the inferior end. Wherever an epiphysis articulates with another bone, a layer of hyaline cartilage, the *articular cartilage,* covers the epiphysis.

The wall of the diaphysis is made of compact bone. The interior of the diaphysis is hollow, forming a space called the **marrow cavity.** This cavity is lined with spongy bone and is a storage site for **marrow**, a loose connective tissue. The marrow in long bones contains a high concentration of lipids and is called **yellow marrow**. A membrane called the **endosteum** (en-DOS-stē-um) lines the marrow cavity. Osteoclasts in the endosteum secrete carbonic acid, which dissolves bone matrix so that it can be replaced with new, stronger bone in a process called **remodeling** or so that minerals stored in the bone can be released into the blood.

Between the diaphysis and either epiphysis is the **metaphysis** (me-TAF-i-sis). In a juvenile's bone, the metaphysis is called the **epiphyseal cartilage** and consists of a plate of hyaline cartilage that allows the bone to grow longer. By early adulthood, the rate of mitosis in the epiphyseal cartilage slows, and *ossification* fuses the epiphysis to the diaphysis. Bone growth stops when all the cartilage in the metaphysis has been replaced by bone. This bony remnant of the growth plate is now called the **epiphyseal line**.

Flat bones, such as the frontal and parietal bones of the skull, are thin bones with no marrow cavity. They are made of a layer of spongy bone sandwiched between superficial layers of compact bone (Figure 6.1b). The compact bone layers, collectively called the *cortex* of the bone and individually called the *external* and *internal tables*, are thick in order to provide strength for the bone. The spongy bone between the tables is called **diploë** (DIP-lō-ē) and is filled with **red marrow**, a type of loose connective tissue made up of stem cells that produces most blood cells.

QuickCheck Questions

1.1 What is the location of the two membranes found in long bones?

1.2 Where is spongy bone found?

1 Materials

- ☐ Preserved or fresh long bone
- ☐ Blunt probe
- ☐ Disposable examination gloves
- ☐ Safety glasses

Procedures

1. Put on the safety glasses and examination gloves before you handle the bone.
2. Examine the bone and locate the periosteum. Does it appear shiny? Are any tendons or ligaments attached to it? _____
3. If the bone has been sectioned, observe the internal bone tissue of the diaphysis. Is this tissue similar in all regions of the sectioned bone? _____
4. Locate an epiphysis and its articular cartilage. What is the function of the cartilage? _____
5. Locate the metaphysis. Most likely the bone has an epiphyseal line, rather than an epiphyseal cartilage. Why? _____ ■

LAB ACTIVITY 2 Histological Organization of Bone

Compact bone contains support columns called either *osteons* or *Haversian systems* (Figure 6.2), and each osteon consists of many rings of calcified matrix called *concentric lamellae*. Between the lamellae, in small *lacunae*, are the mature osteocytes, which maintain the mineral and protein components of bone matrix. Bone requires a substantial supply of nutrients and oxygen. Nerves, blood vessels, and lymphatic vessels all pierce the periosteum and enter the bone in a **perforating canal** that is oriented perpendicular to the osteons. This canal interconnects with the *central canal* at the center of each osteon. Radiating outward from each central canal are the smaller *canaliculi* that facilitate nutrient, gas, and waste exchange with the blood.

In order to maintain its strength and weight-bearing ability, bone tissue is continuously being remodeled in a process that leaves distinct structural features in compact bone. Old osteons are partially removed, and the concentric rings of lamellae are fragmented, which results in **interstitial lamellae** between intact osteons. Typically, the distal end of a bone is extensively remodeled throughout life, whereas areas of the diaphysis may never be remodeled. Other lamellae occur underneath the periosteum and wrap around the entire bone. These **circumferential lamellae** are added as a bone grows in diameter.

Unlike compact bone, spongy bone is not organized into osteons; instead, it forms a lattice, or meshwork, of bony struts called **trabeculae** (tra-BEK-ū-le). Each trabecula is composed of layers of lamellae that are intersected with canaliculi. Red marrow fills the spaces between the trabeculae. Spongy bone is always sealed with a thin outer layer of compact bone.

QuickCheck Questions

2.1 What are the three types of lamellae found in bone and their characteristics?

2.2 How is blood supplied to an osteon?

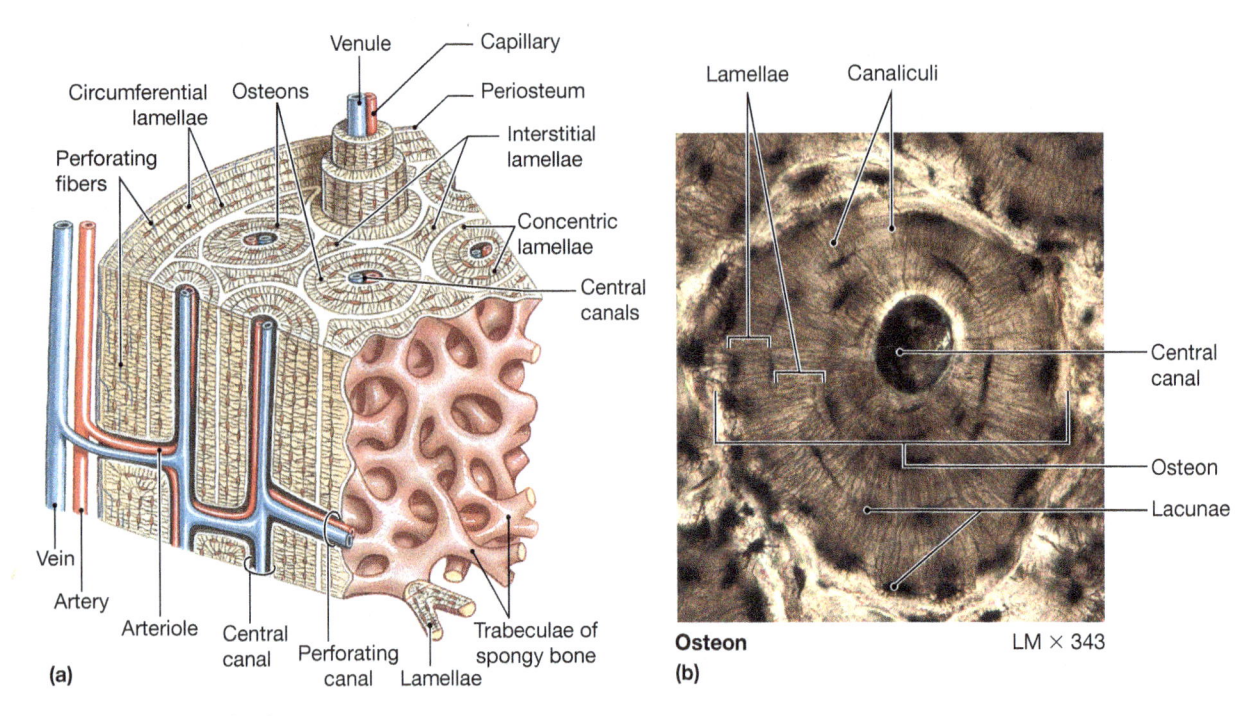

80 EXERCISE 6 Organization of the Skeletal System

Figure 6.2 Bone Histology
(a) The organization of compact and spongy bone. (b) A thin section through compact bone. The intact matrix and central canals appear white, and the lacunae and canaliculi appear black. (LM × 343)

2 Materials

- ☐ Compound microscope
- ☐ Bone model
- ☐ Bone tissue slide (transverse section)

Procedures

1. Examine a bone model and locate each structure that is shown in Figure 6.2. Your bone tissue slide is most probably a transverse section through bone that is ground very thin. This preparation process removes the bone cells but leaves the bone matrix intact for detailed studies.

2. At low magnification, observe the overall organization of the bone tissue. How many osteons can you locate? _____

3. Select an osteon and observe it at medium or high magnification. Identify the central canal, canaliculi, and lacunae. What is the function of the canaliculi? _____

4. Locate an area of interstitial lamellae. How do these lamellae differ from the concentric lamellae? _____ ■

LAB ACTIVITY 3   The Skeleton

The adult skeletal system, shown in Figure 6.3, consists of 206 bones. Each bone is an organ and includes bone tissue, cartilage, and other connective tissues. The skeleton is organized into the axial and appendicular divisions. The **axial division**, which is comprised of 80 bones, includes the **skull**, **vertebrae**, **sternum**, **ribs**, and **hyoid bone**. The **appendicular division** (126 bones) consists of the **pectoral girdle**, **upper limbs**, **pelvic girdle**, and **lower limbs**. Each girdle attaches its respective limbs to the axial skeleton and allows the limbs mobility at the points of attachment.

Each side of the pectoral girdle includes a **scapula** (shoulder blade) and a **clavicle** (collar bone). Each upper limb consists of arm, forearm, wrist, and hand. The **humerus** is the arm bone, and the **ulna** and **radius** together form the forearm. The eight wrist bones, called **carpal bones**, articulate with the elongated **metacarpal bones** of the palm. The individual bones of the fingers are the **phalanges**.

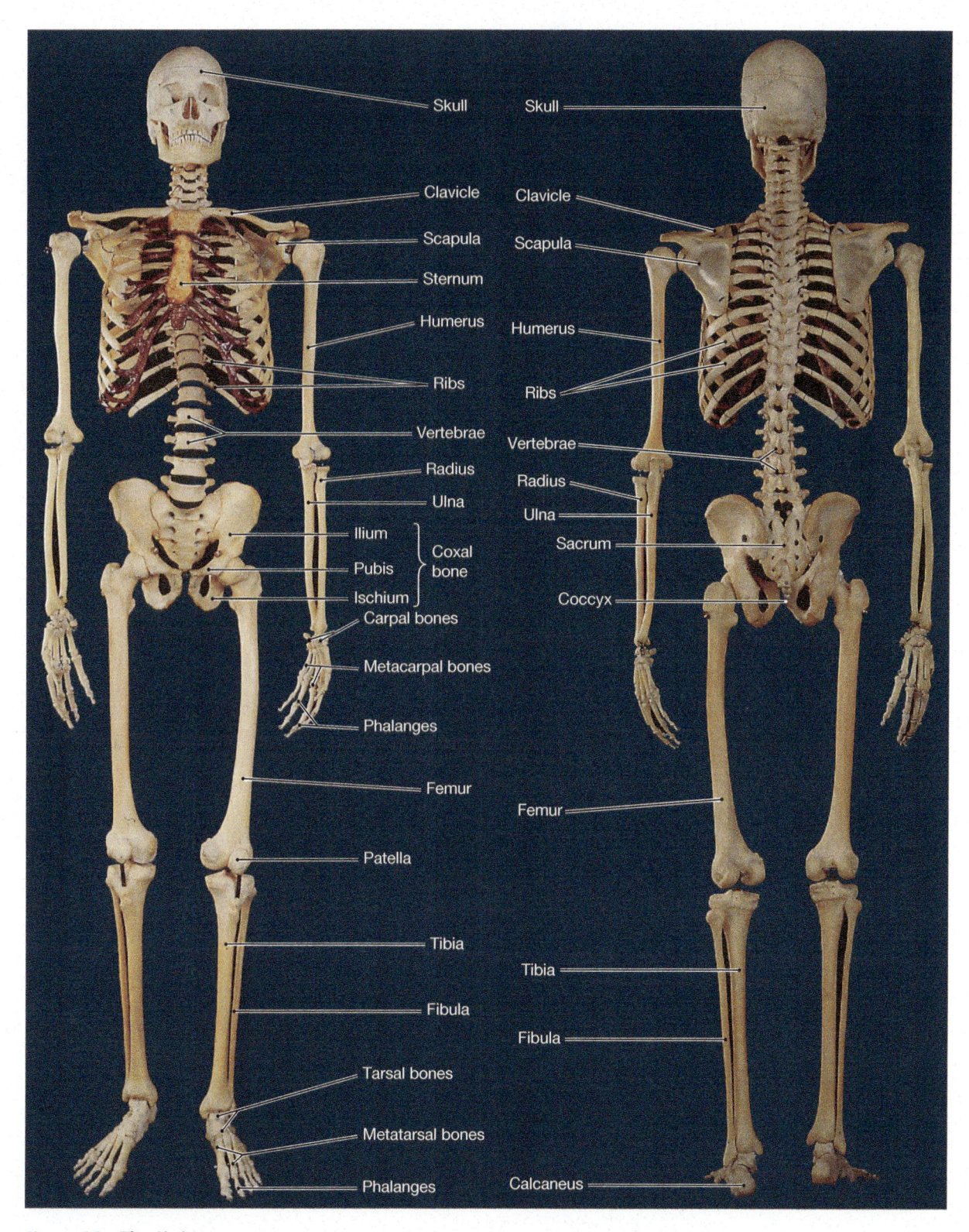

Figure 6.3 The Skeleton
Left, anterior view; right, posterior view.

81

The pelvic girdle is fashioned from two **ossa coxae** (the singular is *os coxae*), each of which is an aggregate of three bones: the superior **ilium** in the hip area, the **ischium** inferior to the ilium, and the **pubis** in the anterior pelvis. Each lower limb is comprised of thigh, kneecap, leg, ankle, and foot. The **femur** is the thighbone and is the largest bone in the body. The two bones of the leg are the medial **tibia**, which bears most of the body weight, and the thin, lateral **fibula**. The **patella** occurs at the articulation between femur and tibia. The seven ankle bones are collectively called the **tarsal bones**. **Metatarsal bones** form the arch of the foot, and **phalanges** form the toes.

Clinical Application   **Osteoporosis**

As we age, our bones change. They become weaker and thinner, and they produce less collagen and are therefore less flexible. Calcium levels in the bone matrix decline, which results in brittle bones. **Osteopenia** is the natural, age-related loss of bone mass that begins as early as 30–40 years of age in some individuals. The loss is a result of a decrease in osteoblast activity while osteoclasts continue to remain active. Bone degeneration beyond normal loss is called **osteoporosis** and affects the epiphyses, vertebrae, and jaws and leads to weak limbs, decrease in height, and loss of teeth. Bone fractures are common as spongy bone becomes more porous and unable to withstand stress. Osteoporosis is more common in women, with 29% of women 45 years or older, but only 18% of males 45 years or older, having osteoporosis.

Osteoporosis is associated with an age-related decline in circulating sex hormones in the blood. Sex hormones stimulate osteoblasts to deposit calcium into new bone matrix. In menopausal women, decreasing levels of estrogen slow osteoblast activity, and one result is bone loss. As men age, hormone levels decline more gradually than in women, and as a result, most men are able to maintain a healthy bone mass.

Exercising more and consuming adequate amounts of calcium can reduce the rate of bone degeneration and occurrence of osteoporosis in both men and women. However, increasing calcium intake is not enough to prevent osteoporosis. New bone matrix must be produced in order to maintain bone density. Hormone replacement therapy is sometimes prescribed to promote new bone growth in postmenopausal women. Unfortunately, many studies link hormone replacement therapy with blood clots in the lungs, uterine cancer, and other clinical complications.

QuickCheck Questions

3.1 What are the two major divisions of the skeleton?

3.2 What are the five bone types that make up the axial skeleton?

3.3 What are the four main bone groups of the appendicular skeleton?

3.4 What is the meaning of *arm* and *leg* to an anatomist?

3 *Materials*   **Procedures**

☐ Articulated skeleton

1. On the articulated skeleton, identify the major bones of the axial division.
2. On the articulated skeleton, identify the major bones of the appendicular division.
3. Count the bones in one upper limb. How many are there? _____
4. What three bones fuse to form the os coxae? _____
5. Count the bones in one lower limb. How many are there? _____

EXERCISE 6 Organization of the Skeletal System 83

LAB ACTIVITY 4 Bone Markings

Each bone has certain anatomical features on its surface, called either **bone markings** or *surface markings*. A particular bone marking may be unique to a single bone or may occur throughout the skeleton. Table 6.1 illustrates the various types of bone markings and organizes them into five groups.

QuickCheck Questions

4.1 What is a foramen?
4.2 What is the neck of a bone?

4 Materials

☐ Articulated skeleton

Procedures

1. Using Table 6.1 as a reference, locate on the skeleton:
   - A *foramen* in the skull. Describe this structure. _____
   - A *fossa* on a humerus. How is the fossa different from a foramen? _____
   - A *head* on a femur and on a humerus. Which other bones have a head? _____
   - A *condyle* on two different bones. Describe this structure. _____
   - A *tuberosity* on a humerus. What is the texture of this structure? _____

2. Locate one instance of each of these markings on the skeleton: process, ramus, trochanter, tubercle, crest, line, spine, neck, trochlea, facet, sulcus, canal, fissure, and sinus. ■

84 EXERCISE 6 Organization of the Skeletal System

Table 6.1 The Terminology of Bone Markings

General Description	Anatomical Term	Definition
Elevations and projections (general)	Process	Any projection or bump
	Ramus	An extension of a bone making an angle with the rest of the structure
Processes formed where tendons or ligaments attach	Trochanter	A large, rough projection
	Tuberosity	A smaller, rough projection
	Tubercle	A small, rounded projection
	Crest	A prominent ridge
	Line	A low ridge
	Spine	A pointed process
Processes formed for articulation with adjacent bones	Head	The expanded articular end of an epiphysis, separated from the shaft by a neck
	Neck	A narrow connection between the epiphysis and the diaphysis
	Condyle	A smooth, rounded articular process
	Trochlea	A smooth, grooved articular process shaped like a pulley
	Facet	A small, flat articular surface
Depressions	Fossa	A shallow depression
	Sulcus	A narrow groove
Openings	Foramen	A rounded passageway for blood vessels or nerves
	Canal	A passageway through the substance of a bone
	Fissure	An elongated cleft
	Sinus or antrum	A chamber within a bone, normally filled with air

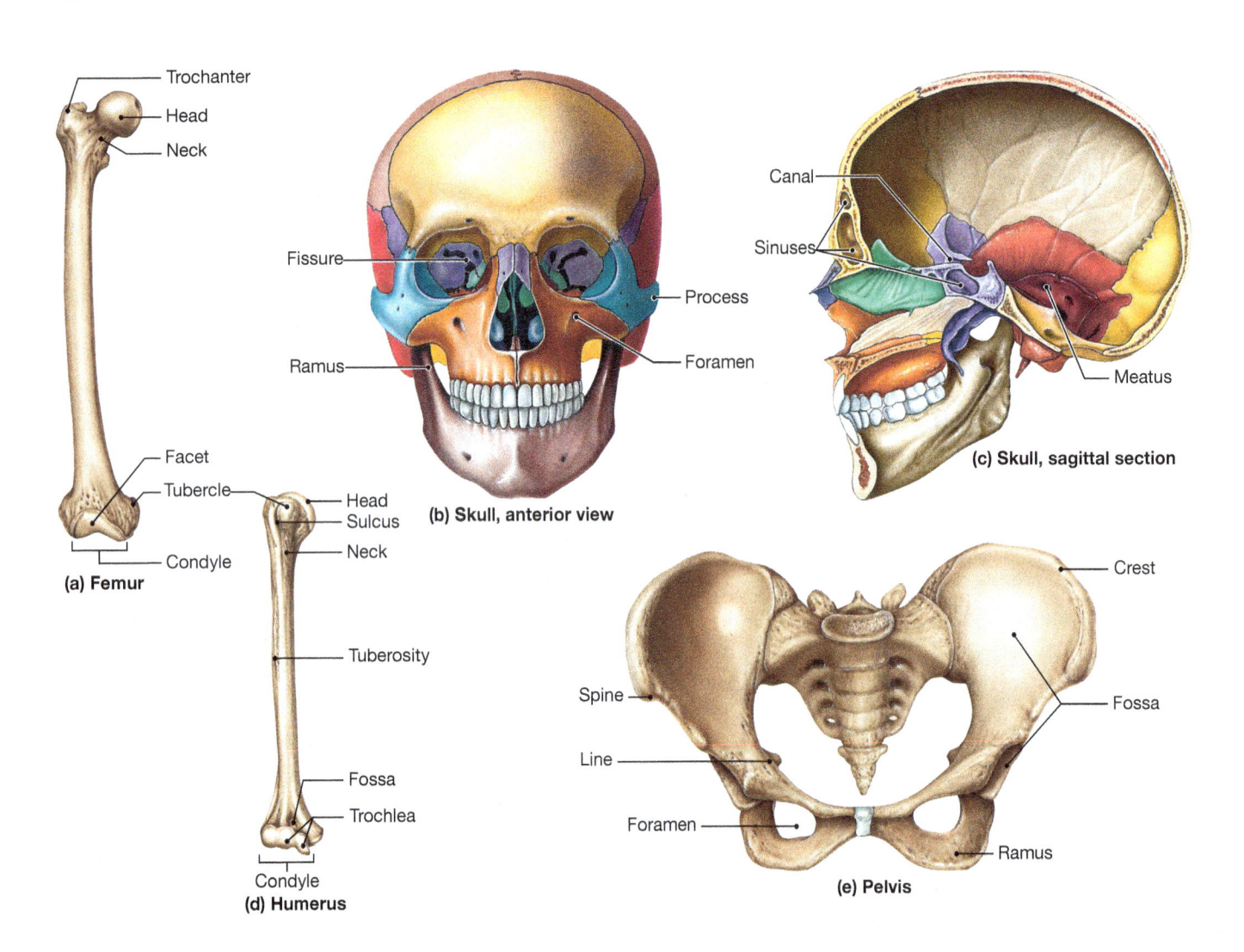

(a) Femur — Trochanter, Head, Neck, Facet, Tubercle, Condyle

(b) Skull, anterior view — Fissure, Ramus, Process, Foramen

(c) Skull, sagittal section — Canal, Sinuses, Meatus

(d) Humerus — Head, Sulcus, Neck, Tuberosity, Fossa, Trochlea, Condyle

(e) Pelvis — Crest, Fossa, Ramus, Foramen, Line, Spine

LAB REPORT

EXERCISE 6

Organization of the Skeletal System

Name _____
Date _____
Section _____

A. Matching

Match each structure in the left column with its correct description from the right column.

_____	1. lacuna	A.	bone shaft
_____	2. trabecula	B.	found in juvenile bones
_____	3. articular cartilage	C.	bony column
_____	4. diaphysis	D.	contains yellow marrow
_____	5. epiphyseal cartilage	E.	bone tip
_____	6. osteon	F.	lines marrow cavity
_____	7. periosteum	G.	found in adult bones
_____	8. interstitial lamellae	H.	forms osteon
_____	9. epiphysis	I.	bony projection
_____	10. endosteum	J.	produces bone matrix
_____	11. marrow cavity	K.	cellular space in bone matrix
_____	12. epiphyseal line	L.	remodeled osteons
_____	13. concentric lamellae	M.	outer membrane of bone
_____	14. osteoblast	N.	cartilage on epiphysis

B. Labeling

Label Figure 6.4.

C. Short-Answer Questions

1. List the major components of the axial skeleton.

2. List the major components of the appendicular skeleton.

85

EXERCISE 6

LAB REPORT

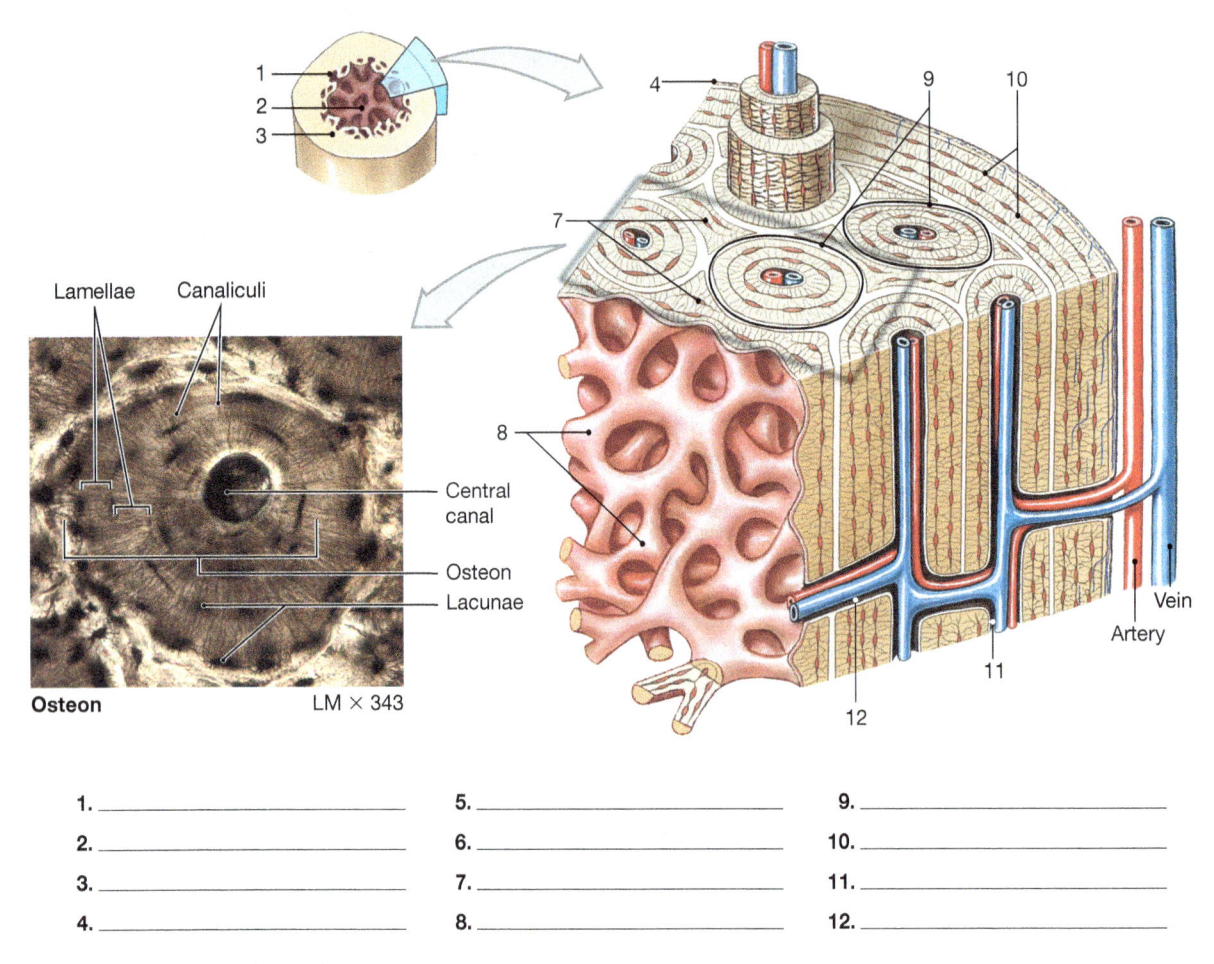

Lamellae **Canaliculi**

Central canal
Osteon
Lacunae

Osteon LM × 343

Vein
Artery

1. _____   5. _____   9. _____
2. _____   6. _____   10. _____
3. _____   7. _____   11. _____
4. _____   8. _____   12. _____

Figure 6.4 **The Histology of Bone**

3. Name the five groups of bone surface markings.

4. Name the six groups used to categorize bones by their shape.

D. **Analysis and Application**

1. Where does spongy bone occur in the skeleton?

2. How are the upper limbs attached to the axial skeleton?

3. Where does growth in length occur in a long bone?

EXERCISE 7

The Axial Skeleton

OBJECTIVES

On completion of this exercise, you should be able to:

- Identify the components of the axial skeleton.
- Identify the cranial and facial bones of the skull and describe their surface features.
- Describe the principal difference between a fetal skull and an adult skull.
- Name the four regions of the vertebral column.
- Identify the features of a typical vertebra.
- Describe the distinguishing features of each of the four types of vertebrae.
- Discuss the way in which the ribs articulate with the thoracic vertebrae.
- Identify the components of the sternum.

LAB ACTIVITIES

1. Cranial Bones   87
2. Facial Bones   95
3. Hyoid Bone   99
4. Sinuses of the Skull   99
5. Fetal Skull   101
6. Vertebral Column   102
7. Thoracic Cage   108

The axial skeleton provides both a central framework for attachment of the appendicular skeleton and protection for the body's internal organs. As noted in Exercise 6, the 80 bones of the axial skeleton include the skull, a thoracic cage made up of ribs, and the sternum, and a flexible vertebral column with 33 vertebrae. Of the 29 bones of the skull, 22 are organized into 14 **facial bones** and eight **cranial bones** that form the **cranium**, which is the space that encases the brain. The seven other bones of the skull, referred to as the *associated bones,* are the six bones of the middle ear (three per ear, described in Exercise 20) and the hyoid bone.

LAB ACTIVITY 1 Cranial Bones

The **frontal bone** of the cranium extends from the forehead posteriorly to the **coronal suture** and articulates (joins) with the two **parietal bones** (Figures 7.1 and 7.2). The parietal bones are joined superiorly by the **sagittal suture**. The **occipital bone** of the cranium meets the parietal bones at the **lambdoid** (LAM-doyd) **suture**, which completes the posterior wall of the cranium. The lambdoid suture is sometimes called the *occipitoparietal suture.*

The two **temporal bones** articulate with the parietal bone at the **squamous suture**. The squamous and coronal sutures are linked by the **sphenoparietal suture**. Inferior to this suture, the **sphenoid** is partly visible in Figure 7.1 as a vertical rectangle of bone. The sphenoid and parts of the frontal, temporal, and occipital bones form the floor of the cranium. Although it spans the floor of the cranium, the sphenoid is visible only on the lateral surface of the skull and in the posterior wall of the eye orbit (Figure 7.2) and in the inferior view of the skull (Figure 7.3). The single **ethmoid** is a small, rectangular bone

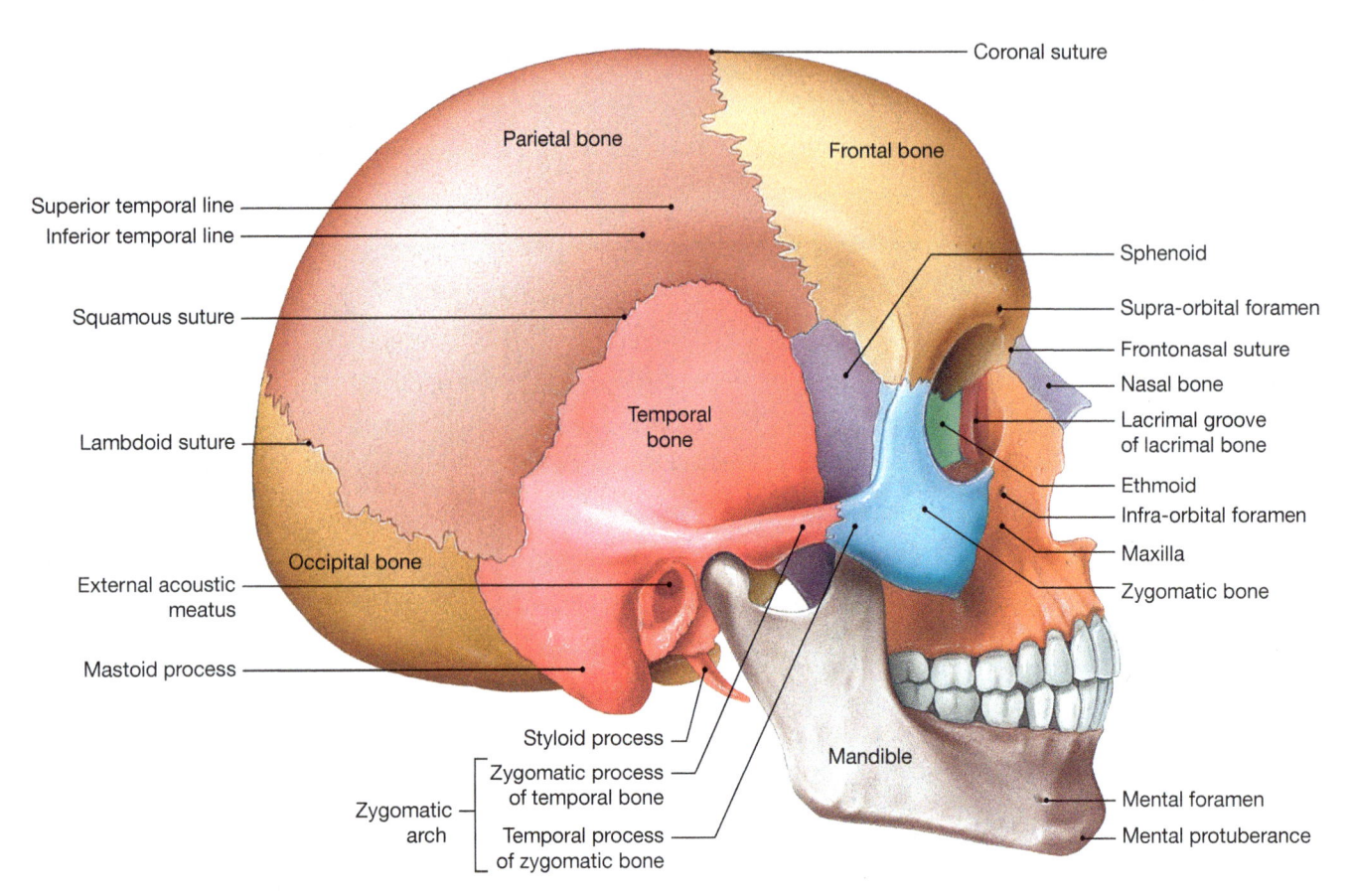

Figure 7.1 Lateral View of Skull

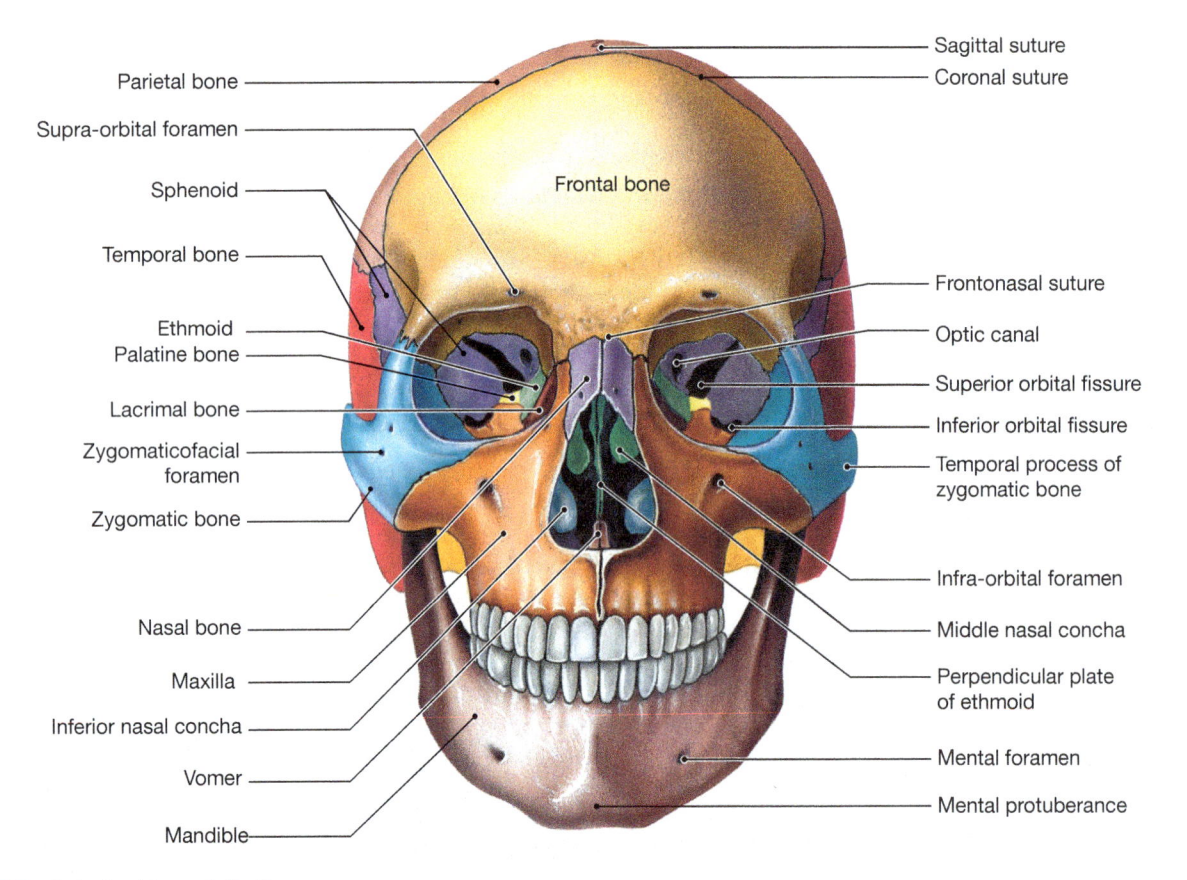

Figure 7.2 Anterior View of Skull

88

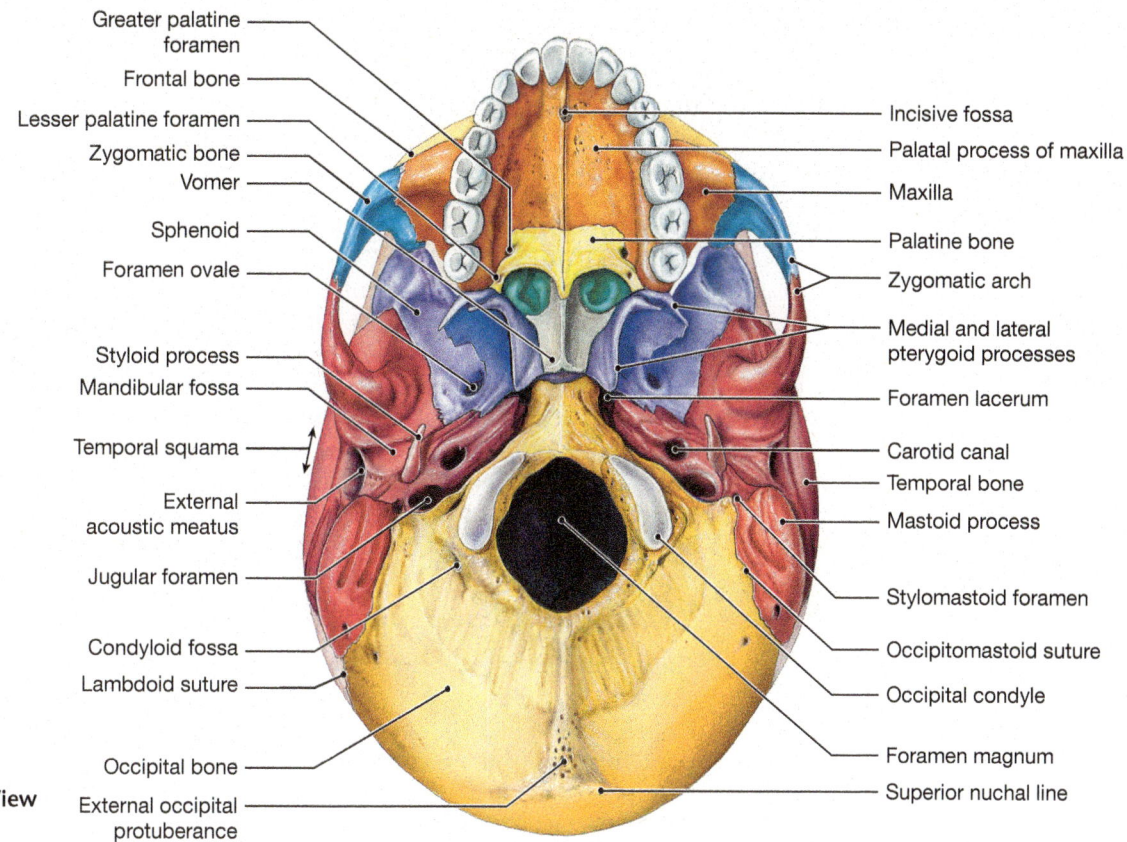

Figure 7.3 Inferior View of Skull

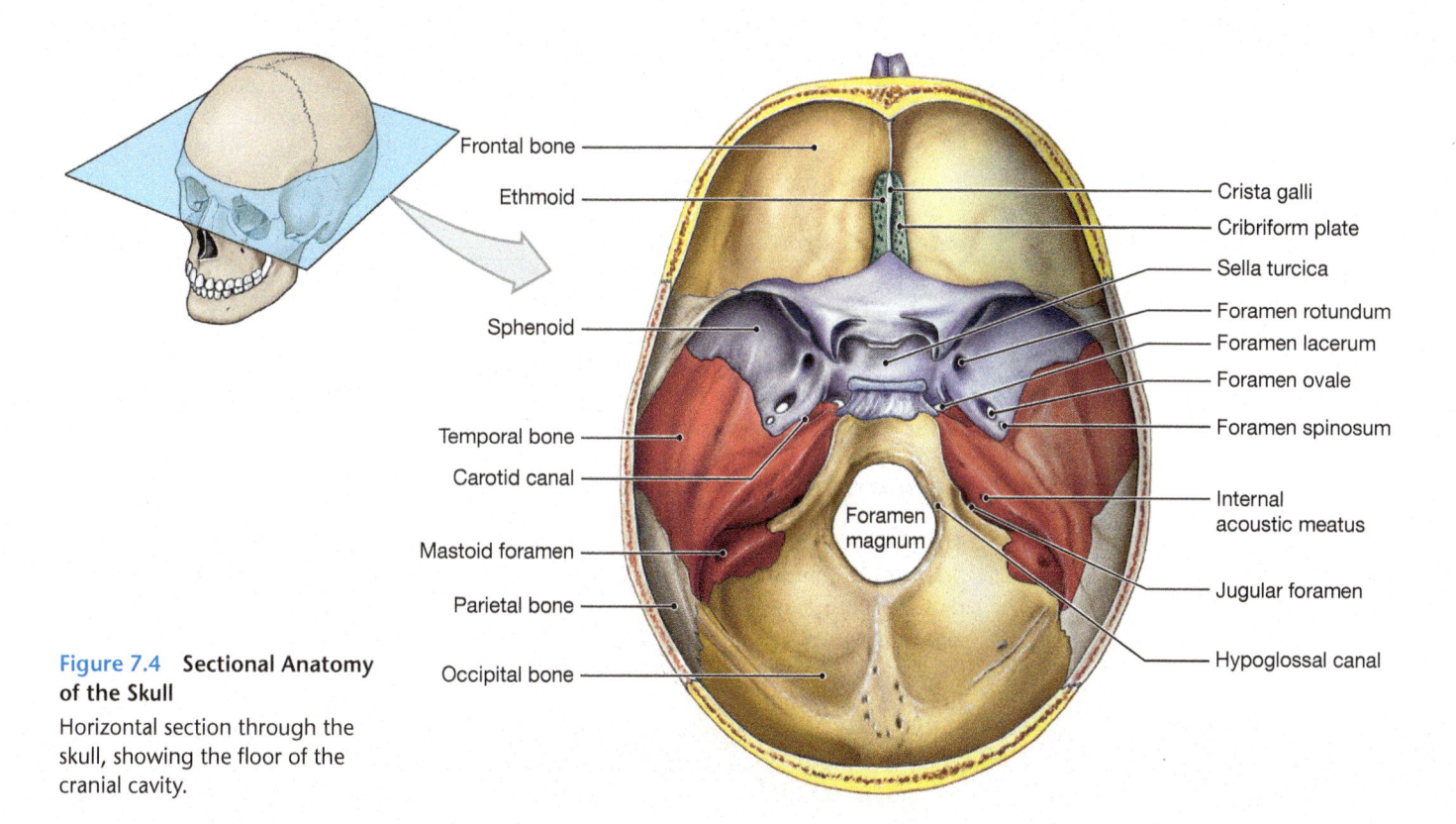

Figure 7.4 Sectional Anatomy of the Skull

Horizontal section through the skull, showing the floor of the cranial cavity.

posterior to the bridge of the nose that forms the posteriomedial wall of both eye orbits.

The floor of the cranium has three depressions called *fossae* (Figure 7.4). The **anterior cranial fossa** is mainly the depression that forms the base of the frontal bone. Small portions of the ethmoid and sphenoid also contribute to the floor of

89

this area. The **middle cranial fossa** is a depressed area extending over the sphenoid and the temporal and occipital bones. The **posterior cranial fossa** is found in the occipital bone.

Frontal Bone

The frontal bone forms the roof, walls, and floor of the anterior cranium and the superior surface of the eye orbit (Figure 7.5). The **frontal squama** is the flattened expanse commonly called the forehead. In the midsagittal plane of the squama is the **frontal** (metopic) **suture**, where the two frontal bones of the fetal skull fuse by eight years of age. By adulthood, the frontal suture is not visible on most skulls. Superior to the eye orbit is the **supra-orbital foramen**, which on some skulls occurs not as a complete hole but rather as a small notch, the **supra-orbital notch**. In the anteriomedial regions of the eye orbit is the **lacrimal fossa**, an indentation for the lacrimal gland, which moistens and lubricates the eye.

Occipital Bone

The occipital bone forms the posterior floor and walls of the skull (Figure 7.6a and b). The most conspicuous structure of the occipital bone is the **foramen magnum**, the large hole where the spinal cord enters the skull and joins the brain. Along the lateral margins of the foramen magnum are flattened **occipital condyles** that articulate with the first vertebra of the spine. Passing under each occipital condyle is the **hypoglossal canal**, a passageway for the hypoglossal nerve, which controls muscles of the tongue and throat.

The occipital bone has many external surface marks that show where muscles and ligaments attach. The **external occipital crest** is a ridge that extends posteriorly from the foramen magnum to a small bump, the **external occipital protuberance**. The **superior** and **inferior nuchal** (NOO-kul) **lines**, surface marks indicating where muscles of the neck attach to the skull, wrap around the occipital bone lateral from the crest and protuberance.

Parietal Bones

The two parietal bones form the posterior crest of the skull and are joined by the sagittal suture, as noted earlier. The bones are smooth and have few surface features. The low ridges of the **superior** and **inferior temporal lines** (Figure 7.6c) are above the squamous suture, where a muscle for chewing attaches. Superior to these lines is the smooth **parietal eminence**.

Temporal Bones

The two temporal bones form the inferior lateral walls of the skull and part of the floor of the middle cranial fossa (Figure 7.7). One of the most distinct features of a temporal bone is its articulation with the zygomatic bone of the face by the **zygomatic arch** (Figures 7.1 and 7.3). This arch is formed by the **zygomatic process** of the temporal bone and the temporal process of the zygomatic bone. Posterior to the zygomatic process is the region of the temporal bone called the **articular tubercle**. Immediately posterior to the articular tubercle is the **mandibular fossa**, a shallow depression where the mandible bone articulates with the temporal bone.

The broad, flattened superior surface of each temporal bone is the **squamous part**. The hole inferior to the squamous part is the **external acoustic meatus**, which conducts sound waves toward the eardrum. Directly posterior to the external acoustic meatus is the conical **mastoid process**, where a muscle tendon that moves the head attaches. Many small interconnected sinuses called **mastoid air cells** are within the mastoid process. The long, needle-like **styloid** (STĪ-loyd; *stylos*,

Figure 7.5 **The Frontal Bone**
(a) Anterior view. (b) Inferior view. (c) Posterior view.

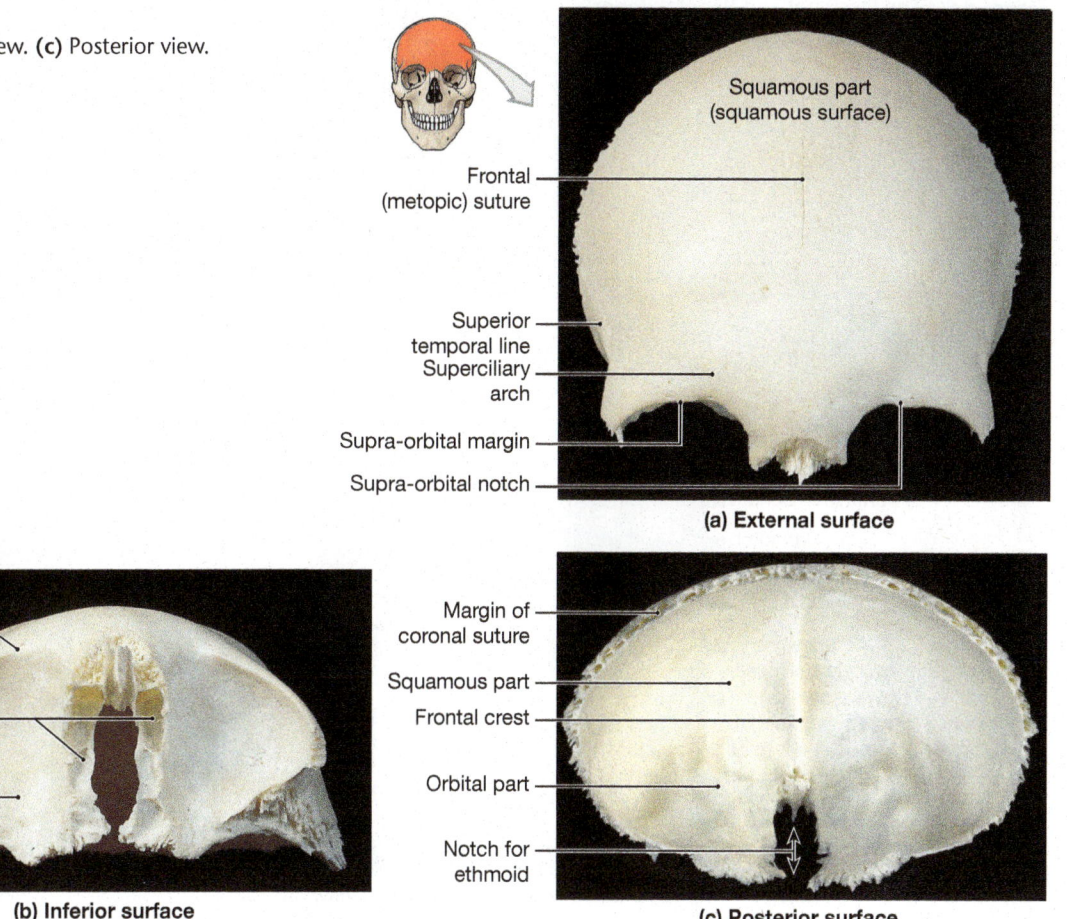

pillar) **process** is located anteriomedial to the mastoid process. Between the styloid and the mastoid processes is a small foramen, the **stylomastoid foramen**, where the facial nerve exits the cranium.

Inside the cranium, the large bony ridge of the temporal bone, is the **petrous part**, which houses the organs for hearing and equilibrium and the tiny bones of the ear. The **internal acoustic meatus** is on the posterior medial surface of the petrous part. The union between the temporal and occipital bones creates an elongated **jugular foramen** that serves as a passageway for cranial nerves and the jugular vein, which drains blood from the brain. On the anterior side of the petrous part is the **carotid canal**, where the internal carotid artery enters the skull to deliver oxygenated blood to the brain.

Sphenoid

On the anterior side the sphenoid contributes to the lateral wall of the eye orbit; on the lateral side it spans the floor of the cranium and braces the walls. The superior surface of the sphenoid is made up of two **lesser wings** and two **greater wings** on either side of the medial line, which give the bone the appearance of a bat (Figure 7.8). Each greater wing has an **orbital surface** that contributes to the wall of the eye orbit. In the center of the sphenoid is the U-shaped **sella turcica** (TUR-si-kuh), commonly called the Turk's saddle. The depression in the sella turcica is the **hypophyseal** (hī-pō-FIZ-ē-ul) **fossa**, which contains the pituitary gland of the brain. The anterior part of the sella turcica is the **tuberculum sellae**; on the posterior wall is the **posterior clinoid** (KLĪ-noyd;

Figure 7.6 The Occipital and Parietal Bones
(a) Occipital bone, inferior view. (b) Occipital bone, superior (internal) view. (c) Right parietal bone, lateral view.

- Foramen magnum
- Occipital condyle
- Hypoglossal canal
- Condyloid fossa
- Inferior nuchal line
- External occipital crest
- Superior nuchal line
- External occipital protuberance

(a) Occipital bone, inferior (external) view

- Foramen magnum
- Jugular notch
- Groove for sigmoid sinus
- Entrance to hypoglossal canal
- Fossa for cerebellum
- Internal occipital crest
- Fossa for cerebrum
- Internal occipital protuberance

(b) Occipital bone, superior (internal) view

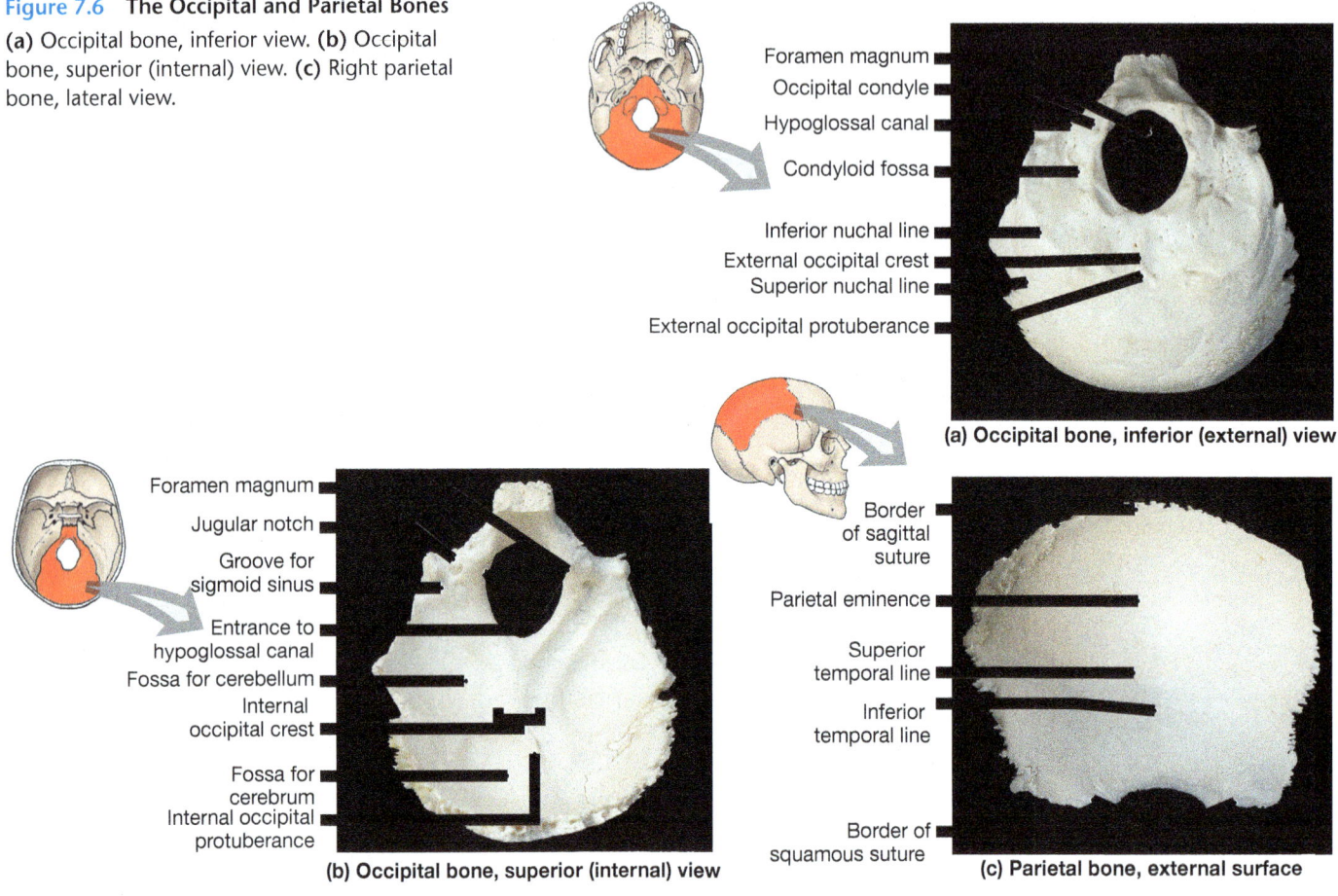

- Border of sagittal suture
- Parietal eminence
- Superior temporal line
- Inferior temporal line
- Border of squamous suture

(c) Parietal bone, external surface

Figure 7.7 The Right Temporal Bone
(a) Lateral view. (b) Medial view.

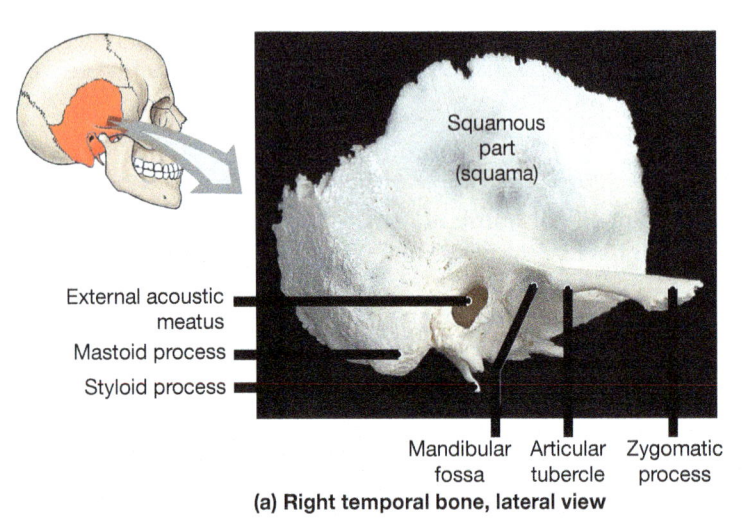

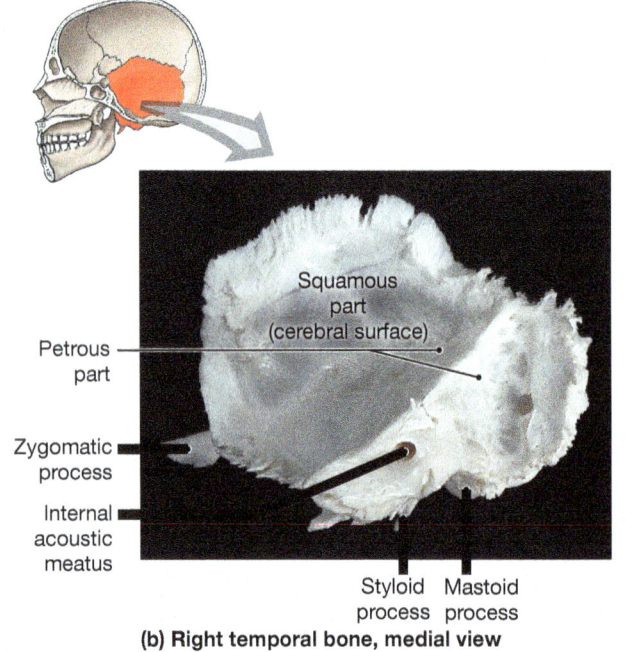

- Squamous part (squama)
- External acoustic meatus
- Mastoid process
- Styloid process
- Mandibular fossa
- Articular tubercle
- Zygomatic process

(a) Right temporal bone, lateral view

- Squamous part (cerebral surface)
- Petrous part
- Zygomatic process
- Internal acoustic meatus
- Styloid process
- Mastoid process

(b) Right temporal bone, medial view

kline, a bed) **process**. The two **anterior clinoid processes** are the horn-like projections on either side of the tuberculum sellae. The **pterygoid** (TER-i-goyd; *pterygion,* wing) **processes** extend vertically from the inferior surface of the sphenoid. Each process divides into a **lateral plate** and a **medial plate**, where muscles of the mouth attach. At the base of each pterygoid process is a small **pterygoid canal** that serves as a passageway for nerves to the soft palate of the mouth.

Four pairs of foramina are aligned on either side of the sella turcica and serve as passageways for blood vessels and nerves. The oval **foramen ovale** (ō-VAH-lē; oval) and, posterior to it, the small **foramen spinosum** are passageways for parts of the trigeminal nerve of the head. The **foramen rotundum**, anterior to the foramen ovale, is the passageway for a major nerve of the face. Directly medial to the foramen ovale, where the sphenoid joins the temporal bone, is the **foramen lacerum** (LA-se-rum; *lacerare,* to tear), where the auditory (eustachian) tube enters the skull. The sphenoid contribution of the foramen lacerum is visible in Figure 7.8a as the notch lateral to the posterior clinoid process. Frequently, the carotid canal merges with the nearby foramen lacerum to form a single passageway.

Superior to the foramen rotundum is a cleft in the sphenoid, the **superior orbital fissure**, where nerves to the ocular muscles pass. The **inferior orbital fissure** is the crevice at the inferior margin of the sphenoid. At the base of the anterior clinoid process is the **optic canal**, where the optic nerve enters the skull to carry visual signals to the brain. Medial to the optic canals is an **optic groove** that lies transverse on the tuberculum sellae.

Making Connections — **Using Foramina as Landmarks**

> Notice the pattern of how the foramina line up along the sphenoid. Use the foramen ovale as a landmark because it is easy to identify by its oval shape. Anterior to the foramen ovale is the foramen rotundum; posterior is the foramen spinosum. Medial to the foramen ovale is the foramen lacerum, with the nearby carotid canal.

Ethmoid

The ethmoid is a rectangular bone that is anterior to the sphenoid (Figure 7.9). It forms the medial walls of the eye orbits, part of the structure that separates the nose into two regions, and the anteriomedial cranial floor. On the superior surface is a vertical crest of bone called the **crista galli** (*crista,* crest; *gallus,* chicken; cock's comb), where membranes that protect and support the brain are attached. At the base of the crista galli is a screen-like **cribriform** (*cribrum,* sieve) **plate** punctured by many small **olfactory foramina** that serve as passageways for branches of the olfactory nerve. The inferior ethmoid has a thin sheet of vertical bone, the **perpendicular plate**, that contributes to the partitioning wall of the nasal cavity. Lateral to the perpendicular plate are **lateral masses** that contain the **ethmoidal labyrinth**, which are full of connected **ethmoidal air cells** (*ethmoidal sinus*) that open into the nasal cavity. Extending inferiorly into the nasal cavity from the lateral masses are the **superior** and **middle nasal conchae**.

QuickCheck Questions

1.1 What are the five main sutures of the adult skull?

1.2 Which two bones form the zygomatic arch?

1.3 What is the difference between the sella turcica and the hypophyseal fossa?

1.4 Which part of the eye orbit is formed by the ethmoid?

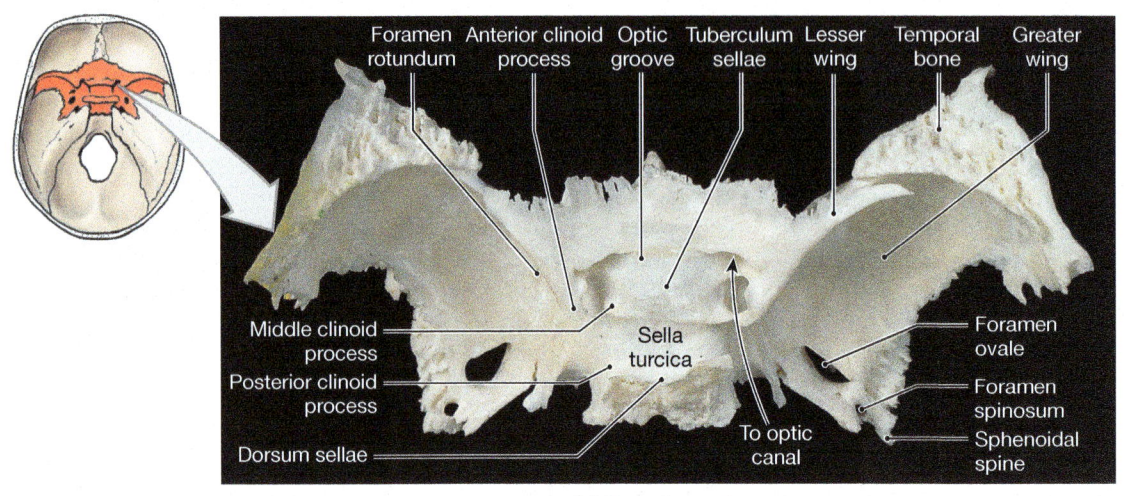

94   EXERCISE 7   The Axial Skeleton

Figure 7.8 The Sphenoid
(a) Superior surface. (b) Anterior surface.

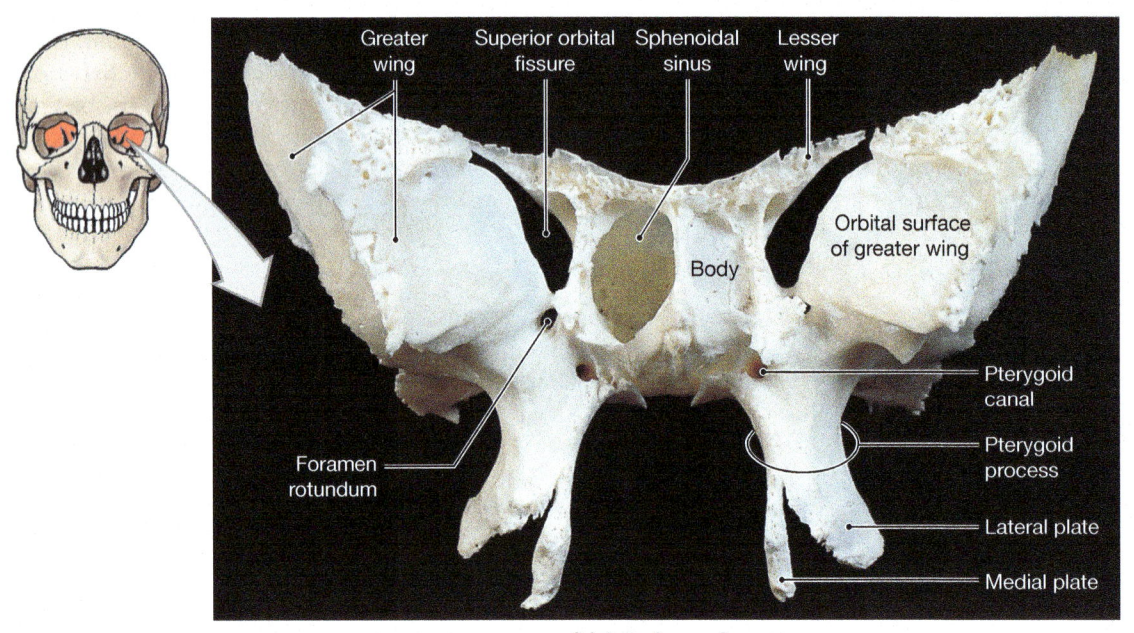

1 Materials

- ☐ Skull sectioned horizontally
- ☐ Disarticulated skull
- ☐ Pipe cleaners

Procedures

1. Identify the eight cranial bones and six sutures on a skull. Touch the general location of each bone on your own head.
2. Use the text descriptions and Figures 7.1–7.9 to identify the features of each cranial bone. Study each bone both individually and in relation to its neighboring bones in the cranium. Select an identifying feature on each bone as a reference structure to facilitate your study.
3. Gently pass pipe cleaners through the foramina to study their entrances and exits. Never force an object into a foramen because doing so may damage the skull. ■

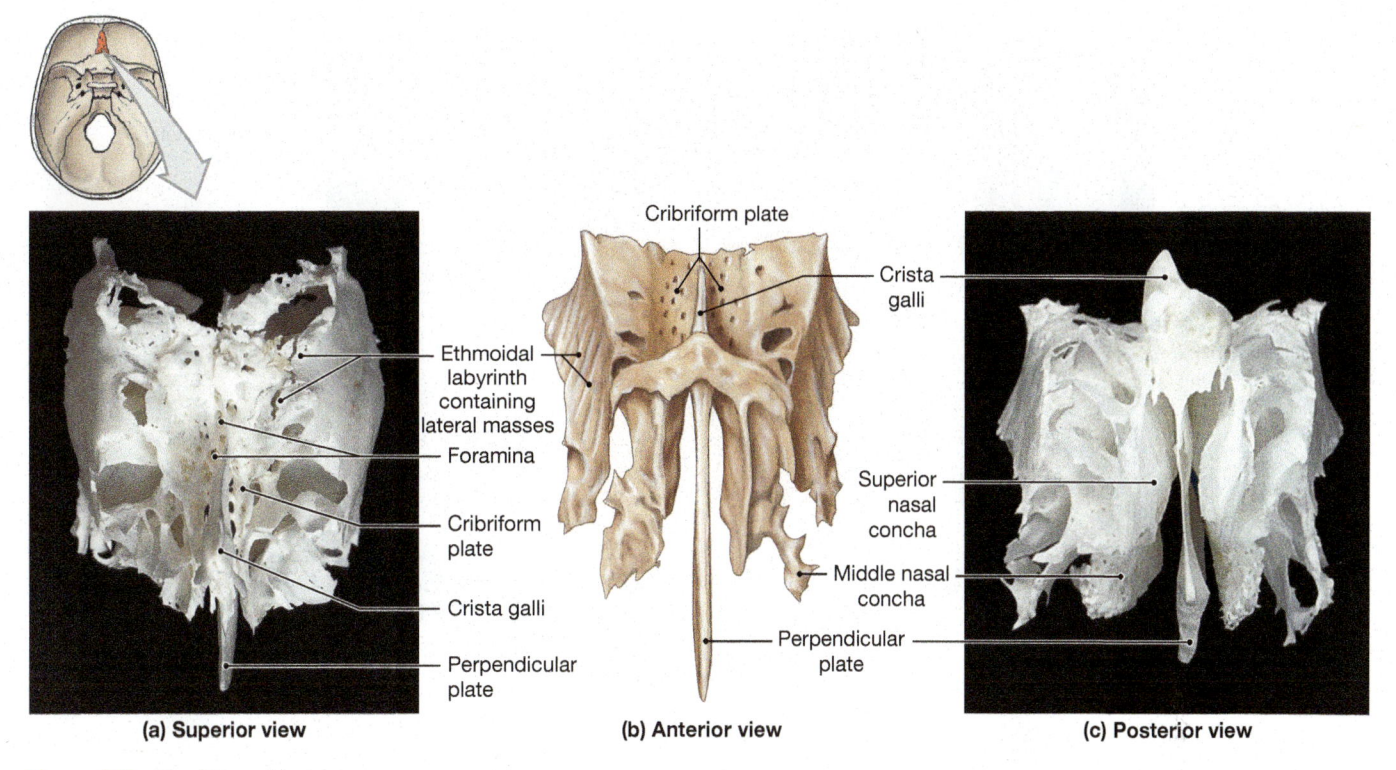

Figure 7.9 The Ethmoid
(a) Superior view. (b) Anterior view. (c) Posterior view.

LAB ACTIVITY 2 — Facial Bones

The face is constructed of 14 bones: two maxillae, two zygomatic, two lacrimal, two inferior nasal conchae, one mandible, two palatine, and one vomer.

The two **nasal bones** join the frontal bone of the cranium at the midline of the face (Figure 7.10). Lateral to the nasal bones are the **maxillae**, which form the floor of the eye orbits and extend inferiorly to form the upper jaw. Below the eye orbits are the **zygomatic bones**, commonly called the cheekbones. At the bridge of the nose, lateral to each maxilla, are the small **lacrimal bones** of the medial eye orbits. Through each lacrimal bone passes a small canal that allows tears to drain into the nasal cavity. The two **inferior nasal conchae** (KONG-kē) are the lower shelves of bone in the nasal cavity. (The middle and superior conchae in the nasal cavity are part of the ethmoid, as described earlier.) The bone of the lower jaw is the **mandible** (Figure 7.11).

On the inferior surface of the skull, the **palatine bones** form the posterior roof of the mouth next to the last molar tooth (Figure 7.11). A thin bone called the **vomer** divides the nasal cavity.

Maxillae

The paired maxillae are the foundation of the face. Inferior to the eye orbit is the **infra-orbital foramen** (see Figure 7.10). The **alveolar process** of each maxilla consists of the U-shaped processes where the upper teeth are embedded in the maxilla. From the inferior aspect, the **palatine process** of the maxilla is visible. This bony shelf forms the anterior hard palate of the mouth. At the anterior margin of the palatine process is the **incisive fossa** (visible in Figure 7.3b).

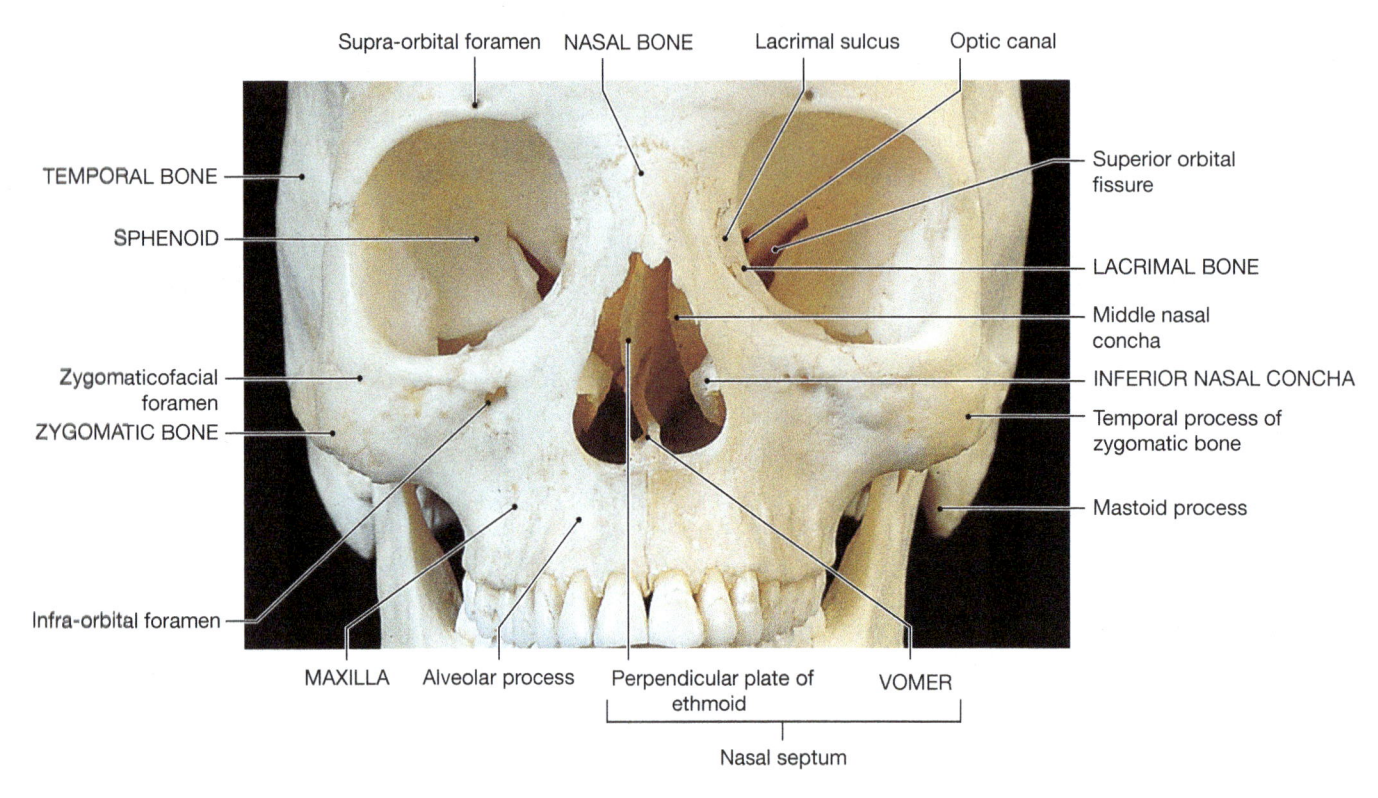

Figure 7.10 Bones of the Face

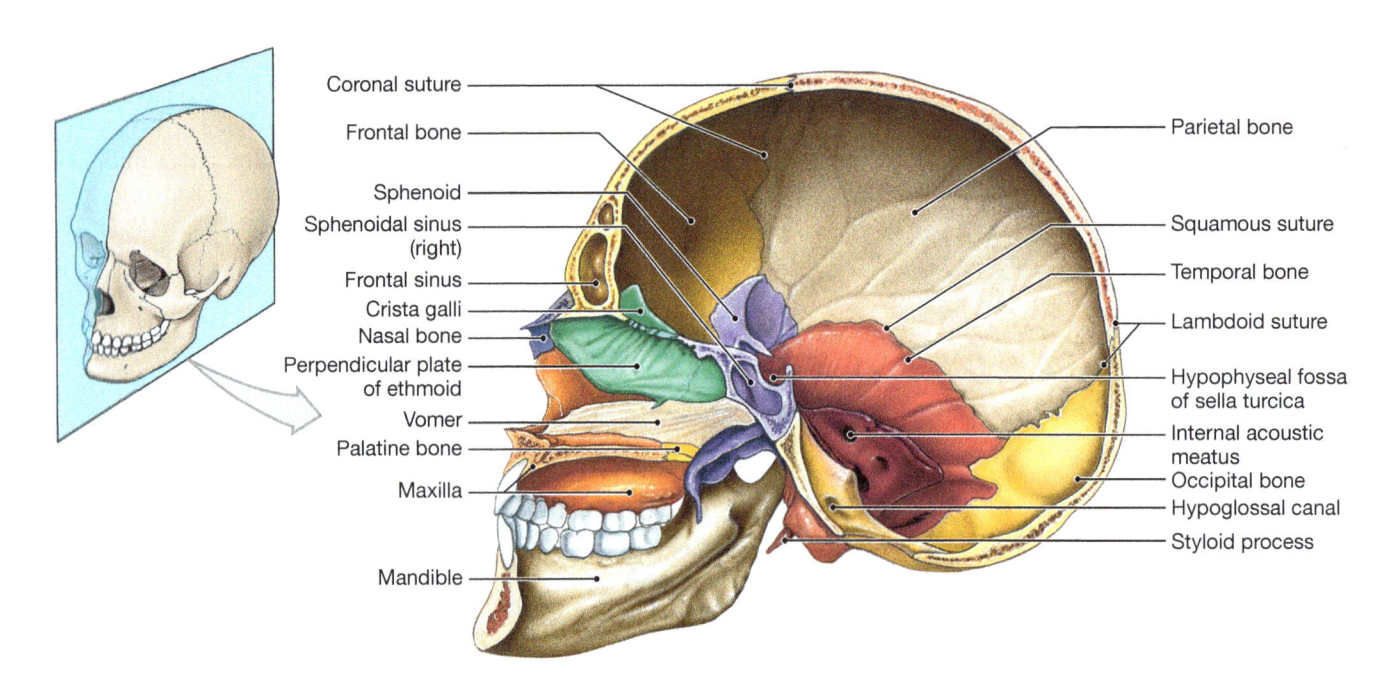

Figure 7.11 Sectional Anatomy of the Skull
Medial view of a sagittal section through the skull. Because the nasal septum (vomer plus perpendicular plate of ethmoid) is illustrated, the right nasal cavity cannot be seen.

Palatine Bones

The palatine bones (see Figures 7.3 and 7.11) are posterior to the palatine processes. The palatine bones and maxillae create the roof of the mouth and separate the oral cavity from the nasal cavity. This separation of cavities allows us to chew and breathe at the same time. Each palatine bone has a **greater palatine foramen** on the lateral margin. Only the inferior portion of each palatine bone is completely visible. The superior surface forms the floor of the nasal cavity and supports the vomer bone.

Zygomatic Bones

The zygomatic bones contribute to the floor and lateral walls of the eye orbit. Lateral and slightly inferior to the orbit is the small **zygomaticofacial foramen**. The posterior margin of each zygomatic bone narrows inferiorly to the **temporal process** of the zygomatic bone, which joins the temporal bone zygomatic process to complete the zygomatic arch (see Figures 7.2a and 7.3).

Lacrimal Bones

The lacrimal bones are the anterior portions of the medial eye orbital wall (see Figure 7.10). Each lacrimal bone is named after the lacrimal glands that produce tears. Tears flow medially across the eye and drain into the inferior lacrimal fossa, which transports them to the nasal cavity.

Nasal Bones

The nasal bones form the bridge of the nose, and the maxilla separates them from the bones of the eye orbit (see Figure 7.10). The superior margin of each nasal bone articulates with the frontal bone; the posterior surface joins the ethmoid deep in the skull.

Inferior Nasal Conchae

The inferior nasal conchae are shelves that extend medially from the lower lateral portion of the nasal wall (see Figure 7.10). They cause inspired air to swirl in the nasal cavity so that the moist mucous membrane that lines the cavity can warm, cleanse, and moisten the air. Similar shelves of bone occur on the lateral walls of the ethmoid bone.

Vomer

The vomer is part of the **nasal septum**, the bony wall that separates the nasal chamber into right and left cavities. The vomer is best viewed from the inferior aspect of the skull looking into the nasal cavities (see Figure 7.3). The midsagittal section of the skull shown in Figure 7.11 illustrates the relationship between the two bones that make up the nasal septum: the perpendicular plate of the ethmoid, which forms the superior portion of the septum, and the vomer, which forms the inferior part.

Mandible

The mandible of the inferior jaw has a horizontal **body** that turns posteriorly at the **angle** to a raised projection, the **ramus**, which terminates at a U-shaped **mandibular notch** (Figure 7.12). Two processes extend upward from the notch, the anterior **coronoid** (kor-Ō-noyd) **process** and the posterior **condylar process**.

98 EXERCISE 7 The Axial Skeleton

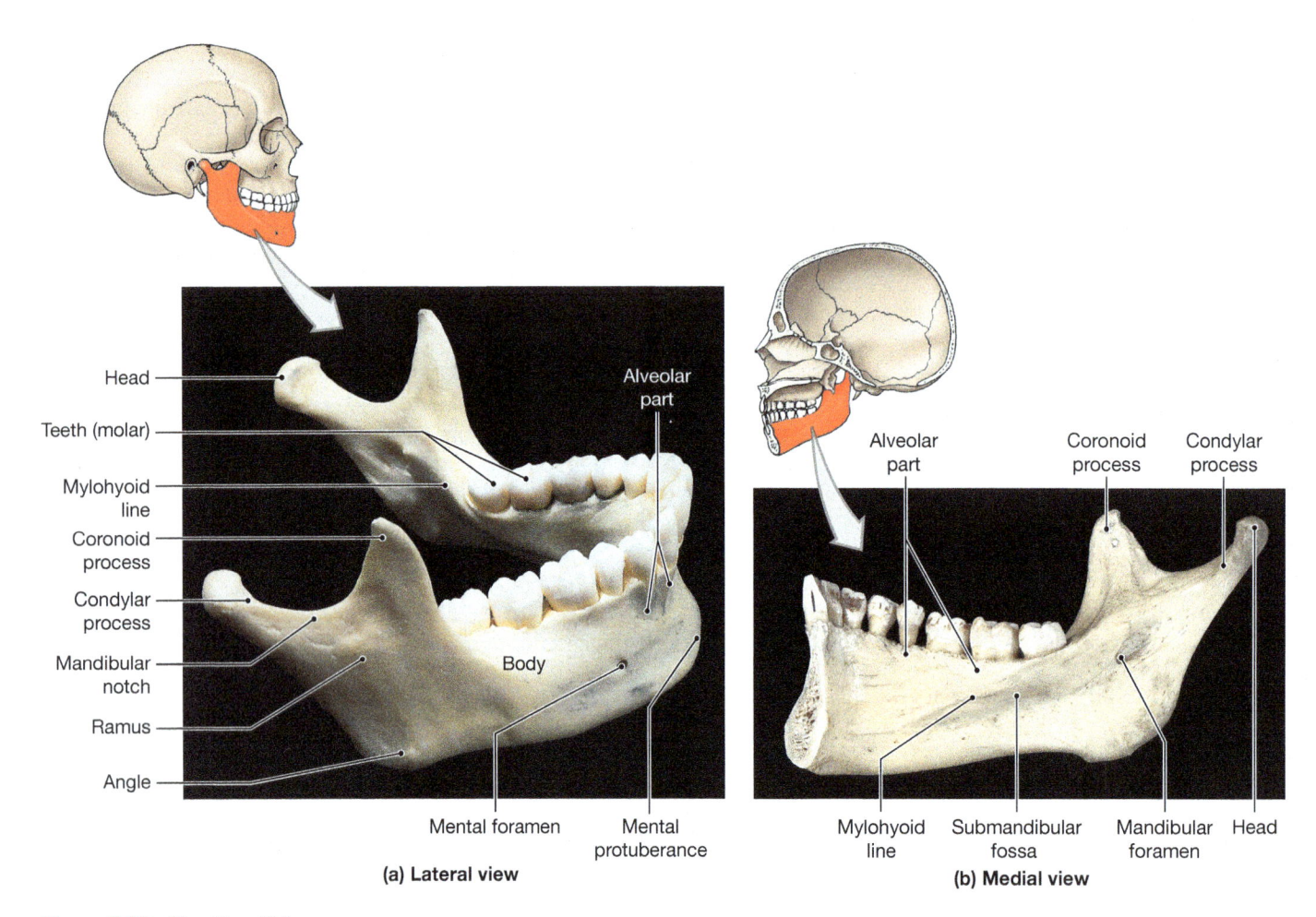

Figure 7.12 The Mandible
(a) Lateral view. (b) Medial view.

Labels (a) Lateral view: Head, Teeth (molar), Mylohyoid line, Coronoid process, Condylar process, Mandibular notch, Ramus, Angle, Alveolar part, Body, Mental foramen, Mental protuberance.

Labels (b) Medial view: Alveolar part, Coronoid process, Condylar process, Mylohyoid line, Submandibular fossa, Mandibular foramen, Head.

The smooth **articular surface** of the condylar **head** articulates in the mandibular fossa on the temporal bone at the **temporomandibular joint**. The **alveolar process** of the mandible is the crest of bone where the lower teeth articulate with the mandible. Lateral to the chin, or **mental protuberance** (*mental,* chin), is the **mental foramen**. The medial mandibular surface features the **submandibular fossa**, where the submandibular salivary gland rests against the bone. At the posterior end of the fossa is the **mandibular foramen**, a passageway for a sensory nerve from the lower teeth and gums. Superior to the fossa is the **mylohyoid line**.

QuickCheck Questions

2.1 Which facial bones contribute to the orbit of the eye?
2.2 Which facial bones form the roof of the mouth?
2.3 How does the mandible articulate with the cranium?

2 Materials

☐ Skull sectioned horizontally
☐ Disarticulated skull

Procedures

1. Identify each facial bone on a skull. Touch the general location of each bone on your own face.
2. Use the text descriptions and Figures 7.1–7.3 and 7.10–7.12 to identify the features of each facial bone. As with your study of the cranial bones, study each bone both individually and in relation to its neighboring bones. Select an identifying feature on each bone as a reference structure to facilitate your study. ■

EXERCISE 7   The Axial Skeleton   99

LAB ACTIVITY 3

Hyoid Bone

The hyoid bone, a U-shaped bone inferior to the mandible (Figure 7.13), is unique because it does not articulate with other bones. The hyoid bone is difficult to palpate because it is surrounded by the ligaments and muscles of the throat and neck. Two horn-like processes for muscle attachment occur on each side of the hyoid bone, an anterior **lesser horn** and a large posterior **greater horn**.

QuickCheck Questions

3.1 Where is the hyoid bone located?

3.2 On which structures of the hyoid bone do muscles attach?

3 Materials

☐ Articulated skeleton

Procedures

1. Examine the hyoid bone on the articulated skeleton.
2. Identify the greater and lesser horns of the hyoid bone.
3. Refer to Figures 11.4–11.7 and note the associations between the hyoid bone and the muscles of the tongue and larynx. ■

LAB ACTIVITY 4

Sinuses of the Skull

The skull contains cavities called **paranasal sinuses** that interconnect with the nasal cavity (Figure 7.14). Like the nasal cavity, the sinuses are lined with a mucous membrane. The **frontal sinus** extends laterally over the orbit of the eyes. The **sphenoidal sinus** is located in the sphenoid directly inferior to the sella turcica. The ethmoid labyrinth houses **ethmoidal air cells** that collectively constitute the **ethmoidal sinus**. Each maxilla contains a large **maxillary sinus** situated lateral to the nasal cavity.

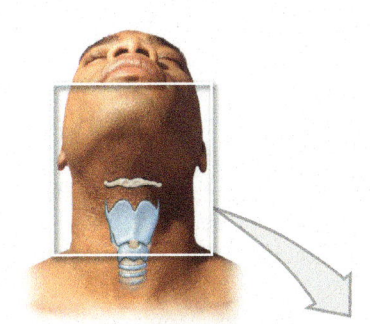

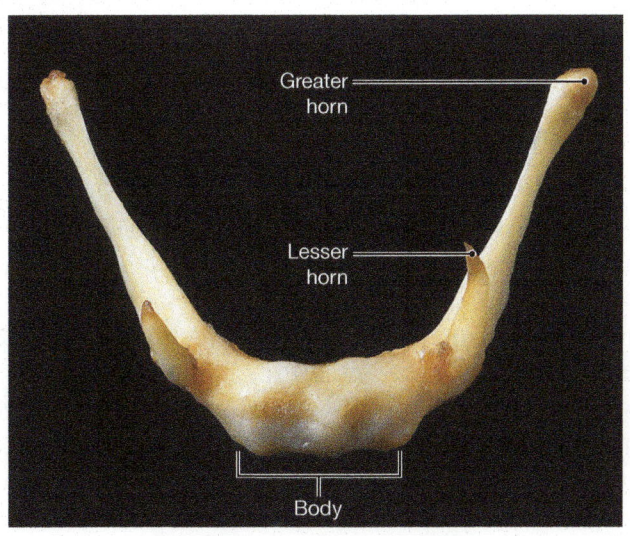

Hyoid bone, anterosuperior view

Figure 7.13 The Hyoid Bone
Anterosuperior view.

100 EXERCISE 7 The Axial Skeleton

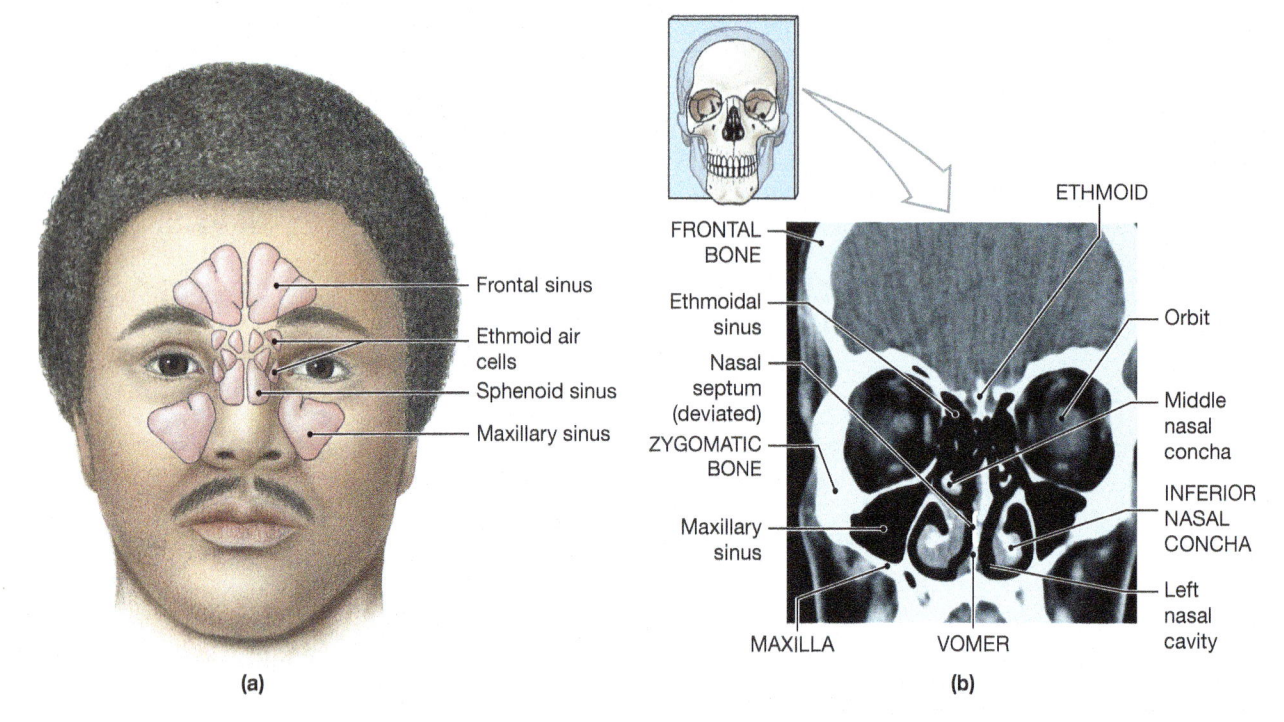

Figure 7.14 Paranasal Sinuses
(a) Location of the four types of paranasal sinuses. (b) An MRI scan that shows a frontal section through the ethmoidal and maxillary sinuses.

Clinical Application   **Sinus Congestion**

In some individuals, allergies or changes in the weather can make the sinus membranes swell and secrete more mucus. The resulting congestion blocks connections with the nasal cavity, and the increased sinus pressure is felt as a headache. The sinuses also serve as resonating chambers for the voice, much like the body of an acoustic guitar amplifies its music, and when the sinuses and nasal cavity are congested, the voice sounds muffled.

QuickCheck Questions
4.1 What are the names of the various paranasal sinuses?
4.2 What is the function of the sinuses?

4 Materials

- Several skulls sectioned midsagittally

Procedures

1. Compare the frontal sinus on several sectioned skulls. Is the sinus the same size on each skull? _____
2. Locate the sphenoidal sinus on a sectioned skull. Under which sphenoidal structure is this sinus located? _____
3. Examine the maxillary sinus on a sectioned skull. How does the size of this sinus compare with the sizes of the other three sinuses? _____
4. Identify the ethmoidal air cells. Which sinus do these cells collectively form? _____

EXERCISE 7 The Axial Skeleton   101

LAB ACTIVITY 5 Fetal Skull

As the fetal skull develops, the cranium must remain flexible to accommodate the enlarging brain. This flexibility is possible because the cranial bones remain incompletely fused until after birth. Between the wide developing sutures are expanses of fibrous connective tissue called **fontanels** (fon-tuh-NELZ). It is these so-called *soft spots* that allow the skull to expand as brain size increases and enable the skull to flex in order to squeeze through the birth canal during delivery. By the age of four to five years, the brain is nearly adult size, the fibrous connective tissue of the fontanels ossifies, and the cranial sutures interlock to securely support the articulating bones.

Four major fontanels are present at birth: the large **anterior fontanel** between the frontal and parietal bones; the **sphenoidal fontanel** at the juncture where the frontal bone, parietal bone, temporal bone, and sphenoid will one day fuse; the **mastoid fontanel** between the parietal and occipital bones; and the **occipital fontanel** between the occipital and parietal bones (Figure 7.15).

QuickCheck Questions

5.1 What does the presence of fontanels allow the fetal skull to do?

5.2 How long are fontanels present in the skull?

5 Materials

☐ Fetal skull
☐ Adult skull

Procedures

1. Examine each fontanel on a fetal skull.
2. Compare the fetal and adult skulls and note the changes due to ossification. ■

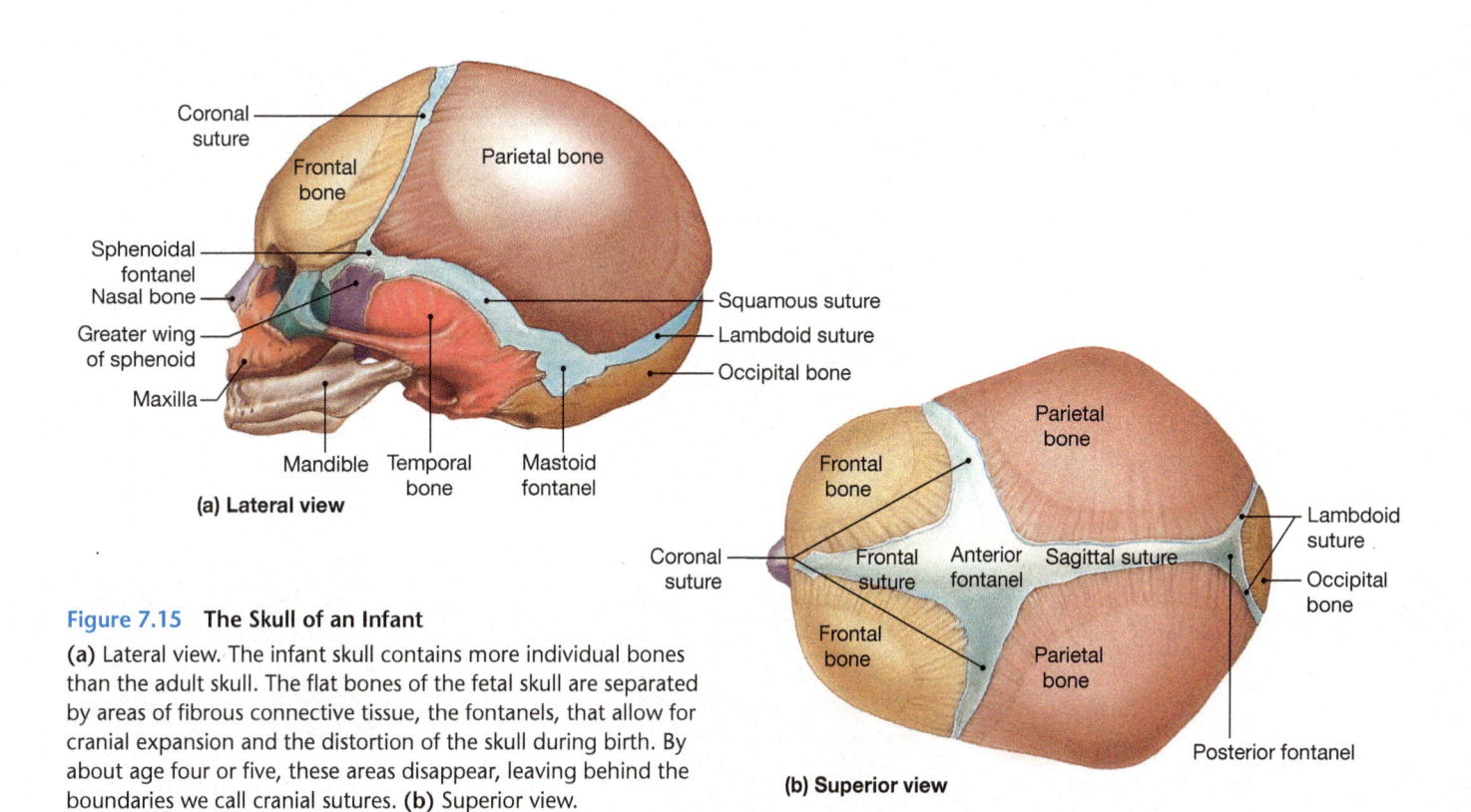

Figure 7.15 The Skull of an Infant
(a) Lateral view. The infant skull contains more individual bones than the adult skull. The flat bones of the fetal skull are separated by areas of fibrous connective tissue, the fontanels, that allow for cranial expansion and the distortion of the skull during birth. By about age four or five, these areas disappear, leaving behind the boundaries we call cranial sutures. **(b)** Superior view.

102   EXERCISE 7   The Axial Skeleton

LAB ACTIVITY 6 — Vertebral Column

The **vertebral column**, or **spine**, is a flexible chain of 33 bones called **vertebrae**. The column articulates with the skull superiorly, with the pelvic girdle inferiorly, and with the ribs laterally. The vertebrae are grouped into five types based on location and anatomical features (Figure 7.16). Starting at the top of the spine, the first seven vertebrae are the **cervical vertebrae** of the neck. Twelve **thoracic vertebrae** articulate with the ribs. The lower back has five **lumbar vertebrae**, and a **sacrum** that joins the hips is comprised of five fused **sacral vertebrae**. The **coccyx** (KOK-siks) is the tailbone portion of the spine and consists of (usually) four fused **coccygeal vertebrae**.

The vertebral column is curved to balance the body weight while standing. Toward the end of gestation, the fetal vertebral column develops **accommodation curves** in the thoracic and sacral regions, curves that provide space for the internal organs in these regions. Because accommodation curves occur first, they are also called **primary curves**. At birth, the accommodation curves are still forming, and the vertebral column is relatively straight. During early childhood, as the individual learns to hold the head up, crawl, and then walk, **compensation (secondary) curves** form in the cervi-

Figure 7.16 The Vertebral Column
Lateral views of the vertebral column. **(a)** The major divisions of the vertebral column, showing the four spinal curves present in an adult. **(b)** Normal vertebral column, lateral view.

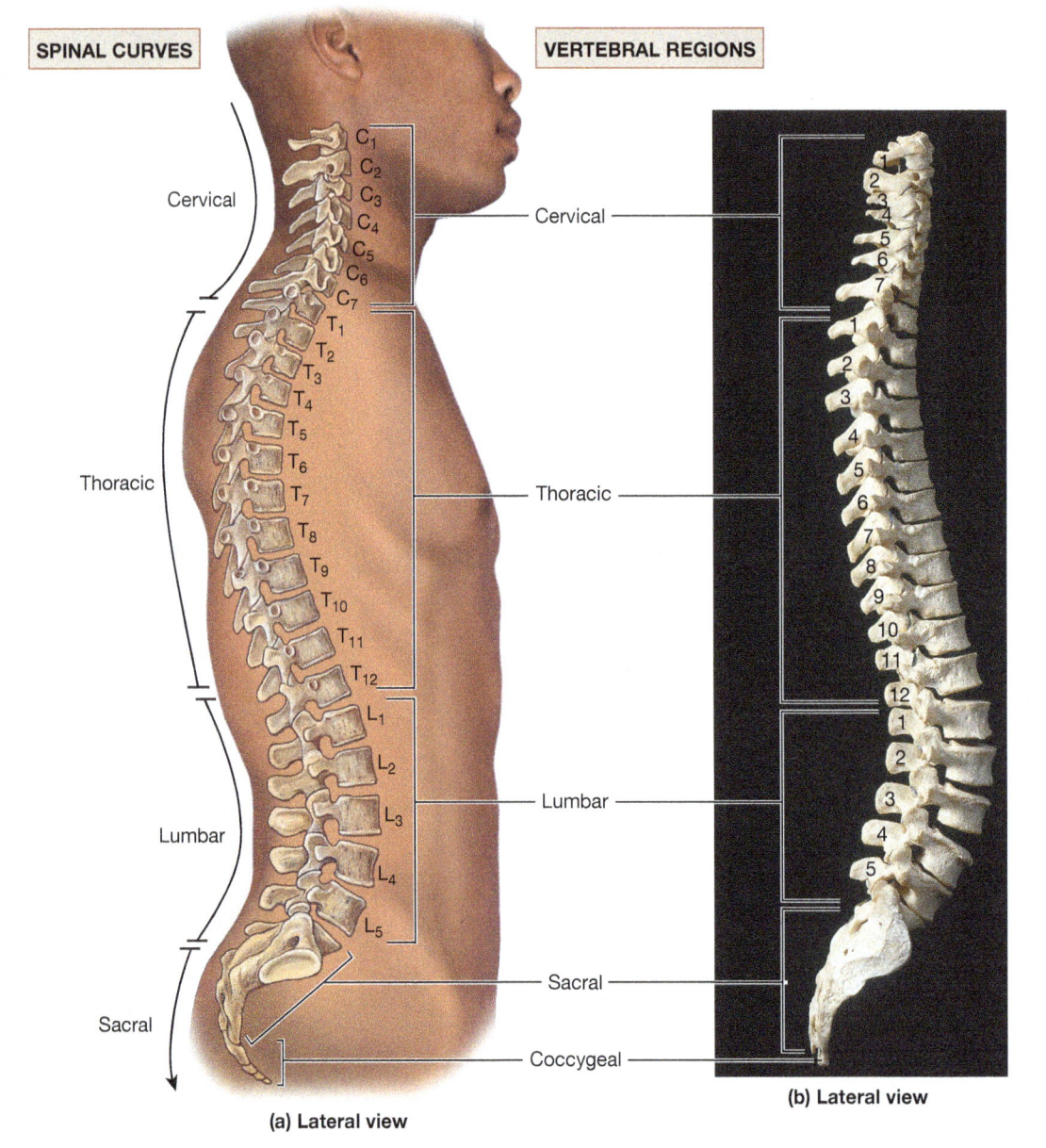

(a) Lateral view

(b) Lateral view

EXERCISE 7 The Axial Skeleton 103

cal and lumbar regions to move the body weight closer to the body's axis for better balance. Once the child is approximately ten years old, the spinal curves are established, and the fully developed column has alternating secondary and primary curves.

In the cervical, thoracic, and lumbar regions are **intervertebral discs**, cushions of fibrocartilage between the articulating vertebrae. Each disc consists of an outer layer of strong fibrocartilage, the **annulus fibrosus**, surrounding an inner mass, the **nucleus pulposus**. Water and elastic fibers in the gelatinous mass of the nucleus pulposus absorb stresses that arise between vertebrae whenever a person is either standing or moving.

Vertebral Anatomy

The anatomical features of a typical vertebra include a large, anterior, disc-shaped **vertebral body** (the *centrum*) and an elongated **spinous process** that extend posteriorly (Figure 7.17). On each side of the vertebral body a **transverse process** ex-

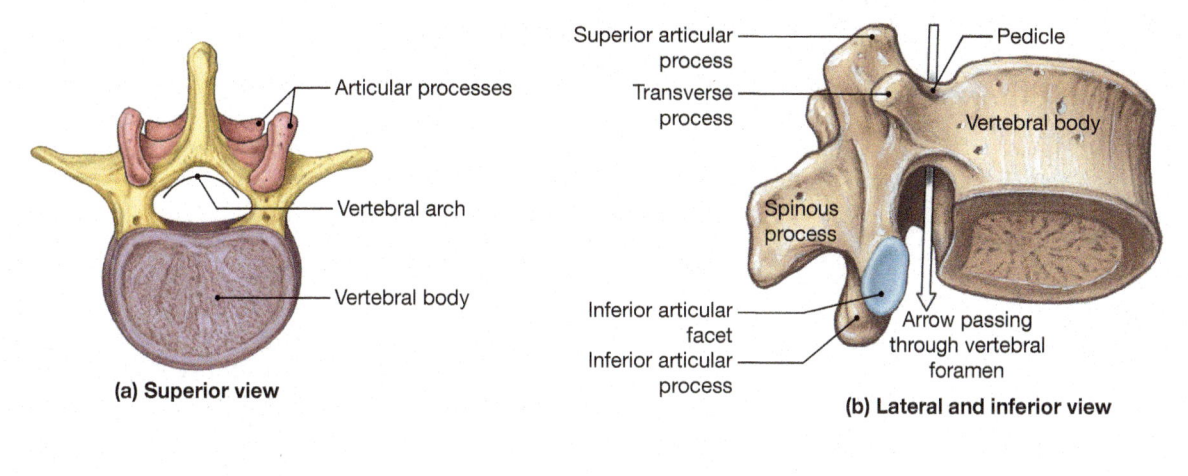

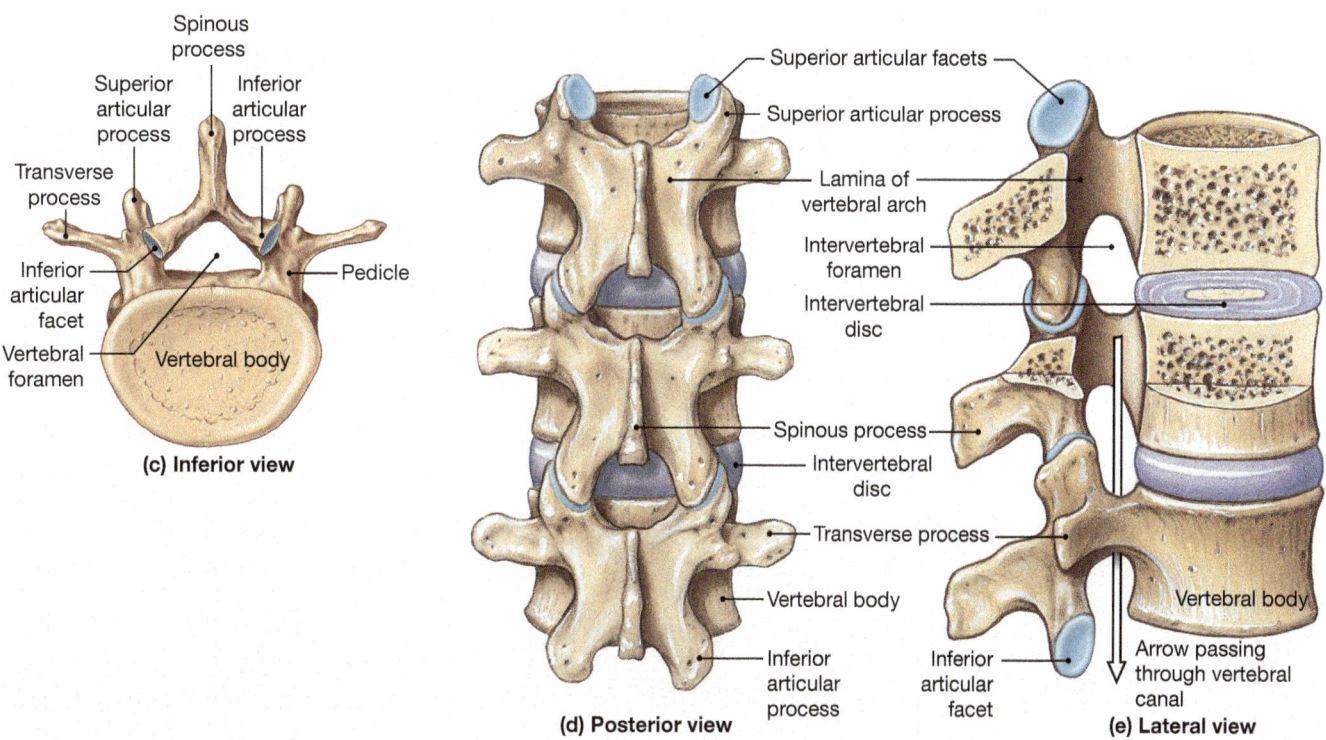

Figure 7.17 Vertebral Anatomy

The anatomy of a typical vertebra and the arrangement of articulations between vertebrae. (a) Superior view. (b) Lateral, slightly inferior view. (c) Inferior view. (d) Posterior view of three articulated vertebrae. (e) Lateral and sectional view of three articulated vertebrae.

tends laterally. The **lamina** (LA-mi-na) is a flat plate of bone between the transverse and spinous processes that forms the curved **vertebral arch**. The **pedicle** (PE-di-kul) is a strut of bone that extends posteriorly from the body to a transverse process. The pedicle and lamina on each side form the wall of the large posterior **vertebral foramen**, which contributes to the spinal cavity where the spinal cord is housed. Inferior to the pedicle is an inverted U-shaped region called the **inferior vertebral notch**. Two articulating vertebrae form an **intervertebral foramen**, in which the inferior vertebral notch of the superior vertebra joins the pedicle of the inferior vertebra. Spinal nerves pass through the intervertebral foramen to access the spinal cord.

The vertebral column moves much like a gooseneck lamp: each joint moves only slightly, but the combination of all the individual movements permits the column a wide range of motion. Joints between adjacent vertebrae occur at smooth articular surfaces called *facets* that project from *articular processes*. The **superior articular process** is on the superior surface of the pedicle of each vertebra and has a **superior articular facet** at the posterior tip. The **inferior articular process** is a downward projection of the inferior lamina wall and has an **inferior articular facet** on the anterior tip. At a vertebral joint, the inferior articular facet of the superior vertebra glides across the superior articular facet of the inferior vertebra. The greatest movement of these joints is in the cervical region for head movement.

Cervical Vertebrae

The seven cervical vertebrae in the neck are recognizable by the presence of a **transverse foramen** on each transverse process (Figure 7.18). The vertebral artery travels up the neck through these foramina to enter the skull. The first two cervical vertebrae are modified for special articulations with the skull. The tip of the spinous process is **bifid** (branched) in vertebrae C_2–C_6. The last cervical vertebra, C_7, is called the **vertebra prominens** because of the broad tubercle at the end of the spinous process. The tubercle can be palpated at the base of the neck.

The first cervical vertebra, C_1, is called the **atlas** (Figure 7.18b), named after the Greek mythological character who carried the world on his shoulders. The atlas is the only vertebra that articulates with the skull. The superior articular facets of the atlas are greatly enlarged, and the occipital condyles of the occipital bone fit into the facets like spoons nested together. When you nod your head, the atlas remains stationary while the occipital condyles glide in the facets. The atlas is unusual in that it lacks a vertebral body and a spinous process and has a very large vertebral foramen formed by the **anterior** and **posterior arches**. A small, rough **posterior tubercle** occurs where the spinous process normally resides. A long spinous process would interfere with occipitoatlas articulation.

The **axis** is the second cervical vertebra, C_2. It is specialized to articulate with the atlas. A peg-like **dens** (DENZ; *dens,* tooth), or *odontoid process,* arises superiorly from the body of the axis (Figure 7.18d). It fits against the anterior wall of the vertebral foramen and provides the atlas with a pivot point for when the head is turned laterally and medially. A **transverse ligament** secures the atlas around the dens.

Thoracic Vertebrae

The 12 thoracic vertebrae, which articulate with the 12 pairs of ribs, are larger than the cervical vertebrae and increase in size as they approach the lumbar region. Most ribs attach to their thoracic vertebra at two sites on the vertebra: on a **transverse costal facet** at the tip of the transverse process and on a **costal**

EXERCISE 7 The Axial Skeleton 105

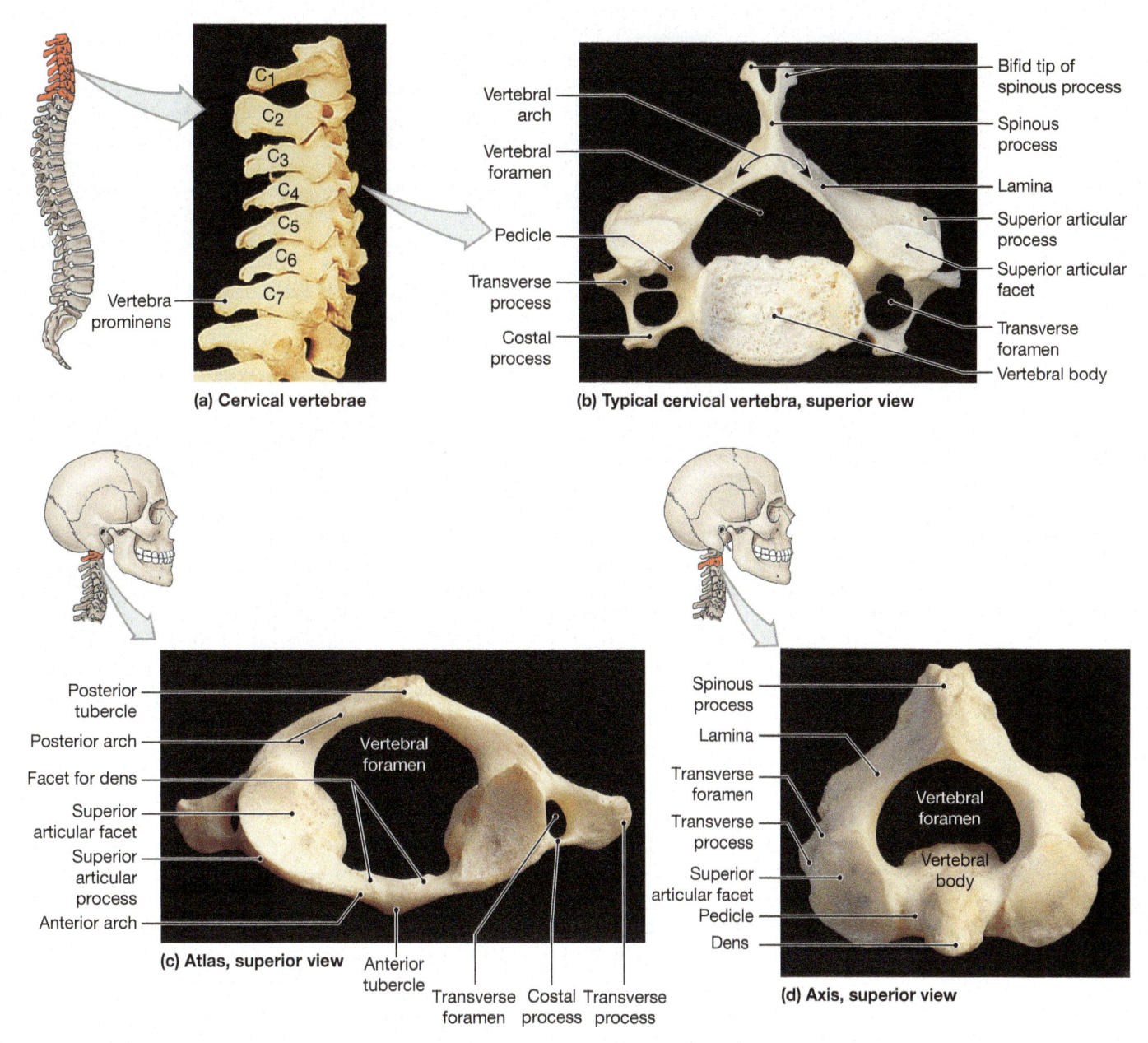

Figure 7.18 The Cervical Vertebrae
(a) Lateral view of the cervical vertebrae. (b) Superior view of a typical (C$_3$–C$_6$) cervical vertebra. Note the characteristic transverse foramen. (c) The first cervical vertebra, the atlas (C$_1$), in superior view. (d) The second cervical vertebra, the axis (C$_2$), in superior view.

facet located on the posterior of the vertebral body (Figure 7.19). Two costal facets usually are present on the same vertebral body, a **superior costal facet** and an **inferior costal facet**. The costal facets are unique to the thoracic vertebrae, and there is variation in where these facets occur on the various thoracic vertebrae.

Lumbar Vertebrae

The five lumbar vertebrae are large and heavy in order to support the weight of the head, neck, and trunk. Compared with thoracic vertebrae, lumbar vertebrae have a wider body, a blunt and horizontal spinous process, and shorter transverse

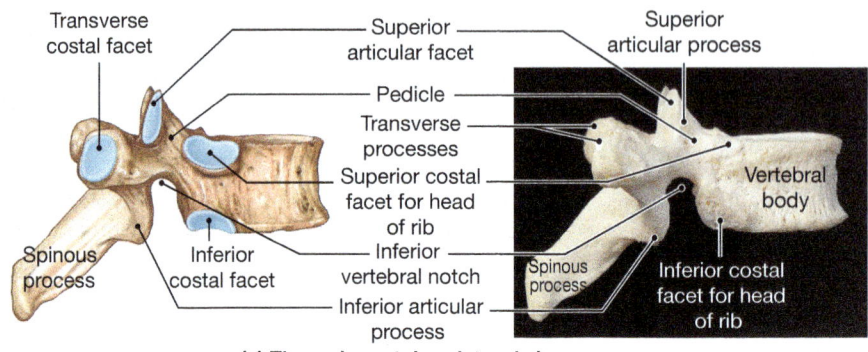

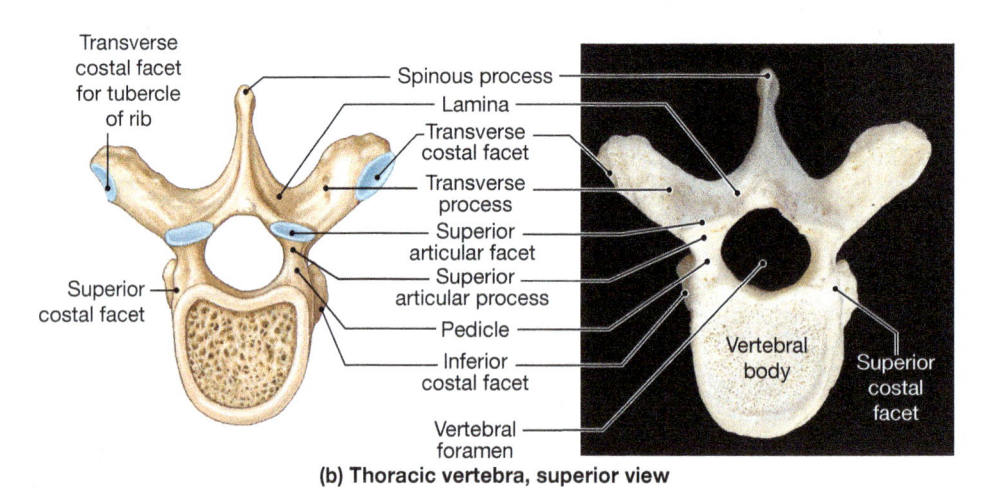

Figure 7.19 The Thoracic Vertebrae
(a) Lateral view. (b) Superior view.

processes (Figure 7.20). The lumbar vertebral foramen is smaller than that in thoracic vertebrae. To prevent the back from twisting when objects are being lifted or carried, the lumbar superior articular process is turned medially and the lumbar inferior articular processes are oriented laterally to interlock the lumbar vertebrae. No facets or transverse foramina occur on the lumbar vertebrae.

Sacral and Coccygeal Vertebrae

As noted earlier, the sacrum is a single bony element composed of five fused sacral vertebrae (Figure 7.21). It articulates with the ilium of the pelvic girdle to form the posterior wall of the pelvis. Fusion of the sacral bones before birth consolidates the vertebral canal into the **sacral canal**. On the fifth sacral vertebra, the sacral canal opens as the **sacral hiatus** (hī-Ā-tus). Along the lateral margin of the fused vertebral bodies are **sacral foramina**. The spinous processes fuse to form an elevation called the **median sacral crest**. A **lateral sacral crest** extends from the lateral margin of the sacrum. The sacrum articulates with each pelvic bone at the large **auricular surface** on the lateral border. Dorsal to this surface is the **sacral tuberosity**, where ligaments attach to support the **sacroiliac joint**.

The coccyx (see Figure 7.21) articulates with the fifth fused sacral vertebra at the **coccygeal cornva** of the first coccygeal vertebra. There may be anywhere from three to five coccygeal bones, but most people have four.

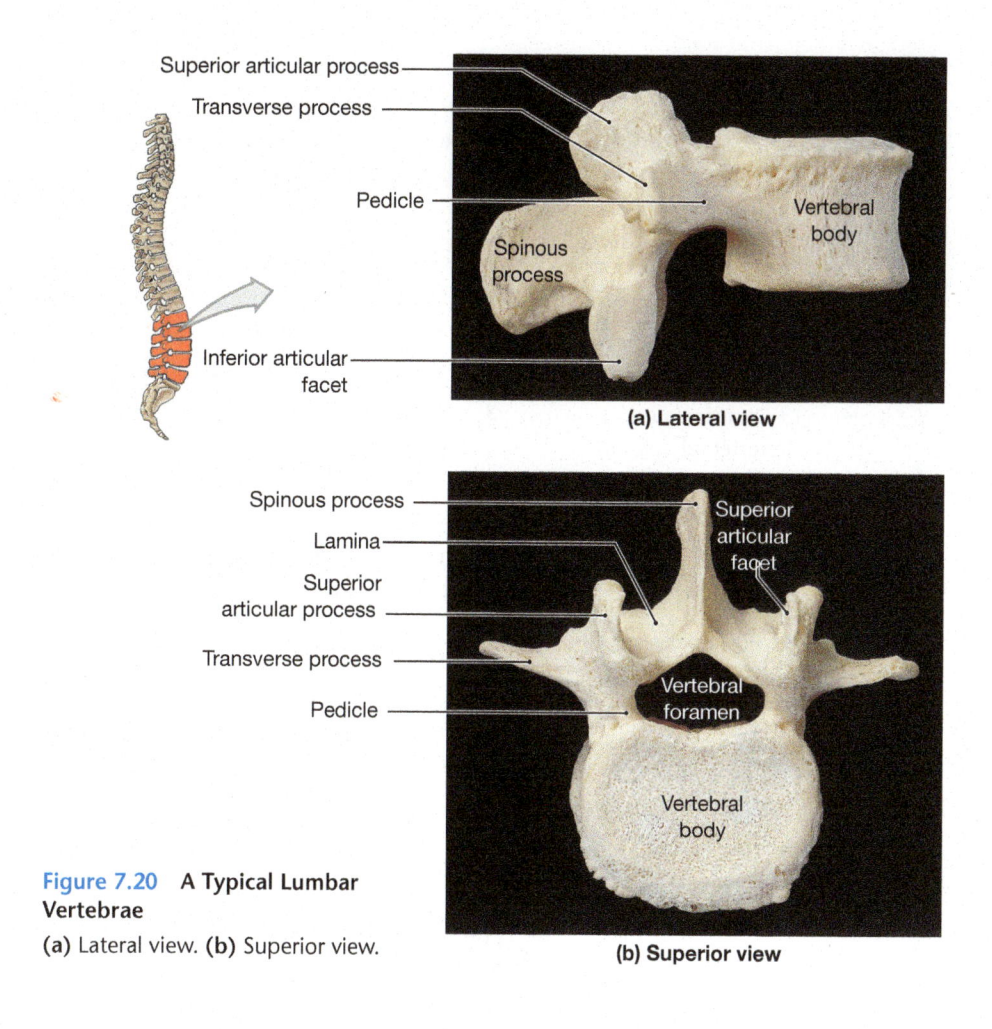

Figure 7.20 A Typical Lumbar Vertebrae
(a) Lateral view. (b) Superior view.

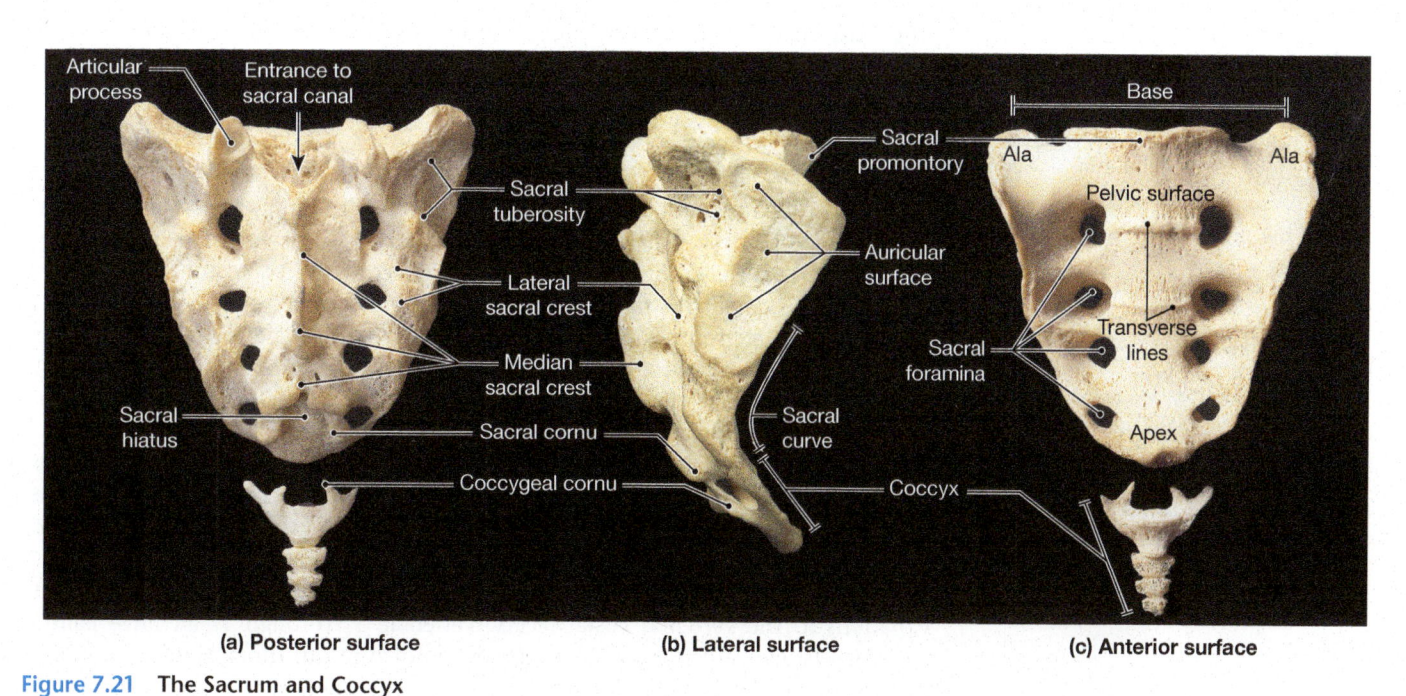

Figure 7.21 The Sacrum and Coccyx
(a) Posterior surface. (b) Lateral surface, right side. (c) Anterior surface.

107

QuickCheck Questions

6.1 What are the five major regions of the vertebral column and the number of vertebrae in each region?

6.2 What are the three features found on all vertebrae?

6 Materials

- ☐ Articulated skeleton
- ☐ Articulated vertebral column
- ☐ Disarticulated vertebral column

Procedures

1. Identify the five regions of the vertebral column on an articulated skeleton and name the type of curve (accommodation or compensation) found in each region.
2. Identify the anatomical features of a typical vertebra on a disarticulated vertebra.
3. Distinguish the anatomical differences among cervical, thoracic, and lumbar vertebrae.
4. Identify the unique features of the atlas and the axis. Examine the articulation of these two vertebrae with the skull.
5. Inspect the sacrum and the coccyx. Note how the sacrum articulates with the bones of the appendicular skeleton. ■

LAB ACTIVITY 7 — Thoracic Cage

The 12 pairs of ribs articulate with the thoracic vertebrae posteriorly and the sternum anteriorly to enclose the thoracic organs in a protective cage. In breathing, muscles move the ribs to increase or decrease the size of the thoracic cavity and cause air to move into or out of the lungs.

Sternum

The **sternum** is the flat bone located anterior to the thoracic region of the vertebral column. It is composed of three bony elements: a superior **manubrium** (ma-NOO-brē-um), a middle **sternal body**, and an inferior **xiphoid** (ZĪ-foyd) **process** (Figure 7.22). The manubrium is triangular and articulates with the first pair of ribs and the clavicle. Muscles that move the head and neck attach to the manubrium. The sternal body is elongated and receives the costal cartilage of ribs 2 through 7. The xiphoid process is shaped like an arrowhead and projects inferiorly off the sternal body. This process is cartilaginous until late adulthood, when it completely ossifies.

Ribs

Ribs, also called **costae**, are classified according to how they articulate with the sternum. The first seven pairs are called either **true ribs** or **vertebrosternal ribs** because their cartilage, the **costal cartilage**, attaches directly to the sternum. Rib pairs 8–12 are called **false ribs** because their costal cartilage does not connect directly with the sternum. Rib pairs 8, 9, and 10 are also called **vertebrochondral ribs** because they do not articulate directly with the sternum but instead fuse with the costal cartilage of rib 7. Rib pairs 11 and 12 are called **floating ribs** because they do not articulate with the sternum.

Each rib has a **head**, or **capitulum** (ka-PIT-ū-lum), and on the head are two **articular facets** for articulating with the costal facets of the rib's thoracic vertebra (Figure 7.22). The **tubercle** of the rib articulates with the transverse costal facet of the rib's vertebra. Between the head and tubercle is a slender **neck**.

Differences in the way ribs articulate with the thoracic vertebrae are reflected in variations in the vertebral costal facets. Vertebrae T_1 through T_8 all have paired costal facets, one superior and one inferior as noted in our thoracic discussion earlier. The first rib articulates with a transverse costal facet of T_1. The second rib articulates with the inferior costal facet of T_1 and the superior costal facet of T_2. Ribs 3

EXERCISE 7 The Axial Skeleton 109

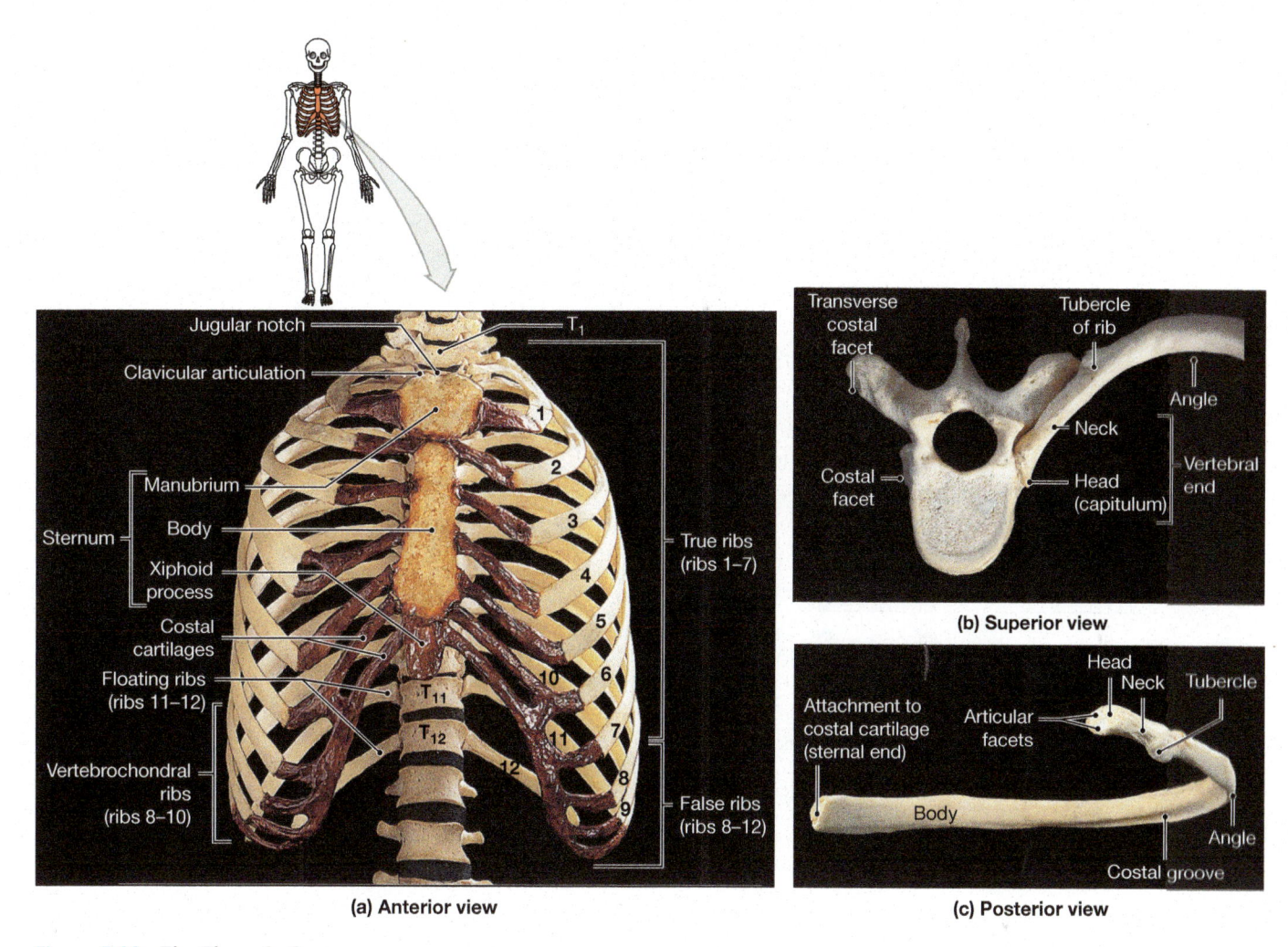

Figure 7.22 The Thoracic Cage
(a) Anterior view of the rib cage and sternum. (b) Superior view of the articulation between a thoracic vertebra and the vertebral end of a right rib. (c) Posterior and medial view of a right rib.

through 9 continue this pattern of articulating with two adjacent costal facets. Vertebrae T_9 through T_{12} have a single costal facet on the vertebral body, and the ribs articulate entirely on the one costal facet. After each rib articulates on the single costal facet, the rib bends laterally and articulates on the transverse costal facet. The eleventh and twelfth pairs of ribs do not articulate on costal facets.

QuickCheck Questions

7.1 Which part of the sternum articulates with the clavicle?
7.2 Which ribs are true ribs, which are false ribs, and which are floating ribs?

7 Materials

- ☐ Articulated skeleton
- ☐ Articulated vertebral column with ribs
- ☐ Disarticulated vertebral column and ribs

Procedures

1. Identify the manubrium, sternal body, and xiphoid process of the sternum.
2. Identify the anatomical features of a typical rib.
3. Observe an articulated male and female skeleton and compare the thoracic cage, noting the number of ribs in each.
4. Examine how each rib articulates with the sternum and note the differences in articulation on the costal facets of the thoracic vertebrae. ■

EXERCISE 7

LAB REPORT

Name _____
Date _____
Section _____

The Axial Skeleton

A. Matching

Match each skull structure in the numbered list with the correct bone in the lettered list. Each term in the lettered list may be used more than once.

_____ 1. sella turcica   A. sphenoid
_____ 2. crista galli   B. maxilla
_____ 3. foramen magnum   C. frontal bone
_____ 4. zygomatic process   D. parietal bone
_____ 5. condylar process   E. occipital bone
_____ 6. lesser wing   F. ethmoid
_____ 7. mandibular fossa   G. nasal bone
_____ 8. coronoid process   H. zygomatic bone
_____ 9. jugular foramen   I. mandible
_____ 10. superior nuchal line   J. temporal bone
_____ 11. superior temporal line
_____ 12. supraorbital foramen

B. Matching

Match each structure of the vertebral column and rib cage in the numbered list with the correct description in the lettered list. Each term in the lettered list may be used more than once.

_____ 1. manubrium   A. all vertebrae
_____ 2. capitulum   B. second cervical vertebra
_____ 3. vertebrosternal ribs   C. thoracic vertebrae
_____ 4. tubercle   D. head of rib
_____ 5. xiphoid process   E. true ribs
_____ 6. vertebral body   F. centrum of vertebra
_____ 7. vertebral foramen   G. articulates with facet
_____ 8. vertebrochondral ribs   H. ribs 8, 9, and 10
_____ 9. axis   I. ribs 11 and 12
_____ 10. dens   J. sternum
_____ 11. floating ribs
_____ 12. costal facet

111

EXERCISE 7 — LAB REPORT

C. Labeling

1. Label Figure 7.23, a lateral view of the skull.

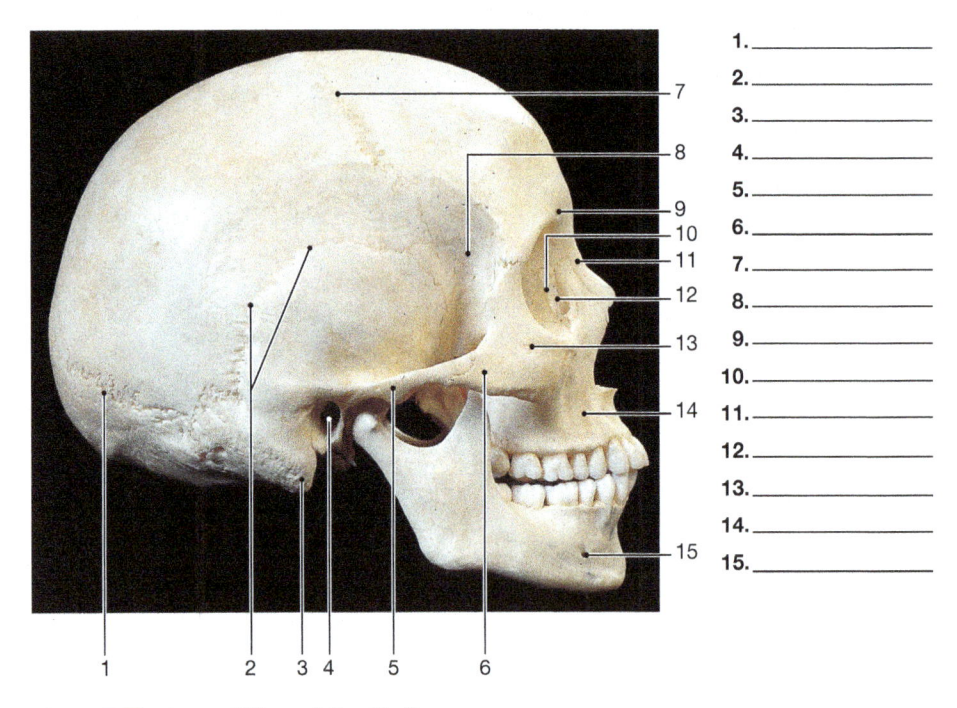

1. _____
2. _____
3. _____
4. _____
5. _____
6. _____
7. _____
8. _____
9. _____
10. _____
11. _____
12. _____
13. _____
14. _____
15. _____

Figure 7.23 Lateral View of the Skull

2. Label Figure 7.24, a typical cervical vertebra.

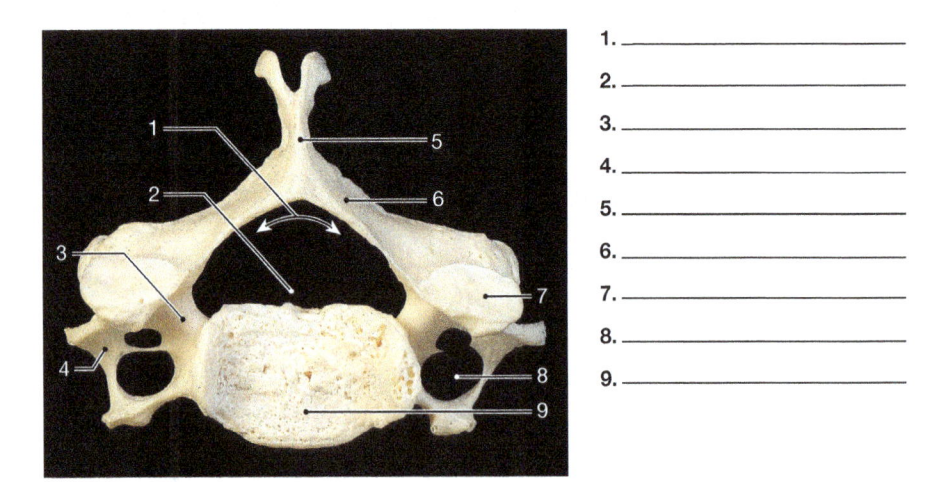

1. _____
2. _____
3. _____
4. _____
5. _____
6. _____
7. _____
8. _____
9. _____

Figure 7.24 Cervical Vertebra, Superior View

112

LAB REPORT — EXERCISE 7

D. Short-Answer Questions

1. List the three primary components of the axial skeleton.

2. How many bones are in the cranium and the face combined?

3. Describe the three cranial fossae and the bones that form the floor of each.

4. List the six primary sutures of the skull and the bones that articulate at each suture.

5. Describe the four regions of the vertebral column.

E. Analysis and Application

1. Name two passageways in the floor of the skull for major blood vessels that supply the brain.

2. Describe the main anatomical features at the point where the vertebral column articulates with the skull.

113

EXERCISE 7

LAB REPORT

3. Compare the articulation on the thoracic vertebrae of rib pair 7 and rib pair 10.

4. List the bones that form the eye orbit.

5. Describe the anatomical features of the nasal cavity.

The Appendicular Skeleton

EXERCISE 8

OBJECTIVES

On completion of this exercise, you should be able to:

- Identify the bones and surface features of the pectoral girdle and upper limb.
- Articulate a clavicle with a scapula.
- Articulate a scapula, humerus, radius, and ulna.
- Identify the bones of the wrist and hand.
- Identify the bones and surface features of the pelvic girdle and lower limb.
- Articulate an os coxae with a sacrum to form a bony pelvis.
- Articulate an os coxae, femur, tibia, and fibula.
- Identify the bones of the ankle and foot.

LAB ACTIVITIES

1. Pectoral Girdle   115
2. Upper Limb   118
3. Pelvic Girdle   123
4. Lower Limb   126
   A Regional Look: Shoulder   132

The appendicular skeleton provides the bony structure of the limbs, permitting us to move and to interact with our surroundings. It is attached to the vertebral column and sternum of the axial skeleton. The appendicular skeleton consists of two pectoral girdles and the attached upper limbs and a pelvic girdle and the attached lower limbs. The pectoral girdles are loosely attached to the axial skeleton, and as a result the shoulder joints have a great range of movement. The pelvic girdle is securely attached to the sacrum of the spine to support the weight of the body.

As you study the appendicular skeleton, keep in mind that each bone is one member of a left/right pair. Carefully observe the orientation of major surface features on the bones and use these features as landmarks for determining whether a given bone is from the left side of the body or from the right side.

LAB ACTIVITY 1   Pectoral Girdle

The **pectoral girdles** consist of four bones: two *clavicles*, commonly called collarbones, and two *scapulae*, the shoulder blades. These four bones are arranged in an incomplete ring that constitutes the bony architecture of the superior trunk. Each scapula rests against the posterior surface of the rib cage and against a clavicle, and provides an anchor for tendons of arm and shoulder muscles. The clavicles are like struts that provide support by connecting the scapulae to the sternum.

Clavicle

The S-shaped **clavicle** (KLAV-i-kul) is the only bony connection between the pectoral girdle and the axial skeleton. The **sternal end** articulates medially with the sternum, and laterally the flat **acromial** (a-KRŌ-mē-al) **end** joins the scapula

116   EXERCISE 8   The Appendicular Skeleton

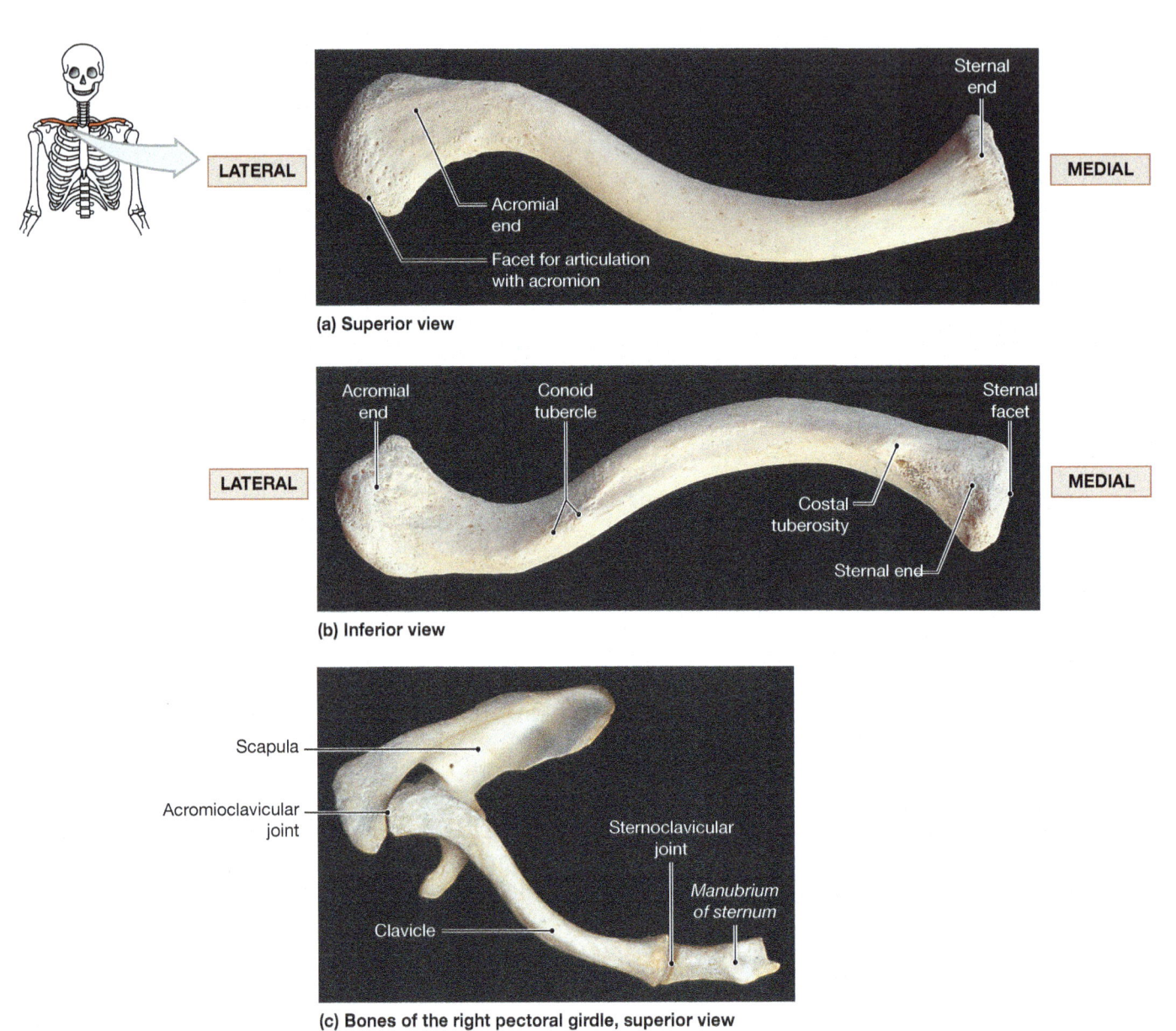

Figure 8.1 The Clavicle
(a) Superior view of right clavicle. (b) Inferior view. (c) The clavicle is the only direct connection of the pectoral girdle and the axial skeleton.

(Figure 8.1). Inferiorly, toward the acromial end, where the clavicle bends, is the **conoid tubercle**, an attachment site for the coracoclavicular ligament. Near the inferior sternal end is the rough **costal tuberosity**.

The sternal end of the clavicle articulates lateral to the jugular notch on the manubrium of the sternum. The point where these two bones articulate is called the **sternoclavicular joint**. From this joint, the clavicle swings posteriorly and articulates with the scapula at the **acromioclavicular joint**.

Scapula

The **scapula** (SKAP-ū-la), shown in Figure 8.2, is composed of a triangular **body**. The long edges of the scapular body are the **superior**, **medial**, and **lateral borders**. The corners where the borders meet are the **superior**, **lateral**, and **inferior**

EXERCISE 8 The Appendicular Skeleton 117

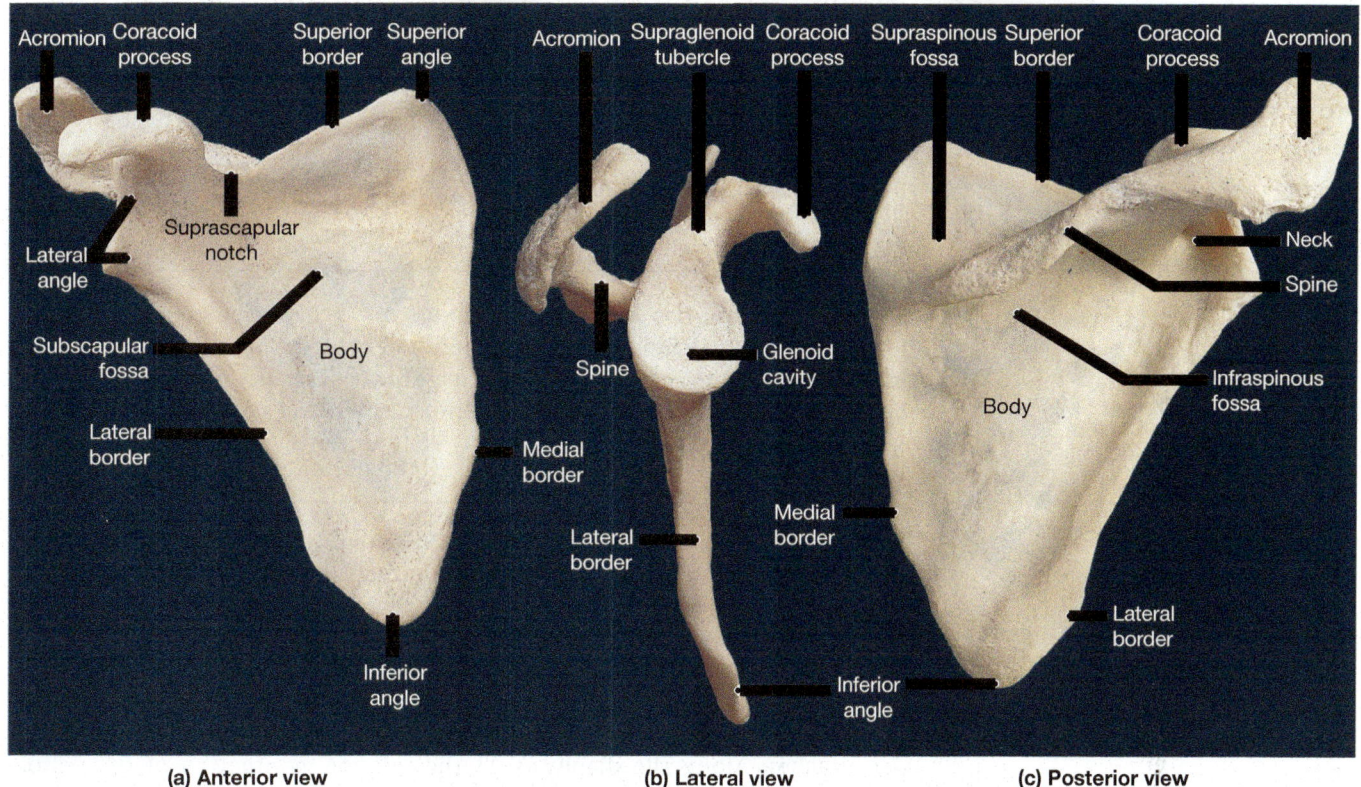

Figure 8.2 The Scapula
(a) Anterior, (b) lateral, and (c) posterior views of right scapula.

angles. An indentation in the superior border is the **suprascapular notch**. The **subscapular fossa** is the smooth, triangular surface where the anterior surface of the scapula faces the ribs.

A prominent ridge, the **spine**, extends across the scapula body on the posterior surface and divides the convex surface into the **supraspinous fossa** superior to the spine and the **infraspinous fossa** inferior to the spine. At the lateral tip of the spine is the **acromion** (a-KRŌ-mē-on), which is superior to the **glenoid cavity** (also called the *glenoid fossa*) where the humerus articulates. Superior and inferior to the glenoid cavity are the **supraglenoid** and **infraglenoid tubercles**, where the biceps brachii and triceps brachii muscles of the arm attach. Superior to the glenoid cavity is the beak-shaped **coracoid** (KOR-uh-koyd) **process**. The **scapular neck** is the ring of bone around the base of the coracoid process and the glenoid cavity.

QuickCheck Questions

1.1 Which bones form the pectoral girdle?
1.2 Which bone of the upper limb articulates with the scapula?

EXERCISE 8 The Appendicular Skeleton

1 Materials

- ☐ Articulated skeleton
- ☐ Disarticulated skeleton

Procedures

1. Review the anatomy of the clavicle from the disarticulated skeleton. Palpate your own shoulder and feel the sternal and acromial ends of one of your clavicles.
2. Review the anatomy of the scapula from the disarticulated skeleton. Palpate to feel the spine and acromion on your own scapula.
3. Articulate a clavicle and scapula from the disarticulated skeleton, being sure both are right bones or both are left bones. Place the clavicle on your appropriate shoulder (left or right) and determine how it articulates with your scapula. Holding the clavicle in place, position the scapula from the disarticulated skeleton over your shoulder and examine the articulation at the acromioclavicular joint. ■

LAB ACTIVITY 2 Upper Limb

Each **upper limb**, also called an *upper extremity*, includes the bones of the arm, forearm, wrist, and hand—a total of 30 bones, all but three of them in the wrist and hand. Note that the correct anatomical usage of the term *arm* is in reference to the **brachium**, the part of the upper limb between elbow and shoulder.

Humerus

The bone of the **arm** (brachium) is the **humerus**, shown in Figure 8.3. The proximal **head** articulates with the glenoid cavity of the scapula. Lateral to the head is the **greater tubercle**, and medial to the head is the **lesser tubercle**: both are sites for muscle tendon attachment. The **intertubercular groove** separates the tubercles. Between the head and the tubercles is the **anatomical neck**; inferior to the tubercles is the **surgical neck**. Inferior to the greater tubercle is the rough **deltoid tuberosity**, where the deltoid muscle of the shoulder attaches. Along the diaphysis, at the inferior termination of the deltoid tuberosity, is the **radial groove**, a depression that serves as the passageway for the radial nerve.

The distal end of the humerus has a specialized **condyle** to accommodate two joints: the hinge-like elbow joint and a pivot joint of the forearm, the latter of which is used when doing such movements as turning a doorknob. The condyle has a round **capitulum** (*capit*, head) on the lateral side and a medial cylindrical **trochlea** (*trochlea*, a pulley). Superior to the trochlea are two depressions, the **coronoid fossa** on the anterior surface and the triangular **olecranon** (ō-LEK-ruh-non) **fossa** on the posterior surface. The **medial** and larger **lateral epicondyles** are to the sides of the condyle.

Ulna

The **forearm** has two parallel bones, the medial **ulna** and the lateral **radius** (Figure 8.4), both of which articulate with the humerus at the elbow. The ulna is the larger forearm bone and articulates with the humerus and radius. A fibrocartilage disc occurs between the ulna and the wrist. The ulna has a conspicuous U-shaped **trochlear notch** (or *semilunar notch*) that is like a C-clamp, with two processes that articulate with the humerus: the superior **olecranon** and the inferior **coronoid process** (Figure 8.4c). Each process fits into its corresponding fossa on the humerus. On the lateral surface of the coronoid process is the flat **radial notch**. Inferior to the notch is the rough **ulnar tuberosity**. At the distal extremity is the **ulnar head** and the pointed **styloid process of the ulna**.

EXERCISE 8   The Appendicular Skeleton   119

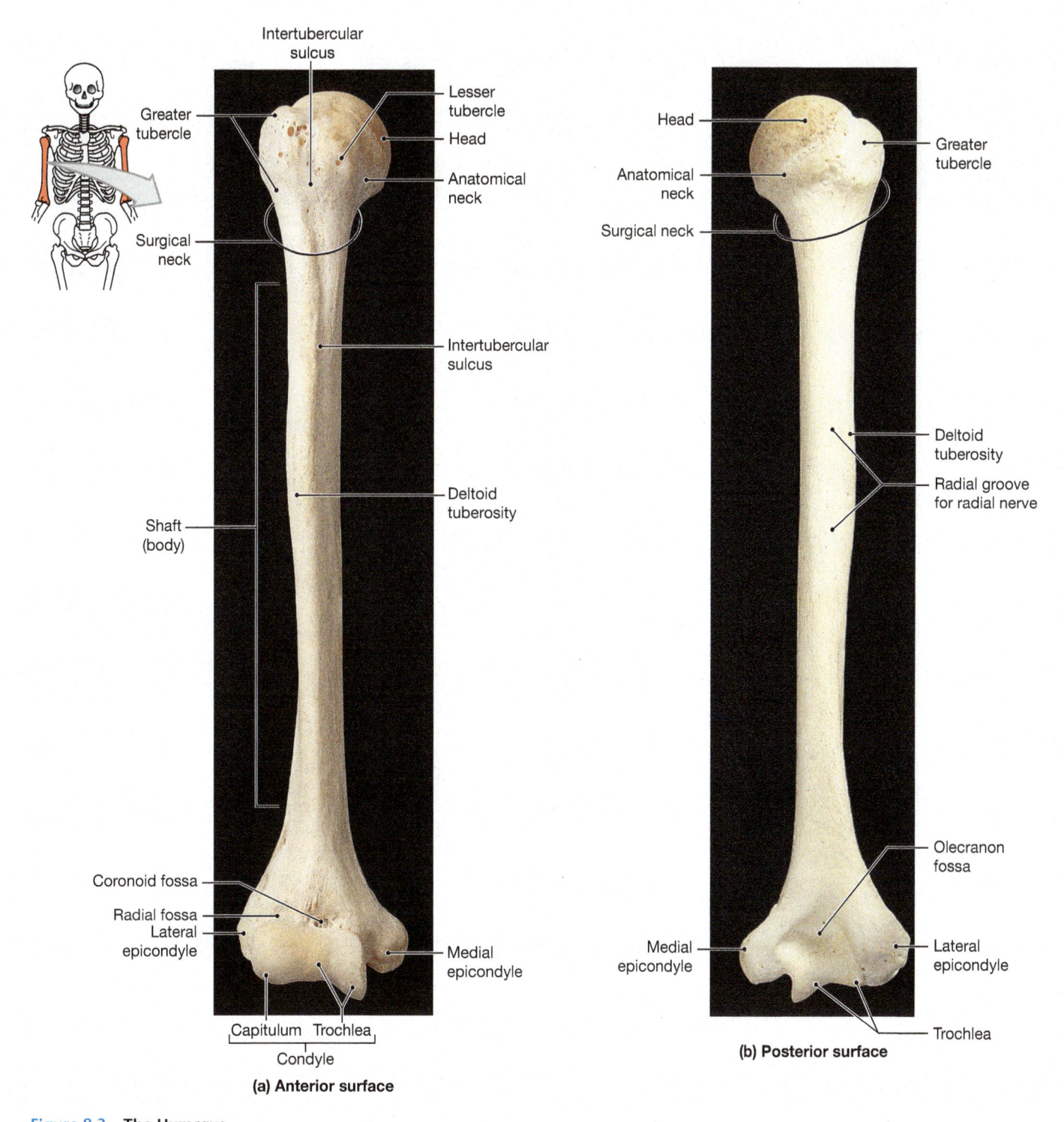

Figure 8.3   The Humerus
(a) Anterior and (b) posterior surfaces of right humerus.

Making Connections   Elbow Terminology

Notice that the terminology of the elbow is consistent in the humerus and ulna. The trochlear notch of the ulna fits into the trochlea of the humerus. The coronoid process and olecranon fit into their respective fossae on the humerus.

120 EXERCISE 8   The Appendicular Skeleton

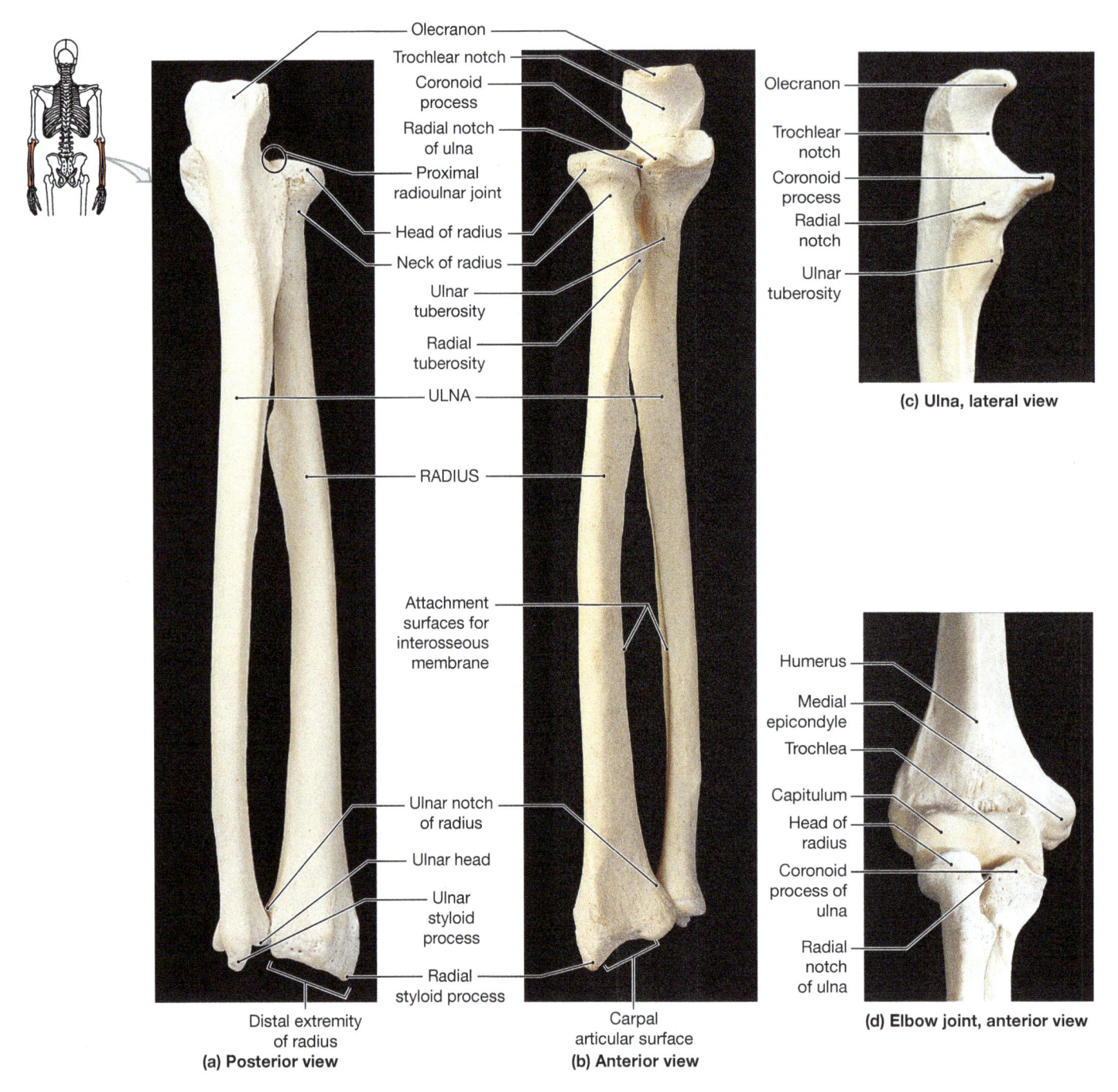

Figure 8.4 The Radius and Ulna
(a) Posterior and (b) anterior views of both bones. (c) Lateral view of right ulna. (d) Anterior view of right elbow joint showing the articulated bones.

Radius

The radius (see Figure 8.4) has a disc-shaped **radial head** that pivots in the radial notch of the ulna at the **proximal radioulnar joint**. The superior surface of the head has a depression where it articulates with the capitulum of the humerus. Supporting the head is the **neck**, and inferior to the neck is the **radial tuberosity**. On the distal portion, the **styloid process of the radius** is larger and not as pointed as the styloid process of the ulna. The **ulnar notch** on the medial surface articulates with the ulna at the **distal radioulnar joint**. The **interosseous membrane** extends between the ulna and radius to support the bones.

Wrist and Hand

The wrist is formed by eight **carpal bones** (KAR-pulz) arranged in two rows of four, the **proximal** and **distal carpal bones**. An easy method to identify the carpal bones is to use the anterior wrist and start with the carpal bone next to the styloid process of the radius (Figure 8.5). From this reference point moving medially, the proximal carpal bones are the **scaphoid bone**, **lunate bone**, **triquetrum**, and small **pisiform bone**. Returning on the lateral side, the four distal carpal bones are the **trapezium**, **trapezoid bone**, **capitate bone**, and **hamate bone**. The hamate bone has a process called the **hook of the hamate**.

The five long bones of the palm are **metacarpal bones**. Each metacarpal bone is numbered with a roman numeral; the metacarpal bone forms the thumb, and is number I.

The 14 bones of the fingers are called **phalanges**. Digits II, III, IV, and V each have a **proximal**, **middle**, and **distal phalanx**. The thumb, or **pollex**, has only proximal and distal phalanges.

Making Connections

Carpal Bone Identification

Here's a mnemonic to help you remember the carpal bones. Also, remember that "zi" comes before "zo," to remind you that trapezium is before trapezoid.

Sam	Scaphoid
Likes	Lunate
To	Triquetrum
Push	Pisiform
The	Trapezium
Toy	Trapezoid
Car	Capitate
Hard	Hamate

QuickCheck Questions

2.1 What are the three bones that constitute the arm and forearm?

2.2 What are the three major groups of bones in the wrist and hand?

2 Materials

- ☐ Articulated skeleton
- ☐ Disarticulated skeleton

Procedures

1. Locate a humerus of the disarticulated skeleton and review its surface features. Begin your observations at the proximal end of the bone and proceed toward the distal end. Compare the bone with the two humerus bones on the articulated skeleton and use the anatomical markings to determine whether your disarticulated bone is a right or left bone. Notice which surface features are good landmarks for distinguishing right from left bones. Palpate the sides of your own elbow to feel the epicondyles on your humerus.

2. Locate an ulna and radius of the disarticulated skeleton and review their surface features from the proximal end to the distal end. Compare each bone with its counterparts on the articulated skeleton and use the anatomical markings to determine whether your disarticulated bones are right or left.

3. Locate a wrist and hand of the disarticulated skeleton. The bones are probably wired together for easy study and to examine the articulations. Examine the proximal and distal carpal bones, metacarpal bones I–V, and the phalanges. Compare the wrist and hand with those of the articulated skeleton to determine whether your disarticulated unit is right or left, using the pollex as a landmark.

4. Articulate the bones of the upper limb with those of the pectoral girdle. Examine how the humerus articulates proximally with the scapula and distally

Figure 8.5 Bones of the Wrist and Hand
(a) Anterior and (b) posterior views of the right wrist and hand.

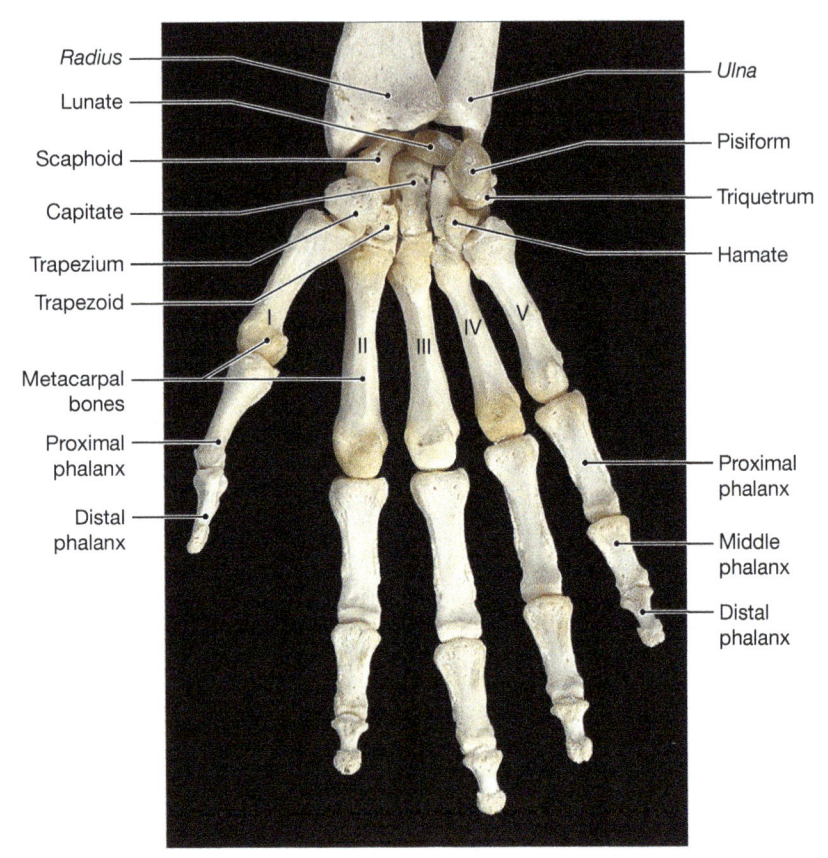

(a) Right wrist and hand, anterior (palmar) view

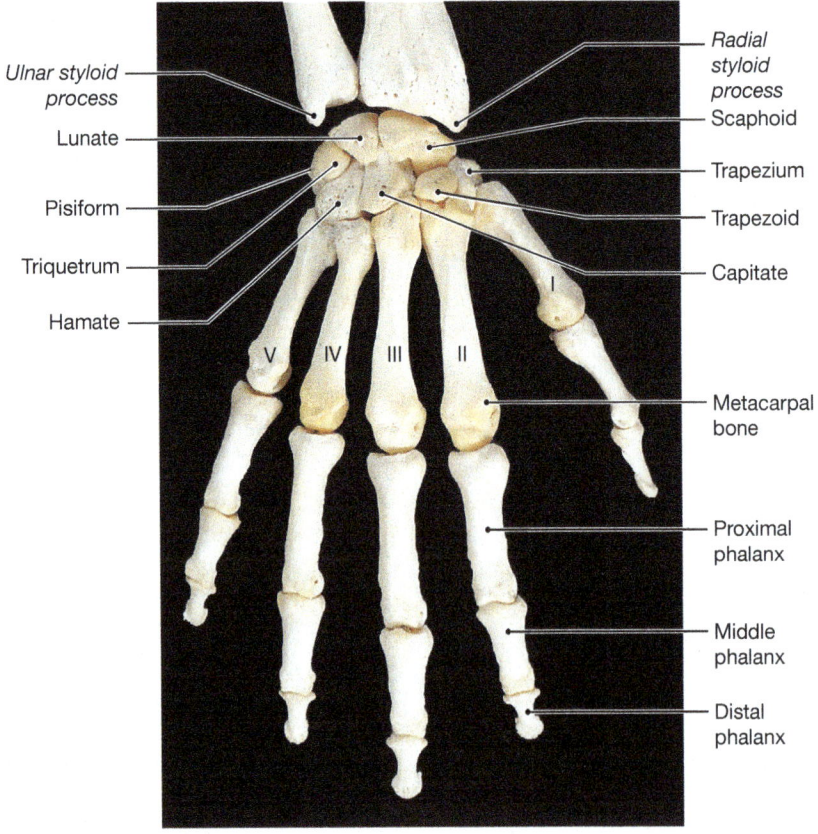

(b) Right wrist and hand, posterior (dorsal) view

122

with the ulna and radius. Observe the joints between the humerus, ulna, and radius. On an articulated skeleton, move the bones of each joint and identify the articulating surfaces. ∎

LAB ACTIVITY 3 Pelvic Girdle

The **pelvic girdle** is made up of the two hipbones, called **ossa coxae** (singular *os coxae*), which articulate with the vertebral column and attach the lower limbs (Figure 8.6). The os coxae, also called a **coxal bone**, is formed by the fusion of three bones: the **ilium** (IL-ē-um), **ischium** (IS-kē-um), and **pubis** (PŪ-bis). By 20–25 years of age, these three bones ossify into a single os coxae, but the three bones are still used to name related structures.

In addition to the pelvic girdle, Figure 8.6 shows the rest of the region called the **pelvis,** defined as the bony ring that encases the pelvic cavity. The pelvis is made up of the girdle (in other words, the two ossa coxae) of the appendicular skeleton and the sacrum and coccyx of the axial skeleton. On the medial surface of each os coxae, the **iliac fossa** forms the wall of the upper pelvis, called the **false pelvis**. The **arcuate line** on this same surface marks where the pelvis narrows into the lower pelvis, called the **true pelvis**. Anteriorly, the pubis bones join each other at the **pubic symphysis,** a strong joint that contains fibrocartilage.

The pelvis of the male differs anatomically from that of the female. The female pelvis has a wider **pelvic outlet**, which is the space between the two ischii. The circle formed by the top of the os coxae, called the **brim** of the pelvis, defines the **pelvic inlet**. This opening is wider and rounder in females. Additionally, the **pubic angle** at the pubis symphysis is wider in the female and more U-shaped. This angle is V-shaped in the male. The wider female pelvis provides a larger passageway for childbirth. Refer to Table 8.1 for more sex differences in the human skeleton.

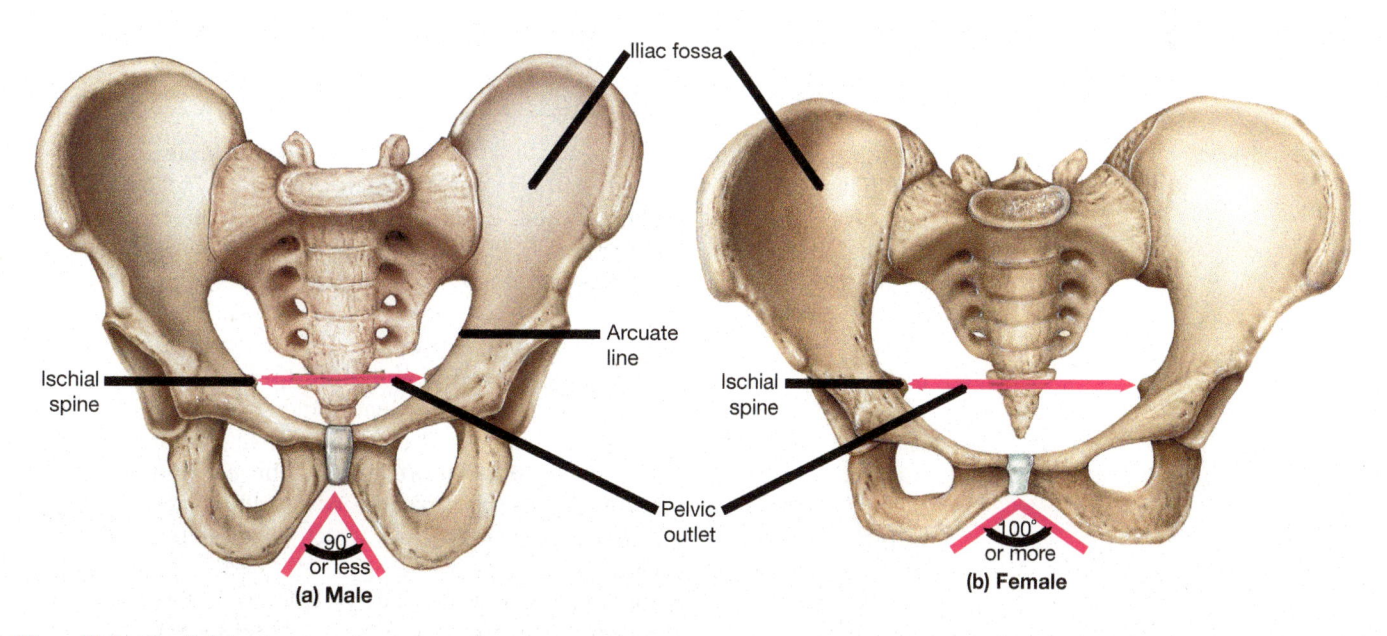

Figure 8.6 The Pelvis
Anatomical differences in the **(a)** male and **(b)** female pelvis. Note the sharper pubic angle in the male pelvis. The female pelvis has a wider pelvic outlet than a male pelvis.

Table 8.1 Sex Differences in the Human Skeleton

Region and Feature	Male (compared with female)	Female (compared with male)
Skull		
General appearance	Heavier, rougher	Lighter, smoother
Forehead	More sloping	More vertical
Sinuses	Larger	Smaller
Cranium	About 10% larger (average)	About 10% smaller
Mandible	Larger, more robust	Smaller, lighter
Teeth	Larger	Smaller
Pelvis		
General appearance	Narrower, more robust, rougher	Broader, lighter, smoother
Pelvic inlet	Heart-shaped	Oval to round
Iliac fossa	Deeper	Shallower
Ilium	More vertical; extends farther superior to sacroiliac joint	Less vertical; less extension superior to sacral articulation
Angle inferior to pubic symphysis	Under 90°	100° or more (see Figure 8.6 123)
Acetabulum	Directed laterally	Faces slightly anteriorly as well as laterally
Obturator foramen	Oval	Triangular
Ischial spine	Points medially	Points posteriorly
Sacrum	Long, narrow triangle with pronounced sacral curvature	Broad, short triangle with less curvature
Coccyx	Points anteriorly	Points inferiorly
Other Skeletal Elements		
Bone weight	Heavier	Lighter
Bone markings	More prominent	Less prominent

Os Coxae

A prominent feature of the os coxae is the deep socket, the **acetabulum** (as-e-TAB-ū-lum), where the head of the femur articulates (Figure 8.7). The smooth inner wall of the acetabulum is the C-shaped **lunate surface**. The center of the acetabulum is the **acetabular fossa**. The anterior and inferior rims of the acetabulum are not continuous; instead, there is an open gap between them, the **acetabular notch**.

The superior ridge of the ilium is the **iliac crest**. It is shaped like a shovel blade, with the **anterior** and **posterior superior iliac spines** at each end. The large indentation below the posterior superior iliac spine is the **greater sciatic** (sī-AT-ik) **notch**. A conspicuous feature on the posterior iliac crest is the rough **auricular surface,** where the **sacroiliac joint** attaches the pelvic girdle to the sacrum of the axial skeleton. On the flat expanse of the ilium are ridges, the **anterior, posterior,** and **inferior gluteal lines,** which are attachment sites for muscles that move the femur.

The ischium is the bone we sit on. The greater sciatic notch terminates at a bony point, the **ischial spine**. Inferior to this spine is the **lesser sciatic notch**. The **ischial tuberosity** is in the most inferior portion of the ischium and is a site for muscle attachment. The **ischial ramus** extends from the tuberosity and fuses with the pubis bone.

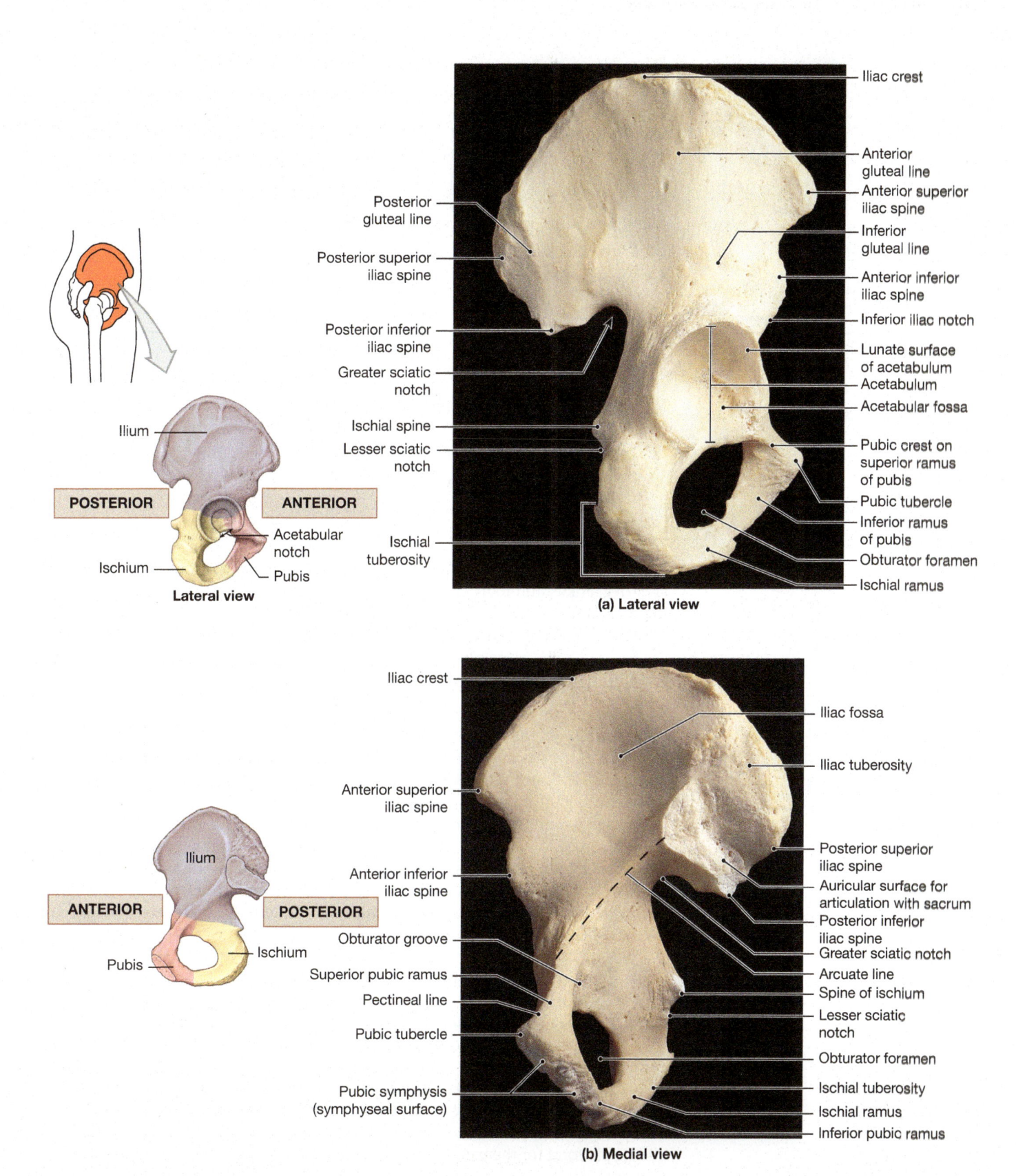

Figure 8.7 The Pelvic Girdle
The pelvic girdle consists of the two ossa coxae. Each os coxae forms as a result of the fusion of an ilium, an ischium, and a pubis. (a) Lateral view and (b) medial view of right os coxae.

125

The pubis forms the anterior portion of the os coxae. The most anterior region of the pubis is the pointed **pubic tubercle**. The **superior ramus** of the pubis is above the tubercle and extends to the ilium. On the medial surface, the superior ramus narrows to a rim called the **pectineal line** of the pubis. The **inferior ramus** joins the ischial ramus, which creates the **obturator** (OB-tū-rā-tor) **foramen**.

QuickCheck Questions

3.1 Which bones make up the pelvic girdle?

3.2 With what structure does the femur articulate at the hip?

3.3 What is the difference between the terms *pelvic girdle* and *pelvis*?

3 Materials

☐ Articulated skeleton
☐ Disarticulated skeleton

Procedures

1. Examine the pelvis of an articulated skeleton and note how the os coxae articulates with the axial skeleton. Examine the pelvis, and determine whether the skeleton is male or female. Articulate the os coxae and sacrum of the skeleton and identify the joints between adjacent bones.

2. Locate an os coxae of the disarticulated skeleton and identify the surface features of the bone, including the boundaries of the ilium, ischium, and pubis. ■

LAB ACTIVITY 4 Lower Limb

Each **lower limb**, also called the *lower extremity*, includes the bones of the thigh, knee, leg, ankle, and foot—a total of 30 bones. Recall that the term *leg* refers not to the entire lower limb but only to the region between the knee and ankle. Superior to the leg is the *thigh*.

Femur

The **femur** is the largest bone of the skeleton (Figure 8.8). It supports the body's weight and bears the stress from the leg. The smooth, round **head** fits into the acetabulum of the os coxae and permits the femur a wide range of movement. The depression on the head is the **fovea capitis**, where the *ligamentum capitis femoris* stabilizes the hip joint during movement. A narrow **neck** joins the head to the proximal shaft. Lateral to the head is a large stump, the **greater trochanter** (trō-KAN-ter); on the inferiomedial surface is the **lesser trochanter**. These large processes are attachment sites for tendons of powerful hip and thigh muscles. On the anterior surface of the femur, between the trochanters, is the **intertrochanteric line**, where the *iliofemoral ligament* inserts to encase the hip joint. Posteriorly, the **intertrochanteric crest** lies between the trochanters.

On the lateral side of the intertrochanteric crest, the **gluteal tuberosity** continues inferiorly and joins with the medial **pectineal line** of the femur as the **linea aspera**, a rough line for thigh muscle attachment. Toward the distal end of the femur, the linea aspera divides into the **medial** and **lateral supracondylar ridges** that encompass a flat triangle called the **popliteal surface**. The medial supracondylar ridge terminates at the **adductor tubercle**.

The **lateral** and **medial condyles** of the femur, which articulate with the tibial head, are the largest condyles of the skeleton. The **intercondylar fossa** separate the condyles posteriorly. A smooth **patellar surface** spans the condyles and serves as a gliding platform for the patella. **Lateral** and **medial epicondyles** are to the sides of the condyles.

Patella

The **patella** is the kneecap and protects the knee joint during movement. It is a sesamoid bone that is encased in the distal tendons of the anterior thigh muscles.

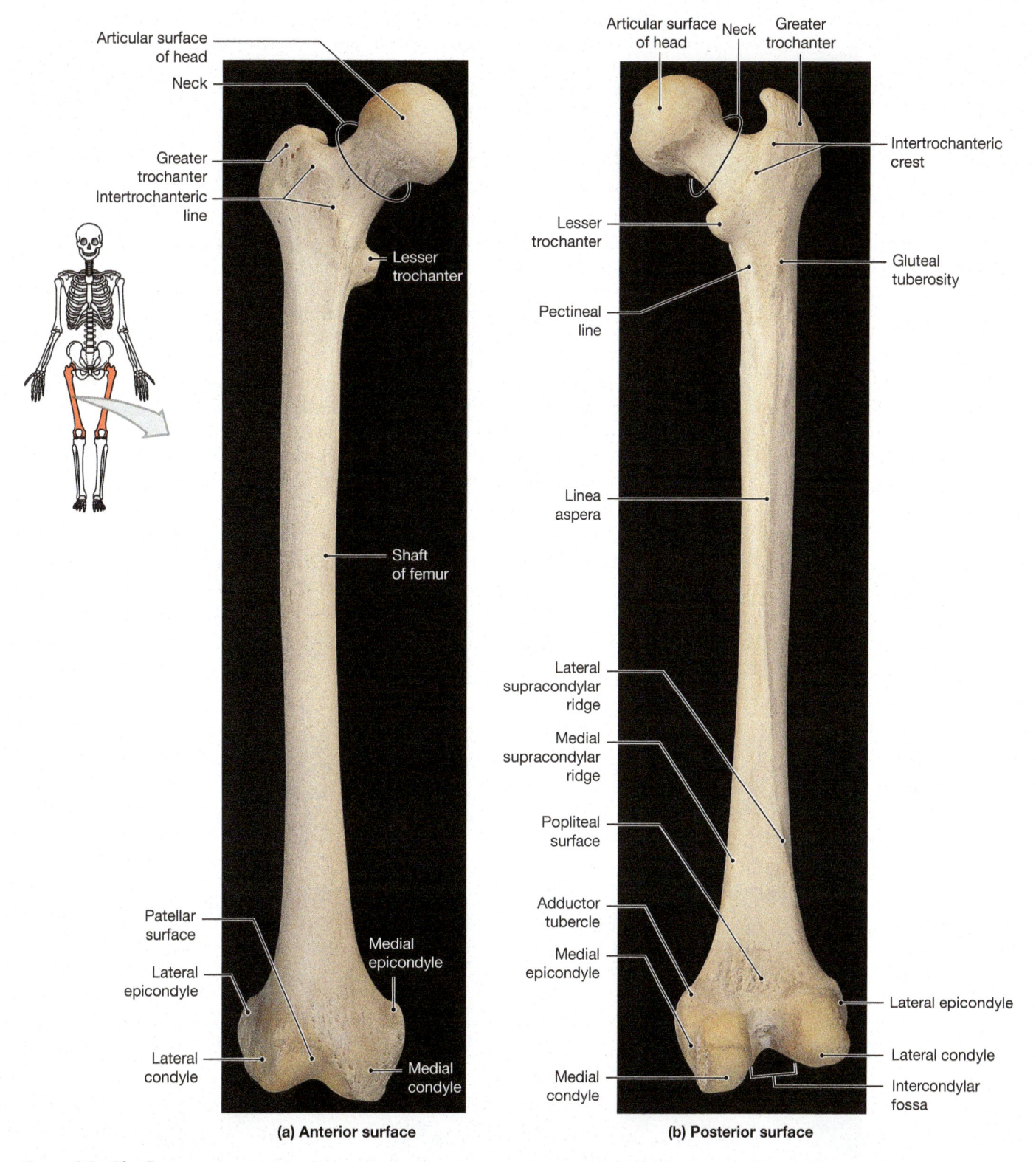

Figure 8.8 The Femur
(a) Anterior surface and (b) posterior surface of right femur.

127

The superior border of the patella is the flat **base**; the **apex** is at the inferior tip (Figure 8.9). Along the base of the patella is the attachment site of the quadriceps muscle tendons that straighten (*extend*) the leg. The patellar ligament joins around the apex of the bone. Tendons attach to the rough anterior surface, and the smooth posterior facets glide over the condyles of the femur. The **medial facet** is narrower than the **lateral facet**.

Tibia

The **tibia** (TIB-ē-uh) is the large medial bone of the leg (Figure 8.10). The proximal portion of the tibia flares to develop the **lateral** and **medial condyles,** which articulate with the corresponding femoral condyles. Separating the tibial condyles is a ridge of bone, the **intercondylar eminence**. This eminence has two projections, the **medial** and **lateral tubercles,** that fit into the intercondylar fossa of the femur. On the anterior surface of the tibia, inferior to the condyles, is the large **tibial tuberosity**, where the patellar ligament attaches. Along most of the length of the anterior shaft is the **anterior margin**, a ridge commonly called the *shin*. The distal tibia is constructed to articulate with the ankle. A large wedge, the **medial malleolus** (ma-LĒ-ō-lus) **of the tibia**, stabilizes the ankle joint. The inferior **articular surface** is smooth so that it can slide over the talus of the ankle. Posteriorly, the proximal tibial shaft has a rough line, the **popliteal line**, where leg muscles attach.

Fibula

The **fibula** (FIB-ū-la) is the slender bone lateral to the tibia (see Figure 8.10). The proximal and distal regions of the fibula appear very similar at first, but closer examination reveals the proximal head to be more rounded (less pointed) than the distal **lateral malleolus of the fibula**. The head of the fibula articulates below the lateral condyle of the tibia at the **superior tibiofibular joint**. The distal articulation creates the **inferior tibiofibular joint**.

Ankle and Foot

The ankle is formed by seven **tarsal bones** (Figure 8.11). One of them, the **talus**, sits on top of the heel bone, the **calcaneus** (kal-KĀ-nē-us), and articulates with the tibia and the lateral malleolus of the fibula. Anterior to the talus is the tarsal bone called the **navicular bone**, which articulates with the **medial, intermediate,** and

(a) Anterior view — Base of patella; Attachment area for quadriceps tendon; Attachment area for patellar ligament; Apex of patella

(b) Posterior view — Medial facet, for medial condyle of femur; Lateral facet, for lateral condyle of femur; Articular surface of patella

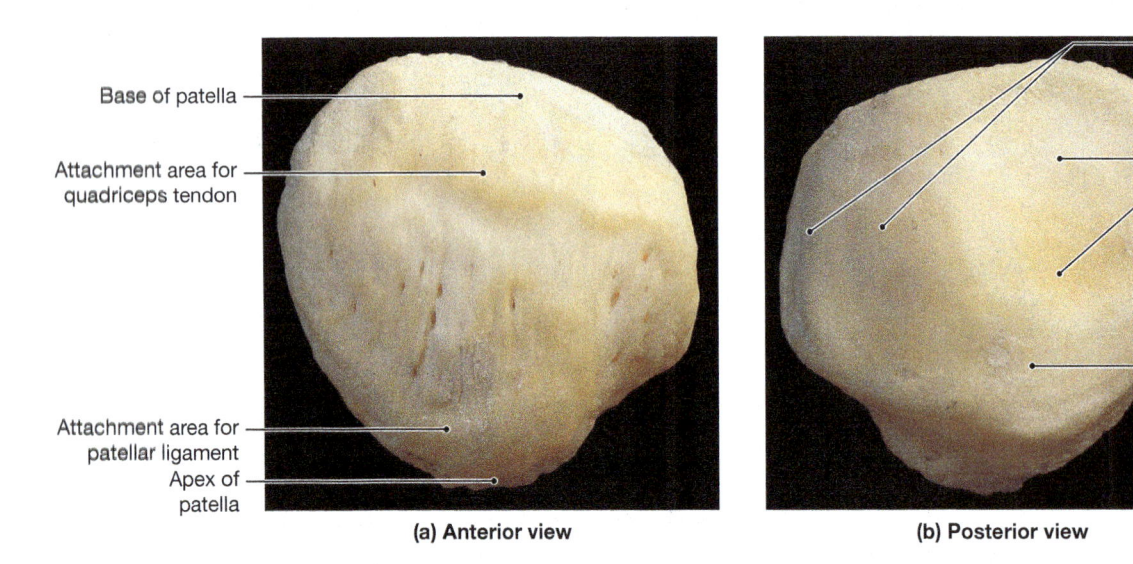

Figure 8.9 The Patella
(a) Anterior surface and **(b)** posterior surface of right patella.

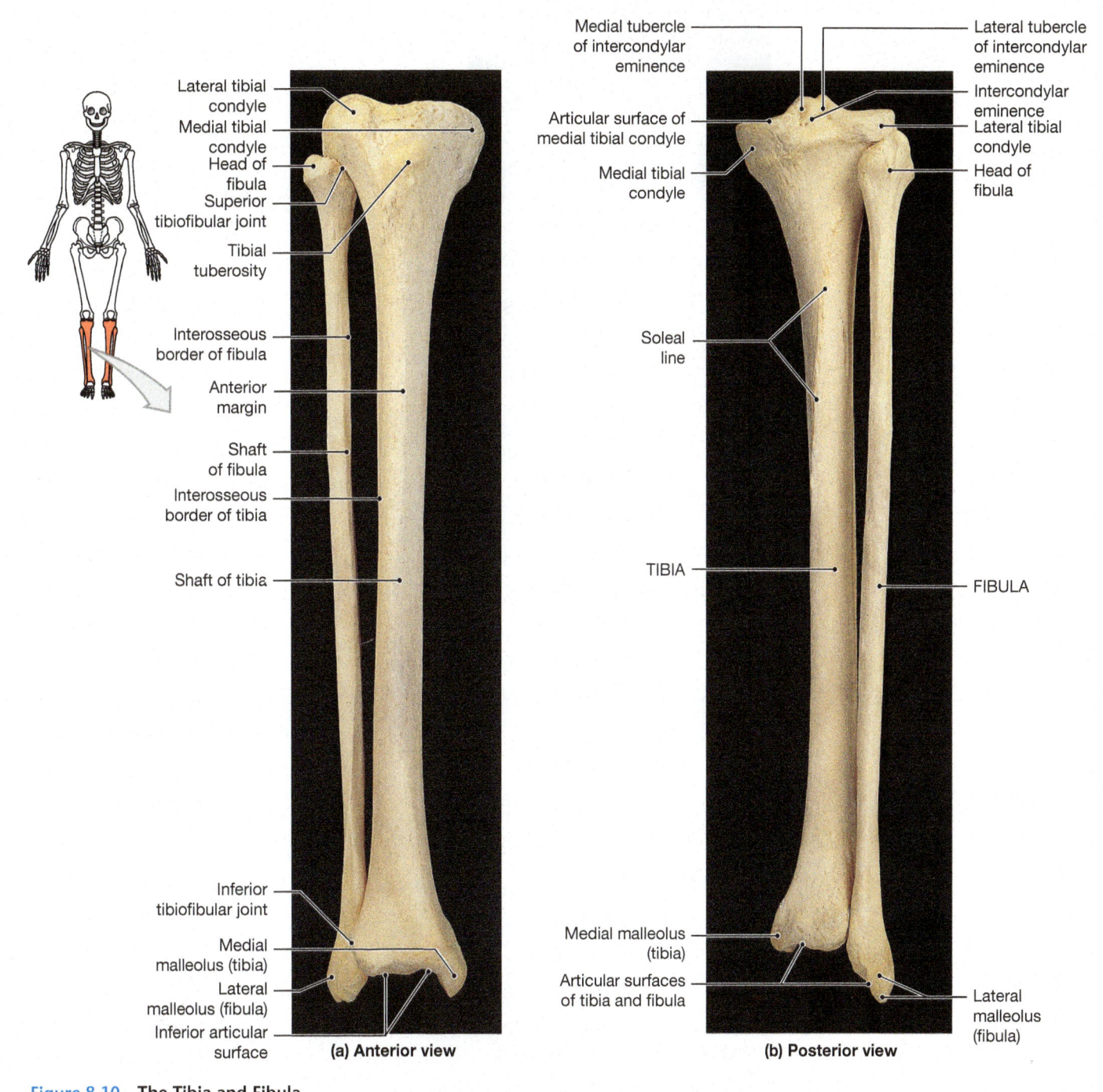

Figure 8.10 The Tibia and Fibula
(a) Anterior view and (b) posterior view of right tibia and fibula.

lateral cuneiform bones. Lateral to the lateral cuneiform bone is the **cuboid bone**, which articulates posteriorly with the calcaneus.

The arch of the foot is formed by five **metatarsal bones** spanning the arch. Each metatarsal bone is named with a roman numeral, with the one forming the big toe being I.

The 14 bones of the toes are called **phalanges**. Like the fingers of the hand, toes II–V have a **proximal**, **middle**, and **distal phalanx**. The big toe, or **hallux**, has only proximal and distal phalanges.

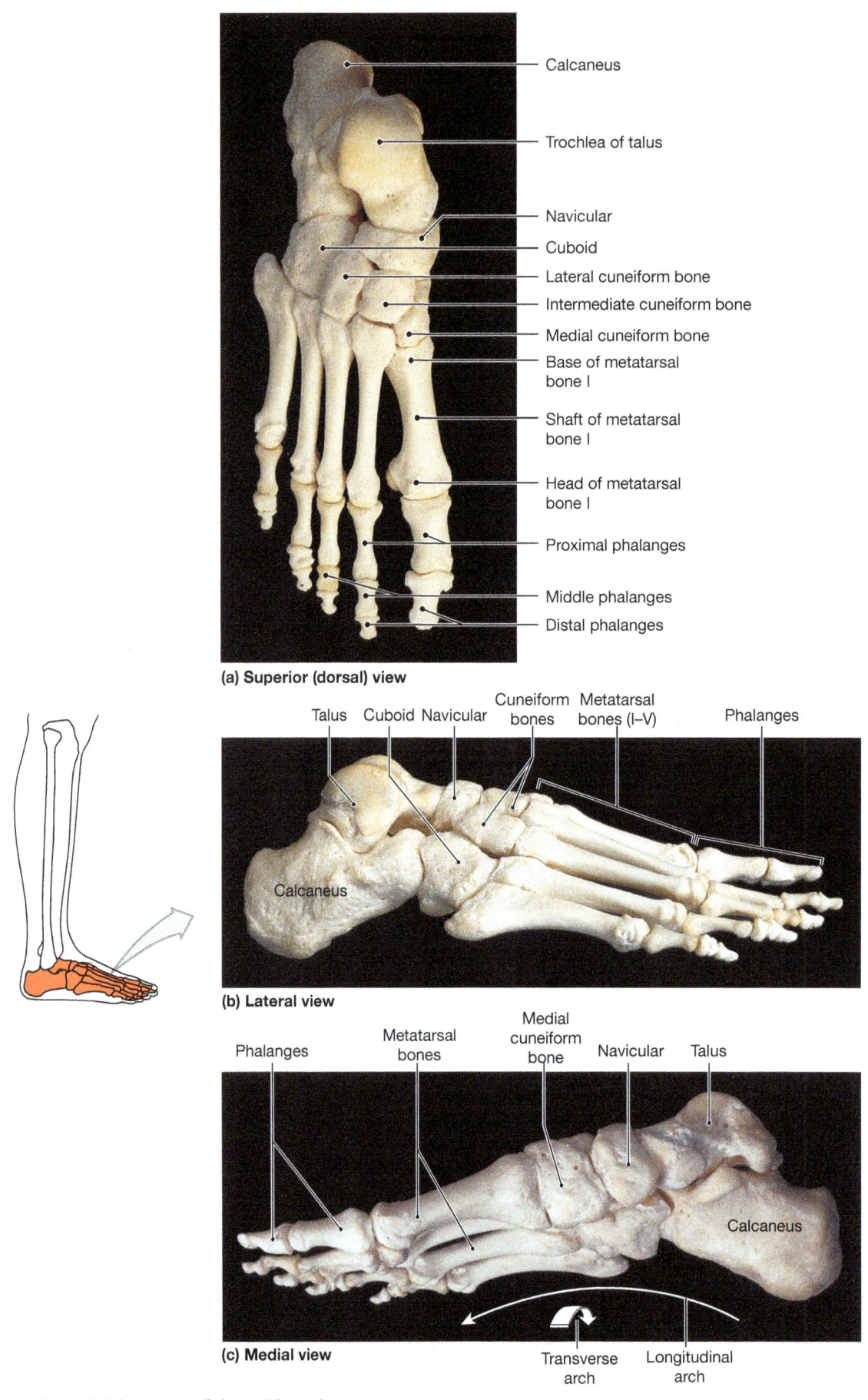

Figure 8.11 Bones of the Ankle and Foot
(a) Superior, (b) lateral, and (c) medial views of right ankle and foot.

QuickCheck Questions

4.1 List the bones of the thigh, knee, and leg.

4.2 What are the three major groups of bones in the ankle and foot?

4 Materials

- ☐ Articulated skeleton
- ☐ Disarticulated skeleton

Procedures

1. Locate a femur of the disarticulated skeleton and review its surface features. Begin your observations at the proximal end of the bone and proceed toward the distal end. Compare the bone with the two femurs on the articulated skeleton and use the anatomical markings to determine whether your disarticulated bone is a right or left bone. Palpate the side of your own hip to feel the greater trochanter of your femur.

2. Locate a tibia and fibula of the disarticulated skeleton and review their surface features from proximal end to distal end. Compare each bone with its counterparts on the articulated skeleton and use the anatomical markings to determine whether your disarticulated bones are right or left.

3. Locate an ankle and foot of the disarticulated skeleton. Like the wrist and hand, the ankle and foot bones are typically strung together in their natural articulations. Examine the seven tarsal bones, metatarsal bones I–V, and the phalanges. Compare the foot with the feet of the articulated skeleton to determine whether your disarticulated foot is right or left, using the hallux as a landmark.

4. Articulate the bones of the lower limb with those of the pelvic girdle. Examine how the femur articulates proximally with the acetabulum of the os coxae and distally with the tibia. Observe the joints between tibia and fibula. On an articulated skeleton, move the bones of each joint and identify the articulating surfaces. ∎

shoulder A Regional Look

The shoulder is the most dynamic area of the appendicular skeleton and exhibits a wide range of motion in a triaxial plane. This is an excellent region to study the anatomical associations among bones, muscles and their tendons, and major blood vessels. To secure the head of the humerus into the glenoid cavity of the scapula, four shoulder muscles form a rotator cuff that wraps around the joint and prevents dislocation during arm movements. Figure 8.12, a horizontal section of a cadaver's shoulder, reveals two of the rotator cuff muscles; the infraspinatus and the subscapularis muscles.

Use Figure 8.12, this manual, and your lecture book to answer the following questions.

1. Follow the tendons of the infraspinatus and subscapularis muscles to their points of attachment on the humerus. How does the position of these muscles and tendons stabilize the shoulder joint?
2. Besides the infraspinatus and subscapularis muscles, which other muscles are part of the rotator cuff?
3. What major artery passes through the shoulder region?
4. Where is the pectoralis major muscle located? Describe how this muscle attaches on the humerus and how the muscle works to move the upper limb.

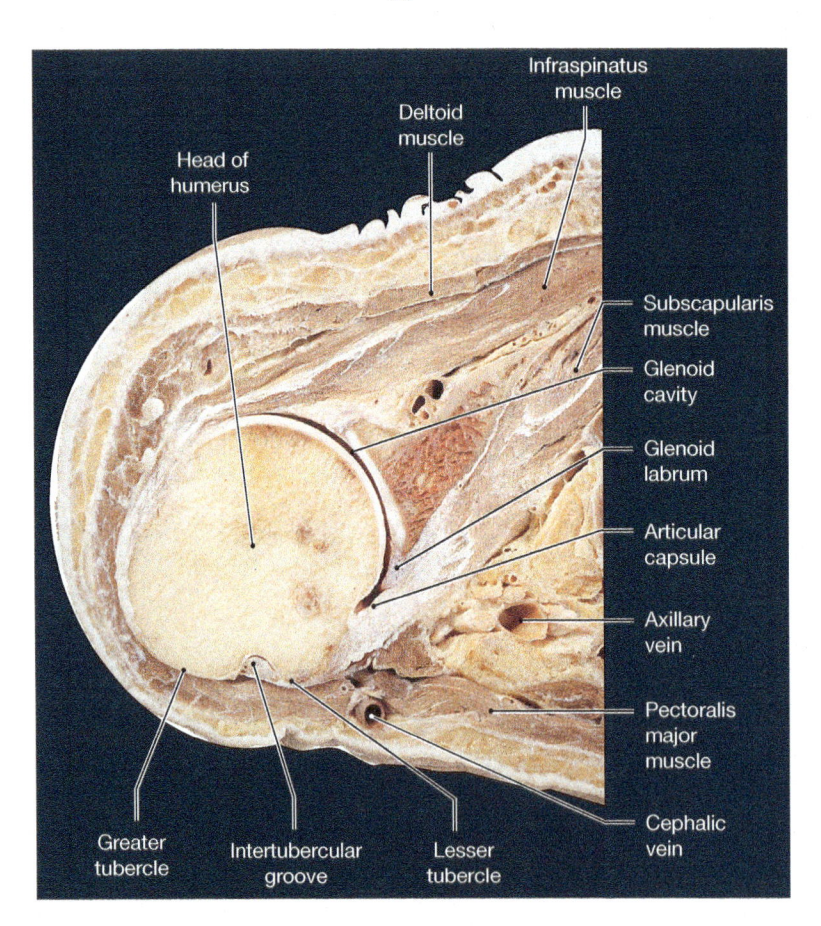

Figure 8.12 Horizontal Section through the Right Shoulder
A cadaver section detailing the regional anatomy of the shoulder.

LAB REPORT

EXERCISE 8

The Appendicular Skeleton

Name _____
Date _____
Section _____

A. Matching

Match each surface feature in the left column with its correct bone from the right column. Each choice from the right column may be used more than once.

_____	1. acromion	A.	clavicle
_____	2. intercondylar fossa	B.	patella
_____	3. trochlea	C.	fibula
_____	4. glenoid cavity	D.	humerus
_____	5. ulnar notch	E.	femur
_____	6. deltoid tuberosity	F.	scapula
_____	7. greater trochanter	G.	tibia
_____	8. sternal end	H.	radius
_____	9. lateral malleolus		
_____	10. linea aspera		
_____	11. capitulum		
_____	12. medial malleolus		
_____	13. intercondylar eminence		
_____	14. base		

B. Labeling

Label the numbered surface features of the bones in Figure 8.13.

C. Short-Answer Questions

1. List the bones of the pectoral girdle and the upper limb.

2. List the bones of the pelvic girdle and the lower limb.

133

EXERCISE 8 — LAB REPORT

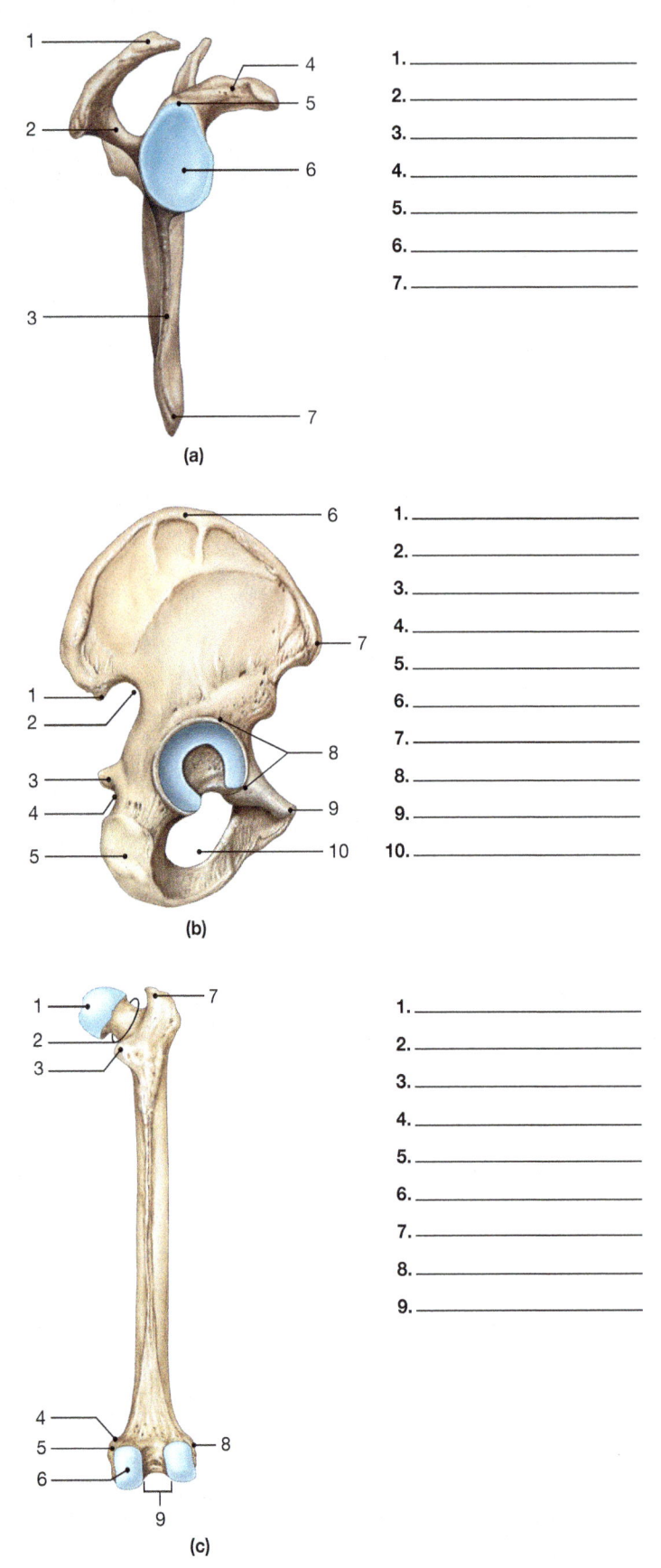

(a)
1. _____
2. _____
3. _____
4. _____
5. _____
6. _____
7. _____

(b)
1. _____
2. _____
3. _____
4. _____
5. _____
6. _____
7. _____
8. _____
9. _____
10. _____

(c)
1. _____
2. _____
3. _____
4. _____
5. _____
6. _____
7. _____
8. _____
9. _____

Figure 8.13
(a) Scapula, (b) os coxae, and (c) femur.

LAB REPORT — EXERCISE 8

3. Compare the male and female pelvis.

4. Which bony process acts like a doorstop to prevent excessive movement of the elbow?

5. On what two structures does the radial head pivot during movements such as turning a doorknob?

6. How are the carpal bones arranged in the wrist?

7. Where is the deltoid tuberosity located?

8. Which appendicular bones have a styloid process?

9. Which bone of the ankle articulates with the tibia?

10. What are the major features of the proximal portion of the femur?

11. Do the toes all have the same number of phalanges?

12. Which bones form the arch of the foot?

EXERCISE 8

LAB REPORT

13. What is the ridge on the tibial head called?

14. Where is the glenoid cavity located?

D. **Analysis and Application**

1. Describe the condyle of the humerus, where the ulna and radius articulate.

2. Compare the bones of the wrist and hand with those of the ankle and foot.

3. Describe how the fibula articulates with the tibia and with the ankle.

4. Name a tuberosity for shoulder muscle attachment and a tuberosity for thigh muscle attachment.

EXERCISE 9

Articulations

OBJECTIVES

On completion of this exercise, you should be able to:

- List the three types of functional joints and give an example of each.
- List the four types of structural joints and give an example of each.
- Describe the three types of diarthroses and the movement each produces.
- Describe the anatomy of a typical synovial joint.
- Describe and demonstrate the various movements of synovial joints.

LAB ACTIVITIES

1. Joint Classification 137
2. Structure of Synovial Joints 140
3. Types of Diarthroses 141
4. Movement at Diarthrotic Joints 141
5. Selected Synovial Joints: Elbow and Knee Joints 147

Arthrology is the study of the structure and function of **joints**; a joint is defined as any location where two or more bones articulate. (In anatomic terminology, a synonym for *joint* is **articulation**.) If you were asked to identify joints in your body, you would most likely name those that allow a large range of movement, such as your knee joint and your hip joint. In large-range joints like these, a cavity between the two bones of the joint permits free movement. In some joints of the body, however, the bones are held closely together, a condition that allows no movement; an example of this type of nonmoving joint is the sutures of the cranium presented in Exercise 7.

Some individuals have more movement in a particular joint than other people and are called "double-jointed." Of course, they do not have two joints; the additional movement is a result of either the anatomy of the articulating bones or the position of tendons and ligaments around the joint.

LAB ACTIVITY 1 Joint Classification

Two classification schemes are commonly used for articulations. The functional scheme groups joints by the amount of movement permitted, and the structural scheme groups joints by the type of connective tissue between the articulating bones. The three kinds of functional joints permit no, some, or free movement of the articulating bones. Four types of structural joints occur: bony fusion, fibrous, cartilagenous, and synovial. Table 9.1 organizes joints into functional groups.

1. **Synarthroses** (sin-ar-THRŌ-sēz; *syn-*, together + *arthros*, joint) are immovable joints in which the bones are either closely fitted together or surrounded by a strong ligament. Three types of synarthroses are found in the skeleton: *sutures*, *synchondroses*, and *synostoses*.

138 EXERCISE 9 Articulations

Table 9.1 Functional Classification of Articulations

Functional Category	Structural Category	Description	Example
Synarthrosis (no movement)	**Fibrous**		
	Suture	Fibrous connections plus interlocking projections	Between the bones of the skull
	Gomphosis	Fibrous connections plus insertion in alveolar process	Between the teeth and jaws
	Cartilaginous		
	Synchondrosis	Interposition of cartilage plate	Epiphyseal plates
	Bony fusion		
	Synostosis	Conversion of other articular form to solid mass of bone	Portions of the skull, epiphyseal lines
Amphiarthrosis (little movement)	**Fibrous**		
	Syndesmosis	Ligamentous connection	Between the tibia and fibula
	Cartilaginous		
	Symphysis	Connected by a fibrocartilage pad	Between right and left halves of pelvis; between adjacent vertebral bodies along vertebral column
Diarthrosis (free movement)	**Synovial**	Complex joint bounded by joint capsule and containing synovial fluid	Numerous; subdivided by range of movement
	Monaxial	Permits movement in one plane	Elbow, ankle
	Biaxial	Permits movement in two planes	Ribs, wrist
	Triaxial	Permits movement in all three planes	Shoulder, hip

- **Fibrous joints**, which are strong synarthroses that have fibrous connective tissue between the bones, are of two types, *sutures* and *gomphoses*. **Sutures** (*sutura,* a sewing together) occur in the skull wherever the bones lock together, and this locking together is the primary characteristic of any suture fibrous joint. The second type of fibrous joint is called a **gomphosis** (gom-FŌ-sis; *gompho,* a peg or nail) and is characterized by the insertion of a conical process into a socket in the alveolar process of the jaw. The gomphosis is the joint between the teeth and the alveolar bone of the jaw. It is lined with a strong periodontal ligament that holds the teeth in place and permits no movement.
- **Synchondroses** (sin-kon-DRŌ-sēz; *syn-,* together + *condros,* cartilage) are synarthroses that have cartilage between the bones that comprise the joints. Two examples of this type of synarthrosis are the epiphyseal plate in a child's long bones and the cartilage between the ribs and sternum.
- **Synostoses** (sin-os-TŌ-sēz; *-osteo,* bone) are synarthroses that are formed by the fusion of two bones, as in the frontal bone, os coxae, and mandible. The joint between the diaphysis and either epiphysis of a mature long bone is also a synostosis.

2. **Amphiarthroses** (am-fē-ar-THRŌ-sēz) are joints that are held together by strong connective tissue; they are capable of only minimal movement. The two types of amphiarthroses are *syndesmoses* and *symphyses*.
   - **Syndesmoses** (sin-dez-MŌ-sēz; *syn-,* together + *desmo-,* band) occur between the radius and ulna of the forearm, and tibia and fibula of the leg. A ligament of fibrous connective tissue forms a strong band that joins the bones. This syndesmosis prevents excessive movement in the joint.
   - **Symphyses** are amphiarthroses that are characterized by the presence of fibrocartilage between the articulating bones. The intervertebral disks, for instance, construct a symphysis between any two articulating vertebrae. Another symphysis in the body is the pubic symphysis, where the ossa coxae

unite at the pubis. This strong joint limits expansion of the pelvis. During childbirth, a hormone softens the fibrocartilage to widen the pelvic bowl.

3. **Diarthroses** (dī-ar-THRŌ-sēz) are joints in which the bones are separated by a small membrane-lined cavity. The cavity allows a wide range of motion, which makes diarthroses freely movable **synovial** (sin-NŌ-vē-ul) **joints**. Movements are classified according to the number of planes through which the bones move. **Monaxial** (mon-AKS-ē-ul) **joints**, like the elbow, move in one plane. **Biaxial** (bī-AKS-ē-ul) **joints** allow movement in two planes; move your wrist up and down and side to side to demonstrate biaxial movement. **Triaxial** (trī-AKS-ē-ul) **joints** occur in the ball-and-socket joints of the shoulder and hip and permit movement in three planes. **Nonaxial joints**, also called **multiaxial joints**, are plane joints in which the articulating bones can move slightly in a variety of directions. The anatomy of a diarthrotic joint is examined in more detail later in this exercise.

Clinical Application

Arthritis

Arthritis, a disease that destroys synovial joints by damaging the cartilage, comes in two forms. Rheumatoid arthritis is an autoimmune disease that occurs when the body's immune system attacks the cartilage and membrane of the joint. As the disease progresses, the joint cavity is eliminated and the articulating bones fuse, which results in painful disfiguration of the joint and loss of joint function. Osteoarthritis is a degenerative joint disease that often occurs due to age and wearing of the joint tissues. The cartilage is damaged, and bone spurs may project into the joint cavity. Osteoarthritis tends to occur in the knee and hip joints, whereas rheumatoid arthritis is more common in the smaller joints of the hand.

QuickCheck Questions

1.1 What is the difference between the functional classification scheme and the structural classification scheme for joints?

1.2 What are the three types of functional joints and the amount of movement each allows?

1.3 What are the four types of structural joints and the type of connective tissue found in each?

1 Materials

☐ Articulated skeleton

Procedures

1. Locate, either on the articulated skeleton or on your body, a joint from each of the three functional groups and one from each of the four structural groups.

2. Give an example of each of the following joint types:
   - Gomphosis _____
   - Syndesmosis _____
   - Synchondrosis _____
   - Synostosis _____
   - Symphysis _____
   - Suture _____

3. Identify two monaxial joints, two biaxial joints, and two triaxial joints on your body:
   - Monaxial _____
   - Biaxial _____
   - Triaxial _____

EXERCISE 9 Articulations

LAB ACTIVITY 2 — Structure of Synovial Joints

The wide range of motion of synovial joints is attributed to the small **joint cavity** between articulating bones (Figure 9.1). When you consider how a door can swing open even though there is only a small space between the metal pieces of the hinges, you can appreciate how a joint cavity permits free movement of a joint. The epiphyses are capped with **articular cartilage,** a slippery gelatinous surface of hyaline cartilage that protects the epiphyses and prevents the bones from making contact across the joint cavity. A membrane called the **synovial membrane** lines the cavity and produces **synovial fluid**. Injury to a joint may cause inflammation of the membrane and lead to excessive fluid production.

A **bursa** (BUR-sa; *bursa,* a pouch) is similar to a synovial membrane except that, instead of lining a joint cavity, the bursa provides padding between bones and other structures. The periosteum of each bone is continuous with the strong **articular capsule** that encases the joint.

QuickCheck Questions

2.1 Where is the synovial membrane located in a joint?
2.2 Where is cartilage located in a synovial joint?

Materials

☐ Fresh beef joint

Procedures

1. Examine the fresh beef joint and locate the articular cartilage and joint cavity.
2. Manipulate the bones and note the movement of the articulating surfaces. Observe how ligaments and the shape of the bones limit movement. ∎

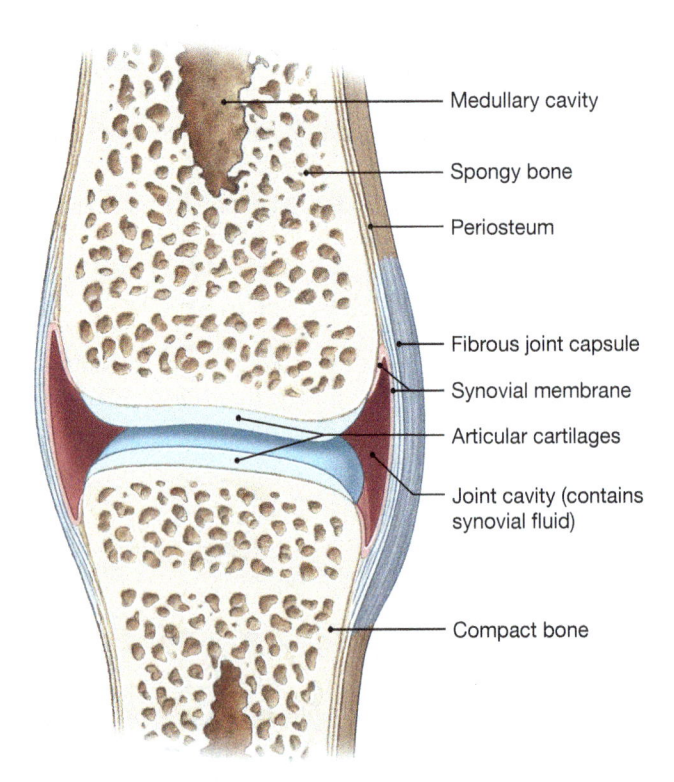

Synovial joint, sagittal section

Figure 9.1 Structure of a Typical Synovial Joint

LAB ACTIVITY 3 — Types of Diarthroses

Six types of diarthroses (synovial joints) occur in the skeleton. Each type permits a certain amount of movement according to the joining surfaces of the articulating bones. Figure 9.2 details each type of joint and includes mechanical representations to show direction of motion.

- **Plane joints** (also called *gliding joints*) are common where flat articular surfaces, such as those in the wrist, slide by neighboring bones. The movement is typically nonaxial and limited due to supporting ligaments. In addition to the wrist, plane joints also occur between sternal bones and between the tarsals. When you place your open hand, palm facing down, on your desktop and press down hard, you can observe the gliding of your wrist bones.
- **Hinge joints** are monaxial, operating like a door hinge, and are located in the elbows, fingers, toes, and knees.
- **Pivot joints** are monaxial joints that permit one bone to rotate around another.
- **Condylar joints**, also called *ellipsoidal joints*, are characterized by a convex surface of one bone that articulates in a concave depression of another bone. This concave-to-convex spooning of articulating surfaces permits biaxial movement.
- The **saddle joint** is a biaxial joint that is found only at the junction between the thumb metacarpus and the trapezium bone of the wrist.
- **Ball-and-socket joints** occur where a spherical head of one bone fits into a cup-shaped fossa of another bone, as in the joint between the humerus and the scapula. This triaxial joint permits a variety of movements, including rotation.

QuickCheck Questions
3.1 What are the six types of diarthroses?
3.2 What type of diarthrosis is a knuckle joint?

Materials
- Articulated skeleton

Procedures
1. Locate each type of synovial joint on the articulated skeleton or on your body. On the skeleton, notice how the structure of the joining bones determines the amount of joint movement.
2. Give an example of each type of synovial joint:
   - Plane _____
   - Hinge _____
   - Pivot _____
   - Condylar _____
   - Saddle _____
   - Ball-and-socket _____

LAB ACTIVITY 4 — Movement at Diarthrotic Joints

The diversity of bone shapes and joint types permits the skeleton to move in a variety of ways. Figure 9.3 illustrates angular movements, which occur either front to back in the anterior/posterior plane or side to side in the lateral plane, and Figure 9.4 highlights rotational movements. For clarity, the figures include a small dot at the joint where a demonstrated movement is described. Tables 9.2, 9.3, and 9.4 summarize joints of the axial and appendicular division of the skeleton.

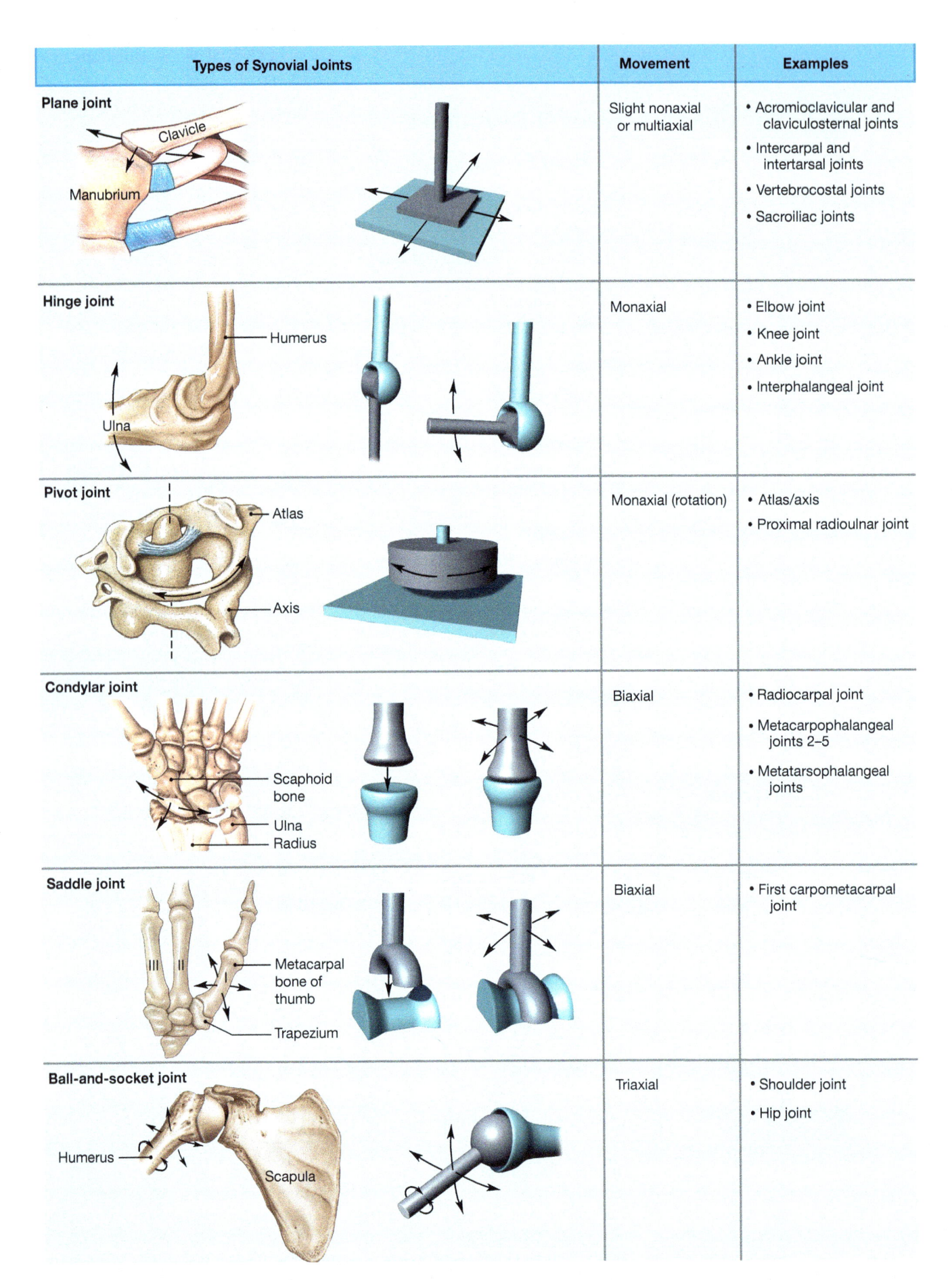

Types of Synovial Joints		Movement	Examples
Plane joint (Clavicle, Manubrium)		Slight nonaxial or multiaxial	• Acromioclavicular and claviculosternal joints • Intercarpal and intertarsal joints • Vertebrocostal joints • Sacroiliac joints
Hinge joint (Humerus, Ulna)		Monaxial	• Elbow joint • Knee joint • Ankle joint • Interphalangeal joint
Pivot joint (Atlas, Axis)		Monaxial (rotation)	• Atlas/axis • Proximal radioulnar joint
Condylar joint (Scaphoid bone, Ulna, Radius)		Biaxial	• Radiocarpal joint • Metacarpophalangeal joints 2–5 • Metatarsophalangeal joints
Saddle joint (Metacarpal bone of thumb, Trapezium)		Biaxial	• First carpometacarpal joint
Ball-and-socket joint (Humerus, Scapula)		Triaxial	• Shoulder joint • Hip joint

Figure 9.2 The Six Types of Synovial Joints
The types of movement permitted are illustrated on the left anatomically, and on the right by a mechanical model.

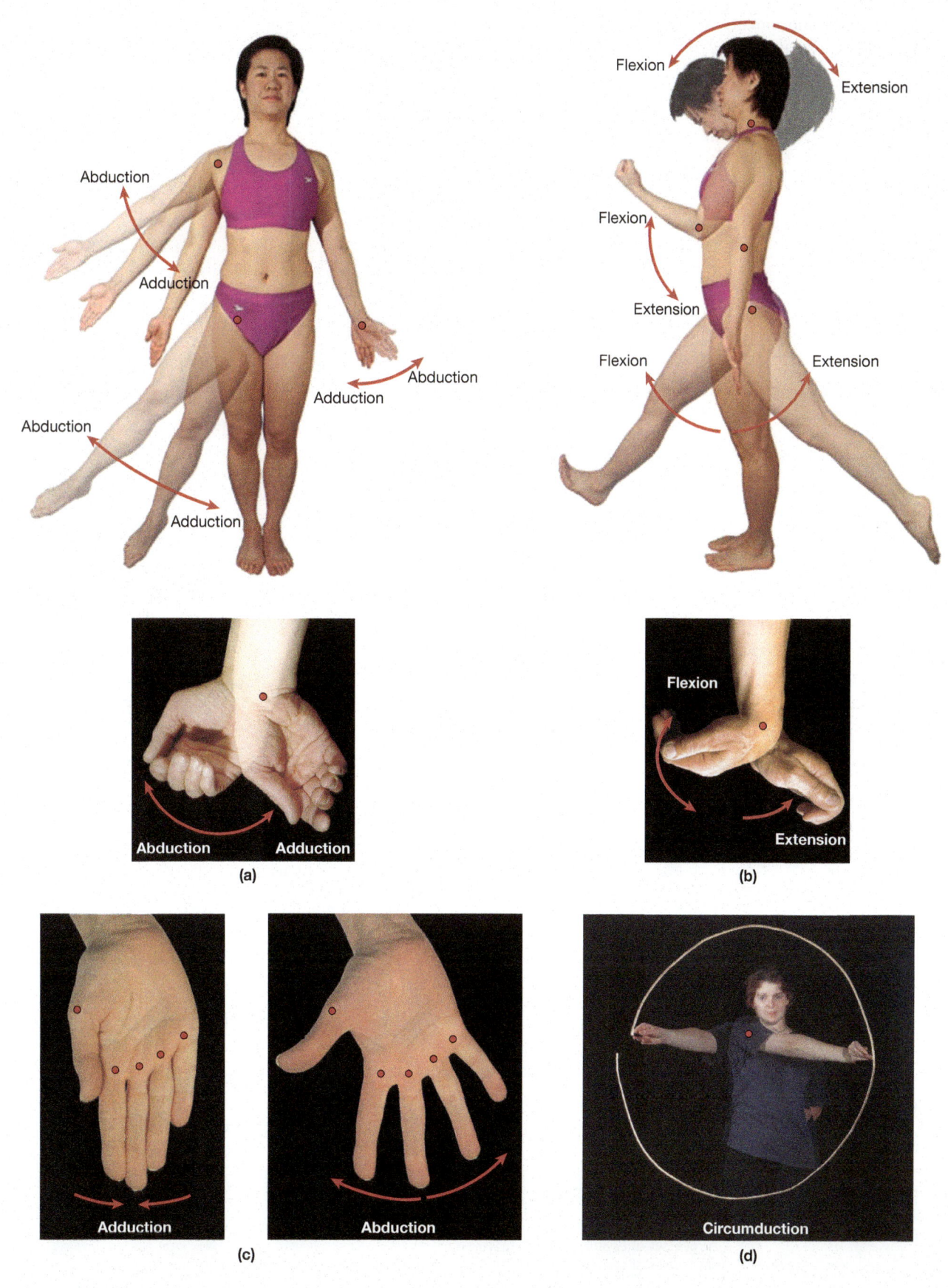

Figure 9.3 Angular Movements
Examples of angular movements that change the angle between the two bones that make up a joint. The red dots indicate the locations of the joints that are involved in the illustrated movement. **(a)** Abduction and adduction in the upper and lower limbs and in the hand. **(b)** Flexion and extension in the upper and lower limbs, in the head, and in the hand. **(c)** Adduction and abduction in the fingers. **(d)** Circumduction.

143

Figure 9.4 Rotational Movements
Examples of motion in which a body part rotates. (a) Lateral and medial rotation of the upper limb and right and left rotation of the head. (b) Supination and pronation of the hand and forearm.

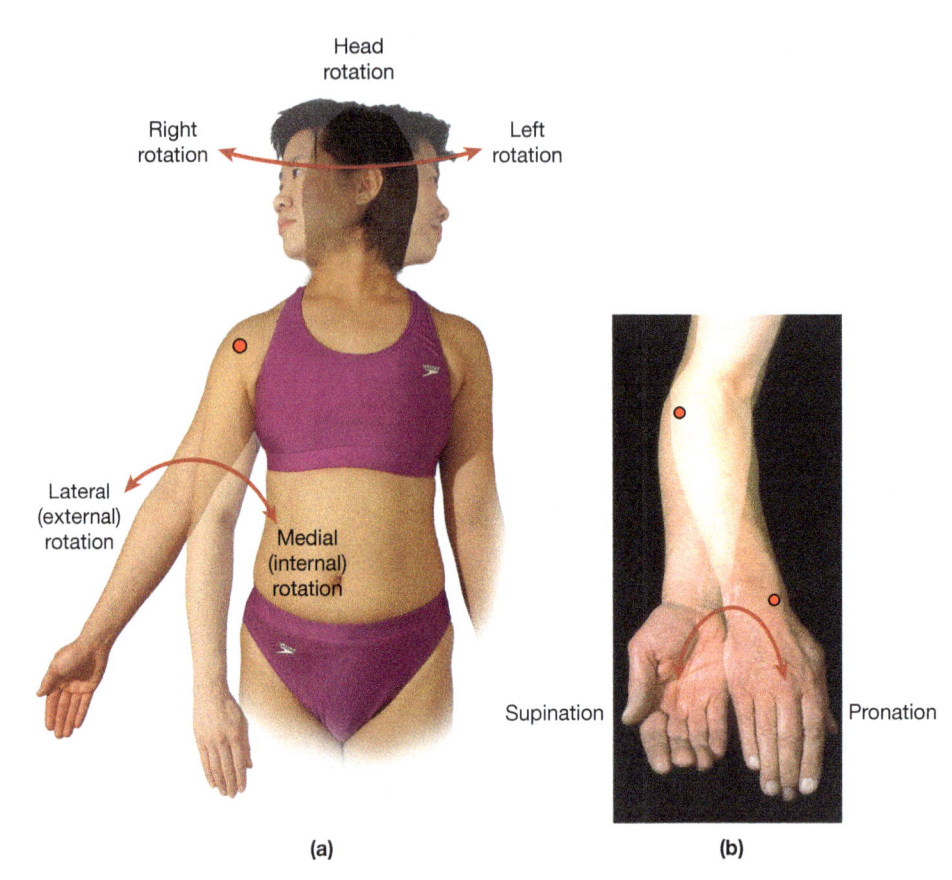

Table 9.2 Articulations of the Axial Skeleton

Element	Joint	Type of Articulation	Movements
Skull			
Cranial and facial bones of skull	Various	Synarthrosis (suture or synostosis)	None
Maxillae/teeth	Alveolar	Synarthrosis (gomphosis)	None
Mandible/teeth	Alveolar	As above	None
Temporal bone/mandible	Temporomandibular	Combined plane joint and hinge diarthrosis	Elevation/depression, lateral gliding, limited protraction/retraction
Vertebral Column			
Occipital bone/atlas	Atlanto-occipital	Condylar diarthrosis	Flexion/extension
Atlas/axis	Atlanto-axial	Pivot diarthrosis	Rotation
Other vertebral elements	Intervertebral (between vertebral bodies)	Amphiarthrosis (symphysis)	Slight movement
	Intervertebral (between articular processes)	Planar diarthrosis	Slight rotation and flexion/extension
Thoracic vertebrae/ribs	Vertebrocostal	Planar diarthrosis	Elevation/depression
Rib/costal cartilage		Synchondrosis	None
Costal cartilage/sternum	Sternocostal	Synchondrosis (rib 1) Planar diarthrosis (ribs 2–7)	None Slight gliding movement
L₅/sacrum	Between body of L₅ and sacral body	Amphiarthrosis (symphysis)	Slight movement
	Between inferior articular processes of L₅ and articular processes of sacrum	Planar diarthrosis	Slight flexion/extension
Sacrum/os coxae	Sacro-iliac	Planar diarthrosis	Slight gliding movement
Sacrum/coccyx	Sacrococcygeal	Planar diarthrosis (may become fused)	Slight movement
Coccygeal bones		Synarthrosis (synostosis)	None

Table 9.3 Articulations of the Pectoral Girdle and Upper Limb

Element	Joint	Type of Articulation	Movements
Sternum/clavicle	Sternoclavicular	Planar diarthrosis (a double "plane joint", with two joint cavities separated by an articular cartilage)	Protraction/retraction, depression/elevation, slight rotation
Scapula/clavicle	Acromioclavicular	Planar diarthrosis	Slight gliding movement
Scapula/humerus	Glenohumeral (shoulder)	Ball-and-scoket diarthrosis	Flexion/extension, adduction/abduction, circumduction, rotation
Humerus/ulna and humerus/radius	Elbow (humeroulnar and humeroradial)	Hinge diarthrosis	Flexion/extension
Radius/ulna	Proximal radioulnar Distal radioulnar	Pivot diarthrosis Pivot diarthrosis	Rotation Pronation/supination
Radius/carpal bones	Radiocarpal	Condylar diarthrosis	Flexion/extension, adduction/abduction, circumduction
Carpal bone/carpal bone	Intercarpal	Planar diarthrosis	Slight gliding movement
Carpal bone/first Metacarpal bone	Carpometacarpal of thumb	Saddle diarthrosis	Flexion/extension, adduction/abduction, circumduction, opposition
Carpal bones/metacarpal bones II–V	Carpometacarpal	Planar diarthrosis	Slight flexion/extension, adduction/abduction
Metacarpal bones/phalanges	Metacarpophalangeal	Condylar diarthrosis	Flexion/extension, adduction/abduction, circumduction
Phalanx/phalanx	Interphalangeal	Hinge diarthrosis	Flexion/extension

Table 9.4 Articulations of the Pelvic Girdle and Lower Limb

Element	Joint	Type of Articulation	Movements
Sacrum/os coxae	Sacro-iliac	Planar diarthrosis	Gliding movements
Os coxae/os coxae	Pubic symphysis	Amphiarthrosis	None*
Os coxae/femur	Hip	Ball-and-socket diarthrosis	Flexion/extension, adduction/abduction, circumduction, rotation
Femur/tibia	Knee	Complex, functions as hinge	Flexion/extension, limited rotation
Tibia/fibula	Tibiofibular (proximal) Tibiofibular (distal)	Planar diarthrosis Planar diarthrosis and amphiarthrotic syndesmosis	Slight gliding movements Slight gliding movements
Tibia and fibula with talus	Ankle, or talocrural	Hinge diarthrosis	Dorsiflexion/plantar flexion
Tarsal bone to tarsal bone	Intertarsal	Planar diarthrosis	Slight gliding movements
Tarsal bones to metatarsal bones	Tarsometatarsal	Planar diarthrosis	Slight gliding movements
Metatarsal bones to phalanges	Metatarsophalangeal	Condylar diarthrosis	Flexion/extension, adduction/abduction
Phalanx/phalanx	Interphalangeal	Hinge diarthrosis	Flexion/extension

*During pregnancy, hormones weaken the symphysis and permit movement important to childbirth (see Exercise 29).

146 EXERCISE 9 Articulations

- **Abduction** is movement away from the midline of the body. **Adduction** is movement toward the midline. Notice how you move your arm at the shoulder for these two motions. Practice this movement first with your shoulder joint and then with your wrist joint.
- **Flexion** is movement that decreases the angle between the articulating bones of a joint, and **extension** is movement that increases the angle between the bones. *Hyperextension* is a forced extension beyond normal extension and may result in injury to the joint, articulating bones, and regional muscles.
- **Circumduction** is circular movement at a ball-and-socket joint.
- **Rotation** is a turning movement of bones at a joint. **Left rotation** or **right rotation** occur when the head is turned, as in shaking to indicate "no". **Lateral rotation** and **medial rotation** of the limbs occur at ball-and-socket joints and at the radioulnar joint. These movements turn the rounded head of one bone in the socket of another bone.
- **Supination** (soo-pi-NĀ-shun) is movement that moves the palm into the anatomical position. **Pronation** (prō-NĀ-shun) is movement that moves the palm to face posteriorly. During these two motions, the humerus serves as a foundation for the radius to pivot around the ulna.

The following specialized motions are illustrated in Figure 9.5:

- **Eversion** (ē-VER-zhun) is lateral movement of the ankle to move the foot so that the toes point away from the body's midline. Moving the sole medially so that the toes point toward the midline is **inversion**; the foot moves "in". Eversion and inversion are commonly mistaken for pronation and supination of the ankle.

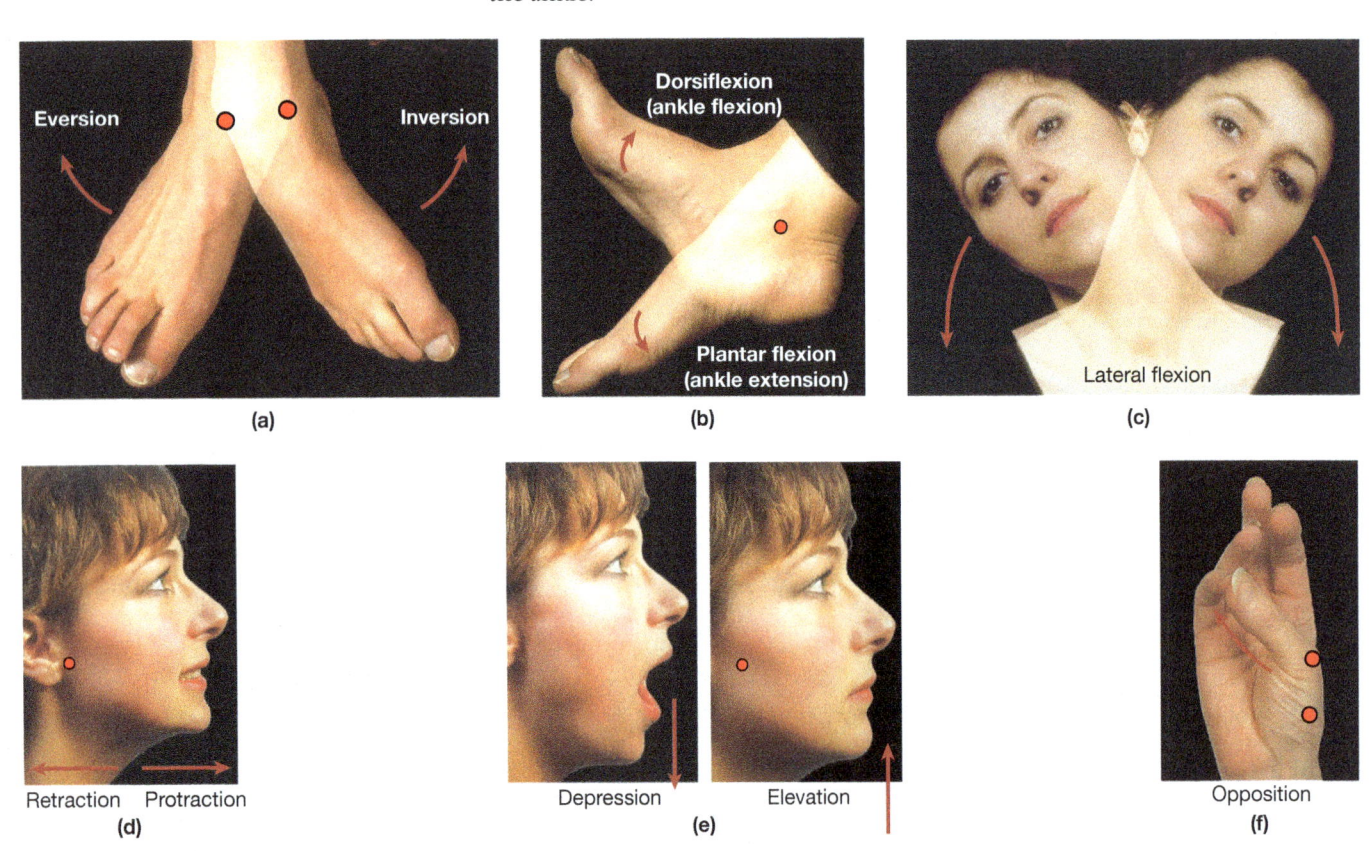

Figure 9.5 Special Movements
(a) Eversion moves the foot so that the toes point away from the body's midline. Inversion points the toes toward the midline.
(b) Dorsiflexion and plantar flexion in the foot. (c) Lateral flexion in the neck. (d) Retraction and protraction in the jaw. (e) Depression and elevation in the jaw. (f) Opposition, which occurs only in the hand.

EXERCISE 9 Articulations 147

- Two other terms that describe ankle movement are dorsiflexion and plantar flexion. **Dorsiflexion** is the joint movement that permits you to walk on your heels, which means the soles of your feet are raised up off the floor and the angle between the ankle and the bones of the leg is decreased. **Plantar flexion** (*plantar*, sole) moves the foot so that you can walk on your tiptoes; here, the angle between the ankle and the tibia/fibula is increased.
- **Opposition** is touching the tips of the thumb and little finger together.
- **Retraction**, which means to take back, moves structures posteriorly out of the anatomical position, as when the mandible is moved posteriorly to demonstrate an overbite. **Protraction** moves a structure anteriorly, as when you jut your mandible forward.
- **Depression** lowers bones. This motion occurs, for instance, when you lower your mandible bone to take a bite of food. Closing your mouth is **elevation** of the mandible bone.
- **Lateral flexion** is the bending of the vertebral column from side to side. Most of the movement occurs in the cervical and lumbar regions.

QuickCheck Questions

4.1 How does flexion differ from extension?
4.2 What are pronation and supination?
4.3 How does dorsiflexion differ from plantar flexion?

4 Materials

☐ Articulated skeleton

Procedures

1. Use the articulated skeleton or your body to demonstrate each of the movements in Figures 9.3, 9.4, and 9.5.
2. Give an example of each of the following movements:
   - Abduction _____
   - Extension _____
   - Left rotation _____
   - Pronation _____
   - Supination _____
   - Depression _____
   - Retraction _____
   - Lateral rotation _____ ■

LAB ACTIVITY 5 — Selected Synovial Joints: Elbow and Knee Joints

The elbow is a hinge joint that involves humeroradial and humeroulnar articulations. (The elbow complex also contains the radioulnar joint, which allows the radius to pivot during supination and pronation.) The morphology of the articulating bones and a strong articular capsule and ligaments result in a strong and highly movable elbow. **Radial** and **ulnar collateral ligaments** reinforce the lateral aspects of the joint, and the **annular ligament** holds the radial head in position to pivot (Figure 9.6).

The knee is a hinge joint that permits flexion and extension of the leg. Most support for the knee is provided by seven bands of ligaments that encase the joint (Figure 9.7). Cushions of fibrocartilage, the **lateral meniscus** (me-NIS-kus; *meniskos*, a crescent) and the **medial meniscus**, pad the area between the condyles of the femur and tibia. Areas where tendons move against the bones in the knee are protected with bursae.

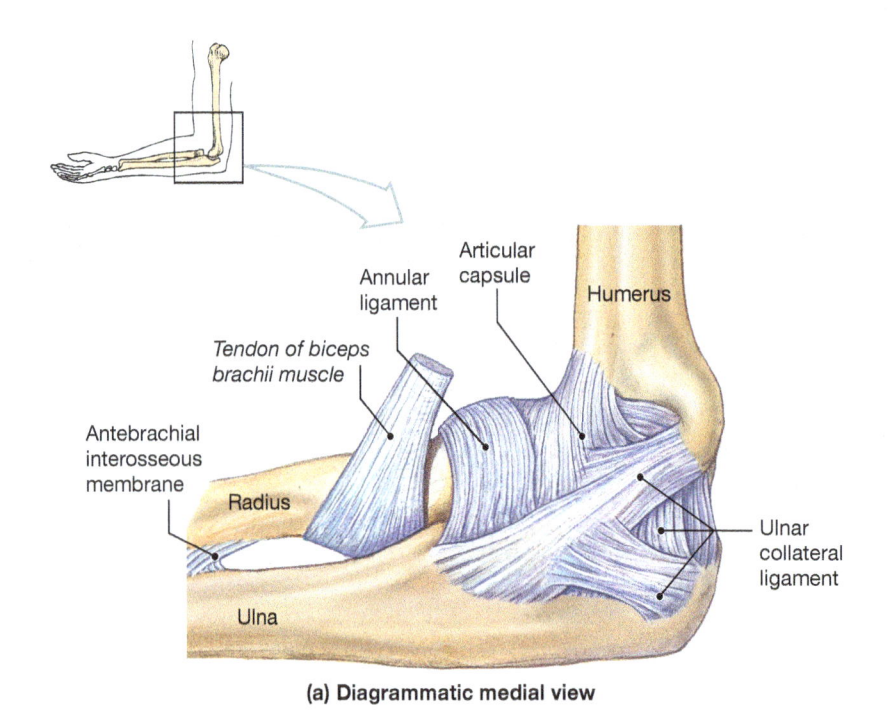

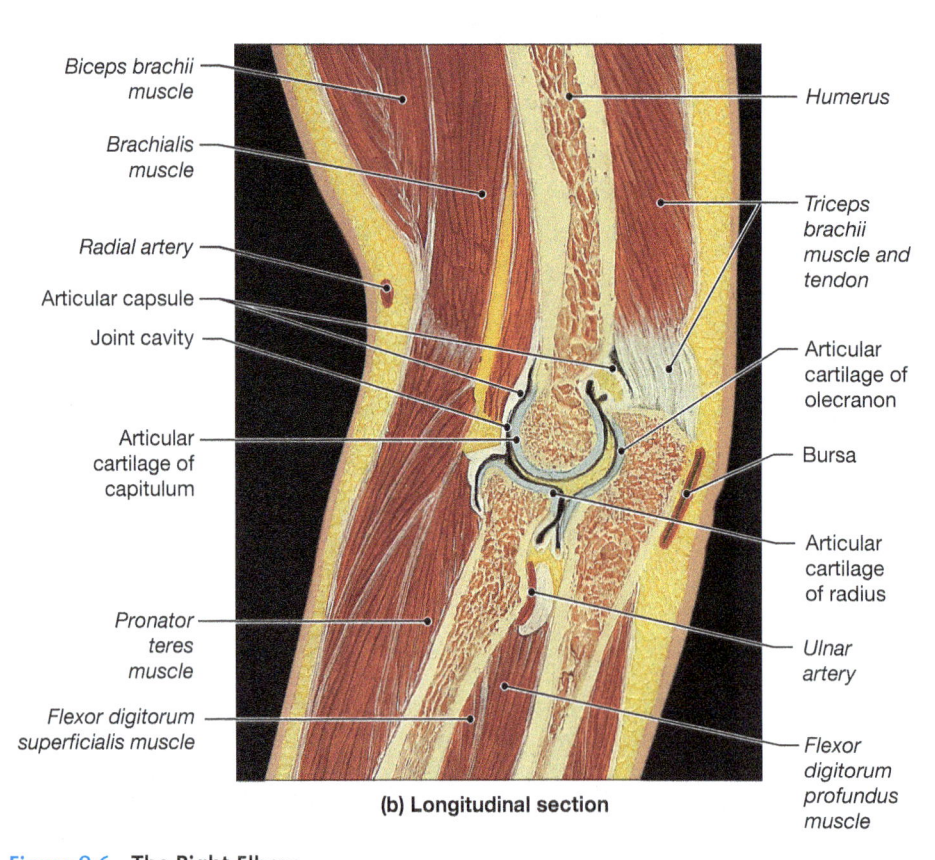

Figure 9.6 The Right Elbow
(a) Diagrammatic medial view. (b) Plastic model based on oblique section through elbow region.

148

The seven ligaments of the knee occur in three pairs and a single patellar ligament. **Tibial** and **fibular collateral ligaments** provide medial and lateral support when a person is standing. Two **popliteal ligaments** extend from the head of the femur to the fibula and tibia to support the posterior of the knee. The **anterior** and **posterior cruciate ligaments** are inside the articular capsule. The cruciate (*cruciate,* a cross) ligaments originate on the tibial head and cross each other as they pass through the intercondylar fossa of the femur. The **patellar ligament** attaches the inferior aspect of the patella to the tibial tuberosity, which adds anterior support to the knee. The large quadriceps tendon is attached to the superior margin of the patella. Cords of ligaments called the **patellar retinaculae** contribute to anterior support of the knee.

QuickCheck Questions

5.1 What structure reinforces the radial head?

5.2 How many ligaments are in the knee?

5.3 What structures cushion the knee?

5 Materials

- ☐ Articulated skeleton
- ☐ Elbow model
- ☐ Knee model

Procedures

1. Examine the elbow of the articulated skeleton and review the skeletal anatomy of the joint.
2. Locate the annular and collateral ligaments on the elbow model.
3. Review the skeletal components of the knee joint on the articulated skeleton.
4. On the knee model, examine the relationship of the ligaments and determine how each supports the knee. Note how the menisci provide a cushion between the bones. ■

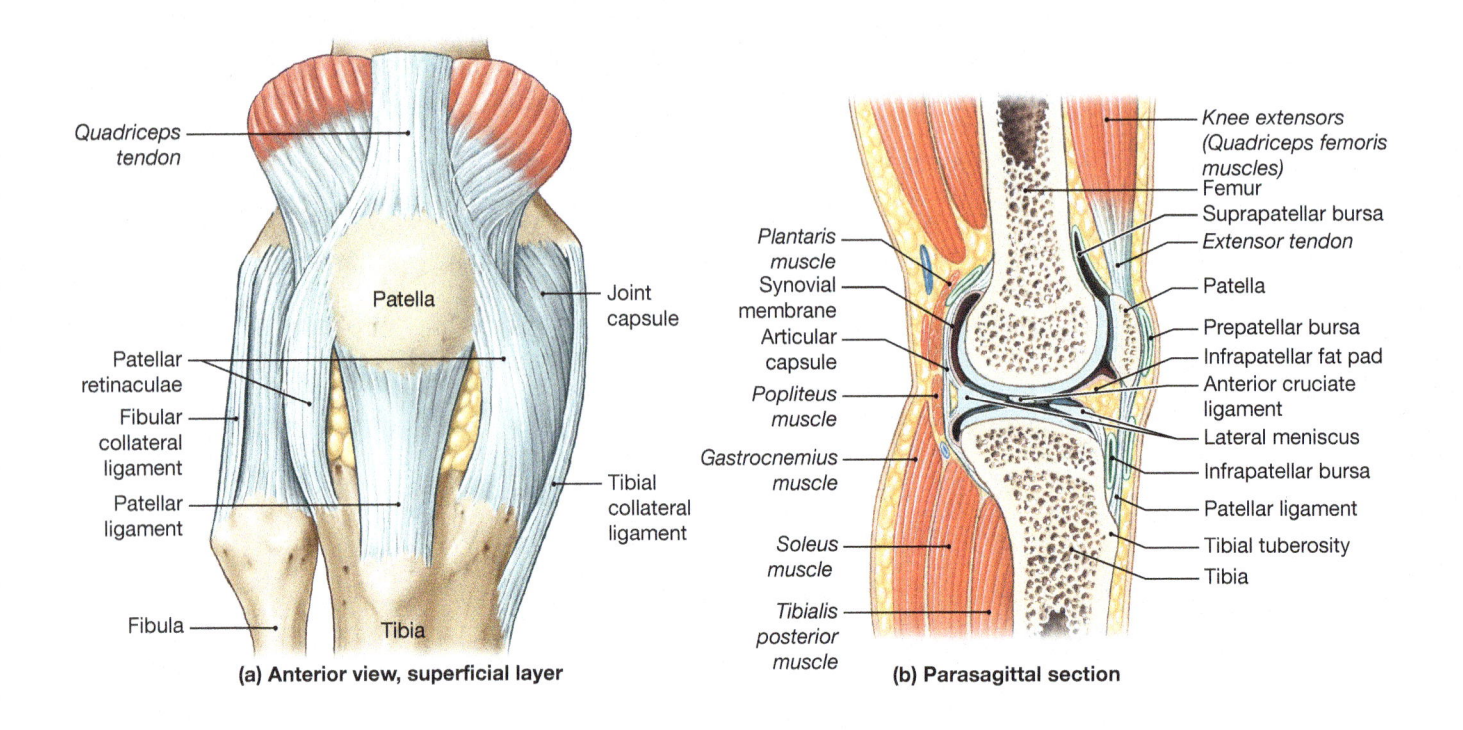

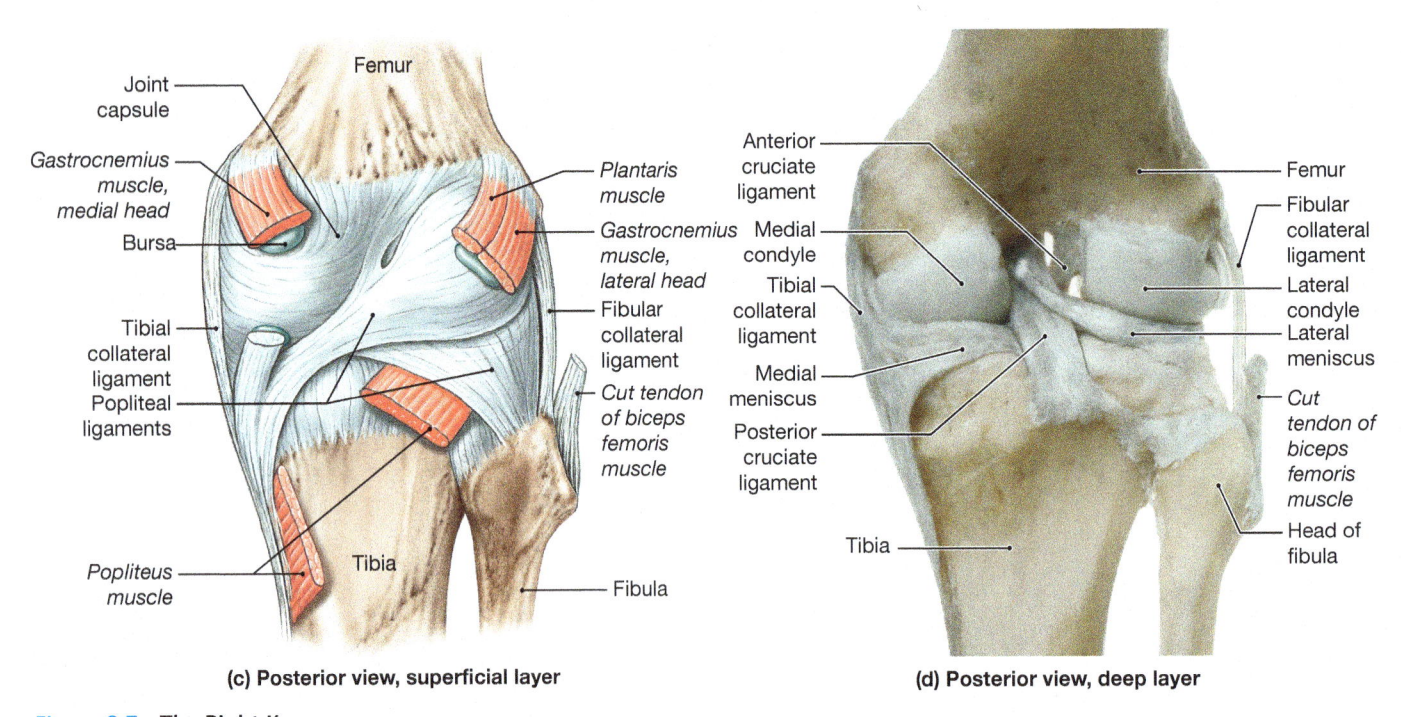

Figure 9.7 The Right Knee
(a) Diagrammatic anterior view of superficial dissection of extended knee. (b) Diagrammatic view of parasagittal view of extended knee. (c) Diagrammatic posterior view of knee dissection, showing the ligaments that support the capsule. (d) Posterior view of the knee at full extension after removal of joint capsule.

LAB REPORT

EXERCISE 9: Articulations

Name _____
Date _____
Section _____

A. Matching

Match each joint in the left column with its correct description from the right column.

_____ 1. pivot
_____ 2. symphysis
_____ 3. ball and socket
_____ 4. gomphosis
_____ 5. hinge
_____ 6. suture
_____ 7. synostosis
_____ 8. syndesmosis
_____ 9. condylar
_____ 10. synchondrosis

A. forearm-to-wrist joint
B. joint between parietal bones
C. rib-to-sternum joint
D. joint between vertebral bodies
E. femur-to-os coxae bone joint
F. phalangeal joint
G. distal tibia-to-fibula joint
H. atlas-to-axis joint
I. fused frontal bones
J. joint that holds a tooth in a socket

B. Matching

Match each movement in the left column with its correct description from the right column.

_____ 1. retraction
_____ 2. dorsiflexion
_____ 3. eversion
_____ 4. inversion
_____ 5. pronation
_____ 6. plantar flexion
_____ 7. protraction
_____ 8. supination
_____ 9. adduction
_____ 10. abduction

A. moves body part away from midline of body
B. turns sole of foot laterally
C. moves palm to face posteriorly
D. moves palm to face anteriorly
E. moves body part out of the anatomical position in posterior direction
F. allows person to stand on tiptoes
G. moves body part out of the anatomical position in anterior direction
H. turns sole of foot medially
I. allows person to stand on heels
J. moves body part toward midline of body

151

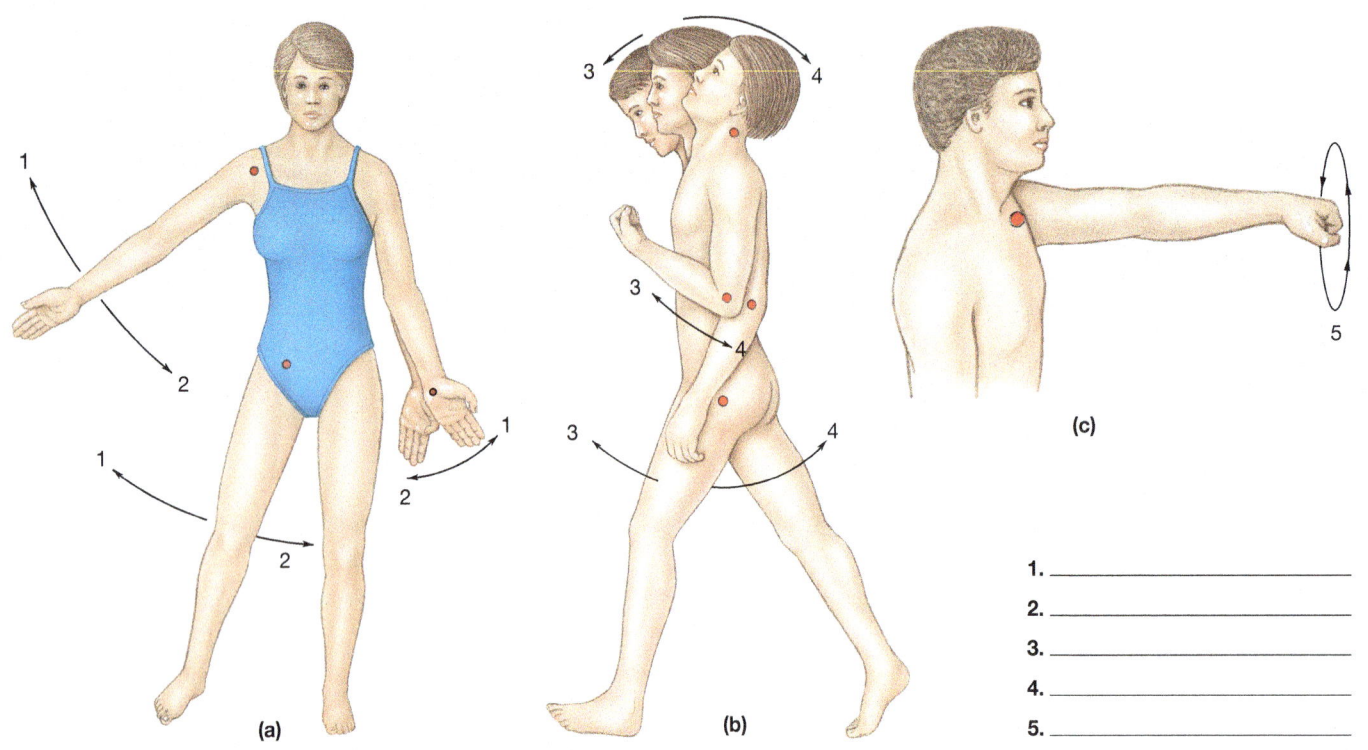

EXERCISE 9

LAB REPORT

Figure 9.8 Angular and Rotational Movements

1. _____
2. _____
3. _____
4. _____
5. _____

C. Labeling

1. Label the five numbered joint movements in Figure 9.8.
2. Label the four numbered joint movements in Figure 9.9.

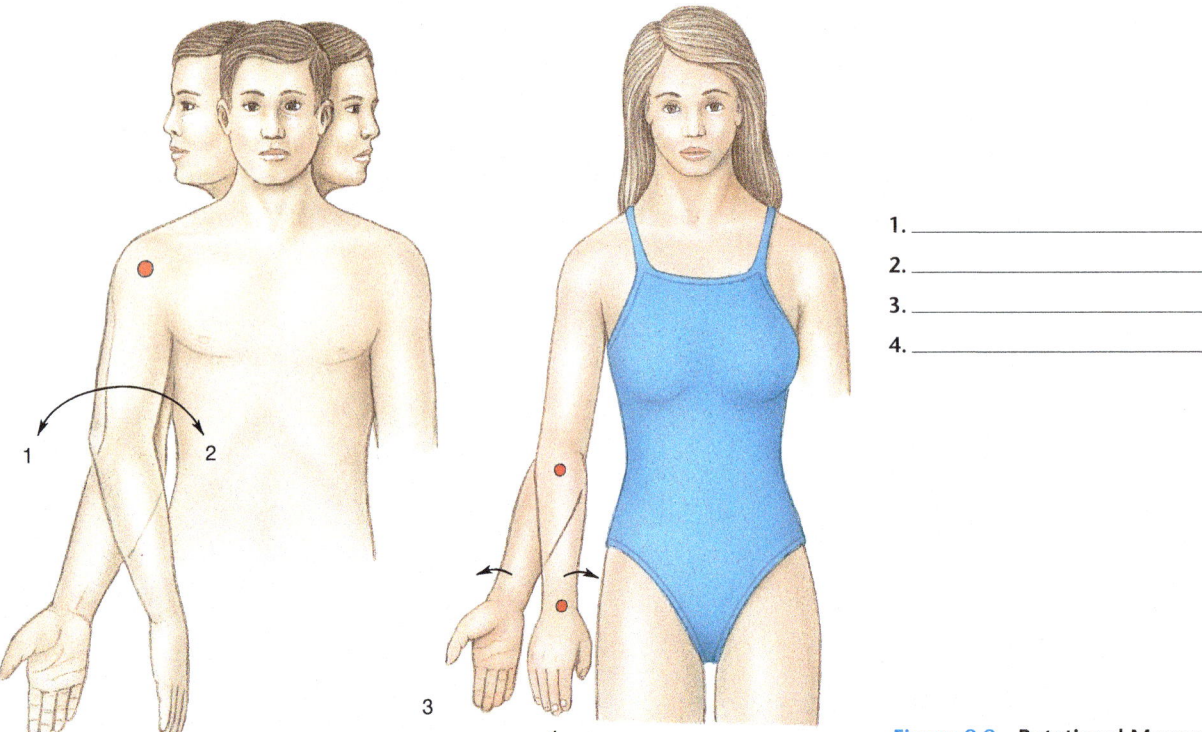

1. _____
2. _____
3. _____
4. _____

Figure 9.9 Rotational Movements

152

LAB REPORT   EXERCISE 9

D. Fill in the Blanks

List the joints and movements involved in:

1. Walking _____
2. Throwing a ball _____
3. Turning a doorknob _____
4. Crossing one lower limb over the other while sitting _____
5. Shaking your head "no" _____
6. Chewing food _____

E. Short-Answer Questions

1. Describe the three types of functional joints.

2. What factors limit the range of movement of a joint?

3. Describe the four types of structural joints.

4. List the seven ligaments of the knee and describe how each supports the joint.

153

EXERCISE 9

LAB REPORT

F. Analysis and Application

1. Describe how the articulating bones of the elbow prevent hyperextension of this joint.

2. What happens to make bones fuse in joints that are damaged by rheumatoid arthritis?

3. Which joint is unique to the hand, and how does this joint move the hand?

4. Which structural feature enables diarthrotic joints to have free movement?

5. Why is the lateral meniscus often associated with a knee injury?

EXERCISE 10

Organization of Skeletal Muscle

OBJECTIVES

On completion of this exercise, you should be able to:

- Describe the basic functions of skeletal muscles.
- Describe the organization of a skeletal muscle.
- Identify the cellular anatomy of a muscle fiber.
- Describe how a muscle fiber is excited by a motor neuron.

LAB ACTIVITIES

1 Skeletal Muscle Arrangement 155
2 Neuromuscular Junction 159

Every time you move some part of your body, either consciously or unconsciously, you use muscles. Recall from Exercises 4 that there are three kinds of muscle tissue: skeletal, smooth, and cardiac. Skeletal muscles, the type we focus on in this exercise, are primarily responsible for **locomotion**, or movement of the body. Such locomotions as rolling your eyes, writing your name, and speaking are the result of highly coordinated skeletal muscle contractions. Other functions of skeletal muscle include maintaining posture and body temperature and supporting soft tissues, as with the muscles of the abdomen.

In addition to the ability to contract, muscle tissue has several other unique characteristics. It is **excitable,** which means that, in response to a stimulus, the tissue produces electrical impulses called **action potentials**. Muscle tissue is **extensible,** which means it can be stretched. When the ends of a stretched muscle are released, it recoils to its original size, like a rubber band. This property is called **elasticity**.

LAB ACTIVITY 1 Skeletal Muscle Arrangement

Connective Tissue Coverings

Connective tissues support and organize skeletal muscles and attach them to bones. Three layers of connective tissue partition a muscle. Superficially, a collagenous connective tissue layer called the **epimysium** (ep-i-MIZ-ē-um; *epi*, on + *mys*, muscle) covers the muscle and separates it from neighboring structures (Figure 10.1). The epimysium folds into the muscle as the **perimysium** (per-i-MIZ-ē-um; *peri-*, around) and divides the muscle cells into groups called **fascicles** (FAS-i-kulz). Connective tissue fibers of the perimysium extend deep into the fascicles as the **endomysium** (en-dō-MIZ-ē-um; *endo-*, inside), which surrounds individual muscle fibers. (Recall from Exercises 4 that *muscle fiber* means a muscle cell.) The parallel, thread-like muscle fibers that make up the fascicles can easily be seen when a muscle is teased apart with a probe.

The central, thicker part of a muscle is called the **belly.** The connective tissues of the muscle interweave and combine as the tendon at the end of the muscle. The fibers

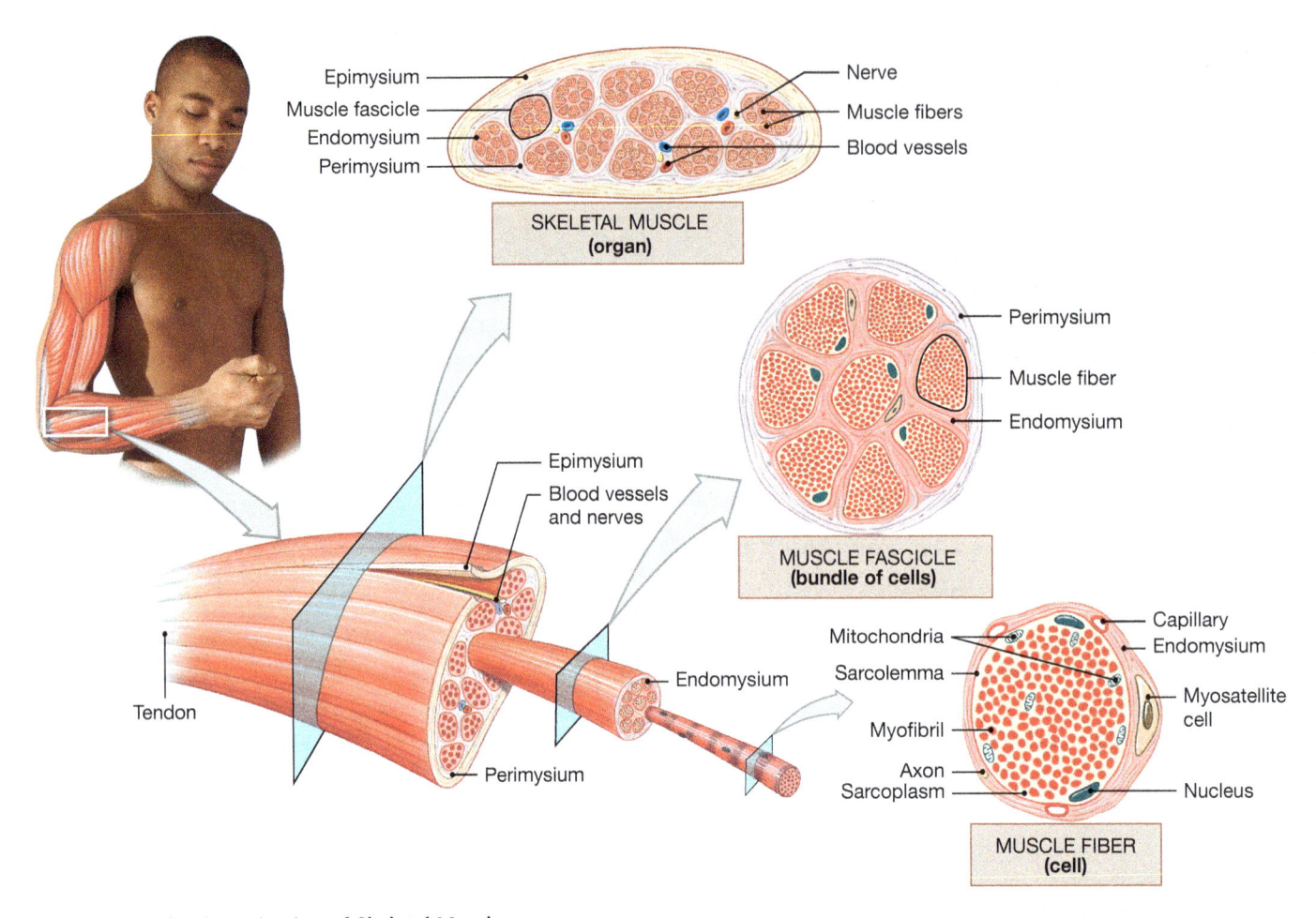

Figure 10.1 The Organization of Skeletal Muscles
A skeletal muscle consists of fascicles (bundles of muscle fibers) enclosed by the epimysium. The fascicles are separated by the connective tissue fibers of the perimysium, and within each fascicle the muscle fibers are surrounded by the endomysium. Each muscle fiber has many superficial nuclei, as well as mitochondria and other organelles.

of the tendon and of the bone's periosteum interlace to firmly attach the tendon to the bone. When the muscle fibers contract and generate tension, they transmit this force through the connective tissue layers to the tendon, which pulls on the associated bone and produces movement.

Structure of a Skeletal Muscle Fiber

Recall from Exercises 4 that the cell membrane of a muscle fiber is called the **sarcolemma** (Figure 10.2). The cytoplasm is called **sarcoplasm**. Many **transverse tubules**, also called *T tubules*, connect the sarcolemma to the interior of the muscle fiber. The function of these tubules is to pass contraction stimuli to deeper regions of the muscle fiber.

Inside the muscle fiber are proteins arranged in thousands of rods, called **myofibrils**, that extend the length of the fiber. Each myofibril is surrounded by the **sarcoplasmic reticulum**, where calcium ions are stored. Branches of the sarcoplasmic reticulum fuse to form large calcium ion storage chambers called **terminal cisternae** (sis-TUR-nē), which lie adjacent to the transverse tubules. A **triad** is a "sandwich" that consists of a transverse tubule plus the terminal cisterna on either side of the tubule. In order for a muscle to contract, calcium ions must be released from the cisternae; the transverse tubules stimulate this ion release. When

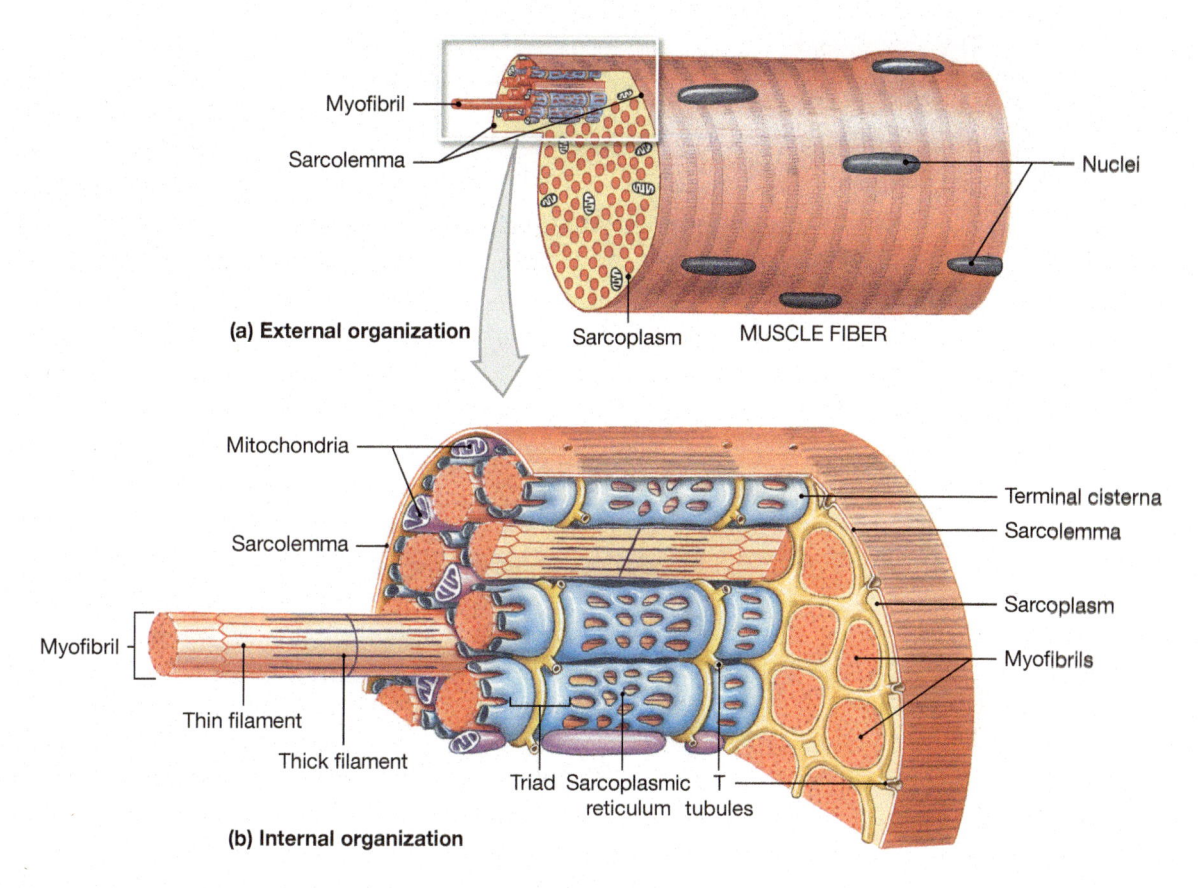

Figure 10.2 The Structure of a Skeletal Muscle Fiber

a muscle relaxes, protein carriers in the sarcoplasmic reticulum transport calcium ions back into the cisternae.

Each myofibril consists of several kinds of proteins that are arranged in thousands of **thin filaments** and **thick filaments** (Figure 10.3). During contraction, the thick and thin filaments interact to produce tension and shorten the muscle. The thin filaments are mostly composed of the protein **actin**, and the thick filaments are made of the protein **myosin**. The filaments are arranged in repeating patterns called **sarcomeres** (SAR-kō-mērz; *sarkos*, flesh + *meros*, part) along a myofibril. The thin filaments connect to one another at the **Z lines** on each end of the sarcomere. Each Z line is made of a protein called **actinin**. Areas near the Z line that contain only thin filaments are **I bands**. Between I bands in a sarcomere is the **A band**, an area that contains both thin and thick filaments. The edges of the A band are the **zone of overlap,** where the thick and thin filaments bind during muscle contraction. The middle region of the A band is the **H zone** and contains only thick filaments. A dense **M line** in the center of the A band attaches the thick filaments. Because the thick and thin filaments do not overlap completely, some areas of the sarcomere appear lighter than others. This organization results in the striated (striped) appearance of skeletal muscle tissue that is visible in Figure 10.3b.

When a muscle fiber contracts, the thin filaments are pulled deep into the sarcomere. As the thin filaments slide inward, the I band and H zone become smaller. Each myofibril consists of approximately 10,000 sarcomeres that are joined end to end. During contraction, the sarcomeres compress and the myofibril shortens and pulls on the sarcolemma, which causes the muscle fiber to shorten as well.

158 EXERCISE 10 Organization of Skeletal Muscle

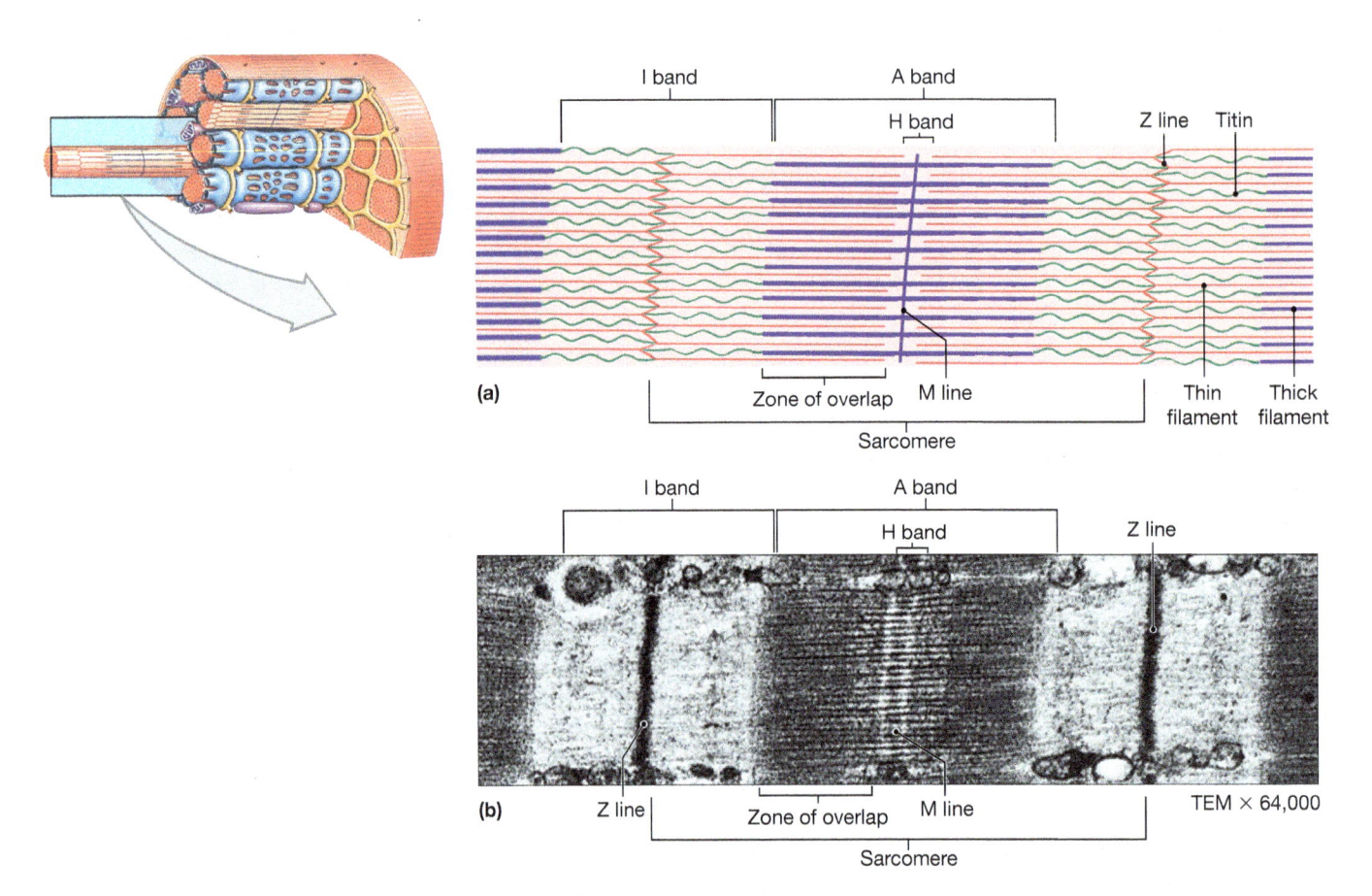

Figure 10.3 Sarcomere Structure
(a) Longitudinal section of a sarcomere. (b) Micrograph of a corresponding view of a sarcomere from a muscle fiber in the gastrocnemius muscle of the calf. (TEM × 64,000)

QuickCheck Questions

1.1 Describe the connective tissue organization of a muscle.

1.2 What is the relationship between myofibrils and sarcomeres?

1.3 Where are calcium ions stored in a muscle fiber?

1 Materials

- ☐ Muscle model
- ☐ Muscle fiber model
- ☐ Round steak or similar cut of meat
- ☐ Preserved muscle tissue
- ☐ Dissecting microscope
- ☐ Microscope slide of skeletal muscle tissue, transverse section
- ☐ Compound microscope

Procedures

1. Review the histology of skeletal muscle fibers in Exercise 4.

2. Examine the muscle fiber model and identify each feature. Describe the location of the sarcoplasmic reticulum, myofibrils, sarcomeres, and filaments. Identify the connective tissue coverings of muscles on the muscle model and on the fiber model.

3. If your instructor has prepared a muscle demonstration from a cut of meat, examine the meat for the various connective tissue components of the muscles.

4. Examine a specimen of preserved muscle tissue by placing the tissue in saline solution and then teasing the muscle apart using tweezers and a probe. Notice how the fascicles appear as strands of muscle tissue. Examine the fascicles under a dissecting microscope.

5. Using the compound microscope, scan the slide of skeletal muscle. Examine the organization of the muscle and each connective tissue layer. ■

EXERCISE 10 Organization of Skeletal Muscle 159

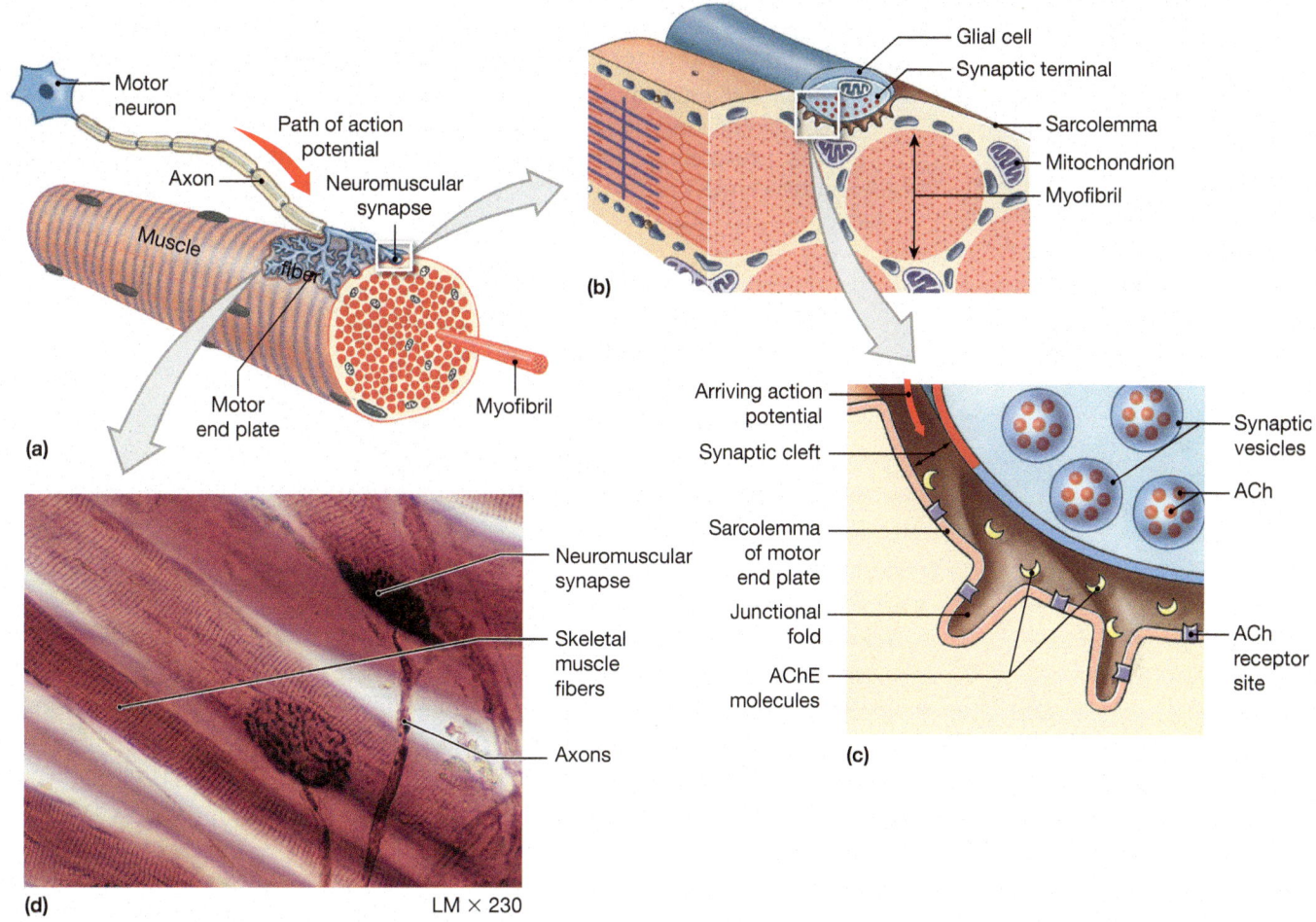

Figure 10.4 Skeletal Muscle Innervation
(a) Diagrammatic view of a neuromuscular junction. (b) Details of the neuromuscular junction. (c) Detail of the synaptic cleft, showing the synaptic vesicles in the synaptic terminal (blue) and the motor end plate of the muscle fiber (beige) innervated by the axon. (d) Micrograph of the neuromuscular junction. (LM × 230)

LAB ACTIVITY 2 — Neuromuscular Junction

Each skeletal muscle fiber is controlled by a nerve cell that is called a **motor neuron** (Figure 10.4). To excite the muscle fiber, the motor neuron releases a chemical called **acetylcholine** (as-ē-til-KŌ-lēn), abbreviated ACh. The motor neuron and the muscle fiber meet at a **neuromuscular junction**, also called a *myoneural junction*. The end of the neuron, called the **axon**, expands to form a branched **synaptic terminal**. In the synaptic terminal are **synaptic vesicles** that contain ACh. A small gap, the **synaptic cleft**, separates the synaptic terminal from a folded area of the sarcolemma that is called the **motor end plate**. At the motor end plate, the sarcolemma releases into the synaptic cleft the enzyme **acetylcholinesterase** (AChE), which prevents overstimulation of the muscle fiber by deactivating ACh.

QuickCheck Questions

2.1 What substance is stored in and released from the synaptic vesicles?

2.2 What substance is released into the sarcoplasm to trigger contraction by the muscle fiber?

160 EXERCISE 10 Organization of Skeletal Muscle

2 Materials

- ☐ Compound microscope
- ☐ Neuromuscular junction slide

Procedures

1. Review the structure of the neuromuscular junction in Figure 10.4.
2. Examine the slide of the neuromuscular junction at low and medium powers. Identify the long, dark, thread-like axons of the motor neurons and the oval disks, which are the neuromuscular junctions. Describe the appearance of the muscle fibers.
3. In the space below, sketch several muscle fibers and their neuromuscular junctions. ■

Muscle fibers and neuromuscular junctions

LAB REPORT

EXERCISE 10

Organization of Skeletal Muscle

Name _____
Date _____
Section _____

A. Matching

Match each term in the left column with its correct description from the right column.

_____ 1. sarcomere
_____ 2. epimysium
_____ 3. perimysium
_____ 4. endomysium
_____ 5. myofibril
_____ 6. striations
_____ 7. sarcolemma
_____ 8. transverse tubule
_____ 9. sarcoplasmic reticulum
_____ 10. actin
_____ 11. myosin
_____ 12. fascicle

A. banding patterns in muscle tissue
B. storage site for calcium ions
C. protein of thin filaments
D. group of muscle fibers
E. cylinder composed of filaments
F. protein of thick filaments
G. carries action potential deep into fiber
H. connective tissue that covers muscle fiber
I. connective tissue that covers whole muscle
J. connective tissue that covers fascicles
K. cell membrane of muscle fiber
L. repeating organization of filaments

B. Labeling

Label the eight numbered muscle fiber features in Figure 10.5.

C. Short-Answer Questions

1. Describe the structure of a fascicle, including the connective tissue covering around, and within, the fascicle.

2. How is a skeletal muscle fiber stimulated by a motor neuron?

3. Describe the structure of a sarcomere.

161

EXERCISE 10   LAB REPORT

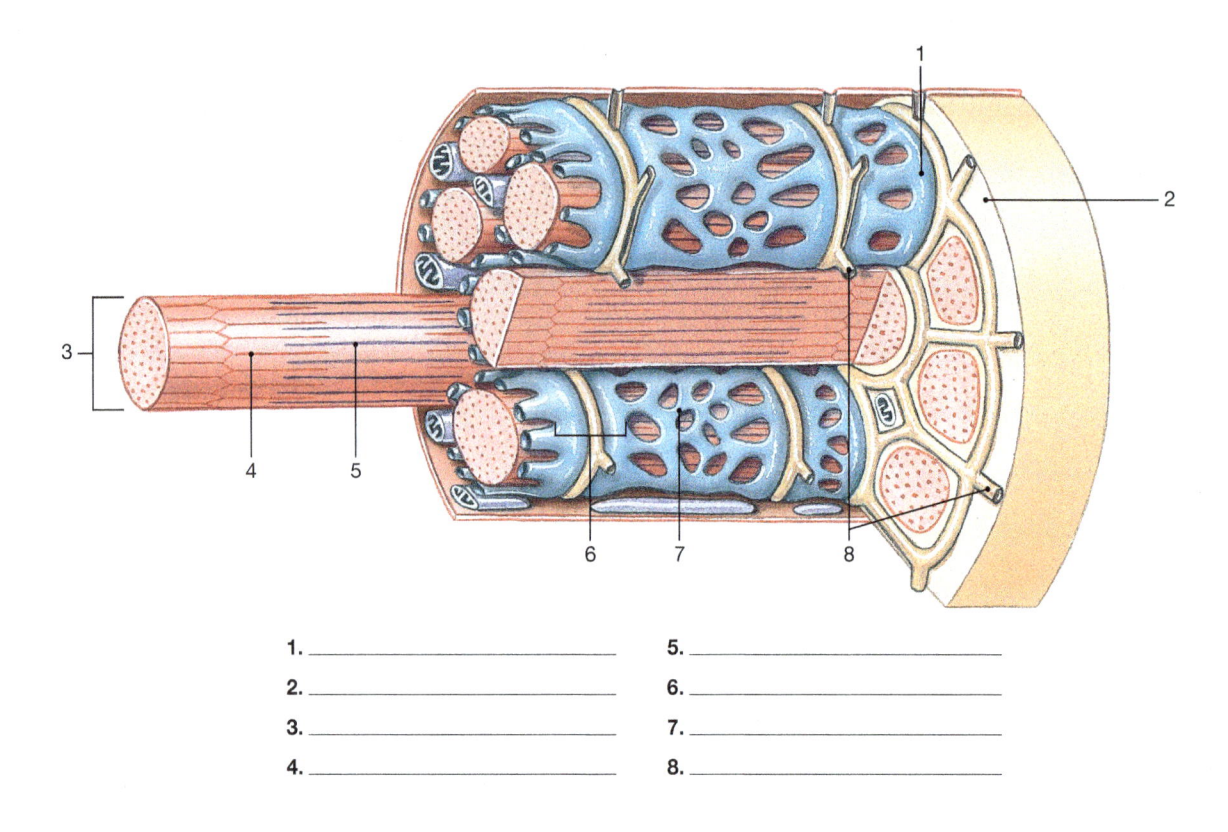

1. _____   5. _____
2. _____   6. _____
3. _____   7. _____
4. _____   8. _____

Figure 10.5   Structure of a Muscle Fiber

D. Analysis and Application

1. Many insecticides contain a compound that is an acetylcholinesterase inhibitor. How would exposure to this poison affect skeletal muscles in a human?

2. What gives skeletal muscle fibers their striations?

3. How are muscles attached to bones?

EXERCISE 11

Axial Muscles

OBJECTIVES

On completion of this exercise, you should be able to:

- Locate the axial muscles on laboratory models and charts.
- Identify on models the origin and insertion of each axial muscle and describe its action.
- Demonstrate the action of each axial muscle.

LAB ACTIVITIES

1. Muscles of Facial Expression   164
2. Muscles of the Eye   167
3. Muscles of Mastication   169
4. Muscles of the Tongue and Pharynx   170
5. Muscles of the Anterior Neck   173
6. Muscles of the Vertebral Column   175
7. Oblique and Rectus Muscles   179
8. Muscles of the Pelvic Region   183

Muscles are organized into the axial and appendicular divisions to reflect their attachment to either the axial or appendicular skeleton. Axial muscles, covered in this exercise, include the muscles of the head, neck, vertebral column, abdomen, and pelvis. Appendicular muscles are discussed in Exercise 12.

Each muscle causes a movement, called the **action**, that depends on many factors, especially the shape of the attached bones. In order for a muscle to produce a smooth, coordinated action, one end of it must serve as an attachment site while the other end moves the intended bone. The relatively stationary part of the muscle is called the **origin**. The opposite end of the muscle, the part that moves the bone, is called the **insertion**. As the muscle contracts, the insertion moves toward the origin to generate a pulling force and cause the muscle's action. Muscles can generate only a pulling force; they can never push. Usually, when one muscle, called an **agonist**, pulls in one direction, an **antagonistic** muscle pulls in the opposite direction to produce resistance and promote smooth movement. **Synergists** are muscles that work together and are often classified together in a **muscle group**, such as the oblique group of the abdomen.

Numerous methods are used to name muscles. The size, shape, origin, and action of a muscle are often reflected in its name.

Making Connections — Muscle Modeling

Your fingers and hands can be used to simulate the origin, insertion, and action of muscles. For example, place the base of your right index finger on your right

164 EXERCISE 11 Axial Muscles

> zygomatic bone and the tip of the index finger at the right corner of your mouth. The finger now represents the zygomatic major muscle, which elevates the edge of the mouth. The base of the finger represents the muscle's origin at the zygomatic bone, and the tip represents the insertion. When you flex your finger and elevate your mouth, you are mimicking the major action of this muscle. Smile! ●

The muscles of the head and neck produce a wide range of motions for making facial expressions, processing food, producing speech, and positioning the head. The names of these muscles usually indicate either the bone to which a muscle is attached or the structure that a muscle surrounds. In this exercise you will identify the major muscles used for facial expression and mastication, the muscles that move the eyes, and those that position the head and neck. As you study each group, attempt to find the general location of each muscle on your body. Contract the muscle and observe its action as your body moves.

LAB ACTIVITY 1 — Muscles of Facial Expression

The muscles of facial expression are those associated with the mouth, eyes, nose, ears, scalp, and neck. These muscles are unique in that one or both attachments are to the dermis of the skin rather than to a bone. Refer to Figure 11.1 and Table 11.1 for details on the origin, insertion, and actions of these muscles.

Figure 11.1 Muscles of Facial Expression
(a) Anterior view.

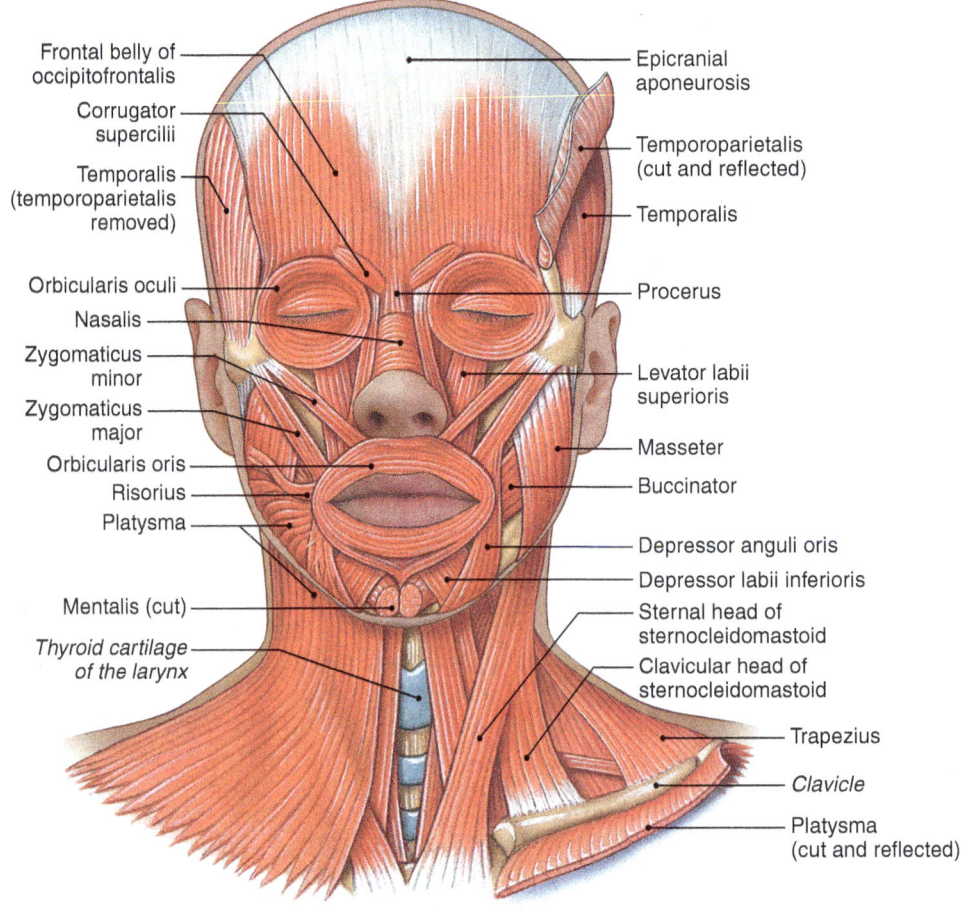

(a) Anterior view

Figure 11.1 **Muscles of Facial Expression**
(b) Anterolateral view of a dissection.

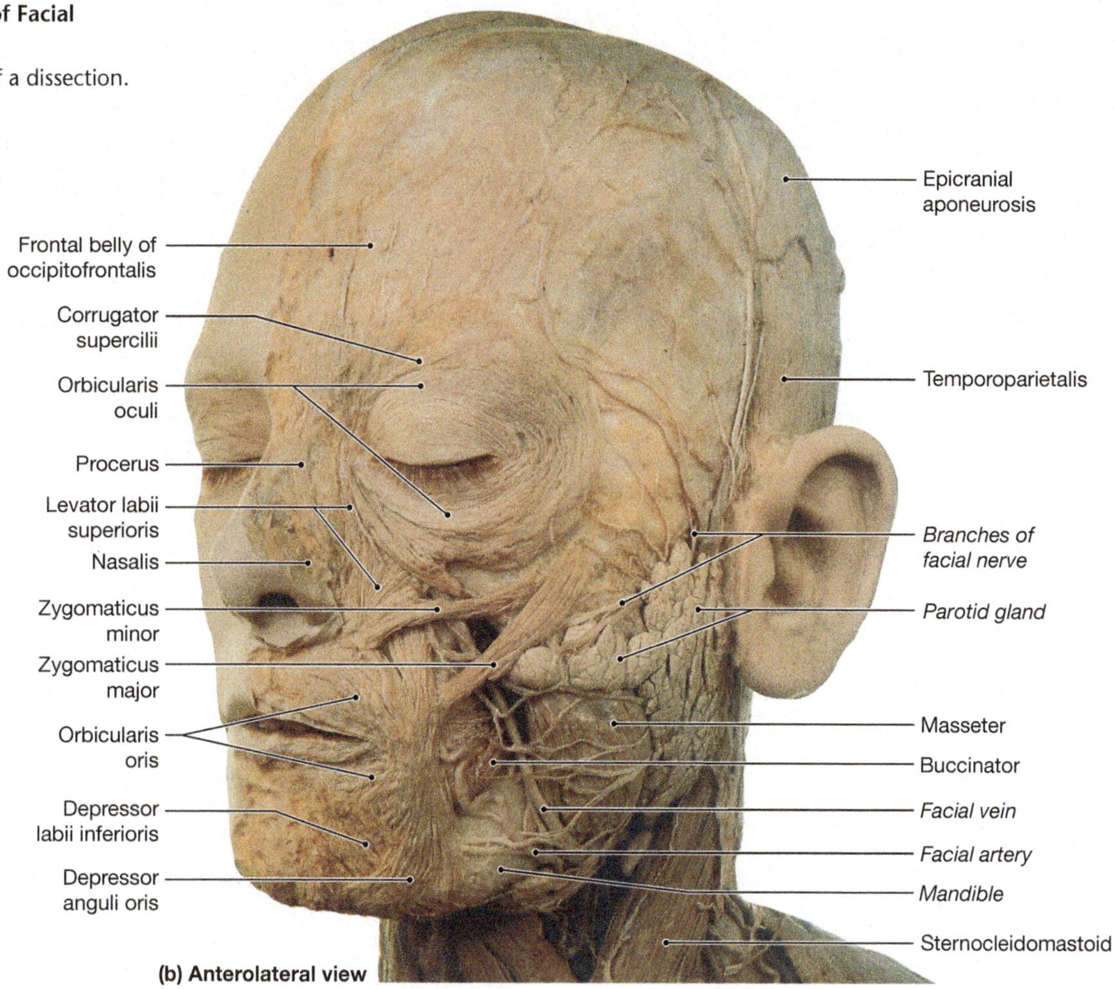

Frontal belly of occipitofrontalis
Corrugator supercilii
Orbicularis oculi
Procerus
Levator labii superioris
Nasalis
Zygomaticus minor
Zygomaticus major
Orbicularis oris
Depressor labii inferioris
Depressor anguli oris

Epicranial aponeurosis
Temporoparietalis
Branches of facial nerve
Parotid gland
Masseter
Buccinator
Facial vein
Facial artery
Mandible
Sternocleidomastoid

(b) Anterolateral view

Mouth

The **buccinator** (BUK-si-nā-tor) **muscle** is the horizontal muscle that spans between the jaws. It compresses the cheeks when you are eating or sucking on a straw. The **orbicularis oris muscle** is a sphincter muscle whose insertion is on the skin that surrounds the mouth. This muscle shapes the lips for a variety of functions, including speech, food manipulation, and facial expressions, and purses the lips together for a kiss. The **levator labii superioris muscle** is lateral to the nose and its insertion is on the superolateral edge of the orbicularis oris muscle. As its name implies, the levator labii superioris muscle elevates the upper lip. Muscles that act on the lower lip are inferior to the mouth. The **depressor anguli oris muscle** inserts on the skin at the angle of the mouth to depress the corners of the mouth. The **depressor labii inferioris muscle** is medial to the anguli muscle and inserts along the edge of the lower lip to depress the lower lip. On the medial chin is the **mentalis muscle**, which elevates and protrudes the lower lip.

The **risorius muscle** is a narrow muscle that inserts on the angle of the mouth. When it contracts, the risorius muscle pulls and produces a grimace-like tensing of the mouth. In the disease tetanus, the risorius is involved in the painful contractions that pull the corners of the mouth back into "lockjaw."

The **zygomaticus major** and **zygomaticus minor muscles** originate on the zygomatic bone and insert on the skin and corners of the mouth. These muscles retract and elevate the corners of the mouth when you smile.

166 EXERCISE 11 Axial Muscles

Table 11.1 *Origins and Insertions* Muscles of Facial Expression

Region/Muscle	Origin	Insertion	Action	Innervation
Mouth				
Buccinator	Alveolar processes of maxilla and mandible	Blends into fibers of orbicularis oris	Compresses cheeks	Facial nerve (N VII)
Depressor labii inferioris	Mandible between the anterior midline and the mental foramen	Skin of lower lip	Depresses lower lip	As above
Levator labii superioris	Inferior margin of orbit, superior to the infraorbital foramen	Orbicularis oris	Elevates upper lip	As above
Mentalis	Incisive fossa of mandible	Skin of chin	Elevates and protrudes lower lip	As above
Orbicularis oris	Maxilla and mandible	Lips	Compresses, purses lips	As above
Risorius	Fascia surrounding parotid salivary gland	Angle of mouth	Draws corner of mouth to the side	As above
Depressor anguli oris	Anterolateral surface of mandibular body	Skin at angle of mouth	Depresses corner of mouth	As above
Zygomaticus major	Zygomatic bone near the zygomaticomaxillary suture	Angle of mouth	Retracts and elevates corner of mouth	As above
Zygomaticus minor	Zygomatic bone posterior to zygomaticotemporal suture	Upper lip	Retracts and elevates upper lip	As above
Eye				
Corrugator supercilii	Orbital rim of frontal bone near frontonasal suture	Eyebrow	Pulls skin inferiorly and anteriorly; wrinkles brow	As above
Levator palpebrae superioris	Tendinous band around optic foramen	Upper eyelid	Elevates upper eyelid	Oculomotor nerve (N III)[a]
Orbicularis oculi	Medial margin of orbit	Skin around eyelids	Closes eye	Facial nerve (N VII)
Nose				
Procerus	Nasal bones and lateral nasal cartilages	Aponeurosis at bridge of nose and skin of forehead	Moves nose, changes position, shape of nostrils	As above
Nasalis	Maxilla and alar cartilage of nose	Bridge of nose	Compresses bridge, depresses tip of nose; elevates corners of nostrils	As above
Scalp (Epicranium)[b]				
Occipitofrontalis Frontal belly	Epicranial aponeurosis	Skin of eyebrow and bridge of nose	Raises eyebrows, wrinkles forehead	As above
Occipital belly	Superior nuchal line	Epicranial aponeurosis	Tenses and retracts scalp	As above
Temporoparietalis	Fascia around external ear	Epicranial aponeurosis	Tenses scalp, moves auricle of ear	As above
Neck				
Platysma	Superior thorax between cartilage of second rib and acromion of scapula	Mandible and skin of cheek	Tenses skin of neck, depresses mandible	As above

[a]This muscle originates in association with the extraocular muscles, so its innervation is unusual.
[b]Includes the epicranial aponeurosis, temporoparietalis, and occipitofrontalis muscles.

Eye

Muscles of the face that surround the eyes wrinkle the brow and move the eyelids. Muscles that move the eyeball are covered in an upcoming activity in this exercise.

The sphincter muscle of the eye is the **orbicularis oculi** (or-bik-ū-LA-ris ok-ū-lī) **muscle**. The muscle acts to close the eye, as during an exaggerated blink. The **corrugator supercilii muscle** is a small muscle that originates on the orbital rim of the frontal bone and inserts on the eyebrow. It acts to pull the skin inferiorly and wrinkles the forehead into a frown. The **levator palpebrae superioris muscle** inserts on and elevates the upper eyelid. (This muscle is not visible in Figure 11.1. See Figure 11.2.)

Nose

The human nose has limited movement, and the related muscles serve mainly to change the shape of the nostrils. The **procerus muscle** has a vertical orientation over the nasal bones; the **nasalis muscle** horizontally spans the inferior nasal bridge.

Ear

The **temporoparietalis muscle** is on the lateral sides of the epicranium (scalp). The muscle is cut and reflected (pulled up) in Figure 11.1 to illustrate deeper muscles of the epicranium. The action of the temporoparietalis is to tense the scalp and move the auricle (flap) of the ear.

Scalp

The **occipitofrontalis muscle** is the major muscle of the epicranium. It consists of two muscle bellies: the **frontal belly** and the **occipital belly**. The frontal belly of the occipitofrontalis muscle is the broad anterior muscle on the forehead that covers the frontal bone. It originates at a sheet of connective tissue called the **epicranial aponeurosis** (ep-i-KRĀ-nē-ul āp-ō-nū-RŌ-sis; *epi-*, on + *kranion*, skull). The actions of the frontal belly include wrinkling the forehead, raising the eyebrows, and pulling the scalp forward. The occipital belly of the occipitofrontalis muscle covers the posterior of the skull. This muscle tenses and retracts the scalp, an action difficult for most people to isolate and perform.

Neck

The **platysma** (pla-TIZ-muh; *platys*, flat) **muscle** is a thin, broad muscle that covers the sides of the neck. The platysma depresses the mandible and the soft structures of the lower face, which results in an expression of horror and disgust.

QuickCheck Questions

1.1 What are two facial muscles that are circular?

1.2 What are the muscles associated with the epicranial aponeurosis?

1 Materials

- Head model
- Muscle chart

Procedures

1. Review the muscles of the head in Figure 11.1 and in Table 11.1.
2. Examine the head model and/or the muscle chart and locate each muscle that is described in the preceding paragraphs.
3. Find the general location of the muscles of facial expression on your face. Practice the action of each muscle and observe how your facial expression changes. ■

LAB ACTIVITY 2 — Muscles of the Eye

The **extraocular muscles** of the eye, also called **extrinsic eye muscles** or **oculomotor muscles**, are those surrounding the eye. (In general, any muscle located outside the structure it controls is called an **extrinsic muscle**, and any muscle inside the structure it controls is called an **intrinsic muscle**.) The extraocular muscles insert on the *sclera*, which is the white, fibrous covering of the eye. **Intrinsic eye muscles** are involved in focusing the eye for vision. These muscles are discussed in Exercise 18.

Six extraocular eye muscles control eye movements (Figure 11.2 and Table 11.2). The **superior rectus**, **inferior rectus**, **medial rectus**, and **lateral rectus muscles**

168 EXERCISE 11 Axial Muscles

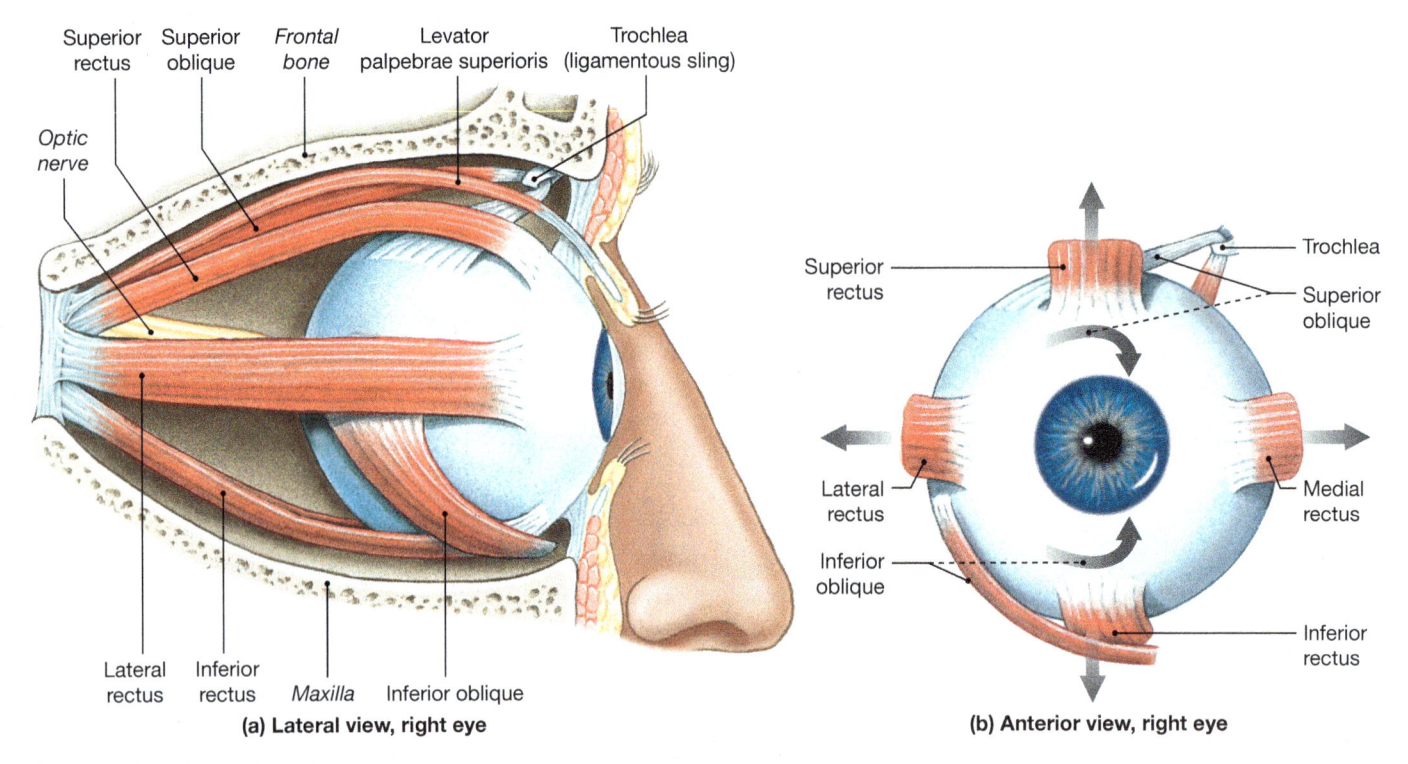

Figure 11.2 Extraocular Muscles
(a) Lateral view of right eye. (b) Anterior view of right eye, showing the orientation of the extraocular muscles and the directions of eye movements produced by contraction of individual muscles.

Table 11.2 *Origins and Insertions* Extraocular Muscles

Muscle	Origin	Insertion	Action	Innervation
Inferior rectus	Sphenoid around optic canal	Inferior, medial surface of eyeball	Eye looks down	Oculomotor nerve (N III)
Medial rectus	As above	Medial surface of eyeball	Eye looks medially	As above
Superior rectus	As above	Superior surface of eyeball	Eye looks up	As above
Lateral rectus	As above	Lateral surface of eyeball	Eye looks laterally	Abducens nerve (N VI)
Inferior oblique	Maxilla at anterior portion of orbit	Inferior, lateral surface of eyeball	Eye rolls, looks up and laterally	Oculomotor nerve (N III)
Superior oblique	Sphenoid around optic canal	Superior, lateral surface of eyeball	Eye rolls, looks down and laterally	Trochlear nerve (N IV)

are straight muscles that move the eyeball in the superior, inferior, medial and lateral directions, respectively. They originate around the optic foramen in the eye orbit and insert on the sclera. The **superior** and **inferior oblique muscles** attach diagonally on the eyeball. The superior oblique muscle has a tendon that passes through a trochlea (pulley) located on the upper orbit. This muscle rolls the eye downward, and the inferior oblique muscle rolls the eye upward.

QuickCheck Questions

2.1 What are the four rectus muscles of the eye?

2.2 Which eye muscle passes through a pulley-like structure?

EXERCISE 11 Axial Muscles   169

2 Materials

☐ Eye model
☐ Eye muscle chart

Procedures

1. Review the muscles of the eye in Figure 11.2 and Table 11.2.
2. Examine the eye model and/or the eye muscle chart, and locate each extraocular eye muscle.
3. Practice the action of each eye muscle by moving your eyeballs. ■

LAB ACTIVITY 3 Muscles of Mastication

The muscles involved in chewing, called mastication, depress and elevate the mandible to open and close the jaws and grind the teeth against the food (Figure 11.3 and Table 11.3). The **masseter** (MAS-se-tur) **muscle** is a short, thick muscle that originates on the zygomatic arch and inserts on the angle and the ramus of the mandible.

(a) Lateral view

(b) Lateral view, pterygoid muscles exposed

Figure 11.3 Muscles of Mastication
The muscles of mastication move the mandible during chewing. (a) The temporalis and masseter muscles are prominent muscles on the lateral surface of the skull. (b) The location and orientation of the pterygoid muscles can be seen after removing the overlying muscles and a portion of the mandible.

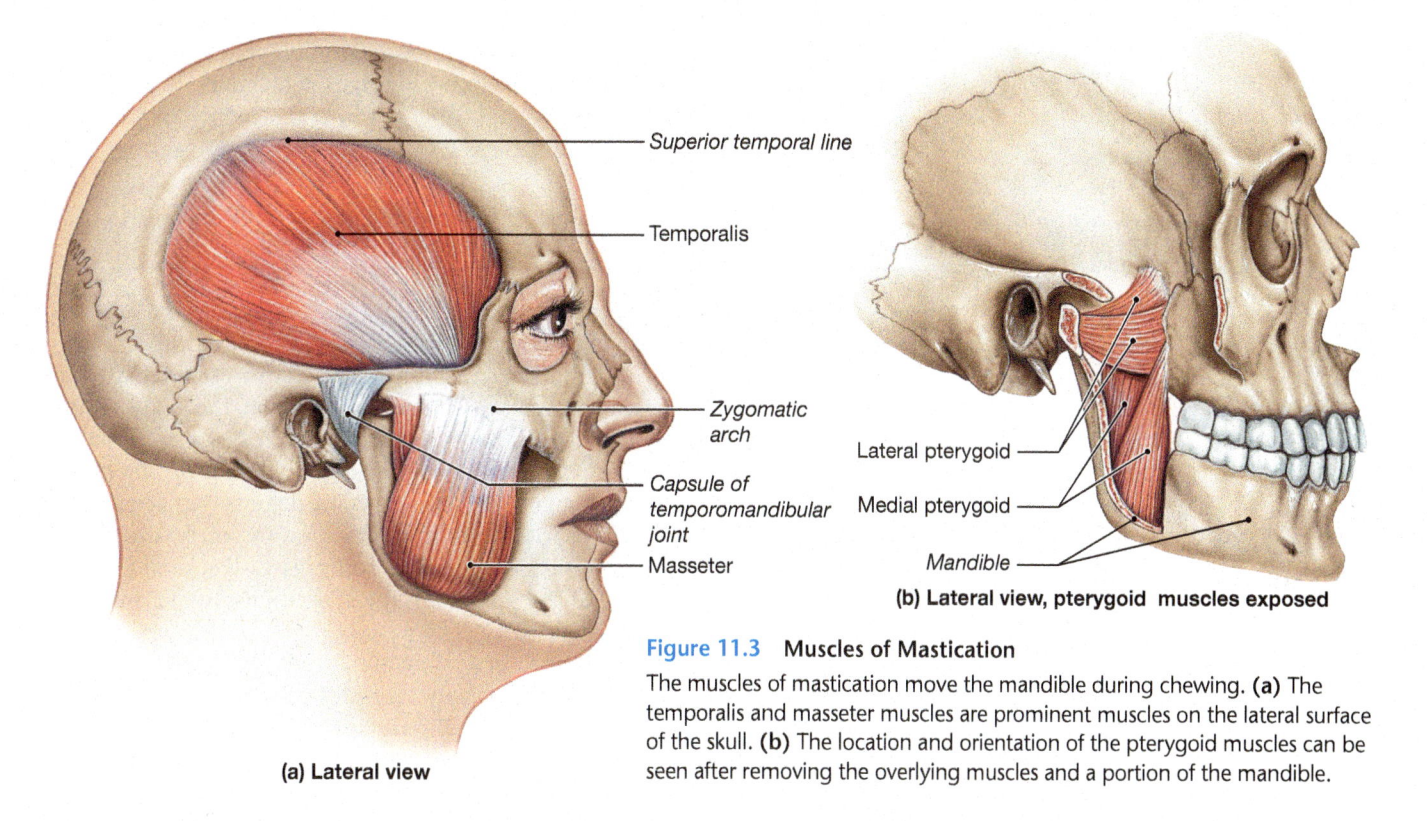

Table 11.3 *Origins and Insertions* Muscles of Mastication

Muscle	Origin	Insertion	Action	Innervation
Masseter	Zygomatic arch	Lateral surface and angle of mandibular ramus	Elevates mandible and closes jaws	Trigeminal nerve (N V), mandibular branch
Temporalis	Along temporal lines of skull	Coronoid process of mandible	As above	As above
Pterygoids	Lateral pterygoid plate	Medial surface of mandibular ramus		
Medial pterygoid	Lateral pterygoid plate and adjacent portions of palatine bone and maxilla	Medial surface of mandibular ramus	Elevates the mandible and closes the jaws, or moves mandible side to side	As above
Lateral pterygoid	Lateral pterygoid plate and greater wing of sphenoid	Anterior part of the neck of the mandibular condyle	Opens jaws, protrudes mandible, or moves mandible side to side	As above

The **temporalis** (tem-pō-RA-lis) **muscle** covers almost the entire temporal fossa. Deep to the masseter and other cheek muscles are the **lateral** and **medial pterygoid** (TER-i-goyd; *pterygoin,* wing) **muscles**, which assist in mastication by elevating and depressing the mandible and moving the mandible from side to side, an action called *lateral excursion*.

QuickCheck Questions

3.1 To which bones do the muscles for mastication attach?

3.2 Which muscles act to move the mandible from side to side?

3 Materials

- ☐ Head model
- ☐ Muscle chart

Procedures

1. Review the mastication muscles in Figures 11.1 and 11.3, and Table 11.3.
2. Examine the head model and/or the muscle chart, and locate each mastication muscle that is described in this activity.
3. Find the general location of the muscles of mastication on your face. Practice the action of each muscle and observe how your mandible moves. For example, put your fingertips at the angle of your jaw and clench your teeth. You should feel the masseter compress as it forces the teeth together. ■

LAB ACTIVITY 4 Muscles of the Tongue and Pharynx

Extrinsic muscles of the tongue constitute the floor of the oral cavity and assist in the complex movements of the tongue for speech, chewing, and initiating swallowing (Figure 11.4 and Table 11.4). Anteriorly, the **genioglossus muscle** originates on the medial mandibular surface around the chin and inserts on the body of the tongue and the hyoid bone. It depresses and protracts the tongue, as in initiating the licking of an ice cream cone. The **hyoglossus muscle** originates on the hyoid bone, inserts on the side of the tongue, and acts to both depress and retract the tongue. The **palatoglossus muscle** arises from the soft palate, inserts on the side of the tongue, elevates the tongue, and depresses the soft palate. The origin of the **styloglossus muscle** is superior to the tongue on the styloid process. This muscle retracts the tongue and elevates its sides.

Muscles of the pharynx are involved in swallowing (Figure 11.5 and Table 11.5). The **superior, middle,** and **inferior constrictor muscles** constrict the pharynx to push food into the esophagus. The **levator veli palatini** and **tensor veli palatini muscles** elevate the soft palate during swallowing. The larynx is elevated by the **palatopharyngeus** (pal-āt-ō-far-IN-jē-us), **salpingopharyngeus** (sal-pin-gō-far-IN-jē-us), **stylopharyngeus muscles**, and by some of the neck muscles.

QuickCheck Questions

4.1 What does the word *glossus* mean?

4.2 Where do the styloglossus and the hyoglossus muscles originate?

4.3 Which muscles constrict the pharynx?

4 Materials

- ☐ Head and neck model
- ☐ Muscle chart

Procedures

1. Review the extrinsic muscles of the tongue in Figure 11.4 and Table 11.4.
2. Examine the head model and/or the muscle chart and identify each muscle of the tongue.
3. Practice the action of each tongue muscle. The ability to curl the sides of your tongue with the styloglossus is genetically controlled by a single gene.

Figure 11.4 Muscles of the Tongue

The left mandibular ramus has been removed to show the extrinsic muscle on the left side of the tongue.

Styloid process
Palatoglossus
Styloglossus
Genioglossus
Hyoglossus
Hyoid bone
Mandible (cut)

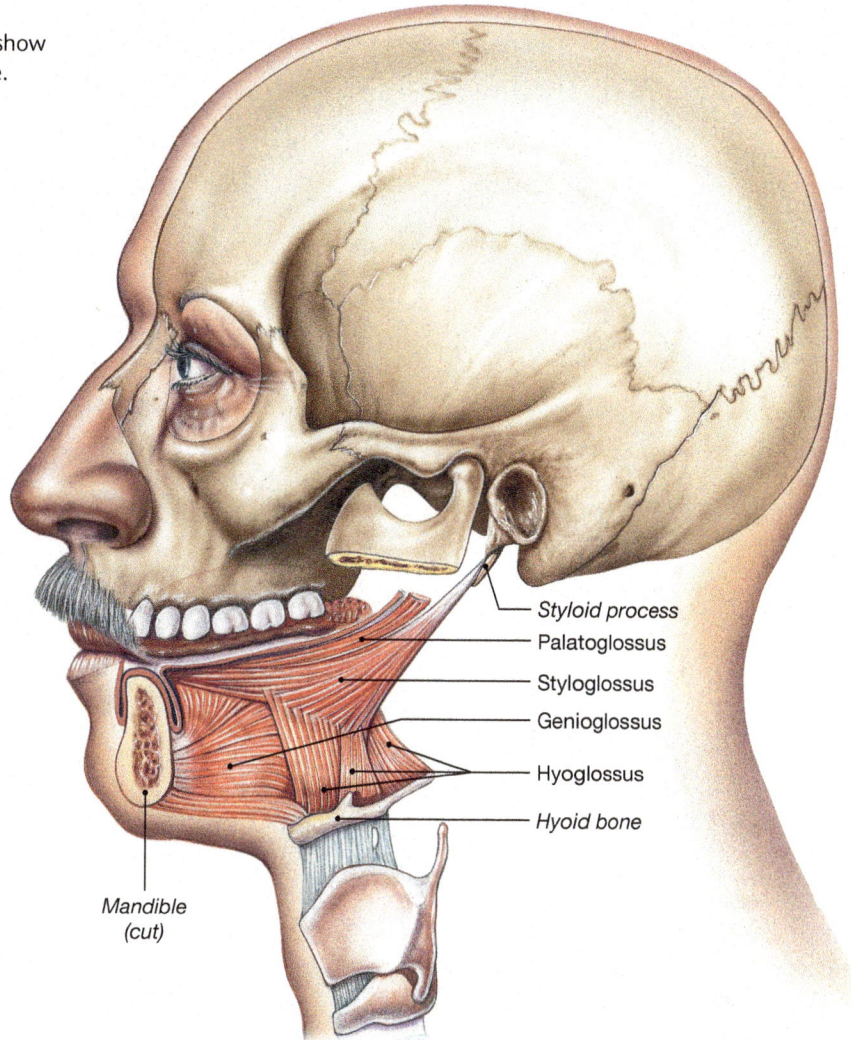

Table 11.4 *Origins and Insertions* **Muscles of the Tongue**

Muscle	Origin	Insertion	Action	Innervation
Genioglossus	Medial surface of mandible around chin	Body of tongue, hyoid bone	Depresses and protracts tongue	Hypoglossal nerve (N XII)
Hyoglossus	Body and greater horn of hyoid bone	Side of tongue	Depresses and retracts tongue	As above
Palatoglossus	Anterior surface of soft palate	As above	Elevates tongue, depresses soft palate	Internal branch of accessory nerve (N XI)
Styloglossus	Styloid process of temporal bone	Along the side to tip and base of tongue	Retracts tongue, elevates sides	Hypoglossal nerve (N XII)

Individuals with the dominant gene are "rollers" and can curl the tongue. Those with the recessive form of the gene are "nonrollers" Are you a "roller" or a "nonroller"?

4. Review the muscles of the pharynx in Figure 11.5 and Table 11.5.
5. Locate each pharyngeal muscle on the head model and/or muscle chart.
6. Place your finger on your larynx (Adam's apple) and swallow. Which muscles caused the larynx to move? ■

EXERCISE 11 Axial Muscles

Figure 11.5 Muscles of the Pharynx
Pharyngeal muscles for swallowing seen in a lateral view.

Lateral view

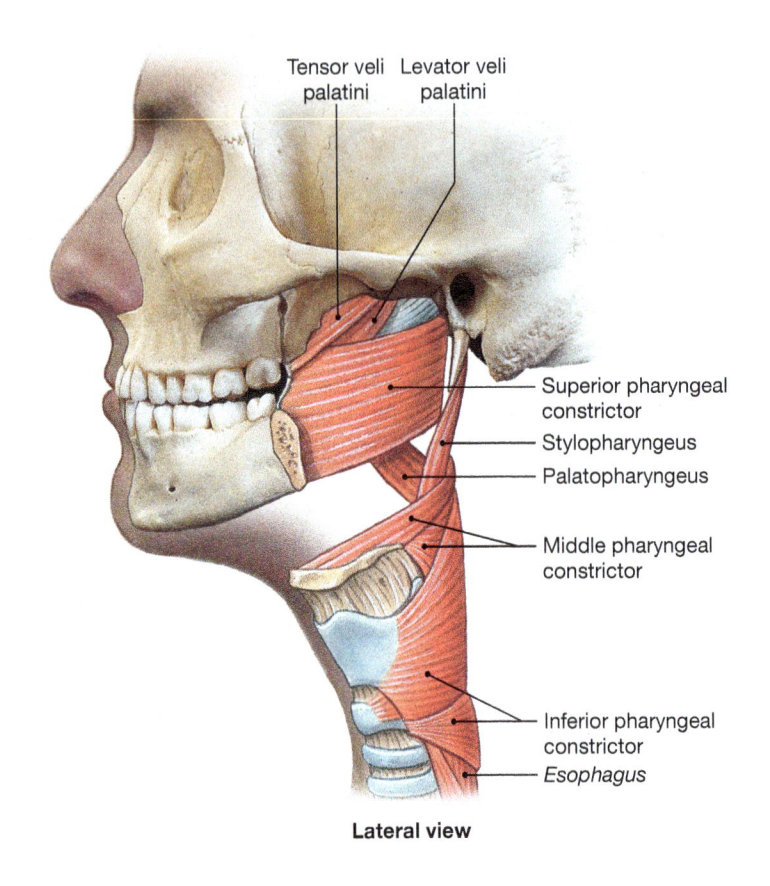

Table 11.5 *Origins and Insertions* Muscles of the Pharynx

Muscle	Origin	Insertion	Action	Innervation
Pharyngeal Constrictors			Constrict pharynx to propel bolus into esophagus	Branches of pharyngeal plexus (N X)
Superior constrictor	Pterygoid process of sphenoid, medial surfaces of mandible	Median raphe attached to occipital bone		N X
Middle constrictor	Horns of hyoid bone	Median raphe		N X
Inferior constrictor	Cricoid and thyroid cartilages of larynx	Median raphe		N X
Laryngeal Elevators*			Elevate larynx	Branches of pharyngeal plexus (N IX & X)
Palatopharyngeus	Soft palate	Thyroid cartilage		N X
Salpingopharyngeus	Cartilage around the inferior portion of the auditory tube	Thyroid cartilage		N X
Stylopharyngeus	Styloid process of temporal bone	Thyroid cartilage		N IX
Palatal Muscles				
Levator veli palatini	Petrous part of temporal bone, tissues around the auditory tube	Soft palate	Elevates soft palate	Branches of pharyngeal plexus (N X)
Tensor veli palatini	Sphenoidal spine and tissues around the auditory tube	Soft palate	As above	N V

*Assisted by the thyrohyoid, geniohyoid, stylohyoid, and hyoglossus muscles, discussed in Tables 11.4 and 11.6.

EXERCISE 11 Axial Muscles 173

LAB ACTIVITY 5 ## Muscles of the Anterior Neck

Muscles of the anterior neck, which stabilize and move the neck, act on the mandible and the hyoid bone (Figures 11.6 and 11.7 and Table 11.6). The principal muscle of the anterior neck is the **sternocleidomastoid** (ster-nō-klī-dō-MAS-toyd) **muscle**. This long, slender muscle occurs on both sides of the neck and is named after its points of attachment on the sternum, clavicle, and mastoid process of the temporal bone.

The *suprahyoid muscles* are a group of neck muscles that originate superior to, and act on, the hyoid bone. The suprahyoid muscle known as the **digastric muscle** has two parts: the **anterior belly** originates on the inferior surface of the mandible near the chin, and the **posterior belly** arises on the mastoid process of the temporal bone. The bellies insert on the hyoid bone and form a muscular swing that elevates the hyoid bone or depresses the mandible. The **mylohyoid muscle** is a wide muscle that is posterior to the anterior belly of the digastric muscle. Deep and medial to the mylohyoid muscle is the **geniohyoid muscle**, which depresses the mandible, elevates the larynx, and can also retract the hyoid bone. The **stylohyoid muscle** originates on the styloid process of the temporal bone, inserts on the hyoid bone, and elevates the hyoid bone and the larynx.

The *infrahyoid muscles* are a group of neck muscles that arise inferior to the hyoid bone, and their actions depress that bone and the larynx. The infrahyoid called the **omohyoid** (ō-mō-HĪ-oyd) **muscle** has two bellies that meet at a central tendon that is attached to the clavicle and the first rib. Medial to the omohyoid muscle is the strap-like **sternohyoid muscle**. Deep to the sternohyoid is the **sternothyroid muscle**. The omohyoid, sternohyoid, and sternothyroid muscles depress the hyoid bone and larynx. The **thyrohyoid muscle** depresses the hyoid and elevates the larynx.

Figure 11.6 Muscles of the Anterior Neck

Muscles of the anterior neck adjust the position of the larynx, mandible, and floor of the mouth and establish a foundation for tongue and pharyngeal muscles.

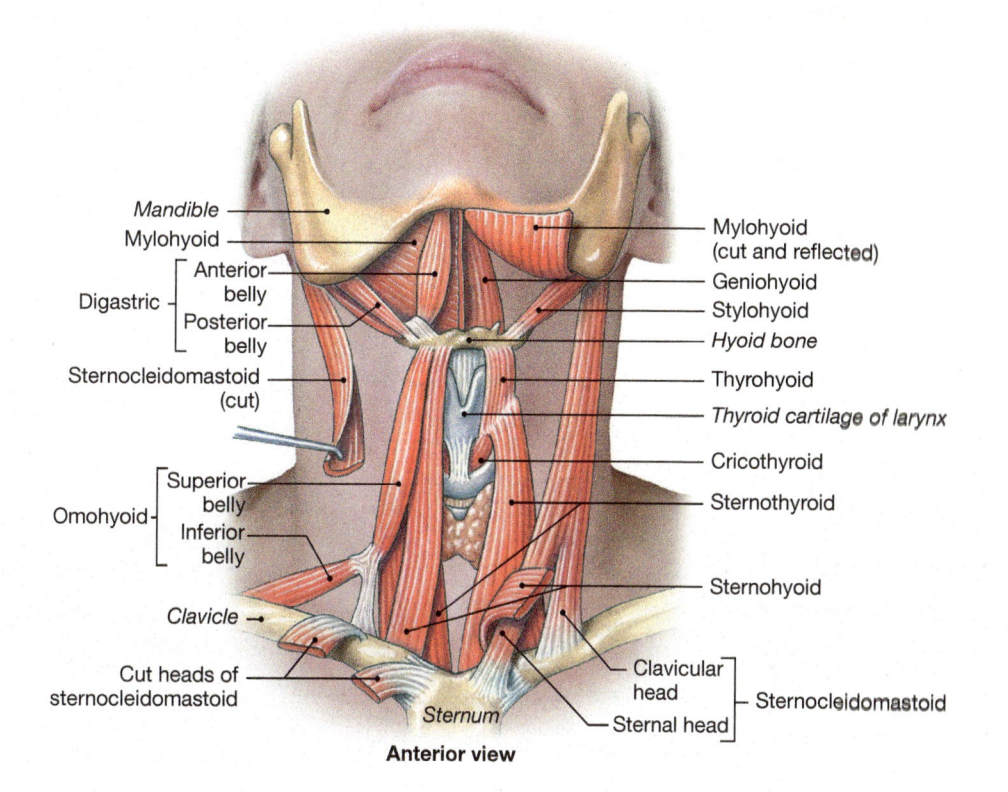

Anterior view

174 EXERCISE 11 Axial Muscles

Table 11.6 *Origins and Insertions* Muscles of the Anterior Neck

Muscle	Origin	Insertion	Action	Innervation
Digastric	Two bellies: *anterior* from inferior surface of mandible at chin; *posterior* from mastoid region of temporal bone	Hyoid bone	Depresses mandible or elevates larynx	*Anterior belly*: Trigeminal nerve (V), mandibular branch *Posterior belly*: Facial nerve (VII)
Geniohyoid	Medial surface of mandible at chin	Hyoid bone	As above and pulls hyoid bone anteriorly	Cervical nerve C_1 via hypoglossal nerve (XII)
Mylohyoid	Mylohyoid line of mandible	Median connective tissue band (raphe) that runs to hyoid bone	Elevates floor of mouth and hyoid bone or depresses mandible	Trigeminal nerve (V), mandibular branch
Omohyoid (superior and inferior bellies united at central tendon anchored to clavicle and first rib)	Superior border of scapula near scapular notch	Hyoid bone	Depresses hyoid bone and larynx	Cervical spinal nerves C_2–C_3
Sternohyoid	Clavicle and manubrium	Hyoid bone	As above	Cervical spinal nerves C_1–C_3
Sternothyroid	Dorsal surface of manubrium and first costal cartilage	Thyroid cartilage of larynx	As above	As above
Stylohyoid	Styloid process of temporal bone	Hyoid bone	Elevates larynx	Facial nerve (VII)
Thyrohyoid	Thyroid cartilage of larynx	Hyoid bone	Elevates thyroid, depresses hyoid bone	Cervical spinal nerves C_1–C_2 via hypoglossal nerve (XII)
Sternocleidomastoid	Two bellies: *clavicular head* attaches to sternal end of clavicle; *sternal head* attaches to manubrium	Mastoid region of skull and lateral portion of superior nuchal line	Together, they flex the neck; alone, one side bends head toward shoulder and turns face to opposite side	Accessory nerve (XI) and cervical spinal nerves (C_2–C_3) of cervical plexus

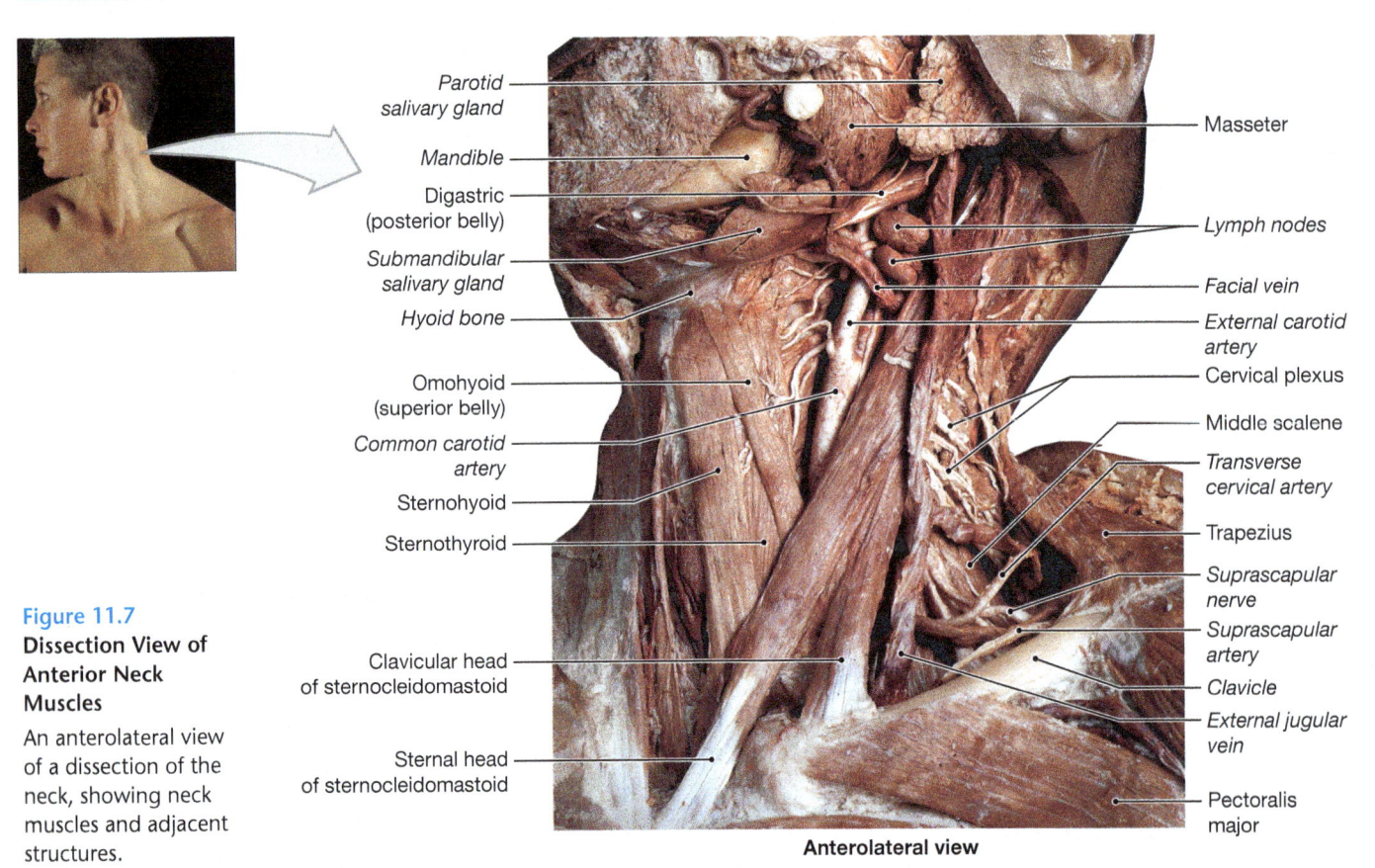

Figure 11.7
Dissection View of Anterior Neck Muscles
An anterolateral view of a dissection of the neck, showing neck muscles and adjacent structures.

Anterolateral view

EXERCISE 11 Axial Muscles   175

> ### QuickCheck Questions
> **5.1** Where does the sternocleidomastoid muscle attach?
> **5.2** What is the suffix in the names of muscles that insert on the hyoid bone?
> **5.3** Where is the digastric muscle located?

5 Materials
- Head-torso model
- Muscle chart

Procedures
1. Review the anterior neck muscles in Figures 11.6 and 11.7, and Table 11.6.
2. Locate each muscle on the head-torso model and/or the muscle chart.
3. Produce the actions of your suprahyoid and infrahyoid muscles and observe how your larynx moves.
4. Locate the sternocleidomastoid muscle on the head-torso model and/or on the muscle chart.
5. Contract your sternocleidomastoid muscle on one side and observe your head movement. Next, contract both sides and note how your head flexes.
6. Rotate your head until your chin almost touches your right shoulder and locate your left sternocleidomastoid muscle just above the manubrium of the sternum. ∎

LAB ACTIVITY 6 Muscles of the Vertebral Column

The muscles of the back are organized into three layers: superficial, intermediate, and deep (Table 11.7). Except for two superficial muscles that act on the appendicular skeleton (the trapezius and latissimus dorsi muscles, discussed in Exercise 12), all the back muscles move the vertebral column. The superficial vertebral muscles move the head and neck. The intermediate muscles are in long bands that stabilize and extend the vertebral column. The deep layer consists of small muscles that connect adjacent vertebrae to each other.

Most of the vertebral muscles are *extensor muscles* that extend the vertebral column and, in doing so, resist the downward pull of gravity. The extensors are located on the back, posterior to the spine. *Flexor muscles* are muscles that cause flexion; the vertebral flexors are positioned lateral to the vertebral column. There are few flexor muscles for the vertebral column because most of the body's mass is positioned anterior to the vertebral column, and consequently the force of gravity naturally flexes the column.

Table 11.7 *Origins and Insertions* Muscles of the Vertebral Column

Group/Muscle	Origin	Insertion	Action	Innervation
Superficial Layer				
Splenius (splenius capitis, splenius cervicis)	Spinous processes and ligaments that connect inferior cervical and superior thoracic vertebrae	Mastoid process, occipital bone of skull, superior cervical vertebrae	The two sides act together to extend neck; either alone rotates and laterally flexes neck to that side	Cervical spinal nerves
Intermediate Layer (Erector Spinae)				
Spinalis group				
Spinalis cervicis	Inferior portion of ligamentum nuchae and spinous process of C_7	Spinous process of axis	Extends neck	As above
Spinalis thoracis	Spinous processes of inferior thoracic and superior lumbar vertebrae	Spinous processes of superior thoracic vertebrae	Extends vertebral column	Thoracic and lumbar spinal nerves

(continued)

Table 11.7 **Origins and Insertions** Muscles of the Vertebral Column *(continued)*

Muscle	Origin	Insertion	Action	Innervation
Longissimus group				
Longissimus capitis	Transverse processes of inferior cervical and superior thoracic vertebrae	Mastoid process of temporal bone	The two sides act together to extend neck; either alone rotates and laterally flexes neck to that side	Cervical and thoracic spinal nerves
Longissimus cervicis	Transverse processes of superior thoracic vertebrae	Transverse processes of middle and superior cervical vertebrae	As above	As above
Longissimus thoracis	Broad aponeurosis and at transverse processes of inferior thoracic and superior lumbar vertebrae; joins iliocostalis	Transverse processes of superior thoracic and lumbar vertebrae and inferior surfaces of lower ten ribs	Extension of vertebral column; alone, each produces lateral flexion to that side	Thoracic and lumbar spinal nerves
Iliocostalis group				
Iliocostalis cervicis	Superior borders of vertebrosternal ribs near the angles	Transverse processes of middle and inferior cervical vertebrae	Extends or laterally flexes neck, elevates ribs	Cervical and superior thoracic spinal nerves
Iliocostalis thoracis	Superior borders of ribs 6–12 medial to the angles	Superior ribs and transverse processes of last cervical vertebra	Stabilizes thoracic vertebrae in extension	Thoracic spinal nerves
Iliocostalis lumborum	Iliac crest, sacral crests, and lumbar spinous processes	Inferior surfaces of ribs 6–12 near their angles	Extends vertebral column, depresses ribs	Inferior thoracic nerves and lumbar spinal nerves
Deep Layer (Transversospinalis)				
Semispinalis group				
Semispinalis capitis	Processes of inferior cervical and superior thoracic vertebrae	Occipital bone, between nuchal lines	Together the two sides extend neck; alone, each extends and laterally flexes neck and turns head to opposite side	Cervical spinal nerves
Semispinalis cervicis	Transverse processes of T_1–T_5 or T_6	Spinous processes of C_2–C_5	Extends vertebral column and rotates toward opposite side	As above
Semispinalis thoracis	Transverse processes of T_6–T_{10}	Spinous processes of C_5–T_4	As above	Thoracic spinal nerves
Multifidus	Sacrum and transverse process of each vertebra	Spinous processes of the third or fourth more superior vertebra	As above	Cervical, thoracic, and lumbar spinal nerves
Rotatores (cervicis, thoracis, and lumborum)	Transverse processes of the vertebrae in each region (cervical, thoracic, and lumbar)	Spinous process of adjacent, more superior vertebra	As above	As above
Interspinales	Spinous process of each vertebra	Spinous processes of more superior vertebra	Extends vertebral column	As above
Intertransversarii	Transverse processes of each vertebra	Transverse process of more superior vertebra	Lateral flexion of vertebral column	As above
Spinal Flexors				
Longus capitis	Transverse processes of cervical vertebrae	Base of the occipital bone	Together the two sides flex the neck; alone each rotates head to that side	Cervical spinal nerves
Longus colli	Anterior surfaces of cervical and superior thoracic vertebrae	Transverse processes of superior cervical vertebrae	Flexes and/or rotates neck; limits hyperextension	As above
Quadratus lumborum	Iliac crest and iliolumbar ligament	Last rib and transverse processes of lumbar vertebrae	Together they depress ribs; alone, each produces lateral flexion of vertebral column; fixes floating ribs (11 and 12) during forced exhalation	Thoracic and lumbar spinal nerves

Many vertebral muscles are named after their insertion to assist with grouping and identification. Muscles that insert on the skull include *capitis* in their name. Muscles that insert on the neck are called *cervicis*, those that insert on the thoracic vertebrae are *thoracis*, and those that insert on the lumbar vertebrae are *lumborum*.

Superficial Layer

The superficial vertebral muscles are the **splenius capitis muscle** (Figure 11.8) and the **splenius cervicis muscle.** When the two left splenius muscles and the two right ones contract in concert, the neck extends. When the splenius capitis and splenius cervicis muscles on only one side of the neck contract, the neck rotates laterally and flexes.

Intermediate Layer

The **erector spinae group** of muscles forms the intermediate layer of the back musculature. This group is made up of three subgroups: *spinalis*, *longissimus*, and *iliocostalis*. The **spinalis cervicis** muscles extend the neck, and the **spinalis thoracis** muscles extend the vertebral column.

The **longissimus capitis** and **longissimus cervicis** muscles act on the neck. When either both longissimus capitis muscles or both longissimus cervicis muscles contract, the neck extends. When only one longissimus capitis or one longissimus cervicis contracts, the neck flexes and rotates laterally. The **longissimus thoracis** muscles extend the vertebral column, and when only one of these muscles contracts, the column flexes laterally.

The **iliocostalis cervicis, iliocostalis thoracis**, and **iliocostalis lumborum** muscles all extend the neck and vertebral column and stabilize the thoracic vertebrae.

Deep Layer

The back muscles that make up the deep layer are collectively called the *transversospinalis muscles*; they interconnect and support the vertebrae. The various types in this layer are the semispinalis group and the multifidus, rotatores, interspinales, and intertransversarii muscles.

The **semispinalis capitis** muscles extend the neck when both of them contract; if only one semispinalis capitis contracts, it extends and laterally flexes the neck and turns the head to the opposite side. The **semispinalis cervicis** muscles extend the vertebral column when both contract and rotate the column to the opposite side when only one contracts. The **semispinalis thoracis** muscles work in the same way.

The **multifidus muscles** are a deep band of muscles that span the length of the vertebral column. Each portion of the band originates either on the sacrum or on a transverse process of a vertebra and inserts on the spinous process of a vertebra that is three or four vertebrae superior to the origin. Between transverse processes are the **rotatores cervicis, rotatores thoracis**, and **rotatores lumborum** muscles, each named after the vertebra of origin. The multifidus and rotatores muscles act with the semispinalis thoracis to extend and flex the vertebral column. Spanning adjacent spinous processes are **interspinales muscles,** which extend the vertebral column. Contiguous transverse processes have **intertransversarii muscles,** which laterally flex the column.

Spinal Flexors

The spinal flexor muscles are located along the lateral and anterior surfaces of the vertebrae (Figure 11.8b). The **longus capitis** muscles are visible as bands along the

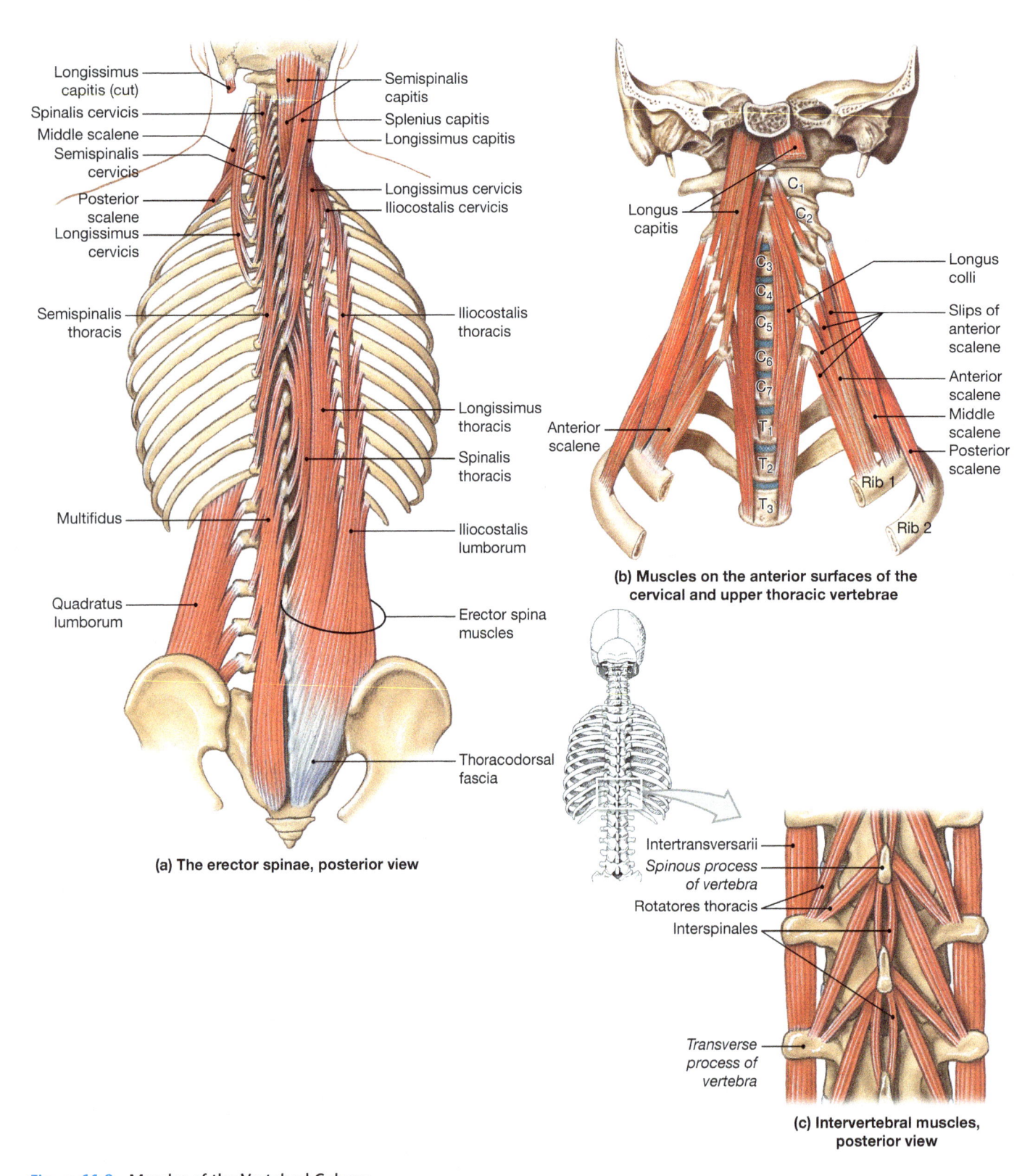

Figure 11.8 Muscles of the Vertebral Column

These muscles adjust the position of the vertebral column, head, neck, and ribs. (a) Posterior view of muscles of the vertebral column. (b) Muscles on anterior surface of the cervical and superior thoracic vertebrae. (c) Posterior view of intertransversarii, rotatores, and interspinales muscles.

178

anterior margin of the vertebral column that insert on the occipital bone and flex the neck; when only one longus capitis contracts, it rotates the head to the side of contraction. The **longus colli** muscles insert on the cervical vertebrae, flex and rotate the neck, and limit extension. The **quadratus lumborum** muscles originate on the iliac crest and the iliolumbar ligament and insert on the inferior border of the twelfth pair of ribs and the transverse processes of the lumbar vertebrae. These muscles flex the vertebral column; when only one quadratus lumborum contracts, the column is flexed laterally toward the side of contraction.

QuickCheck Questions

6.1 What are the three muscles of the longissimus group and their action?

6.2 Which muscle inserts on the twelfth pair of ribs?

6 Materials

- ☐ Torso model
- ☐ Muscle chart

Procedures

1. Review the muscles of the vertebral column in Figure 11.8 and Table 11.7.
2. Examine the back of the torso model and identify the superficial vertebral muscles. Note the insertion of each muscle.
3. Distinguish among the various erector spinae muscles in the intermediate layer of vertebral muscles. Note the insertion of each muscle group.
4. Identify the transversospinalis muscles associated with the individual vertebrae on the torso model and/or muscle chart.
5. Locate the flexor muscles of the vertebral column on the torso model and/or muscle chart.
6. Extend and flex your vertebral column and consider the muscles that produce each action. ■

LAB ACTIVITY 7 Oblique and Rectus Muscles

Muscles between the vertebral column and the anterior midline are grouped into either the oblique (slanted) or rectus muscle groups. As the names imply, the oblique muscles are slanted relative to the body's vertical central axis and the rectus muscles are oriented either parallel or perpendicular to this axis. Both muscle groups are in the cervical, thoracic, and abdominal regions (Table 11.8). All these muscles support the vertebral column, provide resistance against the erector spinae muscles, move the ribs during respiration, and constitute the abdominal wall. Another major action of these muscles is to increase intraabdominal pressure during urination, defecation, and childbirth.

Oblique Muscles

The oblique muscles of the neck, collectively called the *scalene group,* are the **anterior**, **middle**, and **posterior scalene muscles** (see Figure 11.8b). Each muscle originates on the transverse process of a cervical vertebra and inserts on a first or second rib. When the ribs are held in position, the scalene muscles flex the neck. When the neck is stationary, they elevate the ribs during inspiration.

Oblique muscles of the thoracic region include the intercostal and transversus thoracis muscles (Figure 11.9). The intercostal muscles are located between the ribs and, along with the diaphragm, change the size of the chest for breathing. The superficial **external intercostal muscles** and the deep **internal intercostal muscles** span the gaps between the ribs. The **transversus thoracis muscle** lines the posterior surfaces of the sternum and the cartilages of the ribs. The muscle is covered by the serous membrane of the lungs (pleura). It depresses the ribs.

Table 11.8 *Origins and Insertions* Oblique and Rectus Muscles

Group/Muscle	Origin	Insertion	Action	Innervation
Oblique Group				
Cervical region				
Scalenes (anterior, middle, and posterior)	Transverse and costal processes of cervical vertebrae C_2–C_7	Superior surface of first two ribs	Elevate ribs and/or flex neck; one side bends neck and rotates head and neck to opposite side	Cervical spinal nerves
Thoracic region				
External intercostals	Inferior border of each rib	Superior border of more inferior rib	Elevate ribs	Intercostal nerves (branches of thoracic spinal nerves)
Internal intercostals	Superior border of each rib	Inferior border of the more superior rib	Depress ribs	As above
Transversus thoracis	Posterior surface of sternum	Cartilages of ribs	As above	As above
Serratus posterior superior	Spinous processes of C_7–T_3 and ligamentum nuchae	Superior borders of ribs 2–5 near angles	Elevates ribs, enlarges thoracic cavity	Thoracic nerves (T_1–T_4)
inferior	Aponeurosis from spinous processes of T_{10}–L_3	Inferior borders of ribs 8–12	Pulls ribs inferiorly; also pulls outward, opposing diaphragm	Thoracic nerves (T_9–T_{12})
Abdominal region				
External oblique	External and inferior borders of ribs 5–12	External oblique aponeuroses extending to linea alba and iliac crest	Compresses abdomen; depresses ribs; flexes, laterally flexes, or rotates vertebral column to the opposite side	Intercostal nerves 5–12, iliohypogastric, and ilioinguinal nerves
Internal oblique	Thoracolumbar fascia and iliac crest	Inferior surfaces of ribs 9–12, costal cartilages 8–10, linea alba, and pubis	As above, but rotates vertebral column to same side	As above
Transversus abdominis	Cartilages of ribs 6–12, iliac crest, and thoracolumbar fascia	Linea alba and pubis	Compresses abdomen	As above
Rectus Group				
Cervical region	*Includes the geniohyoid, oamohyoid, sternohyoid, sternothyroid, and thyrohyoid muscles in Table 11.6*			
Thoracic region				
Diaphragm	Xiphoid process, ribs 7–12 and associated costal cartilages, and anterior surfaces of lumbar vertebrae	Central tendinous sheet	Contraction expands thoracic cavity, compresses abdominopelvic cavity	Phrenic nerves (C_3–C_5)
Abdominal region				
Rectus abdominis	Superior surface of pubis around symphysis	Inferior surfaces of cartilages (ribs 5–7) and xiphoid process of sternum	Depresses ribs, flexes vertebral column and compresses abdomen	Intercostal nerves (T_7–T_{12})

The serratus posterior muscles insert on the ribs and assist the intercostal muscles to move the rib cage. The **superior serratus posterior muscle** elevates the ribs, and the **inferior serratus posterior muscle** (Figure 11.9b) pulls the rib inferiorly and opposes the diaphragm.

The abdomen has layers of oblique and rectus muscles that are organized in crossing layers, much like the laminar structure of a sheet of plywood (Figures 11.9 and 11.10). On the lateral abdominal wall is the thin, membranous **external oblique muscle**. This muscle originates on the external and inferior borders of ribs

EXERCISE 11 Axial Muscles 181

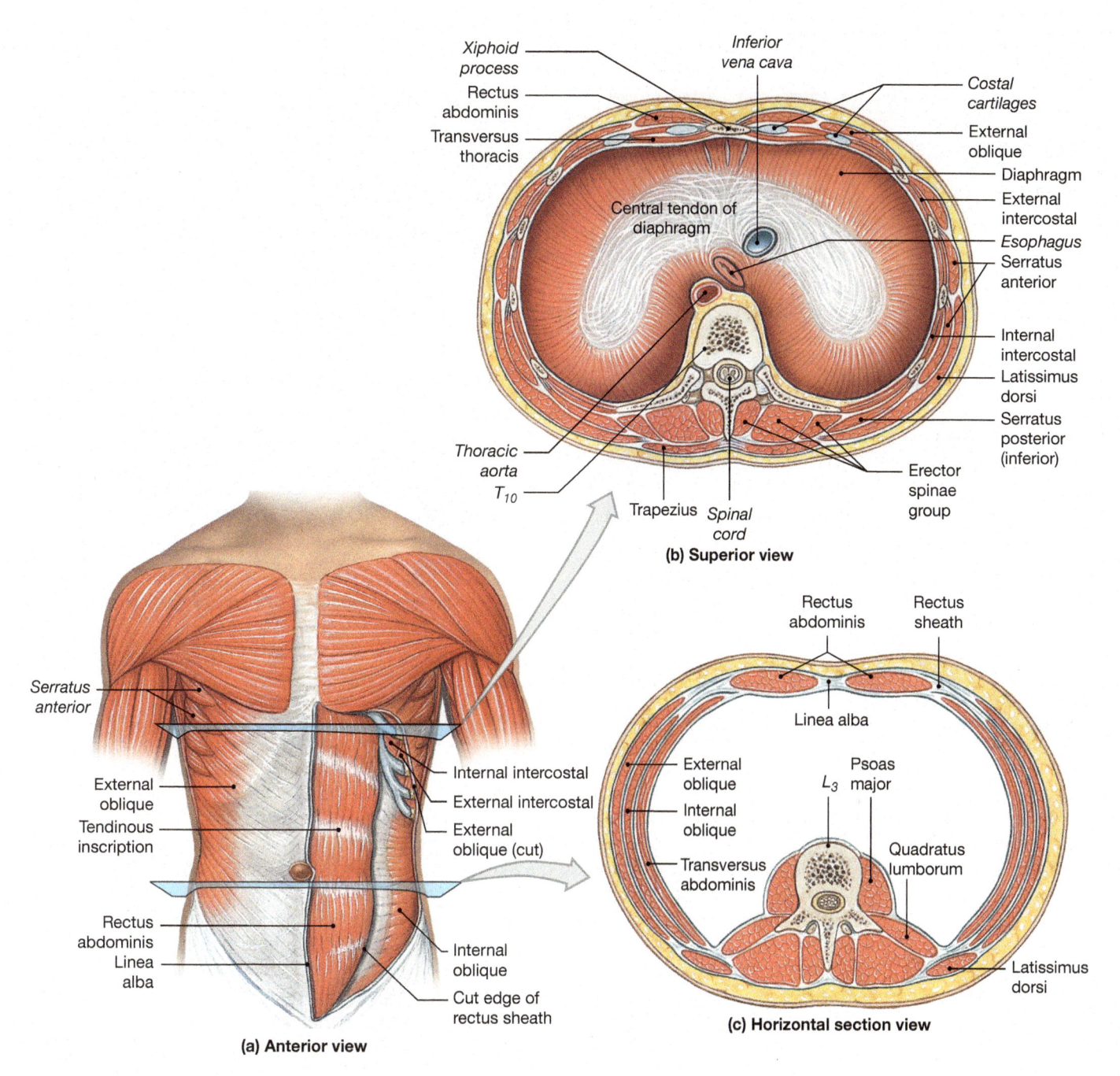

Figure 11.9 Oblique and Rectus Muscles and the Diaphragm
Oblique muscles compress underlying structures between the vertebral column and the ventral midline; rectus muscles are flexors of the vertebral column. **(a)** Anterior view of the trunk, showing superficial and deep muscles of the oblique and rectus groups. **(b)** The diaphragm, shown in superior view, is a muscular sheet that separates the thoracic cavity from the abdominopelvic cavity. **(c)** Horizontal section through the abdominal region near the level of the umbilicus.

5–12 and inserts on the external oblique aponeurosis that extends to the iliac crest and to a midsagittal fibrous line called the **linea alba**. The **internal oblique muscle** lies deep and at a right angle to the external oblique muscle. Both the external and internal oblique muscles compress and flex the abdomen, depress the ribs, and rotate the vertebral column.

The **transversus abdominis muscle** is located deep to the internal oblique muscle. It contracts with the other abdominal muscles to compress the abdomen.

182 EXERCISE 11 Axial Muscles

(a) Anterior view, superficial

(b) Deep dissection of the anterior abdominal wall

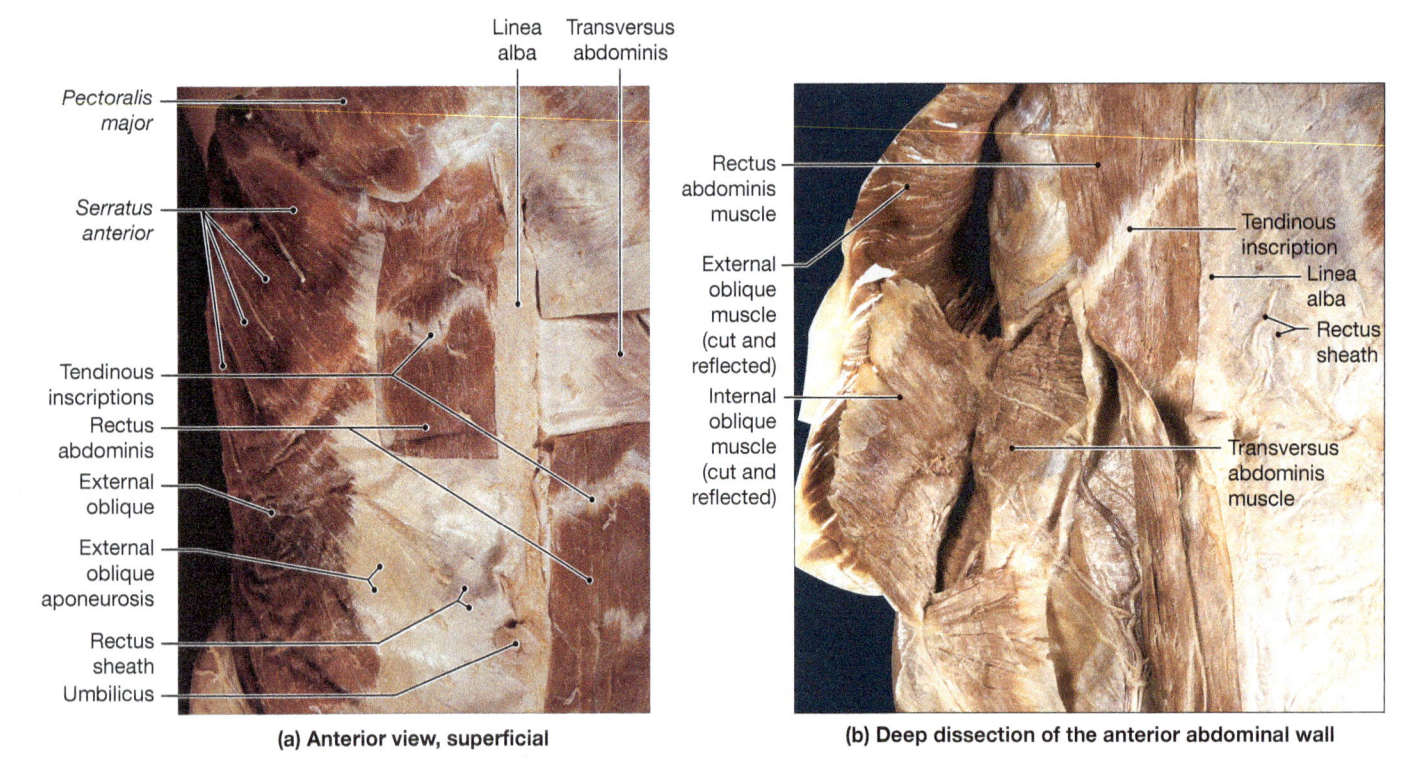

Figure 11.10 Dissection View of Muscles of the Trunk
(a) Cadaver, anterior superficial view of the abdominal wall. **(b)** Deep dissection of the anterior abdominal wall.

Rectus Muscles

Rectus muscles are found in the cervical, thoracic, and abdominal regions of the body. Those of the cervical region are the suprahyoid and infrahyoid muscles which were presented in Lab Activity 5 of this exercise.

The **diaphragm** is a sheet of muscle that forms the thoracic floor and separates the thoracic cavity from the abdominopelvic cavity (Figure 11.9). The diaphragm originates at many points along its edges, and the muscle fibers meet at a central tendon. Contracting the diaphragm to expand the thoracic cavity is the muscular process by which air is inhaled into the lungs.

The **rectus abdominis muscle** is the vertical muscle along the midline of the abdomen between the pubic symphysis and the xiphoid process of the sternum. This muscle is divided by the linea alba. A well-developed rectus abdominis muscle has a washboard appearance because transverse bands of collagen called **tendinous inscriptions** separate the muscle into many segments. Contraction of the rectus abdominis flexes and compresses the vertebral column and depresses the ribs for forced exhalation that occurs during exercise.

Making Connections

Fiber Orientation

Find the external oblique and internal oblique muscles on a muscle model, and notice the difference in the way the muscle fibers are oriented. The fibers of the external oblique muscle flare laterally as they are traced from bottom to top, whereas those of the internal oblique muscle are directed medially. This tip is also useful in examining the external and internal intercostal muscles between the ribs. By the way, the intercostal muscles of beef and pork are the barbecue "ribs" that you might enjoy. ●

QuickCheck Questions

7.1 What are all the muscles of the abdomen wall and their characteristics?

7.2 Why is the rectus abdominis muscle nicknamed the "six-pack"?

7.3 How do the intercostal muscles move the thoracic cage during respiration?

7.4 Where is the diaphragm located?

7.5 What is the basic difference between muscles classified as oblique and those classified as rectus?

7 Materials

- ☐ Head and neck model
- ☐ Torso model
- ☐ Muscle chart

Procedures

1. Review the oblique and rectus muscles in Figures 11.9 and 11.10, and Table 11.8.
2. Examine the head and neck model and distinguish among each muscle of the scalene group.
3. Locate the intercostal muscles on the torso model and note differences in the orientation of the fibers of each muscle.
4. Identify each abdominal muscle on the torso model and/or on the muscle chart.
5. Locate the general position of each oblique and rectus muscles on the torso model. ■

LAB ACTIVITY 8 Muscles of the Pelvic Region

The pelvic floor and wall form a bowl that supports the organs of the reproductive and digestive systems. The floor mainly consists of the **coccygeus muscle** and the **levator ani muscle**, peritoneal muscles, and muscles associated with the reproductive organs (Figure 11.11 and Table 11.9). The coccygeus muscle originates on the ischial spine, passes posteriorly, and inserts on the lateral and inferior borders of the sacrum. The levator ani muscle, which is anterior to the coccygeus muscle, originates on the inside edge of the pubis and the ischial spine and inserts on the coccyx.

These two muscles together form the muscle group called the **pelvic diaphragm**. The action of this group is to flex the coccyx muscle and tense the pelvic floor. During pregnancy, the expanding uterus bears down on the pelvic floor, and the pelvic diaphragm supports the weight of the fetus.

The **external anal sphincter** originates on the coccyx and inserts around the anal opening. This muscle closes the anus and is consciously relaxed for defecation. After the external anal sphincter depresses and protrudes during defecation, the levator ani muscle elevates and retracts the anus.

QuickCheck Questions

8.1 What are the muscles of the pelvic floor?

8.2 Which muscle surrounds the anus?

8 Materials

- ☐ Torso model
- ☐ Muscle chart

Procedures

1. Review the muscles of the pelvic region in Figure 11.11.
2. Locate each muscle of the pelvic region on the torso model and/or muscle chart. ■

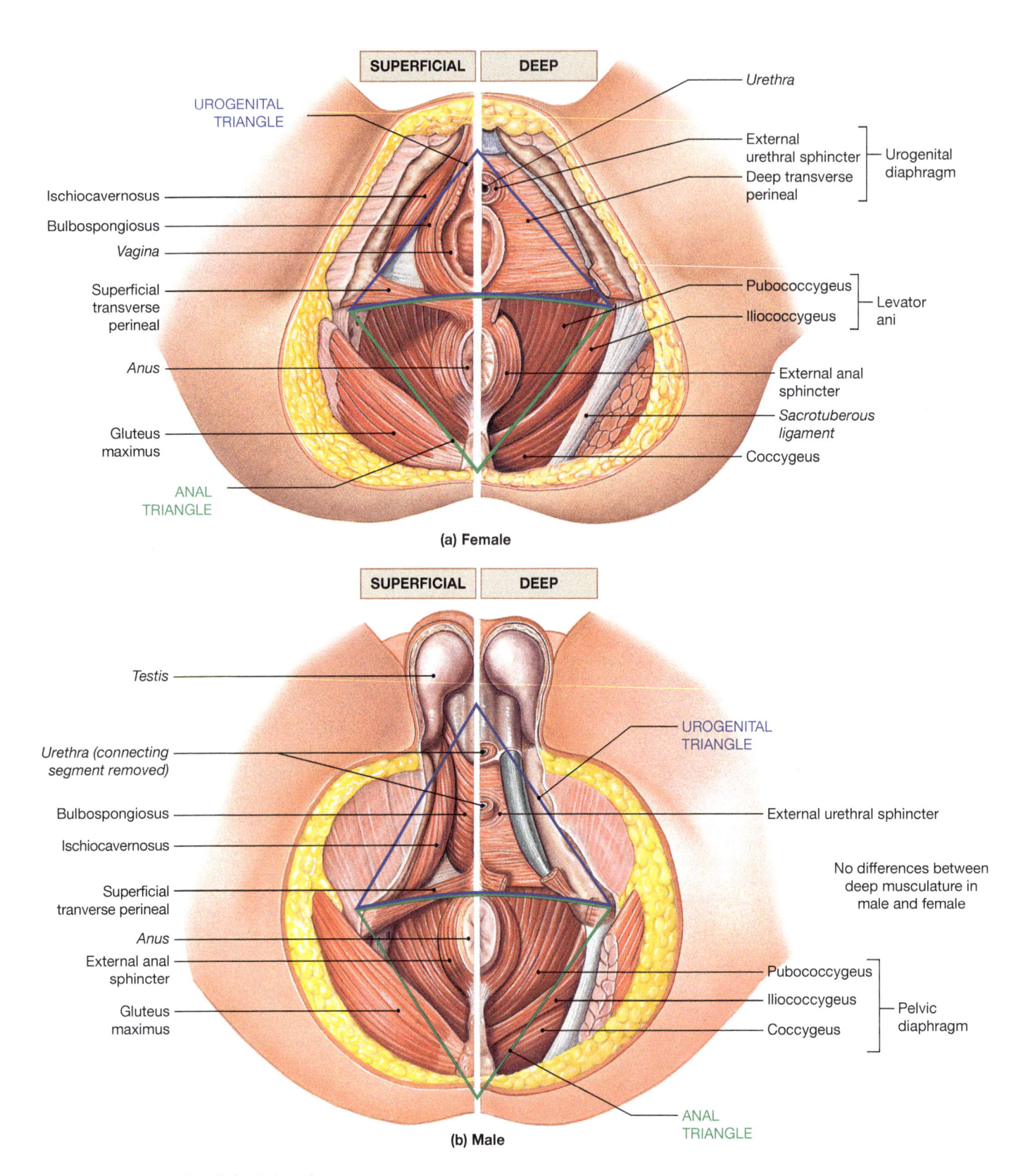

Figure 11.11 Muscles of the Pelvic Floor
The muscles of the pelvic floor form the urogenital triangle and the anal triangle. They support organs of the pelvic cavity, flex the sacrum and coccygx, and control material movement through the urethra and anus. (a) Inferior view, female. (b) Inferior view, male.

184

Table 11.9 *Origins and Insertions* Muscles of the Pelvic Floor

Group/Muscle	Origin	Insertion	Action	Innervation
Urogenital Triangle				
Superficial Muscles				
Bulbospongiosus				
Male	Collagen sheath at base of penis; fibers cross over urethra	Median raphe and central tendon of perineum	Compresses base, stiffens penis, ejects urine or semen	Pudendal nerve, perineal branch (S_2–S_4)
Female	Collagen sheath at base of clitoris; fibers run on either side of urethral and vaginal openings	Central tendon of perineum	Compresses and stiffens clitoris, narrows vaginal opening	As above
Ischiocavernosus	Ramus and tuberosity of ischium	Symphysis pubis anterior to base of penis or clitoris	Compresses and stiffens penis or clitoris, helping to maintain erection	As above
Superficial transverse perineal	Ischial ramus	Central tendon of perineum	Stabilizes central tendon of perineum	As above
Deep muscles				
Deep transverse perineal	Ischial ramus	Median raphe of urogenital diaphragm	As above	As above
External urethral sphincter				
Male	Ischial and pubic rami	To median raphe at base of penis; inner fibers encircle urethra	Closes urethra; compresses prostate and bulbourethral glands	As above
Female	Ischial and pubic rami	To median raphe; inner fibers encircle urethra	Closes urethra; compresses vagina and greater vestibular glands	As above
Anal Triangle				
Coccygeus	Ischial spine	Lateral, inferior borders of the sacrum and coccyx	Flexes coccygeal joints; elevates and supports pelvic floor	Inferior sacral nerves (S_4–S_5)
Levator ani				
Iliococcygeus	Ischial spine, pubis	Coccyx and median raphe	Tenses floor of pelvis, supports pelvic organs, flexes coccygeal joints, elevates and retracts anus	Pudendal nerve (S_2–S_4)
Pubococcygeus	Inner margins of pubis	As above	As above	As above
External anal sphincter	Via tendon from coccyx	Encircles anal opening	Closes anal opening	Pudendal nerve; hemorrhoidal branch (S_2–S_4)

LAB REPORT

EXERCISE 11

Axial Muscles

Name _____
Date _____
Section _____

A. Matching

Match each term in the left column with its correct description from the right column.

_____ 1. orbicularis oculi muscle
_____ 2. splenius capitis muscle
_____ 3. zygomaticus minor muscle
_____ 4. external intercostal muscle
_____ 5. occipital belly of occipitofrontalis muscle
_____ 6. stylohyoid muscle
_____ 7. linea alba
_____ 8. corrugator supercilii muscle
_____ 9. risorius muscle
_____ 10. buccinator muscle
_____ 11. digastric muscle
_____ 12. rectus abdominis muscle
_____ 13. external oblique muscle
_____ 14. masseter muscle
_____ 15. erector spinae
_____ 16. platysma muscle

A. retracts scalp
B. elevates mandible
C. small muscle used in smiling
D. elevates larynx
E. one of two superficial neck muscles
F. compresses cheeks
G. closes eye
H. wrinkles forehead
I. tenses angle of mouth laterally
J. thin muscle that covers sides of neck, depresses jaw
K. superficial lateral muscle of abdomen
L. vertical abdominal muscle
M. found between ribs; elevates rib cage
N. muscle group that extends vertebral column
O. extends neck
P. fibrous line located along midline of trunk

187

EXERCISE 11   LAB REPORT

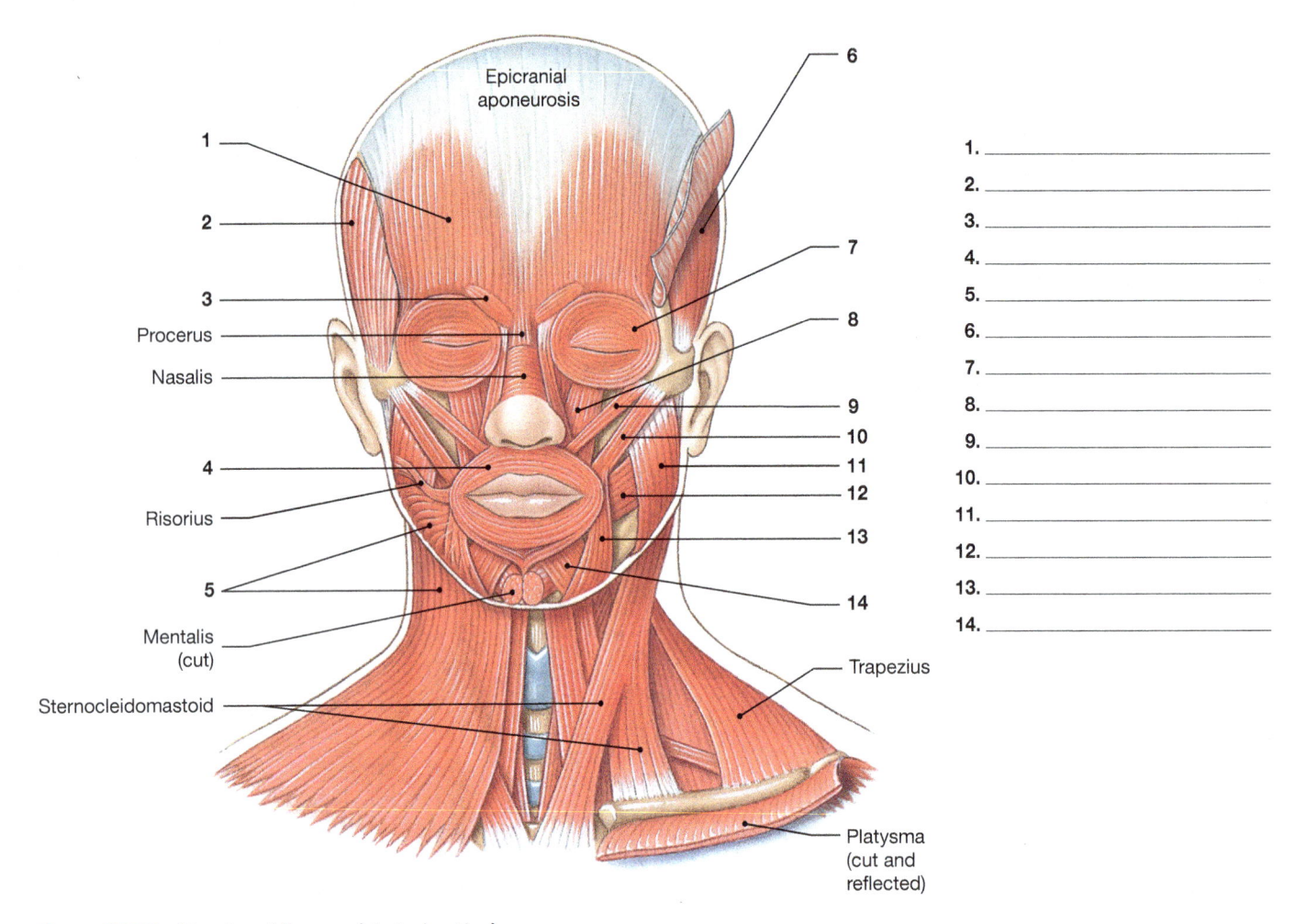

Figure 11.12 Muscles of Face and Anterior Neck

1. _____
2. _____
3. _____
4. _____
5. _____
6. _____
7. _____
8. _____
9. _____
10. _____
11. _____
12. _____
13. _____
14. _____

B. **Labeling**

Label the muscles of the head in Figure 11.12.

C. **Short-Answer Questions**

Describe the location of each of these muscles:

1. Nasalis muscle

2. Masseter muscle

188

LAB REPORT EXERCISE 11

3. Sternocleidomastoid muscle

4. Omohyoid muscle

5. Sternohyoid muscle

6. Styloglossus muscle

7. Multifidus muscles

8. Quadratus lumborum

9. Scalene group of muscles

10. Coccygeus muscle

EXERCISE 11

LAB REPORT

D. Analysis and Application

1. Describe the movement produced by each extrinsic tongue muscle.

2. Explain how the muscles of the tongue and anterior neck are named.

3. Describe the actions of the digastric muscle.

4. The anterior abdominal wall lacks bone. Given this fact, on what structure do the abdominal muscles insert?

5. Which muscle groups work as synergists to extend the vertebral column and neck?

6. Describe the muscles involved in smiling and grimacing.

EXERCISE 12

Appendicular Muscles

OBJECTIVES

On completion of this exercise, you should be able to:

- Locate major muscles of the pectoral girdle and describe each muscle's origin, insertion, and action.
- Locate major muscles of the upper limb and describe each muscle's origin, insertion, and action.
- Locate major muscles of the pelvic girdle and describe each muscle's origin, insertion, and action.
- Locate major muscles of the lower limb and describe each muscle's origin, insertion, and action.
- Describe the muscle compartments of the upper and lower limbs.

LAB ACTIVITIES

1. Muscles That Move the Pectoral Girdle 191
2. Muscles That Move the Arm 195
3. Muscles That Move the Forearm, Wrist, Hand, and Fingers 197
4. Muscles That Move the Thigh 204
5. Muscles That Move the Leg 208
6. Muscles That Move the Ankle, Foot, and Toes 213

   A Regional Look: Compartments 218

The appendicular musculature supports and moves the pectoral girdle and upper limb and the pelvic girdle and lower limb. Many of the muscles of the pectoral and pelvic girdles are on the body trunk but move appendicular bones. For example, the largest muscle that moves the arm, the latissimus dorsi, is located on the lower back.

LAB ACTIVITY 1 Muscles That Move the Pectoral Girdle

The muscles of the shoulder support and position the scapula and help maintain the articulation between the humerus and scapula. The shoulder joint is the most movable and least stable joint of the body, and many of the surrounding muscles help keep the humerus articulated in the scapula. Origin, insertion, action, and innervation for these muscles are detailed in Table 12.1.

The large, diamond-shaped muscle of the upper back is the **trapezius** (tra-PĒ-zē-us) **muscle**. It spans the gap between the scapulae and extends from the lower thoracic vertebrae to the back of the head (Figure 12.1). The superior portion of the trapezuis originates at three places: on the occipital bone; on the **ligamentum nuchae** (li-guh-MEN-tum NOO-kē; *nucha,* nape), which is a ligament that extends from the cervical vertebrae to the occipital bone; and on the spinous processes of thoracic vertebrae. It inserts on the clavicle and on the

192 EXERCISE 12 Appendicular Muscles

Table 12.1 *Origins and Insertions* Muscles That Position the Pectoral Girdle

Muscle	Origin	Insertion	Action	Innervation*
Levator scapulae	Transverse processes of first four cervical vertebrae	Vertebral border of scapula near superior angle	Elevates scapula	Cervical nerves C_3–C_4 and dorsal scapular nerve (C_5)
Pectoralis minor	Anterior-superior surfaces of ribs 3–5	Coracoid process of scapula	Depresses and protracts shoulder; rotates scapula so glenoid cavity moves inferiorly (downward rotation); elevates ribs if scapula is stationary	Medial pectoral nerve ($C_{8,1}$)
Rhomboid major	Spinous processes of superior thoracic vertebrae	Vertebral border of scapula from spine to inferior angle	Adducts scapula and performs downward rotation	Dorsal scapular nerve (C_5)
Rhomboid minor	Spinous processes of vertebrae C_7–T_1	Vertebral border of scapula near spine	As above	As above
Serratus anterior	Anterior and superior margins of ribs 1–8 or 1–9	Anterior surface of vertebral border of scapula	Protracts shoulder; rotates scapula so glenoid cavity moves superiorly (upward rotation)	Long thoracic nerve (C_5–C_7)
Subclavius	First rib	Clavicle (inferior border)	Depress and protracts shoulder	Nerve to subclavius (C_5–C_6)
Trapezius	Occipital bone, ligamentum nuchae, and spinous processes of thoracic vertebrae	Clavicle and scapula (acromion and scapula spine)	Depends on active region and state of other muscles; may (1) elevate, retract, depress, or rotate scapula upward, (2) elevate clavicle, or (3) extend neck	Accessory nerve (N XI) and cervical spinal nerves (C_3–C_4)

*Where appropriate, spinal nerves involved are given in parentheses.

acromion and scapular spine of the scapula. Because the trapezius has origins superior and inferior to its insertion, it may elevate, depress, retract, or rotate the scapula and/or the clavicle upward. The trapezius also can extend the neck. Deep to the trapezius are the **rhomboid major muscle** and **rhomboid minor muscle,** which extend between the upper thoracic vertebrae and the scapula. The rhomboid muscles adduct the scapula and rotate it downward. The **levator scapulae muscle** originates on cervical vertebrae 1–4 and inserts on the superior border of the scapula. As its name implies, it elevates the scapula.

On the anterior of the trunk, the **subclavius** (sub-KLĀ-vē-us) **muscle** is inferior to the clavicle (Figure 12.2). It arises from the first rib, inserts on the underside of the clavicle, and depresses and protracts the clavicle. The **serratus anterior muscle** appears as wedges on the side of the chest. This arrangement gives the muscle a sawtooth appearance similar to that of a bread knife with its *serrated* cutting edge. The muscle protracts the shoulder and rotates the scapula upward.

The **pectoralis minor muscle** is a deep muscle of the anterior trunk. Its origin is the anterior surfaces and superior margins of ribs 3–5, and it inserts on the coracoid process of the scapula (Figure 12.2). The function of this muscle is to pull the top of the scapula forward and depress the shoulders. It also elevates the ribs during forced inspiration, as during strenuous exercise.

QuickCheck Questions

1.1 What are the actions of the trapezius muscle?
1.2 What is the action of the rhomboid muscles?

EXERCISE 12 Appendicular Muscles   193

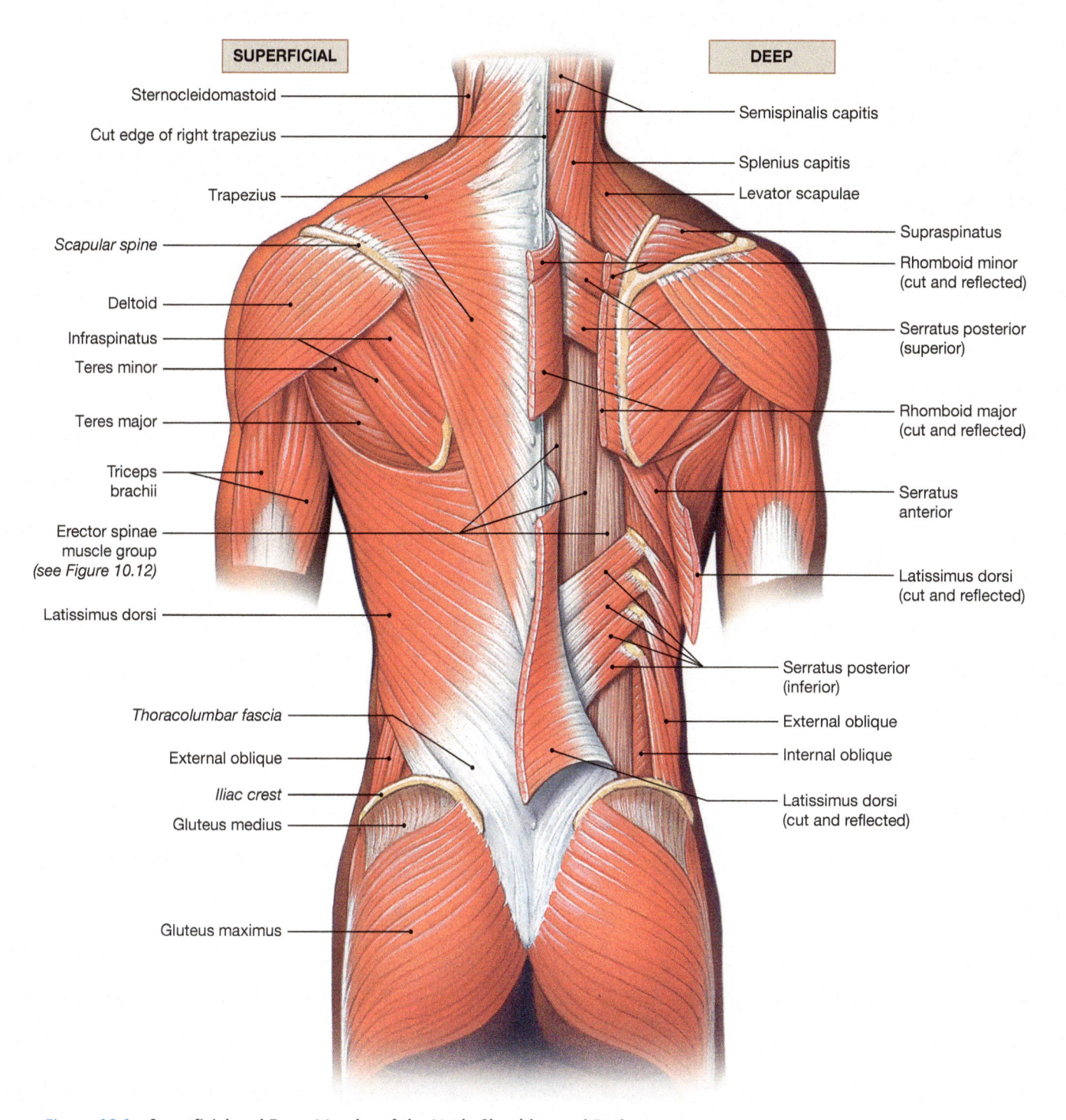

Figure 12.1 Superficial and Deep Muscles of the Neck, Shoulder, and Back
A posterior view of many of the important muscles of the neck, trunk, and proximal portion of the upper limbs.

1 Materials

- ☐ Torso model
- ☐ Articulated skeleton
- ☐ Muscle chart

Procedures

1. Review the posterior and anterior muscles of the chest in Figures 12.1 and 12.2.
2. Identify each muscle on the torso model and the muscle chart.
3. Examine an articulated skeleton and note the origin, insertion, and action of the major muscles that act on the shoulder.
4. Locate the position of these muscles on your body and practice each muscle's action. ■

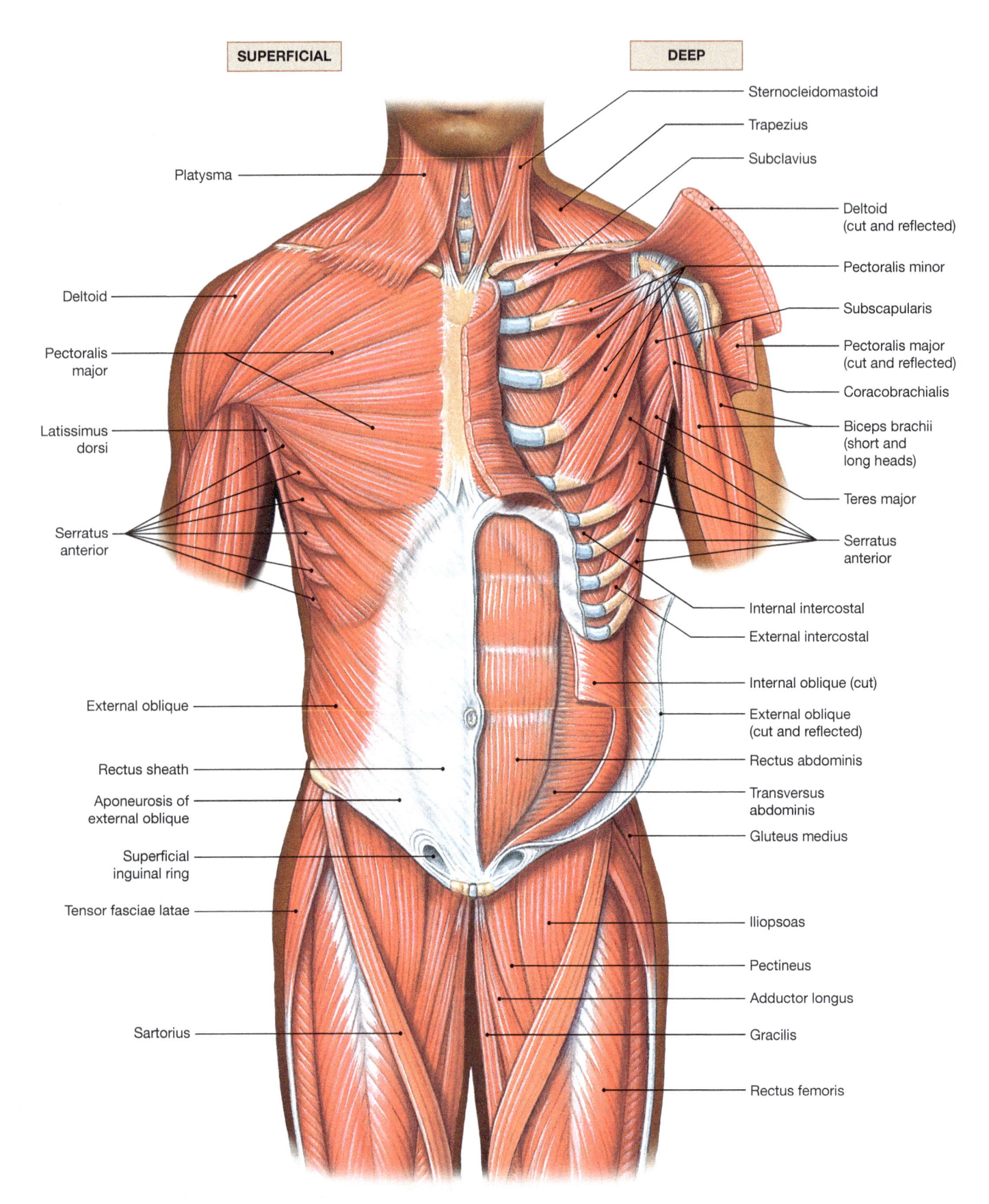

Figure 12.2 **Superficial and Deep Muscles of the Trunk and Proximal Limbs**
Anterior view of the axial muscles of the trunk and the appendicular musculature associated with the pectoral and pelvic girdles and the proximal portion of the upper and lower limbs.

LAB ACTIVITY 2 — **Muscles That Move the Arm**

Muscles that move the arm originate either on the scapula or on the vertebral column, span the ball-and-socket joint of the shoulder, and insert on the humerus to abduct, adduct, flex, or extend the arm. Refer to Table 12.2 for details on origin, insertion, action, and innervation for these muscles.

The **coracobrachialis** (KOR-uh-kō-brā-kē-A-lis) **muscle** is a small muscle that originates on the coracoid process of the scapula and adducts and flexes the shoulder (see Figure 12.2). The **pectoralis** (pek-to-RA-lis; *pectus*, the breast) **major muscle**, which covers most of the upper rib cage on the two sides of the chest, is one of the main muscles that move the arm. It originates on the clavicle, on the body of the sternum, and on costal cartilages for ribs 2–6 and inserts on the humerus at the greater tubercule and lateral surface of the intertubercular groove. This muscle flexes, adducts, and medially rotates the arm. In females, the breasts cover the inferior part of the pectoralis major muscle. Lateral to the pectoralis major muscle is the **deltoid muscle**, the triangular muscle of the shoulder. It originates on the anterior edge of the clavicle, on the inferior margins of the scapular spine, and on the acromion process of the scapula. The deltoid inserts on the deltoid tuberosity and is the major abductor of the humerus.

The **subscapularis muscle** is deep to the scapula next to the posterior surface of the rib cage (see Figure 12.2). It originates on the subscapular fossa, inserts on the lesser tubercle of the humerus, and medially rotates the shoulder.

Table 12.2 *Origins and Insertions* Muscles That Move the Arm

Muscle	Origin	Insertion	Action	Innervation*
Deltoid	Clavicle and scapula (acromion and adjacent scapular spine)	Deltoid tuberosity of humerus	*Whole muscle*: adbuction at shoulder; *anterior part*: flexion and medial rotation; *posterior part*: extension and lateral rotation	Axillary nerve (C_5–C_6)
Supraspinatus	Supraspinous fossa of scapula	Greater tubercle of humerus	Abduction at the shoulder	Suprascapular nerve (C_5)
Subscapularis	Subscapular fossa of scapula	Lesser tubercle of humberus	Medial rotation at shoulder	Subscapular (nerves (C_5–C_6)
Teres major	Inferior angle of scapula	Passes medially to reach the medical lip of intertubercular groove of humerus	Extension, adduction, and medial rotation at shoulder	Lower subscapular nerve (C_5–C_6)
Infraspinatus	Infraspinous fossa of scapula	Greater tubercle of humerus	Lateral rotation at shoulder	Suprascapular nerve (C_5–C_6)
Teres minor	Lateral border of scapula	Passes laterally to reach the greater tubercle of humerus	Lateral rotation at shoulder	Axillary nerve (C_5)
Coracobrachialis	Coracoid process	Medial margin of shaft of humerus	Adduction and flexion at shoulder	Musculocutaneous nerve (C_5–C_7)
Pectoralis major	Cartilages of ribs 2–6, body of sternum, and inferior, medial portion of clavicle	Crest of greater tubercle and lateral lip of intertubercular groove of humerus	Flexion, adduction, and medial rotation at shoulder	Pectoral nerves (C_5–T_1)
Latissimus dorsi	Spinous processes of inferior thoracic and all lumbar vertebrae, ribs 8–12, and lumbodorsal fascia	Floor of intertubercular groove of the humerus	Extension, adduction, and medial rotation at shoulder	Thoracodorsal nerve (C_6–C_8)
Triceps brachii (long head)	See Table 12.3			

*Where appropriate, spinal nerves involved are given in parentheses.

The **latissimus dorsi** (la-TIS-i-mus DOR-sē; *lati*, broad) **muscle** is the large muscle that wraps around the lower back (see Figure 12.1). This muscle has a broad origin from the sacral and lumbar vertebrae up to the sixth thoracic vertebra and sweeps up and inserts on the humerus. The latissimus dorsi muscle extends, adducts, and medially rotates the arm.

Making Connections — **Insertion Determines Action**

> Always notice whether a muscle passes in front of or behind the bone on which it inserts. This detail determines how the bone moves. For example, the latissimus dorsi passes in front of the humerus before inserting on the intertubercular groove. This allows medial rotation of the humerus. If the muscle passed behind the humerus, the resulting action would be lateral rotation.

The **supraspinatus muscle** originates on the supraspinous fossa, the depression that is located superior to the scapular spine (see Figure 12.1). It abducts the shoulder. The **infraspinatus muscle** arises from the infraspinous fossa of the scapula and inserts on the greater tubercule of the humerus to laterally rotate the humerus at the shoulder. The **teres major muscle** is a thick muscle that arises on the inferior angle of the posterior surface of the scapula. The muscle converges up and laterally into a flat tendon that ends on the anterior side of the humerus. On the lateral border of the scapula is the small and flat **teres minor muscle**. The teres major muscle extends, adducts, and medially rotates the humerus; the teres minor muscle laterally rotates the humerus at the shoulder.

Clinical Application — **Rotator Cuff Injuries**

> Four shoulder muscles—the supraspinatus, infraspinatus, teres minor, and subscapularis muscles—all act to seat the head of the humerus firmly in the glenoid fossa to prevent dislocation of the shoulder. Collectively these muscles are called the **rotator cuff**. Remember the acronym SITS (supraspinatus, infraspinatus, teres minor, subscapularis) for the rotator cuff muscles. Although it is part of the rotator cuff, the supraspinatus is not itself a rotator; rather, it is an abductor. You may be familiar with rotator cuff injuries if you are a baseball fan. The windup and throw of a pitcher involve circumduction of the humerus. This motion places tremendous stress on the shoulder joint and on the rotator cuff—stress that can cause premature degeneration of the joint. To protect the shoulder joint and muscles, bursal sacs are interspersed between the tendons of the rotator cuff muscles and the neighboring bony structures. Repeated friction on the bursae may result in an inflammation called *bursitis*.

QuickCheck Questions

2.1 Where are the two spinatus and two teres muscles located?

2.2 Which muscles adduct the arm?

2.3 Which muscle flexes the arm?

2 Materials

- ☐ Torso model
- ☐ Upper limb model
- ☐ Articulated skeleton
- ☐ Muscle chart

Procedures

1. Review the muscles that move the arm in Figures 12.1 and 12.2.
2. Locate each muscle that moves the arm on the torso model, upper limb model, and muscle chart.
3. Examine the articulated skeleton and note the origin, insertion, and action of the major muscles that act on the arm.
4. Locate the general position of each arm muscle on your body. Contract each muscle and observe how your arm moves.

EXERCISE 12 Appendicular Muscles

LAB ACTIVITY 3

Muscles That Move the Forearm, Wrist, Hand, and Fingers

Muscles that move the forearm arise on the humerus, span the elbow, and insert on the ulna and/or radius. The wrist is positioned by muscles from the forearm that insert on the carpal bones. Refer to Tables 12.3, 12.4, and 12.5 for details on the origin, insertion, action, and innervation for muscles that move the forearm, wrist, and hand.

The **biceps brachii muscle** (Figure 12.3) is the superficial muscle of the anterior brachium that flexes the forearm at the elbow. The term *biceps* refers to the presence of two origins, or "heads". The **short head** of the biceps brachii muscle begins on the coracoid process of the scapula as a tendon that expands into the muscle belly. The **long head** arises on the superior lip of the glenoid fossa at the supraglenoid tubercle. A tendon passes over the top of the humerus into the intertubercular groove and blends into the muscle. The tendon of the long head is enclosed in a protective covering called the *intertubercular synovial sheath*. The two heads of the biceps brachii muscle fuse and constitute most of the mass of the anterior brachium.

The **brachialis** (brā-kē-ā-lis) **muscle** also flexes the elbow. It is located under the distal end of the biceps brachii muscle. You can feel a small part of the brachialis muscle when you flex your arm and palpate the area just lateral to the tendon of the biceps brachii muscle.

The **triceps brachii muscle** on the posterior arm is the principal antagonist to the biceps brachii and brachialis muscles (Figure 12.4) and extends the elbow. The muscle arises from three heads—called the **long**, **lateral**, and **medial heads**—that merge into a common tendon that begins near the middle of the muscle and inserts on the olecranon process of the ulna. At the posterior lateral humerus is the small **anconeus** (an-kō-nē-us) **muscle,** which assists the triceps brachii muscle to extend the elbow.

The flexor muscles that move the wrist and hand are on the anterior forearm. At the wrist, the long tendons of the flexors are supported and stabilized by a wide sheath called the **flexor retinaculum** (ret-i-NAK-ū-lum; Figure 12.5).

The superficial **brachioradialis muscle** is easily felt on the lateral side of the anterior surface of your forearm (see Figure 12.5). This muscle is a good anatomical landmark and divides the flexor and extensor muscles of the forearm. Medial to the brachioradialis is the **flexor carpi radialis muscle**, the flexor muscle that is closest to the radius. The fibers of this muscle blend into a long tendon that inserts on the second and third metacarpals. The **palmaris longus muscle,** which is medial to the flexor carpi radialis muscle, is continuous with the palmar fascia. Medial to the palmaris longus muscle is the **flexor carpi ulnaris muscle**. This muscle rests on the ulnar side of the forearm and inserts on the pisiform and hamate bones of the carpus and on the base of metacarpal IV. The **flexor digitorum superficialis muscle** is located deep to the superficial flexors of the hand. It has four tendons that insert on the midlateral surface of the middle phalanges of fingers 2 through 5. Deeper flexors are also shown Figure 12.5.

Clinical Application

Carpal Tunnel Syndrome

The tendons of the flexor digitorum superficialis muscle pass through a narrow valley, the *carpal tunnel*, that is bounded by carpal bones. A protective synovial sheath lubricates the tendons in the tunnel, but repeated flexing of the hand and fingers, such as with prolonged typing or piano playing, causes the sheath to swell and compress the median nerve. Pain and numbness occur in the palm during flexion, a condition called *carpal tunnel syndrome*.

The **pronator teres muscle** is a thin muscle that is inferior to the elbow and medial to the brachioradialis muscle (see Figure 12.5). Proximal to the wrist joint is the **pronator quadratus muscle,** which is on the anterior surface of the forearm. This muscle acts as a synergist to the pronator teres muscle in pronating the forearm

198 EXERCISE 12 Appendicular Muscles

(a) Superficial muscles, anterior view

(b) Anterior view

(c) Anterior view of forearm, deep muscles

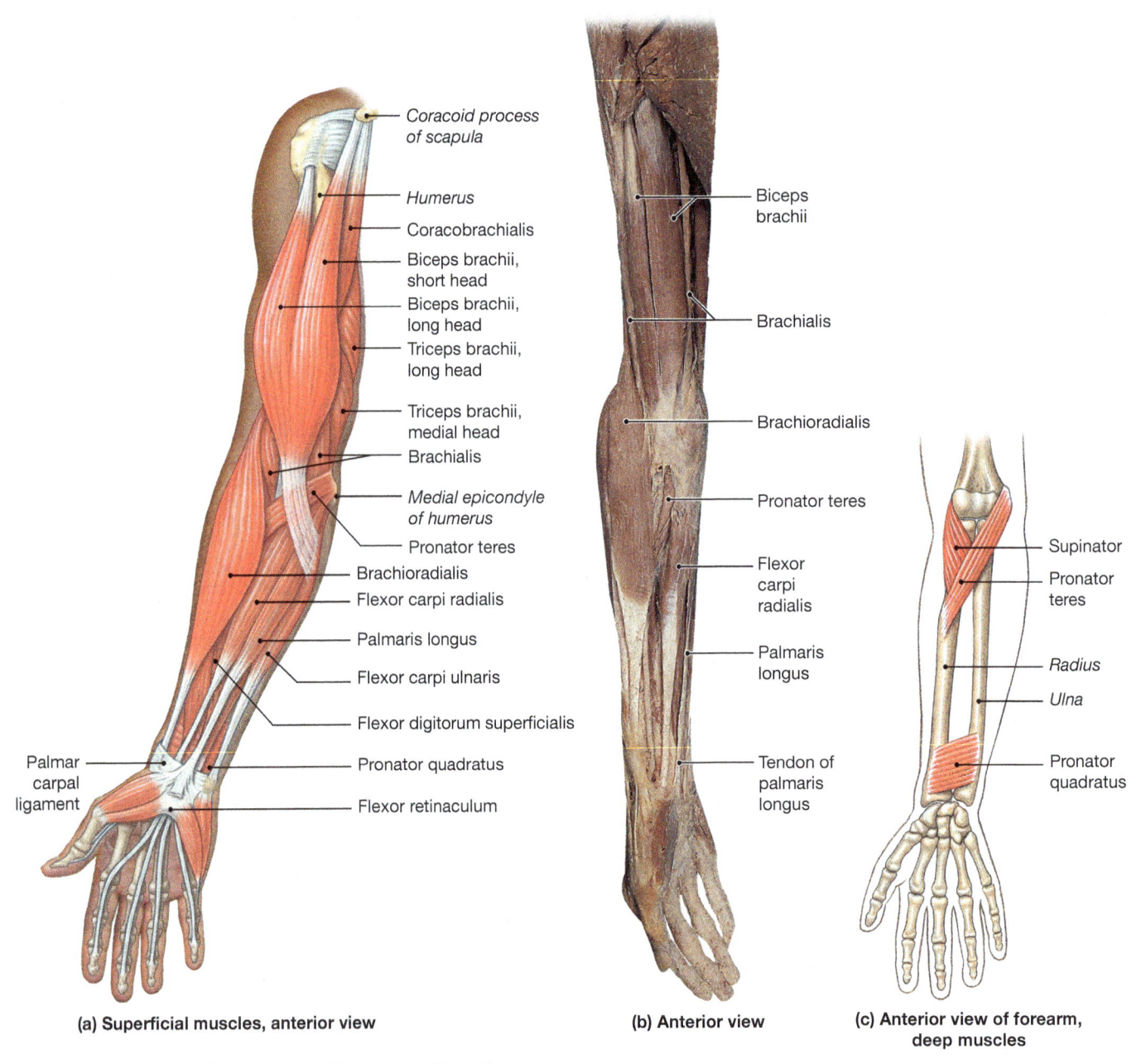

Figure 12.3 Muscles That Move the Forearm and Hand
Relationships among the muscles of the right upper limb are shown. **(a)** Superficial muscles, anterior view. **(b)** Anterior view of dissection. The palmaris longus and flexor carpi muscles (radialis and ulnaris) have been partially removed, and the flexor retinaculum has been cut. **(c)** Anterior view of the deep muscles of the supinated forearm.

and can also cause medial rotation of the forearm. The **supinator muscle** is found on the lateral side of the forearm deep to the brachioradialis muscle. It contracts and rotates the radius into a position parallel to the ulna, which results in supination of the forearm.

Making Connections   Deducing Action

Rather than memorizing the action for each muscle, use the position of the muscle over the joint to determine the action. Muscles on the lateral side of a joint, such as the deltoid muscle, are abductors and move the body away from the midline. Adductor muscles, such as the pectoralis major muscle, are located on the medial

Table 12.3 *Origins and Insertions* Muscles That Move the Forearm, Wrist, and Hand

Muscle	Origin	Insertion	Action	Innervation*
Action at the Elbow				
Flexors				
Biceps brachii	Short head from the coracoid process; long head from the supraglenoid tubercle (both on the scapula)	Tuberosity of radius	Flexion at elbow and shoulder; supination	Musculocutaneous nerve (C_5–C_6)
Brachialis	Anterior, distal surface of humerus	Tuberosity of ulna	Flexion at elbow	As above and radial nerve (C_7–C_8)
Brachioradialis	Ridge superior to the lateral epicondyle of humerus	Lateral aspect of styloid process of radius	As above	Radial nerve (C_5–C_6)
Extensors				
Anconeus	Posterior, inferior surface of lateral epicondyle of humerus	Lateral margin of olecranon on ulna	Extension at elbow	Radial nerve (C_7–C_8)
Triceps brachii				
lateral head	Superior, lateral margin of humerus	Olecranon of ulna	As above	Radial nerve (C_6–C_8)
long head	Infraglenoid tubercle of scapula	As above	As above, plus extension and adduction at the shoulder	As above
medial head	Posterior surface of humerus inferior to radial groove	As above	Extension at elbow	As above
Pronators/Supinators				
Pronator quadratus	Anterior and medial surfaces of distal portion of ulna	Anterolateral surface of distal portion of radius	Pronation	Median nerve (C_8–T_1)
Pronator teres	Medial epicondyle of humerus and coronoid process of ulna	Midlateral surface of radius	As above	Median nerve (C_6–C_7)
Supinator	Lateral epicondyle of humerus, annular ligament, and ridge near radial notch of ulna	Anterolateral surface of radius distal to the radial tuberosity	Supination	Deep radial nerve (C_6–C_8)
Action at the Hand				
Flexors				
Flexor carpi radialis	Medial epicondyle of humerus	Bases of second and third metacarpal bones	Flexion and abduction at wrist	Median nerve (C_6–C_7)
Flexor carpi ulnaris	Medial epicondyle of humerus; adjacent medial surface of olecranon and anteromedial portion of ulna	Pisiform bone, hamate bone, and base of fifth metacarpal bone	Flexion and adduction at wrist	Ulnar nerve (C_8–T_1)
Palmaris longus	Medial epicondyle of humerus	Palmar aponeurosis and flexor retinaculum	Flexion at wrist	Median nerve (C_6–C_7)
Extensors				
Extensor carpi radialis longus	Lateral supracondylar ridge of humerus	Bases of second metacarpal bone	Extension and abduction at wrist	Radial nerve (C_6–C_7)
Extensor carpi radialis brevis	Lateral epicondyle of humerus	Bases of third metacarpal bone	As above	As above
Extensor carpi ulnaris	Lateral epicondyle of humerus; adjacent dorsal surface of ulna	Bases of fifth metacarpal bone	Extension and adduction at wrist	Deep radial nerve (C_6–C_8)

*Where appropriate, spinal nerves involved are given in parentheses.

200 EXERCISE 12 Appendicular Muscles

Figure 12.4 Muscles That Move the Forearm, Wrist, and Hand—Another View
Relationships among the muscles of the right upper limb are shown. **(a)** Superficial muscles, posterior view. **(b)** Posterior view of a superficial dissection of the forearm. **(c)** Relationships among deeper muscles are best seen in this sectional view.

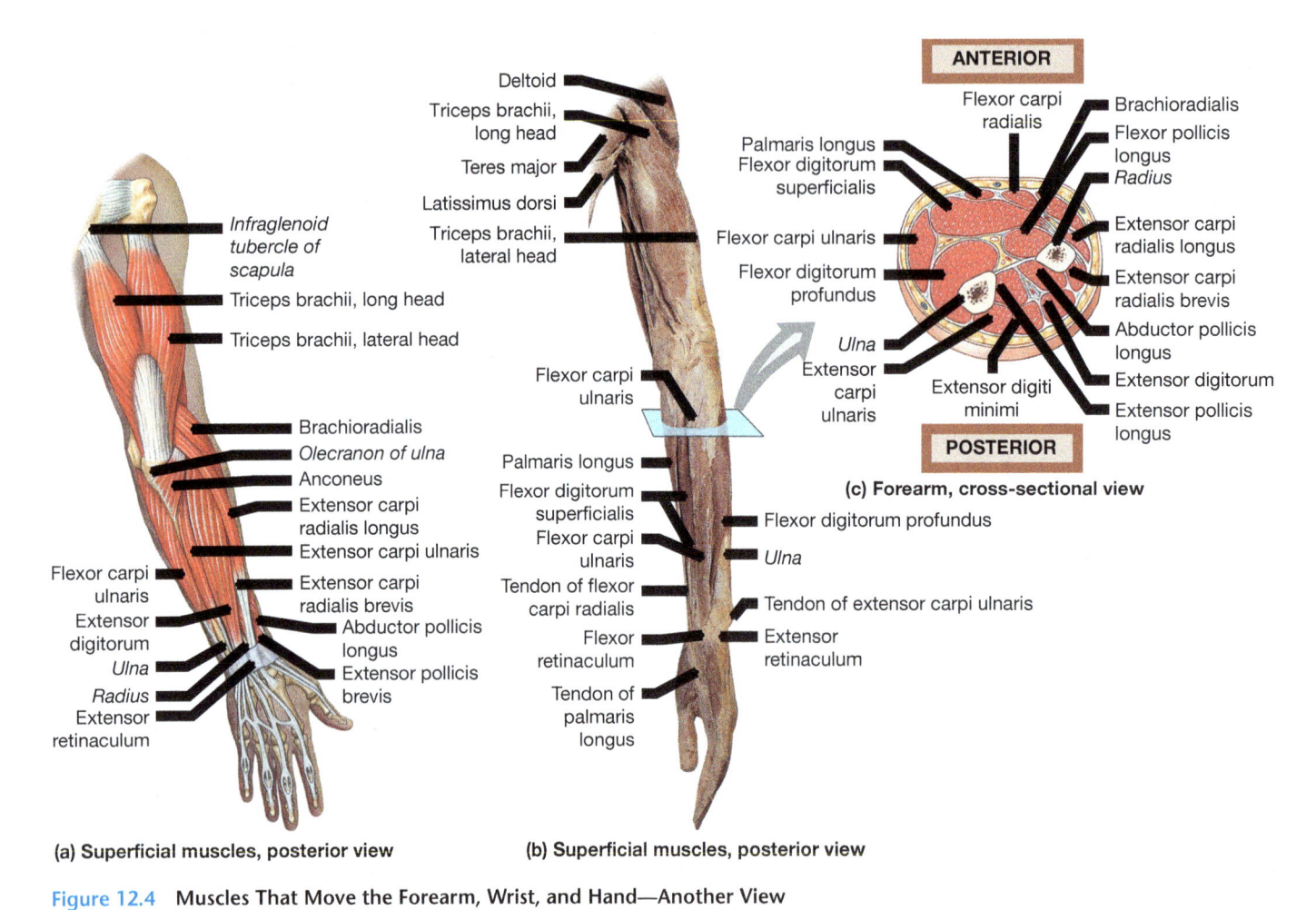

side of a joint and move the body toward the midline. Flexor muscles are positioned on the inner surface of a joint and face either anteriorly or posteriorly. Extensor muscles are found on the outer surface of these joints. For example, the biceps brachii muscle spans the anterior of the elbow joint and flexes the elbow. •

The extensor muscles of the hand and fingers are located on the posterior forearm. All except one of these muscles originate from a common tendon that is attached to the lateral epicondyle of the humerus. Tendons of the extensor muscles are secured by a transverse fibrous band across the posterior aspect of the wrist that is called the **extensor retinaculum** (see Figure 12.5).

Posterior to the brachioradialis muscle, the long **extensor carpi radialis longus muscle** is the only extensor that does not originate on a tendon that is attached to the humerus lateral epicondyle. Instead, it arises from the humerus just proximal to the lateral epicondyle, although a few fibers do extend from the common tendon. Inferior to the longus muscle is the **extensor carpi radialis brevis muscle**. The carpi muscles extend and abduct the wrist. The **extensor digitorum muscle** is medial to the extensor carpi radialis muscles. It has three or four tendons that insert on the posterior surface of the phalanges of fingers 2 through 5. Lateral to the digitorum muscle is the **extensor carpi ulnaris muscle**. Deeper extensor muscles are shown in Figure 12.5.

Table 12.4 *Origins and Insertions* Muscles That Move the Wrist, Hand, and Fingers

Muscle	Origin	Insertion	Action	Innervation*
Abductor pollicis longus	Proximal dorsal surfaces of ulna and radius	Lateral margin of first metacarpal bone	Abduction at joints of thumb and wrist	Deep radial nerve (C_6–C_7)
Extensor digitorum	Lateral epicondyle of humerus	Posterior surfaces of the phalanges, fingers 2–5	Extension at finger joints and wrist	Deep radial nerve (C_6–C_8)
Extensor pollicis brevis	Shaft of radius distal to origin of adductor pollicis longus	Base of proximal phalanx of thumb	Extension at joints of thumb; abduction at wrist	Deep radial nerve (C_6–C_7)
Extensor pollicis longus	Posterior and lateral surfaces of ulna and interosseous membrane	Base of distal phalanx of thumb	As above	Deep radial nerve (C_6–C_8)
Extensor indicis	Posterior surface of ulna and interosseous membrane	Posterior surface of phalanges of index finger (2), with tendon of extensor digitorum	Extension and adduction at joints of index finger	As above
Extensor digiti minimi	Via extensor tendon to lateral epicondyle of humerus and from intermuscular septa	Posterior surface of proximal phalanx of little finger (5)	Extension at joints of little finger	As above
Flexor digitorum superficialis	Medial epicondyle of humerus; adjacent anterior surface of ulna and radius	Midlteral surfaces of middle phalanges of fingers 2–5	Flexion at proximal interphalangeal, metacarpophalangeal, and wrist joints	Median nerve (C_7–T_1)
Flexor digitorum profundus	Medial and posterior surfaces of ulna, medial surface of coronoid process, and interosseus membrane	Bases of distal phalanges of fingers 2–5	Flexion at distal interphalangeal joints and, to a lesser degree, proximal interphalangeal joints and wrist	Palmar interosseous nerve, from median nerve, and ulnar nerve (C_8–T_1)
Flexor pollicis longus	Anterior shaft of radius, interosseous membrane	Base of distal phalanx of thumb	Flexion at joints of thumb	Median nerve (C_8–T_1)

*Where appropriate, spinal nerves involved are given in parentheses.

The masses of tissue at the base of the thumb and along the medial margin of the hand are called **eminences**. See Tables 12.4 and 12.5 for details on the origins, insertions, and actions of these muscles. The **thenar eminence** of the thumb consists of several muscles (Figure 12.6). The most medial of the thenar muscles is the **flexor pollicis brevis muscle**, which flexes and adducts the thumb. Lateral to this flexor is the **abductor pollicis brevis muscle**, which abducts the thumb. The most lateral thenar muscle is the **opponens pollicis muscle**, which opposes the thumb toward the little finger.

The **adductor pollicis muscle** is often not considered part of the thenar eminence, because it is found just medial to the flexor pollicis brevis muscle and deep in the web of tissue between the thumb and palm. This muscle adducts the thumb and opposes the action of the abductor pollicis brevis muscle.

The **hypothenar eminence** is a fleshy mass on the medial side of the palm at the base of the little finger and consists of three muscles (see Figure 12.6). The most lateral is the **opponens digiti minimi muscle**, which opposes the little finger toward the thumb. Medial to this muscle is the **flexor digiti minimi brevis muscle**, which flexes the little finger. The most medial muscle of the hypothenar eminence is the **abductor digiti minimi muscle**, which abducts the little finger.

You should note that no muscles originate on the fingers. Instead, the phalanges of the fingers serve as insertion points for muscles whose origins are more proximal.

(a) Anterior view, superficial

(b) Anterior view, middle

(c) Anterior view, deep

(d) Posterior view, superficial

(e) Posterior view, middle

(f) Posterior view, deep

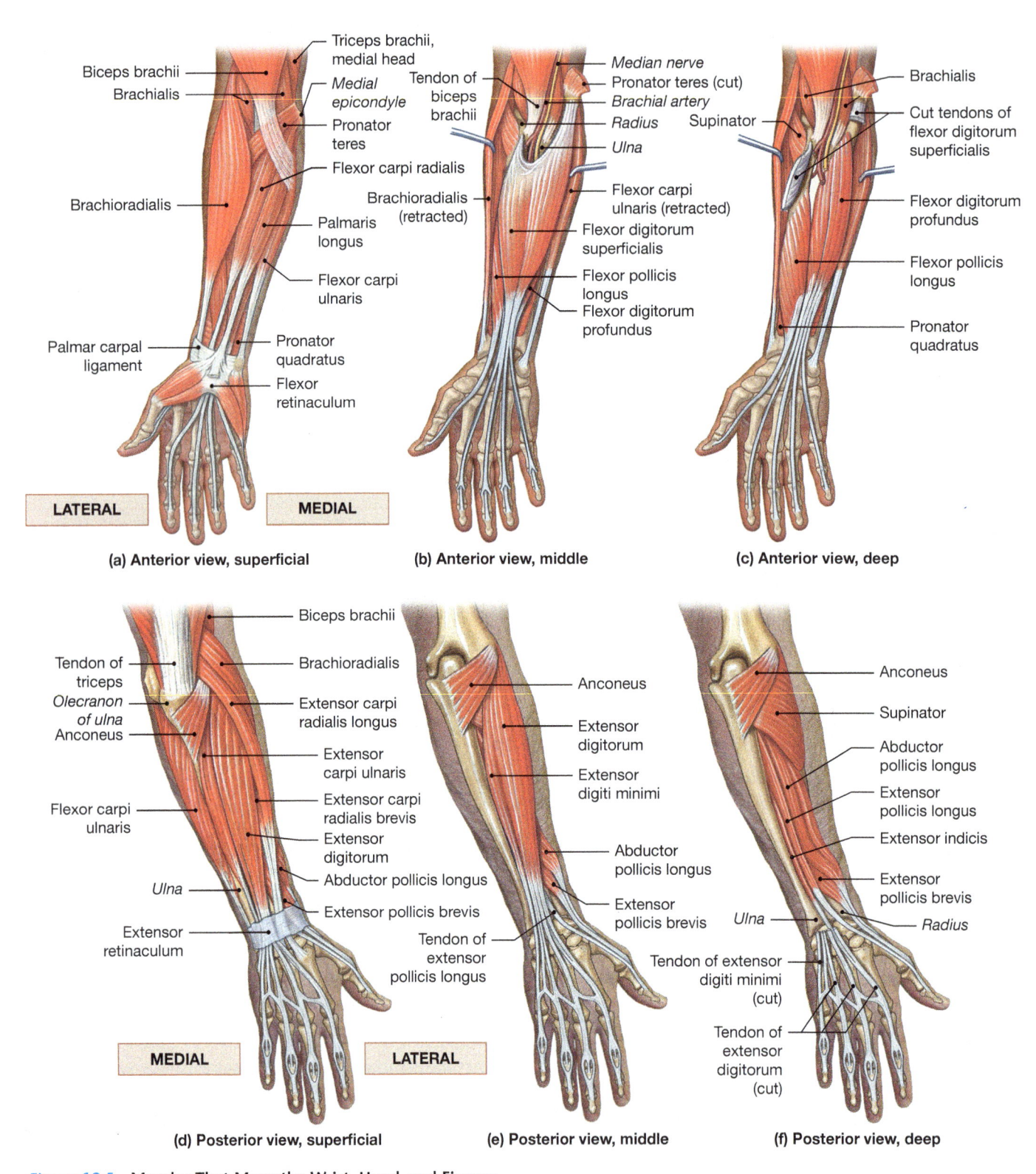

Figure 12.5 Muscles That Move the Wrist, Hand, and Fingers

Middle and deep muscle layers of the right forearm. (a) Anterior view, superficial layer, (b) Anterior view, middle layer. The flexor carpi radialis muscle and palmaris longus muscle have been removed. (c) Anterior view, deep layer. (d) Posterior view, superficial layer. (e) Posterior view, middle layer. (f) Posterior view, deep layer.

Table 12.5 *Origins and Insertions* Muscles That Move the Hand

Muscle	Origin	Insertion	Action	Innervation*
Adductor pollicis	Metacarpal and carpal bones	Proximal phalanx of thumb	Adduction of thumb	Ulnar nerve, deep branch (C_8–T_1)
Opponens pollicis	Trapezium and flexor retinaculum	First metacarpal bone	Opposition of thumb	Median nerve (C_6–C_7)
Palmaris brevis	Palmar aponeurosis	Skin of medial border of hand	Moves skin on medial border toward midline of palm	Ulnar nerve, superficial branch (C_8)
Abductor digiti minimi	Pisiform bone	Proximal phalanx of little finger	Abduction of little finger and flexion at its metacarpophalangeal joint	Ulner nerve, deep branch (C_8–T_1)
Abductor pollicis brevis	Transverse carpal ligament, scaphoid bone, and trapezium	Radial side of base of proximal phalanx of thumb	Abduction of thumb	Median nerve (C_6–C_7)
Flexor pollicis brevis	Flexor retinaculum, trapezium, capitate bone, and ulnar side of first metacarpal bone	Radial and ulnar sides of proximal phalanx of thumb	Flexion and adduction of thumb	Branches of median and ulnar nerves
Flexor digiti minimi brevis	Hamate bone	Proximal phalanx of little finger	Flexion at joints of little finger	Ulnar nerve, deep branch (C_8–T_1)
Opponens digiti minimi	As above	Fifth metacarpal bone	Opposition of fifth metacarpal bone	As above
Lumbrical (4)	Tendons of flexor digitorum profundus	Tendons of extensor digitorum to digits 2–5	Flexion at metacarpophalangeal joint 2–5; extension at proximal and distal interphalangeal joints, digits 2–5	No. 1 and no. 2 by median nerve; no. 3 and no. 4 by ulnar nerve, deep branch
Dorsal interosseus (4)	Each orginates from opposing faces of two metacarpal bones (I and II, II and III, III and IV, IV and V)	Bases of proximal phalanges of fingers 2–4	Abduction at metacarpophalangeal joints of fingers 2 and 4; flexion at metacarpophalangeal joints; extension at interphalangeal joints	Ulnar nerve, deep branch (C_8–T_1)
Palmar interosseus† (3–4)	Sides of metacarpal bones II, IV, and V	Bases of proximal phalanges of fingers 2, 4, and 5	Adduction at metacarpophalangeal joints of fingers 2, 4, and 5; flexion at metacarpophalangeal joints; extension at interphalangeal joints	As above

*Where appropriate, spinal nerves involved are given in parentheses.

†The deep, medial portion of the flexor pollicis brevis that originates on the first metacarpal bone is sometimes called the *first palmar interosseus muscle*; it inserts on the ulnar side of the phalanx and is innervated by the ulnar nerve.

QuickCheck Questions

3.1 Which muscles are antagonistic to the triceps brachii?

3.2 Which muscles are involved when you turn a doorknob?

3.3 What is the general action of the muscles on the posterior forearm?

3.4 What are the muscles of the thenar eminence?

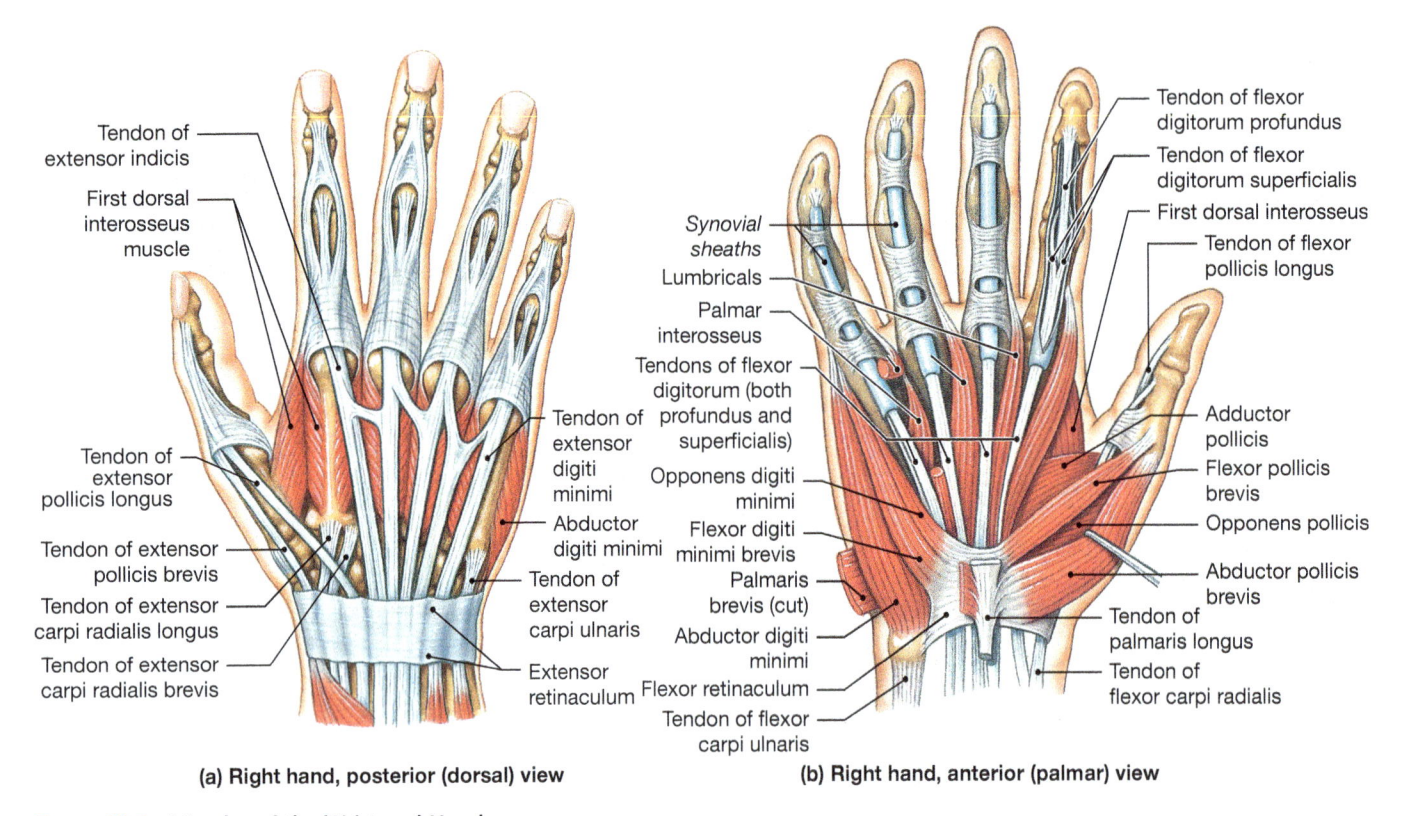

Figure 12.6 Muscles of the Wrist and Hand
Anatomy of the right wrist and hand. (a) Anterior (palmar) view. (b) Posterior view.

3 Materials

- Torso model
- Upper limb model
- Articulated skeleton
- Muscle chart

Procedures

1. Review the muscles of the forearm, wrist, and hand in Figures 12.3 through 12.6 and Tables 12.3, 12.4, and 12.5.
2. Identify each muscle on the torso and upper limb models and on the muscle chart.
3. Examine the articulated skeleton and note the origin, insertion, and action of the muscles that act on the forearm, wrist, and hand.
4. On your body, locate the general position of each muscle that is involved with movement of the forearm, wrist, and hand. Contract each muscle and observe the action of your upper limb. ■

LAB ACTIVITY 4 Muscles That Move the Thigh

Unlike the articulations between the axial skeleton and the pectoral girdle, which give this region great mobility, the articulations between the axial skeleton and the pelvic girdle limit movement of the hips. Axial muscles that move the pelvic girdle are discussed in Exercise 11. Muscles that move the thigh insert on the femur and cause movement at the ball-and-socket joint. These muscles are organized into four groups: gluteal, lateral rotator, adductor, and iliopsoas. Refer to Table 12.6 and Figure 12.7 for details on these muscles.

Gluteal Group

The posterior muscles that originate on the ilium of the pelvis are the three gluteal muscles that constitute the buttocks. The most superficial and prominent is the

Table 12.6 *Origins and Insertions* Muscles That Move the Thigh

Group and Muscle(s)	Origin	Insertion	Action	Innervation*
Gluteal Group				
Gluteus maximus	Iliac crest, posterior gluteal line, and lateral surface of ilium; sacrum, coccyx, and lumbodorsal fascia	Iliotibial tract and gluteal tuberosity of femur	Extension and lateral rotation at hip	Inferior gluteal nerve (L_5–S_2)
Gluteus medius	Anterior iliac crest of ilium, lateral surface between posterior and anterior gluteal lines	Greater trochanter of femur	Abduction and medial rotation at hip	Superior gluteal nerve (L_4–S_1)
Gluteus minimus	Lateral surface of ilium between inferior and anterior gluteal lines	As above	As above	As above
Tensor fasciae latae	Iliac crest and lateral surface of anterior superior iliac spine	Iliotibial tract	Flexion and medial rotation at hip; tenses fascia lata, which laterally supports the knee	As above
Lateral Rotator Group				
Obturators (externus and internus)	Lateral and medial margins of obturator foramen	Trochanteric fossa of femur (externus); medial surface of greater trochanter (internus)	Lateral rotation at hip	Obturator nerve (externus: (L_3–L_4) and special nerve and from sacral plexus (internus: (L_5–S_2)
Piriformis	Anterolateral surface of sacrum	Greater trochanter of femur	Lateral rotation and abduction at hip	Branches of sacral nerves (S_1–S_2)
Gemelli (superior and inferior)	Ischial spine and tuberosity	Medial surface of greater trochanter with tendon of obturator internus	Lateral rotation at hip	Nerves to obturator internus and quadratus femoris
Quadratus femoris	Lateral border of ischial tuberosity	Intertrochanteric crest of femur	As above	Special nerve from sacral plexus (L_4–S_1)
Adductor Group				
Adductor brevis	Inferior ramus of pubis	Linea aspera of femur	Adduction, flexion, and medial rotation at hip	Obturator nerve (L_3–L_4)
Adductor longus	Inferior ramus of pubis anterior to adductor brevis	As above	As above	As above
Adductor magnus	Inferior ramus of pubis posterior to adductor brevis and ischial tuberosity	Linea aspera and adductor tubercle of femur	Adduction at hip; superior part produces flexion and medial rotation; inferior part produces extension and lateral rotation	Obturator and sciatic nerves
Pectineus	Superior ramus of pubis	Pectineal line inferior to lesser trochanter of femur	Flexion, medial rotation, and adduction at hip	Femoral nerve (L_2–L_4)
Gracilis	Inferior ramus of pubis	Medial surface of tibia inferior to medial condyle	Flexion at knee; adduction and medial rotation at hip	Obturator nerve (L_3–L_4)
Iliopsoas Group				
Iliacus	Iliac fossa of ilium	Femur distal to lesser trochanter; tendon fused with that of psoas major	Flexion at hip	Femoral nerve (L_2–L_3) (L_2–L_3)
Psoas major	Anterior surfaces and transverse processes of vertebrae (T_{12}–L_5)	Lesser trochanter in company with iliacus	Flexion at hip or lumbar intervertebral joints	Branches of the lumbar 3plexus (L_2–L_3)

*Where appropriate, spinal nerves involved are given in parentheses.

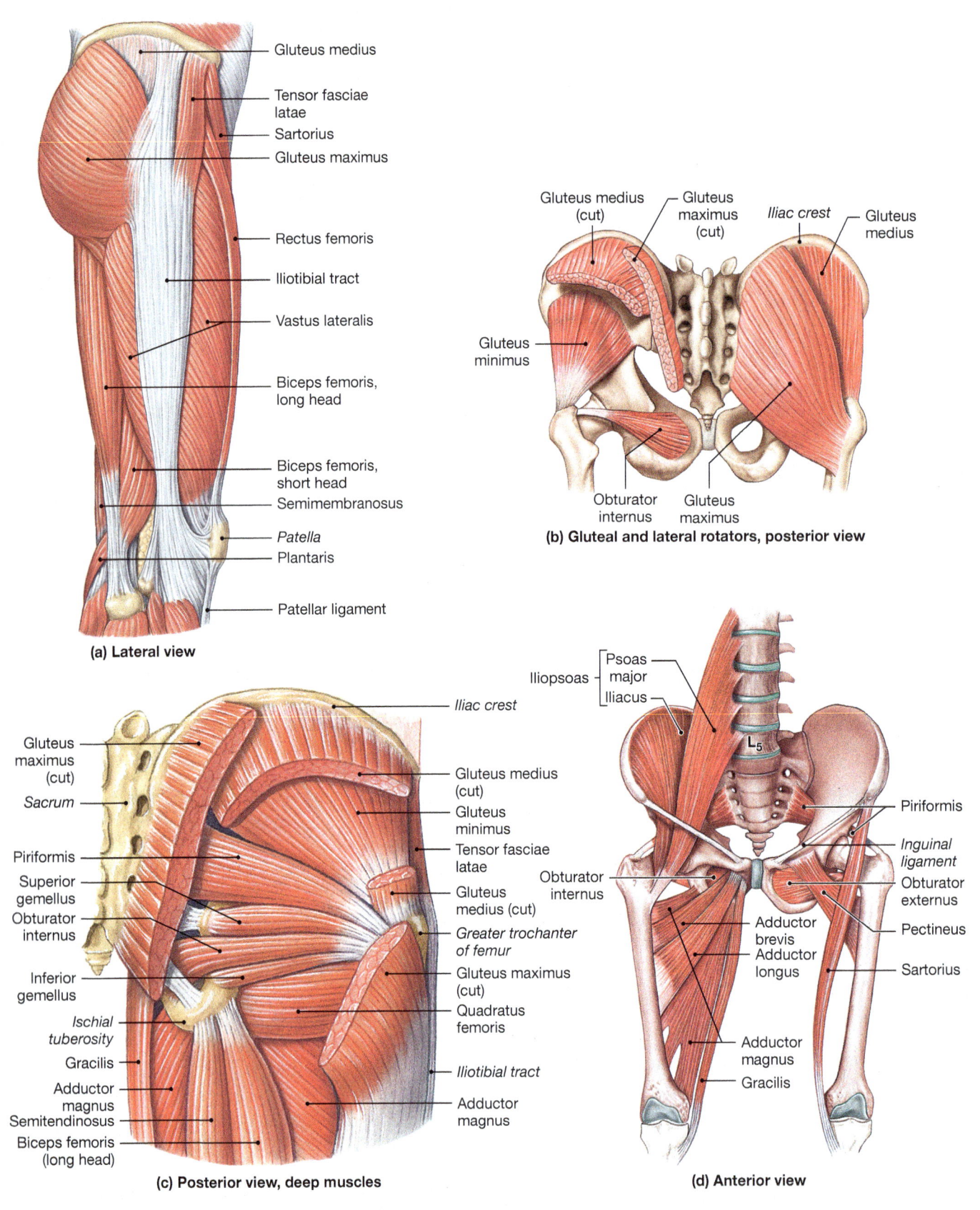

Figure 12.7 Anterior Muscles That Move the Thigh
The gluteal and lateral rotator muscle groups of the right hip. (a) Lateral view of the right thigh. (b) Posterior view of the pelvis with gluteus maximus muscle removed. (c) Posterior view of the pelvis that shows deep dissection of the muscles of the gluteal and lateral rotator groups. (d) The iliopsoas and adductor muscle groups.

gluteus maximus muscle (see Figure 12.7). It is a large, fleshy muscle and is easily located as the major muscle of the buttocks. Its muscle fibers pass inferiorolateral and insert on a thick tendon called the **iliotibial** (il-ē-ō-TIB-ē-ul) **tract** that attaches to the lateral condyle of the tibia.

The **gluteus medius muscle** originates on the iliac crest and on the lateral surface of the ilium and gathers laterally into a thick tendon that inserts posteriorly on the greater trochanter of the femur. The **gluteus minimus muscle** begins on the lateral surface of the ilium, tucked under the origin of the gluteus medius muscle. The fibers of the gluteus minimus muscle also pass laterally to insert on the anterior surface of the greater trochanter. Both the gluteus medius muscle and the gluteus minimus muscle abduct and medially rotate the thigh.

The **tensor fasciae latae** (TEN-sor FASH-ē-ē LĀ-tē) **muscle** originates on the iliac crest and on the outer surface of the anterior superior iliac spine. It is a gluteal muscle because it shares its insertion on the iliotibial tract with the gluteus maximus. As the name implies, the tensor fasciae latae muscle tenses the fascia of the thigh and helps stabilize the pelvis on the femur. The muscle also abducts and medially rotates the thigh.

Lateral Rotator Group

The lateral rotator group consists of the obturator internus and externus muscles and the piriformis, gamellus, and quadratus femoris muscles (see Figure 12.7). All of these muscles rotate the thigh laterally, and the piriformis muscle also abducts the thigh. Both the **obturator internus muscle** and the **obturator externus muscle** originate along the medial and lateral edges of the obturator foramen of the os coxae and insert on the trochanteric fossa, a shallow depression on the medial side of the greater trochanter of the femur.

The **piriformis** (pir-i-FOR-mis) **muscle** arises from the anterior and lateral surfaces of the sacrum and inserts on the greater trochanter of the femur. Inferior to the piriformis is the **quadratus femoris muscle**. Its origin is on the lateral surface of the ischial tuberosity and inserts on the femur between the greater and lesser trochanters.

The **superior gemellus muscle** and **inferior gemellus muscle** are deep to the gluteal muscles. These small rotators originate on the ischial spine and ischial tuberosity and insert on the greater trochanter with the tendon of the obturator internus. Both muscles rotate the thigh laterally.

Adductor Group

Muscles that adduct the thigh are organized into the adductor group and the pectineus and gracilis muscles. Three adductor muscles originate on the inferior pubis and insert on the posterior femur and are powerful adductors of the thigh (see Figure 12.7). They also allow the thigh to flex and rotate medially. The **adductor magnus muscle** is the largest of the adductor muscles. It arises on the inferior ramus of the pubis and the ischial tuberosity and inserts along the length of the linea aspera of the femur. If the superficial muscles are removed it is easy to observe on a leg model. Superficial to the adductor magnus is the **adductor longus muscle**. The **adductor brevis muscle** is superior and posterior to the adductor longus muscle.

The **pectineus** (pek-TI-nē-us) **muscle** is another superficial adductor muscle of the medial thigh (see Figure 12.7d). It is located next to the iliacus muscle. It originates along the superior ramus of the pubic bone and inserts on the pectineal line of the femur.

The **gracilis** (GRAS-i-lis) **muscle** is the most superficial of the thigh adductors. It arises from the superior ramus of the pubic bone, near the symphysis, extends inferiorly along the medial surface of the thigh, and inserts just medial to the insertion of the sartorius near the tibial tuberosity. Because it passes over both the hip and knee joints, it acts to adduct and medially rotate the thigh and flex the knee.

Iliopsoas Group

The iliopsoas (il-ē-ō-SŌ-us) group consists of two muscles, the psoas major and the iliacus (see Figure 12.7d). The **psoas** (SŌ-us) **major muscle** originates on the body and transverse processes of vertebrae T_{12} through L_5. The muscle sweeps inferiorly, passes between the femur and the ischial ramus, and inserts on the lesser trochanter of the femur. The **iliacus** (il-Ē-ah-kus) **muscle** originates on the iliac fossa on the medial portion of the ilium and joins the tendon of the psoas major muscle. The psoas major and iliacus muscles work together to flex the thigh, bringing its anterior surface toward the abdomen.

QuickCheck Questions

4.1 Where are the abductors of the thigh located?

4.2 What is the iliotibial tract?

4.3 Name two muscles that rotate the thigh.

4 Materials

- ☐ Torso model
- ☐ Lower limb model
- ☐ Articulated skeleton

Procedures

1. Review the pelvic and gluteal muscles in Figure 12.7 and Table 12.6.
2. Identify each muscle on the torso and lower limb models and on the muscle chart.
3. On the lower limb model, observe how the gluteal muscles and the tensor fasciae latae muscles insert on the lateral portion of the femur.
4. Locate as many of your own thigh muscles as possible. Practice the actions of the muscles and observe how your lower limb moves.
5. Examine the articulated skeleton and note the origin, insertion, and action of the major muscles that act on the thigh. ∎

LAB ACTIVITY 5 Muscles That Move the Leg

Muscles that flex and extend the leg at the knee joint are on the posterior and anterior sides of the femur. Refer to Table 12.7 for details on these muscles. Some of these muscles originate on the pelvis and cross both the hip and the knee joints and can therefore also move the thigh.

The major muscles of the posterior thigh are collectively called the **hamstrings**. They all have a common origin on the ischial tuberosity and flex the knee. The **biceps femoris muscle** is the lateral muscle of the posterior thigh (Figure 12.8). It has two heads and two origins, one on the ischial tuberosity and a second on the linea aspera of the femur. The two heads merge to form the belly of the muscle and insert on the lateral condyle of the tibia and the head of the fibula. Because this muscle spans both the hip and knee joints, it can extend the thigh and flex the knee. Medial to the biceps femoris muscle is the **semitendinosus** (sem-ē-ten-di-NŌ-sus) **muscle** (Figures 12.8 and 12.9). It is a long muscle that passes the posterior knee to insert on the proximomedial surface of the tibia near the insertion of the grascilis. The **semimembranosus** (sem-ē-mem-bra-NŌ-sus) **muscle** is medial to the semitendinosus muscle and inserts on the medial tibia. These muscles cross both the hip joint and the knee joint and extend the thigh and flex the knee. The hamstrings are therefore antagonists to the quadriceps muscles. When the thigh is flexed and drawn up toward the pelvis, the hamstrings extend the thigh.

The **sartorius muscle** is a thin, ribbon-like muscle that originates on the anterior superior iliac spine and passes inferiorly, and cuts obliquely across the thigh (Figure 12.10a). It is the longest muscle in the body. It crosses the knee

(a) Hip and thigh, posterior view

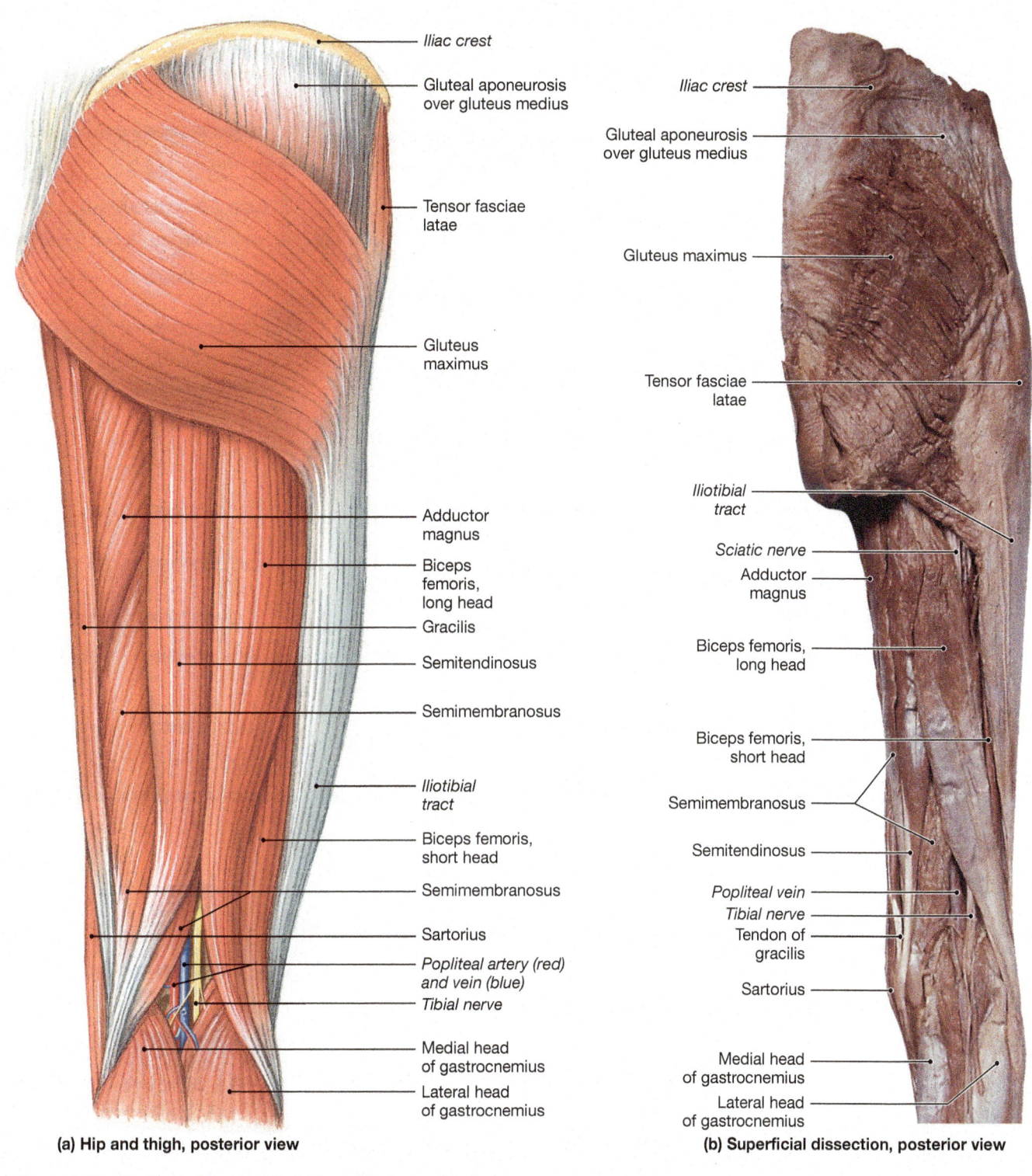

(b) Superficial dissection, posterior view

Figure 12.8 Posterior Muscles That Move the Lower Limb
(a) Posterior view of superficial muscles of the thigh. (b) Posterior view of a dissection of the right thigh muscles.

209

Medial view

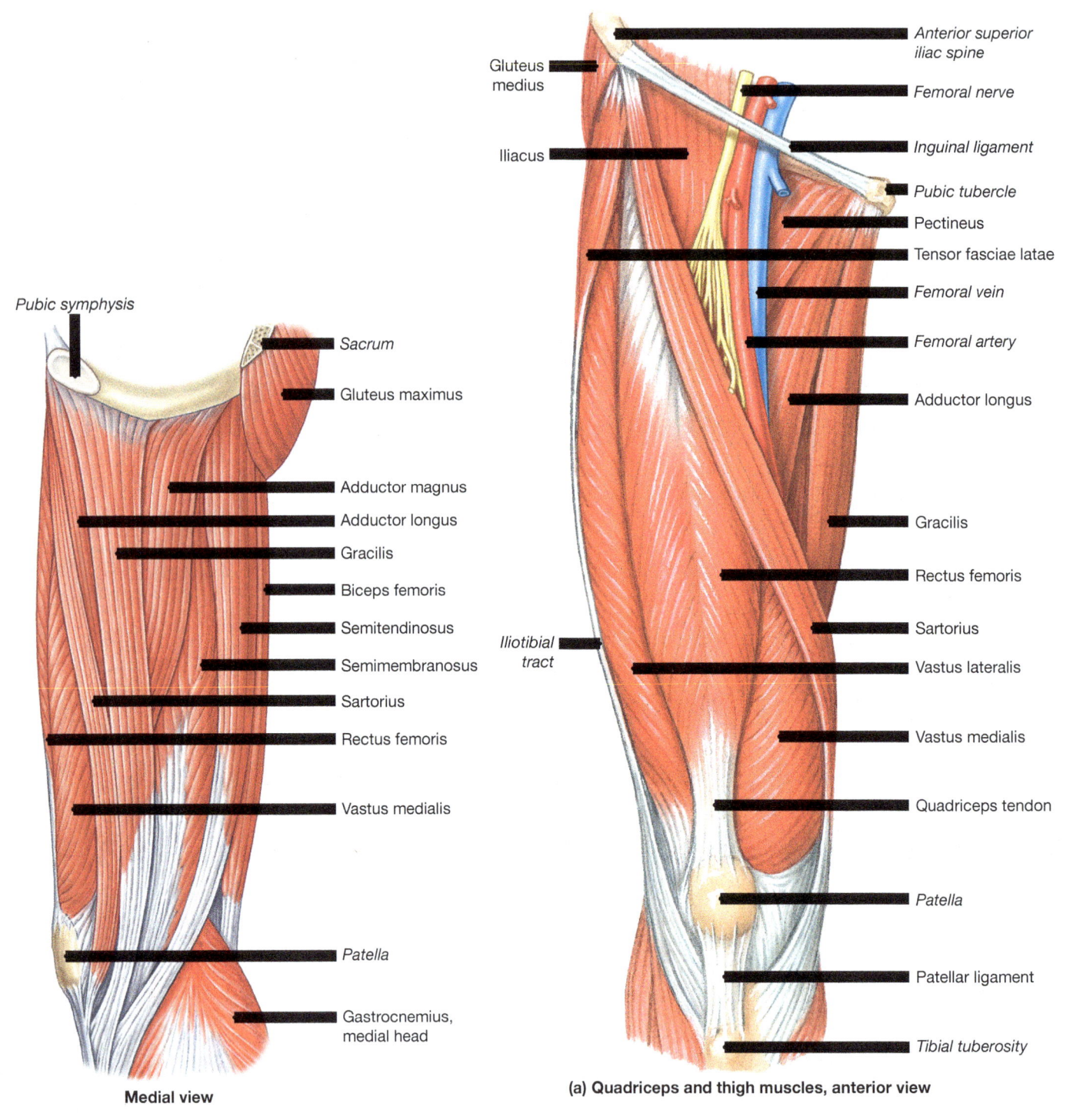

Figure 12.9 Medial Muscles That Move the Leg
Medial view of the muscles of the right thigh.

(a) Quadriceps and thigh muscles, anterior view

Figure 12.10 Muscles That Move the Leg
(a) Anterior view of superficial muscles of the right thigh.

210

(c) **Thigh, transverse section**

(b) **Anterior view**

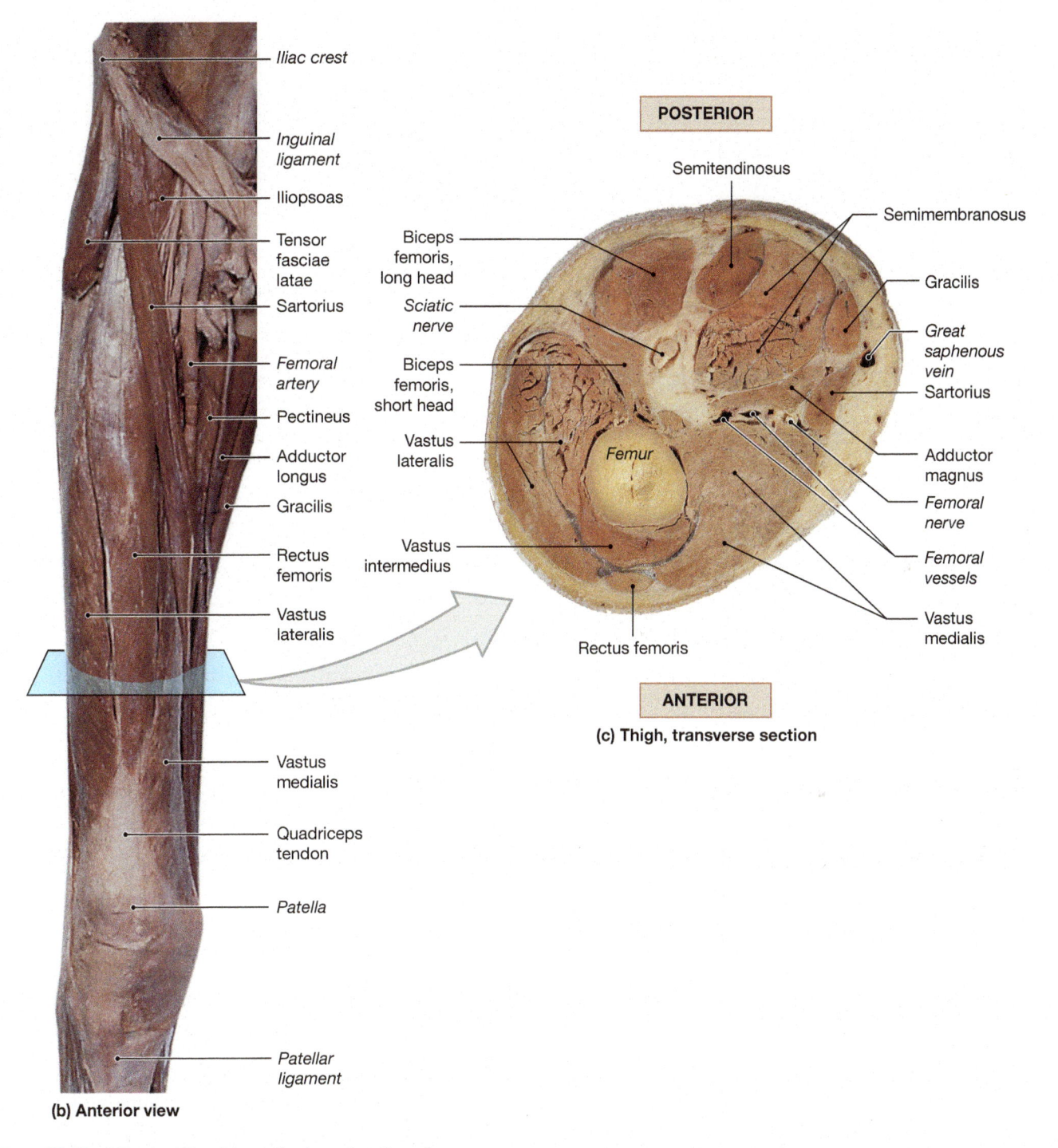

Figure 12.10 Muscles That Move the Leg *(continued)*
(b) Anterior view of a dissection of the right thigh muscles. **(c)** Transverse section of the right thigh.

211

Table 12.7 *Origins and Insertions* Muscles That Move the Leg

Muscle	Origin	Insertion	Action	Innervation*
Flexors of the Knee				
Biceps femoris	Ischial tuberosity and linea aspera of femur	Head of fibula, lateral condyle of tibia	Flexion at knee; extension and lateral rotation at hip	Sciatic nerve; tibial portion (S_1–S_3; to long head) and common fibular branch (L_3–S_2; to short head)
Semimembranosus	Ischial tuberosity	Posterior surface of medial condyle of tibia	Flexion at knee; extension and medial rotation at hip	Sciatic nerve (tibial portion; (L_5–S_2)
Semitendinosus	As above	Proximal, medial surface of tibia near insertion of gracilis	As above	As above
Sartorius	Anterior superior iliac spine	Medial surface of tibia near tibial tuberosity	Flexion at knee; flexion and lateral rotation at hip	Femoral nerve (L_2–L_3)
Popliteus	Lateral condyle of femur	Posterior surface of proximal tibial shaft	Medial rotation of tibia (or lateral rotation of femur); flexion at knee	Tibial nerve (L_4–S_1)
Extensors of the Knee				
Rectus femoris	Anterior inferior iliac spine and superior acetabular rim of ilium	Tibial tuberosity via patellar ligament	Extension at knee; flexion at hip	Femoral nerve (L_2–L_4)
Vastus intermedius	Anterolateral surface of femur and linea aspera (distal half)	As above	Extension at knee	As above
Vastus lateralis	Anterior and inferior to greater trochanter of femur and along linea aspera (proximal half)	As above	As above	As above
Vastus medialis	Entire length of linea aspera of femur	As above	As above	As above

*Where appropriate, spinal nerves involved are given in parentheses.

joint to insert on the medial surface of the tibia near the tibial tuberosity. This muscle is a flexor of the knee and thigh and a lateral rotator of the thigh. The **popliteus** (pop-LI-tē-us) **muscle** (see Figure 12.11d) crosses from its origin on the lateral condyle of the femur to insert on the posterior surface of the tibial shaft.

The extensors of the leg are collectively called either the **quadriceps muscles** or the **quadriceps femoris.** They make up the bulk of the anterior mass of the thigh and are consequently easy to locate. One quadriceps muscle, the **rectus femoris muscle** (see Figure 12.10), is located along the midline of the anterior surface of the thigh. Covering almost the entire medial surface of the femur is the **vastus medialis muscle**. The **vastus lateralis muscle** is located on the lateral side of the rectus femoris muscle, and the **vastus intermedius muscle** is directly deep to the rectus femoris muscle. The quadriceps muscles converge on a patellar tendon and insert on the tibial tuberosity. Because the rectus femoris muscle crosses two joints, the hip and knee, it allows the hip to flex and the leg to extend.

QuickCheck Questions

5.1 What are the names of all the quadriceps muscles?

5.2 What are the names of all the hamstrings?

5 Materials

☐ Torso model
☐ Lower limb model
☐ Articulated skeleton
☐ Muscle chart

Procedures

1. Review the muscles that move the leg in Figures 12.8, 12.9, and 12.10.
2. On the torso and lower limb models and the muscle chart, identify the muscles that move the leg, and categorize each muscle as being a flexor, extensor, adductor, or abductor.
3. Flex your knee and feel the tendons of the semimembranosus and semitendinosus muscles, which are located just above the posterior knee on the medial side. Similarly, on the lateral side of the knee, just above the fibular head, palpate the tendon of the biceps femoris muscle.
4. Examine the articulated skeleton and note the origin, insertion, and action of the major muscles that act on the thigh, knee, and leg. ■

LAB ACTIVITY 6 Muscles That Move the Ankle, Foot, and Toes

Muscles that move the ankle arise on the leg and insert on the tarsal bones. Muscle that move the foot and toes originate either on the leg or in the foot. Details for origin, insertion, action, and innervation for the muscles that move the ankle, foot, and toes are in Tables 12.8 and 12.9.

The **tibialis** (tib-ē-A-lis) **anterior muscle** is located on the anterior side of the leg (see Figure 12.11). This muscle is easy to locate as the lateral muscle mass of the shin on the anterior edge of the tibia. Its tendon passes over the dorsal surface of the foot, and the muscle dorsiflexes and inverts the foot.

Two extensor muscles arise on the anterior leg and insert on the various phalanges of the foot. The **extensor hallucis longus muscle** is lateral and deep to the tibialis anterior muscle. Lateral to the extensor hallucis longus muscle is the **extensor digitorum longus muscle** with four tendons that spread on the dorsal surface of toes 2 through 5. On the lateral side of the leg are the **fibularis longus** and **fibularis brevis muscles**, also called the *peroneus* muscles. These muscles insert on the foot to evert the foot by laterally turning the sole to face outward.

The calf muscles of the posterior leg are the **gastrocnemius** (gas-trok-NĒ-mē-us) and the **soleus** (SŌ-lē-us) **muscles**. These muscles share the calcaneal (Achilles) tendon, which inserts on the calcaneus of the foot. The **plantaris muscle** is a short muscle of the lateral popliteal region, deep to the gastrocnemius muscle. The plantaris muscle has a long tendon that inserts on the posterior of the calcaneus. The gastrocnemius, soleus, and plantaris muscles plantar flex the ankle; the soleus is also a postural muscle for support while standing.

Deep to the soleus muscle is the **tibialis posterior muscle** (Figure 12.12), which adducts and inverts the foot and plantar flexes the ankle. Its tendon passes medially to the calcaneus and inserts on the plantar surface of the navicular and cuneiform bones and metatarsals II, III, and IV. The **flexor hallucis longus muscle** begins lateral to the origin of the tibialis posterior muscle on the fibular shaft. Its tendon runs parallel to that of the tibialis posterior muscle, passes medial to the calcaneus, and inserts on the plantar surface of the distal phalanx of the hallux, or great toe. The **flexor digitorum longus muscle** originates on the posterior tibia and inserts on the distal phalanges of toes 2 through 5. The flexor hallucis longus muscle flexes the joints of the great toe; the flexor digitorum longus flexes the joints of toes 2 through 5. Both of these flexor muscles also dorsiflex the ankle and evert the foot.

The **extensor digitorum brevis muscle** is located on the dorsal surface of the foot and passes obliquely across the foot with four tendons that insert into the dorsal surface of the proximal phalanges of toes 1 through 4 (Figure 12.13). The **flexor digitorum brevis muscle** on the plantar surface inserts tendons on the phalanges of toes 2 through 5. The **abductor hallucis muscle** (see Figure 12.14) is found on the inner margin of the foot on the plantar side of the calcaneus. The **flexor hallucis**

Table 12.8 *Origins and Insertions* Muscles That Move the Ankle, Foot, and Toes

Muscle	Origin	Insertion	Action	Innervation*
Action at the Ankle				
Flexors (Dorsiflexors)				
Tibialis anterior	Lateral condyle and proximal shaft of tibia	Base of first metatarsal bone and medial cuneiform bone	Flexion (dorsiflexion) at ankle; inversion of foot	Deep fibular nerve (L_4–S_1)
Extensors (Plantar flexors)				
Gastrocnemius	Femoral condyles	Calcaneus via calcaneal tendon	Extension (plantar flexion) at ankle; inversion of foot; flexion at knee	Tibial nerve (S_1–S_2)
Fibularis brevis	Midlateral margin of fibula	Base of fifth matatarsal bone	Eversion of foot and extension (plantar flexion) at ankle	Superficial fibular nerve (L_4–S_1)
Fibularis longus	Lateral condyle of tibia, head, and proximal shaft of fibula	Base of first metatarsal bone and medial cuneiform bone	Eversion of foot and extension (plantar flexion) at ankle; supports longitudinal arch	As above
Plantaris	Lateral supracondylar ridge	Posterior portion of calcaneus	Extension (plantar flexion) at ankle; flexion at knee	Tibial nerve (L_4–S_1)
Soleus	Head and proximal shaft of fibula and adjacent posteromedial shaft of tibia	Calcaneus via calcaneal tendon (with gastrocnemius)	Extension (plantar flexion) at ankle	Sciatic nerve, tibial branch (S_1–S_2)
Tibialis posterior	Interosseous membrane and adjacent shafts of tibia and fibula	Tarsal and metatarsal bones	Adduction and inversion of foot; extension (plantar flexion) at ankle	As above
Action at the Toes				
Digital flexors				
Flexor digitorum longus	Posteromedial surface of tibia	Inferior surfaces of distal phalanges, toes 2–5	Flexion at joints of toes 2–5	Sciatic nerve, tibial branch (L_5–S_1)
Flexor hallucis longus	Posterior surface of fibula	Inferior surface, distal phalanx of great toe	Flexion at joints of great toe	As above
Digital extensors				
Extensor digitorum longus	Lateral condyle of tibia, anterior surface of fibula	Superior surface of phalanges, toes 2–5	Extension at joints of toes 2–5	Deep fibular nerve (L_4–S_1)
Extensor hallucis longus	Anterior surface of fibula	Superior surface, distal phalanx of great toe	Extension at joints of great toe	As above

*Where appropriate, spinal nerves involved are given in parentheses.

brevis muscle originates on the plantar surface of the cuneiform and cuboid bones of the foot and splits into two heads, one medial and one lateral. Each head sends a tendon to the base of the first phalanx of the hallux, to either the lateral or the medial side. The **abductor digiti minimi muscle** of the little toe is located on the outer margin of the foot and originates on the plantar and lateral surfaces of the calcaneus. It inserts on the lateral side of the proximal phalanx of the little toe.

QuickCheck Questions

6.1 What are the names and characteristics of the muscles of the calf?

6.2 Which muscles move the great toe?

6.3 What are the names and characteristics of the insertions of the muscles that plantar flex the foot?

6.4 What does the name *flexor hallucis brevis* mean?

Table 12.9 *Origins and Insertions* Muscles of the Foot

Muscle	Origin	Insertion	Action	Innervation*
Extensor digitorum brevis	Calcaneus (superior and lateral surfaces)	Dorsal surfaces of toes 1–4	Extension at metatarsophalangeal joints of toes 1–4	Deep fibular nerve (L_5–S_1)
Abductor hallucis	Calcaneus (tuberosity on inferior surface)	Medial side of proximal phalanx of great toe	Abduction at metatarsophalangeal joint of great toe	Medial plantar nerve (L_4–L_5)
Flexor digitorum brevis	As above	Sides of middle phalanges, toes 2–5	Flexion at proximal interphalangeal joints of toes 2–5	As above
Abductor digiti minimi	As above	Lateral side of proximal phalanx, toe 5	Abduction at metatarsophangeal joint of toe 5	Lateral plantar nerve (L_4–L_5)
Quadratus plantae	Calcaneus (medial inferior surfaces)	Tendon of flexor digitorum longus	Flexion at joints of toes 2–5	As above
Lumbrical (4)	Tendons of flexor digitorum longus	Insertions of extensor digitorum longus	Flexion at metatarsophalangeal joints; extension at proximal interphalangeal joints of toes 2–5	Medial plantar nerve (1), lateral plantar nerve (2–4)
Flexor hallucis brevis	Cuboid and lateral cuneiform bones	Proximal phalanx of great toe	Flexion at metatarsophalangeal joint of great toe	Medial plantar nerve (L_4–L_5)
Adductor hallucis	Bases of metatarsal bones II–IV and plantar ligaments	As above	Adduction at metatarsophangeal joint of great toe	Lateral plantar nerve (S_1–S_2)
Flexor digiti minimi brevis	Base of metatarsal bone V	Lateral side of proximal phalanx of toe 5	Flexion at metatarsophalangeal joint of toe 5	As above
Dorsal interosseus (4)	Sides of metatarsal bones	Medial and lateral sides of toe 2; lateral sides of toes 3 and 4	Abduction at metatarsophalangeal joints of toes 3 and 4	As above
Plantar interosseus (3)	Bases and medial sides of metatarsal bones	Medial sides of toes 3–5	Adduction at metartarsophalangeal joints of toes 3–5	As above

*Where appropriate, spinal nerves involved are given in parentheses.

6 Materials

- ☐ Torso model
- ☐ Lower limb model
- ☐ Foot model
- ☐ Articulated skeleton
- ☐ Muscle chart

Procedures

1. Review the muscles of the leg in Figure 12.11.
2. On the lower limb model and muscle chart, identify each muscle on the leg.
3. Review the muscles of the foot in Figure 12.12 and identify each muscle on the foot model and muscle chart.
4. Locate as many leg and foot muscles on your own lower limb as possible. Practice the actions of the muscles and observe how your leg and foot move.
5. Examine the articulated skeleton and note the origin, insertion, and action of the major muscles that act on the ankle, foot, and toes. ■

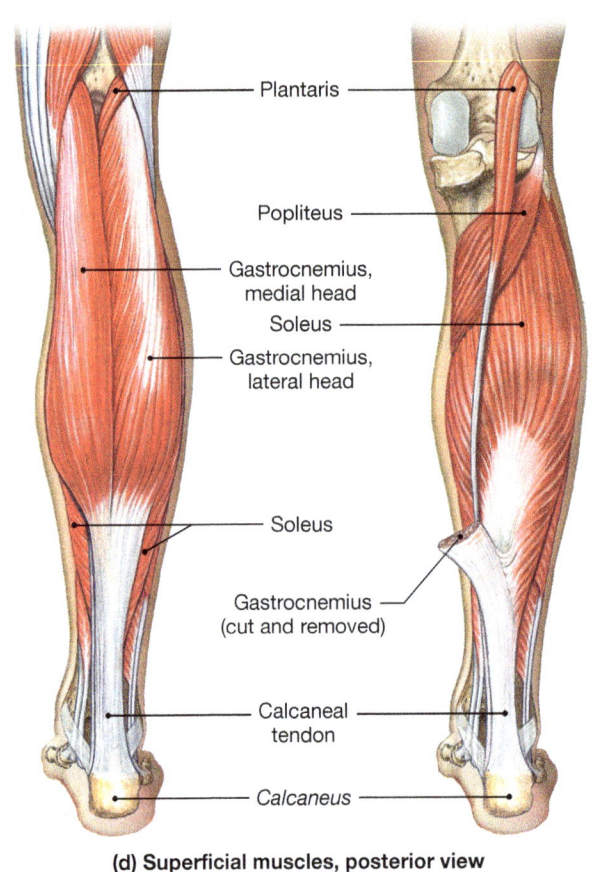

Figure 12.11 Muscles That Move the Ankle, Foot, and Toes

Muscles of the leg. (a) Anterior view. (b) Lateral view. (c) Dissection of superficial muscles, right leg. (d) Superficial muscles, posterior view. (e) Deep muscles, posterior view of right leg.

216

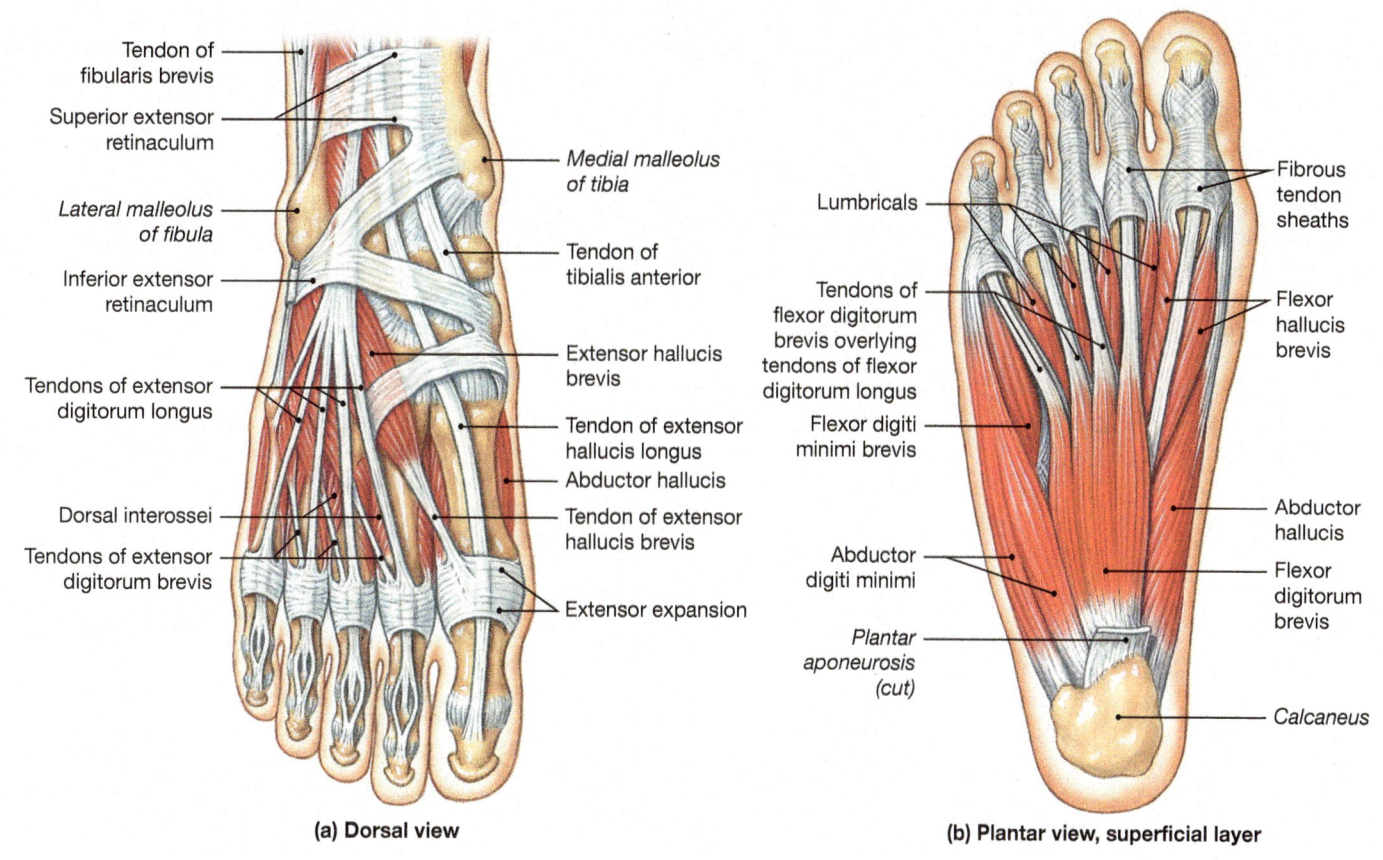

Figure 12.12 Muscles of the Foot
(a) Dorsal view of right foot. (b) Plantar view of right foot, superficial layer.

217

compartments A Regional Look

Muscles on the upper and lower limbs are surrounded by the deep fascia and isolated in sac-like muscle **compartments**. These compartments separate the various muscles into anterior, posterior, lateral, and deep groups that have similar muscle actions. Within a muscle compartment are arteries, veins, nerves, and other structures.

Lower limb compartments are illustrated in Figure 12.13, and the muscles, blood vessels, and nerves in each compartment are listed in Table 12.10. Examine the figure and observe how superficial and deep muscles of the leg are in different compartments. Also study Figure 12.10c, a transverse section of the thigh.

Clinical Application — Compartment Syndrome

Treating a limb injury includes watching for blood trapped in a muscle compartment. Bleeding increases pressure in the compartment and causes compression of local nerves and blood vessels. If the compression persists beyond four to six hours, permanent damage to nerve and muscle tissue may occur, a condition called *compartment syndrome*. To prevent compartment syndrome, drains are inserted into wounds to remove blood and other liquids both from the muscle and from the compartment.

Table 12.10 Compartments of the Lower Limb

Compartment	Muscles	Blood Vessels	Nerves
Thigh			

Anterior compartment	Iliopsoas
	Iliacus
	  Psoas major
	  Psoas minor
	Quadriceps femoris
	  Rectus femoris
	  Vastus intemedius
	  Vastus lateralis
	  Vastus medialis
	Pectineus
	Sartorius
	Tensor fasciae latae
		Femoral artery
		Femoral vein
		Deep femoral artery
		Lateral circumflex femoral artery
			Femoral nerve
			Saphenous nerve
			Superior gluteal nerve

Medial compartment	Adductor brevis Adductor longus Adductor magnus Gracilis Obturator externus	Obturator artery Obturator vein Deep femoral artery Deep femoral vein	Obturator nerve Sciatic nerve
Posterior compartment	Biceps femoris Semimembranosus Semitendinosus	Deep femoral artery Deep femoral vein	Common fibular nerve Sciatic nerve
Leg			
Anterior compartment	Extensor digitorum longus Extensor hallucis longus Fibularis tertius Tibialis anterior	Anterior tibial artery Anterior tibila vein	Deep fibular nerve
Lateral compartment	Fibularis brevis Fibularis longus	Fibular artery Fibular vein	Superficial fibular nerve
Posterior compartment Superficial	Gastrocnemius Plantaris Soleus		
Deep	Flexor digitorum longus Flexor hallucis longus Popliteus Tibialis posterior	Posterior tibial artery Posterior tibial vein	Tibial nerve

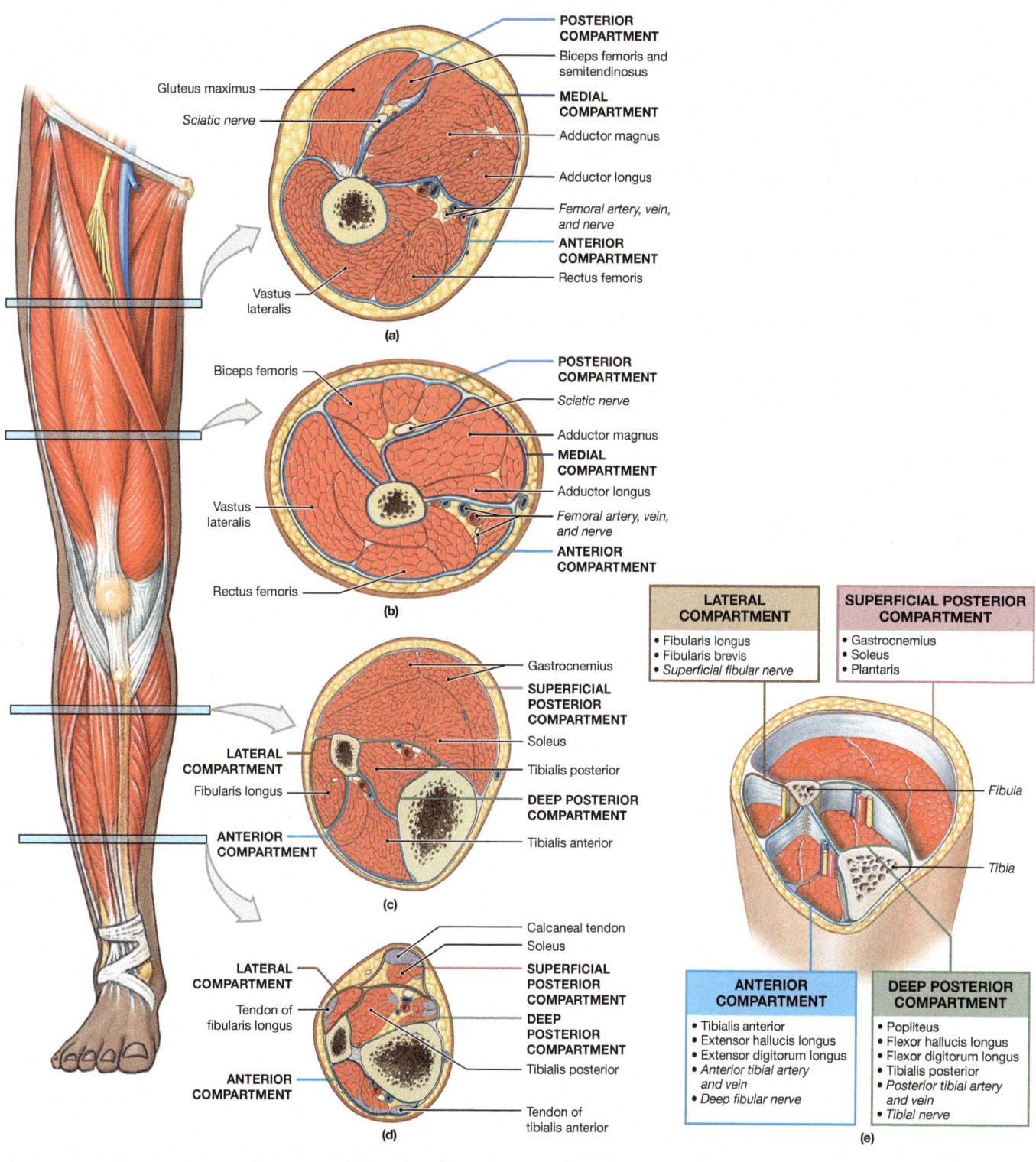

Figure 12.13 Muscle Compartments of the Lower Limb
(a) and (b) Diagrammatic horizontal sections through the proximal and distal portions of the right thigh, showing selected muscles.
(c) and (d) Diagrammatic horizontal sections through the proximal and distal portions of the right leg, showing selected muscles.
(e) The three-dimensional arrangement of septa and fascial boundaries in the right leg.

LAB REPORT

EXERCISE 12

Appendicular Muscles

Name _____
Date _____
Section _____

A. Matching

Match each term in the left column with its correct description from the right column.

_____ 1. triceps brachii muscle
_____ 2. extensor carpi radialis longus muscle
_____ 3. biceps brachii muscle
_____ 4. pronator teres muscle
_____ 5. flexor carpi ulnaris muscle
_____ 6. brachialis muscle
_____ 7. coracobrachialis muscle
_____ 8. deltoid muscle
_____ 9. opponens digiti minimi muscle
_____ 10. extensor digitorum muscle

A. small muscle; has common origin with biceps brachii
B. major pronator of forearm
C. flexes and adducts wrist
D. brings little finger toward thumb
E. flexes, extends, and abducts arm
F. major extensor of forearm
G. major flexor of forearm
H. major flexor and supinator of forearm
I. extends fingers
J. extends and abducts wrist

B. Matching

Match each term in the left column with its correct description from the right column.

_____ 1. sartorius muscle
_____ 2. fibularis longus muscle
_____ 3. gastrocnemius muscle
_____ 4. adductor magnus muscle
_____ 5. tibialis anterior muscle
_____ 6. extensor digitorum longus muscle
_____ 7. semimembranosus muscle
_____ 8. biceps femoris muscle
_____ 9. vastus intermedius muscle
_____ 10. tibialis posterior muscle

A. lateral calf muscle; everts and plantar flexes foot
B. quadriceps muscle; located deep to rectus femoris
C. hamstring; has two heads
D. muscle that crosses anterior thigh
E. hamstring; inserts on medial condyle of tibia
F. largest adductor of femur
G. on anterior leg; dorsiflexes foot
H. calf muscle deep to soleus
I. originates on anterior leg; extends toes
J. superficial calf muscle; plantar flexes and inverts foot

EXERCISE 12 — LAB REPORT

C. Labeling

Label each numbered muscle in Figures 12.14 and 12.15.

D. Short-Answer Questions

1. Describe the muscles involved in rotating the hand, as when twisting a doorknob back and forth.

2. Name the muscles responsible for flexing the arm. Which muscles are antagonists to these flexors?

3. Name a muscle for each movement of the wrist: flex, extend, abduct, and adduct.

4. Describe how the hamstrings move the leg.

5. Which muscle group is the antagonist to the hamstrings?

6. Describe the action of the abductor and adductor muscles of the thigh.

E. Analysis and Application

1. Why would a dislocated shoulder potentially result in injury to the rotator cuff?

2. A brace placed on your wrist to treat carpal tunnel syndrome prevents which type of wrist action? What do you accomplish by limiting this action?

3. Which leg muscles serve a function similar to the function of the arm's rotator cuff muscles?

4. Describe the origin, insertion, and action of the muscles that invert and evert the foot.

5. How can pressure increase around injured muscles, and what effect does this have on the regional anatomy?

LAB REPORT   EXERCISE 12

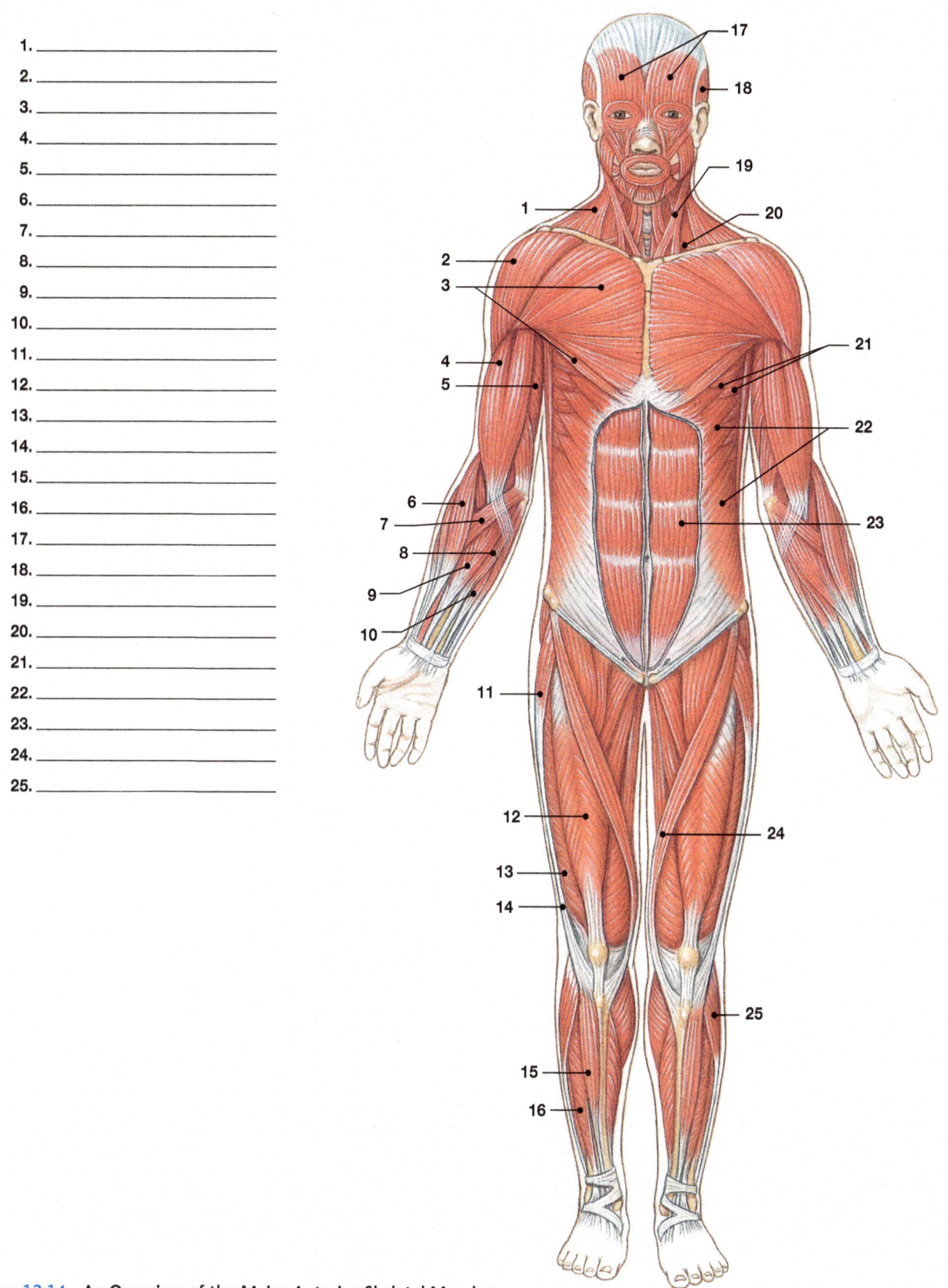

1. _____
2. _____
3. _____
4. _____
5. _____
6. _____
7. _____
8. _____
9. _____
10. _____
11. _____
12. _____
13. _____
14. _____
15. _____
16. _____
17. _____
18. _____
19. _____
20. _____
21. _____
22. _____
23. _____
24. _____
25. _____

Figure 12.14   An Overview of the Major Anterior Skeletal Muscles

223

EXERCISE 12 — LAB REPORT

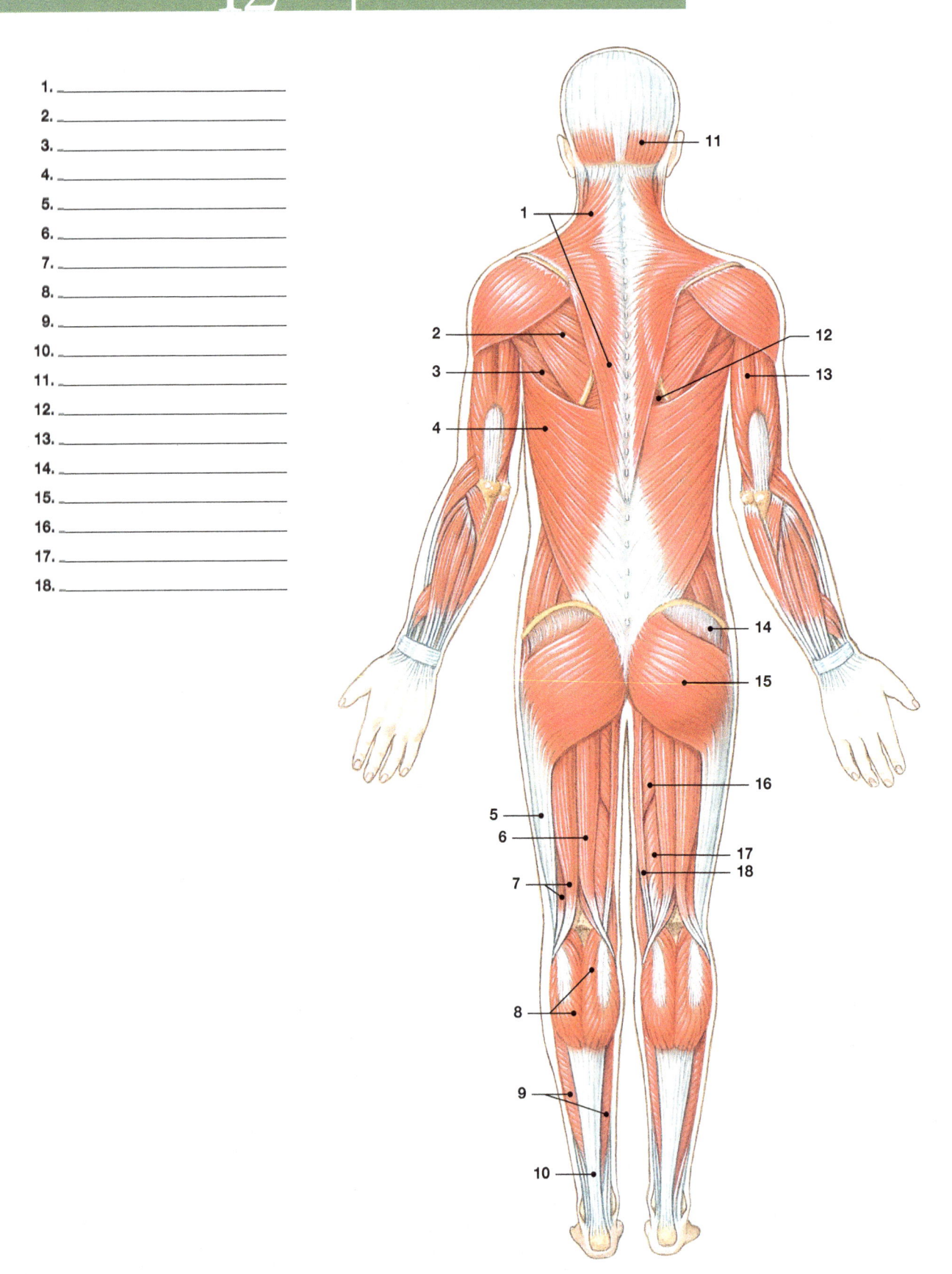

1. _____
2. _____
3. _____
4. _____
5. _____
6. _____
7. _____
8. _____
9. _____
10. _____
11. _____
12. _____
13. _____
14. _____
15. _____
16. _____
17. _____
18. _____

Figure 12.15 An Overview of the Major Posterior Skeletal Muscles

Organization of the Nervous System

EXERCISE 13

LAB ACTIVITIES

1. Histology of the Nervous System 227
2. Anatomy of a Nerve 231
3. Autonomic Nervous System 232

OBJECTIVES

On completion of this exercise, you should be able to:

- Identify six types of glial cells and describe a basic function of each type.
- Describe and identify the cellular anatomy of a neuron.
- Describe and identify the organization of a peripheral nerve.
- Describe the organization of the central and peripheral nervous systems.
- Compare the location of preganglionic outflow in the sympathetic and parasympathetic divisions of the central nervous system.
- Trace a sympathetic pathway and a parasympathetic pathway to the heart.
- Compare the body's response to sympathetic and parasympathetic stimulation.

The nervous system orchestrates body functions to maintain homeostasis. To accomplish this control, the nervous system must perform three vital tasks.

1. It must detect changes in and around the body. For this task, sensory receptors monitor environmental conditions and encode information about environmental changes as electrical impulses.
2. It must process incoming sensory information and generate an appropriate motor response to adjust the activity of muscles and glands.
3. It must orchestrate and integrate all sensory and motor activities so that homeostasis is maintained.

The nervous system is divided into two main components (Figure 13.1): the **central nervous system** (**CNS**), which consists of the brain and spinal cord, and the **peripheral nervous system** (**PNS**), which communicates with the CNS by way of cranial and spinal nerves, collectively called *peripheral nerves*. A **nerve** is a bundle of neurons plus any associated blood vessels and connective tissue. The PNS is responsible for providing the CNS with information concerning changes inside the body and changes in the surrounding environment. Sensory information is sent along PNS nerves that join the CNS in either the spinal cord or the brain. The CNS evaluates the sensory data and determines whether muscle and gland activities should be modified in response to the changes. Motor commands from the CNS are then relayed to PNS nerves that carry the commands to specific muscles and glands.

The PNS is divided into afferent and efferent divisions. The **afferent division** receives sensory information from **sensory receptors**, which are the cells and organs that detect changes in the body and the surrounding environment, and then sends that information to the CNS for interpretation. The CNS decides the appropriate response to the sensory information and sends motor commands to the PNS **efferent division**, which controls the activities of **effectors**, the general term for all the muscles and glands of the body. **Somatic effectors** are skeletal muscles, and **visceral effectors** are cardiac muscle, smooth muscle, and glands.

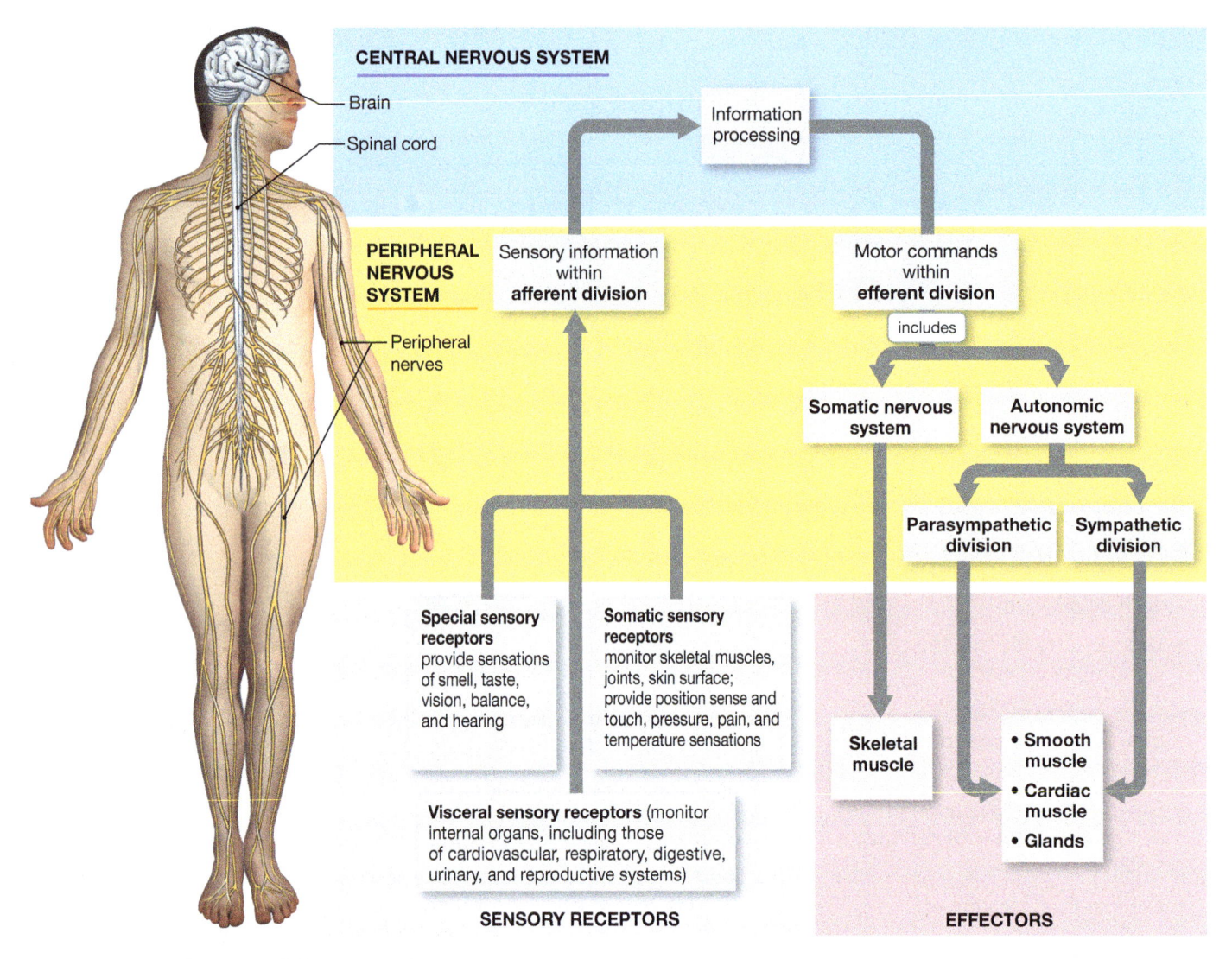

226 EXERCISE 13 Organization of the Nervous System

Figure 13.1 An Overview of the Nervous System
This diagram shows the relationship between the central and peripheral nervous systems and the function and components of the afferent and efferent divisions of the latter.

The efferent division is divided into two parts. One part, the **somatic nervous system,** conducts motor responses to skeletal muscles. The other part, the **autonomic nervous system**, consists of the **sympathetic** and **parasympathetic divisions**, both of which send commands to smooth muscles, cardiac muscles, and glands.

The nerves of the PNS are divided into two groups according to the part of the CNS with which they communicate. **Cranial nerves** communicate with the brain and pass into the face and neck through foramina in the skull. **Spinal nerves** join the spinal cord at intervertebral foramina and pass either into the upper and lower limbs or into the body wall. There are 12 pairs of cranial nerves and 31 pairs of spinal nerves, and each pair transmits specific information between the CNS and the PNS. Functionally, all spinal nerves are **mixed nerves**, which means they carry both sensory signals and motor signals. Cranial nerves are either entirely sensory or mixed. Although a cranial or spinal nerve may transmit both sensory and motor impulses, a single neuron within the nerve transmits only one type of signal.

LAB ACTIVITY 1 ## Histology of the Nervous System

As discussed in Exercise 4, two types of cells populate the nervous system: glial cells and neurons. Glial cells have a supportive role in protecting and maintaining nerve tissue. Neurons are the communication cells of the nervous system and are capable of propagating and transmitting electrical impulses to respond to the ever-changing needs of the body.

Glial Cells

Glial cells, which collectively make up a network called the **neuroglia** (noo-RŌG-lē-a; *glia*, glue), are the most abundant cells in the nervous system. They protect, support, and anchor neurons in place. In the CNS, glial cells are involved in the production and circulation of the cerebrospinal fluid that circulates in the ventricles of the brain and in the central canal of the spinal cord. In both the CNS and the PNS, glial cells isolate and support neurons with myelin.

The CNS has four types of glial cells (Figure 13.2). **Astrocytes** (AS-trō-sīts), shown in Figure 13.3, hold neurons in place and isolate one neuron from another. They also wrap foot-like extensions around blood vessels, which creates a blood-brain barrier that prevents certain materials from passing out of the blood and into nerve tissue. The CNS glial cells known as **oligodendrocytes** (o-li-gō-DEN-drō-sīts)

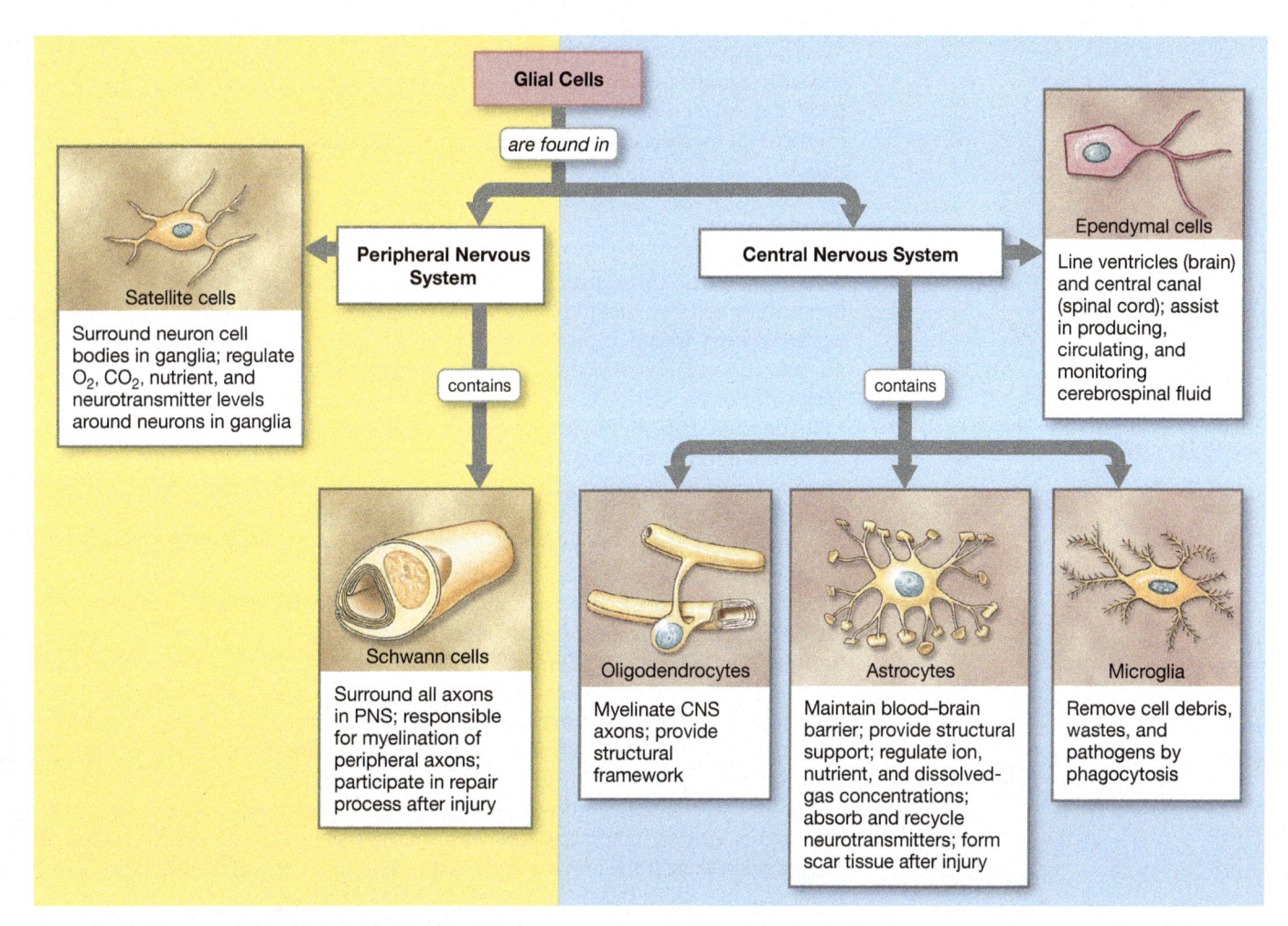

Figure 13.2 **The Classification of Glial Cells**
The categories and functions of the various glial cell types in the CNS and the PNS.

Figure 13.3 Astrocytes
Micrograph of astrocytes that shows the many cellular extensions of this type of glial cell. (LM × 400)

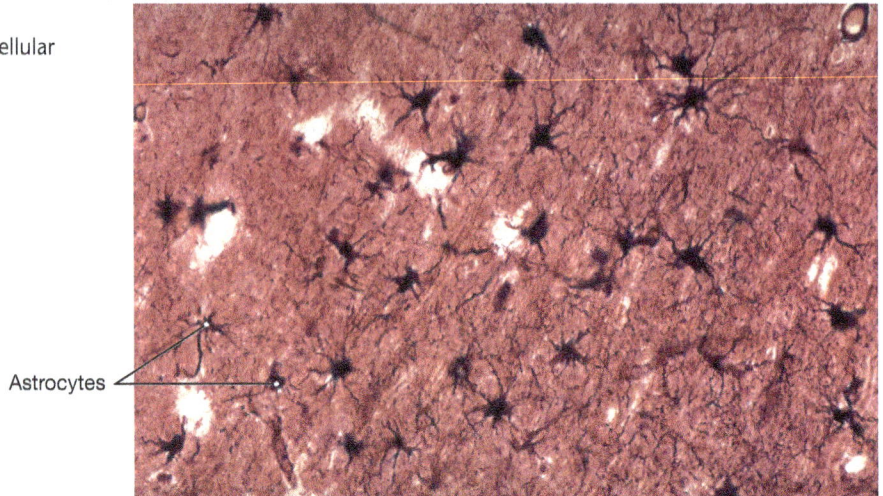

Astrocytes

wrap around axons and form a fatty **myelin sheath**. **Microglia** (mī-KROG-lē-uh) are phagocytic glial cells that remove microbes and cellular debris from CNS tissue. **Ependymal** (e-PEN-dĭ-mul) **cells** line the ventricles of the brain and the central canal of the spinal cord; these glial cells contribute to the production of the cerebrospinal fluid.

The PNS has two types of glial cells (See Figure 13.2). Where neuron cell bodies cluster in groups called *ganglia,* the glial cells called **satellite cells** encase each cell body and isolate it from the interstitial fluid to regulate the neuron's chemical environment. **Schwann cells** surround and myelinate PNS axons in spinal and cranial nerves.

Neurons

A neuron has the three distinguishable features, which are described in Exercise 4: dendrites, a cell body (soma), and an axon (Figures 13.4 and 13.5). The numerous dendrites carry information into the large, rounded cell body, which contains the nucleus and organelles of the cell. The **perikaryon** (per-i-KAR-ē-on), which is the entire area of the cell body that surrounds the nucleus, contains such organelles as mitochondria, free ribosomes, and fixed ribosomes. Also found in the perikaryon are chromatophilic substances, also called **Nissl bodies**, which are groups of free ribosomes and rough endoplasmic reticulum. Nissl bodies account for the dark regions that are clearly visible in a sagittal section of the brain.

The first part of the axon of a neuron, the **initial segment**, extends from a narrow part of the cell body that is called the **axon hillock**. The axon may divide into several **collateral branches** that subdivide into smaller branches called **telodendria** (tel-ō-DEN-drē-uh). At the distal tip of each telodendrion of the neuron is a **terminal bouton**, also called *synaptic terminal* or *synaptic knob,* that houses **synaptic vesicles** which are full of **neurotransmitter** molecules. These molecules are released by the neuron and are the means by which it communicates with another cell, either another neuron or a muscle or gland effector cell. The terminal bouton is the transmitting part of the **synapse,** which is the general term for the neural communication site. At any given synapse, the neuron-releasing neurotransmitter is called the *presynaptic neuron*. If this neuron communicates with another neuron, the latter is called the *postsynaptic neuron.* If the presynaptic neuron communicates with a muscle or gland effector cell, that cell is called the *postsynaptic cell.* A small gap called the **synaptic cleft** separates the presynaptic neuron from the postsynaptic neuron or postsynaptic cell.

As described previously, axons are myelinated by glial cells. In the PNS, a Schwann cell wraps around and encases a small section of axon in multiple layers of the Schwann cell's membrane. Any region of an axon covered in this membrane

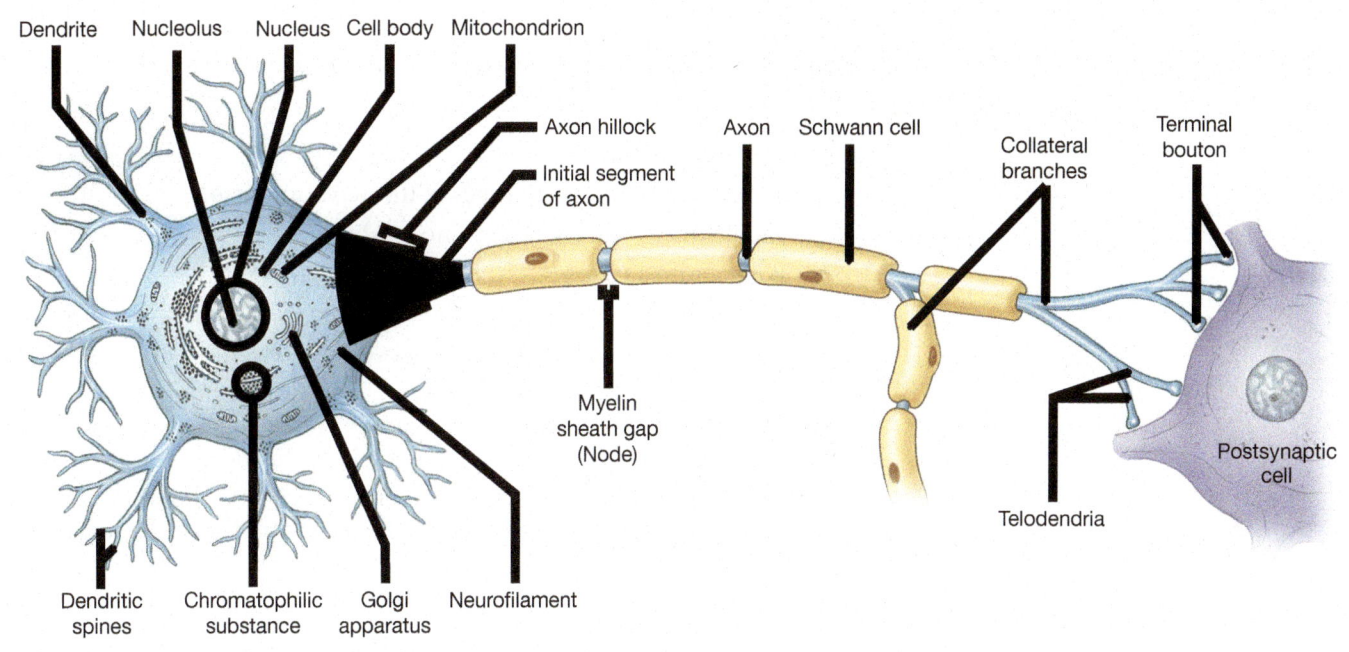

Figure 13.4 The Anatomy of a Representative Neuron
A neuron has a cell body (soma), some branching dendrites, and a single axon. The region of the cytoplasm around the nucleus is the perikaryon. The neuron in this illustration has a myelin sheath that covers the axon.

Figure 13.5 A Representative Motor Neuron
Micrograph of a neuron that has multiple dendrites along with its single axon. (LM × 400)

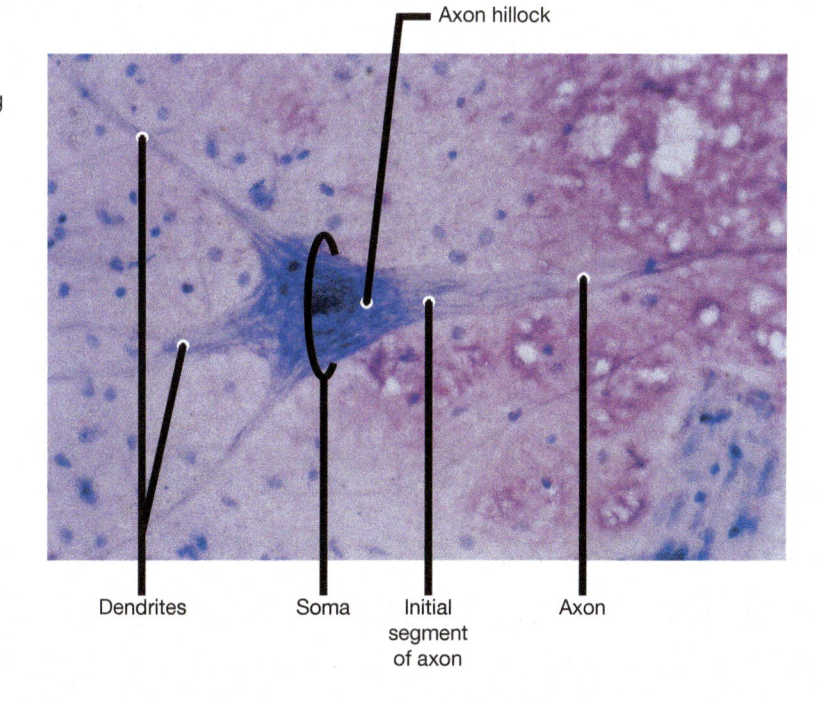

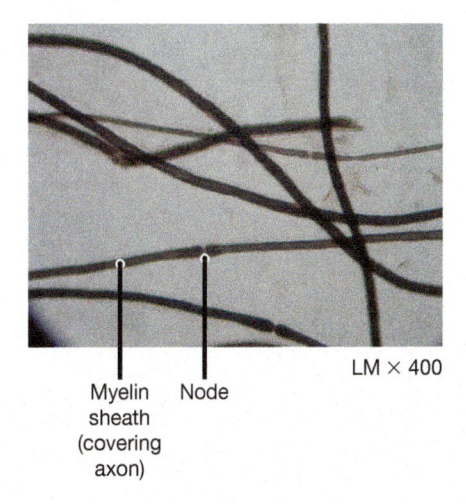

Figure 13.6 Myelinated Neuron
Micrograph of myelinated neurons stained to show the myelin sheath and nodes. (LM × 400)

is called a **myelinated internode.** Between the internodes are myelin sheath gaps, or **nodes** (also called *nodes of Ranvier*) and are shown in Figures 13.4 and 13.6. The membrane of the axon, the **axolemma**, is exposed at the nodes, and this exposure permits a nerve impulse to arc rapidly from node to node. The **neurilemma** (noo-ri-LEM-uh), or outer layer of the Schwann cell, covers the axolemma at the myelinated internodes.

Any regions of the PNS and CNS that contain large numbers of myelinated neurons are called *white matter* because of the white color of the myelin. Regions that contain mostly unmyelinated neurons are called *gray matter* because without any myelin present, gray is the predominant color due to the dark color of the neuron's organelles. White and gray matter are clearly visible in sections of the brain and spinal cord.

QuickCheck Questions

1.1 What are the two major types of cells in the nervous system?

1.2 What are the three main regions of a neuron?

1.3 What is a node on an axon?

1 Materials

- ☐ Compound microscope
- ☐ Prepared slides of:
  - Astrocytes
  - Neurons
  - Myelinated nerve tissue (teased)

Procedures

1. Scan the astrocytes slide at low magnification and locate a group of glial cells. Examine a single astrocyte at high magnification and note the numerous cellular extensions. Draw an astrocyte in the space provided.

2. Examine the neurons slide, which is a smear of neural tissue from the CNS and has many neurons, each made up of numerous dendrites and a single unmyelinated axon. Scan the slide at low magnification and locate several neurons. Select a single neuron, increase the magnification, and identify its cellular anatomy. Draw and label a neuron in the space provided.

3. The myelinated nerve slide is a preparation from a nerve that has been teased apart to separate the individual myelinated axons. Use Figure 13.6 as a reference and examine the slide at each magnification; then, identify the myelin sheath and the nodes. Draw and label a sketch of your observations in the space provided. ■

Astrocyte

Neuron

Myelinated axon

LAB ACTIVITY 2 — **Anatomy of a Nerve**

Cranial and spinal nerves are protected and organized by three layers of connective tissue in much the same way a skeletal muscle is organized (Figure 13.7). The nerve is wrapped in an outer covering called the **epineurium**. Beneath this layer is the **perineurium,** which separates the axons into bundles called *fascicles*. Inside a fascicle, the **endoneurium** surrounds each axon and isolates it from neighboring axons.

QuickCheck Questions

2.1 What are the three connective tissue layers that organize a nerve?

2.2 How are these connective tissue arranged in a typical spinal nerve?

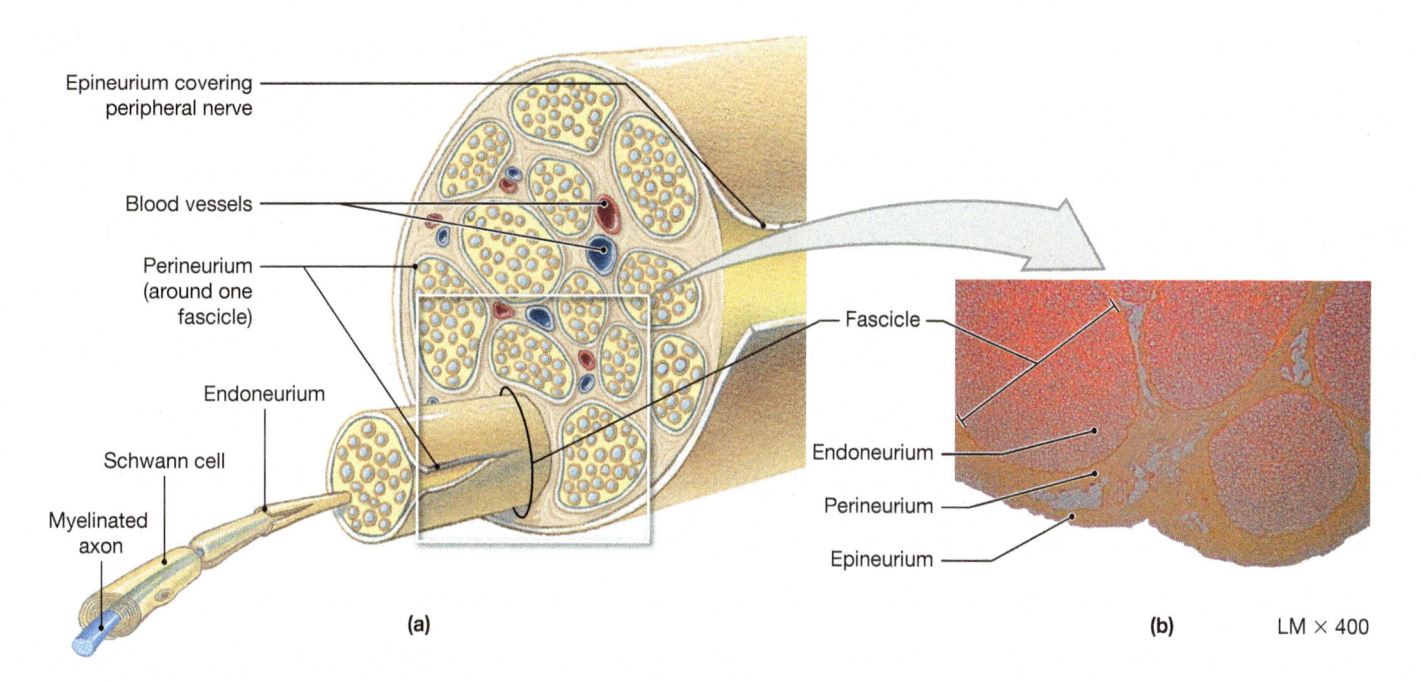

Figure 13.7 **Anatomy of a Spinal Nerve**
A spinal nerve consists of an outer epineurium that encloses a variable number of fascicles (bundles of neurons). The fascicles are wrapped by the perineurium, and within each fascicle the individual axons are encased by the endoneurium. Schwann cells encompass the axons and create a myelin sheath over them. (a) A typical spinal nerve and its connective tissue wrappings. (b) A light micrograph of a spinal nerve showing the fascicles and the three connective tissue layers. (LM × 400)

2 Materials

- ☐ Spinal cord laboratory model
- ☐ Spinal cord chart
- ☐ Compound microscope
- ☐ Prepared slide of spinal nerve

Procedures

1. Examine the spinal nerve slide at low magnification and locate the nerve section. Identify the epineurium and note how it encases the nerve.
2. Examine a single fascicle and distinguish between the perineurium and the epineurium. Locate the individual axons inside a fascicle.
3. Draw and label the nerve in the space provided. ■

Spinal nerve

LAB ACTIVITY 3 — Autonomic Nervous System

The autonomic nervous system controls the motor and glandular activity of the visceral effectors. Most internal organs have **dual innervation** and are innervated by both sympathetic and parasympathetic nerves. Thus the two divisions of the ANS share the role of regulating autonomic function. Typically, one division stimulates a given effector, and the other division inhibits that same effector. Autonomic motor pathways originate in the brain and enter the cranial and spinal nerves.

An autonomic pathway consists of two groups of neurons, both of which have names that reflect the fact that they synapse with one another in bulb-like PNS structures called **ganglia**. An autonomic neuron between the CNS and a sympathetic or parasympathetic ganglion is called a **preganglionic neuron**; an autonomic neuron between the ganglion and the target muscle or gland is a **ganglionic neuron** (Figure 13.8). Preganglionic axons, called **preganglionic fibers**, synapse with ganglionic neurons in the ganglion. Ganglionic axons, called **ganglionic fibers**, synapse with smooth muscles, the heart, and glands.

The preganglionic neurons of both divisions release acetylcholine (ACh) into a ganglion, but the ganglionic neurons of the two divisions release different neurotransmitters to the target effector cells. During times of excitement, emotional stress, and emergencies, sympathetic ganglionic neurons release norepinephrine (NE) to effectors and cause a **fight-or-flight response** that increases overall alertness. Heart rate, blood pressure, and respiratory rate all increase, sweat glands secrete, and digestive and urinary functions cease. Parasympathetic ganglionic neurons release ACh, which slows the body for normal, energy-conserving homeostasis. This **rest-and-repose response** decreases cardiovascular and respiratory activity and increases the rate at which food and wastes are processed.

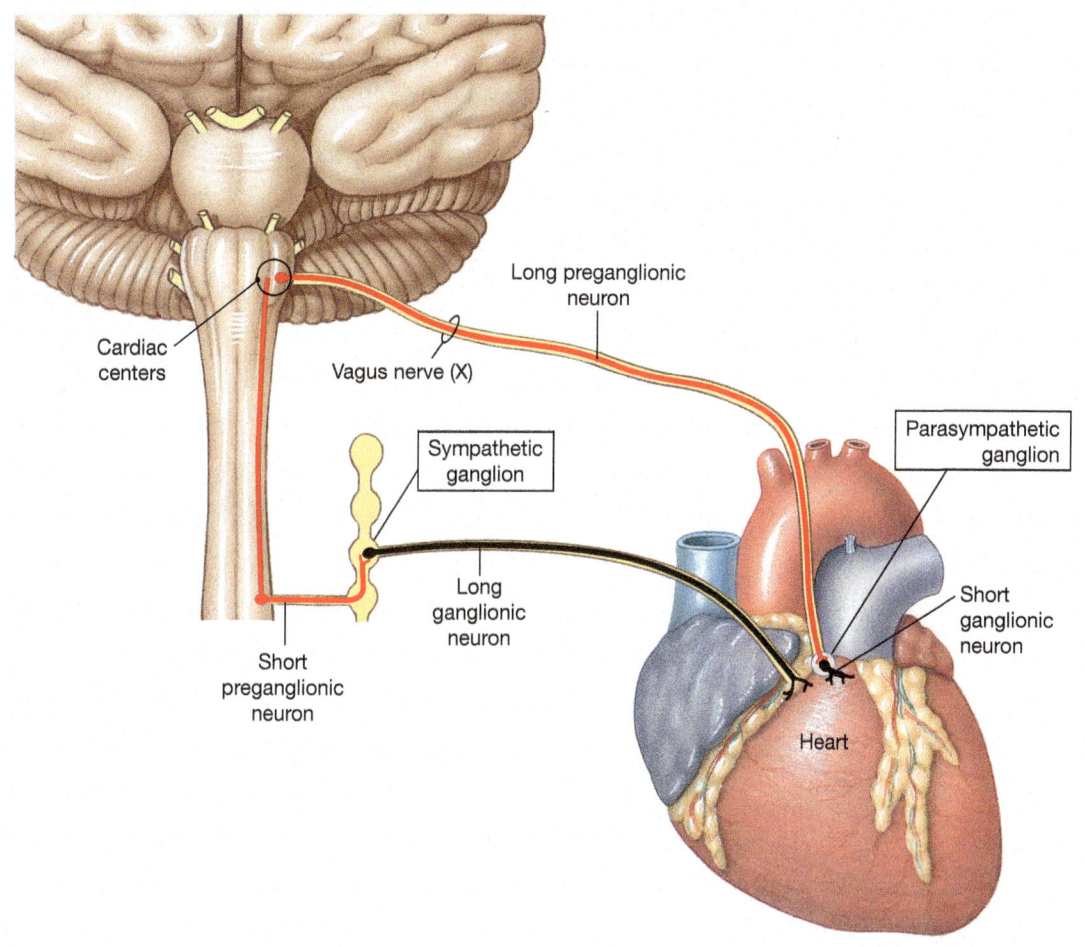

Figure 13.8 An Overview of ANS Pathways

Sympathetic pathways consist of short preganglionic neurons that release acetylcholine (ACh) in sympathetic ganglia. They synapse with long ganglionic neurons that release norepinephrine (NE) at an effector. The sympathetic response is generalized as a fight-or-flight response. Parasympathetic pathways have long preganglionic neurons that exit the CNS either directly from the brain (shown) or by passing down the spinal cord to the sacral region (not shown). They release ACh in terminal and intramural ganglia that are located in or near the effector organ. Preganglionic parasympathetic neurons synapse with short ganglionic neurons that also release ACh. The general parasympathetic response is a rest-and-repose response.

There are two major anatomical differences between the sympathetic and parasympathetic subdivisions of the ANS:

1. *Location of Preganglionic Exit Points from CNS.* Sympathetic preganglionic neurons exit the spinal cord at segments T_1 through L_2 to enter the thoracic and first two lumbar spinal nerves. Because of this nerve distribution, the sympathetic division is also called the **thoracolumbar division**. In the parasympathetic division, the efferent neurons that originate in the brain either exit the cranium in certain cranial nerves or descend the spinal cord and exit at the sacral level. The parasympathetic division is also called the **craniosacral** (krā-nē-ō-SĀ-krul) **division.**

2. *Location of Autonomic Ganglia in PNS.* All autonomic ganglia are in the PNS, but their proximity to the CNS provides another difference between the sympathetic and parasympathetic divisions. Sympathetic ganglia are located close to the spinal cord. This location results in short sympathetic preganglionic neurons and long sympathetic ganglionic neurons. Parasympathetic ganglia are

234   EXERCISE 13   Organization of the Nervous System

located either near or within the visceral effectors. With the ganglia farther away from the CNS, parasympathetic preganglionic neurons are long and parasympathetic ganglionic neurons are short. In Figure 13.8, notice both the difference in the locations of the sympathetic and parasympathetic ganglia and the difference in the preganglionic and ganglionic lengths.

Sympathetic (Thoracolumbar) Division

The organization of the sympathetic division of the ANS is diagrammed on the left in Figure 13.9.

Three types of sympathetic ganglia occur in the body: sympathetic chain ganglia, collateral ganglia, and ganglia in the adrenal medulla. **Sympathetic chain ganglia** are located lateral to the spinal cord and are also called **paravertebral**

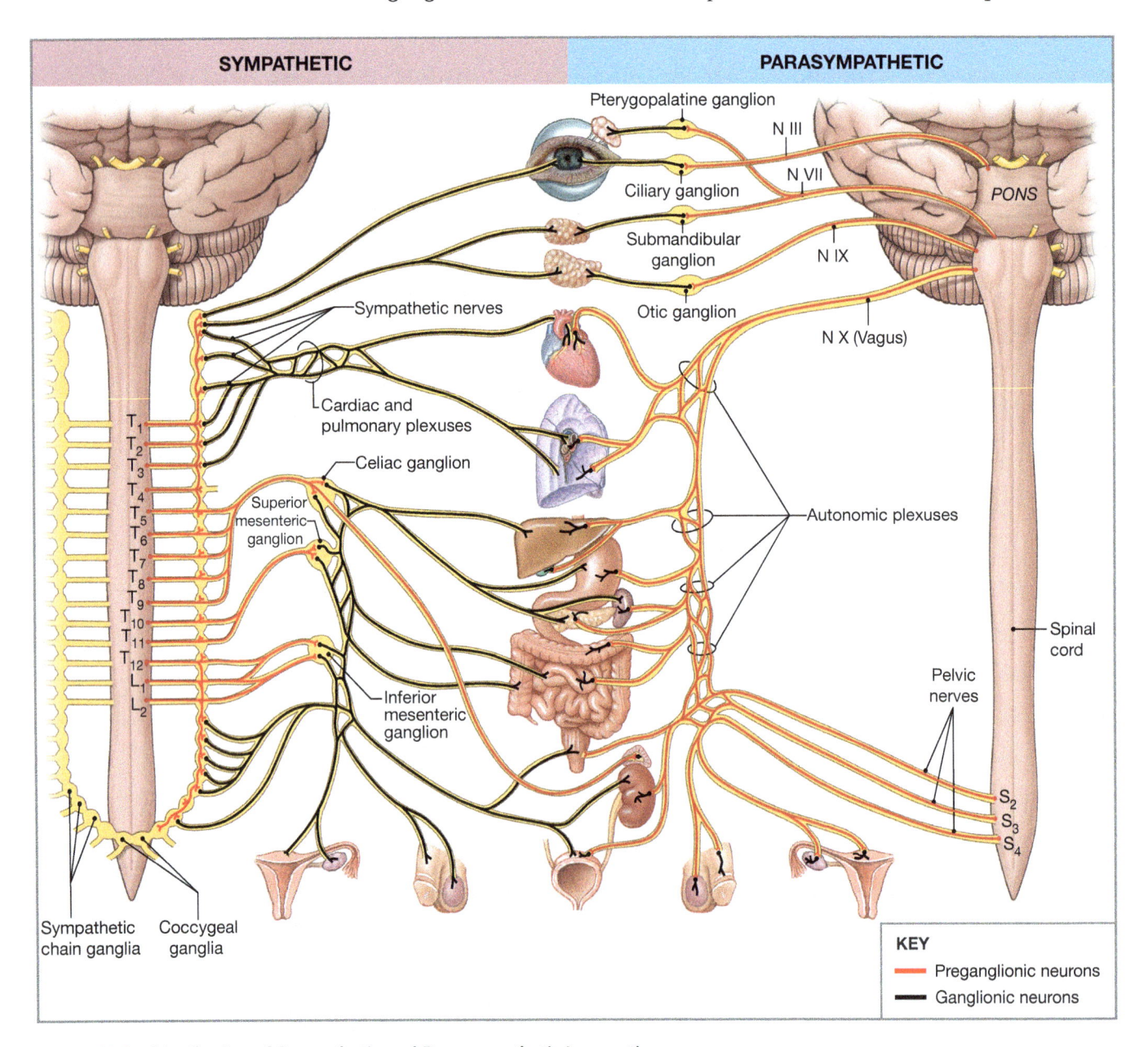

Figure 13.9   Distribution of Sympathetic and Parasympathetic Innervation
A diagram that shows the outflow of ANS nerves from the CNS to effector organs. The sympathetic and parasympathetic divisions are shown separated only for clarity. Naturally, your body is not organized this way!

ganglia. Ganglionic neurons that exit sympathetic chain ganglia innervate the effectors of the thoracic cavity, head, body wall, and limbs.

Collateral ganglia are located anterior to the vertebral column and contain ganglionic neurons that lead to organs in the abdominopelvic cavity. The preganglionic fibers that are associated with collateral ganglia pass through the sympathetic chain ganglia without synapsing and join to form a network called the **splanchnic** (SPLANK-nik) **nerves**. This network divides and sends branches into the collateral ganglia, where the preganglionic fibers synapse with ganglionic neurons. The ganglionic fibers then synapse with abdominopelvic effectors. The collateral ganglia are named after the adjacent blood vessels. The **celiac** (SĒ-lē-ak) **ganglion** supplies the liver, gallbladder, stomach, pancreas, and spleen. The **superior mesenteric ganglion** innervates the small intestine and parts of the large intestine. The **inferior mesenteric ganglion** controls most of the large intestine, the kidneys, the bladder, and the sex organs.

The third type of sympathetic ganglion is associated with the adrenal glands, which are positioned on top of the kidneys. Each adrenal gland has an outer cortex layer that produces hormones and an inner region called the **adrenal medulla.** It is this region that contains sympathetic ganglia and ganglionic neurons. During sympathetic stimulation, the ganglionic neurons in the medulla, like other sympathetic ganglionic neurons, release epinephrine into the bloodstream and contribute to the fight-or-flight response.

Clinical Application

Stress and the ANS

Stress stimulates the body to increase sympathetic commands from the ANS. Appetite may decrease while blood pressure and general sensitivity to stimuli may increase. The individual may become irritable and have difficulty sleeping and coping with day-to-day responsibilities. Prolonged stress can lead to disease. Coronary diseases, for example, are more common in individuals who are employed in stressful occupations or live in stressful environments.

Parasympathetic (Craniosacral) Division

There are two main types of parasympathetic ganglia: terminal and intramural. **Terminal ganglia** are located near the eye and salivary glands; **intramural** (*intra*, within; *mura*, walls) **ganglia** are embedded in the walls of effector organs. In the brain, parasympathetic preganglionic neurons branch into four cranial nerves: oculomotor, facial, glossopharyngeal, and vagus (See Figure 13.9). For the first three of these nerves, there is a separate terminal ganglion for each one. The oculomotor nerve (N III) to the eyes enters the **ciliary ganglion**, the facial nerve (N VII) passes into the **pterygopalatine** and **submandibular ganglia**, and the glossopharyngeal nerve (N IX) includes the **otic ganglion**. Intramural ganglia receive preganglionic neurons in the vagus nerve (N X), which exits the brain, travels down the musculature of the neck, enters the ventral body cavity, and spreads into the intramural ganglia of the internal organs. The sacral portion of the parasympathetic division contains preganglionic neurons in sacral segments S_2, S_3, and S_4. The preganglionic fibers remain separate from spinal nerves and exit from spinal segments S_2 through S_4 as **pelvic nerves**. The organization of the parasympathetic division of the ANS is diagrammed in the right half of Figure 13.9.

Networks of preganglionic neurons, called *autonomic plexuses,* occur between the vagus nerve and the pelvic nerves. In these plexuses, sympathetic preganglionic neurons and parasympathetic preganglionic neurons intermingle as they pass to their respective autonomic ganglia.

QuickCheck Questions

3.1 What are the two main divisions of the ANS?

3.2 How do the heart, lungs, and digestive tract respond to sympathetic stimulation?

3.3 How do the heart, lungs, and digestive tract respond to parasympathetic stimulation?

3 Materials

- Nervous system chart

Procedures

1. Review the anatomy and pathways that are presented in Figures 13.8 and 13.9.
2. On the nervous system chart, identify a sympathetic chain ganglion, a collateral ganglion, and an adrenal gland.
3. On the chart, identify the terminal ganglia of the head and the intramural ganglia in the organs of the ventral body cavity.
4. On the chart, compare the distribution of sympathetic and parasympathetic nerves that run to the heart. ∎

Name _____

Date _____

Section _____

LAB REPORT

EXERCISE

13

Organization of the Nervous System

A. Matching

Match each structure in the left column with its correct description from the right column.

_____ 1. dendrite
_____ 2. axon
_____ 3. collateral branches
_____ 4. bouton terminal
_____ 5. axon hillock
_____ 6. telodendria
_____ 7. astrocyte
_____ 8. myelinated internode
_____ 9. neurilemma
_____ 10. synaptic vesicles
_____ 11. axolemma
_____ 12. cell body

A. main branches of axon
B. outer Schwann cell membrane
C. contains neurotransmitters
D. fine branches of axon
E. axon region associated with Schwann cell
F. forms blood-brain barrier
G. directs impulses to cell body
H. also called soma
I. enlarged end of axon
J. connects cell body and axon
K. membrane of axon
L. conducts impulses away from cell body

B. Short-Answer Questions

1. Compare the CNS and the PNS.

2. Describe the microscopic appearance of an astrocyte.

237

EXERCISE 13

LAB REPORT

3. Describe the three layers of connective tissue that organize a spinal nerve.

4. List four responses to sympathetic stimulation and four responses to parasympathetic stimulation.

C. Analysis and Application

1. How would an injury to the afferent neurons in the left leg affect the victim's sensory and motor functions?

2. While observing a microscopic specimen of nerve tissue from the brain, you notice an axon that is encased by a different cell. Describe the covering over the axon and identify the cell that has surrounded the axon.

3. As a child, you might have been told to wait for up to an hour after eating before going swimming. Suggest a possible rationale for this statement.

4. Compare the outflow of sympathetic and parasympathetic pathways from the CNS into PNS.

5. Compare the lengths of preganglionic and ganglionic neurons in the sympathetic division with the lengths of these neurons in the parasympathetic division and explain the basis of the differences.

EXERCISE 14

The Spinal Cord and Spinal Nerves

OBJECTIVES

On completion of this exercise, you should be able to:

- Identify the major surface features of the spinal cord, including the spinal meninges.
- Describe the internal anatomy of the spinal cord.
- Describe the organization and distribution of spinal nerves.

LAB ACTIVITIES

1. Gross Anatomy of the Spinal Cord   239
2. Spinal Meninges   243
3. Spinal Nerves   243
4. Dissection of the Spinal Cord   251

The **spinal cord** is the long, cylindrical portion of the central nervous system that is located in the spinal cavity of the vertebral column. It connects the peripheral nervous system (PNS) with the brain. Sensory information from the PNS enters the spinal cord and ascends to the brain. Motor signals from the brain descend the spinal cord and exit the spinal cord to reach the effectors. The spinal cord is more than just a conduit to and from the brain, however. It also processes information and produces **spinal reflexes**. A classic example of a spinal reflex is the stretch reflex that occurs when the tendon over the patella is struck; the spinal cord responds to the tap by stimulating the extensor muscles of the leg in the well-known "knee-jerk" reflex.

LAB ACTIVITY 1 Gross Anatomy of the Spinal Cord

The spinal cord is continuous with the inferior portion of the brain stem. It passes through the foramen magnum, descends approximately 45 cm (18 in.) down the spinal canal of the vertebral column, and terminates between lumbar vertebrae L_1 and L_2. In young children, the spinal cord extends through most of the spine. After the age of four, the spinal cord stops lengthening, but the spine continues to grow. By adulthood, therefore, the spinal cord is shorter than the spine and descends only to the level of the upper lumbar vertebrae.

As Figure 14.1 shows, the diameter of the spinal cord is not constant along its length. Two enlarged regions occur where the spinal nerves of the limbs join the spinal cord. The **cervical enlargement** in the neck supplies nerves to the upper limbs. The **lumbar enlargement** occurs near the distal end of the cord, where nerves supply the pelvis and lower limbs. Inferior to the lumbar enlargement, the spinal cord narrows and terminates at the **conus medullaris**. Spinal nerves fan out from the conus medullaris in a group called the **cauda equina** (KAW-duh ek-WĪ-nuh), or the "horse's tail." A thin thread of fibrous tissue, the **filum terminale**, extends past the conus medullaris to anchor the spinal cord in the sacrum.

The spinal cord is organized into 31 segments. Each segment is attached to two spinal nerves, one on each side of the segment as shown in Figure 14.1. Each of the

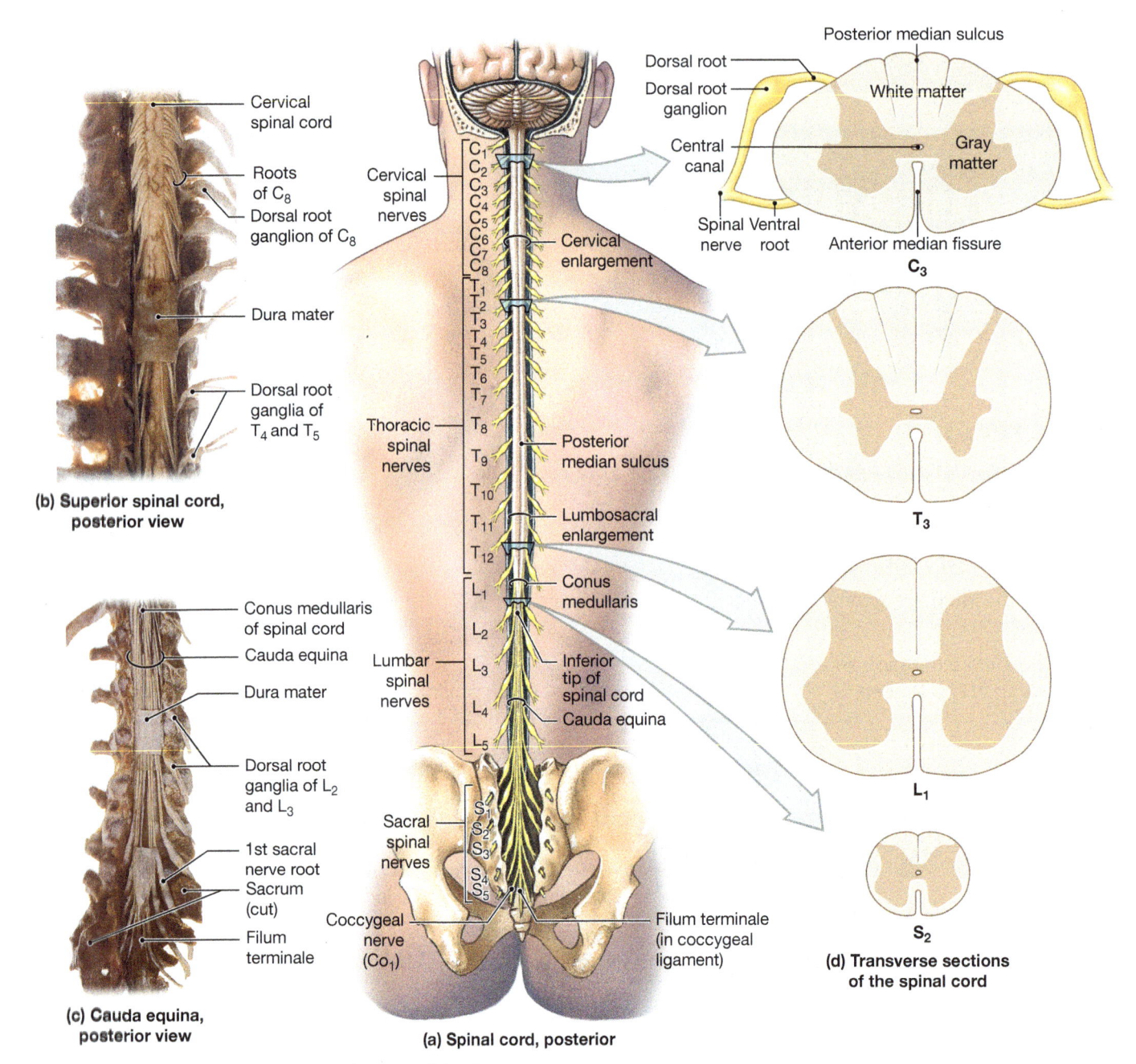

240 EXERCISE 14 The Spinal Cord and Spinal Nerves

Figure 14.1 Gross Anatomy of the Adult Spinal Cord
(a) The superficial anatomy and orientation of the adult spinal cord. The numbered letters on the left identify the spinal nerves and indicate where the nerve roots leave the vertebral canal. The spinal cord extends from the brain only to the level of vertebrae L_1–L_2; the spinal segments found at representative locations are indicated in the cross-sections. (b) Dissection of the superior portion of spinal cord shown in posterior view. (c) Dissection of inferior portion of spinal cord shown in posterior view. (d) Inferior views of cross-sections through representative segments of the spinal cord that show the arrangement of gray matter and white matter.

two spinal nerves on a given cord segment is formed by joining two lateral extensions of the segment. One of these extensions, the **dorsal root**, contains sensory neurons that enter the spinal cord from sensory receptors. The dorsal root swells at the **dorsal root ganglion**, which is where cell bodies of sensory neurons cluster. The other extension, the **ventral root**, consists of motor neurons that exit the CNS and lead to effectors. The two roots join to form the spinal nerve. Each spinal nerve is therefore a *mixed nerve*, as noted in Exercise 13, and carries both sensory and motor information. (The first spinal nerve does not have a dorsal root and is therefore a motor nerve.)

EXERCISE 14 The Spinal Cord and Spinal Nerves   241

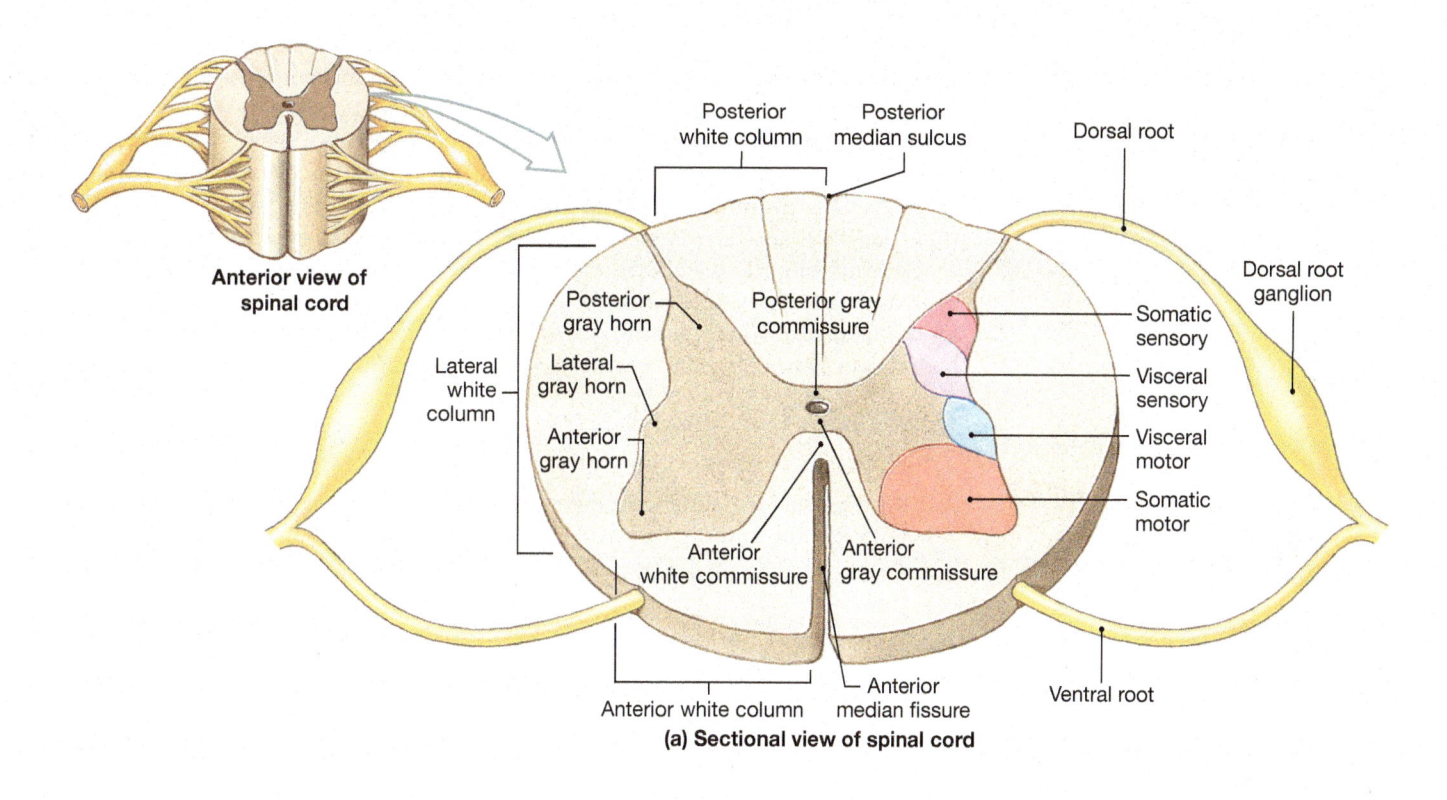

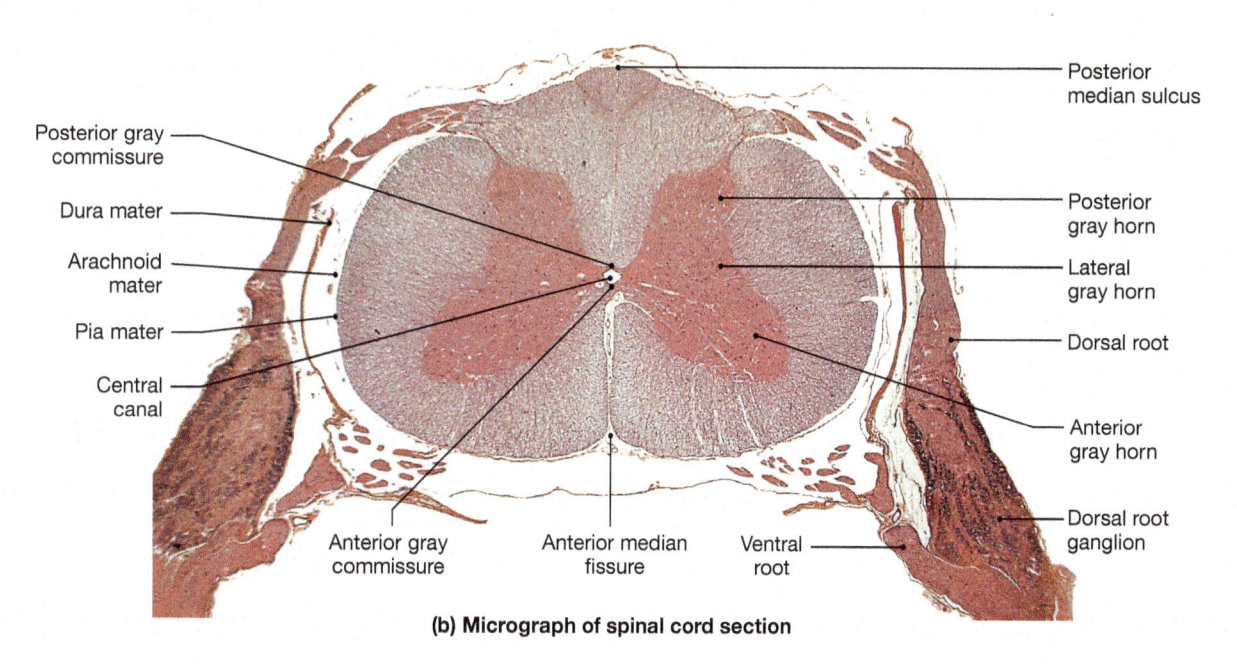

Figure 14.2 Organization of the Spinal Cord
(a) The left half of this cross-sectional view shows important anatomical landmarks in the gray and white matter. The right half indicates the functional organization of the gray matter in the anterior, lateral, and posterior gray horns. (b) A micrograph of a section through the spinal cord that shows major landmarks.

Figure 14.2 illustrates the spinal cord in transverse section to show the internal anatomy, also called the *sectional anatomy*. The cord is divided by the deep and conspicuous **anterior median fissure** and by the shallow **posterior median sulcus**. An H-shaped area called the **gray horns** contains many glial cells and neuron cell bodies. Each horn contains a specific type of neuron. The **posterior gray horns** carry sensory neurons into the spinal cord, and the **anterior gray horns** carry somatic motor neurons out of the cord and to skeletal muscles. In the sacral region, the anterior gray horns have preganglionic neurons of the parasympathetic nervous system. The **lateral gray horns** occur in spinal segments T_1–L_2 and consist of visceral motor neurons. Axons may cross to the opposite side of the spinal cord at the crossbars of the horns, called the **anterior** and **posterior gray commissures**. Between the gray commissures is a hole, called the **central canal,** that contains cerebrospinal fluid. The central canal is continuous with the fluid-filled ventricles of the brain. Collectively, all these structures are sometimes referred to as the spinal cord's **gray matter.**

Surrounding the gray horns are six masses of white matter: the **posterior, lateral**, and **anterior white columns**. The two anterior white columns are connected by the **anterior white commissure**. Within each white column, the myelinated axons form distinct bundles of neurons, which are called either **tracts** or **fascicles**.

QuickCheck Questions

1.1 How is the white and gray matter of the spinal cord organized?

1.2 Which structure is useful in determining which portion of a spinal cord cross-section is the anterior region?

1.3 Why is the spinal cord shorter than the vertebral column?

Materials

- Spinal cord model
- Spinal cord chart
- Dissection microscope
- Compound microscope
- Prepared slide of transverse section of spinal cord

Procedures

1. Review Figures 14.1 and 14.2.
2. Locate each surface feature of the spinal cord on the spinal cord model and chart.
3. Review the internal anatomy of the spinal cord on the spinal cord model.
4. Examine the microscopic features of the spinal cord in transverse section by following this sequence:
   - View the slide at low magnification with the dissection microscope. Identify the anterior and posterior regions.
   - Transfer the slide to a compound microscope. Move the slide around to survey the preparation at low magnification, again identifying the posterior and anterior aspects.
   - Examine the central canal and gray horns. Can you distinguish among the posterior, lateral, and anterior gray horns? Locate the gray commissures.
   - Examine the white columns. What is the difference between gray and white matter in the CNS?
   - Draw a spinal cord cross-section in the space provided here. ∎

Spinal cord cross-section

LAB ACTIVITY 2 Spinal Meninges

The spinal cord is protected within three layers of **spinal meninges** (men-IN-jēz). The outer layer, the **dura mater** (DOO-ruh MĀ-ter), is composed of tough, fibrous connective tissue (Figure 14.3). The fibrous tissue attaches to the bony walls of the spinal canal and supports the spinal cord laterally. Superficial to the dura mater is the **epidural space,** which contains adipose tissue that pads the spinal cord. The **arachnoid** (a-RAK-noyd) **mater** is the second meningeal layer. A small cavity called the **subdural space** separates the dura mater from the arachnoid mater. Deep to the arachnoid mater is the **subarachnoid space**, which contains cerebrospinal fluid that protects and cushions the spinal cord. The **pia mater** is the thin inner meningeal layer that lies directly over the spinal cord. Blood vessels that supply the spinal cord are held in place by the pia mater. The pia mater extends laterally on each side of the spinal cord as the **denticulate ligament,** which joins the dura mater for lateral support to the spinal cord. Another extension of the pia mater, the filum terminale (see Figure 14.1a), supports the spinal cord inferiorly.

Clinical Application

Epidural Injections and Spinal Taps

During childbirth, the expectant mother may receive an **epidural block**, a procedure that introduces anesthesia in the epidural space. A thin needle is inserted between two lumbar vertebrae, and the anesthetic drug is injected into the epidural space. The anesthetic numbs only the spinal nerves of the pelvis and lower limbs and reduces the discomfort the woman feels during the powerful labor contractions of her uterus.

A **spinal tap** is a procedure in which a needle is inserted into the subarachnoid space to withdraw a sample of cerebrospinal fluid. The fluid is then analyzed for the presence of microbes, wastes, and metabolites. To prevent injury to the spinal cord, the needle is inserted into the lower lumbar region inferior to the cord.

QuickCheck Questions

2.1 What are the three layers of spinal meninges?

2.2 Where does cerebrospinal fluid circulate in the spinal cord?

2 Materials

- ☐ Spinal cord model
- ☐ Spinal cord chart
- ☐ Compound microscope
- ☐ Prepared slide of transverse section of spinal cord

Procedures

1. Review Figure 14.3.
2. Locate the spinal meninges on the spinal cord model and chart.
3. Use the compound microscope to examine the spinal meninges in transverse section. Move the slide around to survey the preparation. Locate the dura mater, arachnoid mater, pia mater, and the associated spaces between the meninges.
4. Add the spinal meninges to the drawing you began in Lab Activity 1.

LAB ACTIVITY 3 Spinal Nerves

Two types of nerves connect PNS sensory receptors and effectors to the CNS: 12 pairs of cranial nerves and 31 pairs of spinal nerves. As their names indicate, cranial nerves connect with the brain and spinal nerves communicate with the spinal cord.

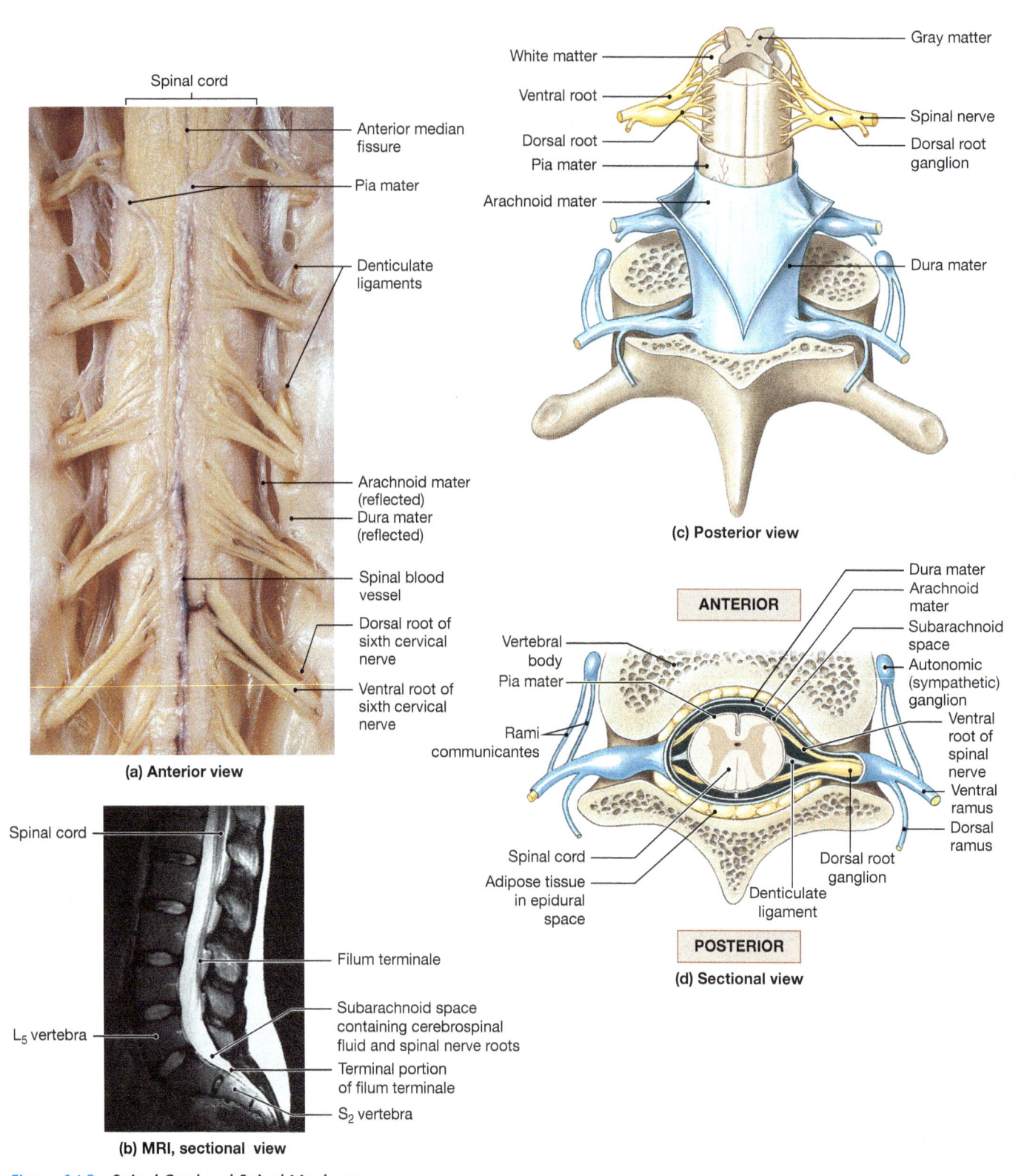

Figure 14.3 Spinal Cord and Spinal Meninges
(a) Anterior view of the spinal cord and spinal nerve roots in the vertebral canal. The dura mater and arachnoid mater have been reflected. (b) MRI of inferior spinal cord in sagittal view. (c) Posterior view of the spinal cord that shows the meningeal layers, superficial landmarks, and distribution of gray and white matter. (d) Sectional view through the spinal cord and meninges that shows the peripheral distribution of the spinal nerves.

244

EXERCISE 14 The Spinal Cord and Spinal Nerves 245

As noted at the opening of this exercise, spinal nerves branch into PNS nerves, and spinal nerves comprise the axons of PNS sensory and motor neurons.

The two spinal nerves on a given spine segment exit the vertebral canal by passing through an intervertebral foramen between two adjacent vertebrae (see Figure 14.3). Each spinal nerve divides into a series of peripheral nerves. The posterior branch is called the **dorsal ramus** and supplies the skin and muscles of the back, and the anterior branch, called the **ventral ramus**, innervates the anterior and lateral skin and muscles. The ventral ramus has additional branches, called the **rami communicantes**, that innervate autonomic ganglions. The rami communicantes consists of two branches: a **white ramus,** which passes ANS preganglionic neurons from the spinal nerve into the ganglion, and a **gray ramus**, which carries ganglionic neurons back into the spinal nerve. Once in the spinal nerve, the ganglionic neurons travel in the ventral or dorsal ramus to their target effector. As their names imply, the white ramus has *myelinated* preganglionic neurons and the gray ramus has *unmyelinated* ganglionic neurons.

There are eight **cervical nerves** (C_1 through C_8), 12 **thoracic nerves** (T_1 through T_{12}), five **lumbar nerves** (L_1 through L_5), five **sacral nerves** (S_1 through S_5), and a single **coccygeal nerve** (Co_1) (Figure 14.4). The cervical nerves exit superior to their corresponding vertebrae, except for C_8, which exits inferior to vertebra C_7.

The thoracic and lumbar spinal nerves are named after the vertebra that is immediately above each nerve, which means that thoracic nerve T_1 is inferior to vertebra T_1. Only the spinal nerves that have autonomic neurons carry visceral motor information. Cervical, some lumbar, and coccygeal spinal nerves do not have ANS neurons.

Groups of spinal nerves join in a network called a **plexus**. As muscles form during fetal development, the spinal nerves that supplied the individual muscles interconnect and create a plexus. There are four of these regions, as shown in Figure 14.4: cervical, brachial, lumbar, and sacral.

Cervical Plexus

The eight cervical spinal nerves supply the neck, shoulder, upper limb, and diaphragm. The various branches of the **cervical plexus** contain nerves C_1 through C_4 and parts of C_5 (Figure 14.5 and Table 14.1). This plexus innervates muscles of the larynx plus the sternocleidomastoid, trapezius, and diaphragm muscles. It also innervates the skin of the neck, shoulder, and upper limb.

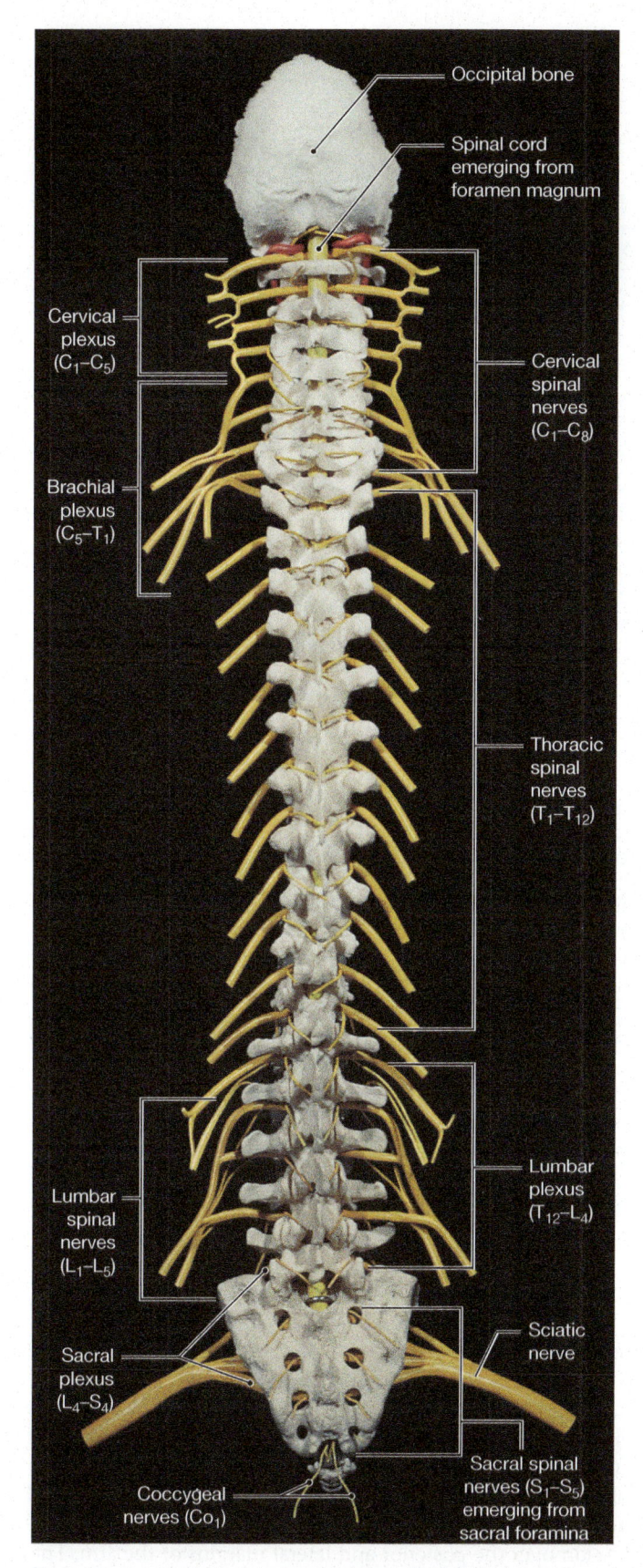

Figure 14.4 Posterior View of the Vertebral Column and Spinal Nerves

The yellow wires represent spinal nerves. The groups of nerves are interwoven into a network called a plexus. There are four plexuses: cervical, thoracic, lumbar, and sacral.

Figure 14.5 Cervical Plexus

The cervical plexus innervates muscles of the neck and branch into the thoracic cavity to control the diaphragm.

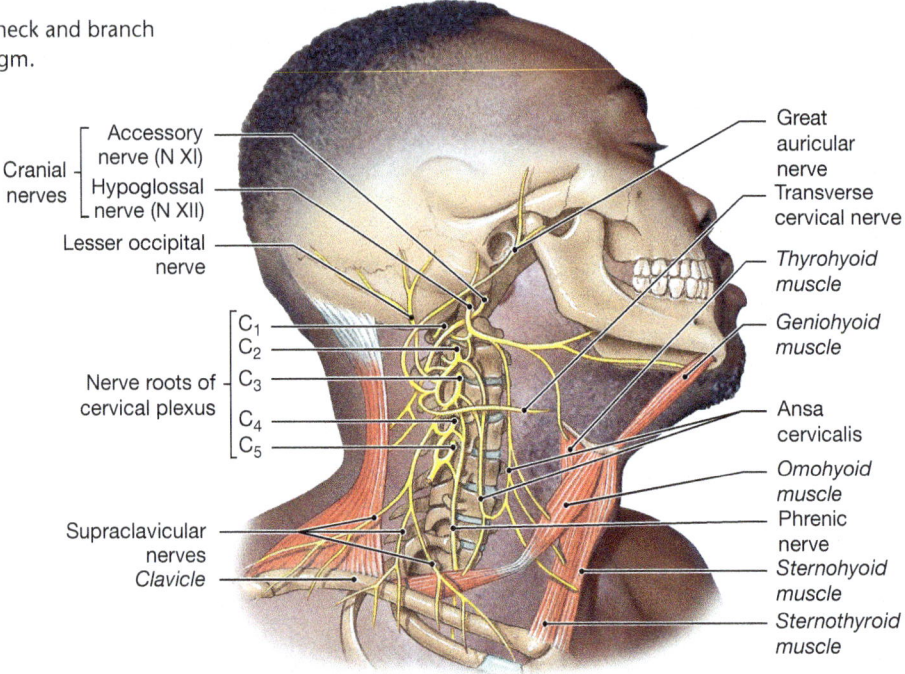

Table 14.1 The Cervical Plexus

Spinal Segments	Nerves	Distribution
C_1–C_4	Ansa cervicalis superior (and inferior branches)	Five of the extrinsic laryngeal muscles (sternothyroid, sternohyoid, omohyoid, geniohyoid, and thyrohyoid) by way of N XII
C_2–C_3	Lesser occipital, transverse cervical, supraclavicular, and great auricular nerves	Skin of upper chest, shoulder, neck, and ear
C_3–C_5	Phrenic nerve	Diaphragm
C_1–C_5	Cervical nerves	Levator scapulae, scalenes, sternocleidomastoid, and trapezius muscles (with N XI)

Brachial Plexus

The **brachial plexus** includes the parts of spinal nerve C_5 that are not involved with the cervical plexus, plus nerves C_6, C_7, C_8, and T_1. This plexus is more complex than the cervical plexus and branches to innervate the shoulder, the upper limb, and some muscles on the trunk (Figure 14.6 and Table 14.2). The major branches of this plexus are the axillary, radial, musculocutaneous, median, and ulnar nerves. The **axillary nerve** (C_5 and C_6) supplies the deltoid and teres minor muscles and the skin of the shoulder. The **radial nerve** (C_5 through T_1) controls the extensor muscles of the upper limb as well as the skin over the posterior and lateral margins of the arm. The **musculocutaneous nerve** (C_5 through C_7) supplies the flexor muscles of the upper limb and the skin of the lateral forearm. The **median nerve** (C_6 through T_1) innervates the flexor muscles of the forearm and digits, the pronator muscles, and the lateral skin of the hand. The **ulnar nerve** (C_8 and T_1) controls the flexor carpi ulnaris muscle of the forearm, other muscles of the hand, and the medial skin of the hand.

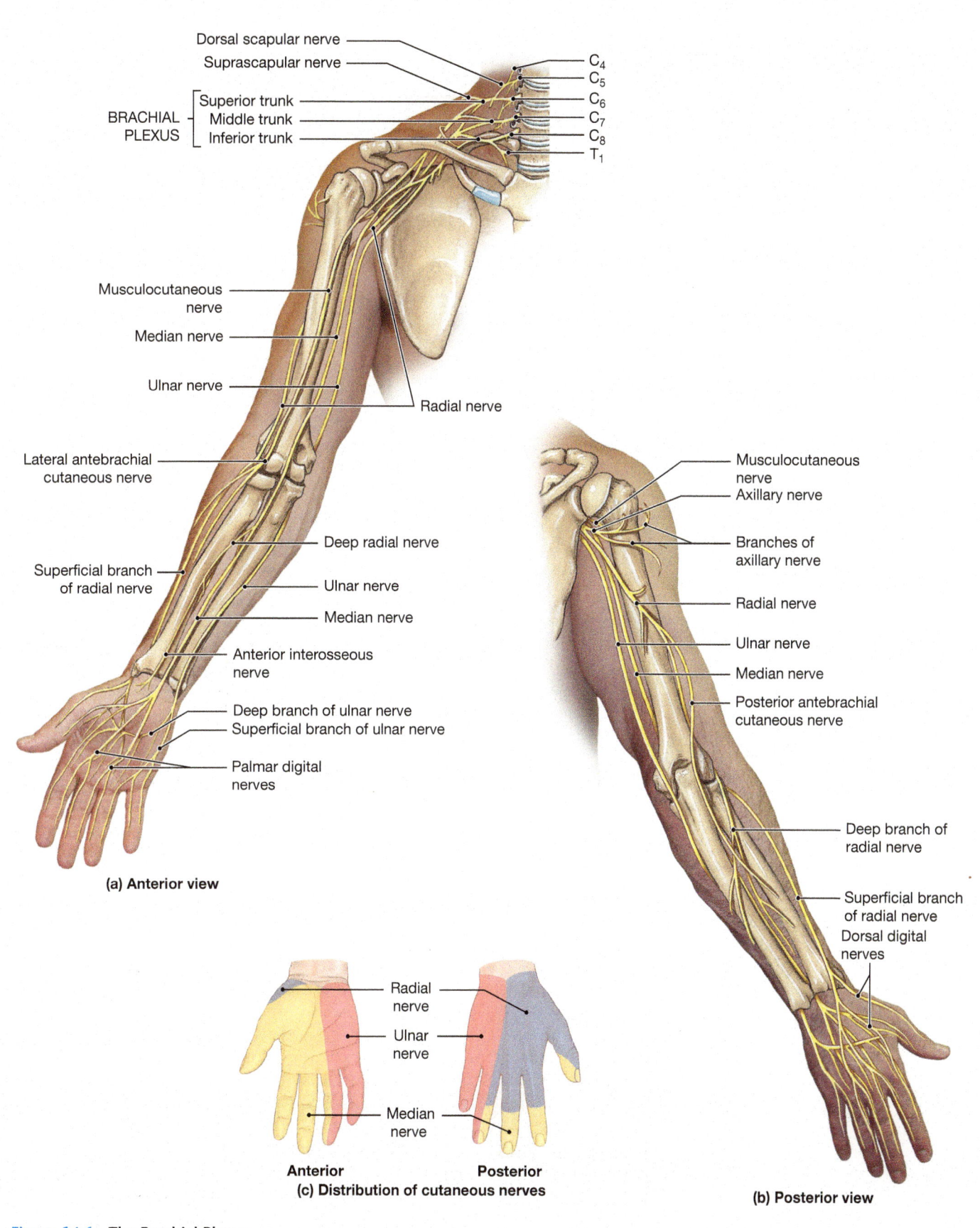

Figure 14.6 The Brachial Plexus

The brachial plexus innervates muscles of the pectoral girdle and upper limb. **(a)** Anterior view of brachial plexus that innervates the right upper limb, showing the peripheral distributions of major nerves. **(b)** Posterior view. **(c)** Area of hands that is serviced by nerves of the right brachial plexus.

247

Table 14.2 The Brachial Plexus

Spinal Segments	Nerve(s)	Distribution
C_4–C_6	Nerve to subclavius	Subclavius muscle
C_5	Dorsal scapular nerve	Rhomboid and levator scapulae muscles
C_3–C_7	Long thoracic nerve	Serratus anterior muscle
C_5, C_6	Suprascapular nerve	Supraspinatus and infraspinatus muscles; sensory from shoulder joint and scapula
C_5–T_1	Pectoral nerves (medial and lateral)	Pectoralis muscles
C_5, C_6	Subscapular nerves	Subscapularis and teres major muscles
C_6–C_8	Thoracodorsal nerve	Latissimus dorsi muscle
C_5, C_6	Axillary nerve	Deltoid and teres minor muscles; sensory from skin of shoulder
C_8–T_1	Medial antebrachial cutaneous nerve	Sensory from skin over anterior and medial surface of arm and forearm
C_5–T_1	Radial nerve	Many extensor muscles on the arm and forearm (triceps brachii, anconeus, extensor carpi radialis, extensor carpi ulnaris, and brachioradialis muscles); supinator muscle, digital extensor muscles, and abductor pollicis muscle via the *deep branch;* sensory from skin over the posterolateral surface of the limb through the *posterior brachial cutaneous nerve* (arm), *posterior antebrachial cutaneous nerve* (forearm), and the *superficial branch* (radial portion of hand)
C_5–C_7	Musculocutaneous nerve	Flexor muscles on the arm (biceps brachii, brachialis, and coracobrachialis muscles); sensory from skin over lateral surface of the forearm through the *lateral antebrachial cutaneous nerve*
C_6–T_1	Median nerve	Flexor muscles on the forearm (flexor carpi radialis and palmaris longus muscles); pronator quadratus and pronator teres muscles; radial half of flexor digitorum profundus muscle, digital flexors (through the *anterior interosseous nerve*); sensory from skin over anterolateral surface of the hand
C_8, T_1	Ulnar nerve	Flexor carpi ulnaris muscle, ulnar half of flexor digitorum profundus muscle, adductor pollicis muscle, and small digital muscles through the *deep branch;* sensory from skin over medial surface of the hand through the *superficial branch*

Notice how overlap occurs in the brachial plexus. For example, spinal nerve C_6 innervates both flexor and extensor muscles. The dissection view of Figure 14.7 shows the four main branches of the brachial plexus.

Lumbar and Sacral Plexuses

The largest nerve network is called the **lumbosacral plexus**. It is a combination of the **lumbar plexus** (T_{12}, L_1–L_4) and the **sacral plexus** (L_4, L_5, S_1–S_4). (Note that thoracic spinal nerves T_2 through T_{11} are not part of any plexus but instead constitute **intercostal nerves** that enter the spaces between the ribs. The intercostal nerves innervate the intercostal muscles and abdominal muscles and receive sensations from the lateral and anterior trunk.) Figures 14.8 and 14.9 and Table 14.3 present the distribution of nerves in this combined plexus. The major nerves of the lumbar plexus innervate the skin and muscles of the abdominal wall, genitalia, and thigh. The **genitofemoral nerve** supplies some of the external genitalia and the anterior and lateral skin of the thigh. The **lateral femoral cutaneous nerve** innervates the

Figure 14.7 Cervical and Brachial Plexuses, Anterior View
This dissection shows the major nerves that arise from the cervical and brachial plexuses.

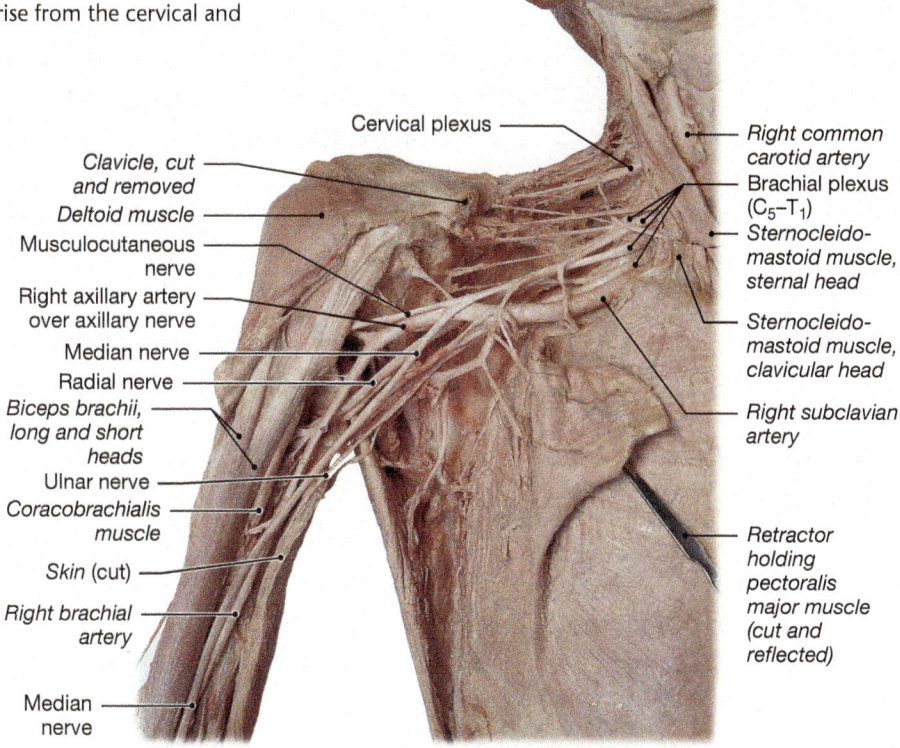

skin of the thigh from all aspects except the medial region. The **femoral nerve** controls the muscles of the anterior thigh and the adductor muscles and medial skin of the thigh.

The sacral plexus consists of two major nerves, the sciatic and the pudendal. The **sciatic nerve** descends the posterior lower limb and sends branches into the posterior thigh muscles and the musculature and skin of the leg. The **pudendal nerve** supplies the muscular floor of the pelvis, the perineum, and parts of the skin of the external genitalia.

QuickCheck Questions

3.1 What are the two groups of nerves in the PNS?
3.2 Which branch of a peripheral nerve innervates the limbs?
3.3 What are the four rami of a spinal nerve, and their characteristics?
3.4 What is a plexus?
3.5 What are the four plexuses in the body?

3 Materials

- ☐ Spinal cord model
- ☐ Spinal cord chart

Procedures

1. Review Figures 14.4–14.9.
2. Locate each nerve plexus on the spinal cord model and chart.
3. Locate the spinal nerves that are assigned by your instructor on the spinal cord model. ∎

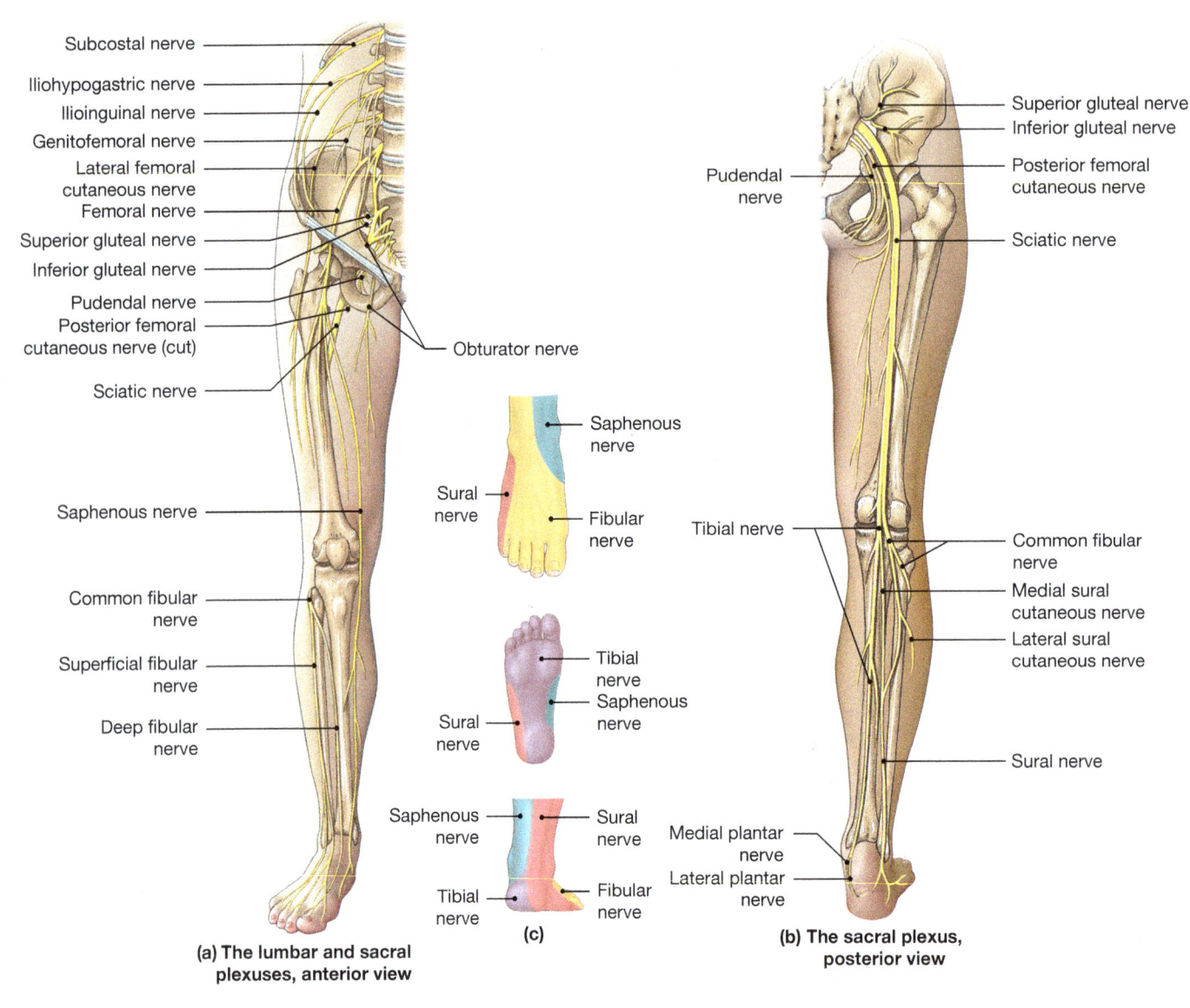

Figure 14.8 Nerves of the Lumbar and Sacral Plexuses

Nerves of the lumbar and sacral plexuses innervate the pelvis and lower limb. Areas of the foot that are serviced by these nerves is shown. (a) The major nerves of the right lumbar plexus. (b) The major nerves of the right sacral plexus. (c) Area of feet that is serviced by nerves of the right lumbar and sacral plexuses.

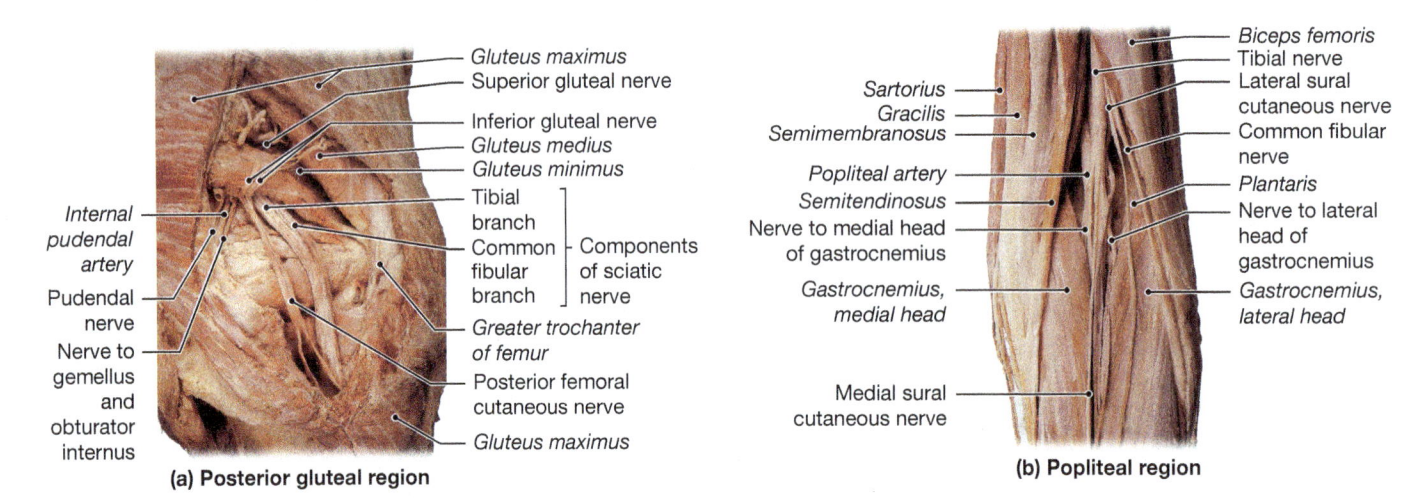

Figure 14.9 Lumbar and Sacral Plexuses, Posterior View

Major nerves are seen in (a) a dissection of the right gluteal region and (b) a dissection of the popliteal fossa of the right lower limb.

Table 14.3 The Lumbar and Sacral Plexuses

Spinal Segment(s)	Nerve(s)	Distribution
Lumbar Plexus		
T_{12}–L_1	Iliohypogastric nerve	Abdominal muscles (external and internal oblique muscles, transverse abdominis muscles); skin over inferior abdomen and buttocks
L_1	Ilioinguinal nerve	Abdominal muscles (with *iliohypogastric nerve*); skin over superior and medial thigh, and portions of external genitalia
L_1–L_2	Genitofemoral nerve	Skin over anteromedial surface of thigh and portions of external genitalia
C_2–L_3	Lateral femoral cutaneous nerve	Skin over anterior, lateral, and posterior surfaces of thigh
L_2–L_4	Femoral nerve	Anterior muscles of thigh (sartorius muscle and quadriceps group); adductors of hip (pectineus and iliopsoas muscles); skin over anteromedial surface of thigh and medial surface of leg and foot
L_2–L_4	Obturator nerve	Adductors of hip (adductors magnus, brevis, and longus); gracilis muscle; skin over medial surface of thigh
L_2–L_4	Saphenous nerve	Skin over medial surface of leg
Sacral Plexus		
L_4–S_2	Gluteal nerves:	
	Superior	Abductors of hip (gluteus minimus, gluteus medius, and tensor fasciae latae)
	Inferior	Extensor of hip (gluteus maximus)
S_1–S_3	Posterior femoral cutaneous nerve	Skin of perineum and posterior surface of thigh and leg
L_4–S_3	Sciatic nerve:	Two of the hamstrings (semimembranosus and semitendinosus); adductor magnus (with *obturator nerve*)
	Tibial nerve	Flexors of knee and extensors (plantar flexors) of ankle (popliteus, gastrocnemius, soleus, and tibialis posterior muscles and long head of the biceps femoris muscle); flexors of toes; skin over posterior surface of leg; plantar surface of foot
	Fibular nerve	Short head of biceps femoris muscle; fibularis (brevis and longus) and tibialis anterior muscles; extensors of toes; skin over anterior surface of leg and dorsal surface of foot; skin over lateral portion of foot (through the *sural nerve*)
S_2–S_4	Pudendal nerve	Muscles of perineum, including urogenital diaphragm and external anal and urethral sphincter muscles; skin of external genitalia and related skeletal muscles (bulbospongiosus and ischiocavernosus muscles)

LAB ACTIVITY 4 Dissection of the Spinal Cord

Dissecting a preserved sheep or cow spinal cord provides you the opportunity to examine the meningeal layers and the internal anatomy.

⚠ Safety Alert: Dissecting the Spinal Cord

You must—repeat, *must*—practice the highest level of laboratory safety while handling and dissecting the spinal cord. Keep the following guidelines in mind during the dissection:

1. Be sure to use only a *preserved* spinal cord for dissection because fresh spinal cords can carry disease.
2. Wear gloves and safety glasses to protect yourself from the fixatives used to preserve the specimen.
3. Do not dispose of the fixative from your specimen. You will later store the specimen in the fixative to keep the specimen moist and to keep it from decaying.
4. Be extremely careful when using a scalpel or other sharp instrument. Always direct cutting and scissor motions away from you to prevent an accident if the instrument slips on moist tissue.

EXERCISE 14 The Spinal Cord and Spinal Nerves

5. Before cutting a given tissue, make sure it is free from underlying and/or adjacent tissues so that they will not be accidentally severed.
6. Never discard tissue in the sink or trash. Your instructor will inform you of the proper disposal procedure. ▲

QuickCheck Questions

4.1 What safety equipment is required for the spinal cord dissection?

4.2 What is the disposal procedure as discussed by your laboratory instructor?

4 Materials

- ☐ Gloves
- ☐ Safety glasses
- ☐ Segment of preserved sheep or cow spinal cord
- ☐ Dissection pan
- ☐ Dissection pins
- ☐ Scissors
- ☐ Scalpel
- ☐ Forceps
- ☐ Blunt probe

Procedures

Put on gloves and safety glasses before opening the container of preserved spinal cord segments or handling one of the segments.

1. Lay the spinal cord on the dissection pan and cut a thin cross-section about 2 cm (0.75 in.) thick. Lay this cross-section flat on the dissection pan and observe the internal anatomy. Use Figure 14.10 as a guide to help locate the various anatomical features of the spinal cord.

2. Identify the gray horns, central canal, and white columns. What type of tissue is found in the gray horns? What type is found in the white columns? How can you determine which margin of the cord is the posterior margin?

3. Locate the spinal meninges by pulling the outer tissues away from the spinal cord with a forceps and blunt probe. Slip your probe between the meninges on the lateral spinal cord. Cut completely through the meninges and gently peel them back to expose the ventral and dorsal roots. How does the dorsal root differ in appearance from the ventral root?

4. Closely examine the meninges. Separate the arachnoid mater from the dura mater with your probe. Attempt to loosen a free edge of the pia mater with a dissection pin. What function does each of these membranes serve?

5. Clean up your work area, wash the dissection pan and tools, and follow your instructor's directions for proper disposal of the specimen. ■

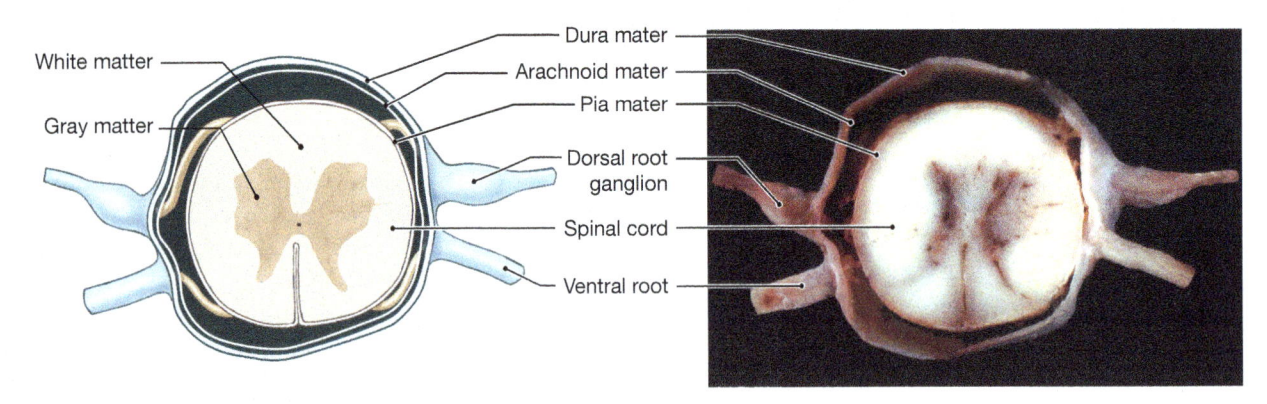

Figure 14.10 Spinal Cord Dissection
Transverse section of a sheep spinal cord that details the spinal meninges and internal organization.

EXERCISE 14

LAB REPORT

The Spinal Cord and Spinal Nerves

Name _____

Date _____

Section _____

A. Matching

Match each term in the left column with its correct description from the right column.

_____ 1. lateral gray horn
_____ 2. posterior median sulcus
_____ 3. bundle of axons
_____ 4. rami communicantes
_____ 5. subarachnoid space
_____ 6. ventral root
_____ 7. dorsal ramus
_____ 8. dorsal root ganglion
_____ 9. conus medullaris
_____ 10. dorsal root

A. site of cerebrospinal fluid circulation
B. sensory branch that enters spinal cord
C. contains visceral motor neurons
D. tapered end of spinal cord
E. fascicle
F. shallow groove of spinal cord
G. posterior branch of a spinal nerve
H. motor branch that exits spinal cord
I. leads to autonomic ganglion
J. contains sensory cell bodies

B. Labeling

Label the numbered regions of the spinal cord cross-section in Figure 14.11.

C. Short-Answer Questions

1. Describe the organization of white and gray matter in the spinal cord.

2. Describe the spinal meninges.

3. Discuss the major nerves of the brachial plexus.

EXERCISE 14   LAB REPORT

Spinal nerve

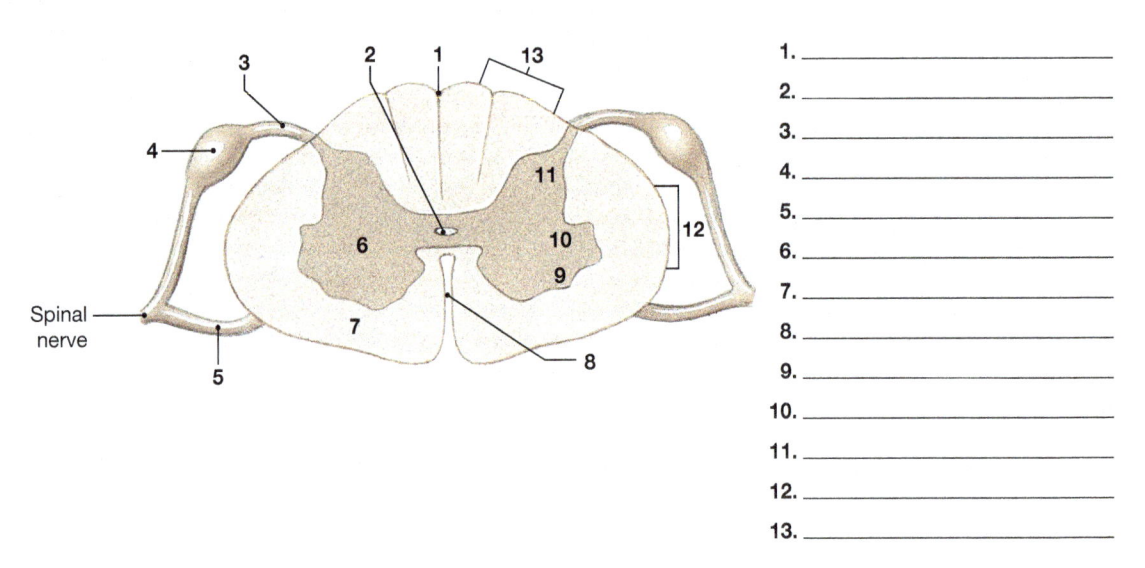

1. _____
2. _____
3. _____
4. _____
5. _____
6. _____
7. _____
8. _____
9. _____
10. _____
11. _____
12. _____
13. _____

Figure 14.11 Anatomy of the Spinal Cord

4. List where sensory, visceral motor, and somatic motor neurons are located in the gray horns.

D. Analysis and Application

1. Trace a sensory pathway from the flexor carpi ulnaris muscle to the spinal cord. Include the names of the peripheral nerve, the spinal nerve, and the various rami and roots involved in the pathway.

2. Starting in the spinal cord, trace a motor pathway to the adductor muscles of the thigh. Include the spinal cord root, spinal nerve, nerve plexus, and specific peripheral nerve involved in the pathway.

3. Suppose someone falls off a roof and injures spinal nerves L_2–L_4. Describe how this injury would affect the sensory and motor functions that are controlled by the injured spinal nerves.

EXERCISE 15

Anatomy of the Brain

OBJECTIVES

On completion of this exercise, you should be able to:

- List the three meninges that cover the brain.
- Describe the extensions of the dura mater.
- Identify the four ventricles of the brain and trace a drop of cerebrospinal fluid through the ventricular system.
- Identify the six major regions of the brain and a basic function of each.
- Identify the surface features of each region of the brain.
- Identify the 12 pairs of cranial nerves.
- Describe the anatomy of a dissected sheep brain.

LAB ACTIVITIES

1. Cranial Meninges and Ventricles of the Brain 256
2. Regions of the Brain 260
3. Cranial Nerves 267
4. Sheep Brain Dissection 271

   A Regional Look: Cranium 276

The brain is one of the largest organs in the body. It weighs approximately three pounds and occupies the cranial cavity. Billions of synapses between neurons form a vast biological circuitry that no electronic computer will ever surpass. Every second, the brain performs a huge number of calculations, interpretations, and visceral-activity coordinations to maintain homeostasis.

The brain is divided into six major regions: cerebrum (ser-Ē-brum or SER-ē-brum), diencephalon (dī-en-SEF-a-lon), mesencephalon, pons, medulla oblongata, and cerebellum. The medulla oblongata, pons, and mesencephalon (midbrain) are collectively called the **brain stem**. Some anatomists include the diencephalon as part of the brain stem.

The **cerebrum** is the largest region of the brain. It is divided into right and left **cerebral hemispheres** by the deep groove known as the **longitudinal fissure**. A left cerebral hemisphere is shown in Figure 15.1a. The hemispheres are covered with a folded **cerebral cortex** (*cortex,* bark or rind) of gray matter, where neurons are not myelinated. Each small fold of the cerebral cortex is called a **gyrus** (JĪ-rus; plural *gyri*), and each shallow groove is called a **sulcus** (SUL-kus; plural *sulci*). Deep in the cerebrum is the brain's white matter, where myelinated neurons that occur in thick bands interconnect the various regions of the brain.

Inferior to the cerebrum are the **thalamus** (THAL-a-mus) and **hypothalamus**, which together make up the **diencephalon** (Figure 15.1b). Inferior to the diencephalon is the **mesencephalon** (midbrain) of the brain stem. The **pons** is the enlarged region of the brain stem just inferior to the mesencephalon, and the **medulla oblongata** is the most inferior part of the brain stem, and connects the brain to the spinal cord. The **cerebellum** is the oval mass posterior to the brain stem.

256 EXERCISE 15 Anatomy of the Brain

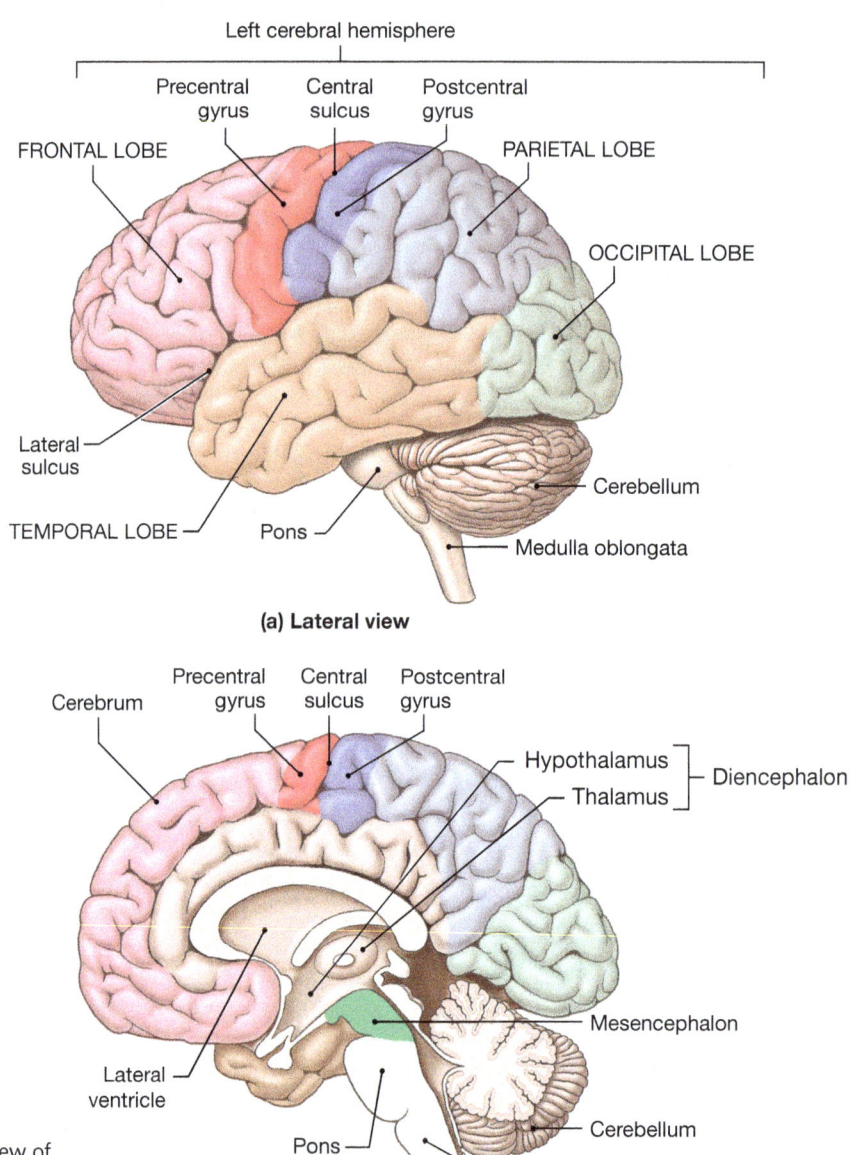

Figure 15.1 The Human Brain
Major regions of the human brain. (a) Lateral view of left side of brain. (b) Midsagittal view of right side of brain.

Making Connections

A Sea Horse's Guide to the Brain

When examining the brain in sagittal section, most people notice how the brain stem and diencephalon form the shape of a sea horse. The pons is the horse's belly, the mesencephalon is the neck, the diencephalon is the head, and the medulla oblongata is the tail. Imagine that the sea horse is wearing the cerebellum as a backpack and the cerebrum as a very large hat. ●

LAB ACTIVITY 1 Cranial Meninges and Ventricles of the Brain

Cranial Meninges

The brain is encased in layers of tough, protective **cranial meninges**. Cerebrospinal fluid (CSF) circulates between certain meningeal layers, cushions the

EXERCISE 15 Anatomy of the Brain 257

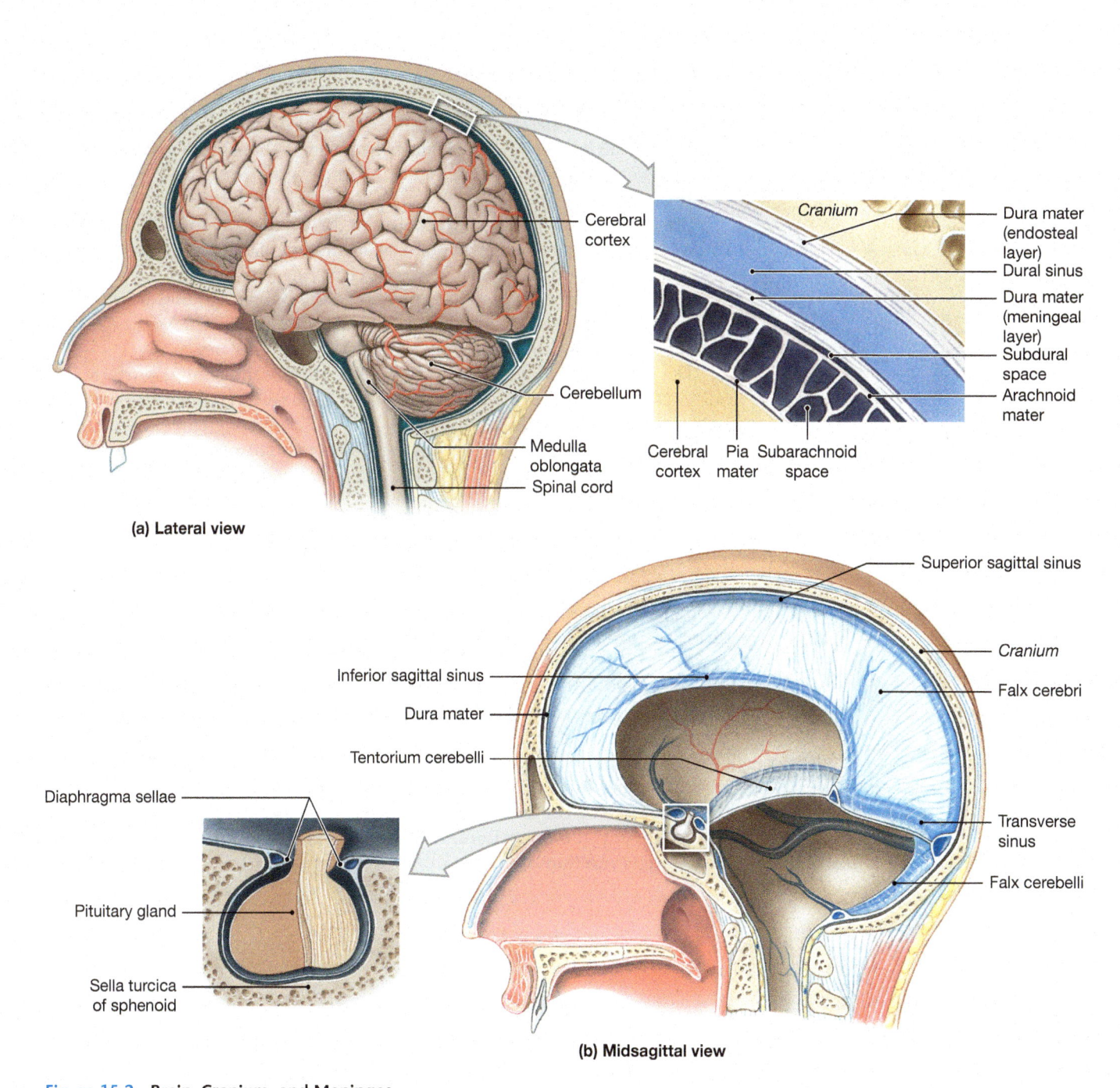

Figure 15.2 Brain, Cranium, and Meninges
(a) A lateral view of the brain that shows its position in the cranium and the organization of the meninges. (b) A diagrammatic view that shows the orientation of the three largest dural folds: falx cerebri, tentorium cerebelli, and falx cerebelli. The location of the pituitary gland is also shown.

brain, and prevents it from contacting the cranial bones during a head injury, much like how a car's airbag prevents a passenger from hitting the dashboard. The cranial meninges are anatomically similar to, and continuous with, the spinal meninges of the spinal cord. Like their spinal counterparts, the cranial meninges consist of three layers: the dura mater, arachnoid mater, and pia mater (Figure 15.2).

The **dura mater** (DOO-ruh MĀ-ter; *dura*, tough + *mater*, mother), the outer meningeal covering, consists of an **endosteal layer** fused with the periosteum of the cranial bones and a **meningeal layer** that faces the arachnoid mater. The endosteal layer is referred to as the *outer dural layer*, and the meningeal layer is referred to as the *inner dural layer*. Between the two layers are large blood sinuses, collectively called **dural sinuses**, that drain blood from cranial veins into the jugular veins. The **superior** and **inferior sagittal sinuses** are large veins in the dura mater between the two hemispheres of the cerebrum. The **transverse sinus** is in the dura mater between the cerebrum and the cerebellum. Between the dura mater and the underlying arachnoid mater is the **subdural space**.

Deep to the dura mater is the **arachnoid** (a-RAK-noyd; *arachno,* spider) **mater**, named after the web-like connection this membrane has with the underlying pia mater. The arachnoid mater forms a smooth covering over the brain.

On the surface of the brain is the **pia** (PĒ-uh; *pia,* delicate) **mater**, which contains many blood vessels that supply the brain. Between the arachnoid mater and pia mater is the **subarachnoid space**, where the CSF circulates.

The dura mater has extensions that help stabilize the brain (Figure 15.2b). A midsagittal fold in the dura mater forms the **falx cerebri** (falks ser-Ē-brē-; *falx,* sickle-shaped) and separates the right and left hemispheres of the cerebrum. Posteriorly, the dura mater folds again as the **tentorium cerebelli** (ten-TOR-ē-um ser-e-BEL-ē-; *tentorium,* a covering) and separates the cerebellum from the cerebrum. The **falx cerebelli** is a dural fold between the hemispheres of the cerebellum.

Ventricles

Deep in the brain are four chambers called **ventricles** (Figure 15.3). Two **lateral ventricles**, one in each cerebral hemisphere, extend deep into the cerebrum, and are horseshoe-shaped chambers. At the midline of the brain, the lateral ventricles are separated by a thin membrane called the **septum pellucidum**. A brain sectioned at the midsagittal plane exposes this membrane. CSF circulates from the lateral ventricles through the **interventricular foramen** (also called the *foramen of Monro*) and enters the **third ventricle**, which is a small chamber in the diencephalon. CSF in the third ventricle passes through the **aqueduct of the midbrain** and enters the **fourth ventricle** between the brain stem and the cerebellum. In the fourth ventricle, two **lateral apertures** and a single **median aperture** direct CSF laterally to the exterior of the brain

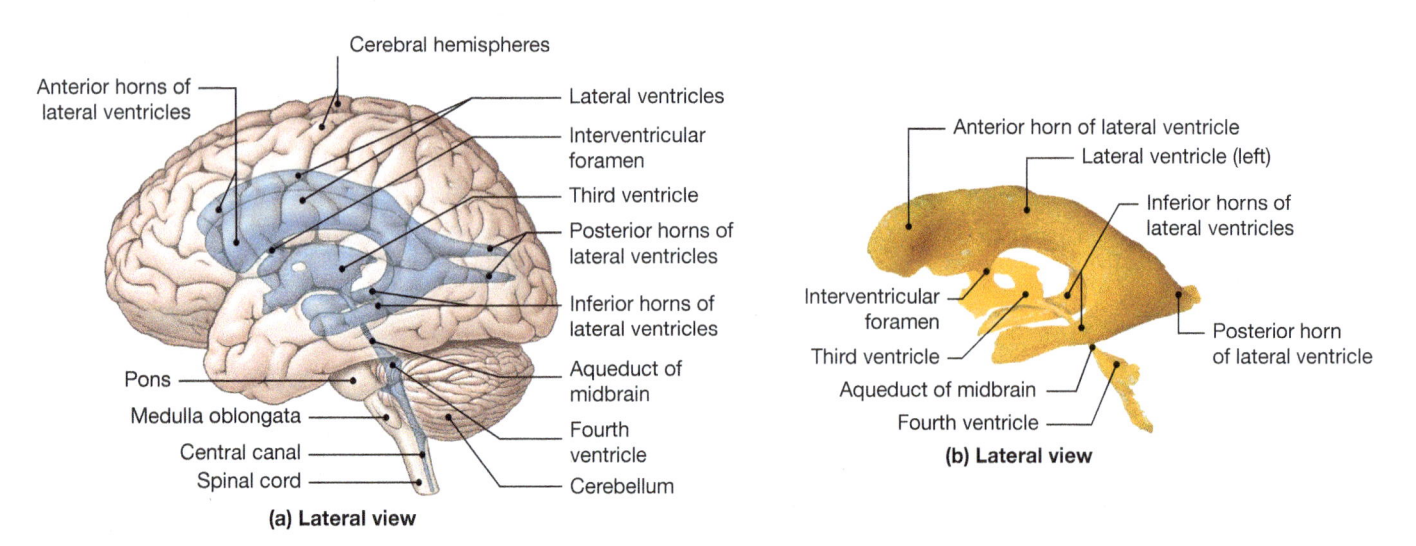

Figure 15.3 Ventricles of the Brain

The ventricles contain cerebrospinal fluid, which transport nutrients, chemical messengers, and waste products. (a) Orientation and extent of the ventricles as seen in a lateral view of a transparent brain. (b) A lateral view of a plastic cast of the ventricles.

EXERCISE 15 Anatomy of the Brain 259

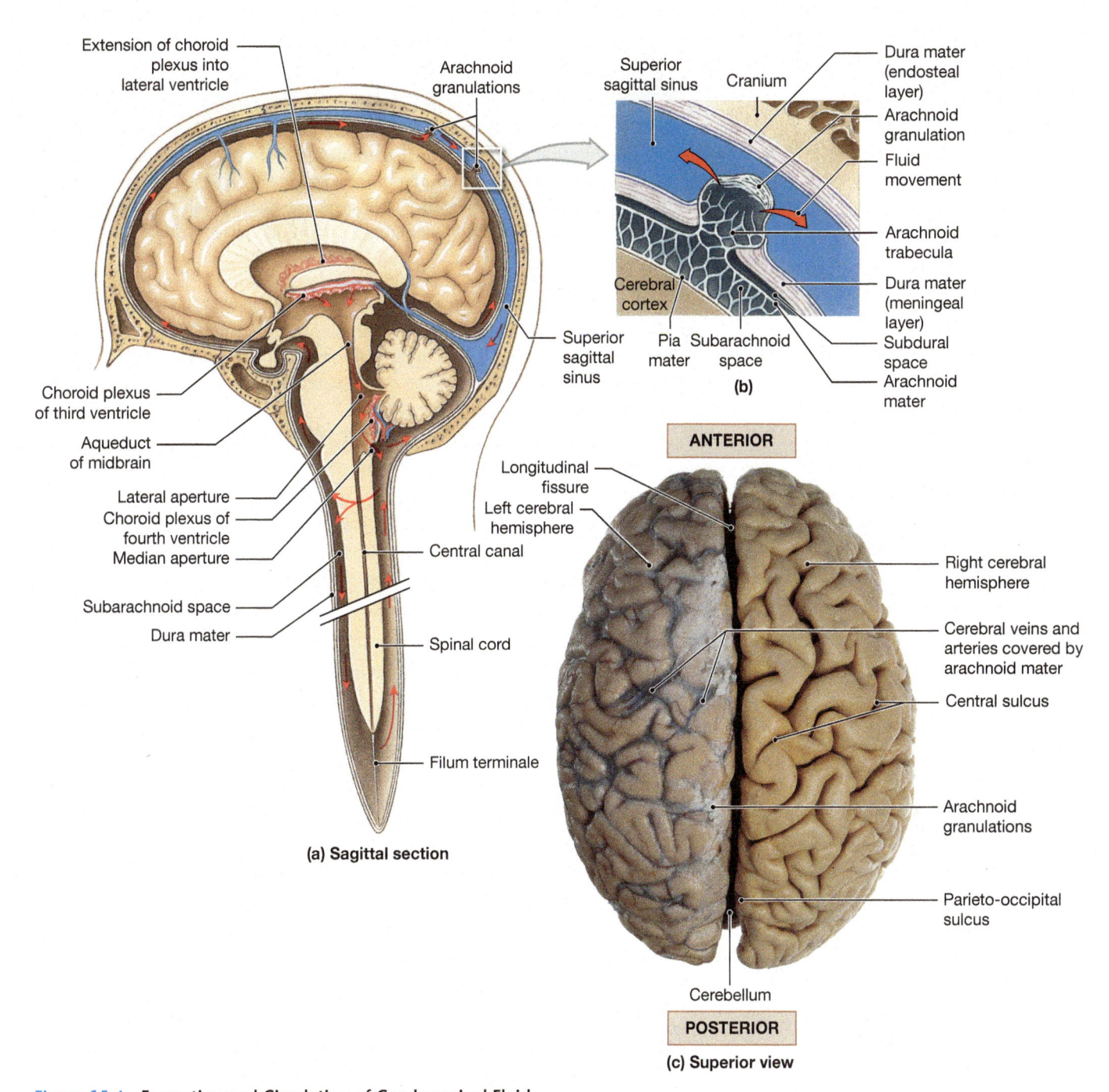

Figure 15.4 Formation and Circulation of Cerebrospinal Fluid
(a) A sagittal section of the central nervous system. Cerebrospinal fluid (*CSF*), formed in the choroid plexus, circulates via the routes indicated by the red arrows. (b) Detail of an arachnoid granulation, where CSF is reabsorbed into the blood. (c) Superior view of the brain that shows arachnoid granulations in the arachnoid mater along the longitudinal fissure.

and spinal cord and into the subarachnoid space. CSF then circulates around the brain and spinal cord and is reabsorbed at **arachnoid granulations**, which project into the veins of the dural sinuses (Figure 15.4).

Inside each ventricle is a specialized capillary called the **choroid plexus** where cerebrospinal fluid is produced. The choroid plexus of the third ventricle has two folds that pass through the interventricular foramen and expand to line the floor of the lateral ventricles. The choroid plexus of the fourth ventricle lies on the posterior wall of the ventricle.

Clinical Application

Hydrocephalus

The choroid plexus of an adult brain produces approximately 500 mL of cerebrospinal fluid daily, and constantly replaces the 150 mL that circulates in the ventricles and subarachnoid space. Because CSF is constantly being made, a volume equal to that produced must be removed from the central nervous system to prevent a buildup of fluid pressure in the ventricles. In an infant, if CSF production exceeds CSF reabsorption, the increase in cranial pressure expands the unfused skull, which creates a condition called *hydrocephalus*. There are two types of hydrocephalus: internal and external. Internal hydrocephalus occurs when CSF accumulates in the ventricles inside the brain. This form of hydrocephalus is almost always fatal because of damaging distortion of the brain tissue. External hydrocephalus is the buildup of CSF in the subdural space which results in an enlarged skull and possible brain damage caused by high fluid pressure on the delicate neural tissues. Surgical treatment of external hydrocephalus involves installation of small tubes called shunts to drain the excess CSF and reduce intracranial pressure.

QuickCheck Questions

1.1 What are the functions of the cranial meninges?
1.2 Between which meningeal layers does CSF circulate?
1.3 What fold separates the cerebellum from the cerebrum?
1.4 Where is CSF produced?
1.5 Where does CSF circulate, and where does it return to the blood?

1 Materials

- ☐ Brain model
- ☐ Brain chart
- ☐ Ventricular system model
- ☐ Preserved and sectioned human brain (if available)

Procedures

1. Review the meningeal anatomy presented in Figure 15.2.
2. Locate the dura mater, arachnoid mater, and pia mater on the ventricular system model.
3. On the brain model or preserved brain, examine the dura mater and identify the falx cerebri, falx cerebelli, and tentorium cerebelli.
4. Review the ventricular system in Figures 15.3 and 15.4.
5. On the brain model, observe how the lateral ventricles extend into the cerebrum. If your model is detailed enough, locate the interventricular foramen. Identify the third ventricle, aqueduct of the midbrain, and fourth ventricle.
6. Starting from one of the two lateral ventricles on the brain model, trace a drop of CSF as it circulates through the brain and then is reabsorbed at an arachnoid granulation.

LAB ACTIVITY 2 Regions of the Brain

Cerebrum

The cerebrum is the most complex part of the brain. Conscious thought, intellectual reasoning, and memory processing and storage all take place in the cerebrum.

Each cerebral hemisphere consists of five lobes, most of which are named for the overlying cranial bone (Figure 15.5). The anterior cerebrum is the **frontal lobe**, and the prominent **central sulcus**, which is located approximately midposterior, separates the frontal lobe from the **parietal lobe**. The **occipital lobe** lies under the occipital bone of the posterior skull. The **lateral sulcus** defines the boundary be-

EXERCISE 15 Anatomy of the Brain   261

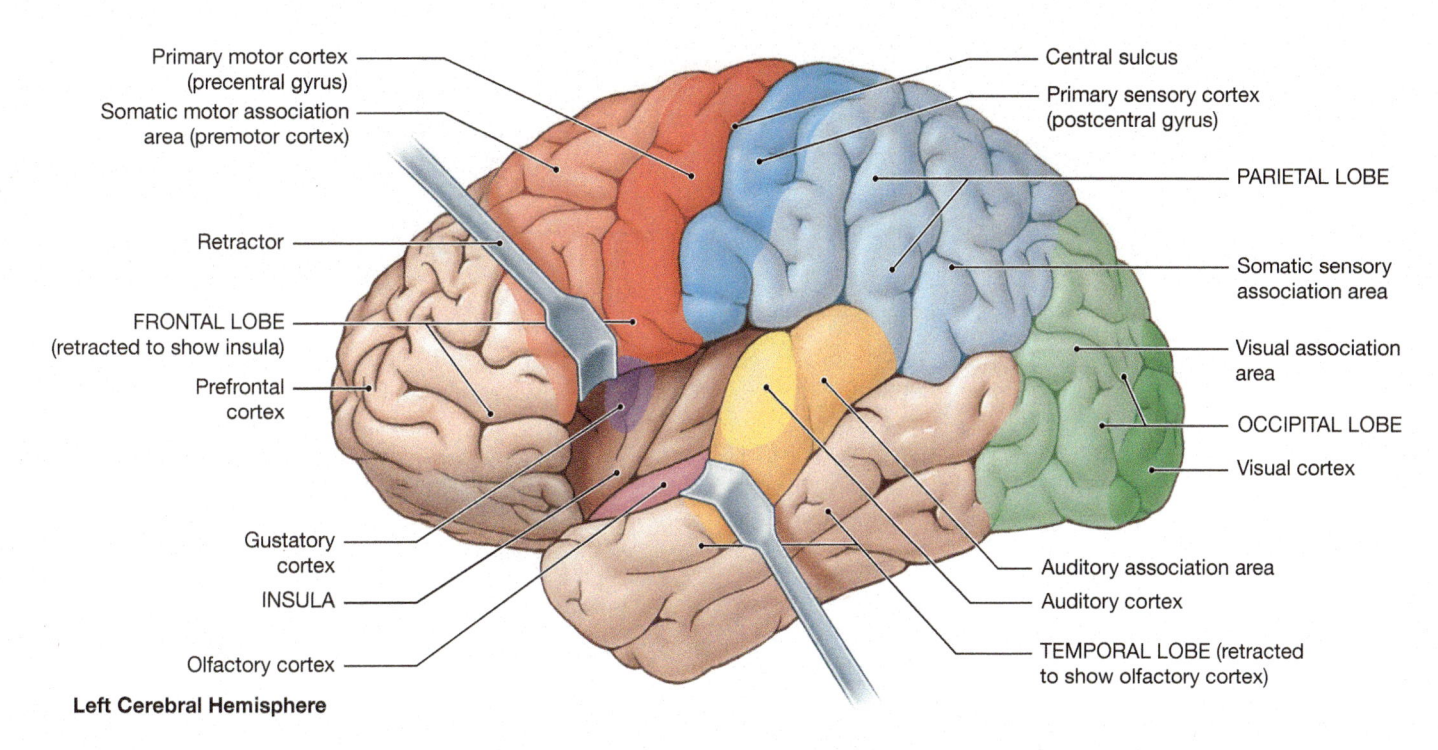

Figure 15.5 Lobes of a Cerebral Hemisphere
Major anatomical landmarks on the surface of the left cerebral hemisphere. Association areas are colored. The lateral sulcus has been pulled open with two retractors to expose the insula.

tween the large frontal lobe and the **temporal lobe** of the lower lateral cerebrum. Cutting into the lateral sulcus and peeling away the temporal lobe reveals a fifth lobe, the **insula** (IN-sū-luh; *insula,* island).

Regional specializations occur in the cerebrum. The central sulcus separates the motor region of the cerebrum (frontal lobe) from the sensory region (parietal lobe). Immediately anterior to the central sulcus is the **precentral gyrus**. This gyrus contains the primary motor cortex, where voluntary commands to skeletal muscles are generated. The **postcentral gyrus**, which is on the parietal lobe, contains the primary sensory cortex, where the general sense of touch is perceived. The other four senses—sight, hearing, smell, and taste—involve the processing of complex information received from many more sensory neurons than the number involved in the sense of touch. These four senses thus require more neurons in the brain to process the sensory signals, and therefore the cerebral cortex areas devoted to processing these messages are larger than the postcentral gyrus of the primary sensory cortex for touch. The occipital lobe contains the visual cortex, where visual impulses from the eyes are interpreted. The temporal lobe houses the auditory cortex and the olfactory cortex.

Figure 15.5 also shows numerous **association areas**, which are regions that either interpret sensory information from more than one sensory cortex or integrate motor commands into an appropriate response. The **premotor cortex** is the somatic motor association area of the anterior frontal lobe. Auditory and visual association areas occur near the corresponding sensory cortex in the occipital lobe.

The cerebral hemispheres are connected by a thick tract of white matter called the **corpus callosum**. This structure, which bridges the two hemispheres at the base of the longitudinal fissure, is easily identified as the curved white structure at the base of the cerebrum (Figures 15.6 and 15.7). The inferior portion of the corpus callosum is the **fornix** (FOR-niks), a white tract that connects deep structures of the limbic system, the "emotional" brain. The fornix narrows anteriorly and meets the **anterior commissure** (kom-MIS-sur), which is another tract of white matter that connects the cerebral hemispheres.

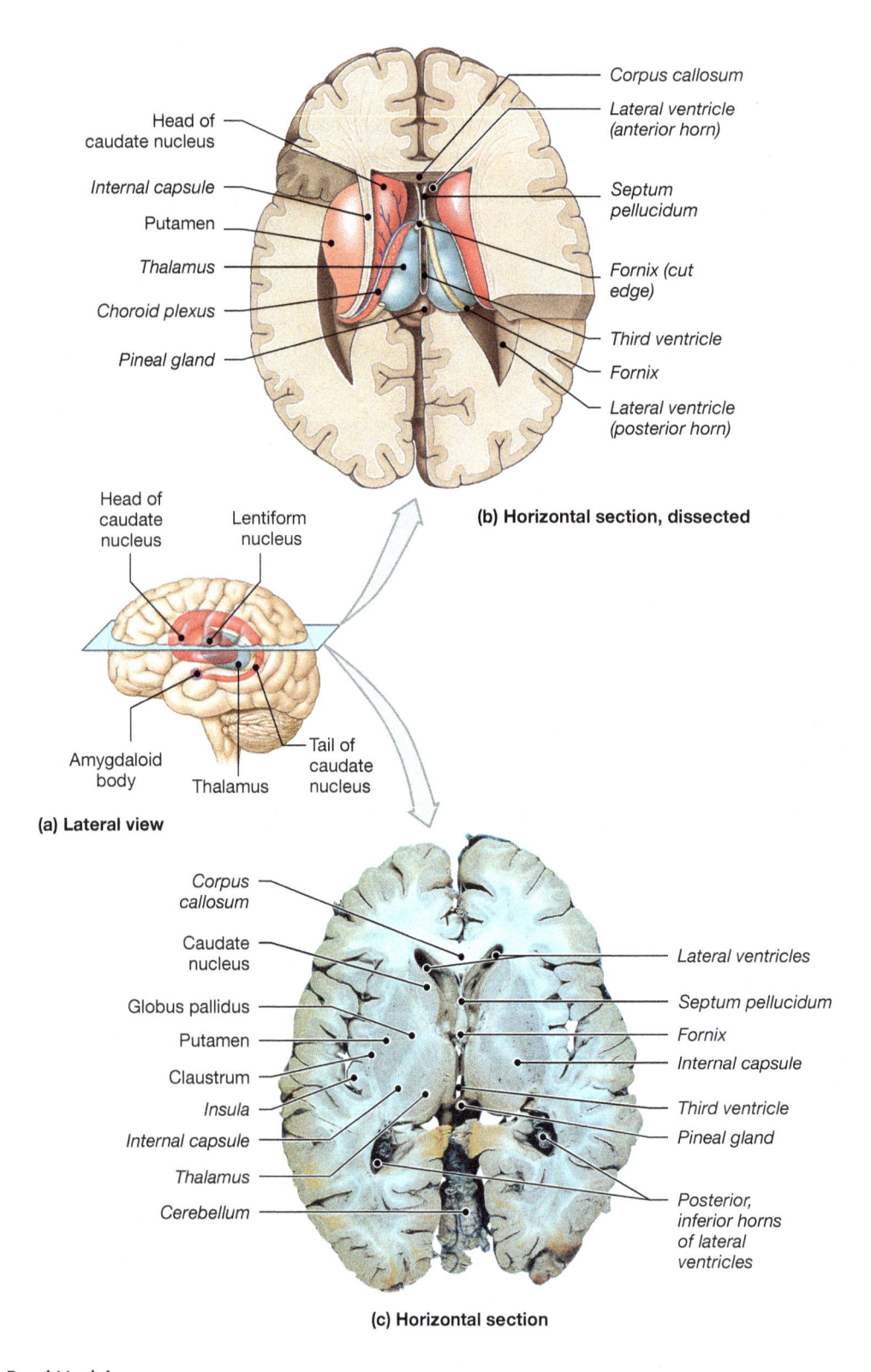

Figure 15.6 The Basal Nuclei

(a) Lateral view showing the relative positions of the basal nuclei. Compare this three-dimensional representation with the horizontal sections (b, c). (b and c) Diagrammatic and dissection views of horizontal sections of the brain.

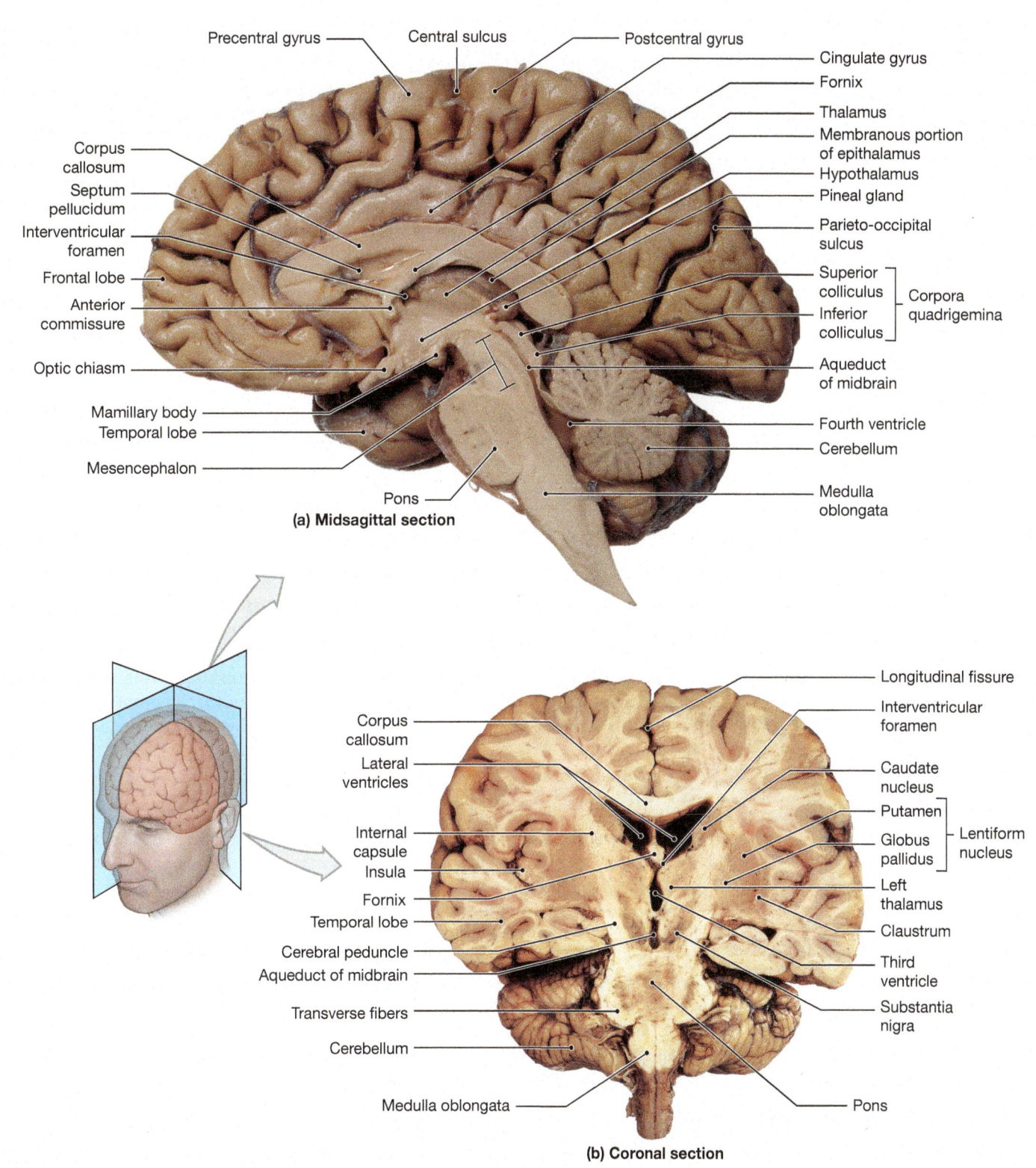

Figure 15.7 Two Views of the Brain
(a) Midsagittal section. (b) Frontal section.

263

Deep structures of the cerebrum and diencephalon are visible when the brain is sectioned on the frontal plane, as in Figure 15.6. Notice in Figure 15.6b how the corpus callosum transverses the brain to connect the two cerebral hemispheres. In each cerebral hemisphere, structures called **basal nuclei** are paired masses of gray matter that are involved in automating voluntary muscle contractions. Each basal nucleus consists of a medial **caudate nucleus** and a lateral **lentiform nucleus**. The latter is made up of two parts: a **putamen** (pū-TĀ-men) and a **globus pallidus**. At the tip of the caudate nucleus is the **amygdaloid** (ah-MIG-da-loyd; *almond*) **body**. Between the caudate nucleus and the lentiform nucleus lies the **internal capsule**, which is a band of white matter that connects the cerebrum to the diencephalon, brain stem, and cerebellum.

Diencephalon

The diencephalon is embedded in the cerebrum and is visible only from the inferior aspect of the brain. The thalamus region of the diencephalon maintains a crude sense of awareness. All sensory impulses except smell and proprioception (the sense that controls muscle tension and position) pass into the thalamus and are relayed to the proper sensory cortex for interpretation. Nonessential sensory data are filtered out by the thalamus and do not reach the sensory cortex.

The hypothalamus is the floor of the diencephalon. On the inferior surface of the brain, a pair of rounded **mamillary** (MAM-i-lar-ē-; *mammilla*, little breast) **bodies** are visible inferior to the hypothalamus (Figure 15.7). These bodies are hypothalamic nuclei that control eating reflexes for licking, chewing, sucking, and swallowing. Anterior to the mamillary bodies is the **infundibulum** (in-fun-DIB-ū-lum; *infundibulum*, funnel), which is the stalk that attaches the **pituitary gland** to the hypothalamus. In sagittal section, the **interthalamic adhesion**, also called the *massa intermedia*, is an oval structure in the diencephalon that connects the right and left sides of the thalamus. The **pineal** (PIN-ē-ul) **gland** is the nipple-like structure that is superior to the mesencephalon positioned between the cerebrum and the cerebellum.

Mesencephalon (Midbrain)

The mesencephalon (Figures 15.7 and 15.8) is posteriorly covered by the cerebrum. The portion of the mesencephalon that is called the **corpora quadrigemina** (KOR-po-ra quad-ri-JEM-i-nuh) is a series of four bulges next to the pineal gland. The two members of the superior pair of bulges are the **superior colliculi** (ko-LIK-ū-lē *colliculus*, small hill), which function as a visual reflex center that moves the eyeballs and the head to keep an object centered on the retina of the eye. The two members of the inferior pair of bulges are the **inferior colliculi**, which function as an auditory reflex center that moves the head to locate and follow sounds. The anterior mesencephalon between the pons and the hypothalamus consists of the **cerebral peduncles** (*peduncles;* little feet), which are a group of white fibers that connect the cerebral cortex with other parts of the brain.

Pons

The pons functions as a relay station that directs sensory information to the thalamus and cerebellum. It also contains certain sensory, somatic motor, and autonomic cranial nerve nuclei.

Medulla Oblongata

The medulla oblongata is the inferior part of the brain stem and is continuous with the spinal cord. Sensory information in ascending tracts in the spinal cord enter the brain at the medulla oblongata, and motor commands in descending tracts pass

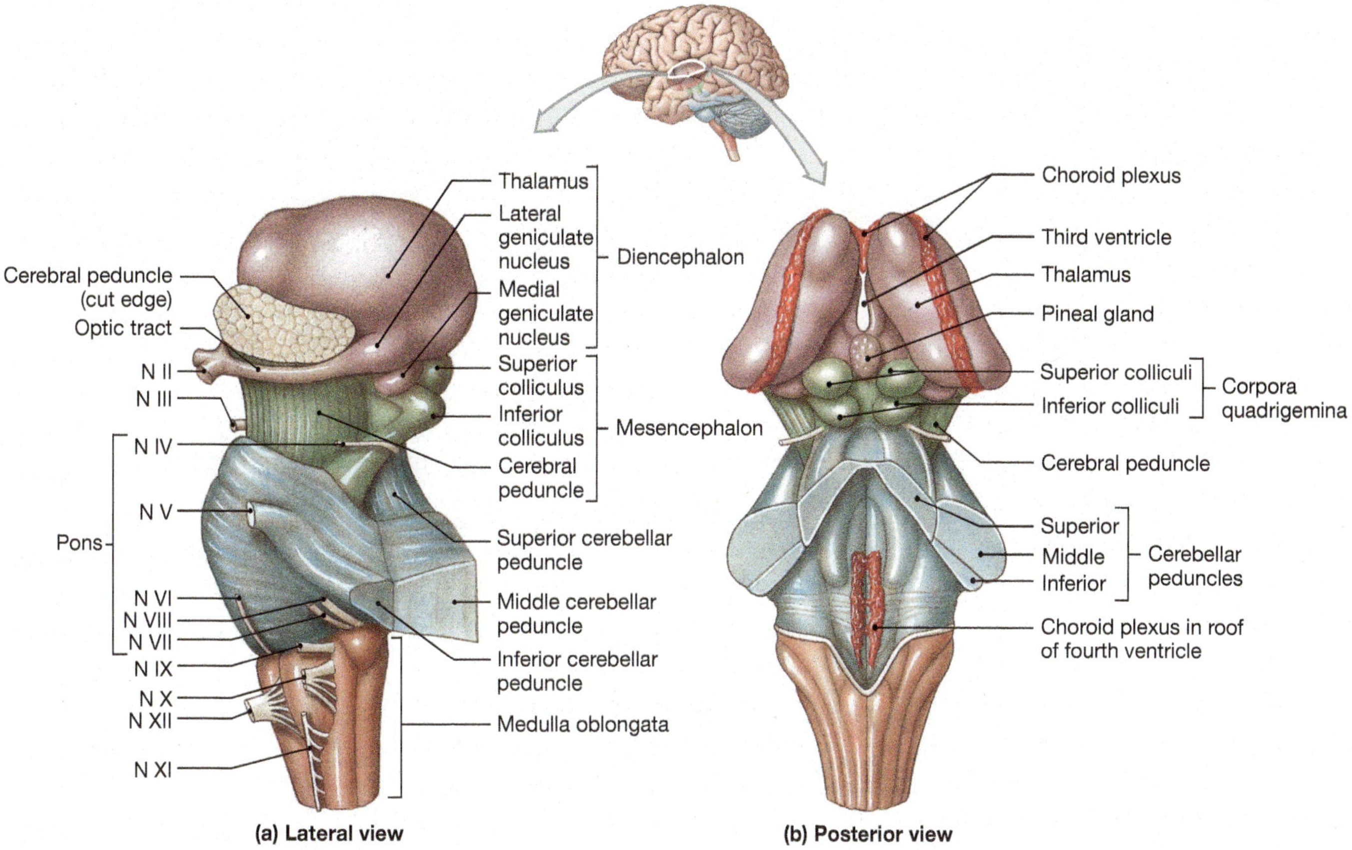

Figure 15.8 The Diencephalon and Brain Stem
(a) Diagrammatic view of the diencephalons and brain stem seen from the left side. (b) Posterior diagrammatic view of the diencephalons and brain stem.

through the medulla oblongata and into spinal cord. The anterior surface of the medulla oblongata has two prominent folds called **pyramids** where some motor tracts cross over, or *decussate,* to the opposite side of the body. The medulla oblongata also functions as an autonomic center for visceral functions. Nuclei in this portion of the brain are vital reflex centers for the regulation of cardiovascular, respiratory, and digestive activities.

Cerebellum

The cerebellum (Figure 15.9) is inferior to the occipital lobe of the cerebrum and is covered by a layer called the **cerebellar cortex**. Small folds on the cerebellar cortex are called **folia** (FŌ-lē-uh; *folia,* leaves; singular *folium*). The cerebellum is divided into right and left **cerebellar hemispheres**, which are separated by a narrow **vermis** (VER-mis; *vermis,* worm). Each cerebellar hemisphere consists of two lobes: a smaller **anterior lobe**, which is directly inferior to the cerebrum, and a **posterior lobe**. The **primary fissure** separates the anterior and posterior cerebellar lobes. In a sagittal section, a smaller **flocculonodular** (flok-ū-lō-NOD-ū-lar) **lobe** is visible where the anterior wall of the cerebellum faces the pons.

In a sagittal section, the white matter of the cerebellum is apparent. Because this tissue is highly branched, it is called the **arbor vitae** (*arbor,* tree; *vitae,* life). In the middle of the arbor vitae are the **cerebellar nuclei**, which regulate involuntary skeletal muscle contraction.

The cerebellum is primarily involved in the coordination of somatic motor functions, primarily skeletal muscle contractions. Adjustments to postural muscles occur when impulses from the cranial nerve of the inner ear pass into the flocculonodular

266 EXERCISE 15 Anatomy of the Brain

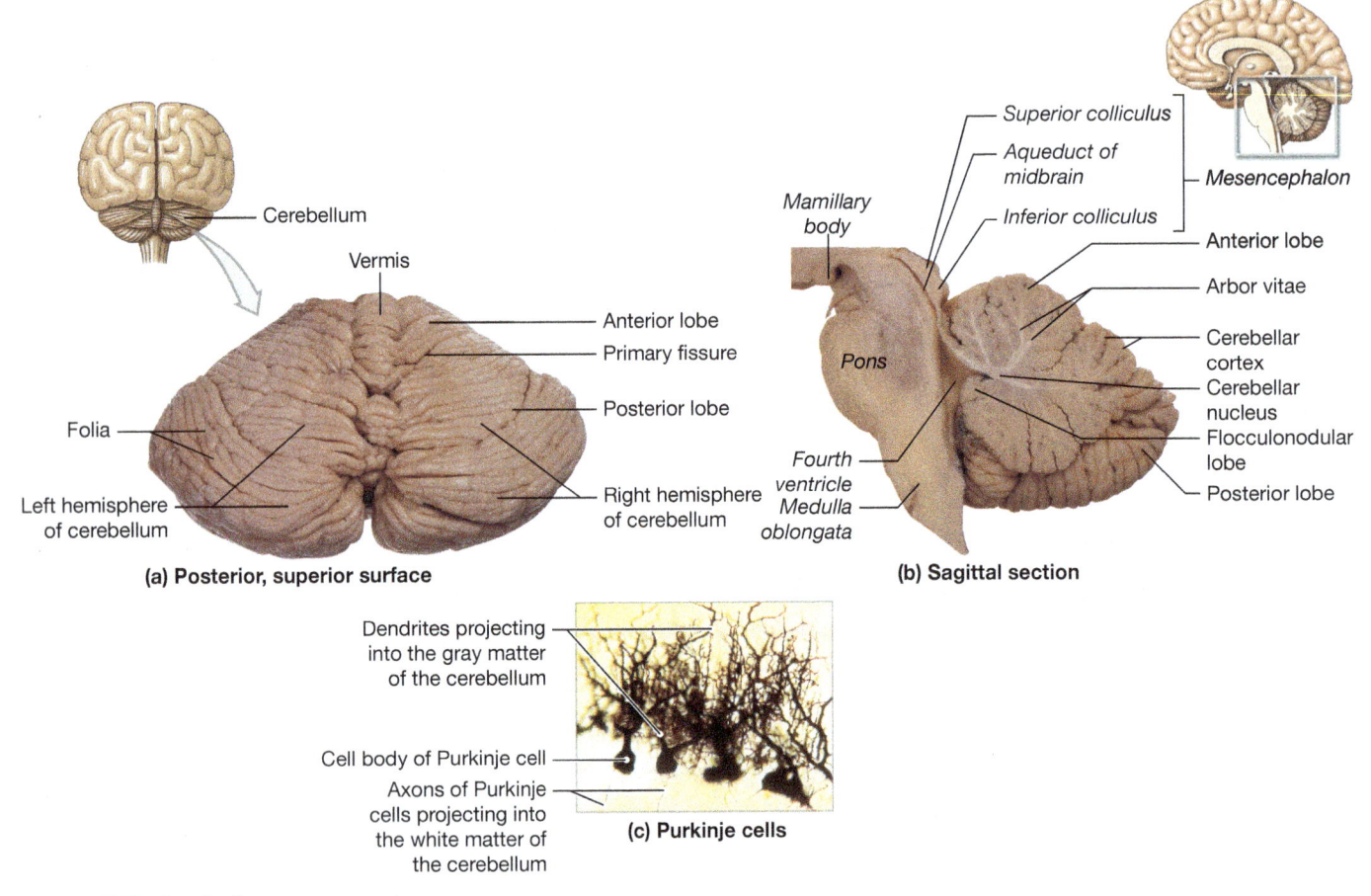

Figure 15.9 Cerebellum
(a) Superior surface of the cerebellum shows major anatomical landmarks and regions. (b) Sagittal view of the cerebellum shows the arrangement of gray matter and white matter. (c) Purkinje cells are seen in the photograph; these large neurons are found in the cerebellar cortex. (LM × 120)

lobe, which is the part of the cerebellum where information about equilibrium is processed. Learned muscle patterns, such as those involved in serving a tennis ball or playing the guitar, are stored and processed in the cerebellum.

QuickCheck Questions

2.1 What are the six major regions of the brain?
2.2 How are the cerebral hemispheres connected to each other?
2.3 Where is the mesencephalon?

2 Materials

- Brain model (midsagittal, frontal, and horizontal sections)
- Brain chart
- Preserved and sectioned human brain (if available)

Procedures

1. Review the brain anatomy in Figures 15.5 through 15.9.
2. On the brain model, identify the following:

   Cerebrum: Note how the longitudinal fissure separates it into two cerebral hemispheres. Identify the five lobes of each hemisphere, along with the central sulcus, precentral gyri, and postcentral gyri. View the brain model in midsagittal section and identify the corpus callosum, fornix, and anterior commissure. In a frontal section and a horizontal section, locate the internal capsule, lentiform nucleus, and caudate nucleus. Distinguish between the putamen and the globus pallidus of the lentiform nucleus.

Diencephalon: In a midsagittal section of the brain model, identify the thalamus, which is recognizable as the lateral wall around the diencephalon, and the wedge-shaped hypothalamus, which is inferior to the thalamus. Observe the third ventricle around the thalamus and the interthalamic adhesion. Identify the infundibulum, which attaches the pituitary gland to the hypothalamus. Locate the mamillary bodies and pineal gland.

Brain stem: Identify the medulla oblongata, pons, and mesencephalon. Locate the two pyramids on the medulla's anterior surface and the cerebral peduncles on the lateral sides of the mesencephalon. Identify the corpora quadrigemina of the mesencephalon, and distinguish between the superior and inferior colliculi.

Cerebellum: Locate the right and left hemispheres and the vermis that separates them. In each hemisphere, identify the primary fissure and the anterior and posterior lobes. In a midsagittal section, locate the arbor vitae and the cerebellar nuclei. ■

LAB ACTIVITY 3 — Cranial Nerves

Cranial nerves emerge from the brain at specific locations and pass through various foramina of the skull to reach the peripheral structures they innervate. Like spinal nerves, cranial nerves occur in pairs; cranial nerves occur in 12 pairs. The nerves are identified by name and are numbered with roman numerals from N I to N XII. The numbers are assigned according to the locations where the nerves contact the brain; N I is most anterior and N XII is most posterior. Some cranial nerves are entirely sensory nerves, but most are mixed. However, the mixed nerves that conduct primarily motor commands are considered motor nerves, even though they have few sensory fibers to inform the brain about muscle tension and position. Figure 15.10 shows the position of each cranial nerve on the inferior surface of the brain. Table 15.1 summarizes the cranial nerves and includes the foramen through which each nerve passes.

Olfactory Nerve (N I)

The **olfactory nerve** is composed of bundles of sensory fibers for the sense of smell and is located in the roof of the nasal cavity. The nerve passes through the cribriform plate of the ethmoid bone and enters an enlarged **olfactory bulb**, which then extends into the cerebrum as the **olfactory tract.**

Optic Nerve (N II)

The **optic nerve** carries visual information. This nerve originates in the retina, which is the neural part of the eye that is sensitive to changes in the amount of light that enters the eye. The nerve is easy to identify as the X-shaped structure at the **optic chiasm** inferior to the hypothalamus. It is at this point that some of the sensory fibers cross to the nerve on the opposite side of the brain. The optic nerve enters the thalamus, which relays the visual signal to the occipital lobe. Some of the fibers enter the superior colliculus for visual reflexes.

Oculomotor Nerve (N III)

The **oculomotor nerve** innervates four extraocular eye muscles—the superior, medial, and inferior rectus muscles, and the inferior oblique muscle—and the levator palpebrae muscle of the eyelid. Autonomic motor fibers also control the intrinsic muscles of the iris and the ciliary body. The oculomotor nerve is located on the ventral mesencephalon just posterior to the optic nerve.

Trochlear Nerve (N IV)

The **trochlear** (TROK-lē-ar) **nerve** supplies motor fibers to the superior oblique muscle of the eye and originates where the mesencephalon joins the pons. The root

268 EXERCISE 15 Anatomy of the Brain

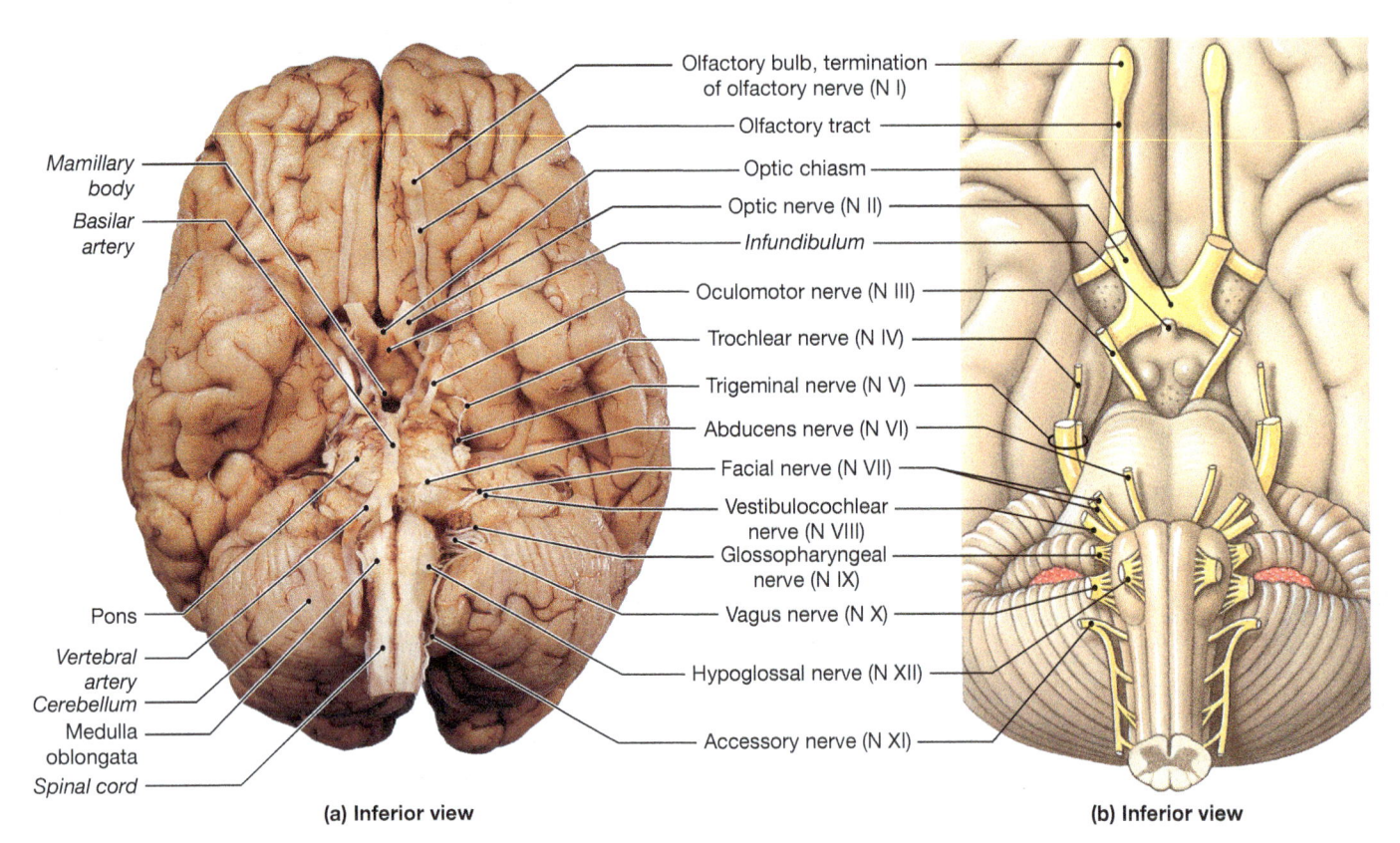

Figure 15.10 Origins of the Cranial Nerves
(a) The inferior surface of the brain as it appears on gross dissection. The roots of the cranial nerves are clearly visible. (b) Diagrammatic inferior view of the human brain that highlights origins of the cranial nerves.

of the nerve exits the mesencephalon on the lateral surface. Because it is easily cut or twisted off during removal of the dura mater, many dissection specimens do not have this nerve intact. The superior oblique eye muscle passes through a trochlea, or "pulley"; hence the name of the nerve.

Trigeminal Nerve (N V)

The **trigeminal** (trī-JEM-i-nal) **nerve** is the largest of the cranial nerves. It is located on the lateral pons near the medulla oblongata and services much of the face. The nerve has three branches: *ophthalmic, maxillary,* and *mandibular*. The ophthalmic branch innervates sensory structures of the forehead, eye orbit, and nose. The maxillary branch contains sensory fibers for structures in the roof of the mouth, including half of the maxillary teeth. The mandibular branch carries the motor portion of the nerve to the muscles of mastication. Sensory signals from the lower lip, gum, muscles of the tongue, and one-third of the mandibular teeth are also part of the mandibular branch.

Abducens Nerve (N VI)

The **abducens** (ab-DŪ-senz) **nerve** controls the lateral rectus extraocular muscle. When this muscle contracts, the eyeball is abducted; hence the name. The nerve originates on the medulla oblongata and is positioned posterior and medial to the trigeminal nerve.

Facial Nerve (N VII)

The **facial nerve** is located on the medulla oblongata, posterior and lateral to the abducens nerve. It is a mixed nerve, with sensory fibers for the anterior two-thirds of the taste buds and somatic and autonomic motor fibers. The somatic motor neurons

Table 15.1   Cranial Nerve Branches and Functions

Cranial Nerve (Number)	Sensory Ganglion	Branch	Primary Function	Foramen	Innervation
Olfactory (I)			Special sensory	Olfactory foramina of ethmoid	Olfactory epithelium
Optic (II)			Special sensory	Optic canal	Retina of eye
Oculomotor (III)			Motor	Superior orbital fissure	Inferior, medial, superior rectus, inferior oblique, and levator palpebrae superioris muscles; intrinsic eye muscles
Trochlear (IV)			Motor	Superior orbital fissure	Superior oblique muscle
Trigeminal (V)	Semilunar		Mixed	Superior orbital fissure	Areas associated with the jaws
		Ophthalmic	Sensory	Superior orbital fissure	Orbital structures, nasal cavity, skin of forehead, upper eyelid, eyebrows, and nose (part)
		Maxillary		Foramen rotundum	Lower eyelid; superior lip, gums, and teeth; cheek, nose (part), palate, and pharynx (part)
		Mandibular		Foramen ovale	*Sensory:* inferior gums, teeth, lips, palate (part), and tongue (part)
					Motor: muscles of mastication
Abducens (VI)			Motor	Superior orbital fissure	Lateral rectus muscle
Facial (VII)	Geniculate		Mixed	Internal acoustic canal to facial canal; exits at stylomastoid foramen	*Sensory:* taste receptors on anterior ⅔ of tongue
					Motor: muscles of facial expression, lacrimal gland, submandibular gland, and sublingual salivary glands
Vestibulocochlear (Acoustic) (VIII)		Cochlear Vestibular	Special sensory	Internal acoustic canal	Cochlea (receptors for hearing) Vestibule (receptors for motion and balance)
Glossopharyngeal (IX)	Superior (jugular) and inferior (petrosal)		Mixed	Jugular foramen	*Sensory:* posterior ⅓ of tongue; pharynx and palate (part); receptors for blood pressure, pH, oxygen, and carbon dioxide concentrations
					Motor: pharyngeal muscles and parotid salivary gland
Vagus (X)	Jugular and nodose		Mixed	Jugular foramen	*Sensory:* pharynx; auricle and external acoustic canal; diaphragm; visceral organs in thoracic and abdominopelvic cavities
					Motor: palatal and pharyngeal muscles and visceral organs in thoracic and abdominopelvic cavities
Accessory (XI)		Internal	Motor	Jugular foramen	Skeletal muscles of palate, pharynx, and larynx (with vagus nerve)
		External	Motor	Jugular foramen	Sternocleidomastoid and trapezius muscles
Hypoglossal (XII)			Motor	Hypoglossal canal	Tongue musculature

innervate the muscles of facial expression, such as the zygomaticus muscle. Visceral motor neurons control the activity of the salivary glands, lacrimal (tear) glands, and nasal mucous glands.

Clinical Application

I'm So Happy I Could Cry!

Why do we cry? Primarily, crying is a protective mechanism that cleans the surface of the eye after some object has touched the eyeball. Why do we cry when we are sad, though, or sometimes even when we are happy? The answer is that strong emotions, such as sorrow and joy, are coordinated by the sympathetic branch of the autonomic nervous system. Sympathetic innervation regulates secretion by glands, including secretion of tears by the lacrimal glands. Thus, such events as receiving a sentimental gift from a loved one, activate sympathetic neurons, which in turn cause the release of tears of joy.

Vestibulocochlear Nerve (N VIII)

The **vestibulocochlear nerve**, also called the *auditory nerve,* is a sensory nerve of the inner ear located on the medulla oblongata near the facial nerve. The vestibulocochlear nerve has two branches. The vestibular branch gathers information about the sense of balance from the vestibule and semicircular canals of the inner ear. The cochlear branch conducts auditory sensations from the cochlea, which is the organ of hearing in the inner ear.

Glossopharyngeal Nerve (N IX)

The **glossopharyngeal** (glos-ō-fah-RIN-jē-al) **nerve** is a mixed nerve of the tongue and throat. It supplies the medulla oblongata with sensory information from the posterior third of the tongue (remember, the facial nerve innervates the anterior two-thirds of the taste buds) and from the palate and pharynx. The glossopharyngeal nerve also conveys barosensory and chemosensory information from the carotid sinus and the carotid body, where blood pressure and dissolved blood gases are monitored, respectively. Motor innervation by the glossopharyngeal nerve controls the pharyngeal muscles involved in swallowing and in the activity of the salivary glands.

Vagus Nerve (N X)

The **vagus** (VĀ-gus) **nerve** is a complex nerve on the medulla oblongata that has mixed sensory and motor functions. Sensory neurons from the pharynx, diaphragm, and most of the internal organs of the thoracic and abdominal cavities ascend along the vagus nerve and synapse with autonomic nuclei in the medulla. The motor portion controls the involuntary muscles of the respiratory, digestive, and cardiovascular systems. The vagus is the only cranial nerve that descends below the neck. It enters the ventral body cavity, but it does not pass to the thorax via the spinal cord; rather, it follows the musculature of the neck. Because this nerve regulates the activities of the organs of the thoracic and abdominal cavities, disorders of the nerve result in systemic disruption of homeostasis.

Parasympathetic fibers in the vagus nerve control swallowing, digestion, heart rate, and respiratory patterns. If this control is compromised, sympathetic stimulation goes unchecked, and the organs respond as during exercise or stress. The cardiovascular and respiratory systems increase their activities, and the digestive system shuts down.

Accessory Nerve (N XI)

The **accessory nerve** is a motor nerve that controls the skeletal muscles involved in swallowing and the sternocleidomastoid and trapezius muscles of the neck. It is the only cranial nerve with fibers that originate from both the medulla oblongata

Table 15.2 Cranial Nerve Tests

Cranial Nerve	Nerve Function Test
I. Olfactory	Hold open container of rubbing alcohol under subject's nose and have subject identify odor. Repeat with open container of wintergreen oil.
II. Optic	Test subject's visual field by moving a finger back and forth in front of subject's eyes. Use eye chart to test visual acuity.
III. Oculomotor	Examine subject's pupils for equal size. Have subject follow an object with eyes.
IV. Trochlear	Tested with oculomotor nerve. Have subject roll eyes downward.
V. Trigeminal	Check motor functions of nerve by having subject move mandible in various directions. Check sensory functions with warm and cold probes on forehead, upper lip, and lower jaw.
VI. Abducens	Tested with oculomotor nerve. Have subject move eyes medially.
VII. Facial	Use sugar solution to test anterior of tongue for sweet taste reception. Observe facial muscle contractions for even muscle tone on each side of face while subject smiles, frowns, and purses lips.
VIII. Vestibulocochlear	Cochlear branch—Hold vibrating tuning fork in air next to ear, and then touch fork to mastoid process for bone-conduction test. Vestibular branch—Have subject close eyes and maintain balance.
IX. Glossopharyngeal	While subject coughs, check position of uvula on posterior of soft palate. Use quinine solution to test posterior of tongue for bitter taste reception.
X. Vagus	While subject coughs, check position of uvula on posterior of soft palate.
XI. Spinal accessory	Hold subject's shoulder while the subject rotates it to test the strength of sternocleidomastoid muscle. Hold head while subject rotates it to test trapezius strength.
XII. Hypoglossal	Observe subject protract and retract tongue from mouth, and check for even movement on two lateral edges of tongue.

3 Materials

- ☐ Brain model
- ☐ Brain chart
- ☐ Isopropyl (rubbing) alcohol
- ☐ Wintergreen oil
- ☐ Eye chart
- ☐ Sugar solution
- ☐ Quinine solution
- ☐ Tuning fork
- ☐ Beaker of ice
- ☐ Cold probes
- ☐ Beaker of warm water
- ☐ Warm probes

and the spinal cord. Numerous thread-like branches from these two regions unite in the spinal accessory nerve.

Hypoglossal Nerve (N XII)

The **hypoglossal** (hī-pō-GLOS-al) **nerve** is located on the medulla oblongata medial to the vagus nerve. This motor nerve supplies motor fibers that control tongue movements for speech and swallowing.

QuickCheck Questions

3.1 What are three cranial nerves that are sensory nerves?

3.2 Which cranial nerve enters the ventral body cavity?

Procedures

1. Review the cranial nerves in Figure 15.10.
2. Locate each cranial nerve on the brain model and chart.
3. Your instructor may ask you to test the function of selected cranial nerves. Table 15.2 lists the basic tests used to assess the general function of each nerve. ■

LAB ACTIVITY 4 Sheep Brain Dissection

The sheep brain, like all other mammalian brains, is similar in structure and function to the human brain. One major difference between the human brain and that of other animals is the orientation of the brain stem relative to the body axis. The human body has a vertical axis, and the brain stem and spinal cord are positioned vertically. In four-legged animals, the body axis is horizontal and the brain stem and spinal cord are also horizontal.

All vertebrate animals—sharks, fish, amphibians, reptiles, birds, and mammals—have a brain stem for basic body functions. These animals can learn through experience, which is a complex neurological process that requires higher-level processing and memory storage, as occurs in the human cerebrum. Imagine the complex motor activity necessary for locomotion in these animals.

Dissecting a sheep brain enhances your study of models and charts of the human brain. Take your time during the dissection and follow the directions carefully. Refer to this manual and its illustrations often during the procedures.

Safety Alert: Brain Dissection

You must—repeat, *must*—practice the highest level of laboratory safety while handling and dissecting the brain. Keep the following guidelines in mind during the dissection:

1. Wear gloves and safety glasses to protect yourself from the fixatives used to preserve the specimen.
2. Do not dispose of the fixative from your specimen. You will later store the specimen in the fixative to keep the specimen moist and to keep it from decaying.
3. Be extremely careful when using a scalpel or other sharp instrument. Always direct cutting and scissor motions away from you to prevent an accident if the instrument slips on moist tissue.
4. Before cutting a given tissue, make sure it is free from underlying and/or adjacent tissues so that they will not be accidentally severed.
5. Never discard tissue in the sink or trash. Your instructor will inform you about the proper disposal procedure. ▲

QuickCheck Questions

4.1 How is the orientation of the sheep brain different from the human brain?

4.2 What type of safety equipment should you wear during the sheep brain dissection?

4.3 How should you dispose of the sheep brain and scrap tissue?

4 Materials

- ☐ Gloves
- ☐ Safety glasses
- ☐ Preserved sheep brain (preferably with dura mater intact)
- ☐ Dissection pan
- ☐ Scissors
- ☐ Blunt probe
- ☐ Large dissection knife

Procedures

Put on gloves and safety glasses before handling the brain.

I. The Meninges

If your sheep brain does not have the dura mater, skip to part II.

1. On the intact dura mater, locate the falx cerebri and the tentorium cerebelli on the overlying dorsal surface of the dura mater. How does the tissue of the falx cerebri compare with the dura mater covering the hemispheres?

2. If your specimen still has the ethmoid, a mass of bone on the anterior frontal lobe, slip a probe between the bone and the dura mater. Carefully pull the bone off the specimen, using scissors to snip away any attached dura mater. Examine the removed ethmoid and identify the crista galli, which is the crest of bone where the meninges attach.

3. Gently insert a probe between the dura mater and the brain and gently work the probe back and forth to separate the two. With scissors, cut completely around the base of the dura mater, and leave the inferior portion intact over the cranial nerves. Make small cuts with the scissors and be careful not to cut or remove any of the cranial nerves. Do not lift the dura too high or the cranial nerves will detach from the brain.

EXERCISE 15 Anatomy of the Brain 273

4. Cut completely through the lateral sides of the tentorium cerebelli and then remove the dura mater in one piece by grasping it with your (gloved) hand and peeling it off the brain.

5. Open the detached dura mater and identify the falx cerebri and tentorium cerebelli. (One difference between the sheep brain and the human brain is that the sheep brain does not have a falx cerebelli.)

II. External Brain Anatomy

1. Examine the cerebrum, and identify the frontal, parietal, occipital, and temporal lobes. The insula is a deep lobe and is not visible externally. Note the longitudinal fissure that separates the right and left cerebral hemispheres. Observe the gyri and sulci on the cortical surface. Examine the surface between sulci for the arachnoid mater and pia mater.

2. Identify the cerebellum and compare the size of the folia with the size of the cerebral gyri. Unlike the human brain, the sheep cerebellum is not divided medially into two lateral hemispheres.

3. To examine the dorsal anatomy of the mesencephalon, position the sheep brain as in Figure 15.11 and use your fingers to gently depress the cerebellum. The mesencephalon will then be visible between the cerebrum and cerebellum. Now identify the four elevated masses of the corpora quadrigemina and distinguish between the superior colliculi and the inferior colliculi. The pineal gland of the diencephalon is superior to the mesencephalon.

4. Turn the brain over to view the ventral surface, as in Figure 15.12. Note how the spinal cord joins the medulla oblongata. Identify the pons and the cerebral peduncles of the mesencephalon. Locate the single mamillary body on the hypothalamus. (Remember that the mamillary body of the human

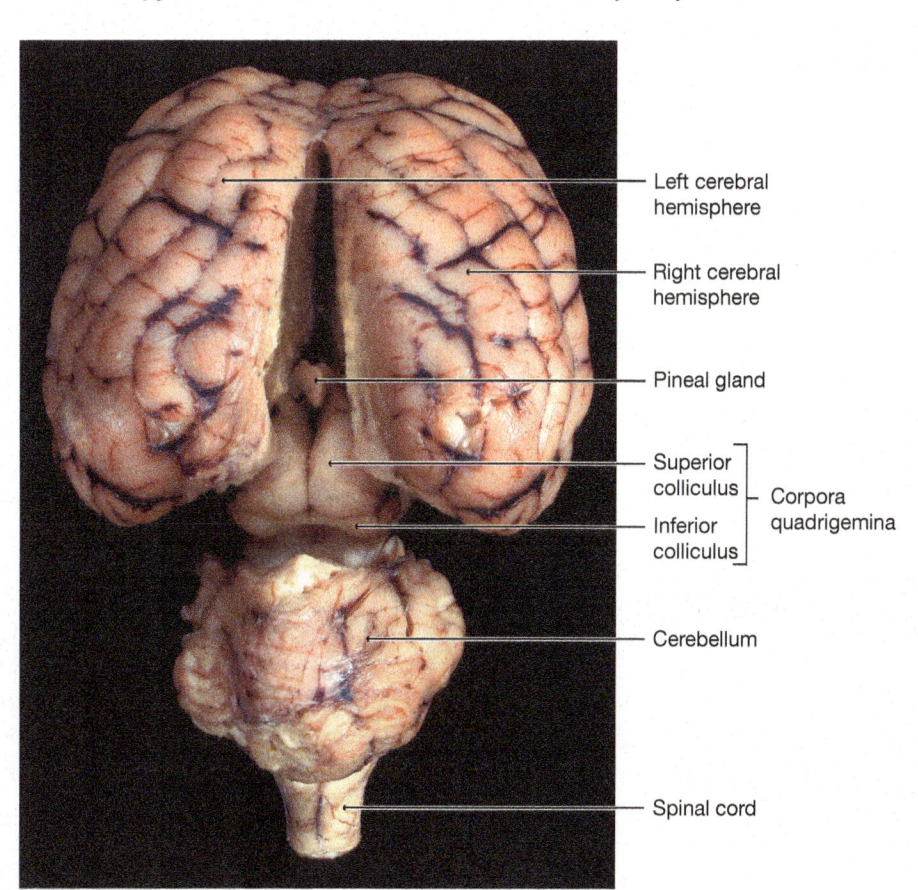

Figure 15.11 Superior View of Sheep Brain
The cerebellum is pushed down to show the location of the corpora quadrigemina.

274 EXERCISE 15 Anatomy of the Brain

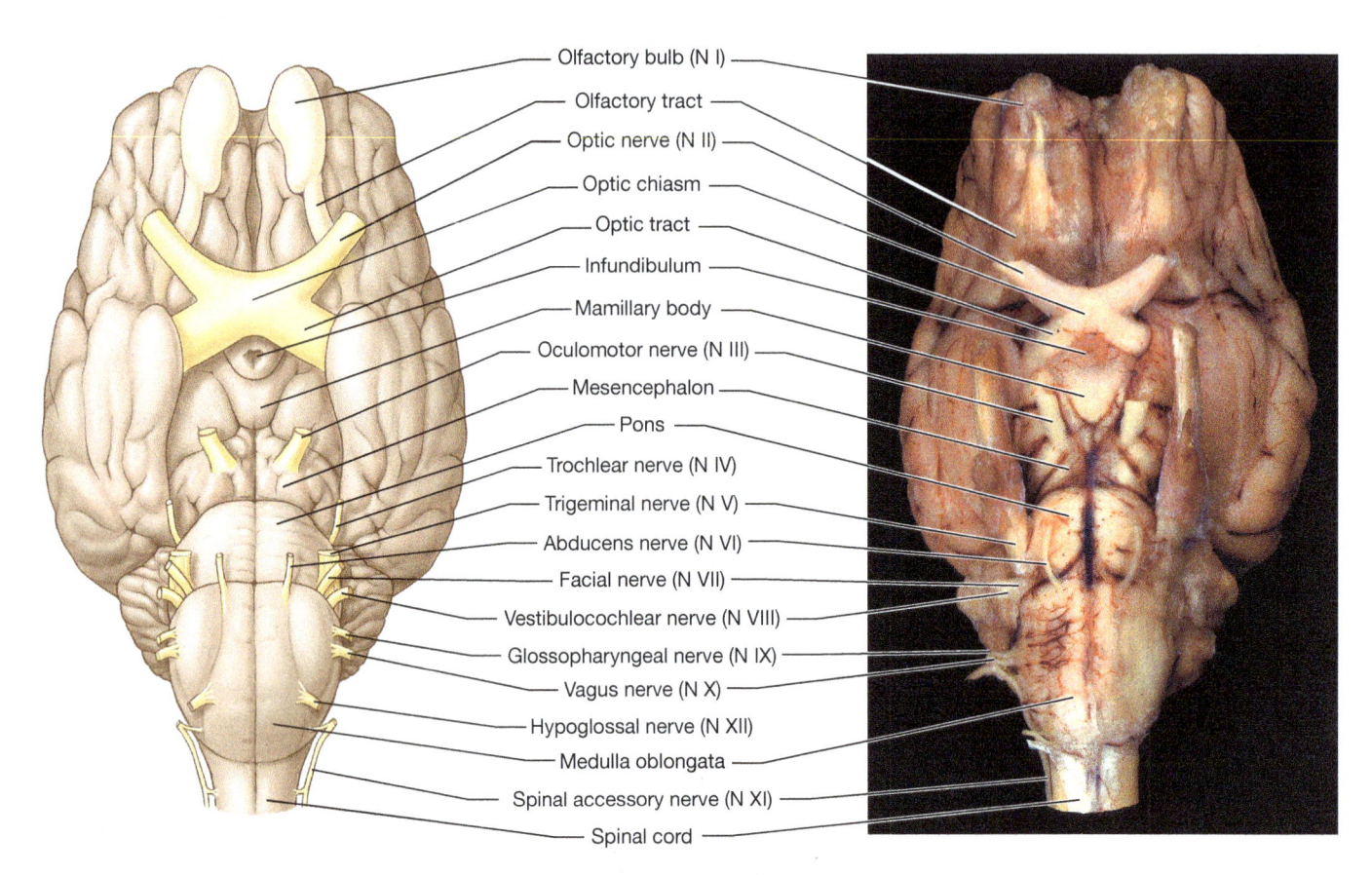

Olfactory bulb (N I)
Olfactory tract
Optic nerve (N II)
Optic chiasm
Optic tract
Infundibulum
Mamillary body
Oculomotor nerve (N III)
Mesencephalon
Pons
Trochlear nerve (N IV)
Trigeminal nerve (N V)
Abducens nerve (N VI)
Facial nerve (N VII)
Vestibulocochlear nerve (N VIII)
Glossopharyngeal nerve (N IX)
Vagus nerve (N X)
Hypoglossal nerve (N XII)
Medulla oblongata
Spinal accessory nerve (N XI)
Spinal cord

Figure 15.12 Ventral View of Sheep Brain
Cranial nerves are clearly visible in the ventral view of the sheep brain.

brain is a *paired* mass.) The pituitary gland has most likely been removed from your specimen; however, you can still identify the stub of the infundibulum that attaches the pituitary to the hypothalamus.

5. Using Figure 15.12 as a guide, identify as many cranial nerves on your sheep brain as possible. Nerves I through III and nerve V are usually intact and easy to identify. Your laboratory instructor may ask you to observe several sheep brains in order to study all the cranial nerves. The three branches of the trigeminal nerve were cut when the brain was removed from the sheep and therefore are not present on any specimen. The glossopharyngeal nerve may have been removed inadvertently when the specimen was being prepared. Even if this nerve is present in your specimen, however, it is difficult to identify on the sheep brain.

III. Internal Brain Anatomy

1. Lay the sheep brain in the dissection pan so that the superior surface faces you, as in Figure 15.11. Place the blade of a large dissection knife in the anterior region of the longitudinal fissure and section the brain by cutting it in half along the fissure. Use as few cutting strokes as possible to prevent damage to the brain tissue.

2. Using Figure 15.13 as a guide, identify the internal anatomical features of the sheep brain. Observe the different regions of white and gray matter.

3. Gently slide a blunt probe between the corpus callosum and fornix and into the lateral ventricle to determine how deep the ventricle extends into the cerebrum. Inside the lateral ventricle, locate the choroid plexus, which appears as a granular mass of tissue.

4. To view deep structures of the cerebrum and diencephalon, put the two halves of the brain together and use a large dissection knife to cut a frontal section

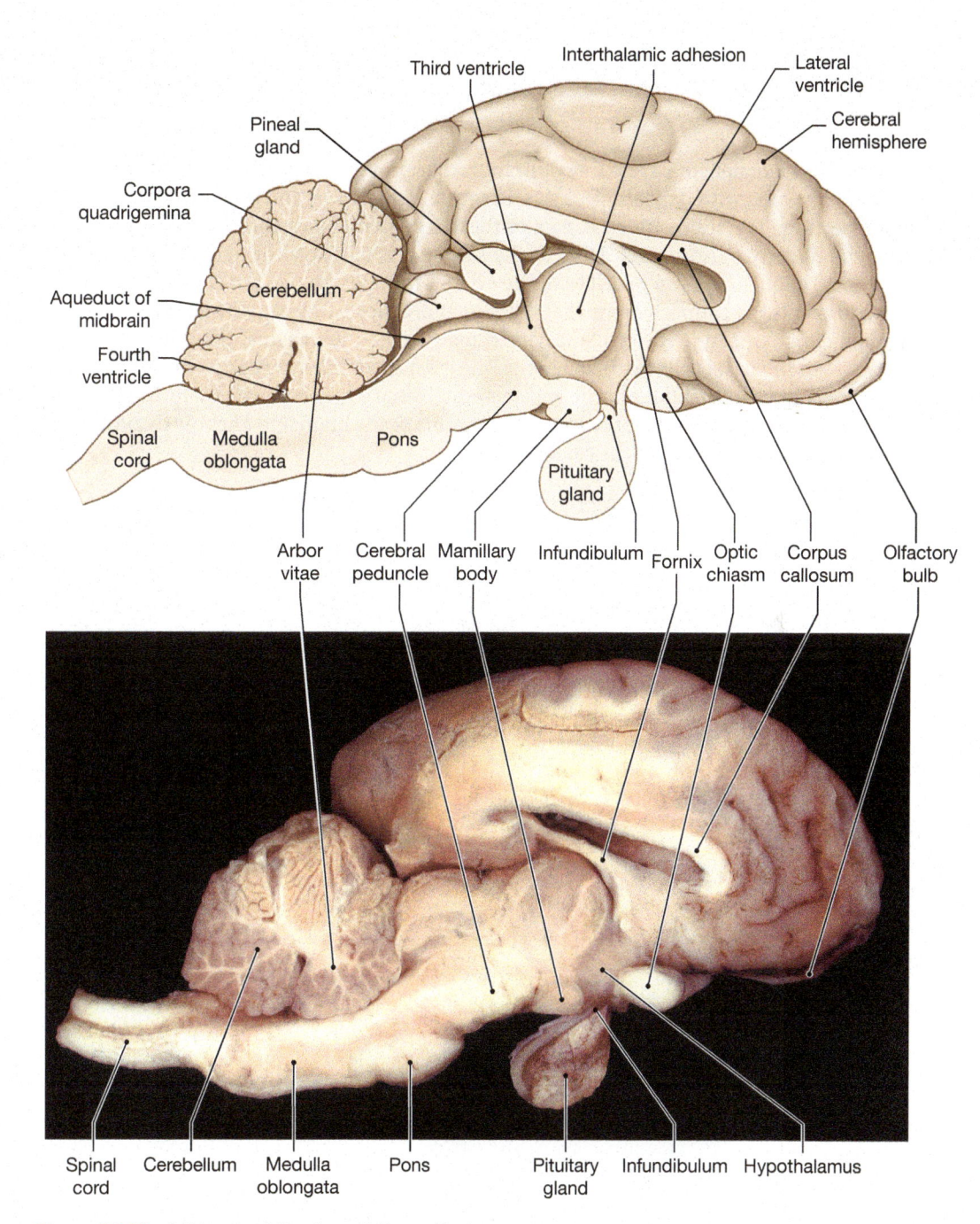

Figure 15.13 Midsagittal Section of Sheep Brain
Internal anatomy of the sheep brain in sagittal section.

through the infundibulum. Make another frontal section just posterior to the first to slice off a thin slab of brain. Lay the slab in the dissection pan with the anterior side up. (The anterior side is the surface where you made your first cut.) Notice the distribution of gray matter and white matter. Observe how the corpus callosum joins each cerebral hemisphere. The gray matter of the basal nuclei is lateral to the ventricles.

5. When finished, store or discard the sheep brain as directed by your laboratory instructor. Proper disposal of all biological waste protects the local environment and is mandated by local, state, and federal regulations. ■

cranium A Regional Look

The head has a lamellate (layered) organization that provides strength, much like a sheet of plywood with its multiple layers. Examine Figure 15.14 and note each layer in the dissection. The scalp, which is part of the integumentary system, encases and protects the underlying structures. Deep to the scalp is a tough layer of tendon called the epicranial aponeurosis where bellies of the occipitofrontalis muscle attach. Deep to the aponeurosis is the connective tissue and periosteum that seal the superficial surface of the cranial bones. Inside the cranial cavity, the dura mater lies against the cranium and attaches to the crista galli of the sphenoid. The arachnoid mater is deep to the dura and contains cerebrospinal fluid in the subarachnoid space to cushion the brain from impacts to the head. The pia mater follows the contours of the brain and anchors blood vessels in place on the surface of the cerebral cortex.

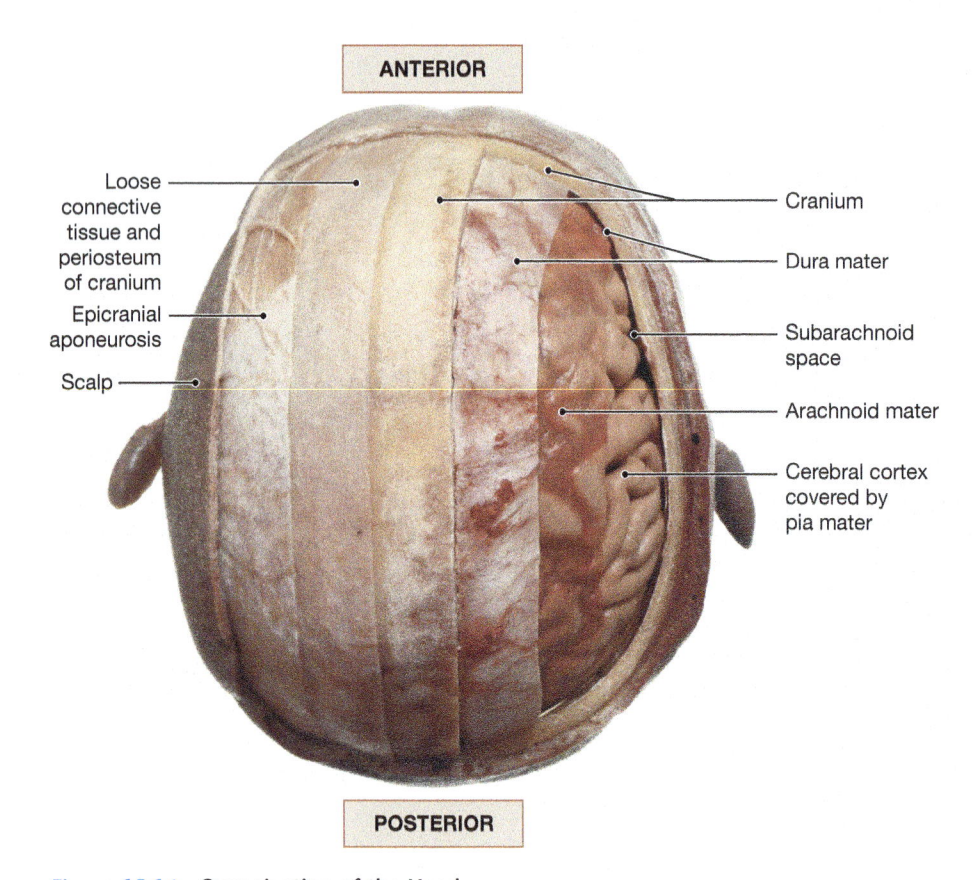

Figure 15.14 Organization of the Head
A superior view of a dissection of the head that shows the cranial meninges and the cerebral cortex.

EXERCISE 15

LAB REPORT

Anatomy of the Brain

Name _____
Date _____
Section _____

A. Matching

Match each structure in the left column with its correct description from the right column.

_____ 1. folia
_____ 2. cerebrum
_____ 3. mamillary body
_____ 4. longitudinal fissure
_____ 5. inferior colliculus
_____ 6. optic chiasm
_____ 7. falx cerebri
_____ 8. hypothalamus
_____ 9. central sulcus
_____ 10. cerebral peduncles
_____ 11. dura mater
_____ 12. vermis
_____ 13. subarachnoid space
_____ 14. pons
_____ 15. tentorium cerebelli

A. area where part of optic nerve crosses to opposite side of brain
B. forms floor of diencephalon
C. part of mesencephalon
D. outer meningeal layer
E. site of cerebrospinal fluid circulation around brain
F. small folds on cerebellum
G. narrow central region of cerebellum
H. separates cerebellum from cerebrum
I. mass posterior to infundibulum
J. area of brain superior to medulla
K. divides motor and sensory cortex
L. tissue between cerebral hemispheres
M. contains five lobes
N. part of corpora quadrigemina
O. cleft between cerebral hemispheres

B. Matching

Match each structure in the left column with its correct description from the right column.

_____ 1. third ventricle
_____ 2. septum pellucidum
_____ 3. thalamus
_____ 4. corpus callosum
_____ 5. pineal gland
_____ 6. superior colliculus
_____ 7. arachnoid granulation
_____ 8. fornix
_____ 9. aqueduct of the midbrain
_____ 10. arbor vitae

A. part of corpora quadrigemina
B. white tract between cerebral hemispheres
C. duct through mesencephalon
D. white matter of cerebellum
E. site of cerebrospinal fluid reabsorption
F. white matter inferior to lateral ventricles
G. chamber of diencephalon
H. nipple-like gland in diencephalon
I. forms lateral walls of third ventricle
J. separates lateral ventricles

EXERCISE 15

LAB REPORT

C. Labeling

Label Figure 15.15, which shows the inferior surface of the human brain.

D. Labeling

Label Figure 15.16, which is a midsagittal close-up view of the human brain.

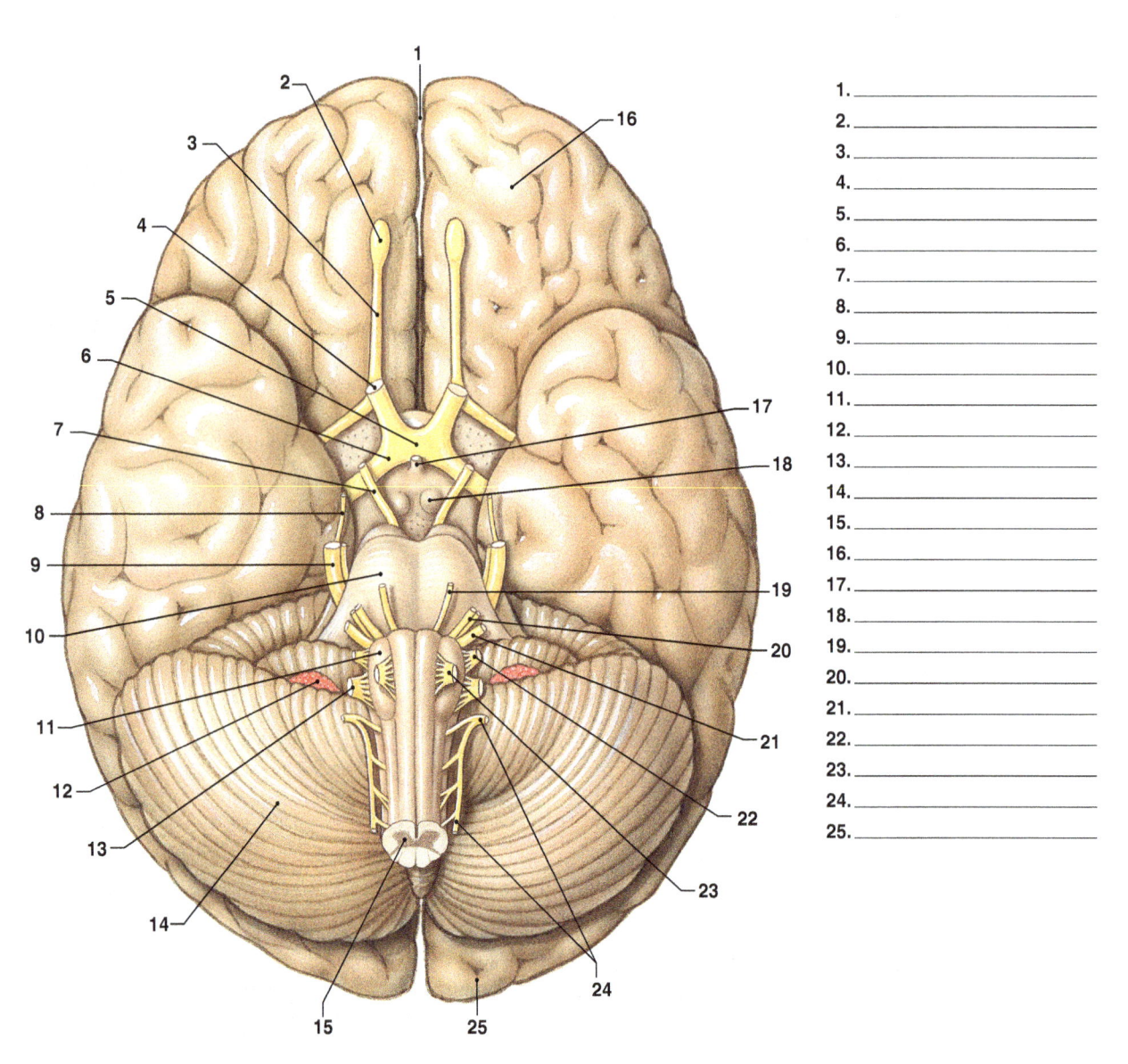

1. _____
2. _____
3. _____
4. _____
5. _____
6. _____
7. _____
8. _____
9. _____
10. _____
11. _____
12. _____
13. _____
14. _____
15. _____
16. _____
17. _____
18. _____
19. _____
20. _____
21. _____
22. _____
23. _____
24. _____
25. _____

Figure 15.15 Inferior Surface of the Human Brain

278

LAB REPORT   **EXERCISE 15**

Figure 15.16 **Detail of Sagittal Section of the Human Brain**

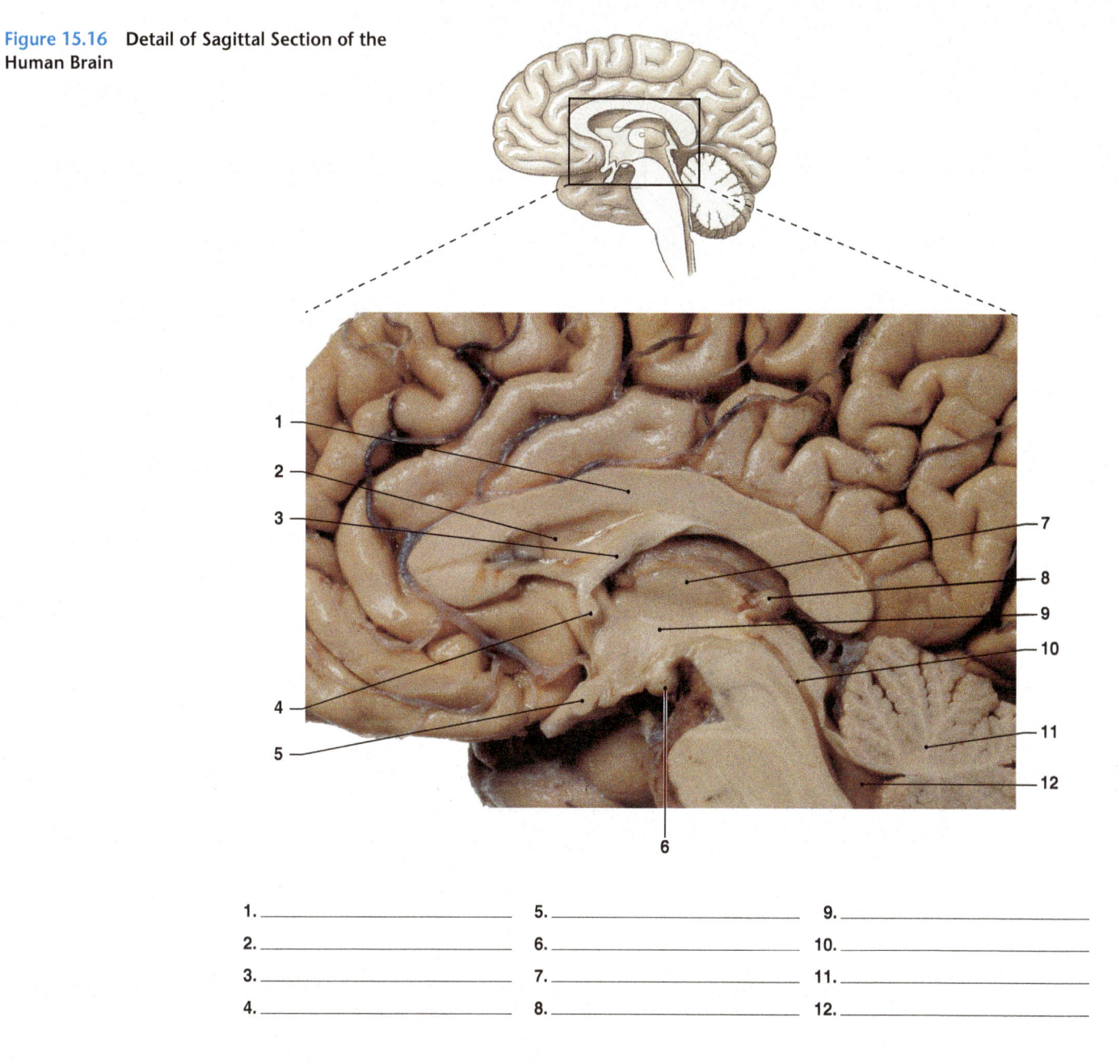

1. _____	5. _____	9. _____
2. _____	6. _____	10. _____
3. _____	7. _____	11. _____
4. _____	8. _____	12. _____

E. **Short-Answer Questions**

1. List the six major regions of the brain.

2. Which cranial nerves conduct the sensory and motor impulses of the eye?

3. Describe the location and anatomical features of the corpora quadrigemina.

279

# EXERCISE 15

LAB REPORT

4. Describe the extensions of the dura mater.

5. What is the function of the precentral gyrus?

6. Trace a drop of CSF from a lateral ventricle to reabsorption at an arachnoid granulation.

F. Analysis and Application

1. Imagine watching a bird fly across your line of vision. What part of your brain is active in keeping an image of the moving bird on your retina?

2. You have just eaten a medium-size pepperoni pizza and you laid down to digest it. Which cranial nerve stimulates the muscular activity of your digestive tract?

3. A child is preoccupied with a large cherry lollipop. What part of the child's brain is responsible for the licking and eating reflexes?

4. Your favorite movie has made you cry yet again. Which cranial nerve is responsible for your tears?

5. A patient is brought into the emergency room with severe whiplash. He is not breathing and has lost cardiac function. What part of the brain has most likely been damaged?

6. A woman is admitted to the hospital with Bell's palsy caused by an inflamed facial nerve. You are the attending physician. What symptoms do you observe? How do you test her facial nerve?

16 EXERCISE

The General Senses

OBJECTIVES

On completion of this exercise, you should be able to:

- List the receptors for the general senses.
- Discuss the distribution of receptors in the integument.
- Describe the two-point discrimination test.

LAB ACTIVITIES

1. General-Sense Receptors 281
2. Two-Point Discrimination Test 284

Changes in the body's internal and external environments are detected by sensory receptors. Most of these receptors are sensitive to a specific stimulus. The taste buds of the tongue, for example, are stimulated only by chemicals dissolved in saliva and not by sound waves or light rays.

The human senses may be grouped into two broad categories: general senses and special senses. The **general senses**, which have simple neural pathways, are touch, temperature, pain, chemical and pressure detection, and body position (proprioception). The **special senses** have complex pathways, and the receptors for these senses are housed in specialized organs. The special senses include gustation (taste), olfaction (smell), vision, audition (hearing), and equilibrium. In this exercise you will study the receptors of the general senses.

A sensory receptor is either a tonic or a phasic receptor. **Tonic receptors** are always active; pain receptors are one example. **Phasic receptors** are usually inactive and are "turned on" with stimulation. These receptors provide information on the rate of change of a stimulus. Some examples are root-hair plexuses, tactile corpuscles, and lamellated corpuscles, which are all phasic receptors for touch.

LAB ACTIVITY 1 General-Sense Receptors

Many kinds of sensory receptors transmit information to the CNS. **Thermoreceptors** are sensors for changes in temperature and have wide distribution in the body, and are found in the dermis, skeletal muscles, and hypothalamus. (The hypothalamus is the body's internal thermostat.) **Chemoreceptors**, which are found in the medulla, in arteries near the heart, and in the heart, monitor changes in the concentrations of various chemicals in body fluids. **Nociceptors** (nō-sē-SEP-turz) are pain receptors in the epidermis.

Mechanoreceptors, which are touch receptors that bend when stimulated, are present in three types: baroreceptors, proprioceptors, and tactile receptors. **Baroreceptors** monitor pressure changes in liquids and gases. These receptors are typically the tips of sensory-neuron dendrites in blood vessels and the lungs. **Proprioceptors** (prō-prē-ō-SEP-turz) are stimulated by changes in body position, such as rotating the head, and convey the information to the cerebellum of the

282 EXERCISE 16 The General Senses

brain so that the CNS knows where we are located in our three-dimensional surroundings. Two types of proprioceptors are **muscle spindles** in muscles, which inform the brain about muscle tension; and **Golgi tendon organs** in tendons near joints, which inform the brain about joint position.

Tactile receptors (Figure 16.1) are located in the skin. They respond to touch and provide us with information regarding texture, shape, size, and location of the

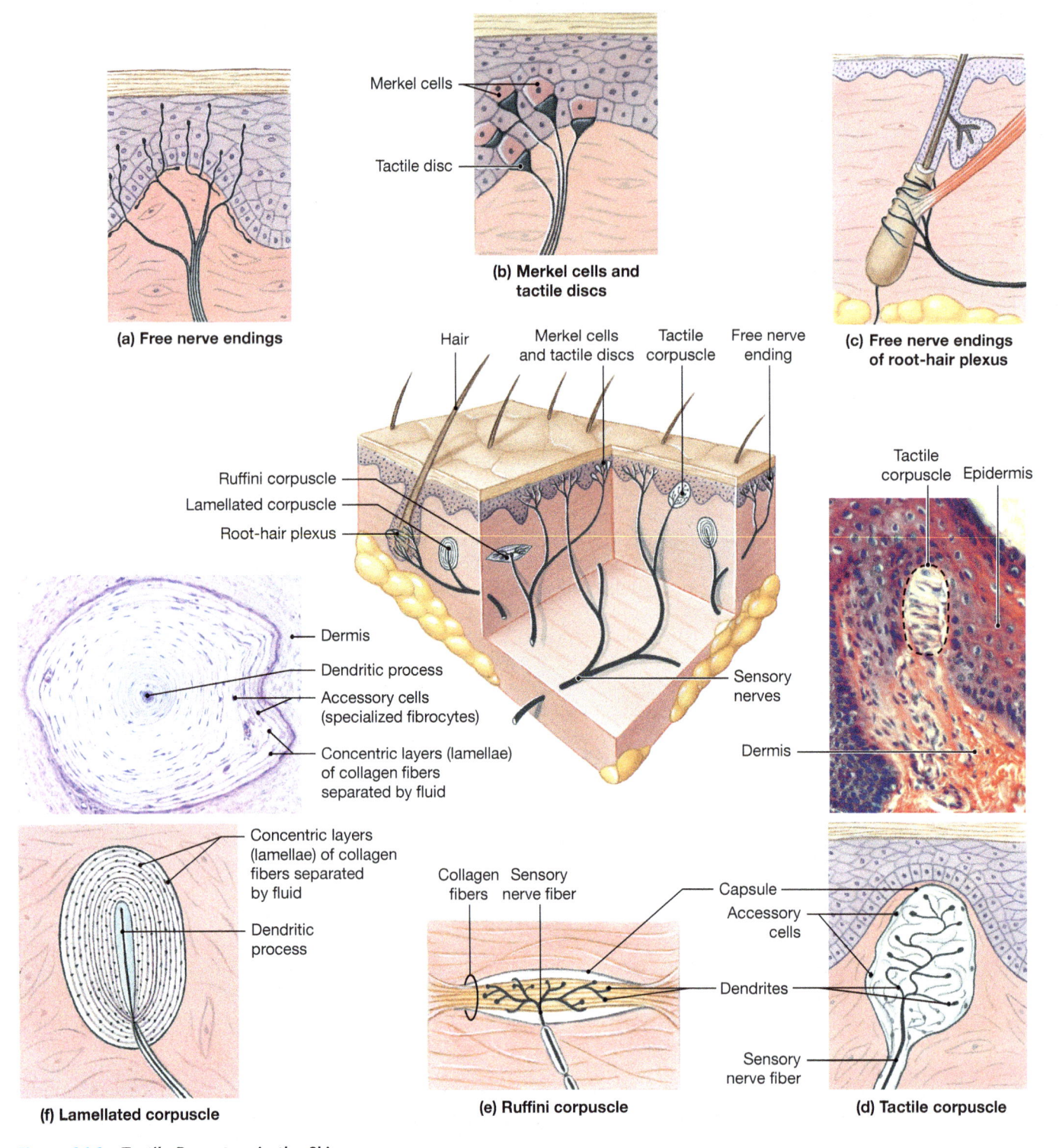

Figure 16.1 Tactile Receptors in the Skin

The location and general appearance of six important tactile receptors. (a) Free nerve endings. (b) Merkel cells and tactile discs. (c) Free nerve endings of root-hair plexus. (d) Tactile corpuscle. (LM × 330) (e) Ruffini corpuscle. (f) Lamellated corpuscle. (LM × .75)

tactile stimulation. The receptor cells may be either unencapsulated or encapsulated with connective tissue. Unencapsulated tactile receptors are very sensitive to touch. The unencapsulated tactile receptors known as **free nerve endings** are simple receptors that are the exposed tips of dendrites in tissues. **Merkel cells** are unencapsulated tactile receptors that are in direct contact with the epidermis and respond to sensory neurons at swollen synapses called **tactile discs**. **Root-hair plexuses** are unencapsulated tactile receptors composed of sensory-neuron dendrites wrapped around hair roots. These receptors are stimulated when an insect, for example, lands on your bare arm and moves one of the hairs there.

Encapsulated tactile receptors have branched dendrites that are covered by specialized cells. **Tactile corpuscles**, also called **Meissner** (MĪS-ner) **corpuscles**, are nerve endings located in the dermal papillae of the skin. The dendrites are wrapped in special Schwann cells and are very sensitive to pressure, change in shape, and touch. Tactile corpuscles are phasic receptors. Deeper in the dermis are encapsulated tactile receptors called either **lamellated** (LAM-e-lā-ted; *lamella,* thin plate) **corpuscles** or **Pacinian** (pa-SIN-ē-an) **corpuscles**. These receptors have a large dendrite encased in concentric layers of connective tissue and respond to deep pressure and vibrations. **Ruffini** (roo-FĒ-nē) **corpuscles** are surrounded by collagen fibers embedded in the dermis. Changes in either the tension or shape of the skin tug on the collagen fibers, and the tugging stimulates Ruffini corpuscles.

Clinical Application

Brain Freeze

Sometimes the body projects a sensation, usually pain, to another part of the body. This phenomenon is called **referred pain**. An excellent example of referred pain is the pain in the forehead some people feel after quickly consuming a cold drink or a bowl of ice cream. The resulting "brain freeze", as it is commonly called, is pain "referred" from the nerves of the throat because these nerves also innervate the forehead.

QuickCheck Questions

1.1 What is chemoreception?
1.2 What is baroreception?
1.3 What is the stimulus for mechanoreceptors?

1 Materials

- ☐ Compound microscope
- ☐ Microscope slide of tactile corpuscles
- ☐ Microscope slide of lamellated corpuscles

Procedures

1. Examine the tactile corpuscle slide at low magnification and locate the junction between the epidermis and dermis. Locate the tactile corpuscles in the papillary region of the dermis, where the dermis folds to attach the epidermis (Figure 16.2). Observe the tactile corpuscles at medium magnification.

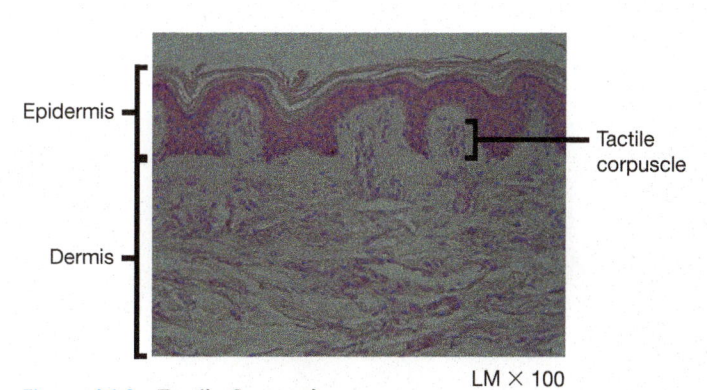

Figure 16.2 **Tactile Corpuscle**
Tactile corpuscles in the dermal papillae.

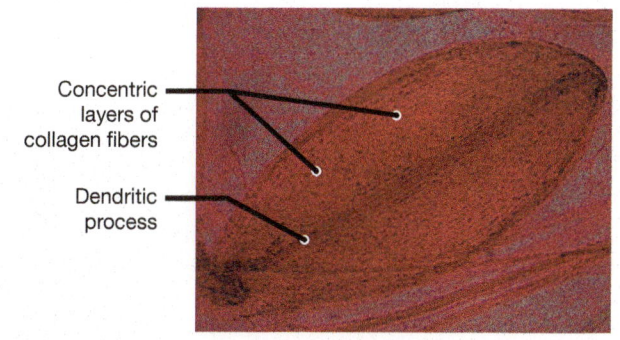

Figure 16.3 **Lamellated Corpuscle**
Lamellated corpuscle with dendritic process visible.

284 EXERCISE 16 The General Senses

2. Using Figure 16.3 for reference, examine the slide of the lamellated corpuscles at low magnification. Note the multiple layers wrapped around the dendritic process of the receptor.

3. Gently touch a hair on your forearm and notice how you suddenly sense the touch. Is the root-hair plexus a phasic receptor or a tonic receptor? ■

LAB ACTIVITY 2 Two-Point Discrimination Test

A sensory receptor monitors a specific region called a **receptive field** (Figure 16.4). Overlap in adjacent receptive fields enables the brain to detect where a stimulus is applied to the body. Sensory receptors are not evenly distributed in the integument. Some areas have a dense population of a particular receptor, whereas other areas have only a few or none of that receptor. This explains why your fingertips, for example, are more sensitive to touch than your scalp. The **two-point discrimination test** is used to map the distribution of touch receptors on the skin. A drawing compass with two points is used to determine the distance between receptors in the skin. The compass points are gently pressed into the skin, and the subject decides if one or two points are felt. If the sensation is that of a single point, only one receptor has been stimulated. By gradually increasing the distance between the points until two distinct sensations are felt, the density of the receptor population in that region can be measured.

QuickCheck Questions

2.1 What does the two-point discrimination test measure?

2.2 Are all parts of the body equally sensitive to touch?

2 Materials

- Drawing compass with millimeter scale

Procedures

1. Push the two points of the compass as close together as they will go, read from the millimeter scale how far apart the points are, and record this distance in the space provided in the column at the far left of Table 16.1, which is in the lab report at the end of this exercise. Gently place the points on the tip of an index finger of your lab partner, and then record whether she or he feels one point or two. Slightly spread the compass points, record the distance apart as read from the millimeter scale, and place them again on the same area of the fingertip; again record whether your partner feels one point or two. Repeat this procedure until your partner feels two distinct points.

2. Reset the compass so that the two points are as close to each other as possible, and repeat the test on the back of the hand, back of the neck, and one side of the nose. Record the data in Table 16.1. ■

Figure 16.4 Receptors and Receptive Fields
Each receptor cell monitors a specific area known as a receptive field.

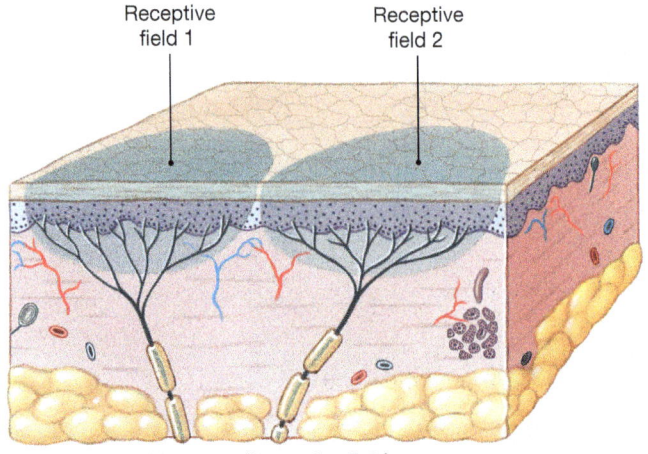

Receptive field 1   Receptive field 2

Receptive fields

EXERCISE 16

LAB REPORT

The General Senses

Name _____

Date _____

Section _____

A. Data Recording and Interpretation

Record your data from the two-point discrimination test in Table 16.1.

Table 16.1 Two-Point Discrimination Test Data

Index Finger		Back of Hand		Back of Neck		Side of Nose	
Distance between points (mm)	1 point or 2 felt?	Distance between points (mm)	1 point or 2 felt?	Distance between points (mm)	1 point or 2 felt?	Distance between points (mm)	1 point or 2 felt?

B. Matching

Match each sense in the left column with its correct receptor from the right column.

_____ 1. very sensitive touch receptor of epidermis

_____ 2. tactile receptor in deep dermis

_____ 3. touch-sensitive receptor in dermal papillae

_____ 4. unencapsulated receptor that consists of dendrite tip

_____ 5. found on hair-covered parts of body

_____ 6. senses muscle tension; one type of proprioceptor

A. lamellated corpuscle
B. tactile corpuscle
C. root-hair plexus
D. muscle spindle
E. tactile disc
F. free nerve ending

EXERCISE 16 — LAB REPORT

C. **Short-Answer Questions**

1. Explain the concept of a receptive field.

2. Describe the appearance of a lamellated corpuscle.

3. What is the sense of proprioception?

D. **Analysis and Application**

1. Evaluate your data from the two-point discrimination test and rank the four body regions tested from most sensitive to least sensitive.

2. What is the difference between a phasic receptor and a tonic receptor?

EXERCISE 17

Special Senses: Olfaction and Gustation

LAB ACTIVITIES

1. Olfaction 287
2. Olfactory Adaptation 289
3. Gustation 290
4. Relationship between Olfaction and Gustation 292

OBJECTIVES

On completion of this exercise, you should be able to:

- Describe the location and structure of the olfactory receptor cells.
- Identify the microscopic features of the olfactory epithelium.
- Describe the location and structure of taste buds and papillae.
- Identify the microscopic features of taste buds.
- Explain how olfaction accentuates gustation.

As mentioned in Exercise 16, the special senses are gustation (taste), olfaction (smell), vision, audition (hearing), and equilibrium. In this exercise you will study the receptors for olfaction and gustation. The receptors for all the special senses are housed in specialized organs, and information from the receptors is processed in dedicated areas of the cerebral cortex. Neural pathways for the special senses are complex, and often branch out to different regions of the brain for integration with other sensory input.

LAB ACTIVITY 1 Olfaction

Olfactory receptor cells are located in the **olfactory epithelium** that lines the roof of the nasal cavity (Figure 17.1a). These receptors are bipolar neurons with many cilia that are sensitive to airborne molecules (Figure 17.1b). Most of the air we inhale passes through the nasal cavity and into the pharynx. Sniffing increases our sense of smell by pulling more air across the olfactory receptor cells.

In addition to olfactory receptors cells, two other types of cells occur in the olfactory epithelium: basal cells and supporting cells. **Basal cells** are stem cells that divide and replace olfactory receptor cells. **Supporting cells**, also called **sustentacular cells**, provide physical support and nourishment to the receptor cells.

The olfactory epithelium is attached to an underlying layer of connective tissue, the lamina propria (Exercise 4). This layer contains **olfactory glands** (also called **Bowman's glands**), which secrete mucus. In order for a substance to be smelled, volatile molecules from the substance must diffuse through the air from the substance to your nose. Once in the nose, the molecules must diffuse through the mucus secreted by the olfactory glands before they can stimulate the cilia of the olfactory receptor cells. The sense of smell is drastically reduced by colds and allergies because mucus production increases in the nasal cavity and keeps molecules from reaching the olfactory receptor cells.

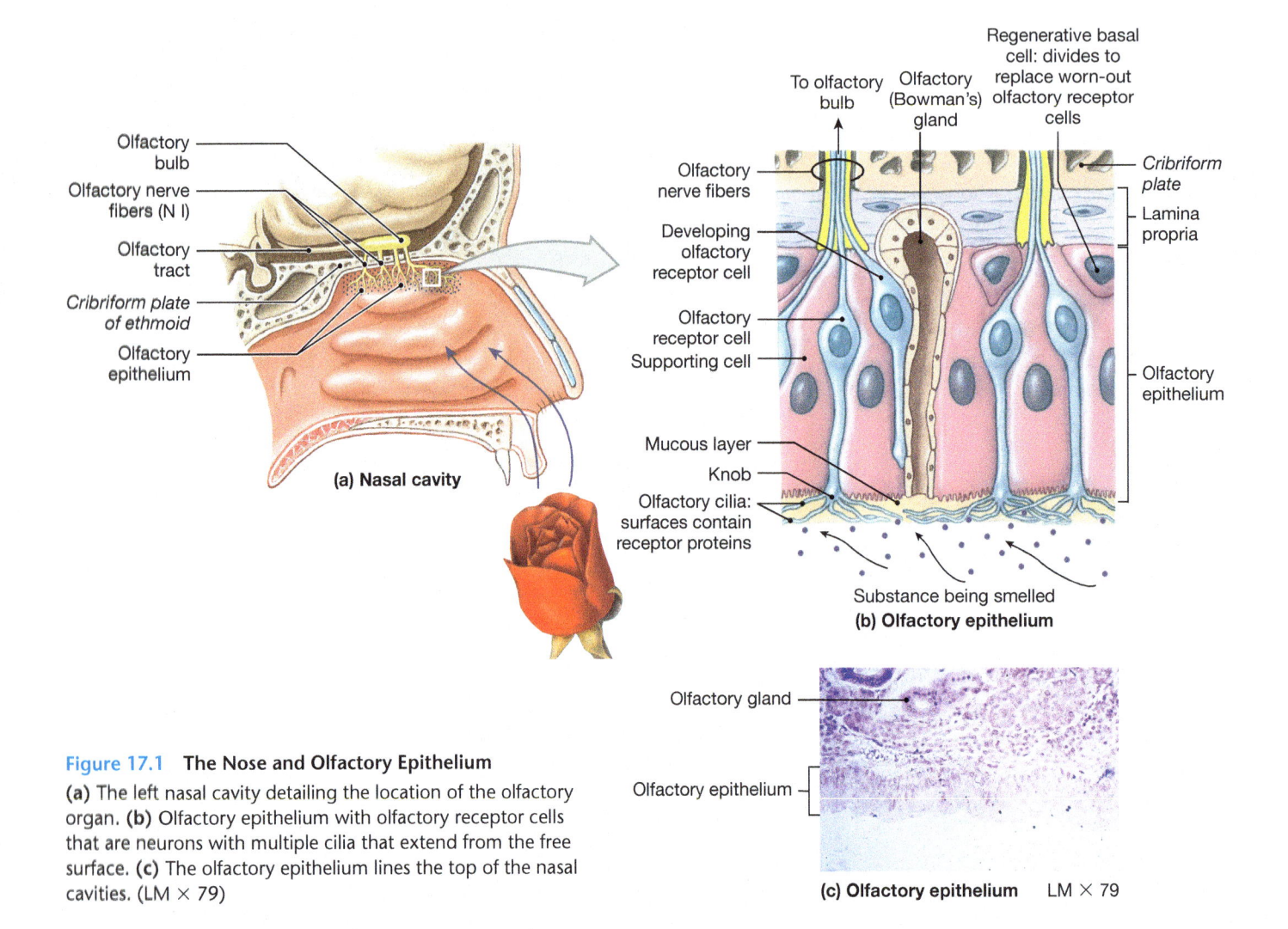

288 EXERCISE 17 Special Senses: Olfaction and Gustation

Figure 17.1 The Nose and Olfactory Epithelium
(a) The left nasal cavity detailing the location of the olfactory organ. (b) Olfactory epithelium with olfactory receptor cells that are neurons with multiple cilia that extend from the free surface. (c) The olfactory epithelium lines the top of the nasal cavities. (LM × 79)

The olfactory nerve, cranial nerve N I, passes through the cribriform plate of the ethmoid and enters the brain at the olfactory bulb (Figure 17.1a). As noted in Exercise 15, the bulb continues as the olfactory tract to the temporal lobe, where the olfactory cortex is located. (Unlike the pathways for many other senses, the olfactory pathway does not have a synapse in the thalamus.) Some branches of the olfactory tract synapse in the hypothalamus and limbic system of the brain, which explains the strong emotional responses associated with olfaction.

QuickCheck Questions

1.1 Where are the olfactory receptor cells located?
1.2 What is the function of the olfactory glands?

1 Materials

- ☐ Compound microscope
- ☐ Prepared slide of olfactory epithelium

Procedures

1. At low magnification, focus on the olfactory epithelium. Identify the supporting cells and olfactory receptor cells, both shown in Figure 17.1.
2. Locate the lamina propria and the olfactory glands.
3. In the space provided, sketch the olfactory epithelium as viewed at medium magnification. ■

Olfactory epithelium

LAB ACTIVITY 2 — Olfactory Adaptation

Adaptation is defined as a reduction in sensitivity to a repeated stimulus. When a receptor is first stimulated, it responds strongly, but then the response declines as the stimulus is repeated. We say the receptor has *adapted* to the stimulus. **Peripheral adaptation** is adaptation that happens because the receptors become desensitized to the stimulus. This type of adaptation occurs rapidly in phasic receptors, and the resulting adaptation reduces the amount of sensory information the CNS must process. **Central adaptation** is adaptation that occurs in the CNS, due to inhibition of sensory neurons along a sensory pathway. Central adaptation allows us to smell a new odor while we have adapted to reduce our awareness of an initial odor.

The olfactory pathway is quick to adapt to a repeated stimulus. A few minutes after you apply cologne or perfume, for instance, you do not smell it as much as you did initially. However, if a new odor is present, the nose is immediately capable of sensing the new scent, which is proof that what is going on is central adaptation and not receptor fatigue. (If the receptors were fatigued instead of adapted, you would not sense new stimuli once the receptors reached exhaustion.) Olfactory adaptation occurs along the olfactory pathway in the brain, not in the receptors.

In this activity, you determine the length of time it takes for your olfactory epithelium to adapt to a particular odor. Use care when smelling the vials. Do not just put the vial right under your nose and inhale. Instead, hold the open vial about six inches in front of your nose and wave your hand over the opening to waft the odor toward your nose.

QuickCheck Questions

2.1 Where does olfactory adaptation occur?

2.2 When does olfactory adaptation occur?

2 Materials

- ☐ Vial that contains oil of wintergreen
- ☐ Vial that contains isopropyl alcohol
- ☐ Stop watch
- ☐ Lab partner

Procedures

1. Hold the vial of wintergreen oil near your face and waft the fumes toward your nose. Ask your partner to start the stop watch.
2. Breathe through your nose to smell the oil. Continue wafting and smelling until you no longer sense the odor. Have your partner stop the stop watch at that instant, and record the time it took for adaptation to occur in Table 17.1, which is in the Lab Report at the end of this exercise.
3. Immediately following loss of sensitivity to the wintergreen oil, smell the vial of alcohol. Explain how you can smell the alcohol but can no longer smell the wintergreen oil.

290　EXERCISE 17　Special Senses: Olfaction and Gustation

4. Wait about 2–3 minutes and then repeat steps 1–3 using the alcohol as the first vial. Is there a difference in adaptation time for the two substances?
5. Repeat steps 1–4; have your partner do the smelling and you do the timing. Record your partner's olfactory adaptation time in Table 17.1.
6. Repeat steps 1–4 with several other classmates and record the times in Table 17.1. ■

LAB ACTIVITY 3　Gustation

Gustation (gus-TĀ-shun) is the sense of taste. The receptors for gustation are **gustatory cells** located in **taste buds** that cover the surface of the tongue, the pharynx, and the soft palate (Figure 17.2). A taste bud can contain up to 100 gustatory cells (Figure 17.2c). The gustatory cells are replaced every 10–12 days by basal cells, which divide and produce transitional cells that mature into the gustatory cells. Each gustatory cell has a small hair, or **microvillus**, that projects through a

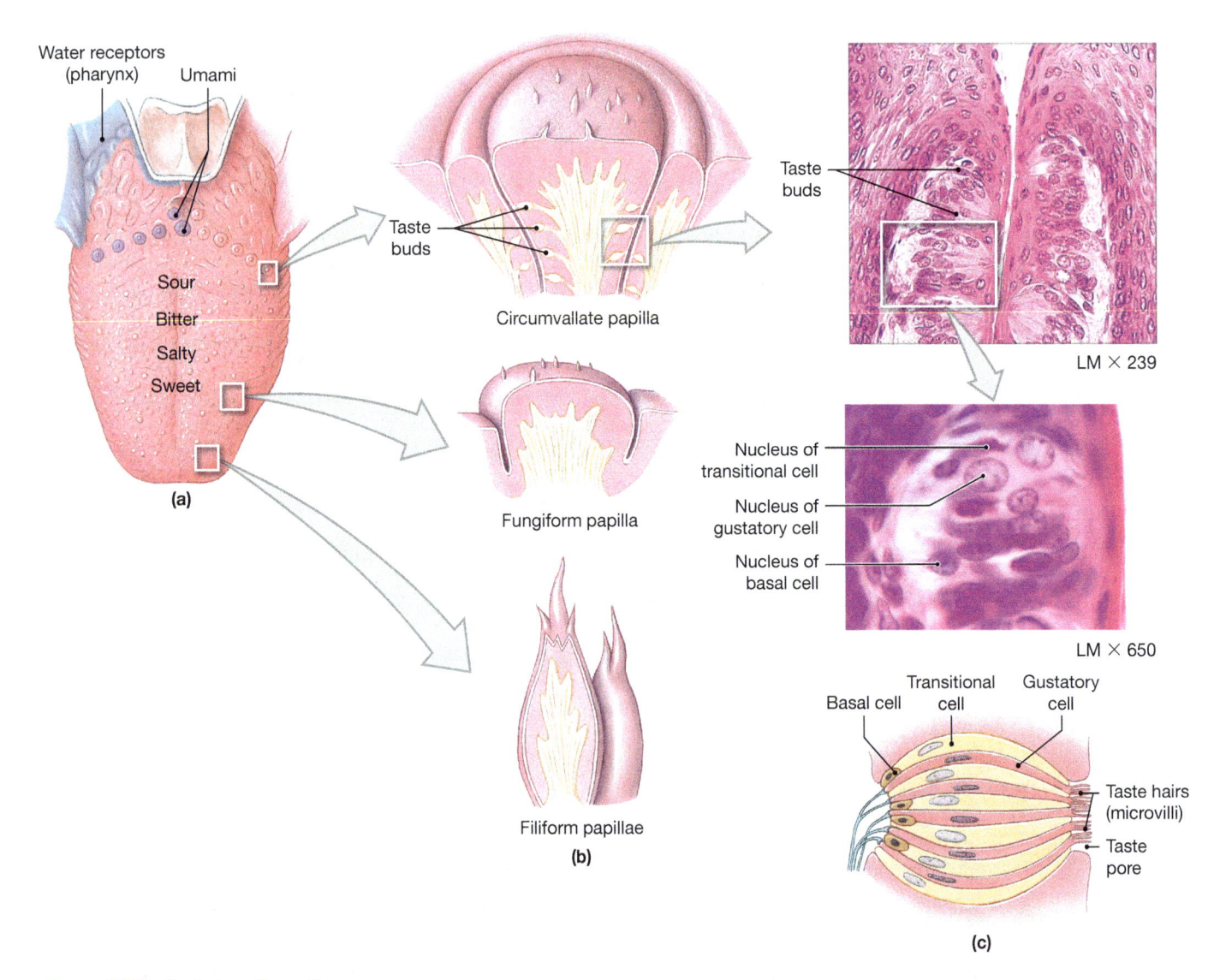

Figure 17.2　Gustatory Reception
(a) Taste receptors are gustatory cells that are located in taste buds, which form pockets in the epithelium of fungiform or circumvallate papillae on the tongue. (b) Taste buds in a circumvallate papilla. (c) A taste bud, with receptor (gustatory) cells, basal cells, and transitional cells. The diagrammatic view shows details of the taste pores that are not visible in the micrograph. (LM × 239) (LM × 650)

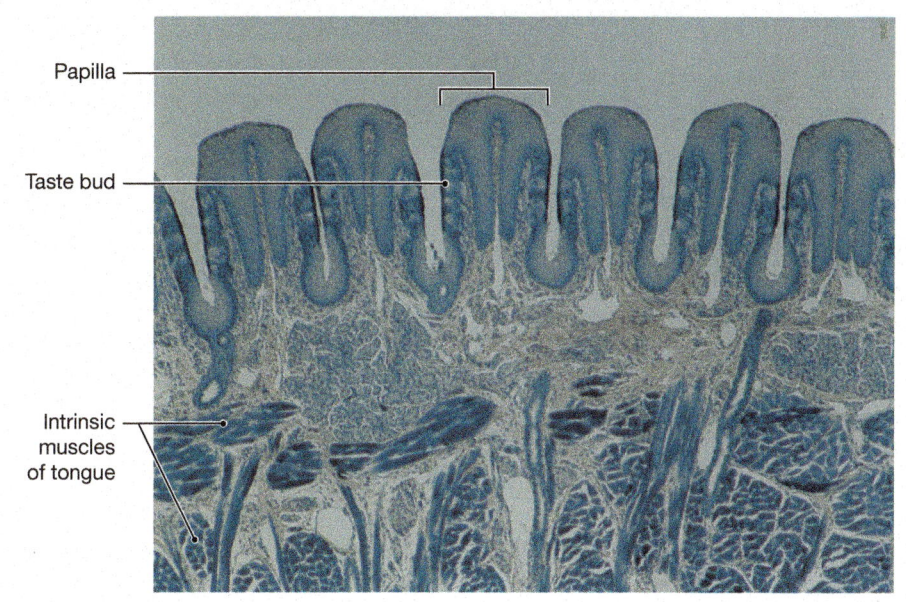

EXERCISE 17 Special Senses: Olfaction and Gustation 291

Papilla

Taste bud

Intrinsic muscles of tongue

LM × 75

Figure 17.3 Taste Buds
Taste buds are visible along the walls of papillae on the tongue surface. (LM × 75)

small **taste pore**. Contact with food that is dissolved in saliva stimulates the microvilli to produce gustatory impulses.

An adult has approximately 10,000 taste buds located inside elevations called **papillae** (pa-PIL-lē), which are detailed in Figure 17.3. The base of the tongue has a number of circular papillae, called **circumvallate** (sir-kum-VAL-āt) **papillae**, arranged in the shape of an inverted V across the width of the tongue. The tip and sides of the tongue contain button-like **fungiform papillae**. Approximately two-thirds of the anterior portion of the tongue is covered with **filiform papillae**, which do not contain taste buds but instead provide a rough surface for the movement of food.

Tastes can be grouped into five categories: sweet, salty, sour, bitter, and **umami** (oo-MAH-mē); umami is the comforting taste of proteins in soup broth. Although water is tasteless, water receptors occur in taste buds of the pharynx. Sensory information from the taste buds is carried to the brain by parts of three cranial nerves: the vagus nerve (N X) that serves the pharynx, the facial nerve (N VII) that serves the anterior two-thirds of the tongue, and the glossopharyngeal nerve (N IX) that serves the posterior one-third of the tongue. Children have more taste buds than adults, and at around the age of 50 the number of taste buds begins to rapidly decline. This difference in taste bud density helps to explain why children might complain that a food is too spicy while a grandparent responds that it tastes bland.

QuickCheck Questions

3.1 Where are taste buds located in the tongue?
3.2 What are the taste receptor cells called?

3 Materials

- ☐ Compound microscope
- ☐ Prepared slide of taste buds
- ☐ PTC taste paper

Procedures

1. Observe the taste bud slide at low and medium magnifications, and use Figure 17.2 as a guide as you identify the three types of papillae and the taste buds.

2. In the space provided, sketch a few papillae with taste buds.

3. The sense of taste is a genetically inherited trait, and consequently two individuals can perceive the same substance in different ways. The chemical phenylthiocarbamide (PTC), for example, tastes bitter to some individuals, sweet to others, and tasteless to still others. Approximately 30 percent of the population are nontasters of PTC.
   a. Place a strip of PTC paper on your tongue and chew it several times. Are you a taster or a nontaster?
   b. If your instructor has each student in the class record her or his taster-or-nontaster results on the chalkboard, calculate the percentage of tasters versus nontasters and complete Table 17.2 in the Lab Report.

Taste buds and papillae

LAB ACTIVITY 4 — Relationship between Olfaction and Gustation

The sense of taste is thousands of times more sensitive when gustatory and olfactory receptor cells are stimulated simultaneously. The following experiment demonstrates the effect of smell on the sense of taste.

QuickCheck Question
4.1 What is this experiment designed to demonstrate?

4 Materials
- Diced onion
- Diced apple
- Paper towels
- Lab partner

Procedures
1. Dry the surface of your partner's tongue with a clean paper towel.
2. Have your partner stand with eyes closed and nose pinched shut with the thumb and index finger.
3. Place a piece of either onion or apple on the dried tongue, and ask if the food can be identified. Record the reply, yes or no, in Table 17.3 in the Lab Report.
4. Have your partner, still with eyes closed and nose pinched shut, chew the piece of food, and again ask if it can be identified. Record the reply in Table 17.3.
5. Have your partner release the nose pinch, and ask one last time if the food can be identified. Record the reply in Table 17.3.

Name _____

Date _____

Section _____

LAB REPORT

EXERCISE

17

Special Senses: Olfaction and Gustation

A. Data Recording and Interpretation

Record your observations and data in the appropriate tables.

Table 17.1 Olfactory Adaptation

Student	Olfactory Adaptation Time for Wintergreen Oil(s)	Olfactory Adaptation Time for Isopropyl Alcohol(s)

Table 17.2 PTC Taste Experiment

Student	Taster	Nontaster
Percentage of class:		

Table 17.3 Gustatory/Olfactory Sensations

Food	Closed Nostrils Dry Tongue	After Chewing	Open Nostrils
Apple	_____	_____	_____
Onion	_____	_____	_____

EXERCISE 17 — LAB REPORT

B. Matching

Match each term in the left column with its correct description from the right column.

_____ 1. cranial nerve for smell
_____ 2. adaptation
_____ 3. filiform papilla
_____ 4. gustation
_____ 5. olfactory gland
_____ 6. basal cell
_____ 7. circumvallate papilla
_____ 8. supporting cell
_____ 9. cranial nerve for taste
_____ 10. olfaction

A. produces mucus
B. sense of smell
C. contains taste buds
D. sense of taste
E. loss of sensitivity due to repeated stimuli
F. provides nourishment to olfactory receptor cell
G. helps manipulate food on tongue surface
H. cranial nerve N I
I. stem cell of olfactory receptor cell
J. cranial nerve N VII

C. Short-Answer Questions

1. Why does a cold affect your sense of smell?

2. Where are the receptors for taste located?

3. Examine the data in Table 17.1. Did adaptation time differ greatly from one individual to another?

D. Analysis and Application

1. Describe an experiment that demonstrates how the senses of taste and smell are linked.

2. Several minutes after applying cologne, you notice that you cannot smell the fragrance. Explain this response by your olfactory receptor cells.

3. Does the class data in Table 17.2 clearly demonstrate which trait, taster or nontaster, is genetically dominant?

EXERCISE 18

Anatomy of the Eye

OBJECTIVES

On completion of this exercise, you should be able to:

- Identify and describe the accessory structures of the eye.
- Explain the actions of the six extraocular eye muscles.
- Describe the external and internal anatomy of the eye.
- Describe the cellular organization of the retina.
- Identify the structures of a dissected cow or sheep eye.

LAB ACTIVITIES

1. External Anatomy of the Eye 295
2. Internal Anatomy of the Eye 298
3. Cellular Organization of the Retina (Neural Tunic) 300
4. Observation of the Retina (Neural Tunic) 303
5. Dissection of the Cow or Sheep Eye 305

The eyes are complex and highly specialized sensory organs, and allow us to view everything from the pale light of stars to the intense, bright blue of the sky. To function in such a wide range of light conditions, the retina of the eye contains two types of receptors, one for night vision and another for bright light and color vision. Because the level and intensity of light are always changing, the eye must regulate the size of the pupil, which allows light to enter the eye. Six oculomotor muscles that surround the eyeball allow it to move. Four of the twelve cranial nerves control the muscular activity of the eyeball and transmit sensory signals to the brain. A sophisticated system of tear production and drainage keeps the surface of the eyeball clean and moist. In this exercise you will examine the anatomy of the eye and dissect the eye of a sheep or a cow.

LAB ACTIVITY 1 External Anatomy of the Eye

The human eyeball is a spherical organ that measures about 2.5 cm (1 in.) in diameter. Only about one-sixth of the eyeball is visible between the eyelids; the rest is recessed in the bony orbit of the skull. The two points where the upper and lower lids meet are the **lateral canthus** (KAN-thus; "corner") and **medial canthus** (Figure 18.1a). A red, fleshy structure in the medial canthus, the **lacrimal caruncle** (KAR-ung-kul; "small soft mass"), contains modified sebaceous and sweat glands. Secretions from the lacrimal caruncle accumulate in the medial canthus during long periods of sleep.

The accessory structures of the eye are the upper eyelid, lower eyelid, eyebrow, eyelashes, lacrimal apparatus, and six extraocular (external) eye muscles. The eyelids, called **palpebrae** (pal-PĒ-brē), are skin-covered muscles that protect the surface

Figure 18.1 Accessory Structures of the Eye

(a) Superficial anatomy of the right eye and its accessory structures.
(b) Diagrammatic representation of a superficial dissection of the right orbit.
(c) Diagrammatic representation of a deeper dissection of the right eye, showing its position in the orbit and its relationship to accessory structures, especially the lacrimal apparatus.

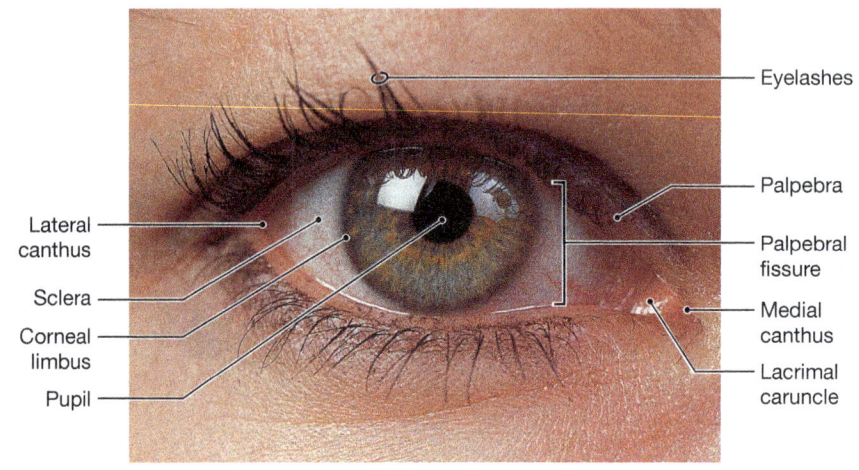

(a) Right eye, accessory structures

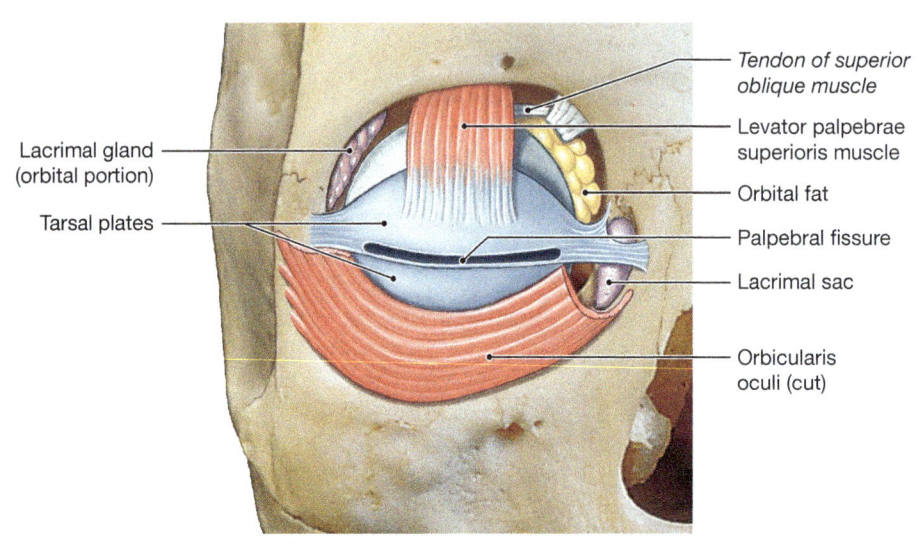

- Eyelashes
- Palpebra
- Palpebral fissure
- Medial canthus
- Lacrimal caruncle
- Lateral canthus
- Sclera
- Corneal limbus
- Pupil

(b) Superficial dissection of right orbit

- Tendon of superior oblique muscle
- Levator palpebrae superioris muscle
- Orbital fat
- Palpebral fissure
- Lacrimal sac
- Orbicularis oculi (cut)
- Lacrimal gland (orbital portion)
- Tarsal plates

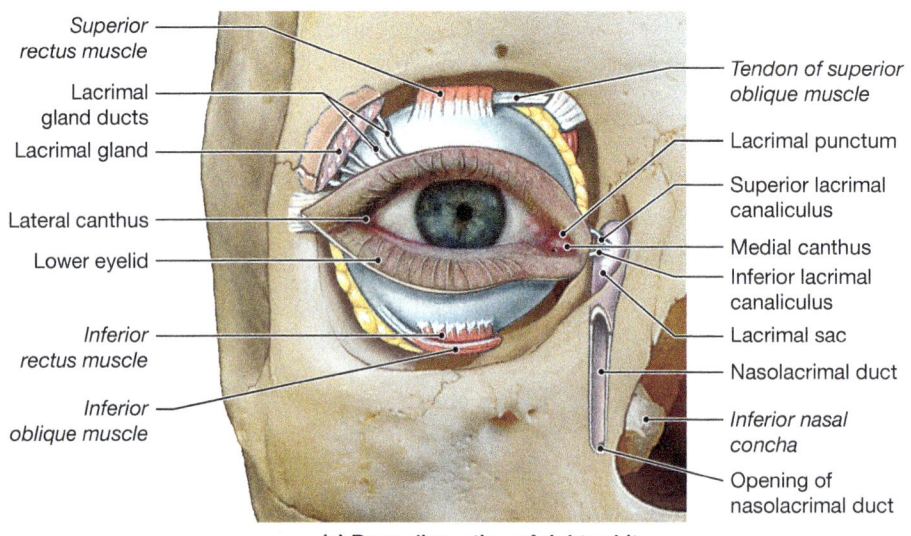

(c) Deep dissection of right orbit

- Superior rectus muscle
- Lacrimal gland ducts
- Lacrimal gland
- Lateral canthus
- Lower eyelid
- Inferior rectus muscle
- Inferior oblique muscle
- Tendon of superior oblique muscle
- Lacrimal punctum
- Superior lacrimal canaliculus
- Medial canthus
- Inferior lacrimal canaliculus
- Lacrimal sac
- Nasolacrimal duct
- Inferior nasal concha
- Opening of nasolacrimal duct

of the eyeball. The **levator palpebrae superioris** muscle raises the upper eyelid, and the **orbicularis oculi** muscle closes the eyelids. Blinking the eyelids keeps the eyeball surface lubricated and clean. A thin mucous membrane called the **palpebral conjunctiva** (kon-junk-TĪ-vuh) covers the underside of the eyelids and reflects over the anterior surface of the eyeball as the **ocular conjunctiva.** This membrane secretes mucus that reduces friction and moistens the eyeball surface. The **eyelashes** are short hairs that project from the border of each eyelid. Eyelashes and the eyebrows protect the eyeball from foreign objects, such as perspiration and dust, and partially shade the eyeball from the sun. Sebaceous **ciliary glands**, located at the base of the hair follicles, help lubricate the eyeball.

Clinical Application

Infections of the Eye

When a ciliary gland is blocked, it becomes inflamed as a sty. Because it is on the tip of the eyelid, the sty irritates the eyeball. **Conjunctivitis** is an inflammation of the conjunctiva that can be caused by bacteria, dust, smoke, or air pollutants; the infected eyeball usually appears red and irritated. The bacterial form of this infection is contagious and spreads easily among young children and individuals who share such objects as tools and office equipment.

The **lacrimal apparatus** consists of the lacrimal glands, lacrimal canals, lacrimal sac, and nasolacrimal duct (Figure 18.1b and c). The **lacrimal glands** are superior and lateral to each eyeball. Each gland contains 6 to 12 excretory **lacrimal ducts** that deliver to the anterior surface of the eyeball a slightly alkaline solution, called either *lacrimal fluid* or *tears*, that cleans, moistens, and lubricates the surface. The lacrimal fluid also contains an antibacterial enzyme called *lysozyme* that attacks any bacteria that may be on the surface of the eyeball. The fluid moves medially across the eyeball surface and enters two small openings of the medial canthus, the **superior** and **inferior lacrimal puncta.** From there, the lacrimal fluid passes into two ducts, the **lacrimal canals**, that lead to an expanded portion of the nasolacrimal duct called the **lacrimal sac.** The **nasolacrimal duct** drains the tears into the nasal cavity.

As discussed in Exercise 11, six extraocular muscles control the movements of the eyeball (Figure 18.2). The **superior rectus**, **inferior rectus**, **medial rectus**, and **lateral rectus** are straight muscles that move the eyeball up and down and side to side. The **superior** and **inferior oblique** muscles attach diagonally on the eyeball. The superior oblique has a tendon that passes through the **trochlea** ("pulley") located on the upper orbit. This muscle rolls the eyeball downward, and the inferior oblique rolls it upward.

QuickCheck Questions

1.1 How are lacrimal secretions drained from the surface of the eyeball?

1.2 Where are the two parts of the conjunctiva located?

1 Materials

- ☐ Dissectible eye model
- ☐ Eyeball chart

Procedures

1. Review the structures of the eye in Figures 18.1 and 18.2.
2. On the eye model and chart, locate the four structures of the lacrimal apparatus and the six accessory structures (count the six extraocular muscles as one accessory structure).
3. On the model, identify the six extraocular muscles. Describe how each one moves the eye.

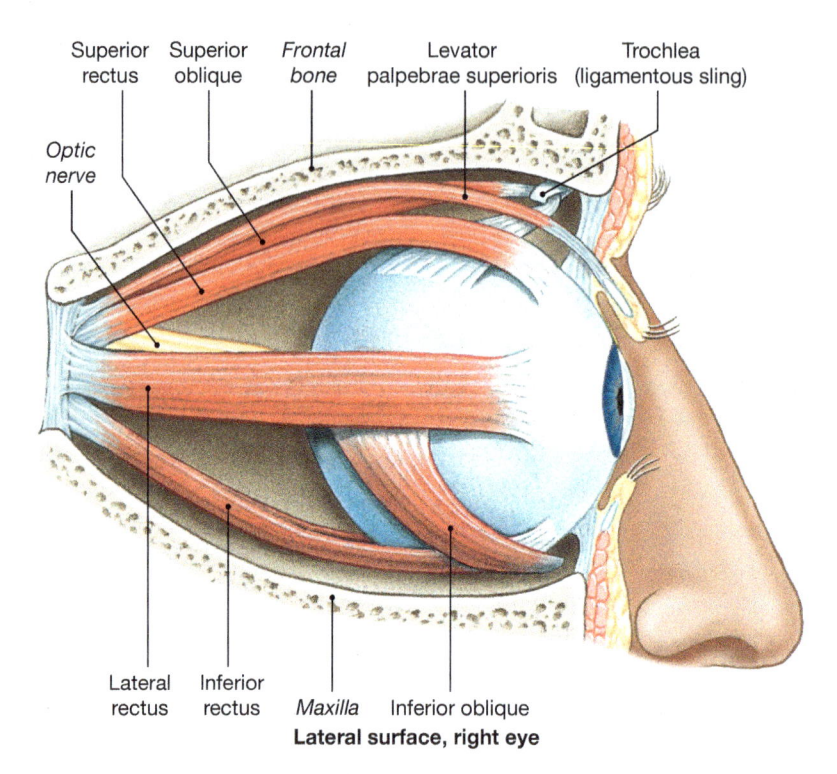

Figure 18.2 Extraocular Eye Muscles
Lateral surface of the right eye, which illustrates the muscles that move the eyeball. The medial rectus muscle is not visible in this view. The levator palpebrae superioris muscle, which raises the upper eyelid, is not classified as one of the six extraocular muscles.

LAB ACTIVITY 2 — Internal Anatomy of the Eye

The eyeball is organized into three layers, each of which are detailed in Figure 18.3a. The outermost layer is called the **fibrous tunic** because of the abundance of dense connective tissue. The **sclera** (SKLER-uh; "hard") is the white part of the fibrous tunic that resists punctures and maintains the shape of the eyeball. It covers the eyeball except at the transparent **cornea** (KOR-nē-uh), which is the region of the fibrous tunic where light enters the eye. The cornea consists primarily of many layers of densely-packed collagen fibers. The **corneal limbus** is the border between the sclera and the cornea. Around the corneal limbus is the **canal of Schlemm** (also called the **scleral venous sinus**), which is a small passageway that drains lacrimal fluid into veins in the sclera.

The second of the eyeball's three layers is the **vascular tunic (uvea)**. The most posterior portion of this layer, the **choroid**, is highly vascularized and contains a dark pigment (melanin) that absorbs light to prevent reflection. Anteriorly, the vascular tunic is modified into the **ciliary body** and the pigmented **iris.** The ciliary body contains two structures, the *ciliary process* and the *ciliary muscle*. The **ciliary process** is a series of folds with thin **suspensory ligaments** that extend to the **lens**, the part of the eye that focuses light. The **ciliary muscle** adjusts the shape of the lens for near and far vision. The front of the iris is pigmented and has a central aperture called the **pupil. Pupillary sphincter muscles** and **pupillary dilator muscles** of the iris change the diameter of the pupil to regulate the amount of light entering the eye. In bright light and for close vision, the pupil constricts as a result of parasympathetic activation that causes the pupillary sphincter muscles to contract and the pupillary dilator muscles to relax. In low light and for distant vision,

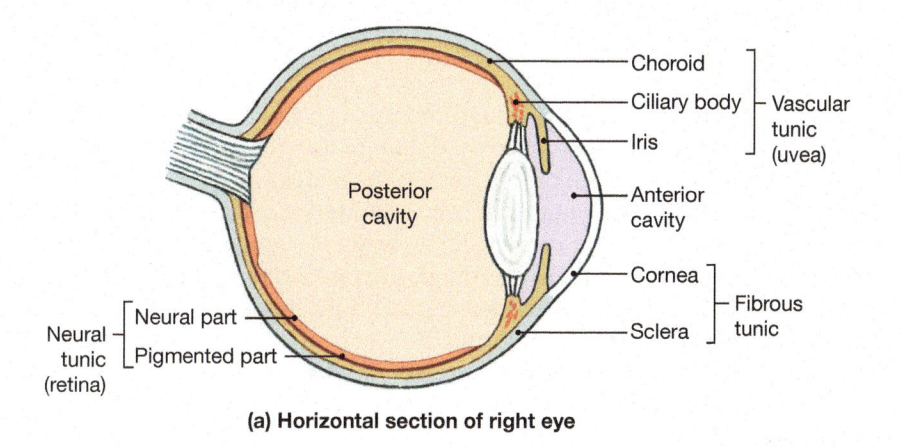

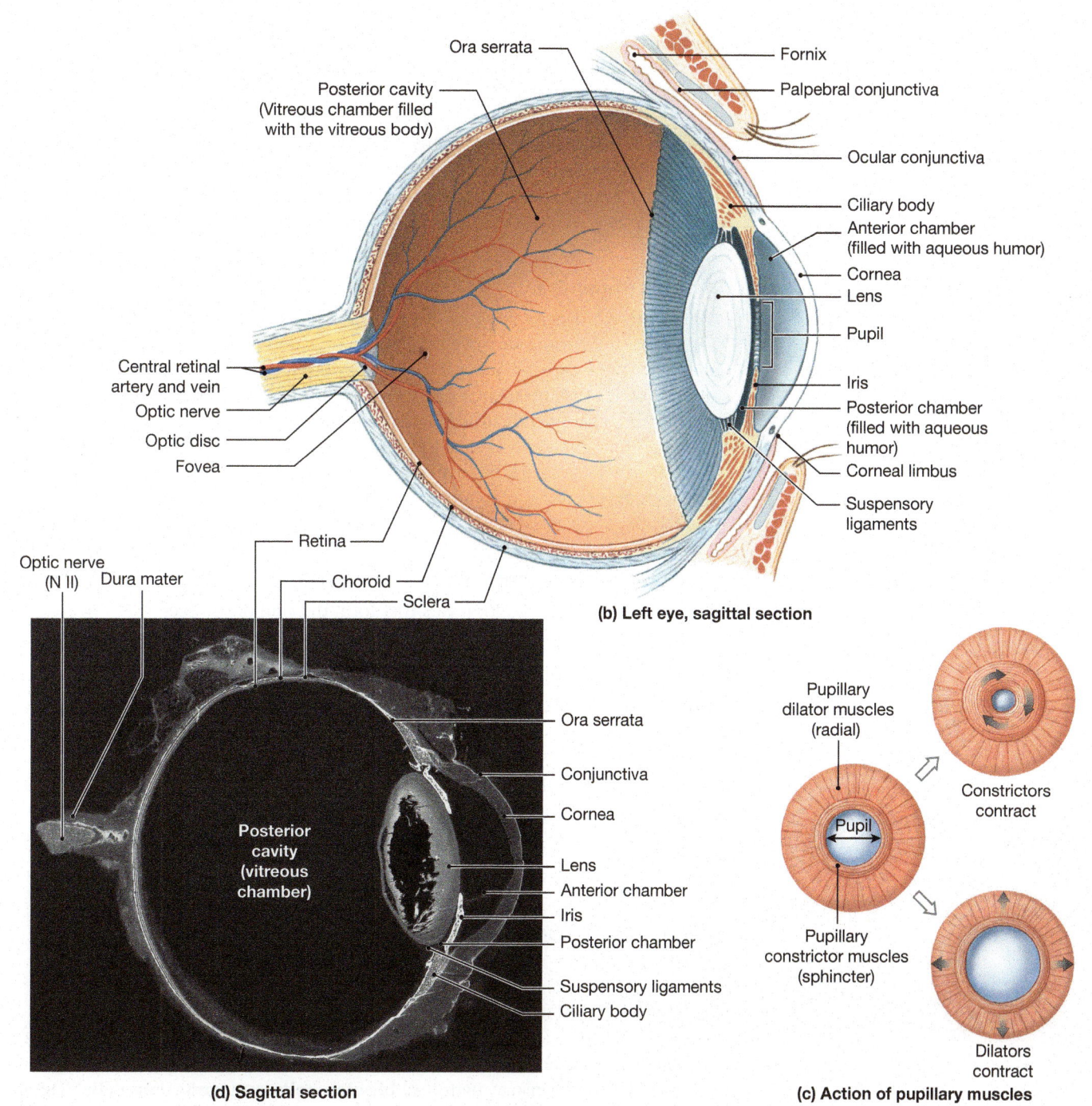

Figure 18.3 Anatomy of the Eye

(a) The three major layers, or tunics, of the eye. (b) Major anatomical landmarks and features viewed in a diagrammatic view of the left eye. (c) The action of the pupillary muscles and changes in pupil diameter. (d) Sagittal section through the eye.

299

sympathetic stimulation causes the dilator muscles to contract and the sphincter muscles to relax; as a result, the pupil expands and more light enters the eye.

The innermost of the eyeball's three layers is the **neural tunic**, usually referred to as the **retina**. This layer contains an outer **pigmented part** that covers the choroid and a **neural part** that contains light-sensitive photoreceptors. The anterior margin of the retina, where the choroid of the vascular tunic is exposed, is the **ora serrata** (Ō-ra ser-RA-tuh; "serrated mouth") and appears as a jagged edge, much like a serrated knife (Figure 18.3b and d).

The lens divides the eyeball into an **anterior cavity**, the area between the lens and the cornea; and a **posterior cavity** (also called *vitreous chamber*), the area between the lens and the retina (Figure 18.3a). The anterior cavity is further subdivided into an **anterior chamber** between the iris and the cornea and a **posterior chamber** between the iris and the lens. Capillaries of the ciliary processes form a watery fluid called **aqueous humor**, which is secreted into the posterior chamber of the anterior cavity. This fluid circulates through the pupil and into the anterior chamber of the anterior cavity, where it is reabsorbed into the blood by the canal of Schlemm. The aqueous humor helps maintain the intraocular pressure of the eyeball and supplies nutrients to the lens and cornea. The posterior cavity, larger than the anterior cavity, contains the **vitreous body**, a clear, jelly-like substance that holds the retina against the choroid and prevents the eyeball from collapsing.

Clinical Application

Diseases of the Eye

Glaucoma is a disease in which the intraocular pressure of the eye is elevated. The increased pressure damages the optic nerve and may eventually result in blindness. If the canal of Schlemm becomes blocked, fluid accumulates in the anterior cavity and intraocular pressure rises. **Diabetic retinopathy** develops in many diabetic individuals. Diabetes causes a proliferation of blood vessels to form over the retina and also causes blood vessels to rupture. These vascular changes occur gradually, but eventually the photoreceptors are damaged and blindness occurs.

QuickCheck Questions

2.1 What are the three major layers that form the wall of the eye?

2.2 Trace a drop of aqueous humor circulating in the eye.

2.3 How does the pupil regulate the amount of light that enters the lens?

2 Materials

- ☐ Dissectible eye model
- ☐ Eyeball chart

Procedures

1. On the eye model and chart, identify the three major layers of the eyeball wall.
2. Identify the sclera and cornea on the eye model. Also locate the corneal limbus and canal of Schlemm.
3. On the model, locate the choroid, and identify the ciliary body and associated structures.
4. Identify the retina (neural tunic), fovea, and ora serrata on the eye model. ∎

LAB ACTIVITY 3 — Cellular Organization of the Retina (Neural Tunic)

The neural part of the retina contains sensory receptors called **photoreceptors** plus two types of sensory neurons: **bipolar cells** and **ganglion cells** (Figure 18.4a). The photoreceptors are stimulated by photons, which are particles of light. Photoreceptors are classified into two types: **rods** and **cones**. Rods are sensitive to

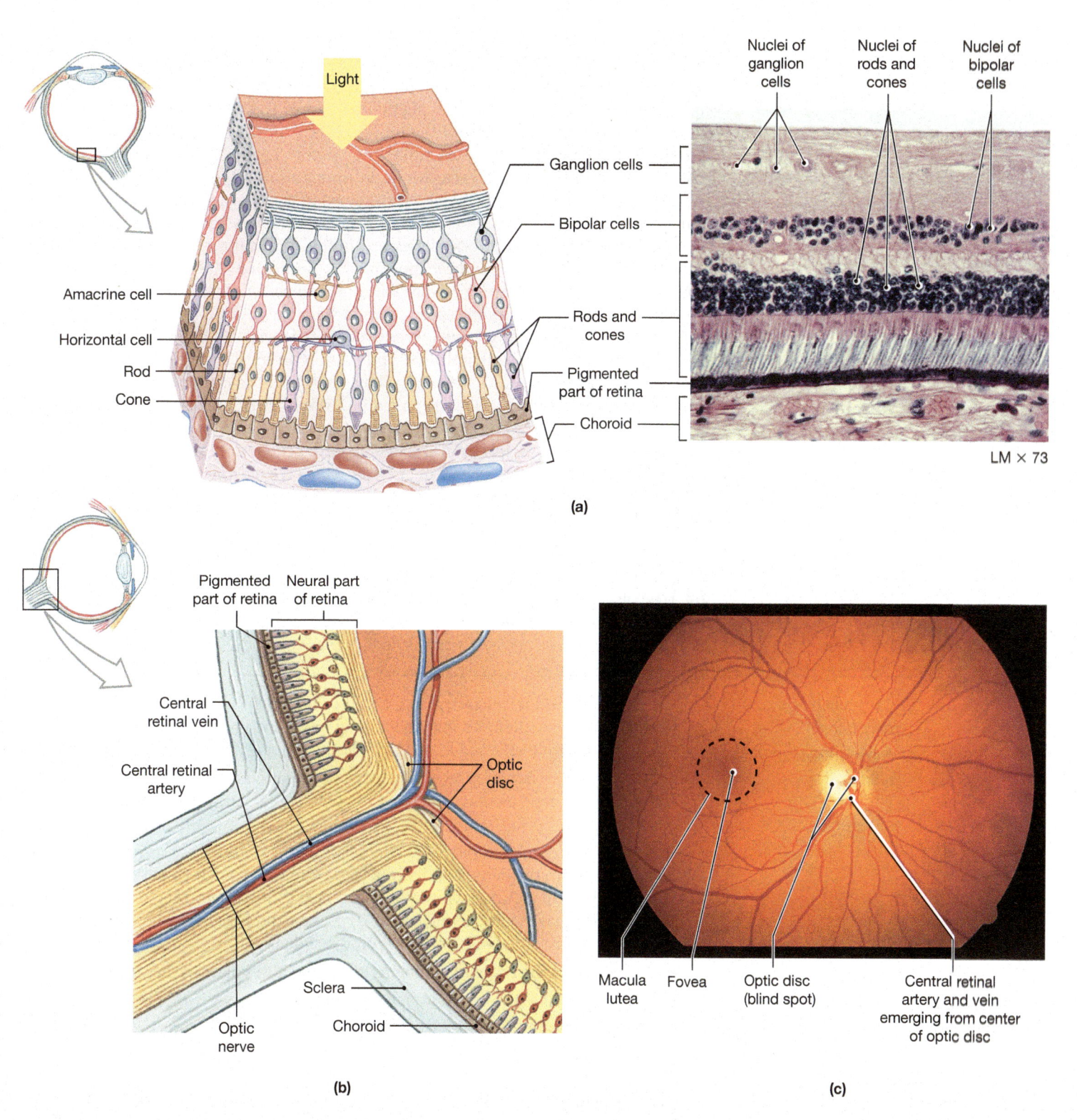

Figure 18.4 Organization of the Retina

(a) Cellular organization of the neural part of the retina. Note that the photoreceptors are located close to the choroid layer of the vascular tunic rather than near the lens. (LM × 73) (b) The optic disc in diagrammatic horizontal section. Note the lack of photoreceptors in the disc. (c) A photograph of the retina as seen through the pupil, which shows the retinal blood vessels, the fovea and macula lutea, the origin of the optic nerve, and the optic disc.

301

low illumination and to motion. They are insensitive to most colors of light, and therefore we see little color at night. Cones are stimulated by moderate or bright light and respond to different colors of light. Our visual acuity is attributed to cones. The rods and cones face the pigmented part of the retina. Light passes all the way though the neural part of the retina, reflects off the pigmented part back into the neural part, and only then strikes the photoreceptors. The photoreceptors pass the signal to the bipolar cells, which in turn pass the signal to the ganglion cells. The axons of the ganglion cells converge at an area of the neural part of the retina called the **optic disc**, where the optic nerve enters the eyeball. Cells called **horizontal cells** form a network that either inhibits or facilitates communication between the photoreceptors and the bipolar cells. **Amacrine** (AM-a-krin) **cells** enhance communication between bipolar and ganglion cells.

The optic disc lacks photoreceptors and is a "blind spot" in your field of vision. Because the visual fields of your two eyes overlap, however, the blind spot is filled in and is not noticeable. Lateral to the optic disc is an area of high cone density called the **macula lutea** (LOO-tē-uh; "yellow spot"). In the center of the macula lutea is a small depression called the **fovea** (FŌ-vē-uh, "shallow depression"). The fovea is the area of sharpest vision because of the abundance of cones. Rods are most numerous at the periphery of the neural part of the retina, and we see best at night by looking out of the corners of our eyes. There are no rods in the fovea.

QuickCheck Questions

3.1 What is the optic disc, and why do you not see a blind spot in your field of vision?

3.2 What are the different types of cells in the neural part of the retina and how are they are organized?

3 Materials

- ☐ Compound microscope
- ☐ Prepared slide of retina

Procedures

1. Focus the slide at low magnification, and use Figure 18.4 as a reference while observing the specimen. Change to medium or high magnification as you examine the neural and pigmented parts of the retina.

2. Locate the thick vascular tunic on the edge of the specimen. Next to the choroid part of the vascular tunic, find the pigmented part of the retina.

3. The three types of cells in the neural part of the retina are clearly visible where the nuclei are grouped into three distinct bands. The photoreceptors—the rods and cones—form the dense band of nuclei next to the pigmented part. The bipolar cells form a thinner cluster of nuclei next to the photoreceptors. The ganglion cells have scattered nuclei and appear on the edge of the neural part of the retina. Locate each layer of cells on the slide.

4. Sketch a view of the two parts of the retina at medium power in the space provided. ■

Two parts of retina

Figure 18.5 An Ophthalmoscope
(a) Front of ophthalmoscope, which shows the view port through which the operator looks. (b) Back of ophthalmoscope, which shows the mirror that shines light into subject's eye.

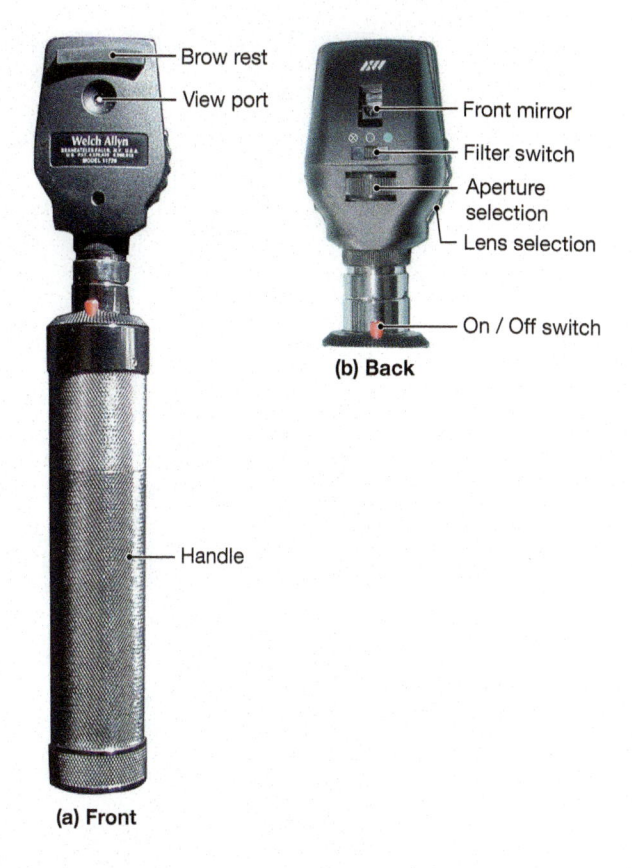

LAB ACTIVITY 4 — Observation of the Retina (Neural Tunic)

The retina is the only location in the body where blood vessels may be directly observed. The retinal blood vessels enter the eyeball by passing through the optic disc and then spread out into the neural part of the retina to provide blood to the photoreceptors and sensory neurons. To observe this vascularization, clinicians use a lighted magnifying instrument called an **ophthalmoscope** (Figure 18.5). The instrument shines a beam of light into the eye while the examiner looks through a lens called a viewing port to observe the retina.

QuickCheck Questions

4.1 An ophthalmoscope is used to observe what part of the eyeball?
4.2 Where do the retinal blood vessels enter the eyeball?

4 Materials

- ☐ Ophthalmoscope
- ☐ Lab partner

Procedures

Important: To protect the subject's eye, make only quick observations with the ophthalmoscope, and move the light beam away from the eye after about two seconds.

1. Before observing the retina, familiarize yourself with the parts of the ophthalmoscope. Use Figure 18.5 as a reference.
2. The examination is best performed in a darkened room. Sit face to face with your partner, the *subject*, who should be relaxed. Be careful not to shine the light from the ophthalmoscope into the eye for longer than one to two seconds at a time. Additionally, ask your partner to look away from the light as needed.
3. Hold the ophthalmoscope in your right hand to examine the subject's right retina. Begin approximately six inches from the subject's right eye and look into the ophthalmoscope with your right eyebrow against the brow rest.

304   EXERCISE 18   Anatomy of the Eye

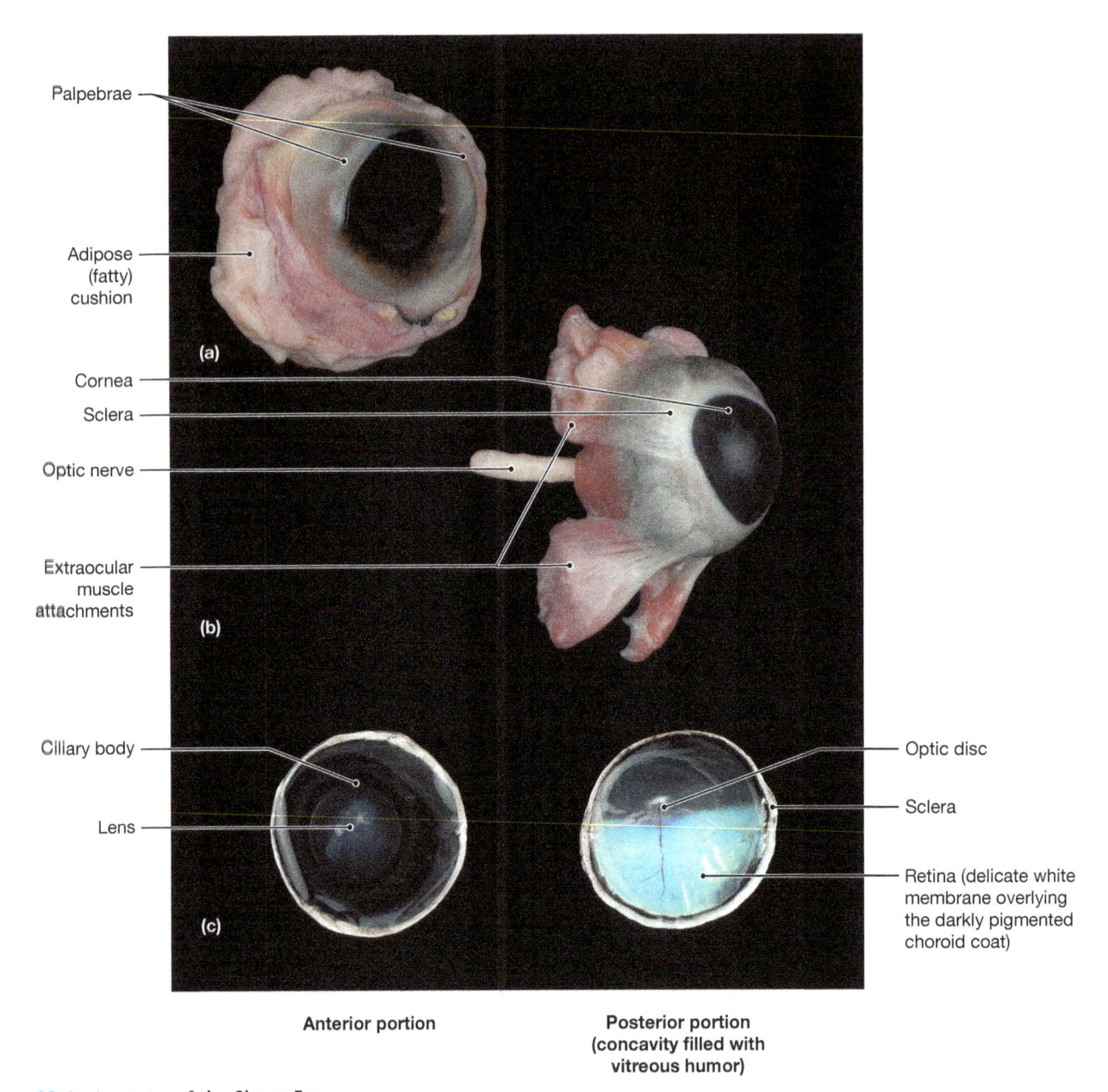

Figure 18.6 Anatomy of the Sheep Eye
(a) Anterior view. **(b)** Lateral view. **(c)** Internal view of frontal sections. The tapetum lucidum is the greenish-blue membrane of the choroid.

Labels (a): Palpebrae; Adipose (fatty) cushion.
Labels (b): Cornea; Sclera; Optic nerve; Extraocular muscle attachments.
Labels (c): Ciliary body; Lens; Optic disc; Sclera; Retina (delicate white membrane overlying the darkly pigmented choroid coat). Anterior portion. Posterior portion (concavity filled with vitreous humor).

4. Move the instrument closer to the subject's eye, and tilt it so that light enters the pupil at an angle. The orange-red image is the interior of the eyeball. The blood vessels should be visible as branched lines as in Figure 18.4c.

5. Observe the macula lutea, and the fovea in its center. Move closer to the subject if you cannot see the fovea. To prevent damage to the fovea, be careful not to shine the light on the fovea for longer than one second.

6. Medial to the macula lutea is the optic disc, the blind spot on the retina. Notice how blood vessels are absent from this area. ■

LAB ACTIVITY 5 | **Dissection of the Cow or Sheep Eye**

The anatomy of the cow eye and sheep eye is similar to that of the human eye (Figure 18.6). Be careful while dissecting the eyeball because the sclera is fibrous and difficult to cut. Use small strokes with the scalpel, and cut away from your fingers. Do not allow your lab partner to hold the eyeball while you dissect.

Safety Alert: Dissecting the Eyeball

You must practice the highest level of laboratory safety while handling and dissecting the eyeball. Keep the following guidelines in mind during the dissection:

1. Wear gloves and safety glasses to protect yourself from the fixatives used to preserve the specimen.
2. Be extremely careful when using a scalpel or other sharp instrument. Always direct cutting and scissor motions away from you to prevent an accident if the instrument slips on moist tissue.
3. Before cutting a given tissue, make sure it is free from underlying and/or adjacent tissues so that they are not accidentally severed.
4. Never discard tissue in the sink or trash. Your instructor will inform you of the proper disposal procedure. ▲

QuickCheck Questions

5.1 What type of safety equipment should you wear during the sheep or cow eye dissection?

5.2 How should you dispose of the eye specimen and scrap tissue?

5 Materials

- ☐ Gloves
- ☐ Safety glasses
- ☐ Fresh or preserved cow or sheep eye
- ☐ Dissection pan
- ☐ Scissors
- ☐ Scalpel
- ☐ Blunt probe
- ☐ Newspaper

Procedures

1. Examine the external features of the cow or sheep eye. Depending on how the eye was removed from the animal, your specimen may have, around the eyeball, adipose tissue, portions of the extraocular muscles, and the palpebrae. If so, note the amount of adipose tissue, which cushions the eyeball. If your specimen lacks these structures, observe them in Figure 18.6.
   - Identify the optic nerve (cranial nerve II) that exits the eyeball at the posterior wall.
   - Examine the remnants of the extraocular muscles and, if present, the palpebrae and eyelashes.
   - Locate the corneal limbus, where the white sclera and the cornea join. The cornea, which is normally transparent, will be opaque if the eye has been preserved.

2. Hold the eyeball securely, and use scissors to remove any adipose tissue and extraocular muscles from the surface. Take care not to remove the optic nerve.

3. Hold the eyeball securely in the dissection pan, and with a sharp scalpel make an incision about 0.6 cm (1/4 in.) back from the cornea. Use numerous, small, downward strokes over the same area to penetrate the sclera.

4. Insert the scissors into the incision, and cut around the circumference of the eyeball. Be sure to maintain the 0.6 cm distance back from the cornea.

5. Carefully separate the anterior and posterior cavities of the eyeball. The vitreous body should stay with the posterior cavity. Examine the anterior portion of the eyeball.
   - Place a blunt probe between the lens and the ciliary processes, and carefully lift the lens up a little. The halo of delicate transparent filaments between the lens and the ciliary processes is formed by the suspensory ligaments. Notice the ciliary body, where the suspensory ligaments originate, and the heavily-pigmented iris with the pupil in its center.
   - Remove the vitreous body from the posterior cavity, set it on a piece of newspaper, and notice how it causes refraction (bending) of light rays.
   - The retina is the tan membrane that is easily separated from the heavily pigmented choroid of the vascular tunic.
   - Examine the optic disc, where the retina attaches to the posterior of the eyeball.
   - The choroid has a greenish-blue membrane, the **tapetum lucidum**, which improves night vision in many animals, including sheep and cows. When headlights shine in a cow's eyes at night, this membrane reflects the light and makes the eyes glow. Humans do not have this membrane, and our night vision is not as good as that of animals that do.
6. Never discard tissue in the sink or trash. Your instructor will inform you of the proper disposal procedure. ■

LAB REPORT

EXERCISE 18

Anatomy of the Eye

Name _____
Date _____
Section _____

A. Matching

Match each term in the left column with its correct description from the right column.

_____	1. transparent part of fibrous tunic	A.	iris
_____	2. thin filaments attached to lens	B.	vitreous body
_____	3. corner of eye	C.	optic disc
_____	4. produces tears	D.	ciliary muscle
_____	5. adjusts lens shape	E.	lacrimal caruncle
_____	6. jelly-like substance of eye	F.	canthus
_____	7. depression in retina	G.	cornea
_____	8. area that lacks photoreceptors	H.	lacrimal gland
_____	9. regulates light that enters eye	I.	suspensory ligaments
_____	10. red structure in medial canthus	J.	fovea

B. Labeling

Label the structures of the eye in Figure 18.7.

C. Short-Answer Questions

1. Name and describe the two major components of the fibrous tunic.

2. Describe the ciliary body region of the vascular tunic.

3. Describe the two cavities and two chambers of the eye and the circulation of aqueous humor through them.

4. How do the two kinds of muscles in the iris respond to high levels and low levels of light that enters the eye?

EXERCISE 18　　　LAB REPORT

1. _____
2. _____
3. _____
4. _____
5. _____
6. _____
7. _____
8. _____
9. _____
10. _____
11. _____
12. _____
13. _____
14. _____
15. _____
16. _____
17. _____
18. _____

Figure 18.7 **Anatomy of the Eye**
A horizontal section of the right eye.

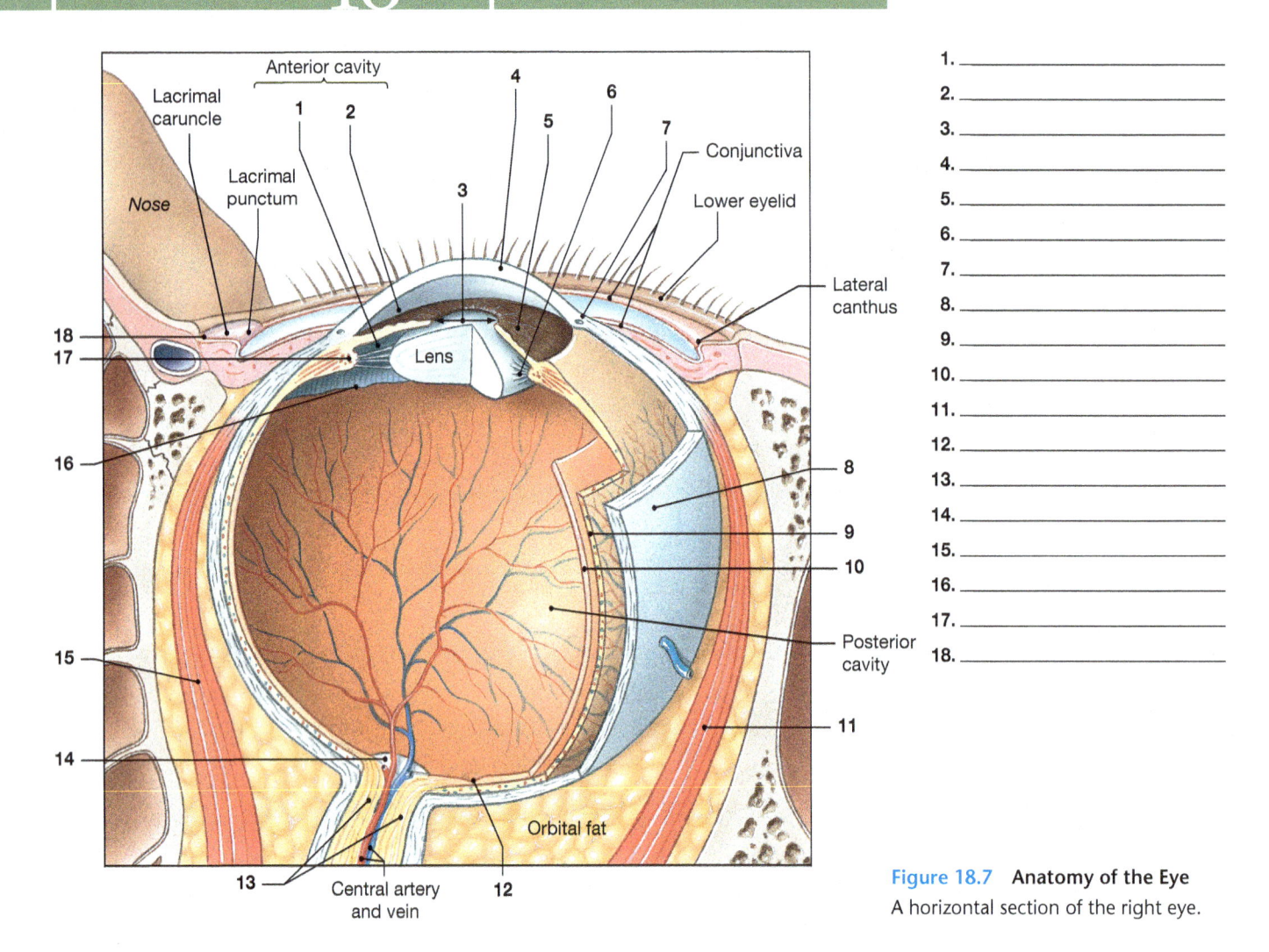

D. Analysis and Application

1. Macular degeneration occurs when cells in the fovea and macula lutea are destroyed. Describe how this disease would change vision.

2. Name the structures of the eye through which light passes; start with the pupil and include the three cell layers of the retina.

3. What causes the pupils to dilate in someone who is excited or frightened?

4. How are lacrimal secretions drained from the surface of the eye into the nasal cavity?

EXERCISE 19

Anatomy of the Ear

OBJECTIVES

On completion of this exercise, you should be able to:

- Identify and describe the components of the external, middle, and inner ear.
- Describe the anatomy of the cochlea.
- Describe the components of the semicircular canals and the vestibule and explain their role in static and dynamic equilibrium.

LAB ACTIVITIES

1. External and Middle Ear 309
2. Inner Ear 311
3. Examination of the Tympanic Membrane 317

The ear is divided into three regions: the **external**, or outer, **ear**; the **middle ear**; and the **inner ear**. The external ear and middle ear direct sound waves to the inner ear for hearing. The inner ear serves two unique functions: balance and hearing. Without a sense of balance, you would not know, at any given moment, where your body is relative to the ground and in three-dimensional space. You would be unable to stand, let alone walk, or even drive a car. Your sense of hearing enables you to enjoy your favorite song while simultaneously conversing with a friend.

In this exercise you will identify the anatomical features of the ear and look at the sensory receptors for equilibrium and hearing.

LAB ACTIVITY 1 External and Middle Ear

The pinna, or **auricle**, is the flap of the outer ear that funnels sound waves into the **external acoustic meatus**, which is a tubular chamber that delivers sound waves to the **tympanic membrane** (*tympanum*), commonly called the *eardrum* (Figure 19.1). The auricle has an inner foundation of elastic cartilage covered with adipose tissue and skin. The tympanic membrane is a thin sheet of fibrous connective tissue stretched across the distal end of the external acoustic meatus and separates the external ear from the middle ear. The meatus contains wax-secreting cells in **ceruminous glands** plus many hairs that prevent dust and debris from entering the middle ear.

The middle ear (Figure 19.2) is a small space inside the petrous part of the temporal bone. It is connected to the back of the upper throat (the nasopharynx) by the **auditory tube**, also called either the *pharyngotympanic tube* or *Eustachian tube*. This tube equalizes pressure between the external air and the cavity of the middle ear. Three small bones of the middle ear, called **auditory ossicles,** transfer vibrations from the external ear to the inner ear. The **malleus** (*malleus,* hammer) is connected on one side to the tympanic membrane and on the other side to the **incus**

310 EXERCISE 19 Anatomy of the Ear

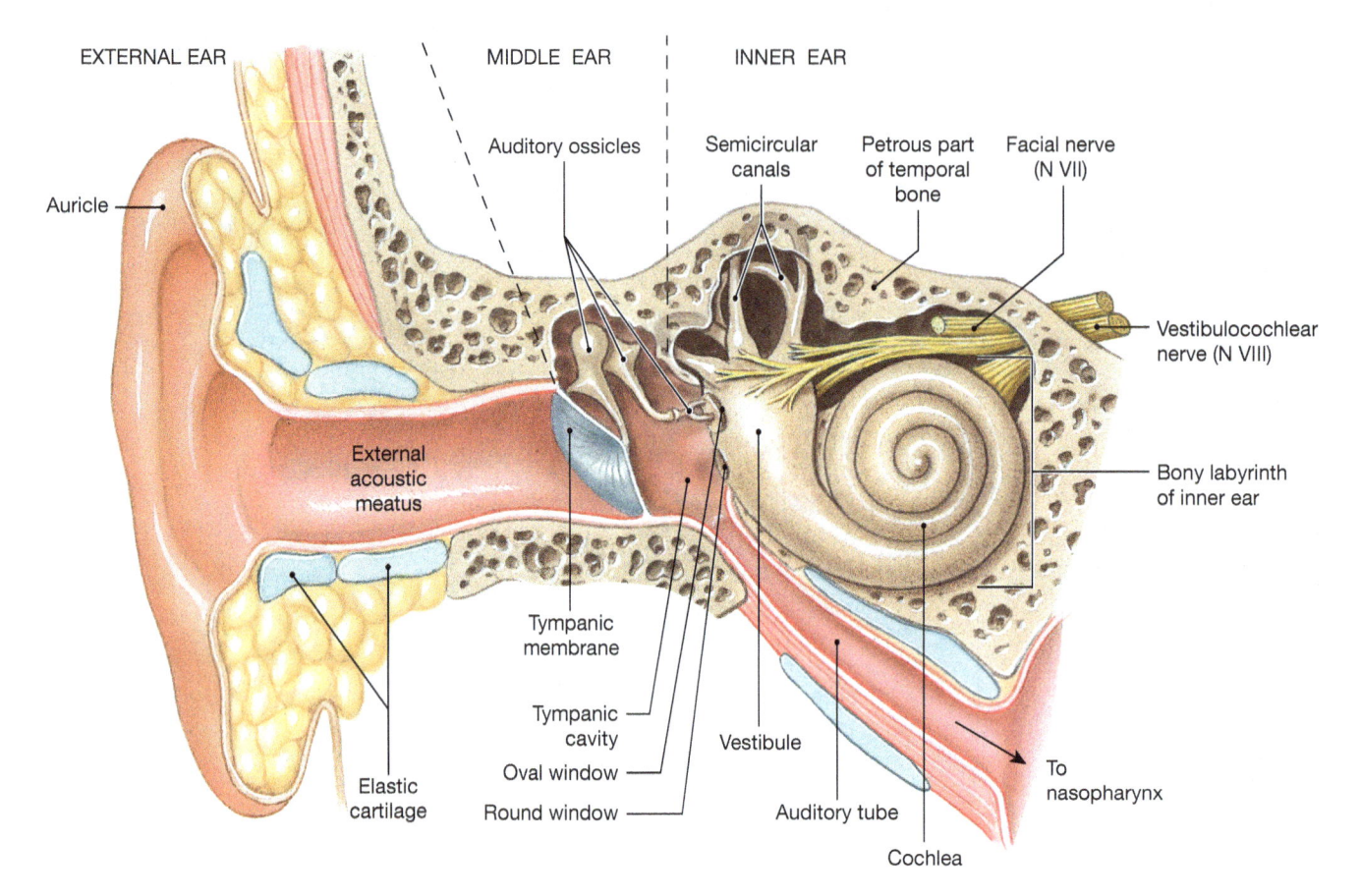

Figure 19.1 Anatomy of the Ear
A general orientation of the external, middle, and inner ear.

(*incus,* anvil), which is in contact with the third auditory ossicle, the **stapes** (*stapes,* stirrup). Vibrations of the tympanic membrane are transferred to the malleus, which then conducts the vibrations to the incus and stapes. The stapes in turn pushes on the **oval window** of the inner ear to stimulate the auditory receptors.

The smallest skeletal muscles of the body are attached to the auditory ossicles. The **tensor tympani** (TEN-sor tim-PAN-ē) **muscle** attaches to the malleus, and the **stapedius** (sta-PĒ-dē-us) **muscle** inserts on the stapes.

Clinical Application — **Otitis Media**

Infection of the middle ear is called **otitis media**. It is most common among infants and children but occurs infrequently in adults. The infection source is typically a bacterial invasion of the throat that migrated to the middle ear by way of the auditory tube. In children, the auditory tube is narrow and more horizontal than in adults. This orientation permits pathogens originally present in the throat to infect the middle ear. Children who frequently get middle-ear infections may have small tubes implanted through the tympanic membrane to drain liquid from the middle ear into the external acoustic meatus.

In severe cases, the microbes infect the air cells of the mastoid process (Exercise 7), which causes a condition known as **mastoiditis**. The passageways between the air cells become congested, and swelling occurs behind the auricle. This condition is serious because it may spread to the brain. Powerful antibiotic therapy is necessary to treat the infection. Otherwise, a mastoidectomy may be necessary, which is a procedure that involves opening and draining the mastoid air cells.

EXERCISE 19 Anatomy of the Ear 311

Temporal bone (petrous part)
Stabilizing ligament
Chorda tympani nerve (cut), a branch of N VII
External acoustic meatus
Tympanic cavity (middle ear)
Tympanic membrane (tympanum)
Malleus
Incus
Base of stapes at oval window
Tensor tympani muscle
Stapes
Round window
Stapedius muscle
Auditory tube

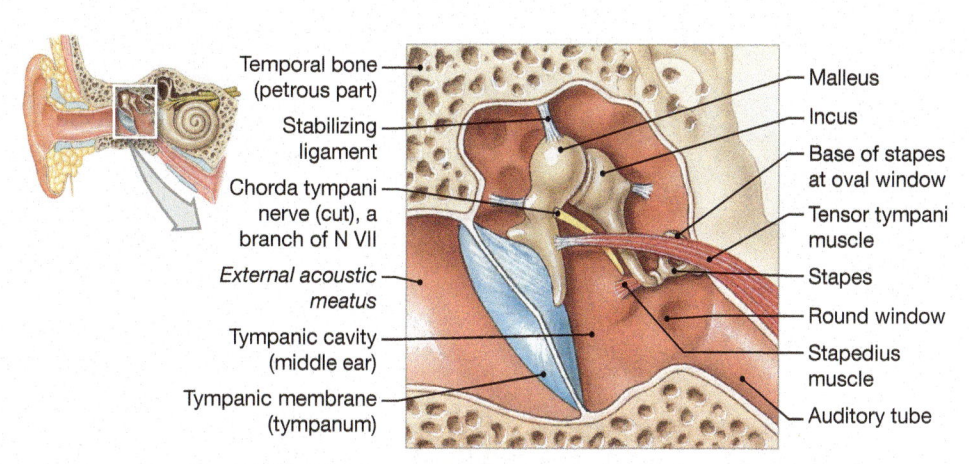

Figure 19.2 The Middle Ear
The structures of the middle ear, which details the auditory ossicles—malleus, incus, and stapes—and associated skeletal muscles.

QuickCheck Questions

1.1 What is the function of the auricle?
1.2 Which two regions of the ear does the tympanic membrane separate?
1.3 What is the function of the auditory tube?

1 Materials

☐ Ear model
☐ Ear chart

Procedures

1. Review the three major regions of the ear in Figure 19.1.
2. Identify the auricle, external acoustic meatus, and tympanic membrane on the ear model and chart.
3. To appreciate how important the auricle is in directing sound into the ear, cup one hand over each of your ears. Do you notice a change in sound? Listen carefully for a moment to the sound, and then experiment by moving your fingers apart. Describe the change in sounds.
4. Identify the three types of auditory ossicle on the ear model and chart. Notice the sequence of articulated structures from the tympanic membrane to the oval window.
5. Identify the auditory tube and the muscles of the middle ear on the ear model and chart. ∎

LAB ACTIVITY 2 Inner Ear

The inner ear consists of three regions. Moving medial to lateral, these regions are a helical **cochlea** (KOK-lē-uh), an elongated **vestibule** (VES-ti-būl), and three **semicircular canals** (Figure 19.3a). The cochlea contains receptors for hearing; the vestibule is receptive to stationary, or static, equilibrium; and the semicircular canals contain receptors for dynamic equilibrium when the body moves. No physical barrier separates one region from the next, and the general internal structure is the same in all three regions. A cross-section of this structure is shown in Figure 19.3b—a "pipe within a pipe" arrangement. The outer pipe, called the **bony labyrinth**, is embedded in the temporal bone and contains a liquid called **perilymph** (PER-i-limf). The inner pipe, the **membranous labyrinth**, is filled with a liquid called **endolymph** (en-dō-limf). The vestibule

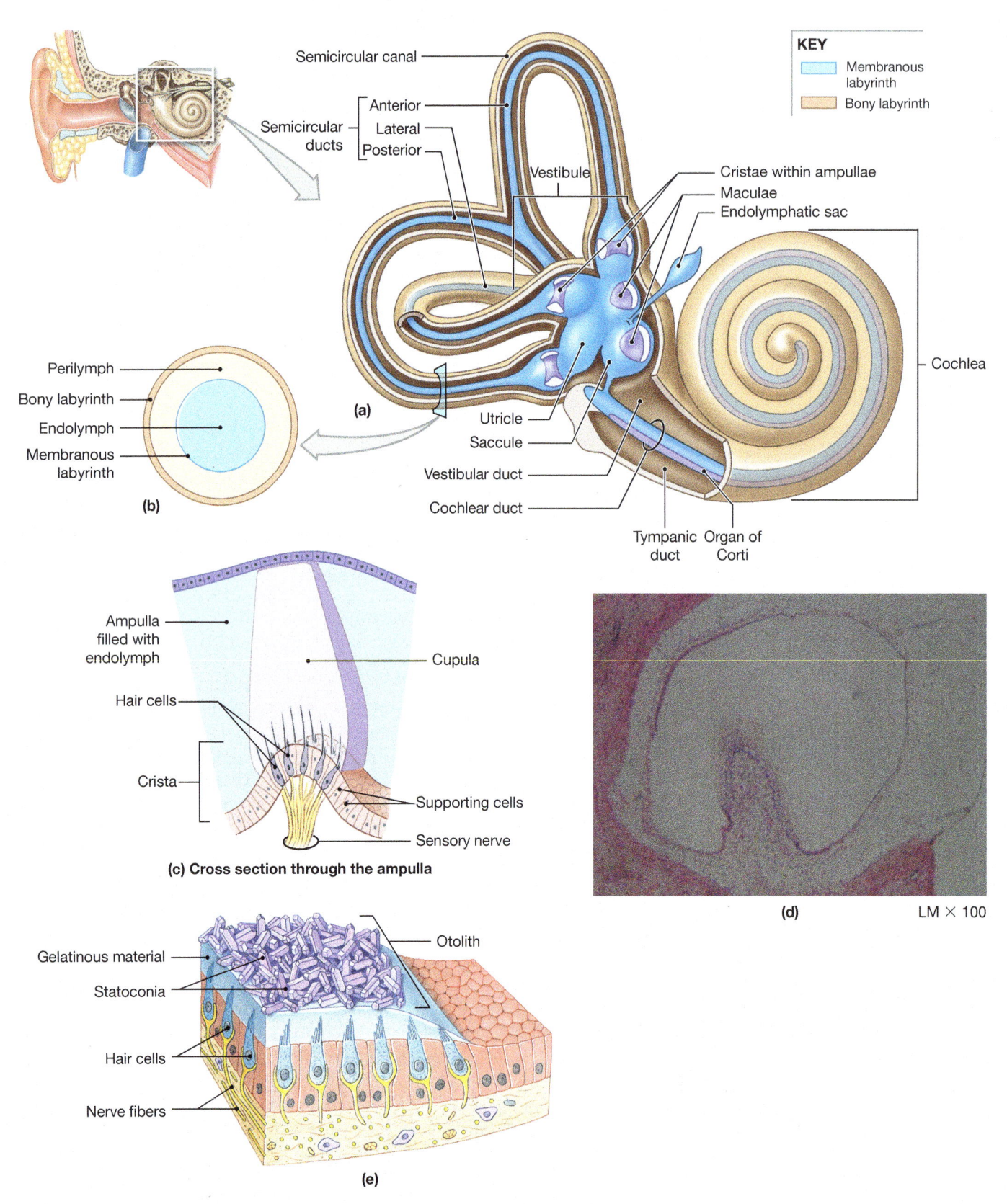

Figure 19.3 The Inner Ear

The inner ear is located in the petrous part of each temporal bone. **(a)** Anterior cutaway view of the semicircular canals and the enclosed semicircular ducts of the membranous labyrinth. **(b)** Cross-section of the inner ear structure. **(c)** Structure of a crista and its overriding cupula. **(d)** Photomicrograph of a crista in an ampulla. (LM × 100) **(e)** Structure of a macula.

312

contains an **endolymphatic duct** that drains endolymph into an **endolymphatic sac**, where the liquid is absorbed into the blood.

The three semicircular canals are oriented perpendicular to one another. Together they function as an organ of dynamic equilibrium and work to maintain equilibrium when the body is in motion. Inside the canals, the membranous labyrinth is called the **semicircular ducts**. At one end of each semicircular duct is a swollen **ampulla** (am-PŪL-la) that houses the balance receptors called **cristae** (Figure 19.3c). Each crista is composed of hair cells and supporting cells, and the cilia of the hair cells extend upward from the crista into a gelatinous material called the **cupula** (KŪ-pū-la). Movement of the head causes the endolymph inside the semicircular ducts to either push or pull on the cupula, so that the embedded hair cells are either bent or stretched.

In the vestibule, the membranous labyrinth contains two sacs, the **utricle** (Ū-tre-kl) and the **saccule** (SAK-ūl), which contain **maculae** (MAK-ū-lē), receptors that work to maintain static equilibrium. Like the cristae of the ampullae, the maculae of the utricle and saccule have hair cells and a **gelatinous material**. Embedded in the gel are calcium carbonate crystals called **statoconia** (Figure 19.3e). The gelatinous material and the statoconia collectively are called an **otolith**, which means "ear stone". When the head is tilted, the otolith changes position, and the hair cells in the utricle and saccule are stimulated. Impulses from the maculae pass to sensory neurons in the vestibular branch of the vestibulocochlear nerve (cranial nerve N VIII).

The cochlea consists of three ducts rolled up together in a spiral formation (Figure 19.4). The **cochlear duct**, also called the **scala media**, contains hair cells that are sensitive to vibrations caused by sound waves. The cochlear duct is part of the membranous labyrinth and so is filled with endolymph. Surrounding the cochlear duct are the **vestibular duct (scala vestibule)** and the **tympanic duct (scala tympani)**. Both of these ducts are part of the cochlea's bony labyrinth and so are filled with perilymph. The floor of the cochlear duct is the **basilar membrane** where the hair cells occur. The **vestibular membrane** separates the cochlear duct from the vestibular duct. These two ducts follow the helix of the cochlea, and the vestibular and tympanic ducts interconnect at the tip of the spiral.

The stapes of the middle ear is connected to the vestibular duct at the oval window. When incoming sound waves make the stapes vibrate against the oval window, the pressure on the window transfers the waves to the ducts of the cochlea. The waves stimulate the hair cells in the cochlear duct and then pass into the tympanic duct, where a second window, the **round window** (Figure 19.4a), stretches to dissipate the wave energy.

The cochlear duct contains the sensory receptor for hearing, called either the **spiral organ** or the **organ of Corti** (Figure 19.5). It consists of hair cells and supporting cells. Extending from the wall of the cochlear duct and projecting over the hair cells is the **tectorial** (tek-TOR-ē-al) **membrane**. Two types of hair cells occur in the spiral organ: **inner hair cells** that rest on the basilar membrane near the proximal portion of the tectorial membrane and **outer hair cells** at the tip of the membrane. The long stereocilia of the hair cells extend into the endolymph and contact the tectorial membrane. Sound waves cause liquid movement in the cochlea, and the hair cells are bent and stimulated as they are pushed against the tectorial membrane. The hair cells synapse with sensory neurons in the cochlear branch of the vestibulocochlear nerve (N VIII), which transmits the impulses to the auditory cortex of the brain. The **spiral ganglia** contain cell bodies of sensory neurons in the cochlear branch of the vestibulocochlear nerve.

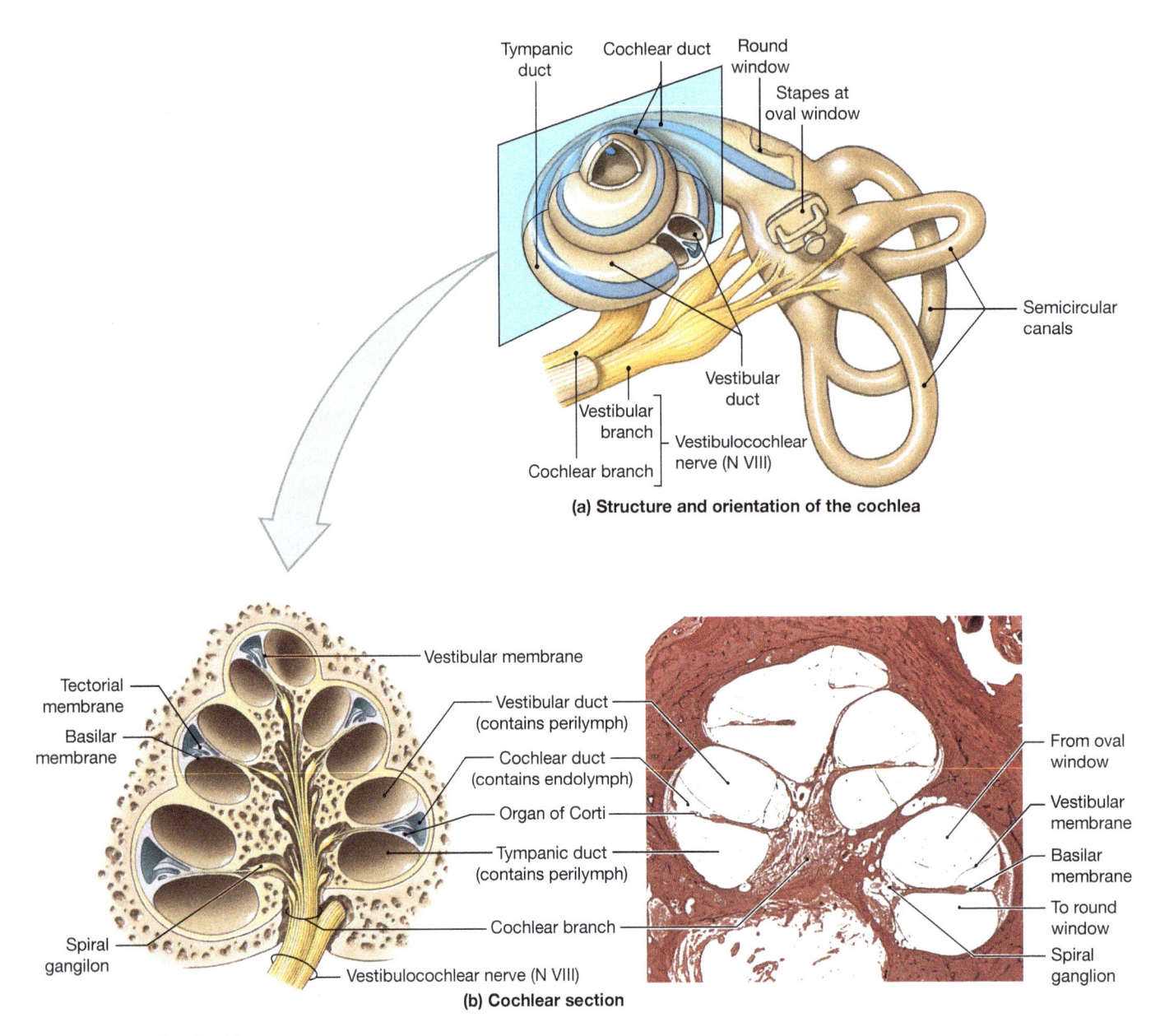

Figure 19.4 The Cochlea

The ducts of the cochlea are coiled approximately 2.5 times. (a) The structure of the cochlea in partial section. (b) Structure of the cochlea in the temporal bone as seen in section, which shows the turns of the vestibular duct, cochlear duct, and tympanic duct. Photomicrograph of the cochlea, showing the ducts as they appear in a sectional view. (LM × 9)

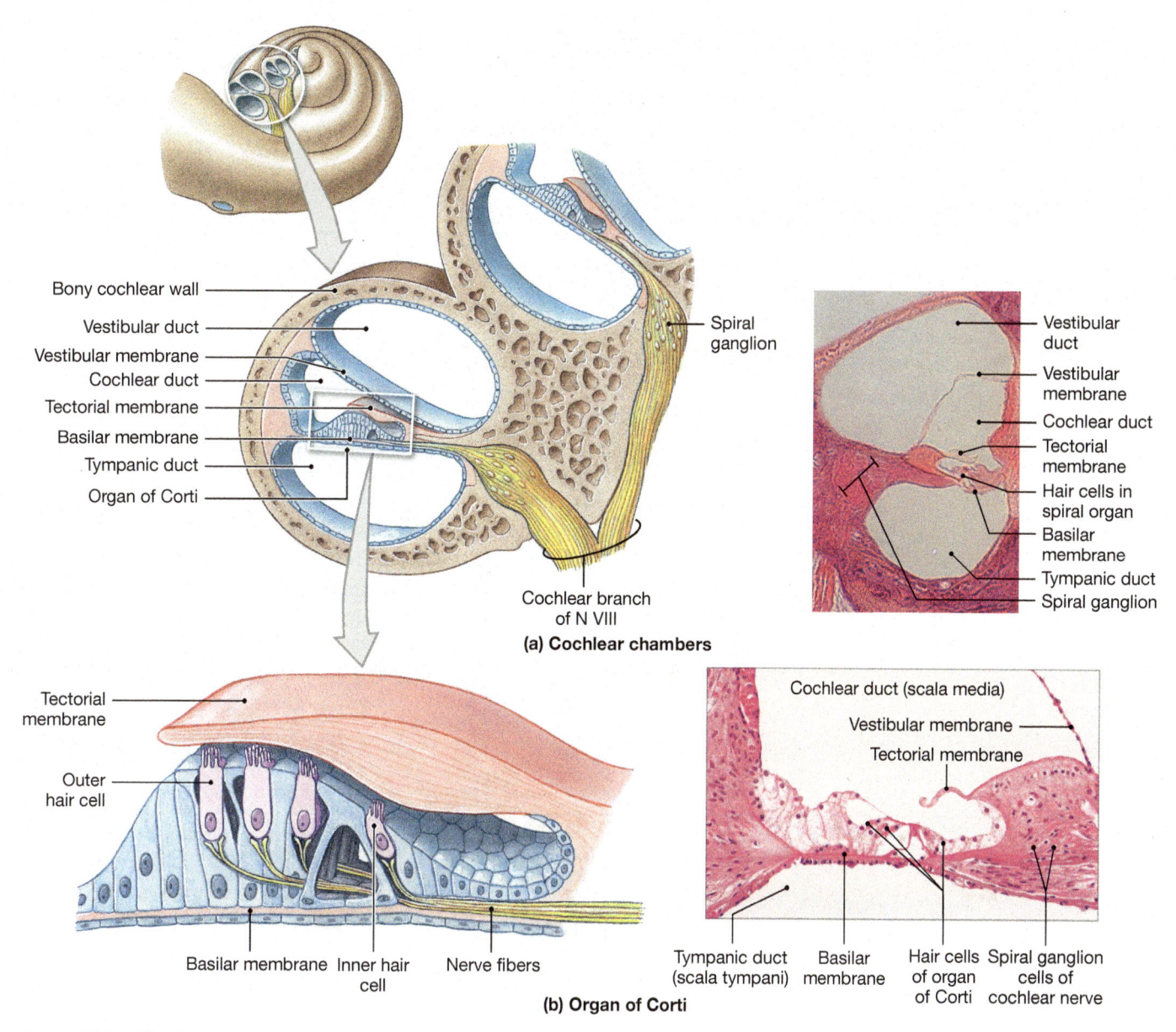

Figure 19.5 The Organ of Corti

The organ of Corti inside the cochlear duct as seen in a sectional view. **(a)** Three-dimensional section that shows the detail of the cochlear chambers, tectorial membrane, and organ of Corti. Photomicrograph of histological section of the cochlea. (LM × 100) **(b)** Diagrammatic and histological sections through the receptor hair cell complex of the organ of Corti. (LM × 1000)

315

EXERCISE 19 Anatomy of the Ear

QuickCheck Questions

2.1 What are the three kinds of sensory receptors of the inner ear, and what is the function of each?

2.2 What is the function of the semicircular canals?

2.3 What is the function of the vestibule?

2 Materials

- Ear model
- Ear chart
- Compound microscope
- Prepared slide of cochlea
- Prepared slide of crista

Procedures

1. Review the anatomy of the inner ear in Figures 19.3 and 19.4.
2. On the ear model and/or chart, distinguish among the anterior, posterior, and lateral semicircular canals. Then locate the ampulla at the base of each canal.
3. On the ear model and/or chart, identify the utricle, saccule, endolymphatic duct, and endolymphatic sac. Note the vestibular branch of the vestibulocochlear nerve (N VIII).
4. Observe the cochlea on the model and/or chart and identify the various ducts and membranes. Examine the organ of Corti and locate the inner and outer hair cells and the tectorial membrane.
5. Examine the cochlea slide and identify the structures shown in Figures 19.4 and 19.5. In the space provided, sketch a cross-section of the cochlea.
6. Examine the crista slide and identify the structures shown in Figure 19.3c. The cochlea is usually presesnt on the same slide as the crista. Search for the cone-shaped crista at the base of the cupula. Observe the crista at medium power and identify the hair cells and the cupula. In the space provided, sketch a cross-section of the crista. ■

Cochlea

Crista

LAB ACTIVITY 3 **Examination of the Tympanic Membrane**

The tympanic membrane that separates the external ear from the middle ear can be examined with an instrument called an **otoscope** (Figure 19.6). The removable tip is the **speculum**, and it is placed in the external acoustic meatus. Light from the instrument illuminates the tympanic membrane, which is viewed through a magnifying lens on the back of the otoscope.

QuickCheck Questions

3.1 What is the name of the instrument used to look at the tympanic membrane?
3.2 What is the removable tip of the instrument called?

3 Materials

- Otoscope
- Alcohol wipes
- Lab partner

Procedures

1. Using Figure 19.6 as a reference, identify the parts of the otoscope.
2. Select the shortest but *largest-diameter* speculum that will fit into your partner's ear.
3. Either wipe the tip clean with an alcohol pad or place a new disposable cover over the speculum.
4. Turn the otoscope light on. Be sure the light beam is strong.
5. Hold the otoscope between your thumb and index finger, and either sit or stand facing one of your partner's ears. Place the tip of your extended little finger against your partner's head to support the otoscope. This finger placement is important to prevent injury by the speculum.

Figure 19.6 Otoscope
The speculum is placed in the external acoustic meatus to examine the tympanic membrane.

6. Carefully insert the speculum into the external acoustic meatus while gently pulling the auricle up and posterolaterally. Neither the otoscope nor the pulling should hurt your partner. If your partner experiences pain, stop the examination.
7. Look into the magnifying lens and observe the walls of the external acoustic meatus. Note if there is any redness in the walls or any buildup of wax.
8. Manipulate the auricle and speculum until you see the tympanic membrane. A healthy membrane appears white. Also notice the vascularization of the region.
9. After the examination, either clean the speculum with a new alcohol wipe or else remove the disposable cover. Dispose of used wipes and covers in a biohazard container. ■

LAB REPORT

EXERCISE 19: Anatomy of the Ear

Name _____
Date _____
Section _____

A. Matching

Match each description in the left column with its correct structure from the right column.

_____ 1. receptors in semicircular canals
_____ 2. coiled region of inner ear
_____ 3. outer layer of inner ear
_____ 4. receptor cells for hearing
_____ 5. receptors in vestibule
_____ 6. attachment membrane for stapes
_____ 7. jelly-like substance in ampulla
_____ 8. contains endolymph
_____ 9. chamber inferior to organ of Corti
_____ 10. membrane that supports organ of Corti

A. basilar membrane
B. membranous labyrinth
C. cupula
D. bony labyrinth
E. tympanic duct
F. cochlea
G. crista
H. organ of Corti
I. oval window
J. maculae

B. Labeling

Label the numbered anatomical features of the ear in Figure 19.7.

C. Short-Answer Questions

1. Describe the components of the external ear.

2. Describe the components of the middle ear.

3. Describe the components of the inner ear.

EXERCISE 19   LAB REPORT

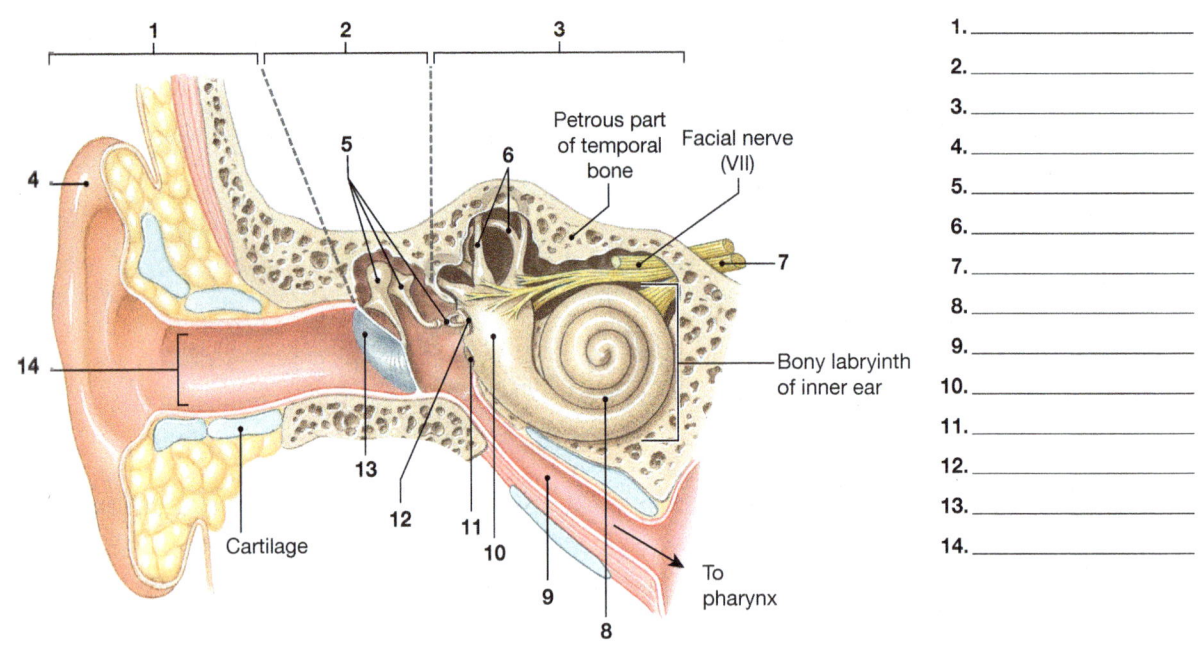

Petrous part of temporal bone
Facial nerve (VII)
Bony labyrinth of inner ear
Cartilage
To pharynx

1. _____
2. _____
3. _____
4. _____
5. _____
6. _____
7. _____
8. _____
9. _____
10. _____
11. _____
12. _____
13. _____
14. _____

Figure 19.7 Anatomy of the Ear

D. Analysis and Application

1. How are sound waves transmitted to the inner ear?

2. How are hair cells stimulated in the inner ear?

3. If the pathway along the vestibular branch of the vestibulocochlear nerve has been disrupted, what symptoms does the patient display?

EXERCISE 20

The Endocrine System

OBJECTIVES

On completion of this exercise, you should be able to:

- Compare the two regulatory systems of the body: the nervous and endocrine systems.
- Identify each endocrine gland on a laboratory model.
- Describe the histological appearance of each endocrine gland.
- Identify each endocrine gland when viewed microscopically.

LAB ACTIVITIES

1. Pituitary Gland   322
2. Thyroid Gland   323
3. Parathyroid Glands   326
4. Thymus Gland   328
5. Suprarenal Glands   329
6. Pancreas   332

A Regional Look: Suprarenal Glands   334

Two kinds of glands are found in the body. **Endocrine glands** produce regulatory molecules called **hormones** that slowly cause changes in the metabolic activities of **target cells**, which are any cells that contain membrane receptors for the hormones. Endocrine glands are commonly called *ductless glands* because they secrete their hormones into the surrounding extracellular fluid instead of secreting into a duct. The other kind of glands, **exocrine glands**, secrete substances into a duct for transport and release onto a free surface of the body. Examples of exocrine glands are the sweat glands and sebaceous glands of the skin.

Two systems regulate homeostasis: the nervous system and the endocrine system. These systems must coordinate their activities to maintain control of internal functions. The nervous system responds rapidly to environmental changes, sending electrical commands that can produce an immediate response in any part of the body. The duration of each electrical impulse is brief, measured in milliseconds. In contrast, the endocrine system maintains long-term control. In response to stimuli, endocrine glands release their hormones, and the hormones then slowly cause changes in the metabolic activities of their target cells. Typically, the effect of a hormone is prolonged and lasts several hours.

The secretion of many hormones is regulated by negative feedback mechanisms. In **negative feedback**, a stimulus causes a response that either reduces or removes the stimulus. An excellent analogy is the operation of an air-conditioner. When a room heats up, the warm air activates a thermostat that then turns on the compressor of the air-conditioner. Cooled air flowing in cools the room and removes the stimulus (the warm air). Once the stimulus is removed, the unit shuts off. Negative feedback is therefore a self-limiting mechanism.

322 EXERCISE 20 The Endocrine System

An example of negative-feedback control of hormonal secretion is the regulation of insulin, a hormone from the pancreas that lowers the concentration of glucose in the blood. When blood glucose levels are high, as they are after a meal, the pancreas secretes insulin. The secreted insulin stimulates the body's cells to increase their glucose consumption and storage, thus lowering the concentration of glucose in the blood. As this concentration returns to normal, insulin secretion stops.

In this exercise, you will study the following glands: pituitary, thyroid, parathyroid, thymus, pancreas, and suprarenal. The ovaries and testes are also important endocrine glands, and they are presented in Exercise 28.

LAB ACTIVITY 1 Pituitary Gland

Anatomy

The **pituitary gland**, or **hypophysis** (hī-POF-i-sis), is located in the sella turcica of the sphenoid of the skull, immediately inferior to the hypothalamus of the brain (Figure 20.1). As noted in Exercise 15, a stalk called the **infundibulum** attaches the pituitary to the hypothalamus. The pituitary gland is organized into two lobes, an **anterior lobe**, also called the **adenohypophysis** (ad-ē-nō-hī-POF-i-sis), and a **posterior lobe**, also called either the **neurohypophysis** (noo-rō-hī-POF-i-sis) or the **pars nervosa**. The main portion of the anterior lobe is the **pars distalis** (dis-TAL-is); the **pars tuberalis** is a narrow portion that wraps around the infundibulum; the **pars intermedia** (in-ter-MĒ-dē-uh) is found in the interior of the gland, and forms the boundary between the anterior and posterior lobes.

The two pituitary lobes are easily distinguished from each other by how they accept stain. The posterior lobe consists mostly of lightly-stained unmyelinated axons from hypothalamic neurons. Darker-stained cells called **pituicytes** are scattered in the lobe and are similar to glial cells in function.

The darker-staining anterior lobe is populated by a variety of cell types that are classified into two main groups determined by their histological staining qualities. **Chromophobes** are light-colored cells that do not react to most stains. **Chromophils**

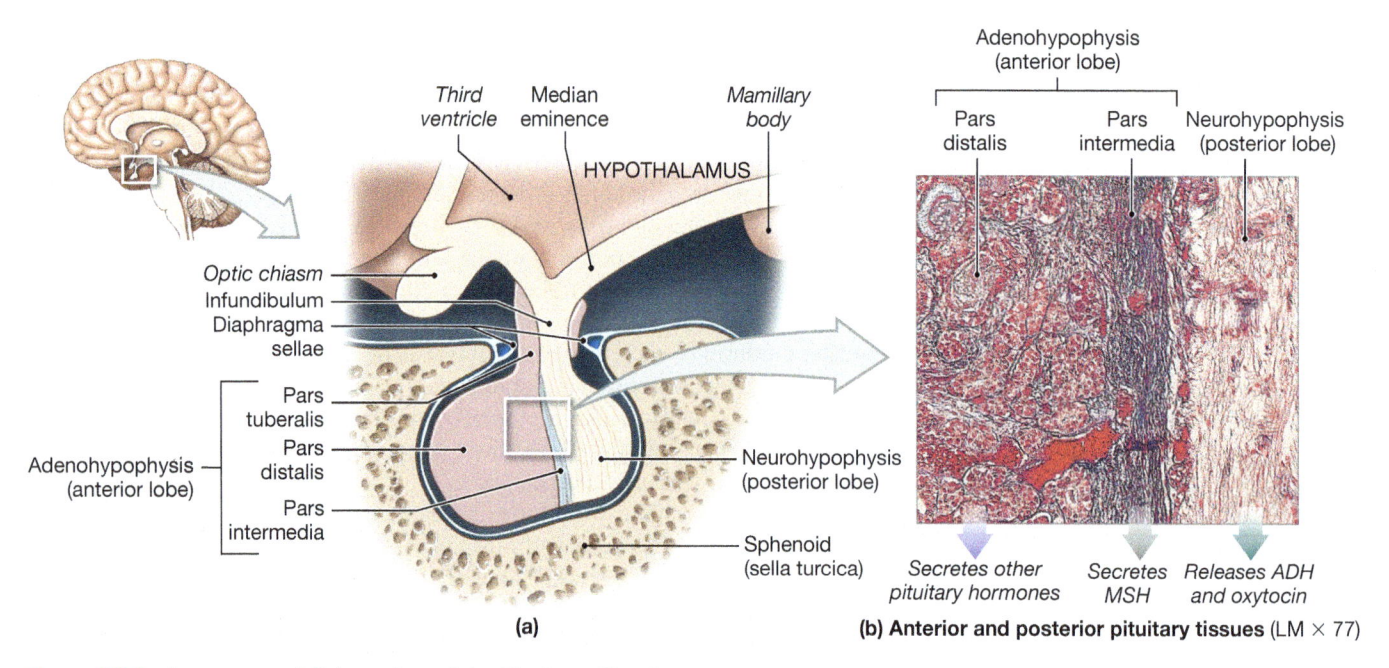

Figure 20.1 Anatomy and Orientation of the Pituitary Gland
(a) Relationship of the pituitary gland to the hypothalamus. (b) Photomicrograph of pituitary gland that shows the anterior and posterior lobes. (LM × 77)

react to histological stains and are darker than chromophobes. Chromophils are subdivided into **acidophils**, which react with acidic stains, and **basophils**, which react with basic stains. In most slide preparations, basophils are stained darker than the more numerous reddish acidophils.

Making Connections — **Histological Stains and Cells**

> Many standard histological stains are mixtures of basic and acidic compounds. When a tissue is stained, some cells may react with the acidic component and turn colorless, other cells may react with the basic component and darken, and still other cells may not react to either component of the stain.

Hormones

The pituitary gland is commonly called the *master gland* because it has a critical role in regulating endocrine function and produces hormones that control the activity of many other endocrine glands. **Regulatory hormones** from the hypothalamus travel down a plexus of blood vessels in the infundibulum and signal the pars distalis to secrete **tropic hormones** that target other endocrine glands, and induces them to produce and secrete their own hormones. The pars intermedia produces a single hormone, melanocyte-stimulating hormone (MSH).

The posterior lobe does not produce hormones. Instead, its function is to store and release antidiuretic hormone (ADH) and oxytocin (OT), which are both produced in the hypothalamus and then passed down the infundibulum to the pituitary gland.

QuickCheck Questions

1.1 Where is the pituitary gland located?
1.2 What is the main staining difference between the anterior and posterior pituitary lobes?

1 Materials

- ☐ Torso model
- ☐ Endocrine chart
- ☐ Dissecting microscope
- ☐ Compound microscope
- ☐ Prepared slide of pituitary gland

Procedures

1. Locate the pituitary gland on the torso model and endocrine chart.
2. Use the dissecting microscope to survey the pituitary gland slide at low magnification. Distinguish between the two lobes of the gland.
3. Examine the slide at low and medium powers with the compound microscope. Identify the anterior and posterior lobes, and note the different cell arrangements in each.

LAB ACTIVITY 2 Thyroid Gland

Anatomy

The **thyroid gland** is located in the anterior aspect of the neck, directly inferior to the thyroid cartilage (Adam's apple) of the larynx and just superior to the trachea (Figure 20.2). This gland consists of two lateral lobes connected by a central mass, the **isthmus** (IS-mus). The thyroid produces two groups of hormones: T_3 and T_4 are associated with the regulation of cellular metabolism and calcitonin for homeostasis.

The thyroid gland is very distinctive. It is composed of spherical **follicles** embedded in connective tissue. Each follicle is composed of a single layer of cuboidal epithelial cells called **follicle cells**, also called T thyrocytes. The lumen of each follicle is filled with a glycoprotein called **thyroglobulin** (thī-rō-GLOB-ū-lin) that stores thyroid hormones. On the superficial margins of the follicles are **C cells**, also called C thyrocytes which are larger and less abundant than the follicle cells. On most slides, the C cells have a light-stained nucleus.

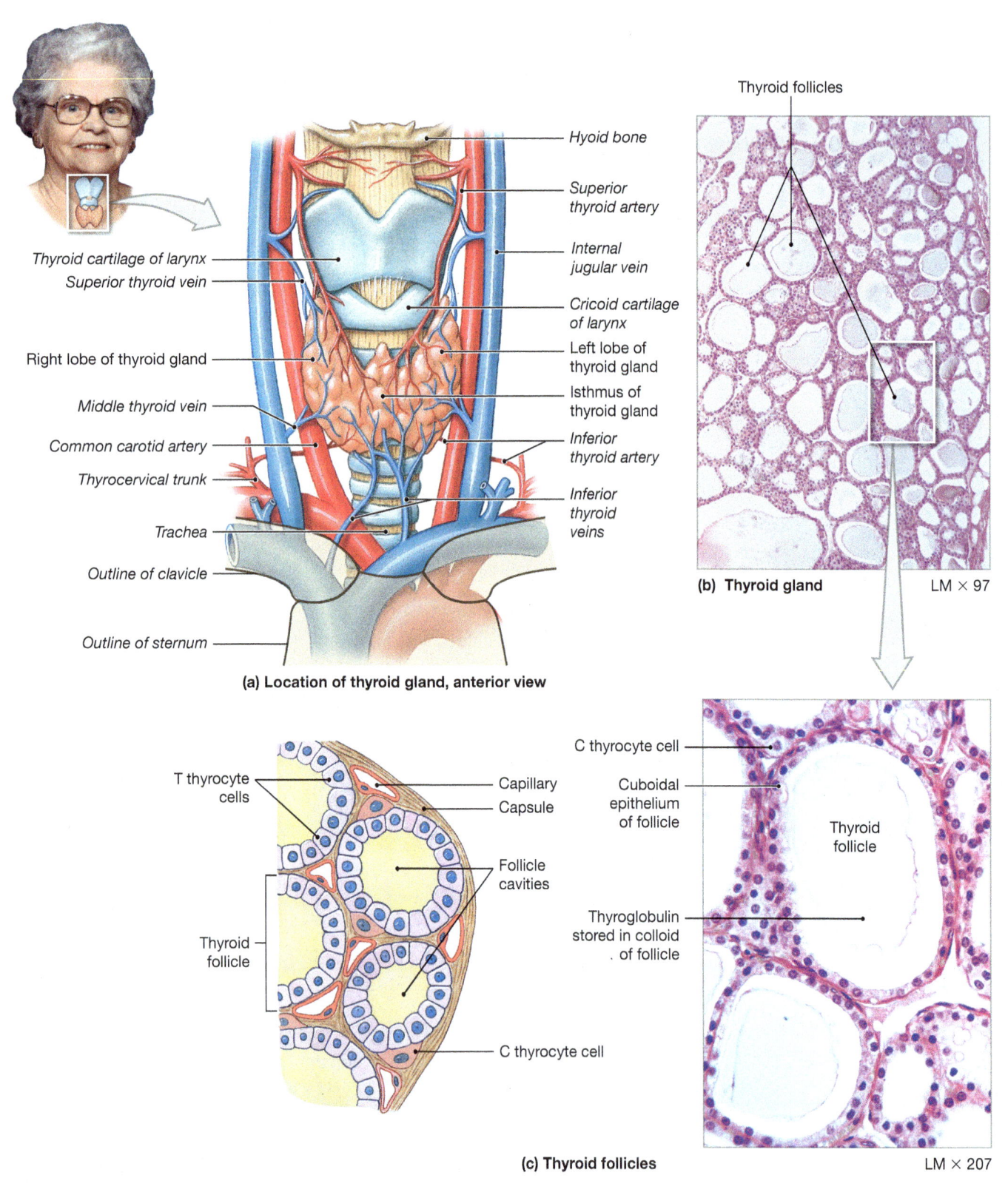

Figure 20.2 Thyroid Gland

(a) Location and anatomy of the thyroid gland. (b) Histological organization of the thyroid gland. (LM × 97) (c) Histological details of the thyroid gland that show thyroid follicles, follicle cells, and C cells. (LM × 207)

324

Hormones

Follicle cells produce the hormones **thyroxine (T_4)** and **triiodothyronine (T_3)**, both of which regulate metabolic rate. These hormones are synthesized in the form of the glycoprotein thyroglobulin. It is secreted into the lumen of the follicles and stored there until needed by the body, at which time it is reabsorbed by the follicle cells and released into the blood.

C cells produce the hormone **calcitonin (CT)**, which decreases blood calcium levels. Calcitonin stimulates osteoblasts in bone tissue to store calcium in bone matrix and lower fluid calcium levels. It also inhibits osteoclasts in bone from dissolving bone matrix and releasing calcium. Calcitonin has a minor role in calcium regulation in humans but is more active in lowering blood calcium in other animals.

Clinical Application

Hyperthyroidism

Hyperthyroidism occurs when the thyroid gland produces too much T_4 and T_3. Because these hormones increase mitochondrial ATP production and increase metabolic rate, individuals with this endocrine disorder are often thin, restless, and emotionally unstable. They fatigue easily because the cells are consuming rather than storing high-energy ATP molecules. *Graves' disease* is a form of hyperthyroidism that occurs when the body has an autoimmune response and produces antibodies that attack the thyroid gland. The gland enlarges to the point that it protrudes from the throat; the enlarged mass is called a *goiter*. Fat tissue is also deposited deep in the eye orbits, which causes the eyeballs to protrude, a condition called *exophthalmos*. Treatment for hyperthyroidism may include partial removal of the gland or destruction of parts of it with radioactive iodine.

QuickCheck Questions

2.1 Where is the thyroid gland located?
2.2 How are the various types of thyroid cells arranged in the gland?

2 Materials

- ☐ Torso model
- ☐ Endocrine chart
- ☐ Dissecting microscope
- ☐ Compound microscope
- ☐ Prepared slide of thyroid gland

Procedures

1. Review the features of the thyroid gland in Figure 20.2.
2. Locate the thyroid gland on the torso model and endocrine chart.
3. Scan the thyroid slide with the dissecting microscope and observe the many thyroid follicles.
4. Use the compound microscope to view the thyroid slide at low and medium powers. Locate a follicle, some follicle cells, thyroglobulin, and C cells.
5. In the space provided, sketch several follicles as observed at medium magnification.

Follicles of a thyroid gland

LAB ACTIVITY 3 — Parathyroid Glands

Anatomy

The **parathyroid glands** are two pairs of oval masses on the posterior surface of the thyroid gland. Each parathyroid gland is isolated from the underlying thyroid tissue by the parathyroid **capsule**. The parathyroid glands are composed mostly of **chief cells**, also called *principal cells*. These cells have a round nucleus, and their cytosol is basophilic and stains pale with basic histological stains (Figure 20.3). The **oxyphil cells** of the parathyroid are larger than the chief cells, and their acidophilic cytosol reacts to acidic stains and turns colorless.

Hormone

The parathyroid glands produce **parathyroid hormone (PTH)**, which is antagonistic to calcitonin from the thyroid gland. Although CT is relatively ineffective in humans, PTH is important in maintaining blood calcium level by stimulating osteoclasts in bone to dissolve small areas of bone matrix and release calcium ions into the blood. PTH also stimulates calcium uptake in the digestive system and reabsorption of calcium from the filtrate in the kidneys.

QuickCheck Questions

3.1 Where are the parathyroid glands located?

3.2 What two types of cells make up the parathyroid glands?

3 Materials

- Torso model
- Endocrine chart
- Dissecting microscope
- Compound microscope
- Prepared slide of parathyroid gland

Procedures

1. Review the parathyroid glands in Figure 20.3.
2. Locate the parathyroid glands on the torso model and endocrine chart.
3. Examine the parathyroid slide with the dissecting microscope. Scan the gland for thyroid follicles that may be on the slide near the parathyroid tissue.
4. Observe the parathyroid slide at low and medium powers with the compound microscope. Locate the dark-stained chief cells and the light-stained oxyphil cells.
5. In the space provided, sketch the parathyroid gland as observed at medium magnification. ■

Parathyroid gland

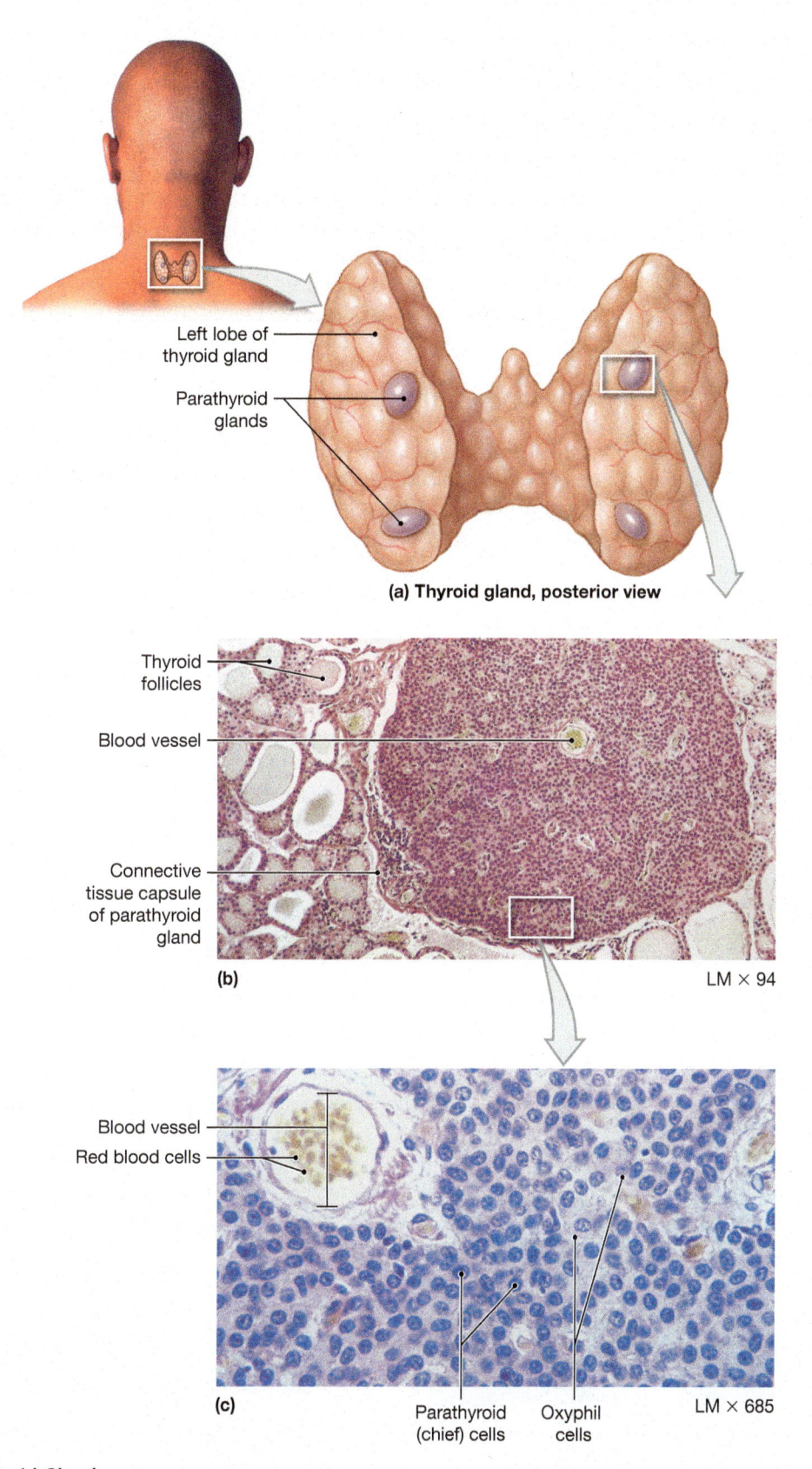

Figure 20.3 Parathyroid Glands

There are usually four separate parathyroid glands bound to the posterior surface of the thyroid gland. **(a)** The location and size of the parathyroid glands on the posterior surfaces of the thyroid lobes. **(b)** A photomicrograph that shows both parathyroid and thyroid tissues. (LM × 94) **(c)** A photomicrograph that shows chief and oxyphil cells of the parathyroid gland. (LM × 685)

328 EXERCISE 20 The Endocrine System

LAB ACTIVITY 4 — Thymus Gland

Anatomy

The **thymus gland** is located inferior to the thyroid gland, in the thoracic cavity posterior to the sternum (Figure 20.4). Because hormones secreted by the thymus gland facilitate development of the immune system, the gland is larger and more active in youngsters than in adults.

The thymus gland is organized into two main lobes, with each lobe made up of many **lobules**, which are very small lobes. The lobules in each lobe of the thymus gland are separated from one another by septae made up of fibrous connective tissue. Each lobule consists of a dense outer **cortex** and a light-staining central **medulla**. The cortex is populated by reticular cells that secrete the thymic hormones. In the medulla, other reticular cells cluster together into distinct oval masses called **thymic corpuscles** (Hassall's corpuscles). Surrounding the corpuscles are developing white blood cells called **lymphocytes** that eventually enter the blood. Adipose and other connective tissues are abundant in an adult thymus because the function and size of the gland decrease after puberty.

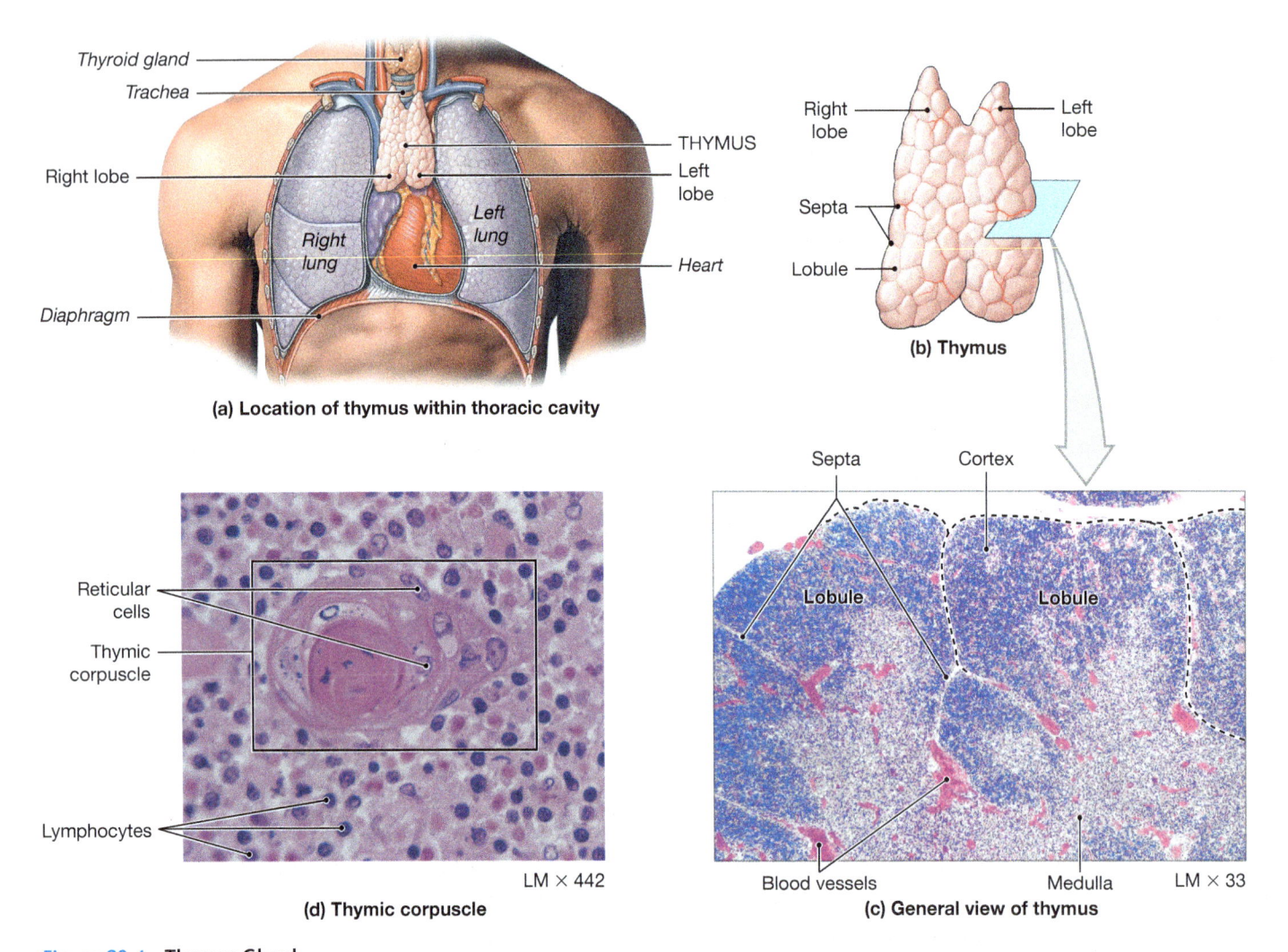

Figure 20.4 Thymus Gland
(a) The appearance and position of the thymus gland in relation to other organs in the chest. (b) Anatomical landmarks on the thymus. (c) A low-power light micrograph of the thymus. (LM × 33) Note the fibrous septa that divide the tissue of the thymus into lobules. (d) At higher magnification, the thymic corpuscles are visible. The small cells are lymphocytes in various stages of development. (LM × 442)

Hormones

Although the reticular cells of the thymus gland produce several hormones, the function of only one, **tymosin**, is understood. Tymosin is essential in the development and maturation of the immune system. Removal of the gland during early childhood usually results in a greater susceptibility to acute infections.

QuickCheck Questions

4.1 Where is the thymus gland located?

4.2 What are the main histological features of the thymus gland?

4 Materials

- ☐ Torso model
- ☐ Endocrine chart
- ☐ Dissecting microscope
- ☐ Compound microscope
- ☐ Prepared slide of thymus gland

Procedures

1. Review the anatomy of the thymus gland in Figure 20.4.
2. Locate the thymus gland on the torso model and endocrine chart.
3. Scan the slide of the thymus gland with the dissecting microscope and distinguish between the cortex and the medulla.
4. Examine the thymus slide with the compound microscope at low magnification to locate a stained thymic corpuscle. Increase the magnification and examine the corpuscle. The cells that surround the corpuscles are lymphocytes.
5. In the space provided, sketch the thymus gland as observed at medium magnification. ■

Thymus gland

LAB ACTIVITY 5 Suprarenal Glands

Anatomy

Superior to each kidney is a **suprarenal gland** (soo-pra-RĒ-nal) also called an adrenal gland (Figure 20.5). A **capsule** protects the gland and attaches it to the kidney. The gland is organized into two major regions: the outer **suprarenal cortex** and the inner **suprarenal medulla**.

The suprarenal cortex is differentiated into three distinct regions, each producing specific hormones. The **zona glomerulosa** (glō-mer-ū-LŌ-suh) is the outermost cortical region. Cells in this area are stained dark and arranged in oval clusters. The next layer, the **zona fasciculata** (fa-sik-ū-LA-tuh), is made up of larger cells organized in tight columns. These cells contain large amounts of lipid, which makes them appear lighter than the surrounding cortical layers (Figure 20.6). The deepest layer of the cortex, next to the medulla, is the **zona reticularis** (re-tik-ū-LAR-is).

330 EXERCISE 20 The Endocrine System

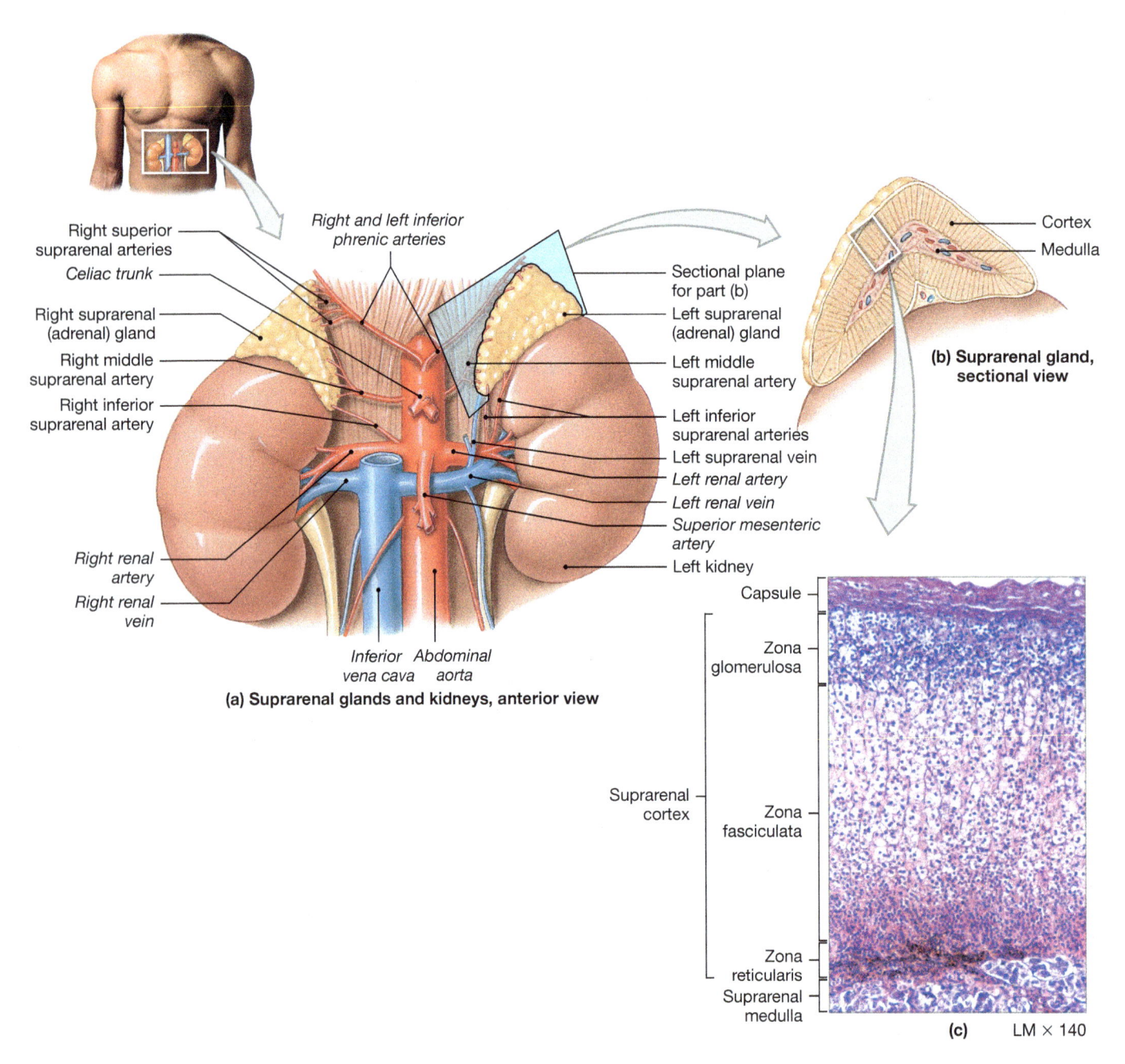

Figure 20.5 Suprarenal Gland
(a) Anterior view of the kidney and suprarenal gland. (b) A suprarenal gland sectioned to show the cortex and medulla.
(c) Photomicrograph that identifies the major regions of the suprarenal gland. (LM × 140)

Cells in this area are small and loosely linked together in chain-like structures. The many blood vessels in the suprarenal medulla give this tissue a dark red color.

Hormones

The suprarenal cortex secretes hormones are collectively called *adrenocorticoids*. These hormones are lipid-based steroids. The zona glomerulosa secretes a group of hormones called **mineralocorticoids** that regulate, as their name implies, mineral or electrolyte concentrations of body fluids. A good example is **aldosterone**, which stimulates the kidney to reabsorb sodium from the liquid being processed into urine. The zona fasciculata produces a group of hormones called **glucocorticoids** that are involved in fighting stress, increasing glucose metabolism, and preventing inflammation. Two of the glucocorticoids, **cortisol** and **corticosterone** (kor-ti-KOS-te-rōn), are commonly found in creams used to treat rashes and allergic responses of the skin. The zona reticularis produces **androgens**, which are male sex hormones. Both males and females produce small quantities of androgens in the zona reticularis.

The suprarenal medulla is regulated by sympathetic neurons from the hypothalamus. In times of stress, exercise, or emotion, the hypothalamus stimulates the suprarenal medulla to release its hormones, the neurotransmitters **epinephrine** (E) and **norepinephrine** (NE) into the blood, which results in a bodywide sympathetic fight-or-flight response.

QuickCheck Questions

5.1 Where are the suprarenal glands located?

5.2 What are the two major regions of the suprarenal gland?

5.3 What are the three layers of the suprarenal cortex?

5 Materials

- Torso model
- Endocrine chart
- Dissecting microscope
- Compound microscope
- Prepared slide of suprarenal gland

Procedures

1. Review the components of the suprarenal gland in Figure 20.5.
2. Locate the suprarenal gland on the torso model and endocrine chart.
3. Examine the slide of the suprarenal gland with the dissecting microscope and distinguish among the capsule, suprarenal cortex, and suprarenal medulla.
4. Observe the suprarenal gland with the compound microscope and differentiate among the three layers of the suprarenal cortex.
5. In the space provided, sketch the suprarenal gland, and show the details of the three cortical layers and the medulla.

Suprarenal gland

332 EXERCISE 20 The Endocrine System

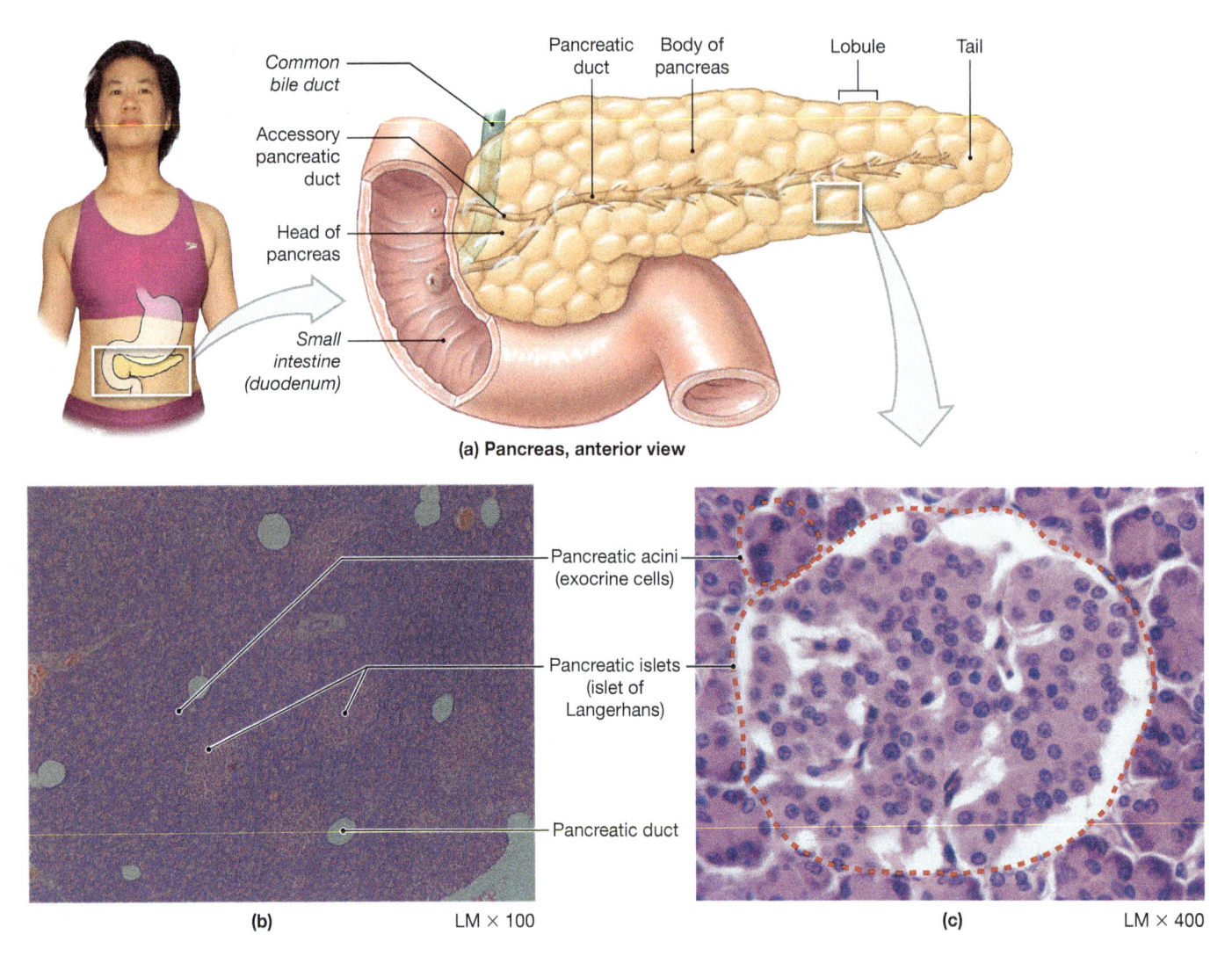

Figure 20.6 Pancreas
The pancreas, which is dominated by exocrine pancreatic acini cells, contains endocrine cells in clusters known as pancreatic islets. **(a)** The gross anatomy of the pancreas. **(b)** A pancreatic islet surrounded by pancreatic acini cells. (LM × 100) **(c)** A high-power photomicrograph of a pancreatic islet and the pancreatic acini cells that surround it. (LM × 400)

LAB ACTIVITY 6 Pancreas

Anatomy

The **pancreas**, a glandular organ that lies posterior to the stomach (Figure 20.6), performs important exocrine and endocrine functions. The exocrine cells secrete digestive enzymes, buffers, and other molecules into a pancreatic duct that empties into the small intestine. The endocrine cells produce hormones that regulate blood sugar metabolism.

The pancreas is densely populated by dark-stained cells called the **pancreatic acini**. These cells make up the exocrine part of the pancreas, and they secrete pancreatic juice, which contains digestive enzymes. Connective tissues and pancreatic ducts are dispersed in the tissue. The endocrine cells of the pancreas occur in isolated clusters of **pancreatic islets**, or *islets of Langerhans* (LAN-ger-hanz), that are scattered throughout the gland. Each islet houses four types of endocrine cells: **alpha cells**, **beta cells**, **delta cells**, and **F cells**. These cells are difficult to distinguish with routine staining techniques and will not be individually examined.

Hormones

Pancreatic hormones regulate carbohydrate metabolism. Alpha cells secrete the hormone **glucagon** (GLOO-ka-gon), which raises blood sugar concentration by catabolizing glycogen to glucose for cellular respiration. This process is called *glycogenolysis*. Beta cells secrete **insulin** (IN-su-lin), which accelerates glucose uptake by cells and also accelerates the rate of glycogenesis, which is the formation of glycogen. Insulin lowers blood sugar concentration by promoting the removal of sugar from the blood.

Clinical Application

Diabetes Mellitus

In **diabetes mellitus**, glucose in the blood cannot enter cells, and blood glucose levels rise above normal levels. In **type I diabetes**, the beta cells in the pancreas do not produce enough insulin, and cells are not stimulated to take in glucose. **Type II diabetes** occurs when the body becomes less responsive to insulin. The pancreas produces adequate amounts of insulin, but the cells are not receptive to it. Diabetes is self-aggravating. Because they are glucose-starved, the alpha cells of the pancreas secrete glucagons and elevate blood glucose concentrations.

QuickCheck Questions

6.1 Where is the pancreas located?
6.2 What is the exocrine function of the pancreas?
6.3 Where are the endocrine cells located in the pancreas?

6 Materials

- ☐ Torso model
- ☐ Endocrine chart
- ☐ Compound microscope
- ☐ Prepared slide of pancreas

Procedures

1. Review the histology of the pancreas in Figure 20.6.
2. Locate the pancreas on the torso model and endocrine chart.
3. Use the compound microscope to locate the dark-stained pancreatic acini cells and the oval pancreatic ducts. Identify the clusters of pancreatic islets, the endocrine portion of the gland.
4. In the space provided, sketch the pancreas. Label the pancreatic islets and the pancreatic acini cells.

Pancreas

suprarenal glands A Regional Look

The endocrine system has an intimate relationship with its distribution system, the blood. This is especially true for the emergency gland of the endocrine system, the suprarenal gland. The gland is positioned on the posterior abdominal wall immediately lateral of the inferior vena cava, the major vein that receives blood from the lower limbs and the abdomen (Figure 20.7). During times of excitement, exercise, emotion, or stress the suprarenal medulla secretes hormones epinephrine (adrenalin) and norepinephrine are absorbed into capillaries in the gland that merge into progressively larger vessels and eventually coalesce as the suprarenal vein which empties into the renal vein which unites with the inferior vena cava (Figure 20.7). Suprarenal medullary hormones target most cells of the body and this vascular association serves to quickly distribute suprarenal hormones into systemic circulation. Consider your body's response when you are suddenly surprised or frightened. It takes only a fraction of a second to feel the "adrenalin rush" of suprarenal hormones.

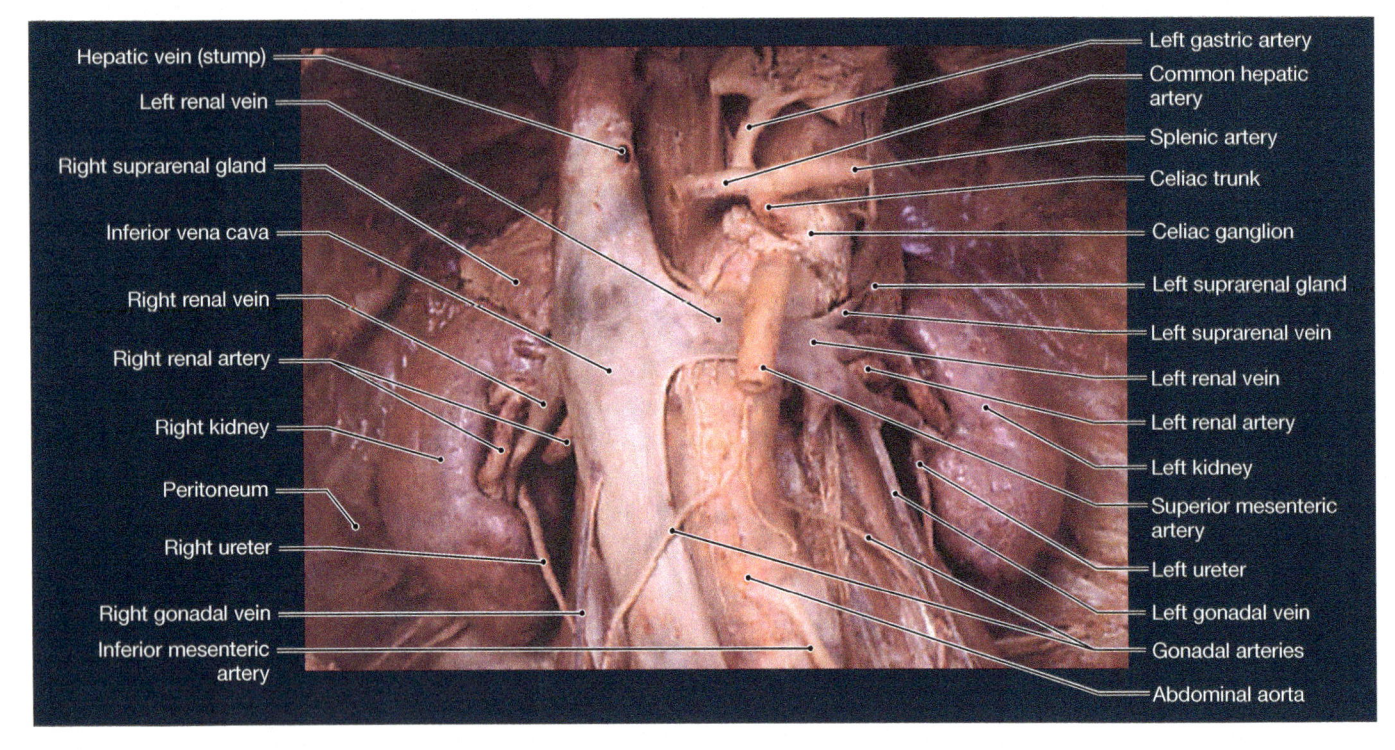

Figure 20.7 **Suprarenal Gland**
Cadaver dissection showing the posterior abdominal wall and the blood vessels serving the suprarenal glands.

334

EXERCISE 20

LAB REPORT

The Endocrine System

Name _____

Date _____

Section _____

A. Matching

Match each endocrine structure in the left column with its correct description from the right column.

_____ 1. thyroid follicle
_____ 2. suprarenal medulla
_____ 3. thymic corpuscle
_____ 4. zona glomerulosa
_____ 5. parathyroid glands
_____ 6. pancreatic acini cells
_____ 7. pituitary gland
_____ 8. master gland
_____ 9. target cell
_____ 10. pancreatic islets
_____ 11. zona reticularis
_____ 12. C cells

A. cell that responds to a specific hormone
B. four oval masses on posterior thyroid gland
C. gland attached to hypothalamus by infundibulum
D. produce insulin
E. cells between thyroid follicles
F. inner cortical layer of suprarenal gland
G. contains thyroglobulin
H. pituitary gland
I. found in thymus gland
J. innermost part of gland above kidney
K. outer cortical layer of suprarenal gland
L. exocrine cells of pancreas

B. Short-Answer Questions

1. Describe how negative feedback regulates the secretion of most hormones.

2. Explain how the pituitary gland functions as the master gland of the body.

335

EXERCISE 20 — LAB REPORT

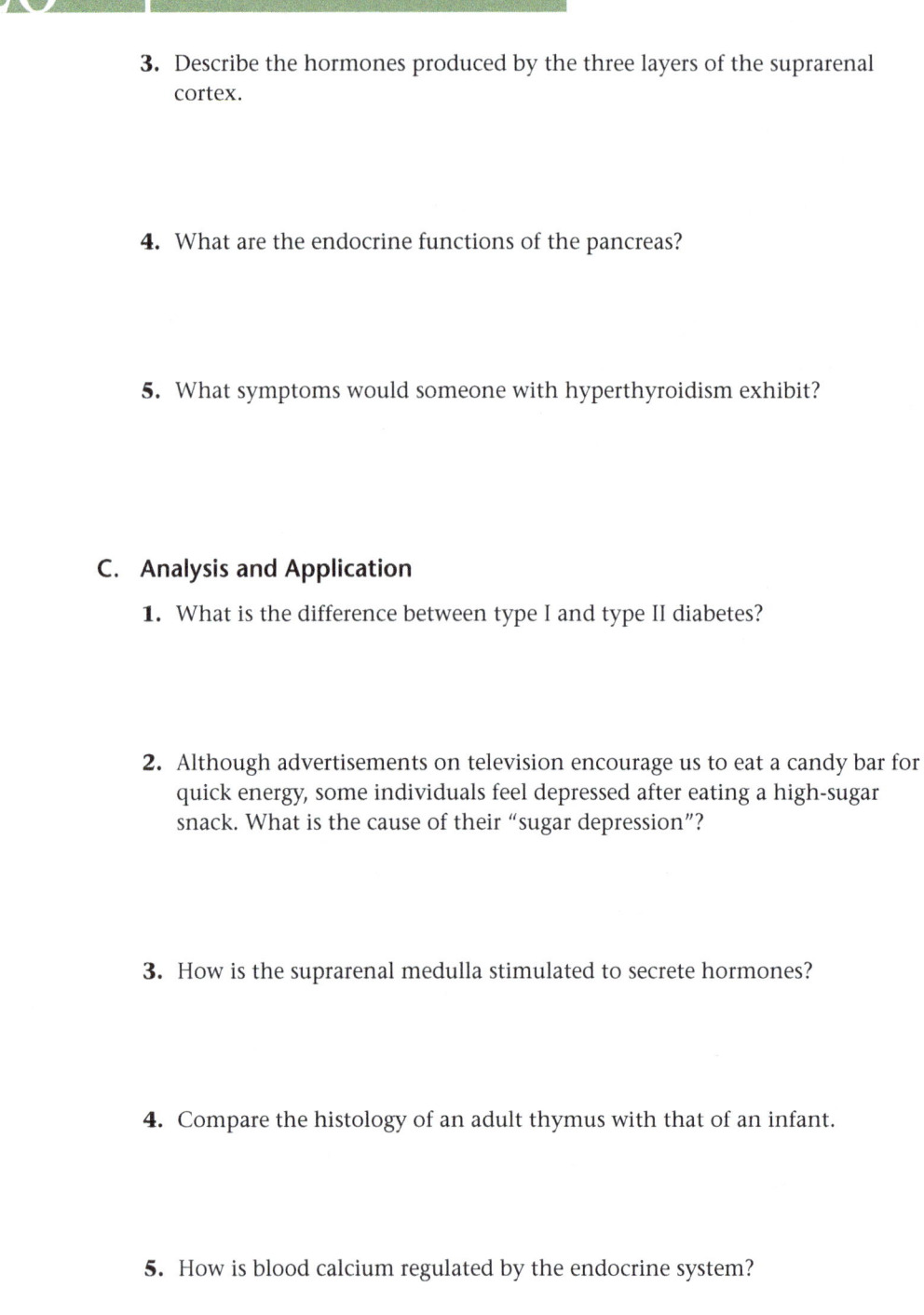

3. Describe the hormones produced by the three layers of the suprarenal cortex.

4. What are the endocrine functions of the pancreas?

5. What symptoms would someone with hyperthyroidism exhibit?

C. **Analysis and Application**

1. What is the difference between type I and type II diabetes?

2. Although advertisements on television encourage us to eat a candy bar for quick energy, some individuals feel depressed after eating a high-sugar snack. What is the cause of their "sugar depression"?

3. How is the suprarenal medulla stimulated to secrete hormones?

4. Compare the histology of an adult thymus with that of an infant.

5. How is blood calcium regulated by the endocrine system?

EXERCISE 21

Blood

OBJECTIVES

On completion of this exercise, you should be able to:

- List the functions of blood.
- Distinguish each type of blood cell on a blood-smear slide.
- Describe the antigen-antibody reactions of the ABO and Rh blood groups.
- Safely collect a blood sample using the blood lancet puncture technique.
- Safely type a sample of blood to determine the ABO and Rh blood types.
- Perform a hematocrit test.
- Describe how to discard blood-contaminated wastes properly.

LAB ACTIVITIES

1. Composition of Whole Blood 337
2. ABO and Rh Blood Groups 342
3. Hematocrit (Packed Red Cell Volume) 346

Blood is a fluid connective tissue that flows through blood vessels of the cardiovascular system. Blood consists of cells and cellular pieces, collectively called the **formed elements**, carried in an extracellular fluid called blood **plasma** (PLAZ-muh). Blood has many functions. It controls the chemical composition of all interstitial fluid by regulating pH and electrolyte levels. It supplies trillions of cells with life-giving oxygen, nutrients, and regulating molecules. Some of its formed elements protect the body from invasion by foreign organisms, such as bacteria, and other formed elements manufacture substances needed for defense against specific biological and chemical threats. In response to injury, blood has the ability to change from a liquid to a gel so as to clot and stop bleeding.

LAB ACTIVITY 1 Composition of Whole Blood

A sample of blood is approximately 55% plasma and 45% formed elements (Figure 21.1). Plasma is 92% water and contains proteins that regulate the osmotic pressure of blood, proteins for clotting, and **antibodies**, the immune system proteins that protect the body from invading pathogens and molecules, collectively referred to as **antigens**. Electrolytes, hormones, nutrients, and some blood gases are transported in the blood plasma. The formed elements are organized into three groups of cells and pieces of cells: red blood cells, white blood cells, and platelets. When stained, each group is easy to identify with a microscope. The reddish cells are erythrocytes, the cells that have visible nuclei are leukocytes, and the small cell fragments between the erythrocytes and leukocytes are platelets.

Red blood cells (RBCs), also called **erythrocytes** (e-RITH-rō-sīts), are red and lack a nucleus. The most abundant of all blood cells, RBCs are biconcave discs that are noticeably thin in the center (Figure 21.2). In a laboratory slide, the thin central section of each disc is not as deeply stained as the surrounding rim. The biconcave shape gives each RBC more surface area than a flat-faced disc would have, an important feature that allows rapid gas exchange between the blood and

338 EXERCISE 21 Blood

PLASMA COMPOSITION

Plasma proteins	7%
Other solutes	1%
Water	92%

Transports organic and inorganic molecules, formed elements, and heat

Components of plasma

Sample of whole blood

consists of

Plasma (46–63%)

+

Formed elements (37–54%)

FORMED ELEMENTS

Platelets	0.1%
White blood cells	
Red blood cells	99.9%

Formed elements of blood

Figure 21.1 The Composition of Whole Blood
Whole blood is composed of a liquid portion, plasma, and a solid portion that comprises three groups of blood cells collectively called the formed elements.

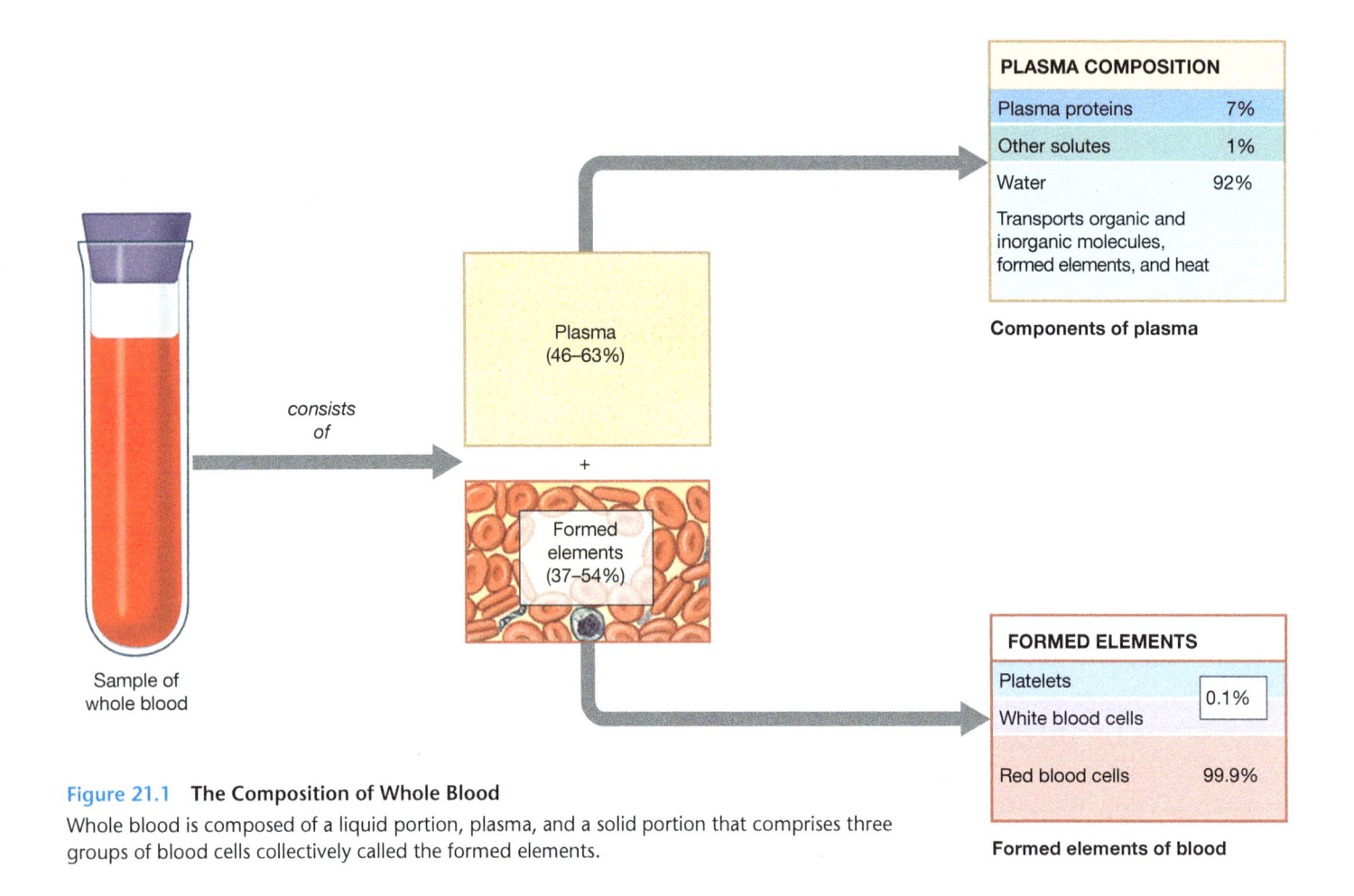

the tissues of the body. Their shape also allows RBCs to flex and squeeze through narrow capillaries.

The major function of RBCs is to transport blood gases. They pick up oxygen in the lungs and carry it to the cells of the body. While supplying the cells with oxygen, the blood acquires carbon dioxide from the cells. The plasma and RBCs convey the carbon dioxide to the lungs for removal during exhalation. To accomplish the task of gas transport, each RBC contains millions of hemoglobin (Hb) molecules. **Hemoglobin** (HĒ-mō-glō-bin) is a complex protein molecule that contains as part of its structure four iron atoms that bind loosely to oxygen and carbon dioxide molecules.

The second type of formed element is **white blood cells** (WBCs), also called **leukocytes** (LOO-kō-sīts). A main feature of WBCs is their nucleus, which takes a very dark stain and is often branched into two or more lobes (Figure 21.3). WBCs lack hemoglobin and therefore do not transport blood gases. They can pass between the endothelial cells of capillaries and enter the interstitial spaces of tissues. Most WBCs are **phagocytes**, scavenger cells that engulf foreign bodies and other unwanted materials circulating in the blood and destroy them, and are therefore part of the immune system.

There are two broad classes of WBCs: granular and agranular. The **granular leukocytes**, also called **granulocytes**, have granules in their cytoplasm and include the neutrophils, eosinophils, and basophils. **Agranular leukocytes**, which include the monocytes and lymphocytes, have few cytoplasmic granules.

Neutrophils (NOO-trō-filz) are the most common leukocytes and account for up to 70% of the WBC population. These granular leukocytes are also called **polymorphonuclear** (pol-ē-mor-fō-NOO-klē-ar) **leukocytes** because the nuclei are complex and branch into two to five lobes. In addition to a dark-staining nu-

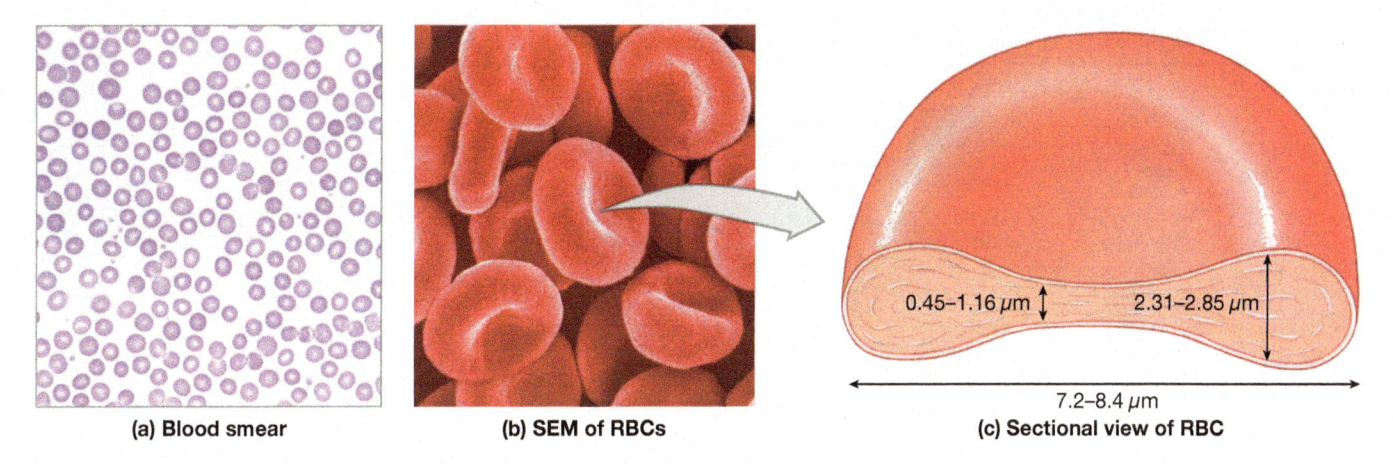

Figure 21.2 The Anatomy of Red Blood Cells
(a) When viewed in a standard blood smear, red blood cells appear two-dimensional because they are flattened against the surface of the slide. (b) A scanning electron micrograph of red blood cells reveals their three-dimensional structure. (SEM × 7000) (c) A sectional view of a mature red blood cell that shows the normal ranges for its dimensions.

cleus, neutrophils have many small cytoplasmic granules that stain pale purple, visible in Figure 21.3b.

Neutrophils are the first leukocytes to arrive at a wound site to begin infection control. They release cytotoxic chemicals and phagocytize (engulf and destroy) invading pathogens. They also release hormones called **cytokines** that attract other phagocytes, such as eosinophils and monocytes, to the site of injury. Neutrophils are short-lived and survive in the blood for up to ten hours. Active neutrophils in a wound may live only 30 minutes until they succumb to the toxins released by the pathogens they have ingested.

About the same size as neutrophils, the granular leukocytes known as **eosinophils** (ē-ō-SIN-ō-filz) are identified by the presence of medium-sized granules that stain orange-red, as shown in Figure 21.3c. The nucleus is conspicuously segmented into two lobes. Eosinophils are phagocytes that engulf bacteria and other microbes that the immune system has coated with antibodies. They also contribute to decreasing the inflammatory response at a wound or site of infection. Approximately 3% of the circulating WBCs are eosinophils.

Basophils (BĀ-sō-filz), the third type of granular leukocyte, constitute less than 1% of the circulating WBCs. They have large cytoplasmic granules that stain dark blue. The granules are so large and numerous that the nucleus is obscured, as illustrated in Figure 21.3d. Smaller than neutrophils and eosinophils, basophils are sometimes difficult to locate on a blood-smear slide because relatively few of them are present. They migrate to injured tissues and release histamines, which cause vasodilation, and heparin, which prevents blood from clotting. Mast cells in the tissue respond to these molecules and induce local inflammation.

The agranular **lymphocytes** (LIM-fō-sīts) are the smallest of the WBCs and are approximately the size of an RBC (Figure 21.3e). The distinguishing feature of any lymphocyte is a large nucleus that occupies almost the entire cell, and leave room for only a small halo of pale blue cytoplasm around the edge of the cell. Lymphocytes are abundant in the blood and compose 20% to 30% of all circulating WBCs. Lymphocytes move freely between the blood and the tissues of the body. As their name suggests, they are the main cells that populate lymph nodes, glands, and other lymphatic tissues.

Although several types of lymphocytes exist, they cannot be individually distinguished with a light microscope. Generally, lymphocytes provide immunity from microbes and defective cells by two methods. **T cells** attach to and destroy foreign cells in a cell-mediated response that involves release of cytotoxic chemicals to

340 EXERCISE 21 Blood

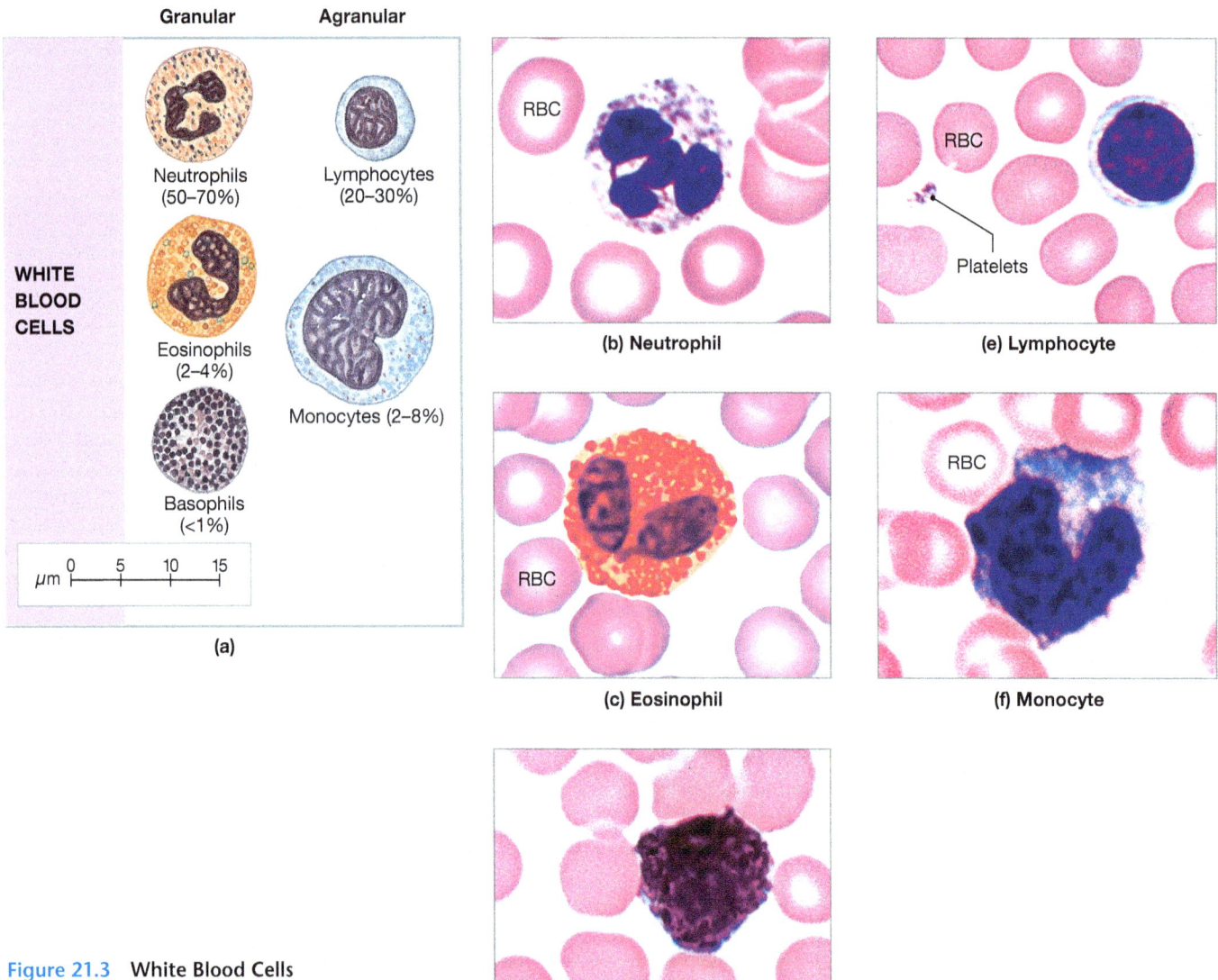

Figure 21.3 White Blood Cells
(a) Comparison of the sizes and abundances of the five types of white blood cells.
(b–f) Enlarged views of the five types.

kill the invaders. The second immunity method uses the lymphocytes known as **B cells**, which become sensitized to a specific antigen, then manufacture and pour antibodies into the blood. The antibodies attach to and help destroy foreign antigens.

Monocytes (MON-ō-sīts) are large, agranular WBCs that contain a dark-staining, kidney-shaped nucleus surrounded by a pale blue cytoplasm (Figure 21.3f). Approximately 2% to 8% of circulating WBCs are monocytes. On a blood-smear slide, monocytes appear roundish and may have small extensions, much like an amoeba. Even though monocytes are agranular leukocytes, materials they ingest, such as phagocytized bacteria and debris, stain and may look like granules under the microscope.

Monocytes are wanderers. They leave the blood by squeezing between the capillary endothelium to patrol the body tissues in search of microbes and worn-out tissue cells. They are second to neutrophils in arriving at a wound site. When neutrophils die from phagocytizing bacteria, the monocytes phagocytize the neutrophils.

Platelets (PLĀT-lets, Figure 21.3e), the third type of formed element, are small cellular pieces produced from the breakdown of **megakaryocytes**, which are large protein-producing cells located in the bone marrow. Platelets lack a nucleus and other organelles. They survive in the blood for a brief time and are involved in blood clotting.

EXERCISE 21 Blood   341

QuickCheck Questions

1.1 What are the three types of formed elements in blood?
1.2 Which is the most abundant type of white blood cell?

1 Materials

- ☐ Compound microscope
- ☐ Immersion oil
- ☐ Human blood-smear slide (Wright's or Giemsa-stained)

Procedures

1. Blood samples are thin and require careful focusing. Bring the sample into focus with the low-power objective. Then use the fine focus knob as you examine individual cells. Notice the abundance of red blood cells. The dark-stained cells are the various white blood cells.

2. Scan the slide at high-dry magnification and locate the different types of WBCs. Note the small platelets between the red and white cells.

3. Use the oil immersion lens to observe the various blood cells. Place a small drop of immersion oil on the coverslip of the slide and gently rotate the oil immersion objective lens so that the tip of the lens becomes covered with the oil. There should be oil, not air, between the lens and the slide. Use the mechanical stage and scan the slide slowly to avoid spreading the oil too thin. When you are finished, it is very important that you correctly clean the oil off the lens and the slide by using a sheet of microscope lens tissue to gently wipe the oil off the lens and slide. Then use a fresh sheet of tissue with a drop of lens cleaner and wipe the lens and slide clean of any remaining oil. To prevent damage to the lens, do not saturate the lens with the cleaner.

4. Sketch each blood cell in the space provided. ■

Neutrophil

Eosinophil

Basophil

Lymphocyte

Monocyte

Platelet

LAB ACTIVITY 2: ABO and Rh Blood Groups

Your blood type is inherited from your parents' genes, and it does not change during your lifetime. Each blood type is a function of the presence or absence of specific antigen molecules on the surface of the red blood cells. [The antigens important in blood types are also called *agglutinogens* (a-gloo-TIN-Ō-jenz), but we shall use the term *antigens*.] The antigens are like cellular nametags that inform your immune system that your red blood cells belong to "self" and are not "foreign."

Blood also contains specialized antibody molecules called *agglutinins* (a-GLOO-ti-ninz). The antibodies and antigens in an individual's blood do not interact with one another, but the antibodies do react with antigens of foreign red blood cells and cause the cells to burst, hence the need for blood type matching prior to a blood transfusion.

More than 50 surface antigens and blood groups occur in the human population. In this activity, you study the two most common, the ABO group and the Rh group. Each blood group is controlled by a different gene, and your ABO blood type does not influence your Rh blood type. Table 21.1 shows the distribution of these two blood groups in the human population.

ABO Blood Group

There are four blood types in the **ABO blood group:** A, B, AB, and O (Figure 21.4). Two surface antigens, A and B, occur in different combinations that determine the blood type. Type A blood has the A surface antigen on its membrane, type B blood has the B surface antigen, type AB blood has both A and B surface antigens, and type O blood has neither. Which antibodies are present in blood depends on type. Type A blood contains anti-B antibodies, which attack red blood cells that carry B surface antigens. Type B blood contains anti-A antibodies to defend against cells that carry A surface antigens. Type AB blood contains no antibodies, and type O contains both anti-A and anti-B antibodies.

The anti-B antibodies in type A blood do not react with the type A surface antigens but do react with the B surface antigens present in blood types B and AB.

Table 21.1 Differences in Blood Group Distribution

Population	O	A	B	AB	Rh⁺
U.S. (Average)	46	40	10	4	85
African-American	49	27	20	4	95
Caucasian	45	40	11	4	85
Chinese-American	42	27	25	6	100
Filipino-American	44	22	29	6	100
Hawaiian	46	46	5	3	100
Japanese-American	31	39	21	10	100
Korean-American	32	28	30	10	100
Native North American	79	16	4	<1	100
Native South American	100	0	0	0	100
Australian Aborigine	44	56	0	0	100

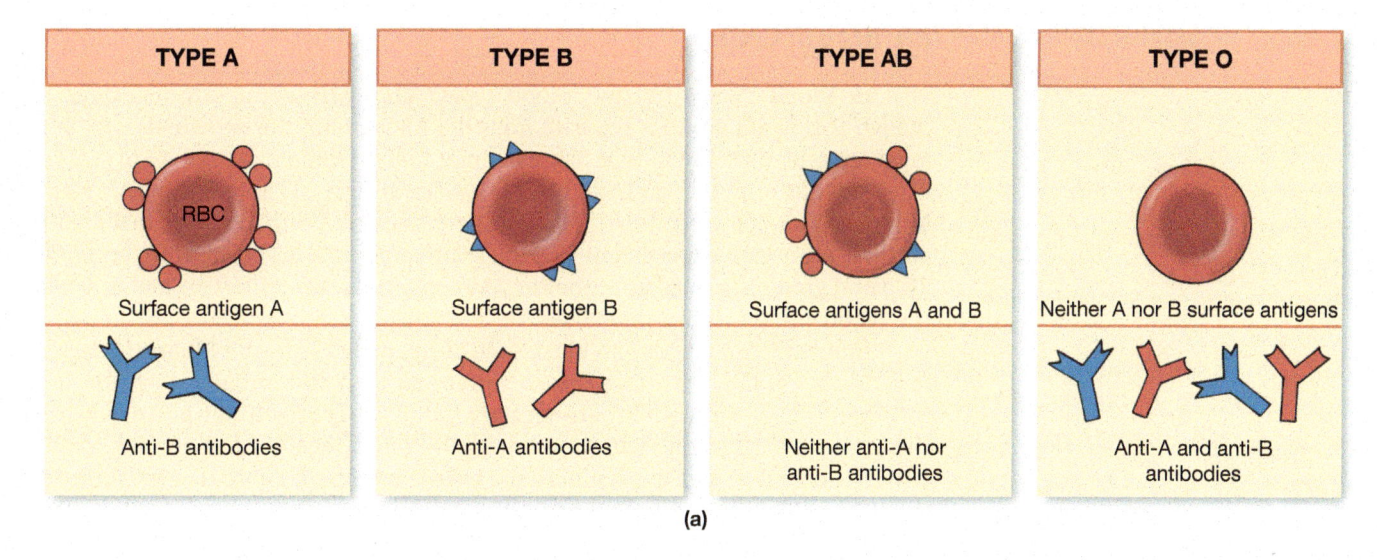

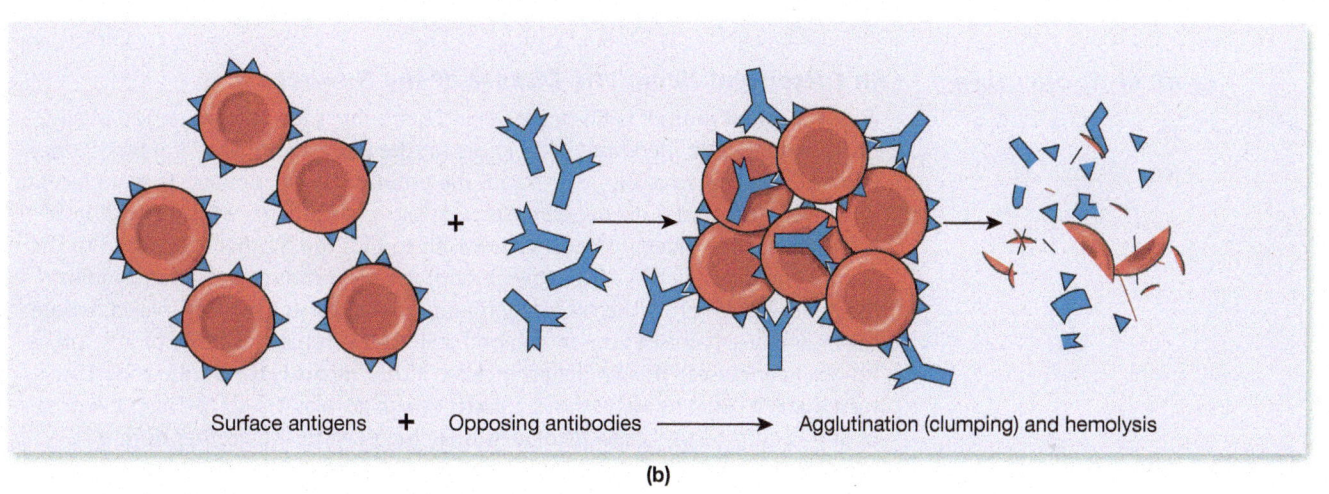

Figure 21.4 Blood Typing and Reactions
Blood type is determined by the kind of surface antigens on the RBC membrane. (a) The plasma contains antibodies that react with foreign surface antigens. (b) Antibodies that encounter their target surface antigens induce agglutination and hemolysis of the affected RBCs.

The same is true for type B blood: the anti-A antibodies do not react with the B surface antigens but do destroy the cells that carry A surface antigens in blood types A and AB. Because AB blood contains neither anti-A antibodies nor anti-B antibodies, it does not react with blood of other types. People with AB blood are called *universal acceptors* because they lack antibodies, and can accept blood of any type in a transfusion. Although surface antigens are absent in type O blood, it has both anti-A and anti-B antibodies. Type O blood has no surface antigens that act as nametags, so it can be transfused to all blood types. People with type O blood are called *universal donors*.

To determine blood type, the presence of antigens is detected by adding to a blood sample drops of **antiserum** that contains either anti-A or anti-B antibodies. The antibodies in the antiserum react with the corresponding surface antigens in the sample. The blood *agglutinates* (forms clumps of solid material that settle out from the plasma) as the antibodies react with the surface antigens.

Rh Blood Group

The **Rh blood group** has two blood types, **Rh-positive** and **Rh-negative**. Although this blood group is separate from the ABO group, the two are usually used together to identify blood type. For example, a blood sample may be A+ or A–. The Rh group has only one antigen, the Rh surface antigen (D antigen), plus a single Rh antibody designated anti-D. The D antigen is present only on RBCs that are Rh-positive; Rh-negative blood cells lack the D antigen. (In case you are wondering, the Rh blood group is named after the rhesus monkey, the laboratory animal in which this blood group was first discovered.)

Rh-positive blood has the Rh surface antigen and lacks the Rh antibody. Rh-negative blood does not have the Rh surface antigen and initially does not have the Rh antibody. However, if Rh-negative blood is exposed to Rh-positive blood, the Rh-negative person's immune system becomes sensitized to the Rh surface antigen and subsequently produces the anti-D antibody. This becomes clinically significant in cases of pregnancy with Rh incompatibilities between mother and fetus.

Clinical Application

Rh Factor and Hemolytic Disease of the Newborn

If an expectant mother is Rh-negative and her baby is Rh-positive, a potentially life-threatening Rh incompatibility exists for the baby. Normally, fetal blood does not mix with maternal blood. Instead, the umbilical cord connects to the placenta, where fetal capillaries exchange gases, wastes, and nutrients with the mother's blood. If internal bleeding occurs, however, so that the mother is exposed to the D antigens in her baby's Rh-positive blood, she will produce anti-D antibodies. These antibodies cross the placental membrane and enter the fetal blood, where they hemolyze (rupture) the fetal blood cells of this fetus and those of any future Rh-positive fetuses. This Rh action is called either **hemolytic disease of the newborn** or **erythroblastosis fetalis** (e-rith-rō-blas-TŌ-sis fē-TAL-is). A dosage of anti-D antibodies, called **RhoGam**, may be given to the mother during pregnancy and after delivery to destroy any Rh-positive fetal cells in her blood. This treatment prevents her from developing anti-D antibodies.

QuickCheck Questions

2.1 What are the two major blood groups used to identify blood type?

2.2 What surface antigens does type A blood have?

Safety Alert: Handling Blood

1. Some infectious diseases are spread by contact with blood. Follow all instructions carefully and protect yourself by wearing gloves and working only with your own blood.
2. Materials contaminated with blood must be disposed of properly. Your instructor will inform you of methods for disposing of lancets, slides, prep pads, and toothpicks.

Your instructor may ask for a volunteer to "donate" blood in order to demonstrate how blood typing is done. Alternatively, many biological supply companies sell simulated blood-typing kits that contain a blood-like solution and antisera. These kits contain no human or animal blood products and safely show the principles of typing human blood.

2 Materials

- ☐ Hand soap
- ☐ Paper towels
- ☐ Gloves
- ☐ Safety glasses
- ☐ Disposable sterile blood lancet
- ☐ Disposable sterile alcohol prep pad
- ☐ Disposable blood typing plate or sterile microscope slide
- ☐ Wax pencil (if using microscope slide)
- ☐ Toothpicks
- ☐ Anti-A, anti-B, and anti-D blood-typing antisera
- ☐ Warming box
- ☐ Biohazardous waste container
- ☐ Bleach solution in spray bottle (optional)

Procedures

Sample Collection

1. If you are using a slide, use the wax pencil to draw three circles across the width of the slide. Label the circles "A", "B", and "D." If you are using a typing plate, label three of the depressions "A", "B", and "D."

2. Wash both hands thoroughly with soap, and then dry them with a clean paper towel. Obtain an additional paper towel on which to place blood-contaminated instruments while collecting a blood sample. Wear gloves and safety glasses while collecting and examining blood. If collecting a sample from yourself, wear a glove on the hand used to hold the lancet.

3. Open a sterile alcohol prep pad, and clean the tip of the index finger from which the blood will be drawn. Be sure to thoroughly disinfect the entire fingertip, including the sides. Place the used prep pad on the paper towel.

4. Open a sterile blood lancet to expose only the sharp tip. Do not use an old lancet, even if it was used on one of your own fingers. Use the sterile tip *immediately* so that there is no time for it to inadvertently become contaminated.

5. With a swift motion, jab the point of the lancet into the lateral surface of the fingertip. Place the used lancet on the paper towel until it can be disposed of in a biohazard container.

6. Gently squeeze a drop of blood either into each depression on the blood-typing plate or into the circles on the slide. If necessary, slowly "milk" the finger to work more blood out of the puncture site. ■

ABO and Rh Typing

1. Add a drop of anti-A antiserum to the sample labeled A, being very careful not to allow blood to touch (and thereby contaminate) the tip of the dropper. Repeat the process by adding a drop of anti-B antiserum to the B sample and a drop of anti-D antiserum to the D sample.

2. Immediately and gently mix each drop of antiserum into the blood with a clean toothpick. To prevent cross-contamination, use a separate, clean toothpick for each sample. Place all used toothpicks on the paper towel until they can be disposed of in a biohazard container.

3. Place the slide or typing plate on the warming box and agitate the samples by rocking the box carefully back and forth for two minutes. *Note:* The anti-D agglutination reaction is often weaker and less easily observed than the anti-A and anti-B agglutination reactions. A microscope may help you observe the anti-D reaction.

4. Examine the drops for any agglutination visible with the unaided eye and compare your samples with Figure 21.5. Agglutination results when the antibodies in the antiserum react with the matching antigen on the red blood cells. For example, if blood agglutinates with the anti-A antiserum and the anti-D antiserum, the blood type is A-positive.

5. Record your results in the first blank row of Table 21.2 in the Lab Report. In each cell of the table, indicate "yes" or "no" for the presence of agglutination.

6. Collect blood-typing data from three classmates to compare agglutination responses among blood types. How does the distribution of blood types in your four-person sample in Table 21.2 compare with the distribution of types given in Table 21.1?

Anti-A	Anti-B	Anti-D	Blood type
			A⁺
			B⁺
			AB⁺
			O⁻

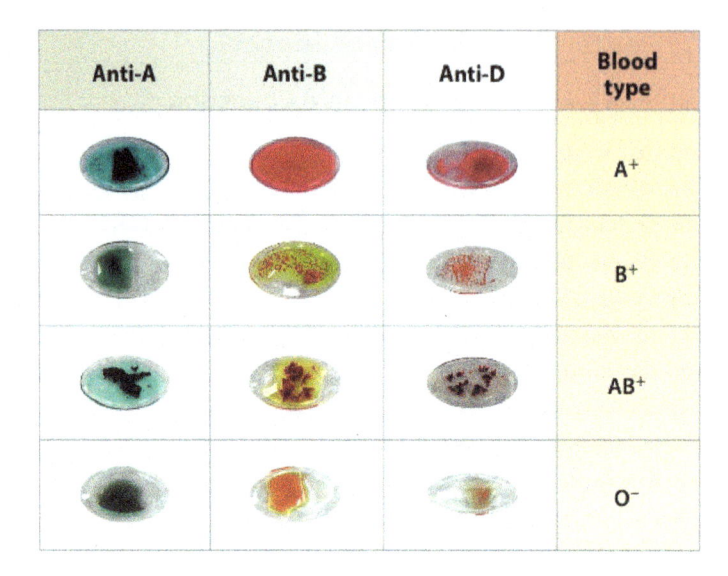

Figure 21.5 Blood Type Testing
Test results for blood samples from four individuals. Drops are taken from the sample at the left and mixed with antiserum solutions that contain antibodies to the surface antigens A, B, and D (Rh). Clumping occurs when the sample contains the corresponding surface antigen(s).

Disposal of Materials and Disinfection of Work Space

1. Dispose of all blood-contaminated materials in the appropriate biohazard box. A box for sharp objects may be available to dispose of the lancets, toothpicks, and microscope slides.
2. Your instructor may ask you to disinfect your workstation with a bleach solution. If so, wear gloves and safety glasses while wiping the surfaces clean.
3. Lastly, remove your gloves and dispose of them in the biohazard box. Remember to wash your hands after disposing of all materials.

LAB ACTIVITY 3 Hematocrit (Packed Red Cell Volume)

The **hematocrit** (he-MA-tō-krit), or packed cell volume (PCV), test measures the volume of packed formed elements in a given volume of blood. Because RBCs far outnumber all the other formed elements, the test mainly measures their volume. Hematocrit results provide information regarding the oxygen-carrying capacity of the blood. A low hematocrit value indicates that the blood has fewer RBCs to transport oxygen. Average hematocrit values range from 40% to 54% in males and from 37% to 47% in females.

QuickCheck Questions

3.1 What does a hematocrit test measure?

3.2 What is the average hematocrit range for males? For females?

3 Materials

- ☐ Hand soap
- ☐ Paper towels
- ☐ Gloves
- ☐ Safety glasses
- ☐ Paper towels
- ☐ Disposable sterile blood lancet
- ☐ Disposable sterile alcohol prep pads
- ☐ Sterile heparinized capillary tubes
- ☐ Seal-easy clay
- ☐ Bleach solution in spray bottle
- ☐ Microcentrifuge
- ☐ Tube reader
- ☐ Biohazardous waste disposal container

Procedures

1. Review the safety tips given in Lab Activity 2.
2. Follow steps 2 through 5 of Lab Activity 2, "Sample Collection", to obtain a blood sample.
3. *Gently* squeeze a drop of blood out of your finger. (Squeeze gently because excess pressure forces interstitial fluid into the blood, and the presence of this fluid may alter your hematocrit reading. If you are having difficulty obtaining a drop, use a clean, sterile lancet to lance your finger again in a different spot.)
4. Place a sterile heparinized capillary tube on the drop of blood. Orient the open end of the tube downward, as shown in Figure 21.6, to allow the blood to flow into the tube. Fill the tube at least two-thirds full with blood.
5. Carefully seal one end of the tube by dipping it into the seal-ease clay as shown in Figure 21.7. Do not force the delicate capillary tube into the clay, for it may break and cause you to jam glass into your hand. Instead, hold the tube the way you hold a pencil for writing, with your thumb and index finger close to the end where the blood has accumulated. Then gently turn the tube while pressing it into the clay. Leave the other end unplugged.
6. Clean any blood off the clay with the bleach solution and a paper towel.
7. Set the tube in the microcentrifuge with the clay end toward the outer margin of the chamber. Because the centrifuge spins at high speeds, the chamber must be balanced by placing tubes evenly in the chamber. Counterbalance your capillary tube by placing another sample directly across from yours. An empty tube sealed at one end with clay may be used if another student's sample is not available.
8. Screw the inner cover on with the centrifuge wrench. Do not overtighten the lid. Close the outer lid and push the latch in.
9. Set the timer to four to five minutes, and allow the centrifuge to spin. Do not attempt to open or stop the centrifuge while it is turning. Always keep loose hair and clothing away from the centrifuge.

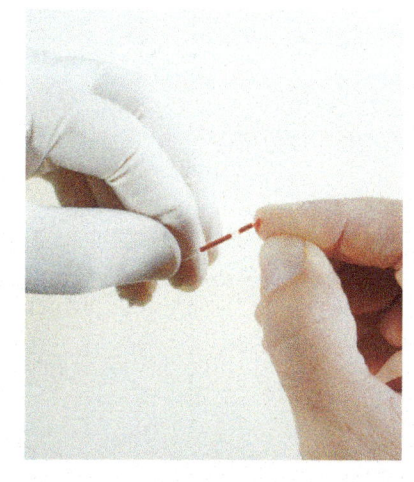

Figure 21.6 Filling a Capillary Tube with Blood
A capillary tube is held slanting downward at the lance site to draw a drop of blood into the tube.

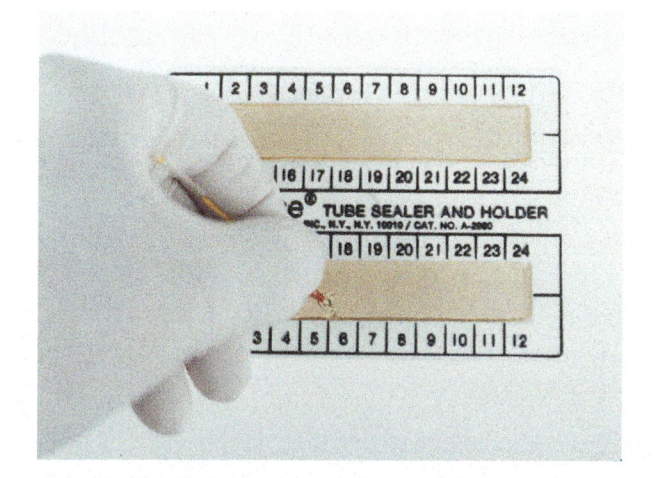

Figure 21.7 Plugging a Capillary Tube with Clay
To avoid breaking the capillary tube, hold the tube at the end nearest the clay and gently press the tip into the clay.

10. After the centrifuge turns off and stops spinning, open the lid and the inner safety cover to remove the capillary tube. Your blood sample should have clear plasma at one end of the tube and packed RBCs at the other end.

11. Place the capillary tube in the tube reader. Because there are a variety of tube readers, your instructor will demonstrate how to use the reader in your laboratory.

12. Record your hematocrit measurement in Table 21.2 in the Lab Report. Is your hematocrit reading within the normal range?

13. Describe the appearance of your blood plasma: _____
    _____
    _____

14. Dispose of all used materials as described in Lab Activity 2, "Disposal of Materials and Disinfection of Work Space." ∎

EXERCISE 21

LAB REPORT

Name _____
Date _____
Section _____

Blood

A. Experimental Data and Observations

Complete Table 21.2 with data collected during the blood typing and hematocrit experiments.

Table 21.2 Blood Typing Data

Student	Anti-A Antiserum Reaction	Anti-B Antiserum Reaction	Anti-D Antiserum Reaction	Blood Type	Hematocrit Reading

B. Matching

Match each term in the left column with its correct description from the right column.

_____ 1. erythrocyte
_____ 2. polymorphonuclear leukocyte
_____ 3. leukocyte
_____ 4. antibody
_____ 5. type A blood
_____ 6. Rh-positive blood
_____ 7. blood cell that stains red-orange
_____ 8. type B blood
_____ 9. Rh-negative blood
_____ 10. antigen

A. eosinophil
B. molecule on RBC surface
C. has surface antigens A and anti-B antibodies
D. has Rh antigens
E. neutrophil
F. lacks Rh antigens
G. red blood cell
H. reacts with a membrane molecule
I. has surface antigens B and anti-A antibodies
J. white blood cell

C. Drawing

Complete each typing slide in Figure 21.8 by indicating with pencil dots where agglutination occurs.

349

EXERCISE 21 — LAB REPORT

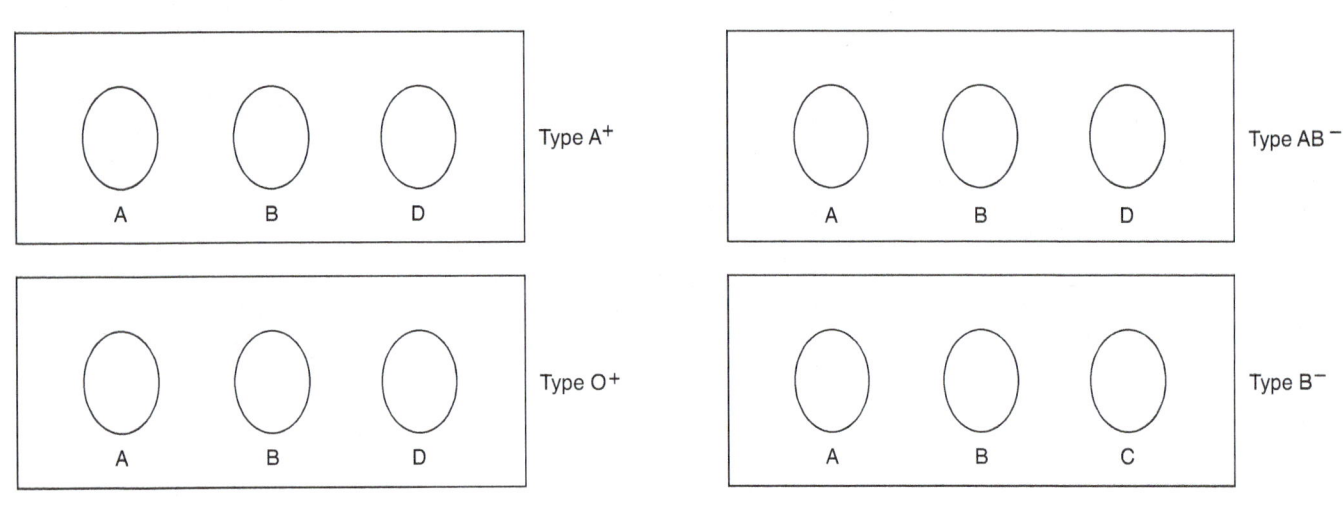

Figure 21.8 Simulated Blood Typing Plates

Type A⁺ — wells A, B, D
Type AB⁻ — wells A, B, D
Type O⁺ — wells A, B, D
Type B⁻ — wells A, B, C

D. Short-Answer Questions

1. What is the main function of RBCs?

2. List the five types of WBCs and describe the function of each.

3. Describe how to do a hematocrit test. What are the average hematocrit values for males and females?

4. Describe how to type blood to detect the ABO and Rh blood groups.

E. Analysis and Application

1. Describe what happens when type A blood is transfused into the blood of someone with type B blood.

2. What happens in the blood of an Rh-negative individual who is exposed to Rh-positive blood?

3. How could you easily determine if two blood samples are compatible?

EXERCISE 22

The Heart

OBJECTIVES

On completion of this exercise, you should be able to:

- Describe the external and internal anatomy of the heart.
- Identify and discuss the function of the valves of the heart.
- Identify the major blood vessels of the heart.
- Trace a drop of blood through the pulmonary circuit and the systemic circuit.
- Identify the vessels of coronary circulation.
- Describe the anatomy of a sheep heart.

LAB ACTIVITIES

1. Heart Wall   351
2. External and Internal Anatomy of the Heart   354
3. Coronary Circulation   359
4. Sheep Heart Dissection   360

The cardiovascular system consists of blood; the heart, which pumps blood through the system; and all the blood vessels through which the blood flows. **Arteries** are the blood vessels that carry blood away from the heart, and **veins** are the blood vessels that return blood to the heart. In addition to arteries and veins, the cardiovascular system also contains small-diameter blood vessels called **capillaries**. It is across the walls of capillaries that gases, nutrients, and cellular waste products enter and exit the blood. The heart beats approximately 100,000 times daily to send blood flowing into thousands of miles of blood vessels, providing the body's cells with nutrients, regulating the amounts of substances and gases in the cells, and removing waste products from them. All organ systems of the body depend on the cardiovascular system. Damage to the heart often results in widespread disruption of homeostasis.

Your laboratory studies in this exercise include the histology of cardiac muscle tissue, external and internal heart anatomy, and circulation of blood through the pulmonary and systemic circuits of the cardiovascular system. The dissection of a sheep heart will reinforce your observations of the human heart.

LAB ACTIVITY 1 Heart Wall

Figure 22.1 illustrates the location of the heart in the mediastinum (mē-dē-as-TĪ-num) of the thoracic cavity. Blood vessels join the heart at the **base**, which is positioned medially in the mediastinum. Because the left side of the heart has more muscle mass than the right side, the **apex** at the inferior tip of the heart is more on the left side of the thoracic cavity. (Note from Figure 22.1a that the heart's base and apex are "upside down" relative to what we usually mean by those words: in the heart, the base is superior to the apex.) Within the mediastinum, the heart is surrounded by the **pericardial** (per-i-KAR-dē-al) **cavity**

352 EXERCISE 22 The Heart

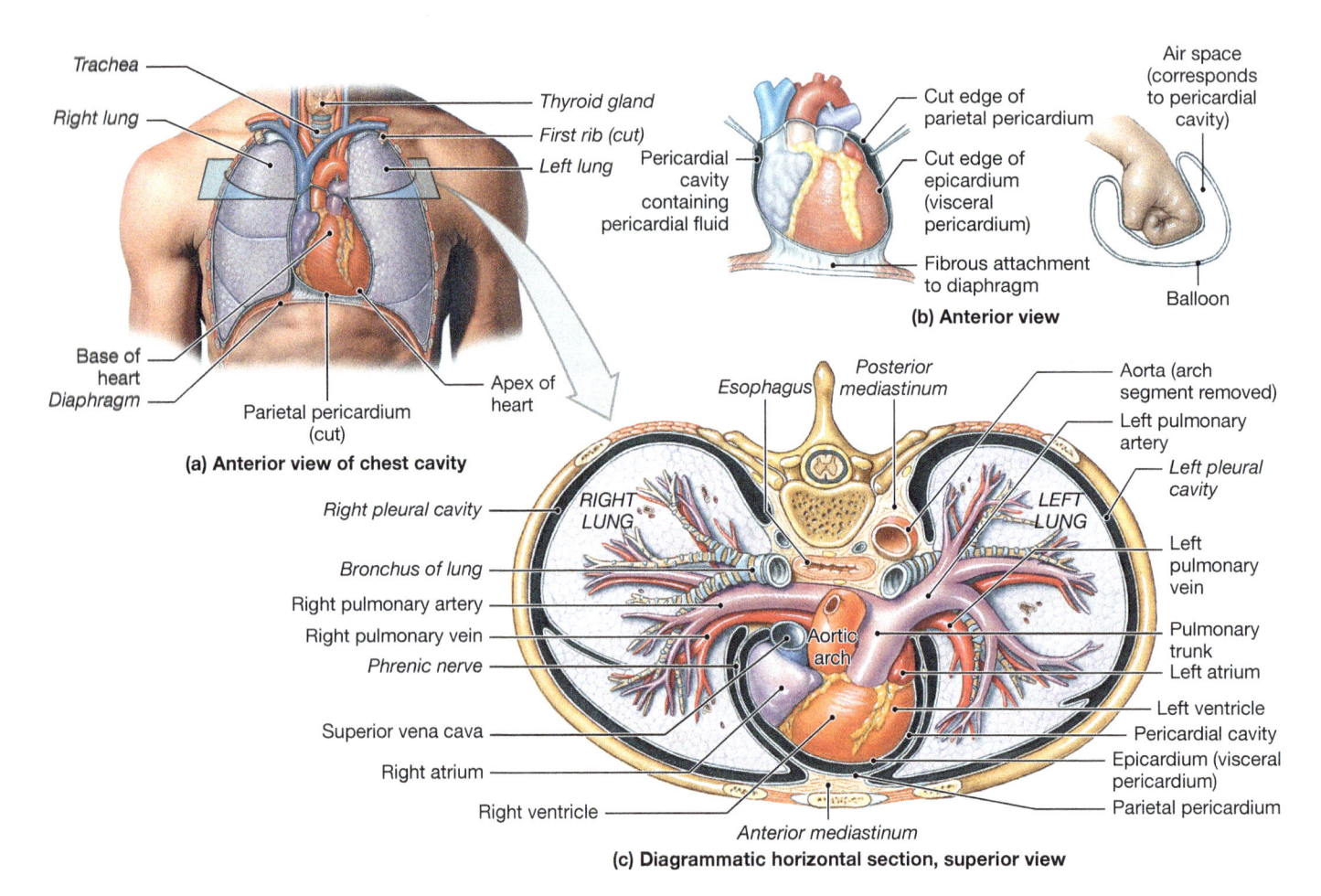

Figure 22.1 Location of the Heart in the Thoracic Cavity
The heart is situated in the anterior part of the mediastinum, immediately posterior to the sternum. (a) An anterior view of the open chest cavity that shows the position of the heart and major vessels relative to the lungs. (b) The relationship between the heart and the pericardial cavity; compare with the fist-and-balloon example. (c) A diagrammatic view of the heart and other organs in the mediastinum with the tissues of the lungs removed to reveal the blood vessels and airways.

formed by the **pericardium**, the serous membrane of the heart. The pericardial cavity contains **serous fluid** to reduce friction during muscular contraction. The superficial **parietal pericardium** attaches the heart in the mediastinum, and the deep **visceral pericardium**, or **epicardium**, covers the heart surface and is considered the outermost layer of the cardiac wall.

The heart wall is organized into three layers: epicardium, myocardium, and endocardium (Figure 22.2). The epicardium is the same structure as the visceral pericardium, as just noted. The **myocardium** constitutes most of the heart wall and is composed of **cardiac muscle cells**, also called **cardiocytes**. Each cardiac muscle cell is **uninucleated** (contains a single nucleus) and branched. Cardiac muscle cells interconnect at their branches via junctions called **intercalated** (in-TER-ka-lā-ted) **discs** (Exercise 4). Deep to the myocardium is the **endocardium**, a thin layer that lines the chambers of the heart. The endocardium is composed of endothelial tissue that rests on a layer of areolar connective tissue.

QuickCheck Questions

1.1 What are the three layers of the heart wall, from superficial to deep?

1.2 How are cardiac muscle cells connected to one another?

EXERCISE 22 The Heart   353

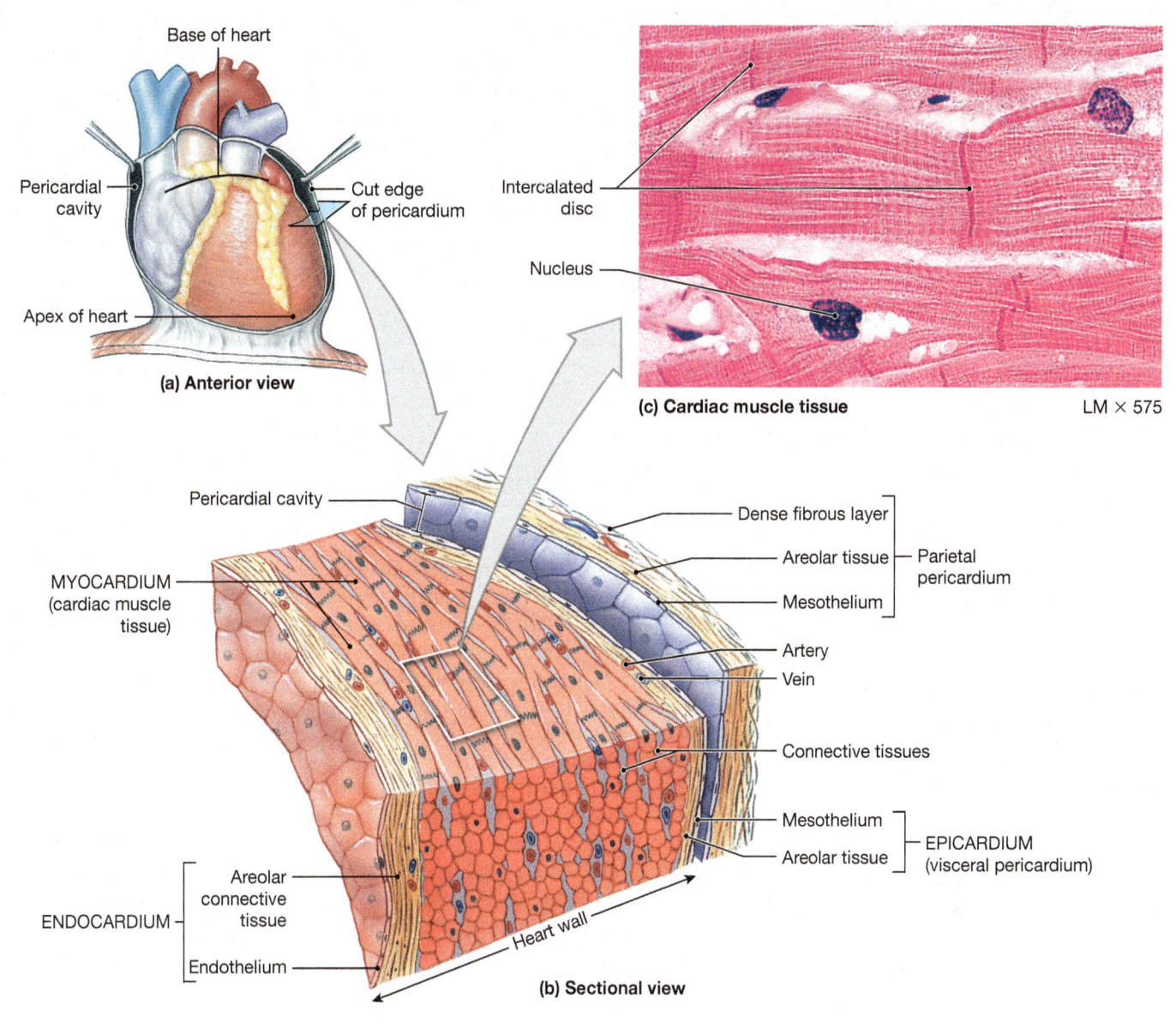

Figure 22.2 The Heart Wall
(a) Anterior view of the heart. (b) Diagrammatic section through the heart wall that shows the relative positions of the epicardium, myocardium, and endocardium. (c) Microphotograph of cardiac muscle tissue that shows intercalated discs. (LM × 575)

1 Materials

- ☐ Heart model and specimens
- ☐ Compound microscope
- ☐ Prepared slide of cardiac muscle

Procedures

1. Review the heart anatomy in Figures 22.1 and 22.2.
2. Identify the layers of the heart wall on the heart model and specimens.
3. With the microscope at low power, examine the microscopic structure of cardiac muscle, using Figure 22.2c for reference. Increase the magnification to high and locate several cardiac muscle cells. Note the single nucleus in each cell and where each cell branches into two arms. Intercalated discs are dark-stained lines where cardiac muscle cells connect together.
4. Sketch several cardiac muscle cells and intercalated discs in the space provided. ■

Cardiac muscle cells

LAB ACTIVITY 2 External and Internal Anatomy of the Heart

The heart is divided into right and left sides, with each side having an upper and a lower chamber (Figure 22.3). The upper chambers are the **right atrium** (A-trē-um; chamber) and the **left atrium**, and the lower chambers are the **right ventricle** (VEN-tri-kul; little belly) and the **left ventricle**. The atria are receiving chambers and fill with blood that is returning to the heart in veins. Blood in the atria flows into the ventricles, the pumping chambers, which squeeze their walls together to pressurize the blood and eject it into two large arteries for distribution to the lungs and body tissues. Most of the blood in the atria flows into the ventricles because of pressure and gravity. Before the ventricles contract, the atria contract and "top off" the ventricles.

For a drop of blood to complete one circuit through the body, it must be pumped by the heart twice—through the **pulmonary circuit**, which directs deoxygenated blood to the lungs; and through the **systemic circuit**, which takes oxygenated blood to the rest of the body (Figure 22.3). Each circuit delivers blood to a series of arteries, then capillaries, and finally veins that drain into the opposite side of the heart.

The right ventricle is the pump for the pulmonary circuit and ejects deoxygenated blood into the large artery called the **pulmonary trunk**. (Remember that, although this blood vessel transports deoxygenated blood, it is an artery because it carries blood away from the heart.) The pulmonary trunk branches into right and left **pulmonary arteries** that enter the lungs and continue to branch ultimately into pulmonary capillaries, where gas exchange occurs to convert the deoxygenated blood to oxygenated blood. The pulmonary circuit ends where four **pulmonary veins** return the oxygenated blood to the left atrium. Not all individuals have four pulmonary veins; some individuals have only three, and others have five.

The myocardium of the left ventricle is thicker than the myocardium of the right ventricle. The thicker left ventricle is the workhorse of the systemic circuit; it ejects oxygenated blood into the **aorta** with enough pressure to deliver blood to the entire body and have it flow back to the heart to complete the pathway. The aorta is the main artery from which all major **systemic arteries** arise. The systemic arteries enter the organ systems, and exchange of gases, nutrients, and waste products occurs in the **systemic capillaries**. **Systemic veins** drain the systemic capillaries and transport the deoxygenated blood to the heart. The systemic veins merge into the two largest systemic veins: the **superior vena cava** (VĒ-na KĀ-vuh) and the **inferior vena cava**, which empty the deoxygenated blood into the right atrium. The cycle of blood flow repeats as the deoxygenated blood enters the right ventricle and is pumped through the pulmonary circuit to the lungs to pick up oxygen for the next journey through the systemic circuit.

Figure 22.3 Generalized View of the Pulmonary and Systemic Circuits

Blood flows through separate pulmonary and systemic circuits, driven by the pumping of the heart. Each circuit begins and ends at the heart and contains arteries, capillaries, and veins. Arrows indicate the direction of blood flow in each circuit.

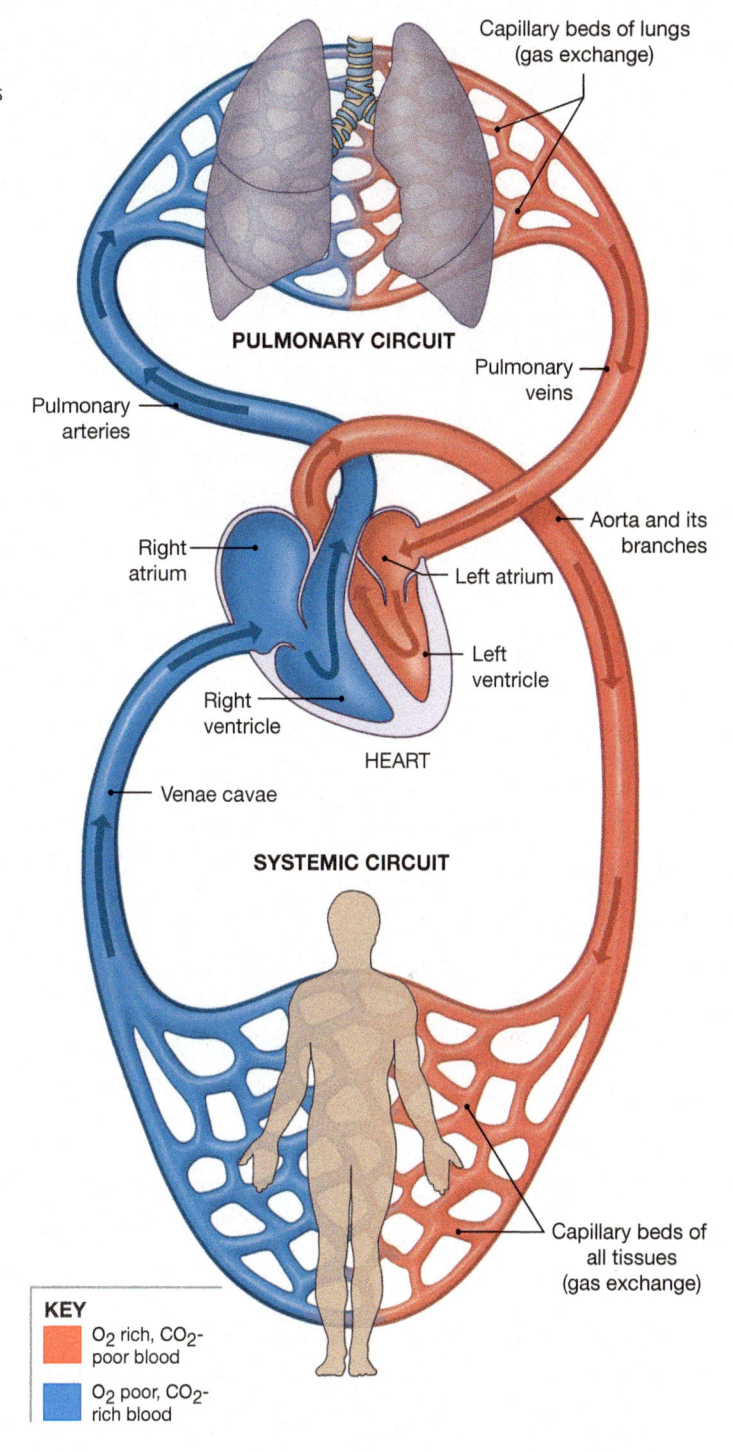

Each atrium is covered by an external flap called the **auricle** (AW-ri-kul; *auris*, ear), which is shown in Figure 22.4a. Adipose tissue and blood vessels occur along grooves in the heart wall. The **coronary sulcus** is a deep groove between the right atrium and right ventricle that extends to the posterior surface. The boundary between the right and left ventricles is marked anteriorly by the **anterior interventricular sulcus** and posteriorly by the **posterior interventricular sulcus**. Coronary blood vessels follow the sulci and branch to the myocardium. At the branch of the pulmonary trunk is the **ligamentum arteriosum**, a relic of a fetal vessel called

356 EXERCISE 22 The Heart

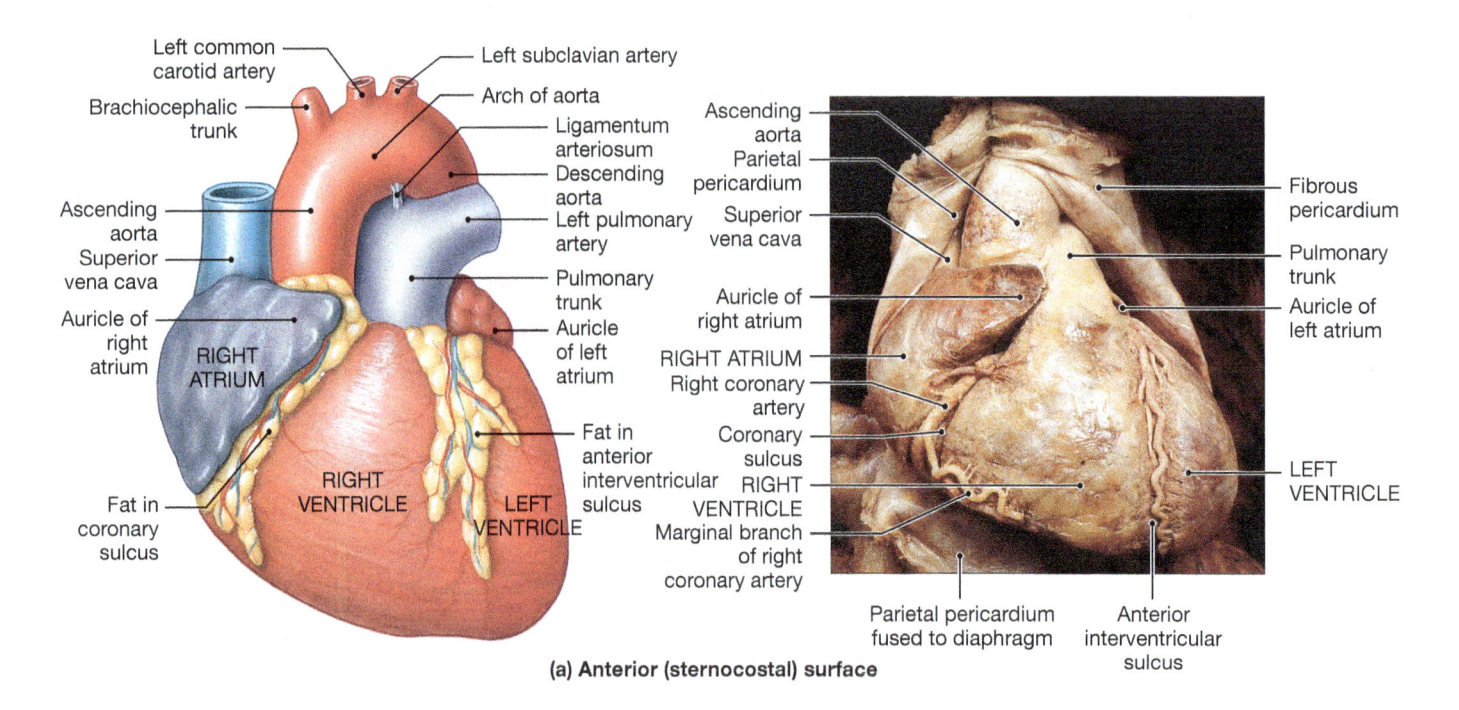

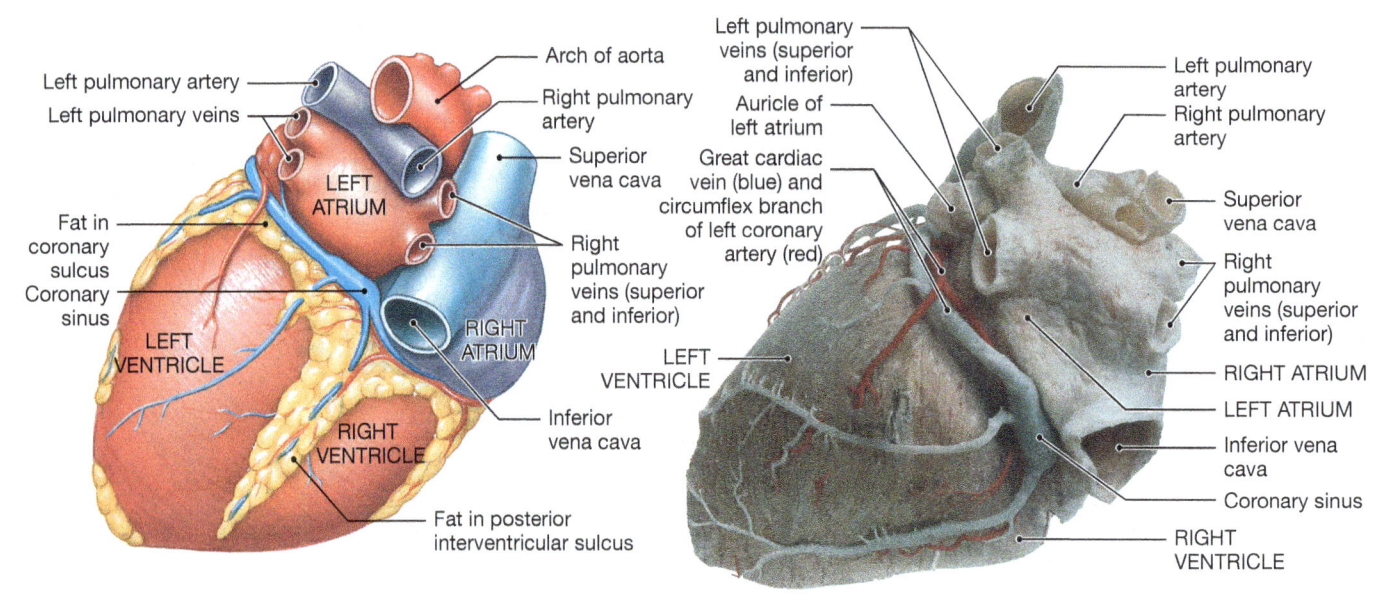

Figure 22.4 External Anatomy of the Heart

(a) Anterior view of the heart and great blood vessels. In the photograph, the pericardial sac has been cut and reflected to expose the heart and great vessels. (b) Posterior view of the heart and great blood vessels. The coronary arteries (which supply the heart) are shown in red; the cardiac veins are shown in blue.

the ductus arteriosus that joined the pulmonary trunk with the aorta. (Fetal circulation is discussed in Exercise 29, Development.)

Figure 22.5 details the internal anatomy of the heart. (Note in Figure 22.5a how much thicker the myocardium is in the left ventricle, as noted above.) The wall between the atria is called the **interatrial septum**, and the ventricles are separated by the **interventricular septum**. In the right atrium, a depression called the **fossa ovalis** is located on the interatrial septum. This is a remnant of fetal circulation, where the foramen ovale allowed blood to bypass the fetal pulmonary circuit. Lining the inside of the right atrium are muscular ridges, the **pectinate** (*pectin,* comb)

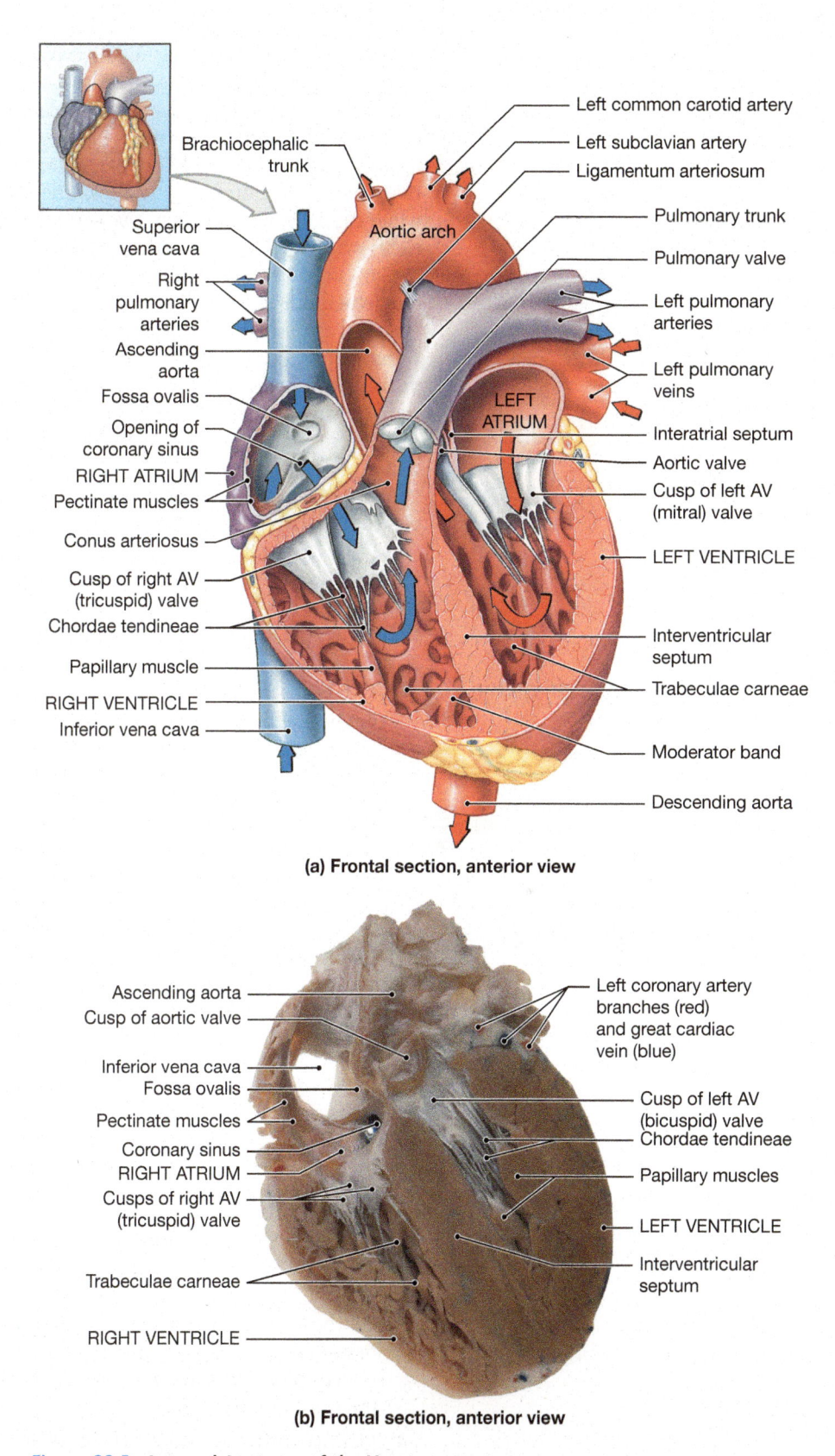

Figure 22.5 Internal Anatomy of the Heart
(a) Diagrammatic frontal section through the heart that shows major landmarks and the path of blood flow (marked by arrows) through the atria, ventricles, and associated blood vessels. (b) Anterior view of a frontal section of the heart.

muscles. Folds of muscle tissue called **trabeculae carneae** (tra-BEK-ū-lē CAR-nē-ē; *carneus*, fleshy) occur on the inner surface of each ventricle. The **moderator band** is a ribbon of muscle that passes electrical signals from the interventricular septum to muscles in the right ventricle.

The heart has two **atrioventricular** (AV) **valves** and two **semilunar valves** that control and direct blood flow. The two pairs generally work in opposition: when the AV valves are open, the semilunar valves are either closed or preparing to close; when the semilunar valves are open, the AV valves are either closed or preparing to close. The two atrioventricular valves prevent blood from re-entering the atria when the ventricles contract. The **right atrioventricular valve**, which joins the right atrium and right ventricle, has three flaps, or cusps, and is also called the **tricuspid** (trī-KUS-pid; *tri*, three; *cuspid*, flap) **valve**. The **left atrioventricular valve** between the left atrium and left ventricle has two cusps and is called either the **bicuspid valve** or the **mitral** (MĪ-tral) **valve**. The cusps of each AV valve have small cords, the **chordae tendineae** (KOR-dē TEN-di-nē-ē; tendon-like cords), which are attached to **papillary** (PAP-i-ler-ē) **muscles** on the floor of the ventricles. When the ventricles contract, the AV valves are held closed by the papillary muscles that pull on the chordae tendineae.

The two semilunar valves are the **aortic valve** and **pulmonary valve** each located at the base of its artery. These valves prevent backflow of blood into the ventricles when the ventricles are relaxed. Each semilunar valve has three small cusps that, when the ventricles relax, fill with blood and close the base of the artery.

Clinical Application

Mitral Valve Prolapse

A common valve problem is **mitral valve prolapse**, a condition in which the left AV valve reverses, like an umbrella in a strong wind. The papillary muscles and chordae tendineae are unable to hold the valve cusps in the closed position, and so the valve inverts. When this happens, the opening between the atrium and ventricle is not sealed shut during ventricular contraction, blood backflows into the left atrium, and cardiac function is diminished.

QuickCheck Questions

2.1 What are the heart chambers associated with the pulmonary circuit and those associated with the systemic circuit?

2.2 What structures separate the walls of the heart chambers?

2.3 What are the four heart valves? What is the function of each valve?

2 Materials

☐ Heart model and specimens

Procedures

1. Review the heart anatomy in Figures 22.3, 22.4, and 22.5.
2. Observe the external features of the heart on the heart model and specimens. Note how the auricles may be used to distinguish the anterior surface. Trace the length of each sulcus, and notice the chambers each passes between.
3. On the heart model, identify each atrium and ventricle. Note which ventricle has the thicker wall. Identify the pectinate muscles in the right atrium and the trabeculae carneae in both ventricles. Locate the moderator band in the inferior right ventricle.
4. Identify the two AV valves, their cusps, and the two semilunar valves.
5. Identify the major arteries and veins at the base of the heart.
6. Starting at the superior vena cava, trace a drop of blood though the heart model, and distinguish between the pulmonary and systemic circuits. ■

LAB ACTIVITY 3 — **Coronary Circulation**

To produce the pressure required for blood to reach throughout the cardiovascular system, the heart can never completely rest. The branch of the systemic circuit known as the **coronary circulation** supplies the myocardium with the oxygen necessary for muscle contraction (Figure 22.6). The right and left **coronary arteries** of the coronary circulation are the first vessels to branch off the base of the ascending aorta and penetrate the myocardium to the outer heart wall. As the right coronary artery (RCA) passes along the coronary sulcus, many atrial branches supply blood to the right atrium and one or more **acute marginal branches** arise to supply the right ventricle. The **posterior interventricular branch** off the RCA supplies adjacent posterior regions of the ventricles.

The left coronary artery (LCA) branches to supply blood to the left atrium, left ventricle, and interventricular septum. The LCA divides into a **circumflex branch** and an **anterior interventricular branch**. The anterior interventricular branch supplies the left ventricle. The circumflex branch follows the left side of the heart; it then has several **diagonal branches** and an **obtuse marginal branch**, then sends the **posterior left ventricular branch** along the posterior coronary sulcus. The posterior branches of the RCA and LCA often unite in the posterior coronary sulcus.

The **cardiac veins** of the coronary circulation collect deoxygenated blood from the myocardium (Figure 22.6). The **great cardiac vein** follows along the anterior interventricular sulcus and curves around the left side of the heart to drain the myocardium that is supplied by the anterior interventricular branch. The **posterior vein of the left ventricle** drains the myocardium that is supplied by the LCA posterior ventricular branch. The **small cardiac vein** drains the superior right area of the heart. The **middle cardiac vein** drains the myocardium that is supplied by the posterior interventricular branch of the RCA. The cardiac veins merge as a large **coronary sinus** situated in the posterior region of the coronary sulcus. The coronary sinus empties deoxygenated blood from the myocardium into the right atrium. As noted previously, the right atrium also receives deoxygenated blood from the venae cavae.

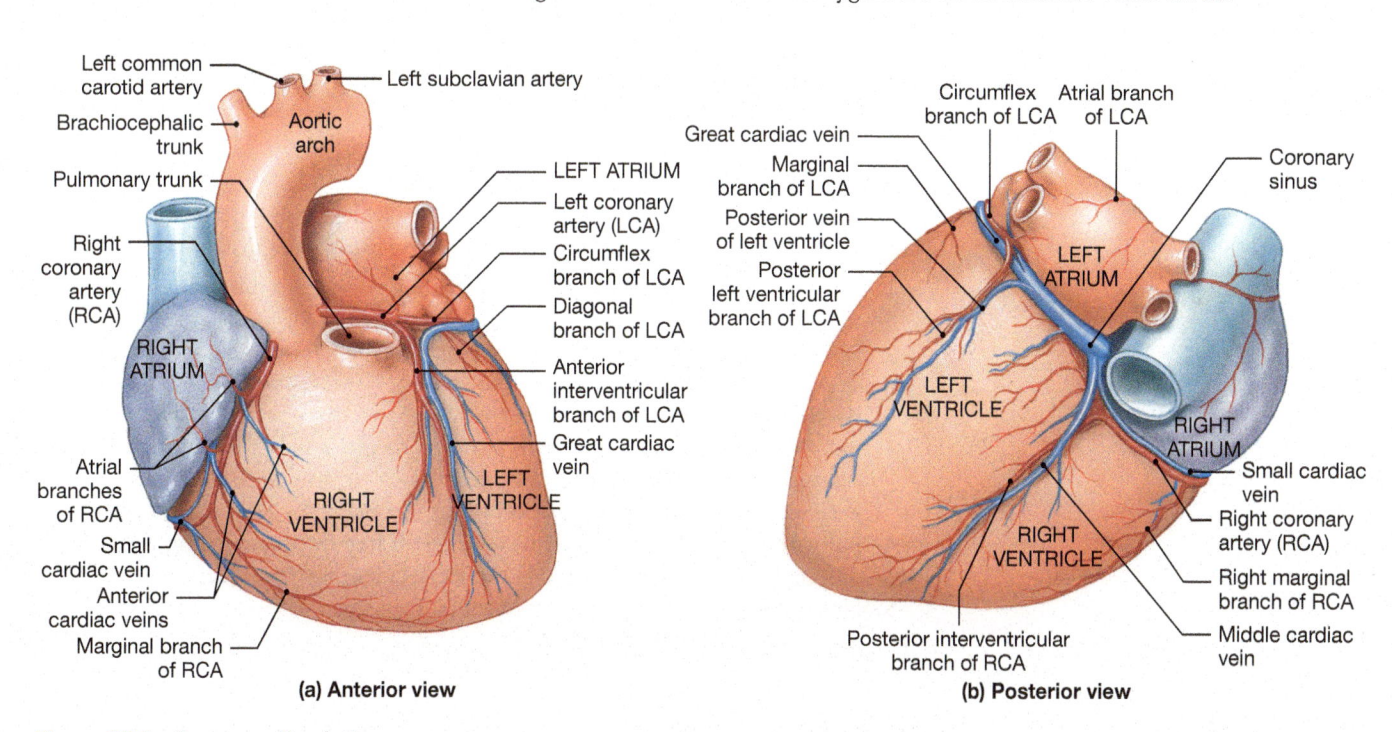

Figure 22.6 Coronary Circulation
Coronary arteries and cardiac veins supply and drain the myocardium of blood. These blood vessels are also shown in Figure 22.4b. **(a)** Blood vessels that supply the anterior surface of the heart. **(b)** Blood vessels that supply the posterior surface of the heart.

Clinical Application: Anastomoses and Infarctions

The interventricular branches connect with one another, as do smaller arteries between the right coronary artery and the circumflex branch of the left coronary artery. These connections, called **anastomoses**, ensure that blood flow to the myocardium remains steady. In coronary artery disease, the arteries become narrower and narrower as fatty plaque is deposited on the interior walls of the vessels. As a result, blood flow is reduced. If enough plaque accumulates in critical areas, blood flow to that part of the heart becomes inadequate, and the heart muscle has an **infarction**, a heart attack.

QuickCheck Questions

3.1 Where do the right and left coronary arteries arise?
3.2 Where do the cardiac veins drain?

3 Materials

☐ Heart model and specimens

Procedures

1. Review the blood vessels of the coronary circulation in Figure 22.6.
2. Follow the RCA and LCA on the heart model and identify their main branches.
3. Identify the cardiac veins and trace them into the coronary sinus. Identify where the coronary sinus drains. ■

LAB ACTIVITY 4 — Sheep Heart Dissection

The sheep heart, like all other mammalian hearts, is similar in structure and function to the human heart. One major difference is in where the great vessels join the heart. In four-legged animals, the inferior vena cava has a posterior connection to the heart instead of the inferior connection that is found in humans. Dissecting a sheep heart will enhance your studies of models and charts of the human heart. Take your time while dissecting and follow the directions carefully.

⚠ Safety Alert: Dissecting the Heart

You *must* practice the highest level of lab safety while handling and dissecting the heart. Keep the following guidelines in mind during the dissection:

1. Wear gloves and safety glasses to protect yourself from the fixatives used to preserve the specimen.
2. Do not dispose of the fixative from your specimen. You will later store the specimen in the fixative to keep the specimen moist and to keep it from decaying.
3. Be extremely careful when using a scalpel or other sharp instrument. Always direct cutting and scissor motion away from you to prevent an accident if the instrument slips on moist tissue.
4. Before cutting a given tissue, make sure it is free from underlying and/or adjacent tissues so that they will not be accidentally severed.
5. Never discard tissue in the sink or trash. Your instructor will inform you of the proper disposal procedure. ▲

QuickCheck Questions

4.1 What type of safety equipment should you wear as you dissect the sheep heart?
4.2 How should you dispose of the sheep heart and scrap tissue?

4 Materials

- ☐ Gloves
- ☐ Safety glasses
- ☐ Dissecting tools
- ☐ Dissecting pan
- ☐ Fresh or preserved sheep heart

Procedures

1. Put on gloves and safety glasses, and clear your workspace before obtaining your dissection specimen.
2. Wash the sheep heart with cold water to flush out preservatives and blood clots. Minimize exposure of your skin and mucous membranes to the preservatives.
3. Carefully follow the instructions in this section. Cut into the heart only as instructed.

External Anatomy

1. Figure 22.7 details the external anatomy of the sheep heart. Examine the surface of the heart to see if the pericardium is present. (Often this serous membrane has been removed from preserved specimens.) Carefully scrape the outer heart muscle with a scalpel to loosen the epicardium.
2. Locate the anterior surface by orienting the heart so that the auricles face you. Under the auricles are the right and left atria. Note the base of the heart above the atria, where the large blood vessels occur. Squeeze gently just above the apex to locate the right and left ventricles. Locate the anterior interventricular sulcus, the fat-laden groove between the ventricles. Carefully remove some of the adipose tissue with the scalpel to uncover some of the coronary blood vessels. Identify two grooves—the coronary sulcus between the right atrium and ventricle and the posterior interventricular sulcus between the ventricles on the posterior surface.
3. Identify the aorta and then the pulmonary trunk anterior to the aorta. If on your specimen the pulmonary trunk was cut long, you may be able to identify the right and left pulmonary arteries that branch off the trunk. The brachiocephalic artery is the first major branch of the aorta and is often intact in preserved material.
4. Follow along the inferior margin of the right auricle to the posterior surface. The prominent vessel at the termination of the auricle is the superior vena cava. At the base of this vessel is the inferior vena cava. Next, examine the posterior aspect of the left atrium and find the four pulmonary veins. You may need to carefully remove some of the adipose tissue around the superior region of the left atrium to locate these veins.

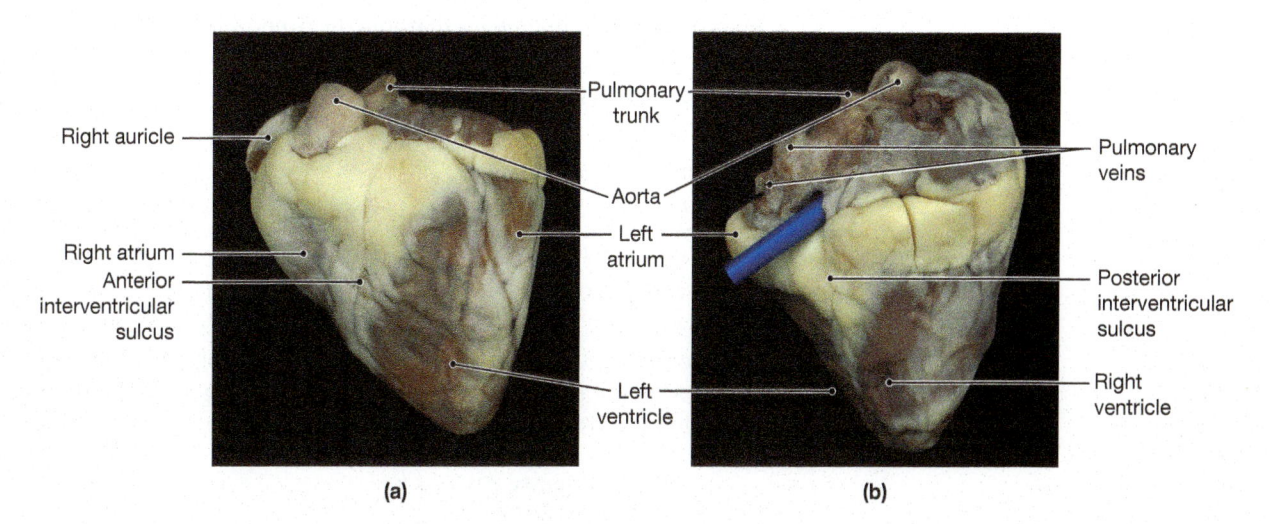

Figure 22.7 External Anatomy of the Sheep Heart
(a) Anterior view. (b) Posterior view.

362   EXERCISE 22   The Heart

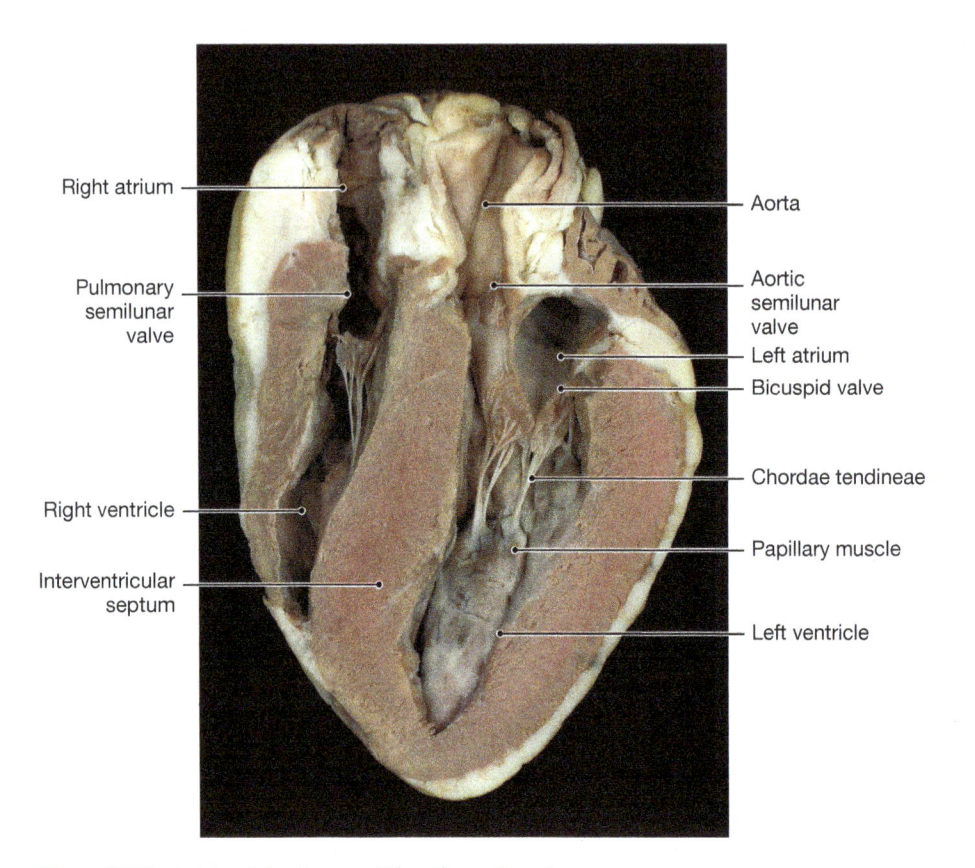

Figure 22.8   Internal Anatomy of the Sheep Heart
The major anatomical features of the sheep heart as shown in a frontal section.

Internal Anatomy

1. Cut a frontal section passing through the aorta. Use Figure 22.8 as a reference to the internal anatomy.
2. Examine the two sides of the heart. Identify the right and left atria, right and left ventricles, and the interventricular septum. Compare the myocardium of the left ventricle with that of the right ventricle. Note the folds of trabeculae carneae along the inner ventricular walls. Examine the right atrium for the comb-like pectinate muscles that line the inner wall.
3. Locate the tricuspid and bicuspid valves. Observe the papillary muscles with chordae tendineae attached.
4. Examine the wall of the left atrium for the openings of the four pulmonary veins.
5. At the entrance of the aorta, locate the small cusps of the aortic valve.
6. At the base of the pulmonary trunk, locate the pulmonary valve.
7. Locate the superior and inferior venae cavae, which drain into the right atrium.
8. Upon completion of the dissection, dispose of the sheep heart as directed by your instructor and wash your hands and dissecting instruments. ■

EXERCISE 22

LAB REPORT

The Heart

Name _____

Date _____

Section _____

A. Matching

Match each heart structure in the left column with its correct description from the right column.

_____ 1. tricuspid valve
_____ 2. superior vena cava
_____ 3. right ventricle
_____ 4. aorta
_____ 5. left ventricle
_____ 6. pulmonary vein
_____ 7. semilunar valve
_____ 8. bicuspid valve
_____ 9. pulmonary trunk
_____ 10. coronary sinus
_____ 11. epicardium
_____ 12. chordae tendineae

A. empties into left atrium
B. left atrioventricular valve
C. pumps blood to body tissues
D. major systemic artery
E. artery that carries deoxygenated blood
F. visceral pericardium
G. drains cardiac veins into heart
H. aortic or pulmonary valve
I. empties into right atrium
J. attached to papillary muscles
K. right atrioventricular valve
L. pumps blood to lungs

B. Labeling

Label the numbered anatomic features of the heart in Figure 22.9.

C. Short-Answer Questions

1. Describe the location of the heart in relation to the other structures in the thoracic cavity.

2. Trace a drop of blood through the pulmonary and systemic circuits.

3. Describe the components of the left atrioventricular valve.

363

EXERCISE 22 — LAB REPORT

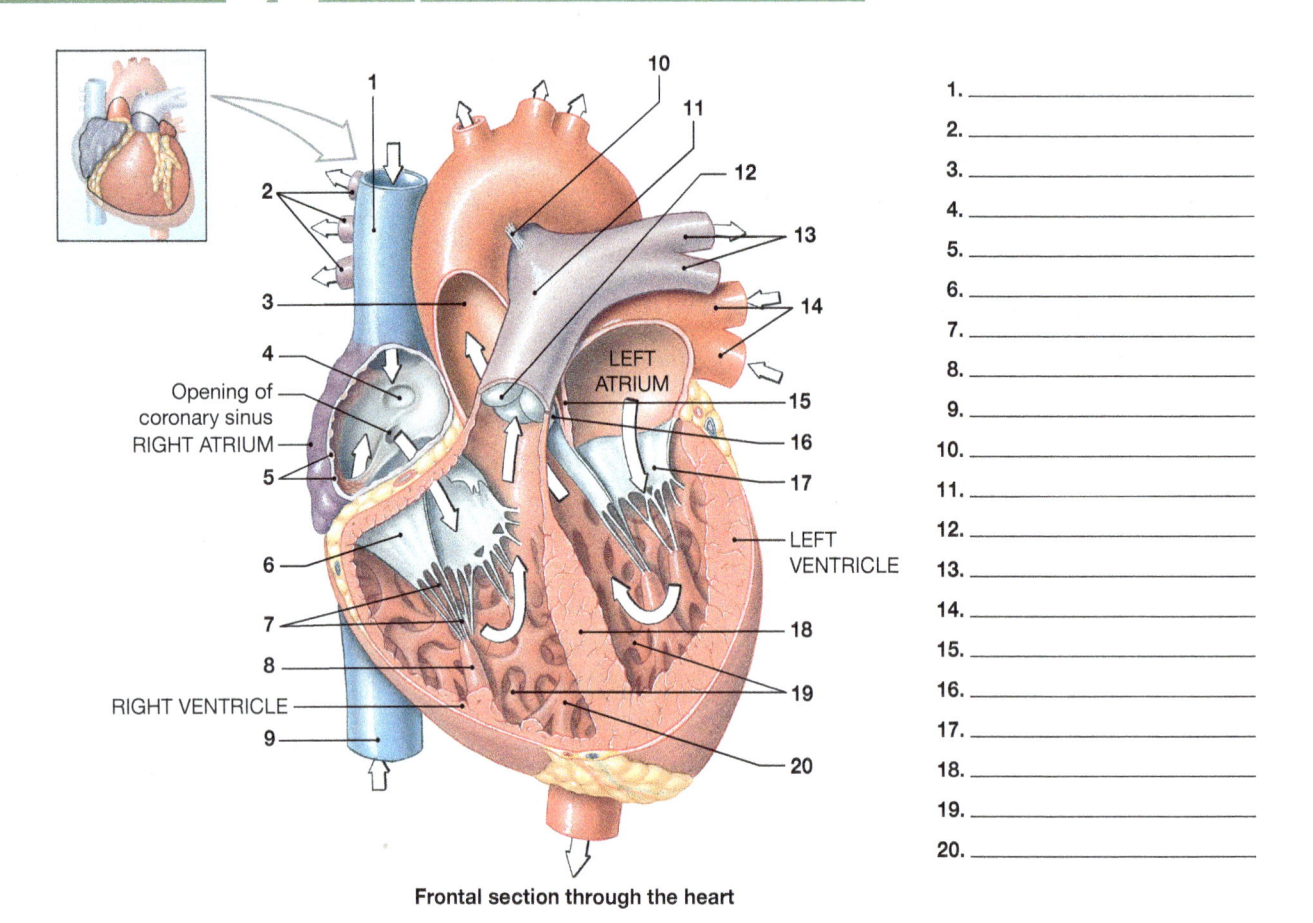

Figure 22.9 Internal Anatomy of the Heart
Diagrammatic frontal section of the heart with arrows that indicate direction of blood flow.

1. _____
2. _____
3. _____
4. _____
5. _____
6. _____
7. _____
8. _____
9. _____
10. _____
11. _____
12. _____
13. _____
14. _____
15. _____
16. _____
17. _____
18. _____
19. _____
20. _____

D. Analysis and Application

1. Explain the difference between the thickness of the myocardium in the right and left ventricles.

2. Suppose a patient has mitral valve prolapse, which is a weakened left atrioventricular valve that does not close properly. How does this defect affect the flow of blood in the heart?

3. Coronary artery disease in the acute marginal branches would affect which part of the myocardium?

EXERCISE 23

The Systemic Circuit

OBJECTIVES

On completion of this exercise, you should be able to:

- Compare the histology of an artery, a capillary, and a vein.
- Describe the difference in the blood vessels that serve the right and left upper limbs.
- Describe the anatomy and importance of the cerebral arterial circle.
- Trace a drop of blood from the ascending aorta to each abdominal organ and to the lower limbs.
- Trace a drop of blood from the foot to the heart.
- Compare circulation in the pulmonary and systemic circuits.

LAB ACTIVITIES

1. Comparison of Arteries, Capillaries, and Veins   365
2. Arteries of the Head, Neck, and Upper Limb   367
3. Arteries of the Abdominopelvic Cavity and Lower Limb   370
4. Veins of the Head, Neck, and Upper Limb   374
5. Veins of the Lower Limb and Abdominopelvic Cavity   378

A Regional Look: Limbs   381

The body contains more than 60,000 miles of blood vessels to transport blood to the trillions of cells in the tissues. Arteries of the systemic circuit distribute oxygen and nutrient-rich blood to microscopic networks of thin-walled vessels called capillaries. At the capillaries, nutrients, gases, wastes, and cellular products diffuse either from blood to cells or from cells to blood. Veins drain deoxygenated blood from the capillaries and direct it toward the heart, which then pumps it into the pulmonary circuit, to be carried to the lungs to pick up oxygen and release carbon dioxide.

Blood vessels are a continuous network of "pipes", and often there is little anatomical difference along the length of a given vessel as it passes from one region of the body to another. To facilitate identification and discussion, however, anatomists assign different names to a given vessel, depending on which part of the body the vessel is passing through. The subclavian artery becomes the axillary artery, for instance, and then the brachial artery. Each name is usually related to the name of a bone or organ adjacent to the vessel; therefore, because they often run parallel to each other, arteries and veins often have the same name.

Refer to Exercise 22 and Figure 22.3 for a review of the pulmonary vessels. In this exercise you will study the major arteries and veins of the systemic circuit.

LAB ACTIVITY 1 Comparison of Arteries, Capillaries, and Veins

The walls of the body's blood vessels have three layers (Figure 23.1). The **tunica externa** is a layer of connective tissue that anchors the vessel to surrounding tissues.

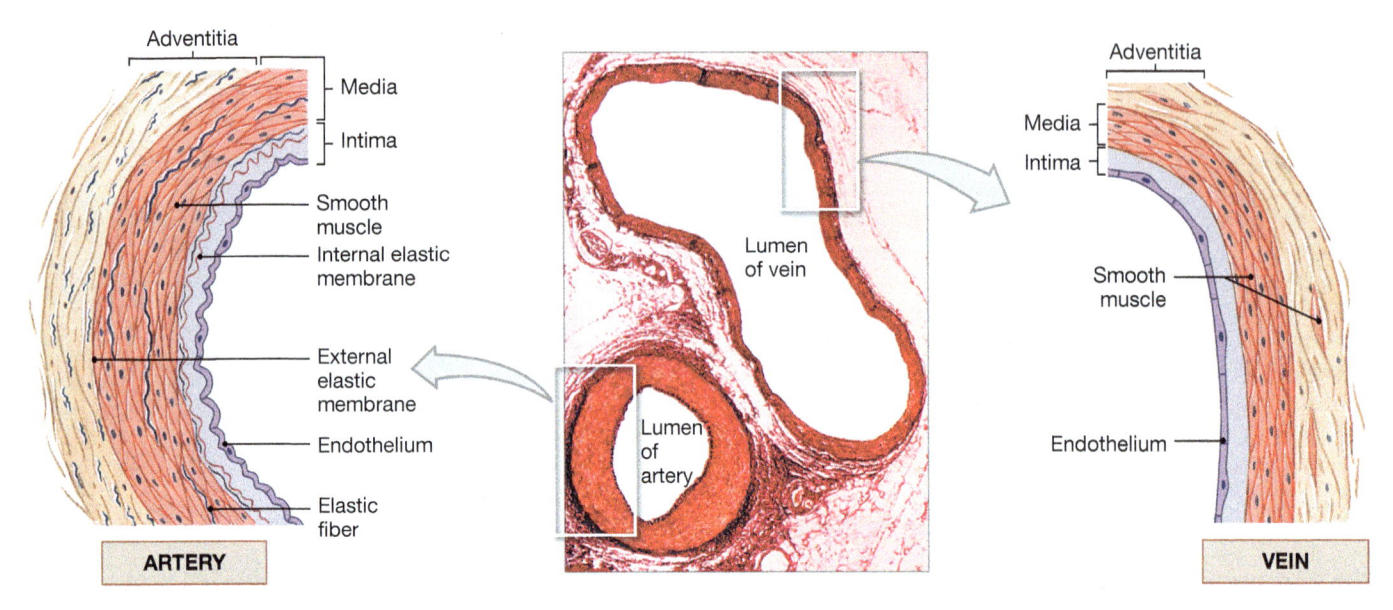

Figure 23.1 Comparison of the Structure of a Typical Artery and Vein (LM × 60)

Collagen and elastic fibers give this layer strength and flexibility. The **tunica media** is a layer of smooth muscle tissue. In the tunica media of arteries are elastic fibers that allow the vessels to stretch and recoil in response to blood pressure changes. In veins, there are fewer elastic fibers; collagen fibers in the tunica media provide strength. Lining the inside of the vessels is the third layer, the **tunica intima**, a thin layer of simple squamous epithelium called **endothelium**. In arteries, the luminal surface of the endothelium has a thick, dark-staining **internal elastic membrane**.

Because blood pressure is much higher in arteries than in veins and also because the pressure fluctuates more in arteries than in veins, the walls of arteries are thicker than those of veins. Notice how the artery cross-section in the micrograph of Figure 23.1 is round and thick-walled, while the adjacent vein is irregularly shaped and thin-walled. In a slide preparation, the tunica intima of an artery may appear pleated because the vessel wall has recoiled due to a loss of pressure. In reality, the luminal surface is smooth and the vessel can expand and shrink to regulate blood flow.

A capillary consists of a single layer of endothelium that is continuous with the tunica intima of the artery and vein that supply and drain the capillary. Capillaries are so narrow that RBCs must line up in single file to squeeze through.

Veins have a thinner wall than arteries. The walls of a vein collapse if the vessel is emptied of blood. Blood pressure is low in veins, and to prevent backflow, the peripheral veins have valves that keep blood flowing in one direction, toward the heart.

QuickCheck Questions

1.1 Describe the three layers in the wall of an artery.

1.2 How do arterial walls differ from venous walls?

1 Materials

- ☐ Compound microscope
- ☐ Prepared slide of artery and vein

Procedures

1. Review the structure of arteries and veins in Figure 23.1.
2. Place the artery/vein slide on the microscope stage and locate the artery and vein at low magnification. Most slide preparations have one artery, an adjacent vein, and a nerve. The blood vessels are hollow and most likely have blood cells in the lumen. The nerve appears as a round, solid structure.
3. Increase the magnification to high and compare each arterial layer with its venous counterpart.

4. Draw and label a cross-section of an artery and a vein in the space provided. Include enough detail in your drawing to show the anatomical differences between the vessels. ■

Artery cross-section

Vein cross-section

LAB ACTIVITY 2   Arteries of the Head, Neck, and Upper Limb

The aorta receives oxygenated blood from the left ventricle of the heart and distributes the blood to the major arteries that arise from the aorta and supply the head, limbs, and trunk. The initial portion of the aorta is curved like an inverted letter U, and the various regions have different names. The **ascending aorta** exits the base of the heart, curves upward and to the left to form the **aortic arch**, and then as the **descending aorta** descends behind the heart (Figure 23.2). At the point where it passes through the diaphragm, the descending aorta becomes the **abdominal aorta**. Arteries that branch off the aortic arch serve the head, neck, and upper limb. Intercostal arteries stem from the thoracic aorta and supply the thoracic wall. Branches off of the abdominal aorta serve the abdominal organs. The abdominal aorta enters the pelvic cavity and divides to send a branch into each lower limb.

The first branch of the aortic arch, the **brachiocephalic** (brā-kē-ō-se-FAL-ik) **trunk**, or **innominate artery**, is short and divides into the **right common carotid artery** and the **right subclavian artery** (Figure 23.2). The right common carotid artery supplies blood to the right side of the head and neck; the right subclavian artery supplies blood to the right upper limb. The second and third branches of the aortic arch are the **left common carotid artery**, which supplies the left side of the head and neck, and **left subclavian artery**, which supplies the left upper limb as well as the shoulder and head. Note that only the right common carotid artery and right subclavian artery are derived from the brachiocephalic trunk. The left common carotid artery and left subclavian artery arise directly from the peak of the aortic arch. A **vertebral artery** branches off each subclavian artery and supplies blood to the brain and spinal cord.

Each common carotid artery ascends deep in the neck and divides at the larynx into an **external carotid artery** and an **internal carotid artery** (Figure 23.3a). The base of the internal carotid swells as the **carotid sinus** and contains baroreceptors that monitor blood pressure. The external carotid artery branches to supply blood to the neck and face. The pulse in the external carotid artery can be felt by placing your fingers lateral to your thyroid cartilage (Adam's apple). The external carotid artery branches into the **facial artery**, **maxillary artery**, and **superficial temporal artery** to serve the external structures of the head. The internal carotid artery ascends to the base of the brain and divides into three arteries: the **ophthalmic artery**, which supplies the eyes, and the **anterior cerebral artery** and **middle cerebral artery**, both of which supply the brain.

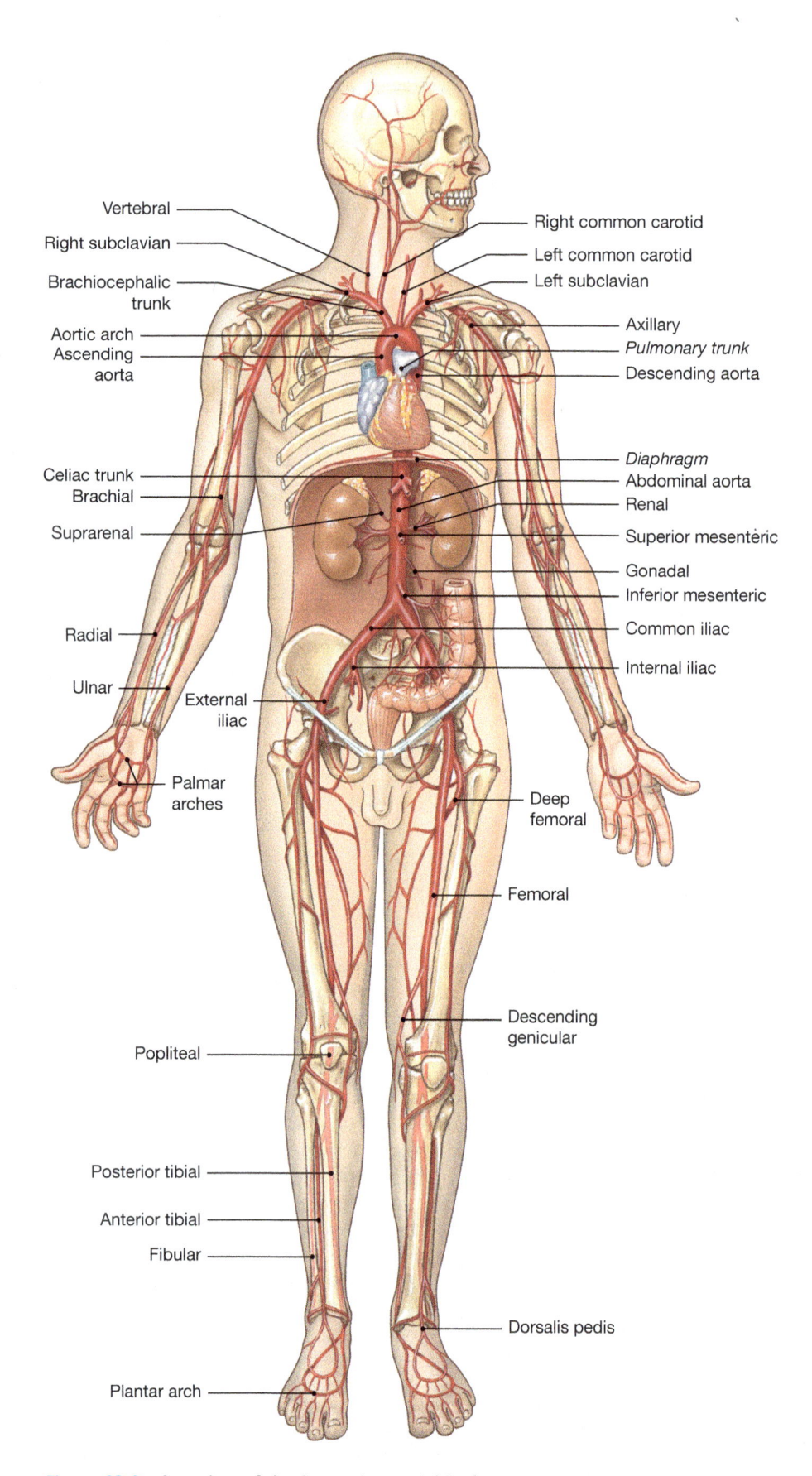

Figure 23.2 Overview of the Systemic Arterial Pathways

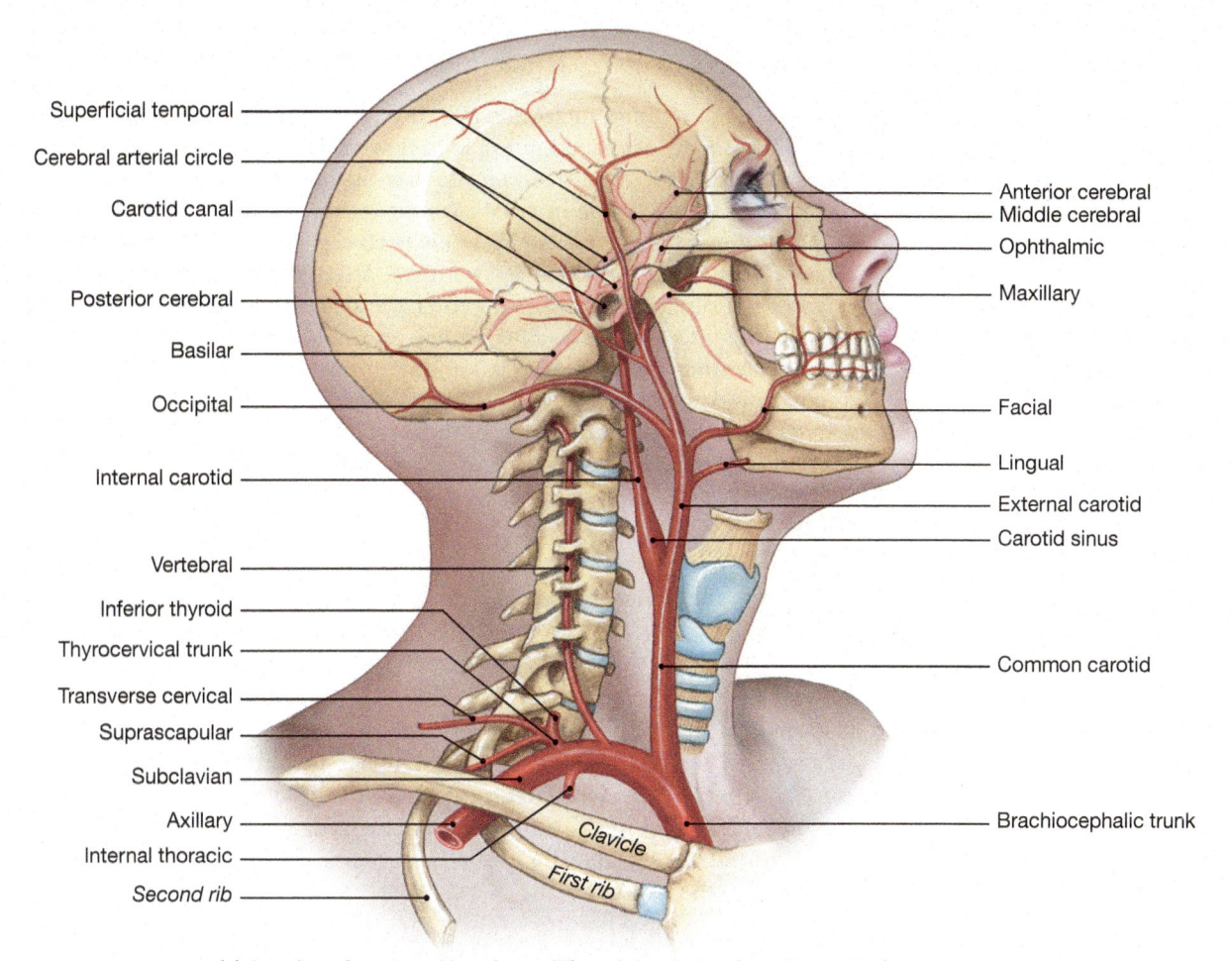

(a) Arteries of neck and head, an oblique lateral view from the right side

(b) Arteries of the brain, inferior view

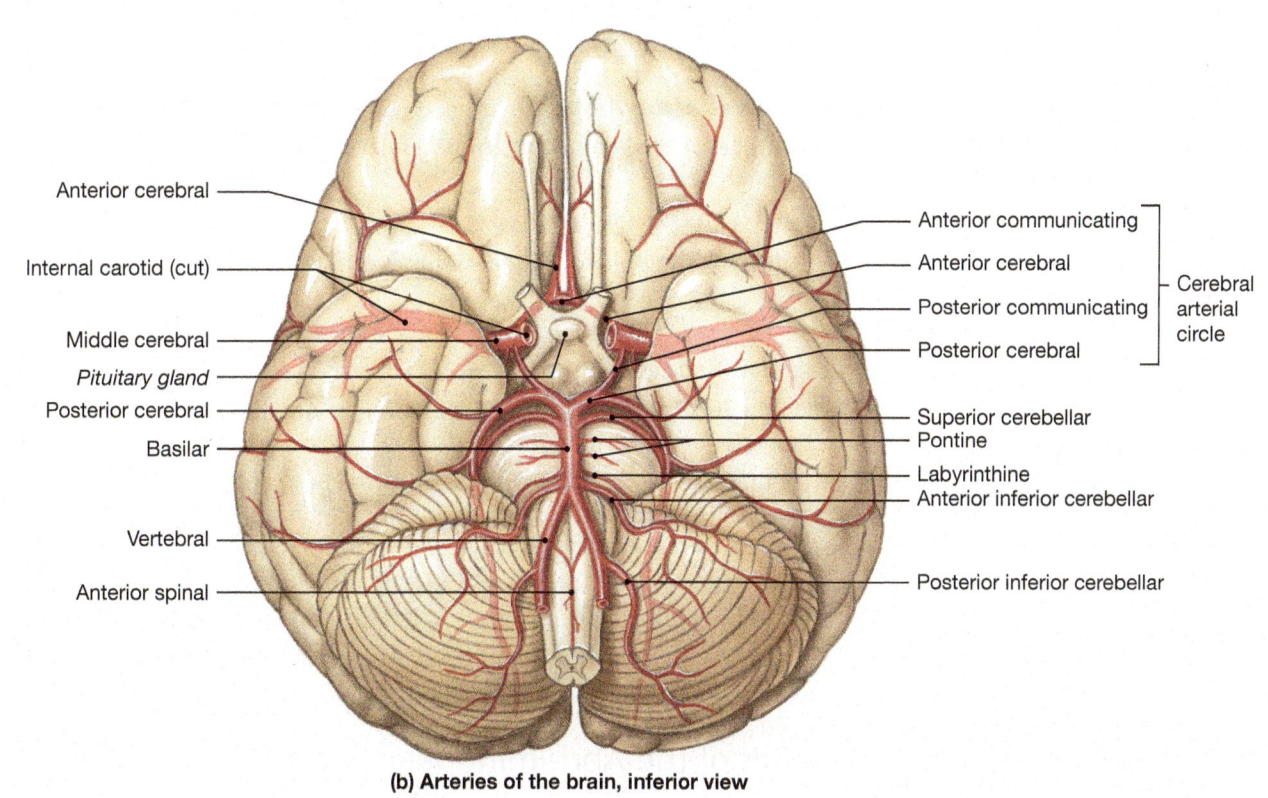

Figure 23.3 Arteries of the Neck, Head, and Brain
(a) Oblique lateral view from right side that shows general circulation pattern of arteries that supply the neck and superficial structures of the head. (b) Inferior view of the brain that shows the distribution of arteries and the cerebral arterial circle.

Because of its high metabolic rate, the brain has a voracious appetite for oxygen and nutrients. A reduction in blood flow to the brain may result in permanent damage to the affected area. To ensure that the brain receives a continuous supply of blood, branches of the internal carotid arteries and other arteries interconnect, or **anastomose**, as the **cerebral arterial circle**, also called the **circle of Willis** (Figure 23.3b). The right and left vertebral arteries ascend in the transverse foramina of the cervical vertebrae and enter the skull at the foramen magnum. These arteries fuse into a single **basilar artery** on the inferior surface of the brain stem. The basilar artery branches into left and right **posterior cerebral arteries** and left and right **posterior communicating arteries**. The right and left anterior cerebral arteries form the anterior portion of the cerebral arterial circle. Between these arteries is the **anterior communicating artery**, which completes the anastomosis.

Making Connections

What's in a Name?

Arteries and veins with the term *common* as part of their name always branch into an external and an internal vessel. The common carotid artery, for example, branches into an external carotid artery and an internal carotid artery. The internal and external iliac veins join as the common iliac vein.

The subclavian arteries supply blood to the upper limbs. Each subclavian artery passes under the clavicle, crosses the armpit as the **axillary artery**, and continues into the arm as the **brachial artery** (Figure 23.4). (Blood pressure is usually taken at the brachial artery.) At the antecubitis (elbow), the brachial artery divides into the lateral **radial artery** and the medial **ulnar artery**, each named after the bone it follows. In the palm of the hand, these arteries are interconnected by the **superficial** and **deep palmar arches**, which send small **digital arteries** to the fingers. Except in the vicinity of the heart, where the right arterial pathway has a brachiocephalic trunk that is absent from the left pathway, the arrangement of the arteries that supply the left and right upper limbs is symmetrical.

QuickCheck Questions

2.1 How does arterial branching in the left side of the neck differ from branching in the right side?

2.2 What is an anastomosis?

2.3 Which arteries in the brain anastomose with one another?

2 Materials

- ☐ Vascular system chart
- ☐ Torso model
- ☐ Head model
- ☐ Upper limb model

Procedures

1. Review the arteries in Figures 23.2, 23.3, and 23.4.
2. On the torso model, examine the aortic arch and identify the three branches that arise from the superior margin of the arch.
3. On the head model, trace the arteries to the head and note the differences between the right and left common carotid arteries. Identify the arteries that converge at the cerebral arterial circle.
4. On the torso model, identify the arteries of the shoulder and limb. Note the difference in origin of the right and left subclavian arteries.
5. Using your index and middle fingers, locate the pulse in your radial and common carotid arteries.

LAB ACTIVITY 3 Arteries of the Abdominopelvic Cavity and Lower Limb

The arteries that stem from the abdominal aorta are shown in Figures 23.2 and 23.5. An easy way to identify the branches of the abdominal aorta is to distinguish between paired arteries, which have right and left branches, and unpaired arteries.

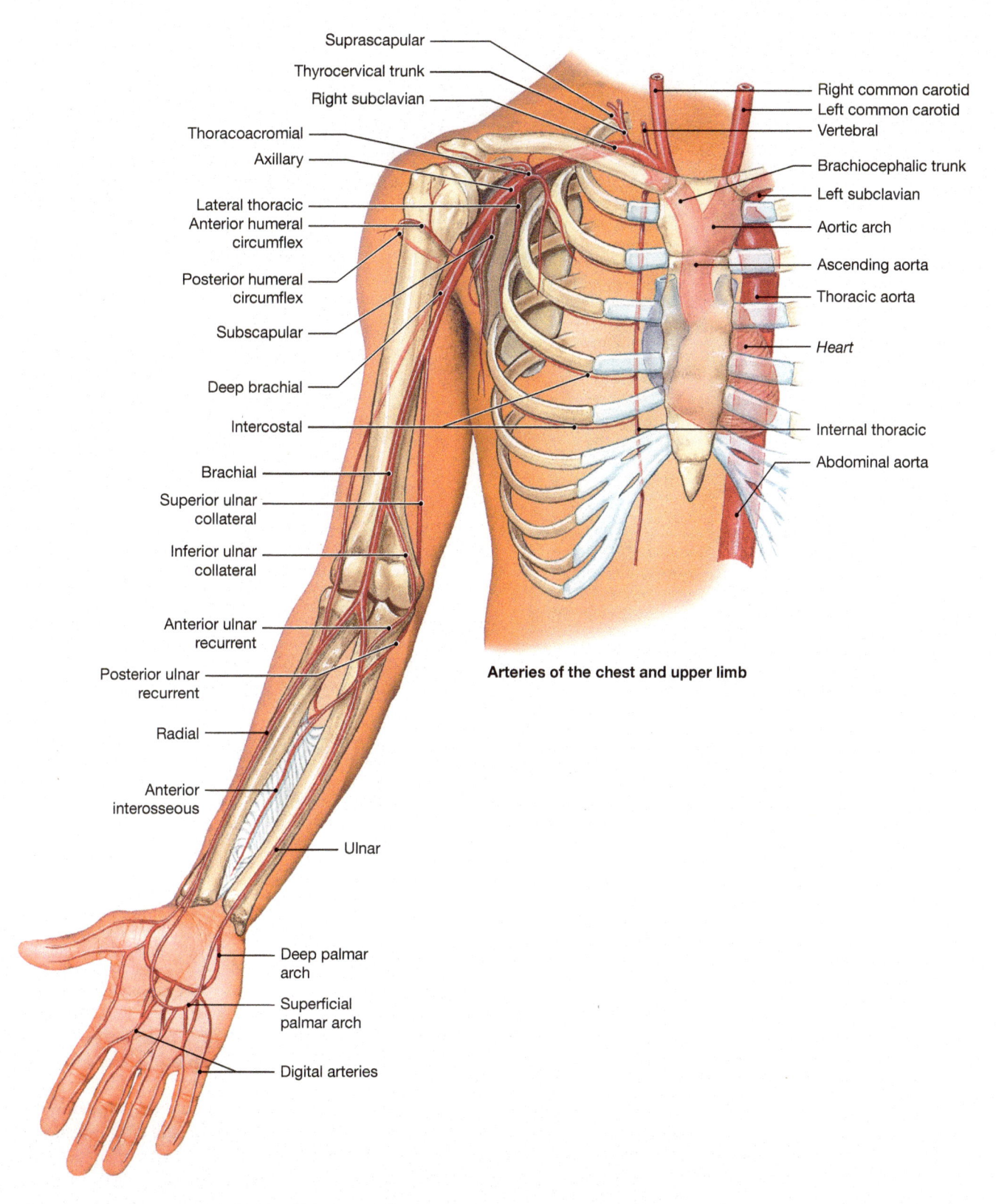

Figure 23.4 Arteries of the Chest and Upper Limb
Arteries that originate along the aortic arch are shown branching into the chest and right upper limb.

371

372 EXERCISE 23 The Systemic Circuit

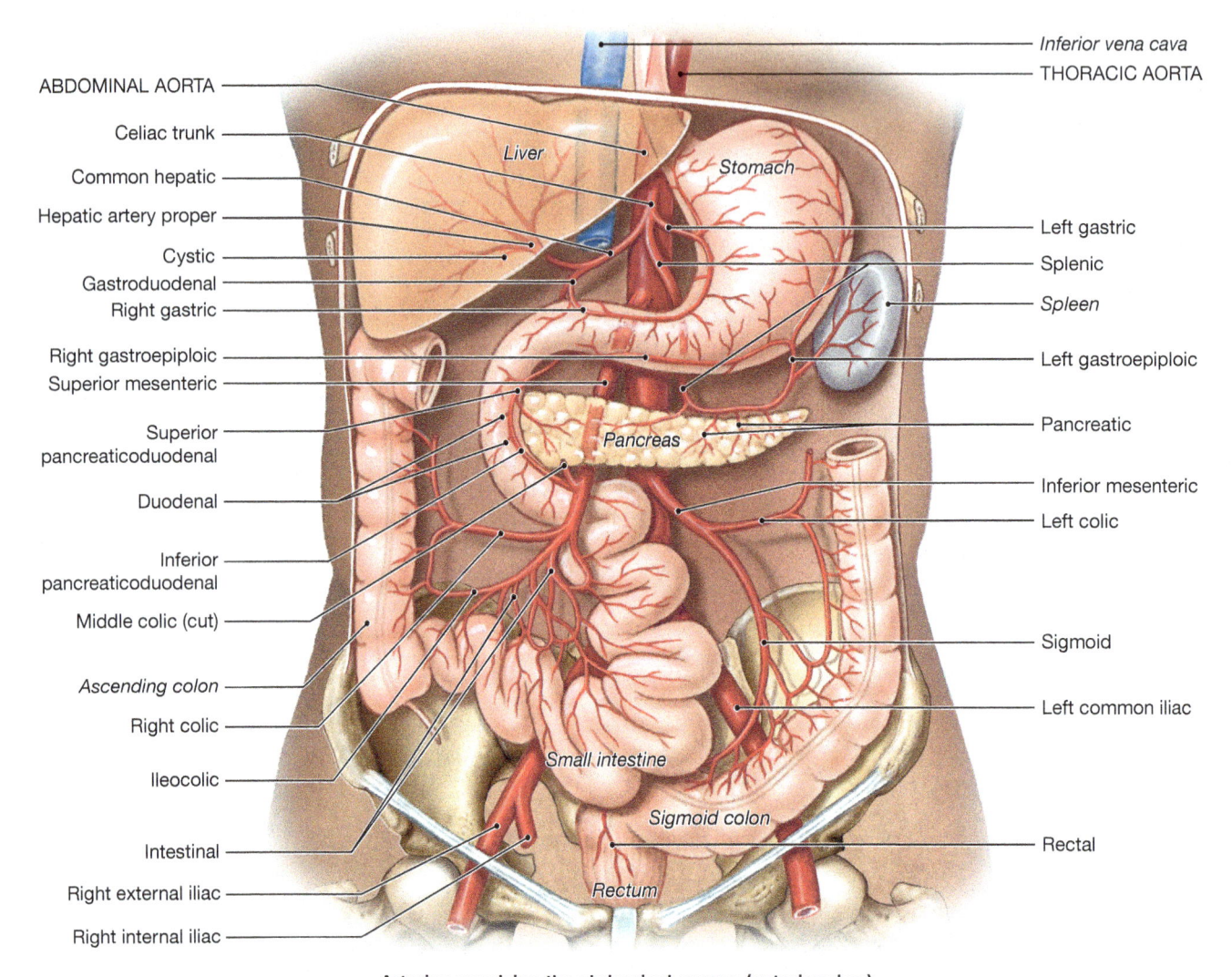

Arteries supplying the abdominal organs (anterior view)

Figure 23.5 Major Arteries of the Abdomen

Three unpaired arteries arise from the abdominal aorta: celiac trunk, superior mesenteric artery, and inferior mesenteric artery. The short **celiac** (SĒ-lē-ak) **trunk** arises inferior to the diaphragm and splits into three arteries. The **common hepatic artery** divides to supply blood to the liver, gallbladder, and part of the stomach. The **left gastric artery** supplies the stomach. The **splenic artery** supplies the spleen, stomach, and pancreas.

Inferior to the celiac trunk is the next unpaired artery, the **superior mesenteric** (mez-en-TER-ik) **artery**. This vessel supplies blood to the large intestine, parts of the small intestine, and other abdominal organs. The third unpaired artery, the **inferior mesenteric artery**, originates before the abdominal aorta divides to enter the pelvic cavity and lower limbs. This artery supplies parts of the large intestine and the rectum.

Four major sets of paired arteries arise off the abdominal aorta (see Figure 23.2). The right and left **suprarenal arteries** arise near the superior mesenteric artery and branch into the suprarenal glands, located on top of the kidneys. The right and left **renal arteries**, which supply the kidneys, stem off the abdominal aorta just inferior to the suprarenal arteries. The right and left **gonadal arteries** arise near the

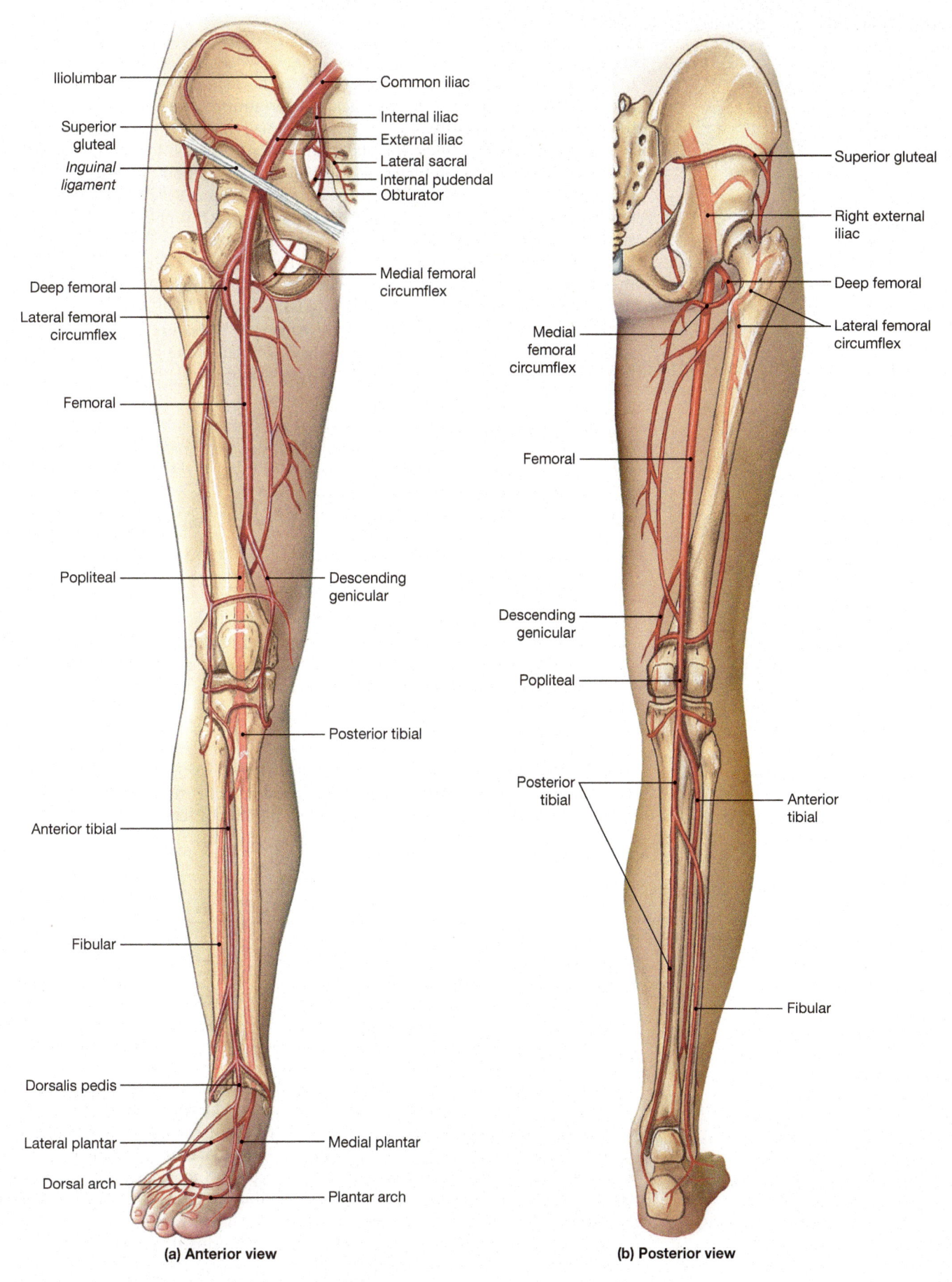

Figure 23.6 Major Arteries of the Lower Limb
(a) Anterior view of arteries that supply the right lower limb. (b) Posterior view.

373

inferior mesenteric artery and bring blood to the reproductive organs. The right and left **lumbar arteries** originate near the terminus of the abdominal aorta and service the lower body wall.

At the level of the hips, the abdominal aorta divides into the right and left **common iliac** (IL-ē-ak) **arteries** (Figure 23.6). Each common iliac artery descends through the pelvic cavity and branches into an **external iliac artery**, which enters the lower limb, and an **internal iliac artery**, which supplies blood to the organs of the pelvic cavity. The external iliac artery pierces the abdominal wall and becomes the **femoral artery** of the thigh (Figure 23.6). A **deep femoral artery** that arises off the femoral artery supplies deep thigh muscles. The femoral artery passes though the posterior knee as the **popliteal** (pop-LIT-ē-al) **artery** and divides into the **posterior tibial artery** and the **anterior tibial artery**, each of which supply blood to the leg. The **fibular artery**, also called the *peroneal artery*, stems laterally off the posterior tibial artery. The arteries of the leg branch into the foot and anastomose at the **dorsal arch** and the **plantar arch**.

QuickCheck Questions

3.1 What are the three branches of the celiac trunk?

3.2 Which arteries supply the intestines?

3.3 What does the external iliac artery become in the lower limb?

3 Materials

- ☐ Vascular system chart
- ☐ Torso model
- ☐ Lower limb model

Procedures

1. Review the arteries in Figures 23.5 and 23.6.
2. On the torso model, locate the celiac trunk and its three branches. Identify the superior and inferior mesenteric arteries and the four sets of paired arteries that stem from the abdominal aorta.
3. On the torso model, observe how the abdominal aorta branches into the left and right common iliac arteries.
4. On the lower limb model, locate the major arteries that supply the lower limb.
5. On your body, trace the location of your abdominal aorta, common iliac artery, external iliac artery, femoral artery, popliteal artery, and posterior tibial artery. ■

LAB ACTIVITY 4 Veins of the Head, Neck, and Upper Limb

Once you have learned the major systemic arteries, identifying the systemic veins is easy because most arteries have a corresponding vein (Figure 23.7). Unlike arteries, many veins are superficial and are easily seen under the skin. Systemic veins are usually painted blue on vascular models and torso models to indicate that they transport deoxygenated blood. When identifying veins, work in the direction of blood flow, from the periphery toward the heart.

Blood in the brain drains into large veins called *sinuses* (Figure 23.8). (Do not confuse this meaning of *sinus* with the more familiar meaning "cavity," as, for instance, the sinuses of the skull presented in Exercise 7.) Small-diameter veins deep inside the brain drain into progressively larger veins that empty into the **superior sagittal sinus** located in the falx cerebri that separates the cerebral hemispheres. Other venous sinuses and the superior sagittal sinus drain into the **internal jugular vein**, which exits the skull via the jugular foramen, descends the neck, and empties into the **brachiocephalic vein**. Superficial veins that drain the face and scalp empty into the **external jugular vein**, which descends the neck to join the **subclavian vein**. The **internal thoracic vein** joins the left brachiocephalic vein and drains the anterior thoracic wall. The right and left brachiocephalic veins merge

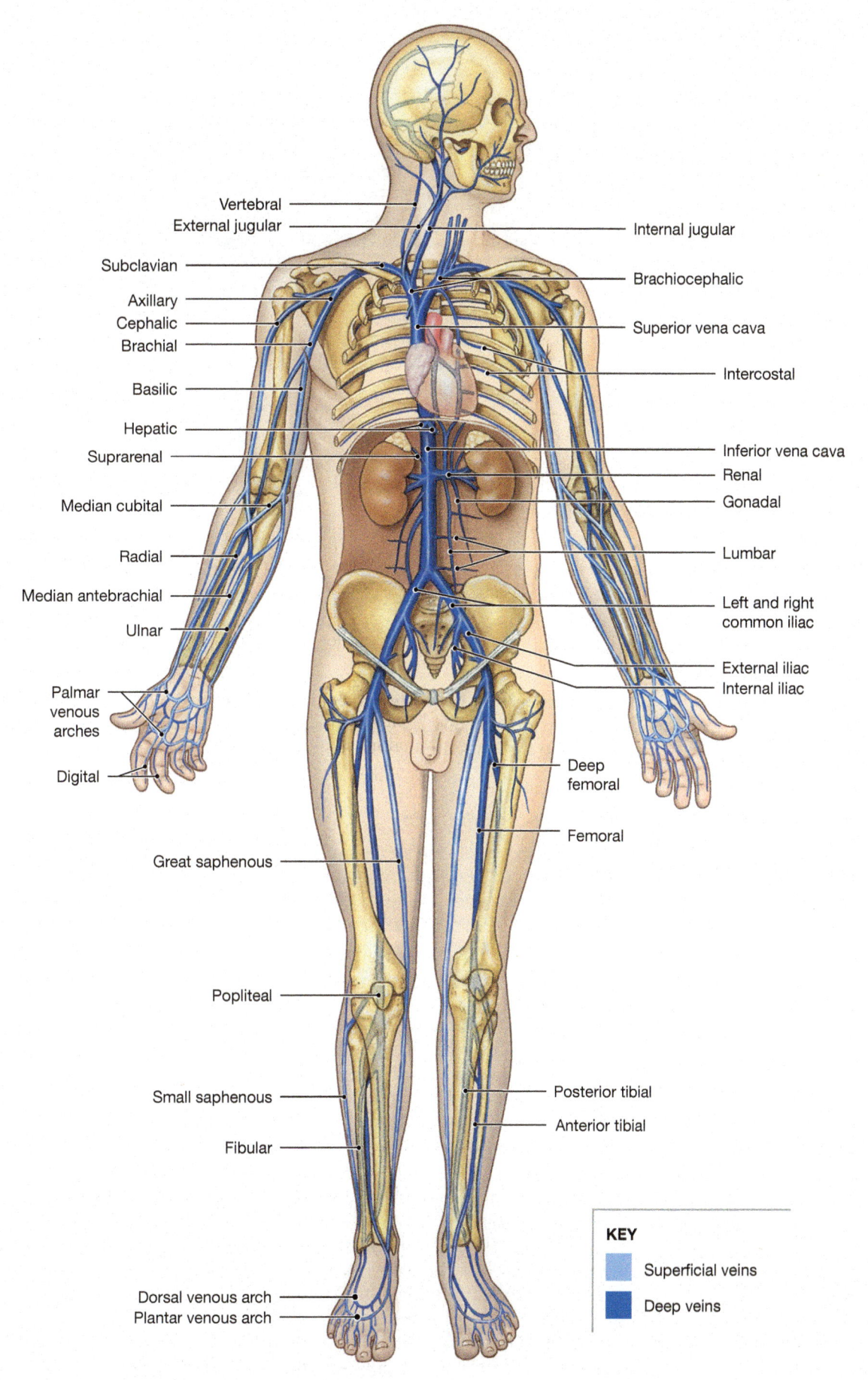

Figure 23.7 Overview of the Systemic Venous Pathways

375

376 EXERCISE 23 The Systemic Circuit

Veins of the head and neck, lateral view

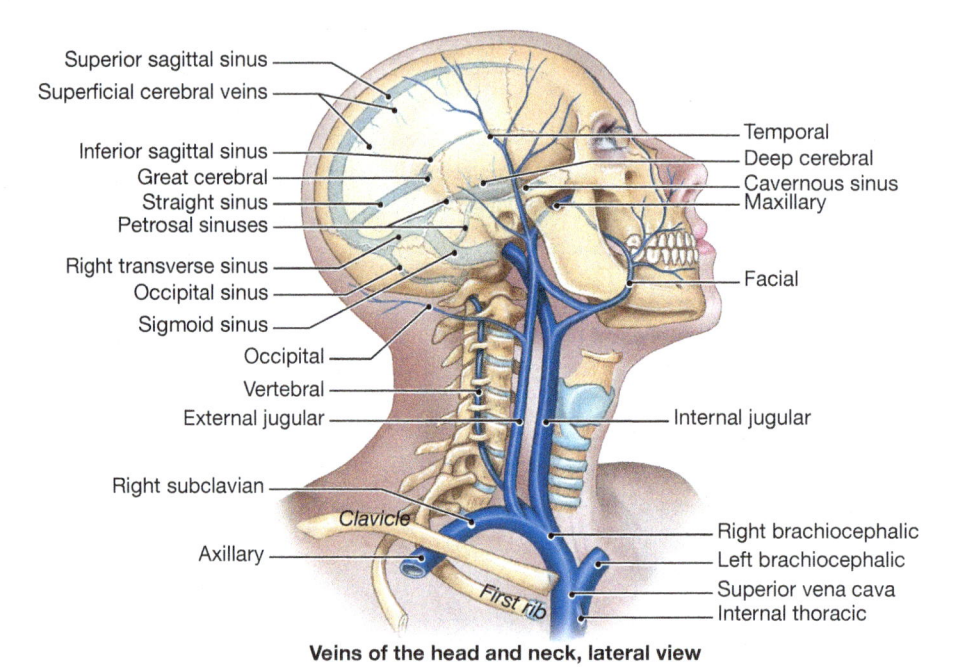

Figure 23.8   Major Veins of the Head, Neck, and Brain
An oblique lateral view of the head and neck that shows the major superficial and deep veins.

at the **superior vena cava** and empty deoxygenated blood into the right atrium of the heart. The blood then enters the right ventricle, which contracts and pumps the blood to the lungs through the pulmonary circuit.

Making Connections   **Brachiocephalic Veins**

> One difference between the systemic arteries and the systemic veins is that the venous pathway has both a right and a left brachiocephalic vein, each formed by the merging of subclavian, vertebral, internal jugular, and external jugular veins. The arterial pathway has a single brachiocephalic trunk that branches into the right common carotid artery and right subclavian artery. On the left side of the body, the common carotid artery and subclavian artery originate directly off the aortic arch, as noted earlier.

Figure 23.9 illustrates the venous drainage of the upper limb, chest, and abdomen. Small veins in the fingers drain into **digital veins** that empty into a network of **palmar venous arches**. These vessels drain into the **cephalic vein**, which ascends along the lateral margin of the arm. The **median antebrachial vein** ascends to the elbow, is joined by the **median cubital vein** that crosses over from the cephalic vein, and becomes the **basilic vein**. The median cubital vein is often used to collect blood from an individual. Also in the forearm are the **radial** and **ulnar veins**, which fuse above the elbow into the **brachial vein**. The brachial and basilic veins meet at the armpit as the **axillary vein**, which joins the cephalic vein and becomes the subclavian vein. The subclavian vein plus veins from the neck and head drain into the brachiocephalic vein, which then empties into the superior vena cava, which empties into the right atrium.

QuickCheck Questions

4.1 Which two veins combine to form the superior vena cava?

4.2 Where is the cephalic vein?

4.3 Where is the superior sagittal sinus?

EXERCISE 23 The Systemic Circuit 377

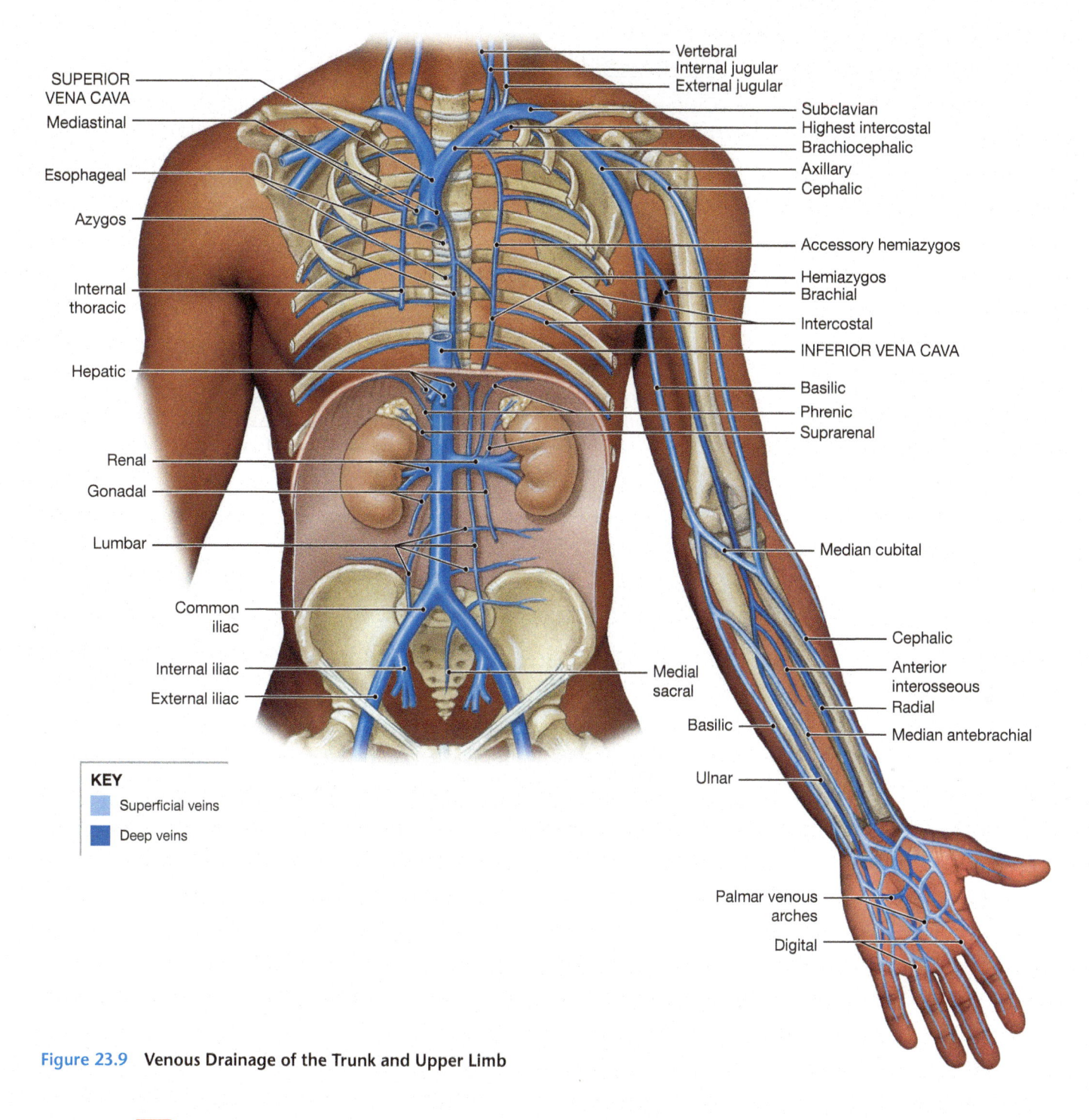

Figure 23.9 Venous Drainage of the Trunk and Upper Limb

4 Materials

- ☐ Vascular system chart
- ☐ Torso model
- ☐ Head model
- ☐ Upper limb model

Procedures

1. Review the head, neck, and upper limb veins in Figures 23.7, 23.8, and 23.9.
2. On the head model, identify the superior sagittal sinus and other veins that drain the head into the external and internal jugular veins.
3. Using the torso and upper limb models, start at one hand and name the veins that drain the limb and shoulder. Notice how the right and left brachiocephalic veins join as the superior vena cava.
4. On your body, trace your cephalic vein, subclavian vein, brachiocephalic vein, and superior vena cava. ■

LAB ACTIVITY 5: Veins of the Lower Limb and Abdominopelvic Cavity

Veins that drain the lower limbs and abdominal organs empty into the **inferior vena cava**, the large vein that pierces the diaphragm and empties into the right atrium of the heart. Figure 23.10 illustrates the venous drainage of the lower limb. Just like the hand, the foot contains digital veins, which in the foot drain into the **plantar venous arch** and the **dorsal venous arch**, which drain into the lateral **fibular vein** (also called *peroneal vein*) and the **anterior tibial vein**, which is located on the medial aspect of the anterior leg. These veins, along with the **posterior tibial vein**, merge and become the **popliteal vein** of the posterior knee. The **small saphenous** (sa-FĒ-nus) **vein**, which ascends from the ankle to the knee and drains blood from superficial veins, also empties into the popliteal vein. Superior to the knee, the popliteal vein becomes the **femoral vein**, which ascends along the femur to the inferior pelvic girdle, where it joins the **deep femoral vein** at the **external iliac vein**. The **great saphenous vein** ascends from the medial side of the ankle to the superior thigh and drains into the external iliac vein. In the pelvic cavity, the external iliac vein and the **internal iliac vein** fuse to form the **common iliac vein**. The right and left common iliac veins merge and drain into the inferior vena cava.

Six major veins from the abdominal organs drain blood into the inferior vena cava. The **lumbar veins** (see Figure 23.7) drain the muscles of the lower body wall and the spinal cord and empty into the inferior vena cava close to the common iliac veins. A pair of **gonadal veins** empty blood from the reproductive organs into the inferior vena cava above the lumbar veins. Pairs of **renal** and **suprarenal veins** drain into the inferior vena cava next to their respective organs. Before entering the thoracic cavity to drain blood into the right atrium, the inferior vena cava collects blood from the **hepatic veins** that drain the liver (Figure 23.11) and the **phrenic veins** from the diaphragm.

Veins leaving the digestive tract are diverted to the liver before continuing on to the heart. The inferior and superior mesenteric veins drain nutrient-rich blood from the digestive tract. These veins empty into the **hepatic portal vein** (Figure 23.11), which passes the blood through the liver, where blood sugar concentration is regulated. Phagocytic cells in the liver cleanse the blood of any microbes that may have entered it through the mucous membrane of the digestive system. Blood from the hepatic arteries and hepatic portal vein mixes in the liver and is returned to the inferior vena cava by the hepatic veins.

QuickCheck Questions

5.1 What are the vessels that drain blood from the lower limb into the inferior vena cava?

5.2 Which veins drain into the hepatic portal vein?

5 Materials

- ☐ Vascular system chart
- ☐ Torso model
- ☐ Lower limb model

Procedures

1. Review the veins in Figures 23.7, 23.9, 23.10, and 23.11.
2. On the lower limb model, identify the veins that drain blood from the ankle to the inferior vena cava.
3. On the torso model, identify the veins that drain the major abdominal organs. Locate where the superior and inferior mesenteric veins drain into the hepatic portal vein.
4. On your body, trace the location of the veins in your lower limb.

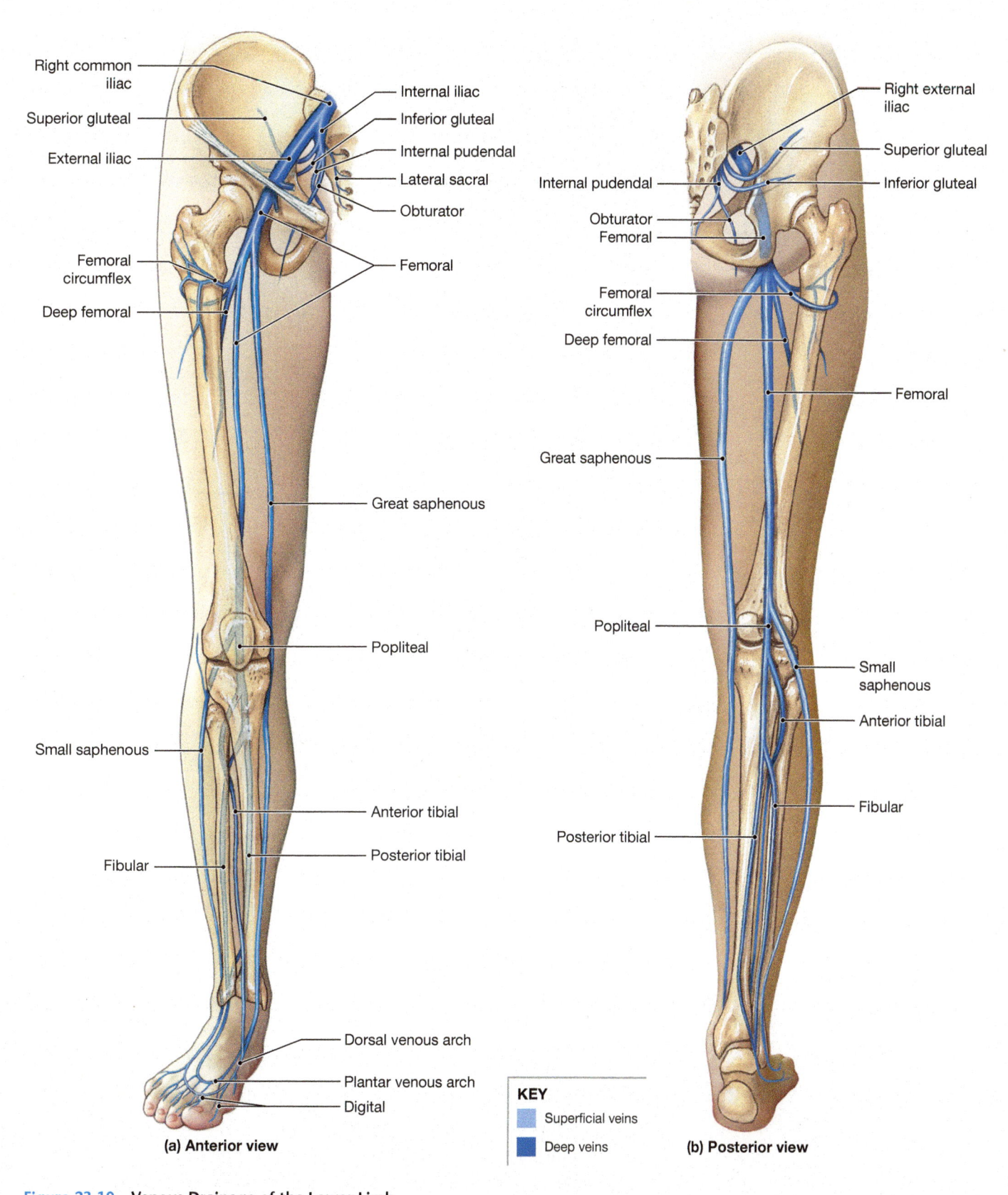

Figure 23.10 Venous Drainage of the Lower Limb
(a) Anterior view of right lower limb. (b) Posterior view.

379

380 EXERCISE 23 The Systemic Circuit

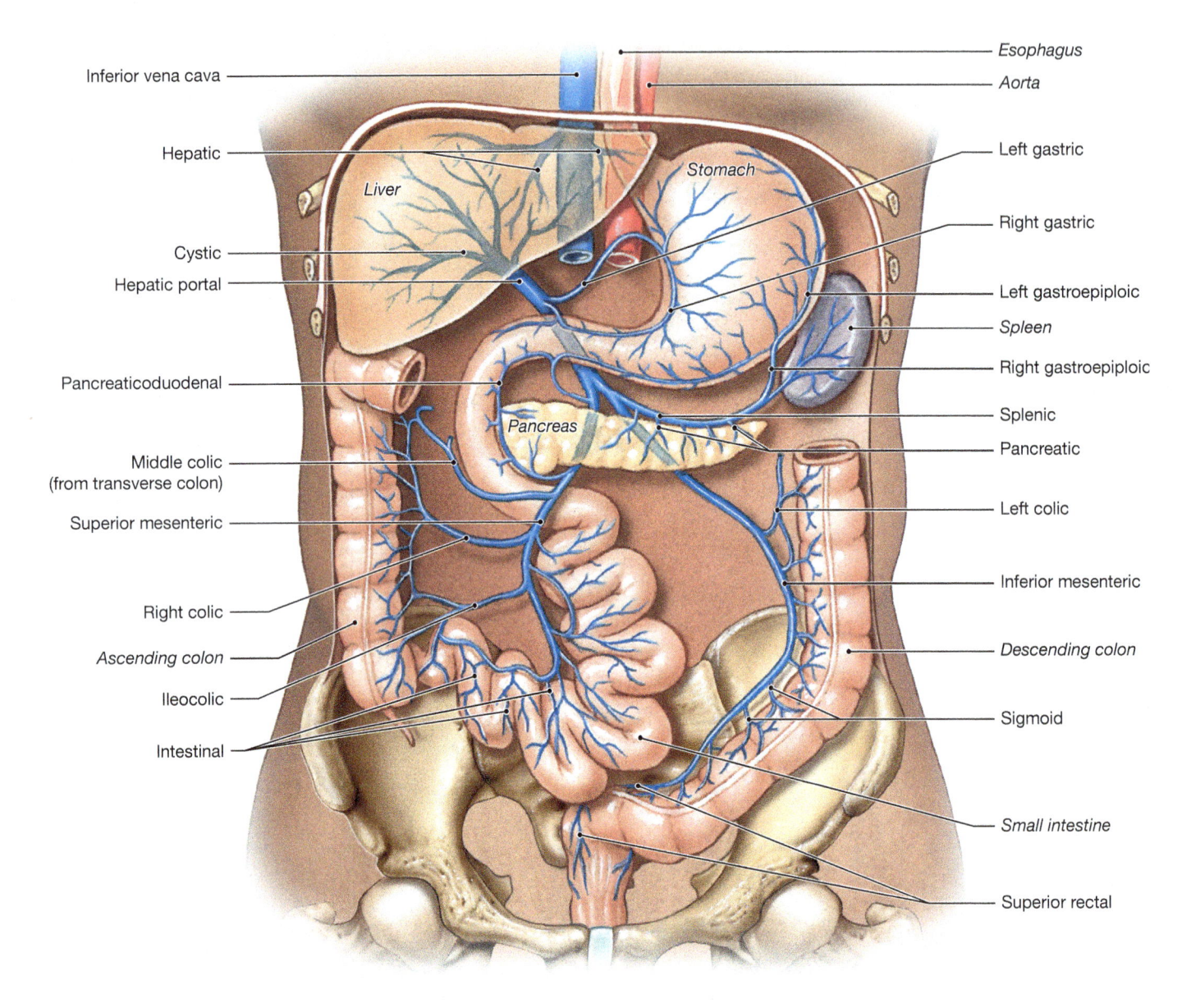

Figure 23.11 The Hepatic Portal System

5. Although you have studied the arterial and venous divisions separately, they are anatomically connected to each other by capillaries. To reinforce this connectedness, practice identifying blood vessels while tracing the following systemic routes:
   a. From the heart through the left upper limb and back to the heart.
   b. From the heart through the brain and back to the heart.
   c. From the heart through the liver and back to the heart.
   d. From the heart through the right lower limb and back to the heart. ■

limbs A Regional Look

Blood vessels and nerves follow similar distributions to and from organs, and it is therefore convenient to use regional terms to name these parallel structures. Figure 23.12 is an anterior view of a deep dissection of the right shoulder, exposing portions of the brachial plexus and the regional vascularization. Observe in the axilla the cords of the brachial plexus and the axillary artery. As these structures enter the brachium, they branch to enter the local muscles. Note the deep brachial artery that supplies blood to the posterior muscles of the arm. Also note the dissection of the biceps brachii muscle's origin to reveal the two heads.

Figure 23.13 is superficial dissection of the anterior groin region. Muscles that flex the ball-and-socket joint of the hip are located in this region and are served by the femoral artery, which similar to function of the brachial artery, gives rise to a deep artery to nourish posterior structures. The femoral nerve innervates the anterior thigh muscles. Following the saphenous vein is the saphenous nerve, which provides sensation to the medial surface of the leg and foot.

Right axillary region, anterior view

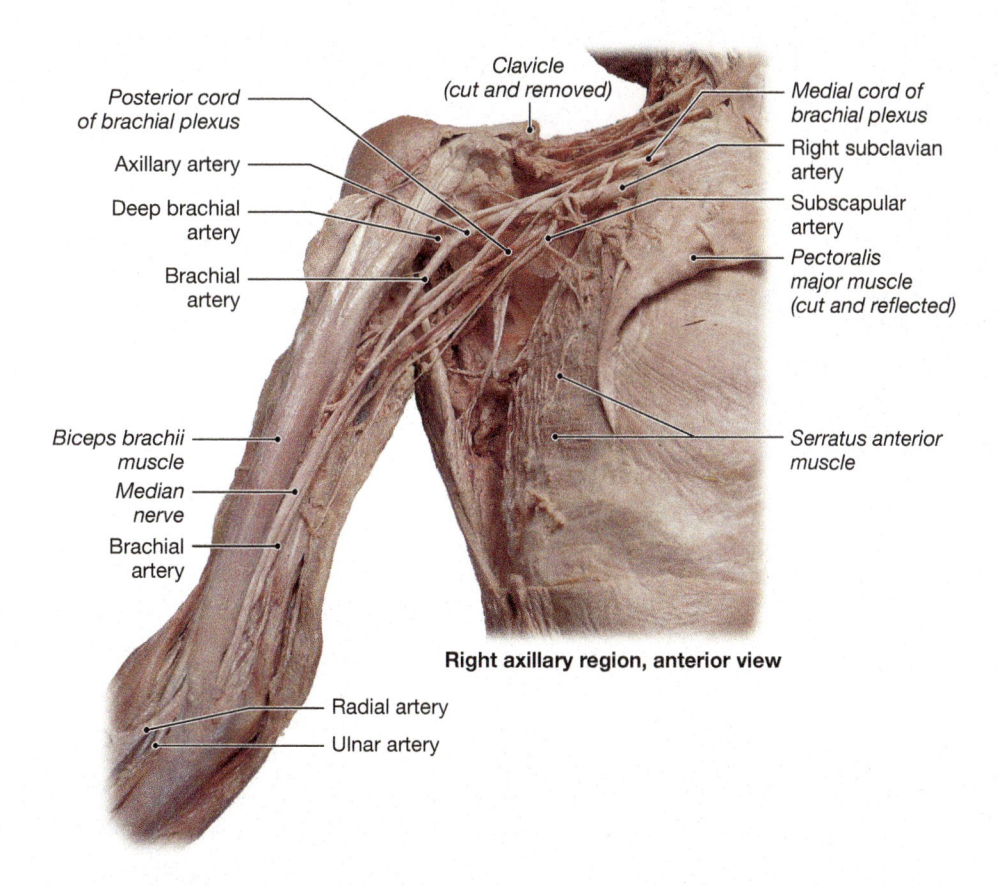

Figure 23.12 Dissection of Right Axilla
Anterior view of the right axillary region dissected to show blood vessels and nerves.

381

Femoral vessels

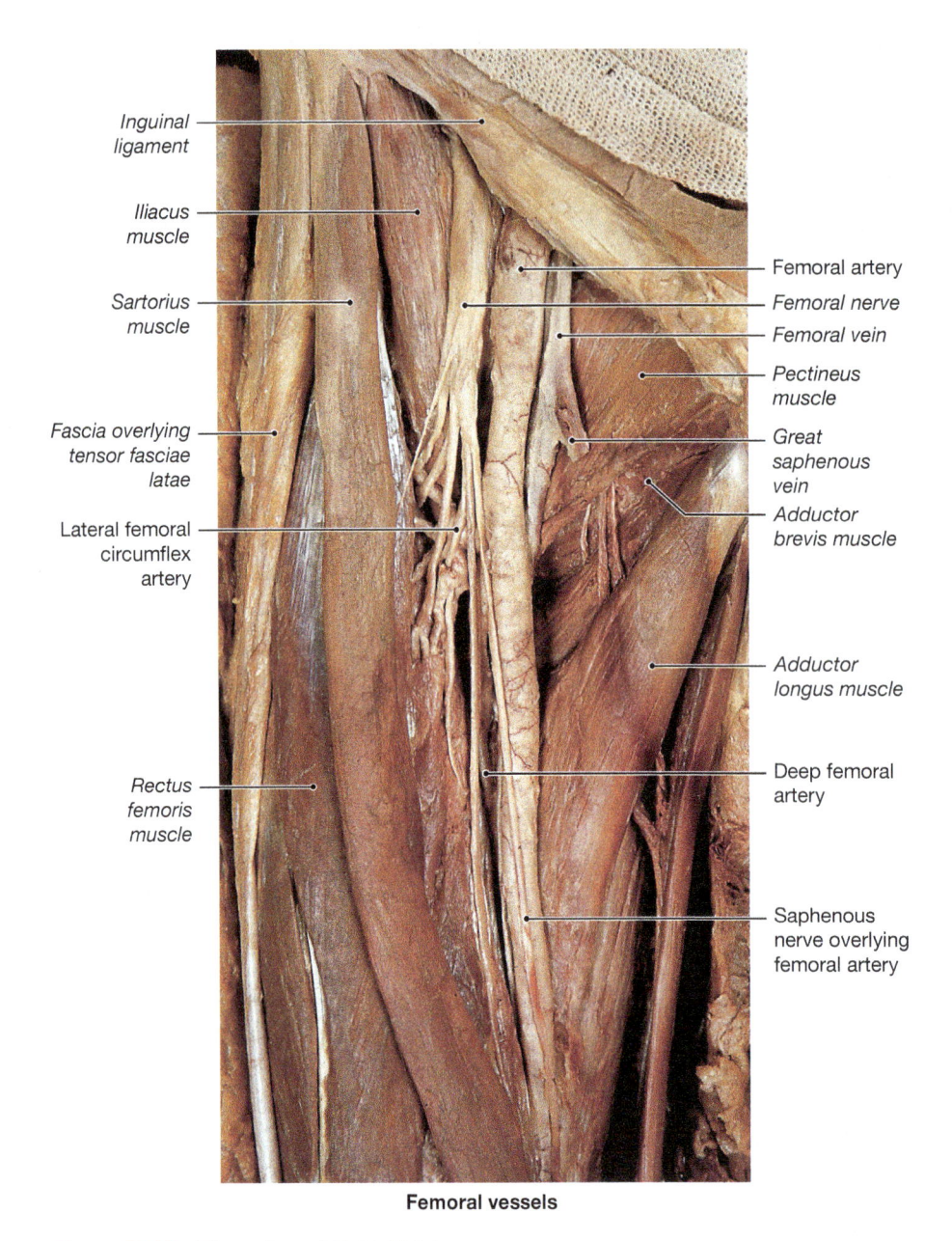

Figure 23.13 Dissection of Right Thigh
Anterior view of the right thigh dissected to show blood vessels and nerves.

EXERCISE 23

LAB REPORT

The Systemic Circuit

Name _____
Date _____
Section _____

A. Matching

Match each term in the left column with its correct description from the right column.

_____ 1. artery in armpit
_____ 2. artery that has three branches
_____ 3. artery on right side only
_____ 4. long vein of lower limb
_____ 5. carries deoxygenated blood to liver
_____ 6. artery to large intestine
_____ 7. cerebral anastomosis
_____ 8. long vein of upper limb
_____ 9. vein in knee
_____ 10. artery to reproductive organ
_____ 11. major artery in neck
_____ 12. found only in veins

A. superior mesenteric
B. popliteal
C. cephalic
D. common carotid
E. gonadal
F. valves
G. axillary
H. great saphenous
I. hepatic portal
J. cerebral arterial circle
K. celiac
L. brachiocephalic trunk

B. Labeling

Label the arteries in Figure 23.14 and the veins in Figure 23.15.

C. Short-Answer Questions

1. Which blood vessel is normally used to obtain blood from a patient?

2. What is the function of valves in the peripheral veins?

3. Describe the major blood vessels that return deoxygenated blood to the right atrium of the heart.

EXERCISE 23

LAB REPORT

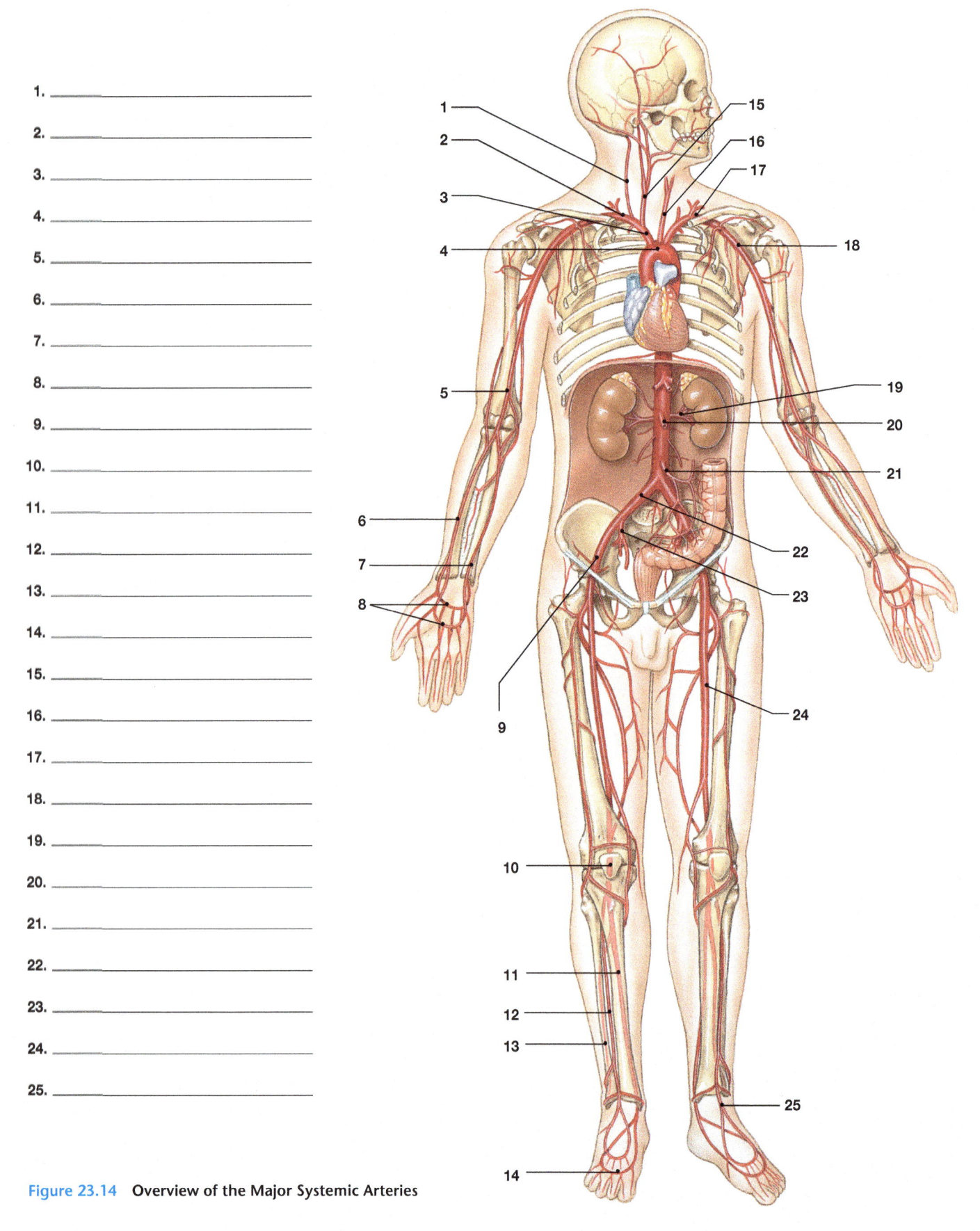

1. _____
2. _____
3. _____
4. _____
5. _____
6. _____
7. _____
8. _____
9. _____
10. _____
11. _____
12. _____
13. _____
14. _____
15. _____
16. _____
17. _____
18. _____
19. _____
20. _____
21. _____
22. _____
23. _____
24. _____
25. _____

Figure 23.14 Overview of the Major Systemic Arteries

384

LAB REPORT EXERCISE 23

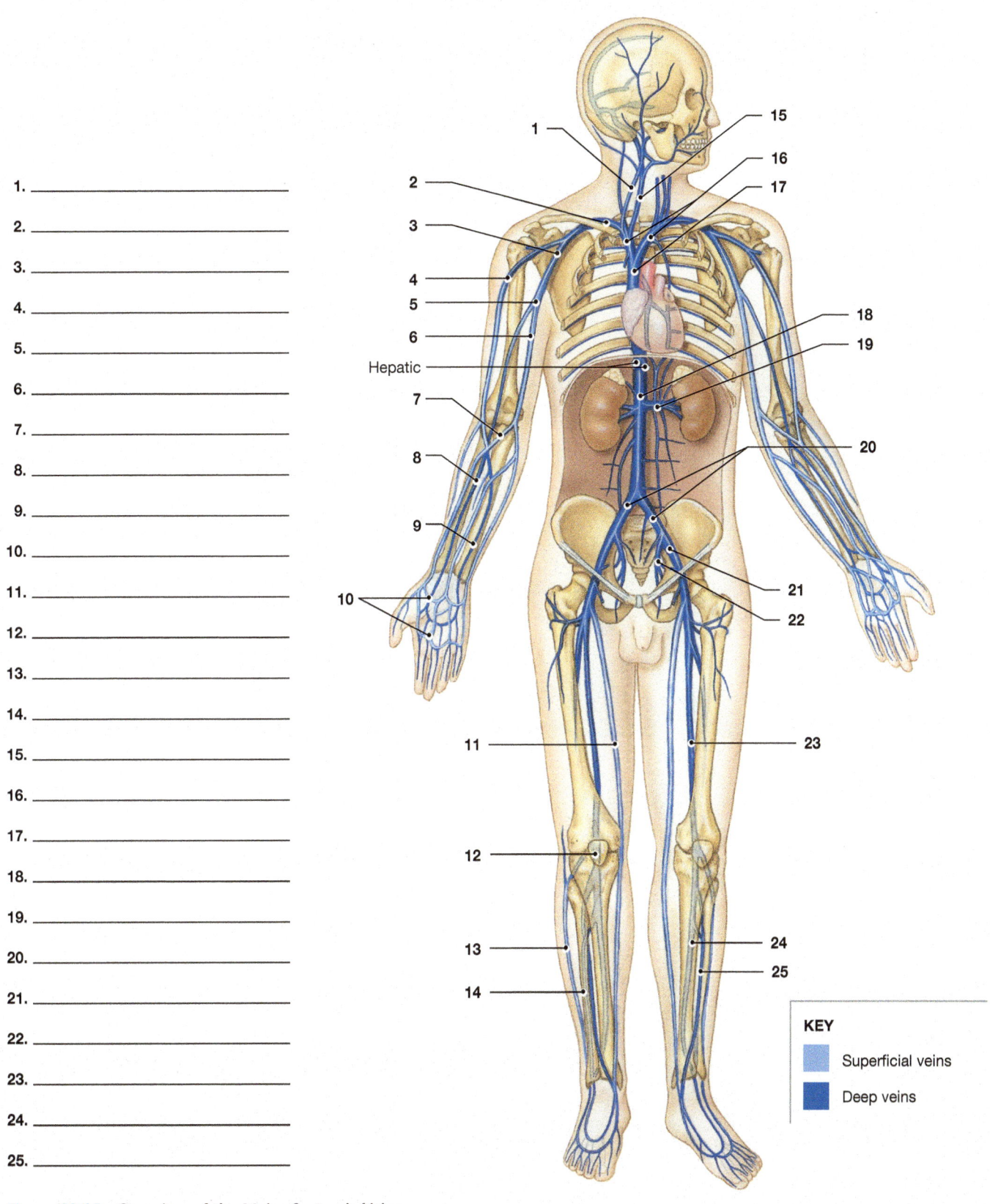

1. _____
2. _____
3. _____
4. _____
5. _____
6. _____
7. _____
8. _____
9. _____
10. _____
11. _____
12. _____
13. _____
14. _____
15. _____
16. _____
17. _____
18. _____
19. _____
20. _____
21. _____
22. _____
23. _____
24. _____
25. _____

Figure 23.15 Overview of the Major Systemic Veins

385

EXERCISE 23 — LAB REPORT

D. Analysis and Application

1. How does the cerebral arterial circle ensure that the brain has a constant supply of blood?

2. How is the venous drainage of the digestive organs organized to help regulate blood sugar concentration?

3. How is the anatomy of the arteries that run from the aorta to the right upper limb different from that of the arteries that run from the aorta to the left upper limb?

EXERCISE 24

Lymphatic System

OBJECTIVES

On completion of this exercise, you should be able to:

- List the functions of the lymphatic system.
- Describe the exchange of blood plasma, interstitial fluid, and lymph.
- Describe the structure of a lymph node.
- Explain how the lymphatic system drains into the vascular system.
- Describe the gross anatomy and basic histology of the spleen.

LAB ACTIVITIES

1. Lymphatic Vessels   389
2. Lymphoid Tissues and Lymph Nodes   390
3. The Spleen   393

   A Regional Look: Female Breast   395

The lymphatic system includes the lymphatic vessels, lymph nodes, tonsils, spleen, and thymus gland (Figure 24.1). **Lymphatic vessels** transport liquid called **lymph** from the extracellular spaces to the veins of the cardiovascular system. Scattered along each lymphatic vessel are **lymph nodes** that contain lymphocytes (one type of white blood cell, Exercise 21) and macrophages (Exercise 4). The macrophages remove invading microbes and other substances from the lymph before the lymph is returned to the blood. Although lymphocytes are classified as a formed element of the blood, they are the main cells of the lymphatic system and colonize dense populations in lymph nodes and the spleen. The antigens present in invading pathogens and other foreign substances cause the lymphocytes to produce antibodies to defend against the antigens. As macrophages capture the antigens in the lymph, lymphocytes exposed to the antigens activate the immune system to respond to the intruding cells. The lymphocytes called B cells produce antibodies that chemically combine with and destroy the antigens.

The thymus gland is involved in the development of the functional immune system in infants. In adults, this gland controls the maturation of lymphocytes. We covered the thymus gland with the endocrine system in Exercise 20.

Pressure in the systemic capillaries forces liquids and solutes out of the capillaries and into the interstitial spaces. This constant renewal of extracellular fluid bathes the cells with nutrients, dissolved gases, hormones, and other materials. After this exchange, osmotic pressure forces most of the extracellular fluid back into the capillaries. Some extracellular fluid drains into lymphatic vessels and becomes lymph. The lymph travels through the lymphatic vessels to the lymph nodes, where macrophages remove abnormal cells and microbes from the lymph.

Two **lymphatic ducts** join with veins near the heart and return the lymph to the blood. Approximately 3 liters of liquid per day is forced out of the capillaries and flows through lymphatic vessels as lymph.

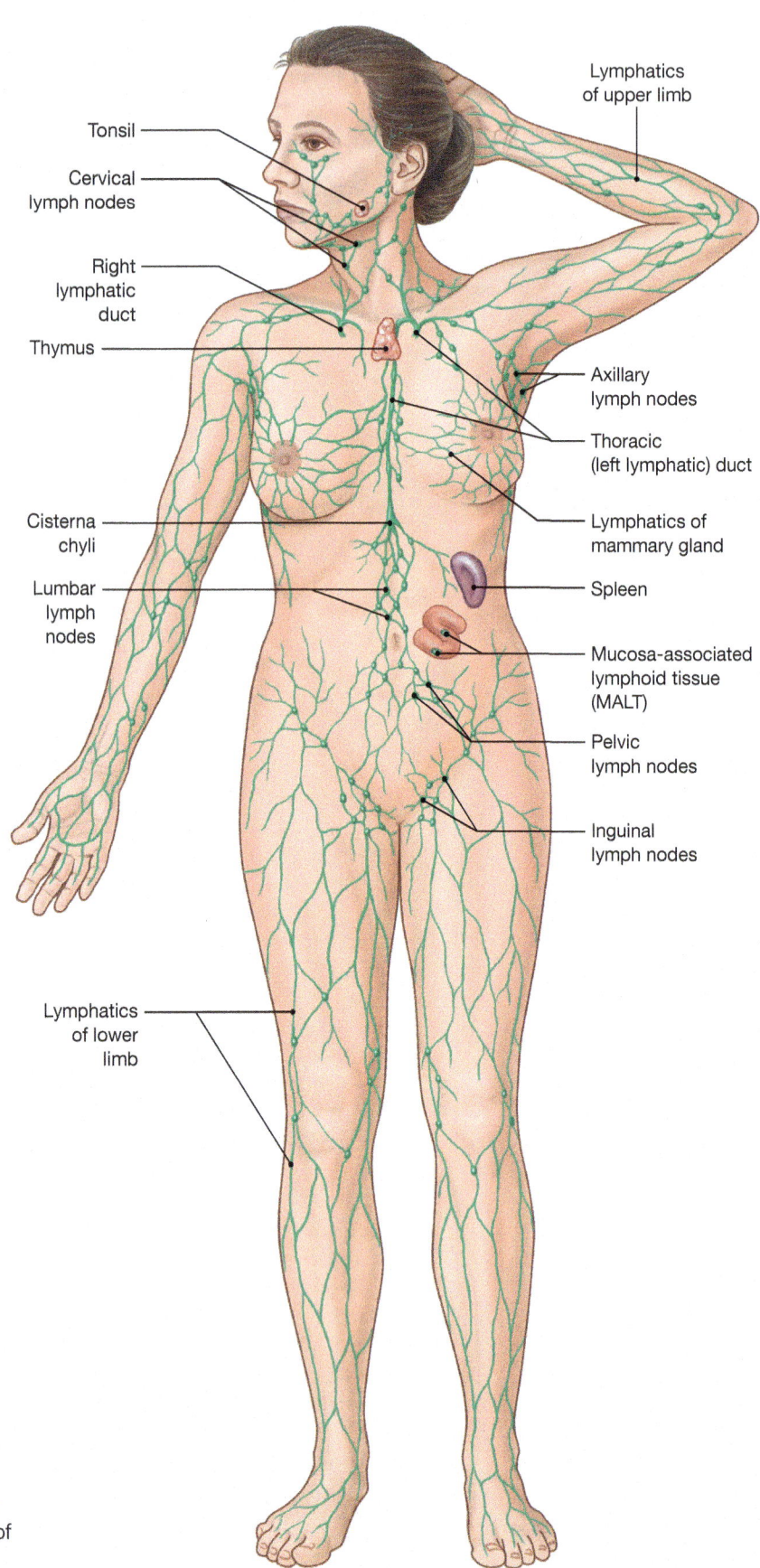

Figure 24.1 **Lymphatic System**
An overview of the distribution of lymphatic vessels, lymph nodes, and the other organs of the lymphatic system.

EXERCISE 24 Lymphatic System   389

LAB ACTIVITY 1 Lymphatic Vessels

Lymphatic vessels, or simply **lymphatics**, occur next to the vessels of the vascular system, as Figure 24.2 shows. Lymphatic vessels are structurally similar to veins. The vessel wall has similar layers and **lymphatic valves** to prevent backflow of liquid. Lymphatic pressures are very low, and the lymphatic valves are close together to keep the lymph circulating toward the body trunk. The lymphatic capillaries, which gradually expand to become the lymphatic vessels, are closed at the ends that lie near the arterial blood capillaries. Lymph enters the lymphatic system at or near these closed ends and then moves into the lymphatic vessels, which conduct the lymph toward the body trunk and into the large-diameter lymphatics that empty into veins near the heart. Smaller lymphatic vessels combine into larger **lymphatic trunks** that eventually converge to empty lymph into the blood.

Two large lymphatic vessels, the **thoracic duct** and the **right lymphatic duct**, return lymph to the venous circulation (Figure 24.1 and 24.3). Most of the lymph is returned to the circulation by the thoracic duct, which commences at the level of the second lumbar vertebra on the posterior abdominal wall behind the abdominal aorta. Lymphatics from the lower limbs, pelvis, and abdomen drain into an inferior sac-like portion of the thoracic duct called the **cisterna chyli** (KĪ-lē; *cistern,* storage well; *chyl,* juice). The thoracic duct ascends the abdomen and pierces the diaphragm. At the base of the heart, the thoracic duct joins with the left subclavian vein to return the lymph to the blood. The only lymph that does not drain into the thoracic duct is from lymphatic vessels in the right upper limb and the right side of the chest, neck, and head. These areas drain into the right lymphatic duct near the right clavicle. This duct empties lymph at or near the junction of the right internal jugular and the right subclavian veins near the base of the heart.

QuickCheck Questions

1.1 What is lymph?
1.2 Where is lymph returned to the vascular system?
1.3 What is the function of lymphocytes?

(a) Association of blood capillaries, tissue, and lymphatic capillaries

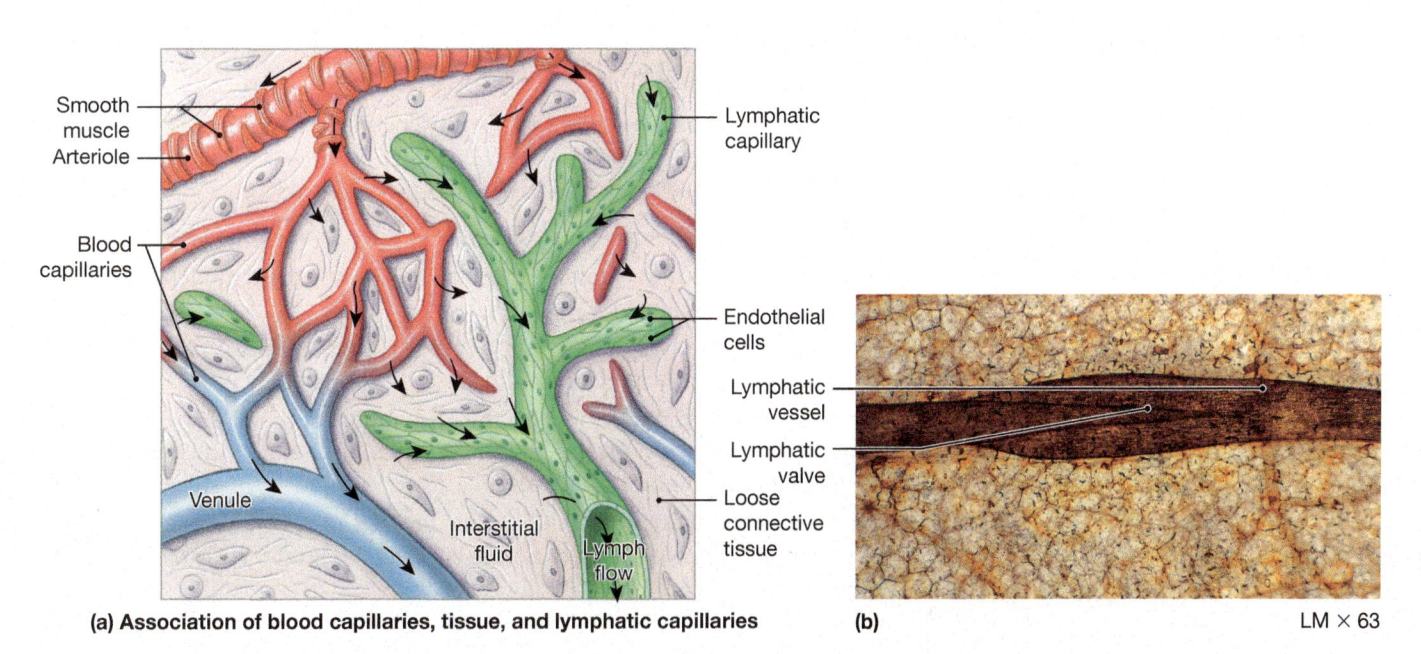

(b) LM × 63

Figure 24.2 Lymphatic Capillaries
(a) A three-dimensional view of the association of blood capillaries, tissue, interstitial fluid, and lymphatic capillaries. Arrows show the direction of interstitial fluid and lymph movement. **(b)** A valve in a small lymphatic vessel. (LM × 63)

390   EXERCISE 24   Lymphatic System

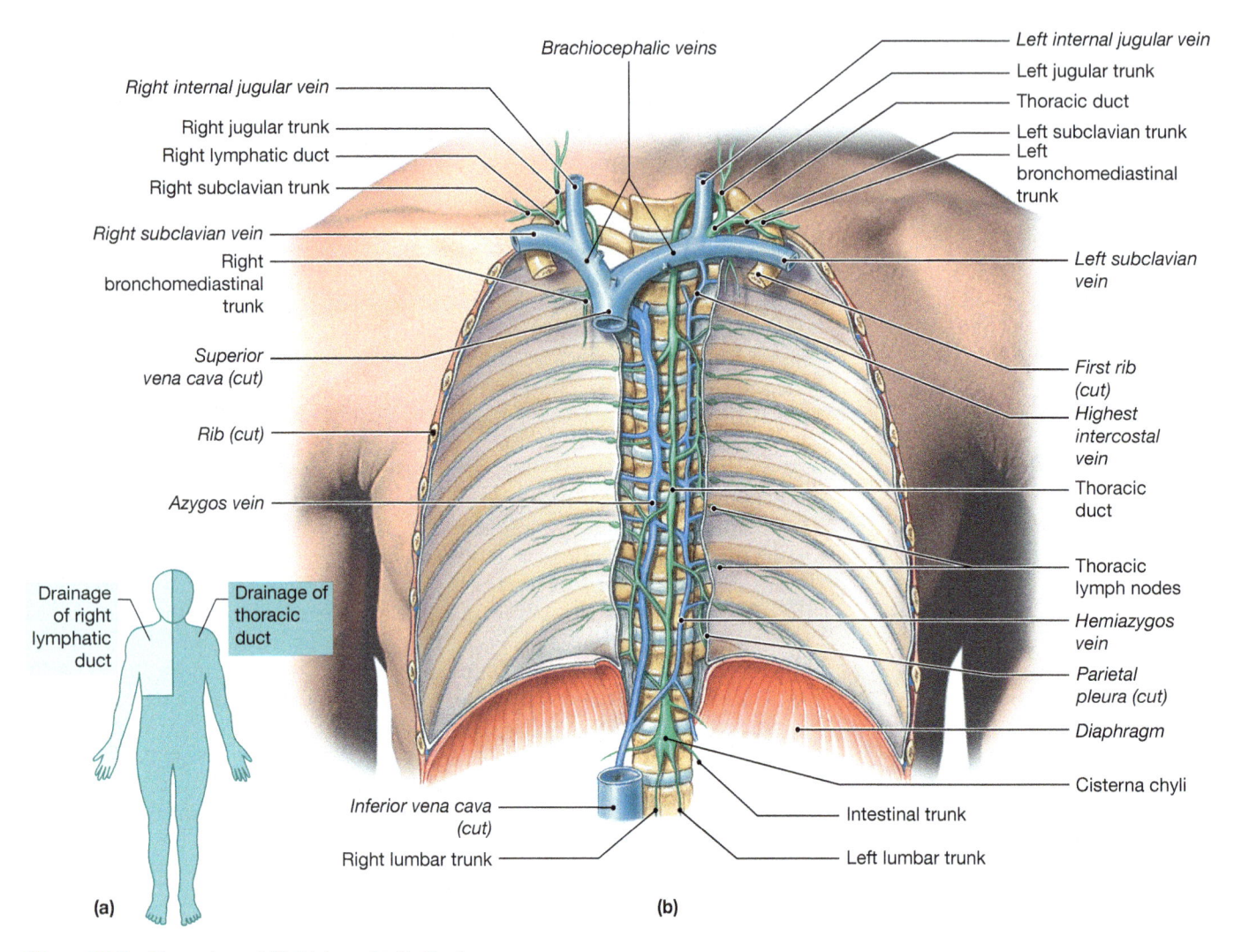

Figure 24.3   Thoracic and Right Lymphatic Ducts
(a) The thoracic duct carries lymph from the entire body inferior to the diaphragm and from the left side of the upper body. The smaller right lymphatic duct services the rest of the body. (b) The lymphatic vessels of the thoracic cavity. The thoracic duct empties into the left subclavian vein. The right lymphatic duct drains into the blood at or near where the right internal jugular and right subclavian veins meet.

1   Materials

- Torso model or lymphatic system chart
- Compound microscope
- Prepared slide of lymphatic vessel

Procedures

1. Locate the thoracic duct and the cisterna chyli on the torso model or lymphatic system chart.
2. Which areas of the body drain lymph into the thoracic duct? Where does this lymphatic vessel return lymph to the blood?
3. Locate the right lymphatic duct on the torso model or lymphatic system chart. Where does this lymphatic vessel join the vascular system? Which regions of the body drain lymph into this vessel?
4. Observe the lymphatic vessel slide at low power and search for a lymphatic valve. Consider in which direction the valve allows lymph to flow. ■

LAB ACTIVITY 2   Lymphoid Tissues and Lymph Nodes

Two major groups of lymphatic structures occur in connective tissues: **encapsulated lymph organs** and **diffuse lymphoid tissues**. The encapsulated lymph organs include lymph nodes, the thymus gland, and the spleen. Each encapsulated organ is separated from the surrounding connective tissue by a fibrous capsule. Diffuse lymphoid tissues do not have a defined boundary that separates them from the connective tissue.

Each lymph node is an oval organ that functions like a filter cartridge. As lymph passes though a node, phagocytes remove microbes, debris, and other antigens from the lymph. Lymph nodes are scattered throughout the lymphatic system, as depicted in Figure 24.1. Lymphatic vessels from the lower limbs pass through a network of inguinal lymph nodes located in the groin region (*inguinal* means "pertaining to the groin"). Pelvic and lumbar nodes filter lymph from the pelvic and abdominal lymphatic vessels. Many lymph nodes occur in the upper limbs and in the axillary and cervical regions. The breasts in women also contain many lymphatic vessels and nodes. Often, infections occur in a lymph node before they spread systemically. A swollen or painful lymph node suggests an increase in lymphocyte abundance and general immunological activity in response to antigens in the lymph nodes.

Figure 24.4 details the anatomy of a lymph node. Each node is encased in a dense connective tissue **capsule**. Collagen fibers from the capsule extend as partitions called **trabeculae** into the interior of the node. The region immediately inside the capsule is the **subcapsular sinus**. Interspersed in the subcapsular sinus are regions of cortical tissue called **outer cortex**, which is rich in B cells. As Figure 24.4 shows, each region of outer cortex surrounds a pale-staining **germinal center**, where lymphocytes are produced. Deep to the ring of germinal centers is the **deep cortex**, here lymphocytes carried into the node by the blood leave the blood and enter the node. The central region of the node is the **medulla**, where **medullary cords** made up of B cells and plasma cells extend into a network of **medullary**

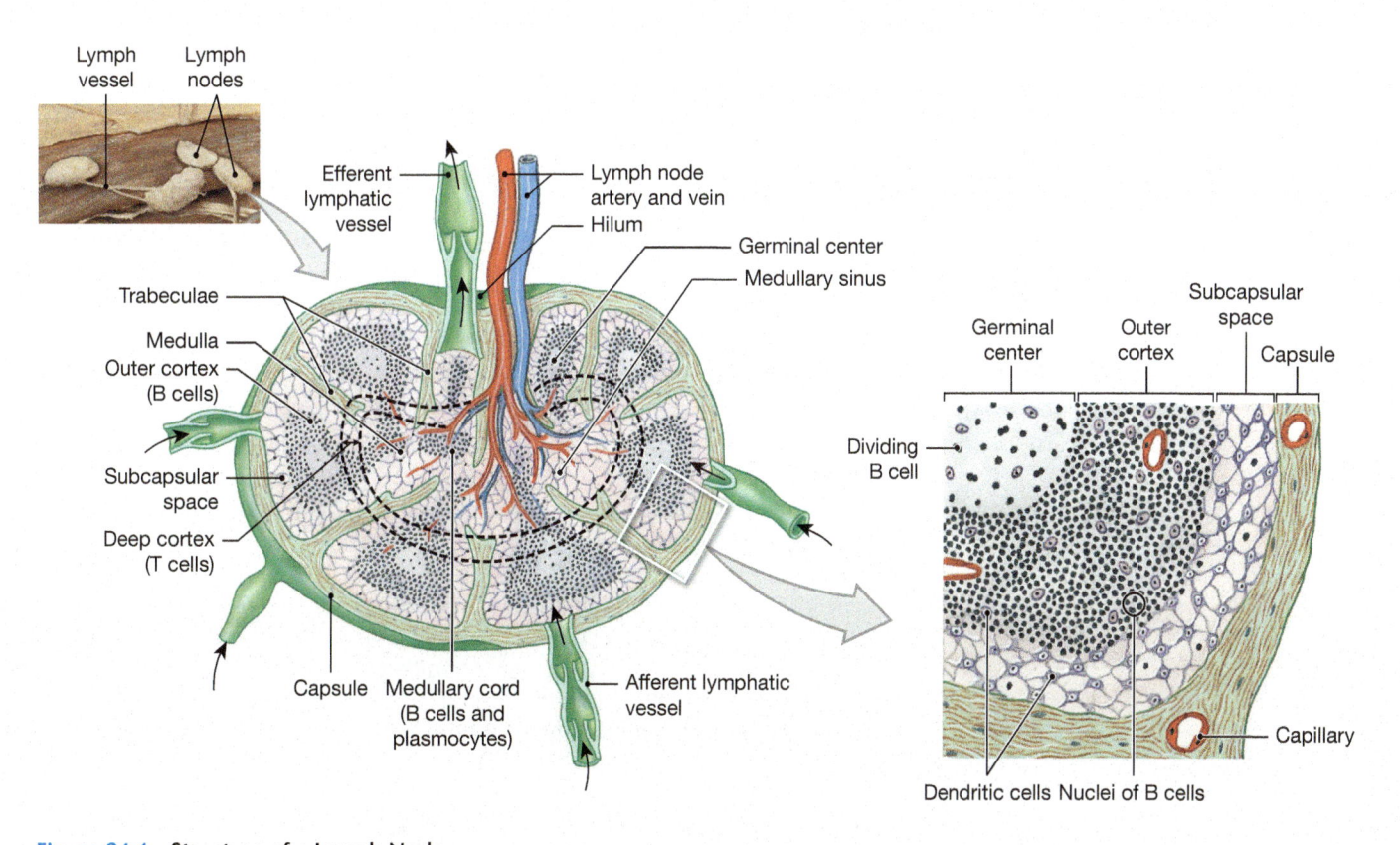

Figure 24.4 Structure of a Lymph Node

Lymph nodes are covered by a capsule of dense, fibrous connective tissue. Lymphatic vessels and blood vessels penetrate the capsule to reach the lymphoid tissue within. Note that a lymph node has several afferent lymphatic vessels but only one efferent vessel.

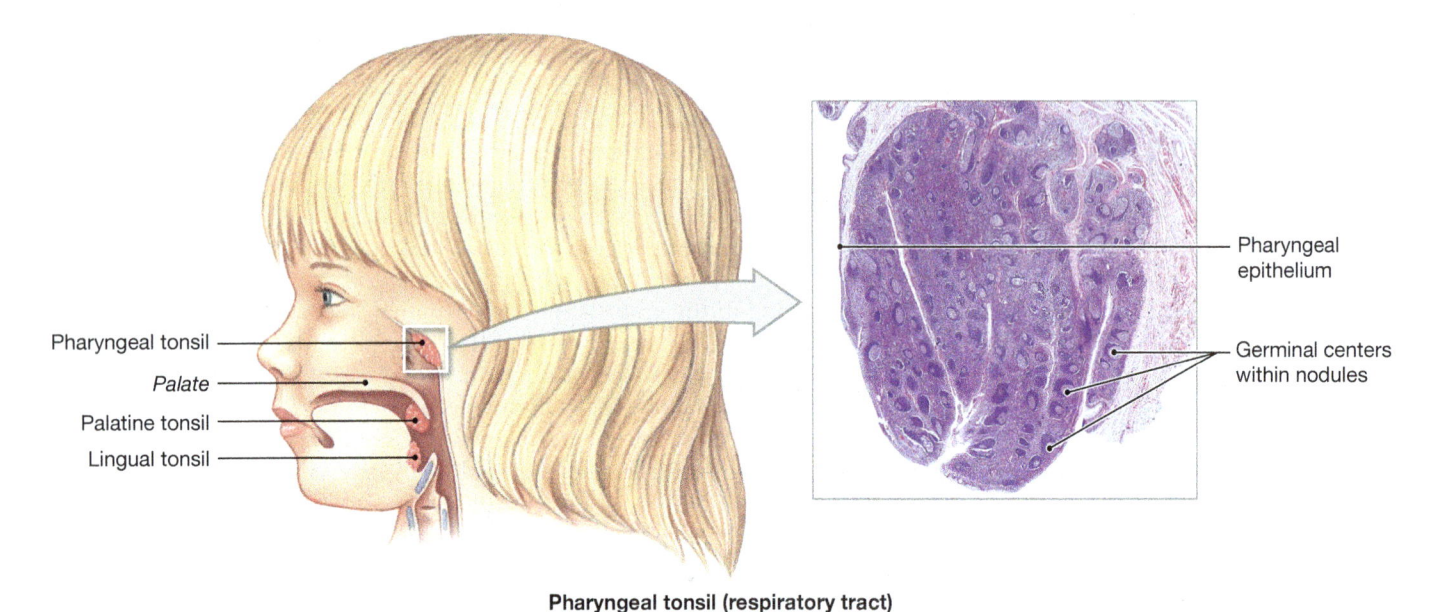

Figure 24.5 Tonsilar Lymphoid Tissue
Tonsils are lymphoid tissue in the mouth and pharynx. The photomicrograph shows the histological organization of a tonsil. (LM × 10)

sinuses. Lymph enters a node via several **afferent lymphatic vessels**. As the lymph flows through the subcapsular and medullary sinuses, macrophages phagocytize abnormal cells, pathogens, and debris. Draining the lymph node is a single **efferent lymphatic vessel**, which exits the node at a slit called the **hilus**.

Lymphoid nodules, which are diffuse lymphoid tissue, are found in connective tissue under the lining of the digestive, urinary, and respiratory systems. Microbes that penetrate the exposed epithelial surface pass into lymphoid nodules and into the lymph, where lymphocytes and macrophages destroy the foreign cells and remove them from the lymph. Some nodules have a germinal center where lymphocytes are produced by cell division.

Tonsils are lymphoid nodules in the mouth and pharynx (Figure 24.5). A pair of **lingual tonsils** sits at the posterior base of the tongue. The **palatine tonsils** are easily viewed because they hang off the posterior arches of the oral cavity. A single **pharyngeal tonsil**, or **adenoid**, is located in the upper pharynx near the opening to the nasal cavity.

Clinical Application   Tonsillitis

The lymphatic system usually has the upper hand in the immunological battle against invading bacteria and viruses. Occasionally, however, microbes manage to populate a lymphoid nodule. When the tonsils are infected, they swell and become irritated. This condition is called **tonsillitis** and is treated with antibiotics to control the infection. If the problem is recurrent, the tonsils are removed in a surgical procedure called a *tonsillectomy*. Usually, the palatine tonsils are removed. If the pharyngeal tonsil is also infected or is abnormally large, it is also removed during the procedure.

QuickCheck Questions

2.1 What are the names of the two types of lymphatic structures in the body?
2.2 Where are lymph nodes located?

EXERCISE 24 Lymphatic System   393

2 Materials

- ☐ Torso model
- ☐ Compound microscope
- ☐ Prepared slide of lymph node

Procedures

1. Review the structure of a lymph node in Figure 24.4.
2. On the torso model, locate the two pairs of tonsils and the single pharyngeal tonsil.
3. Examine the lymph node slide at low magnification and identify the capsule and trabeculae.
4. Change the microscope to high magnification and examine a germinal center inside the outer cortex. Identify the cells produced in the germinal center. ■

LAB ACTIVITY 3   The Spleen

The **spleen**, the largest lymphatic organ in the body, is located lateral to the stomach (Figure 24.6). A capsule surrounds the spleen and protects the underlying tissue of **red pulp** and **white pulp**. The color of the red pulp is due to the blood that

(a) Abdomen, transverse section

(b) Visceral surface of spleen

(c) Histological appearance of spleen LM × 38

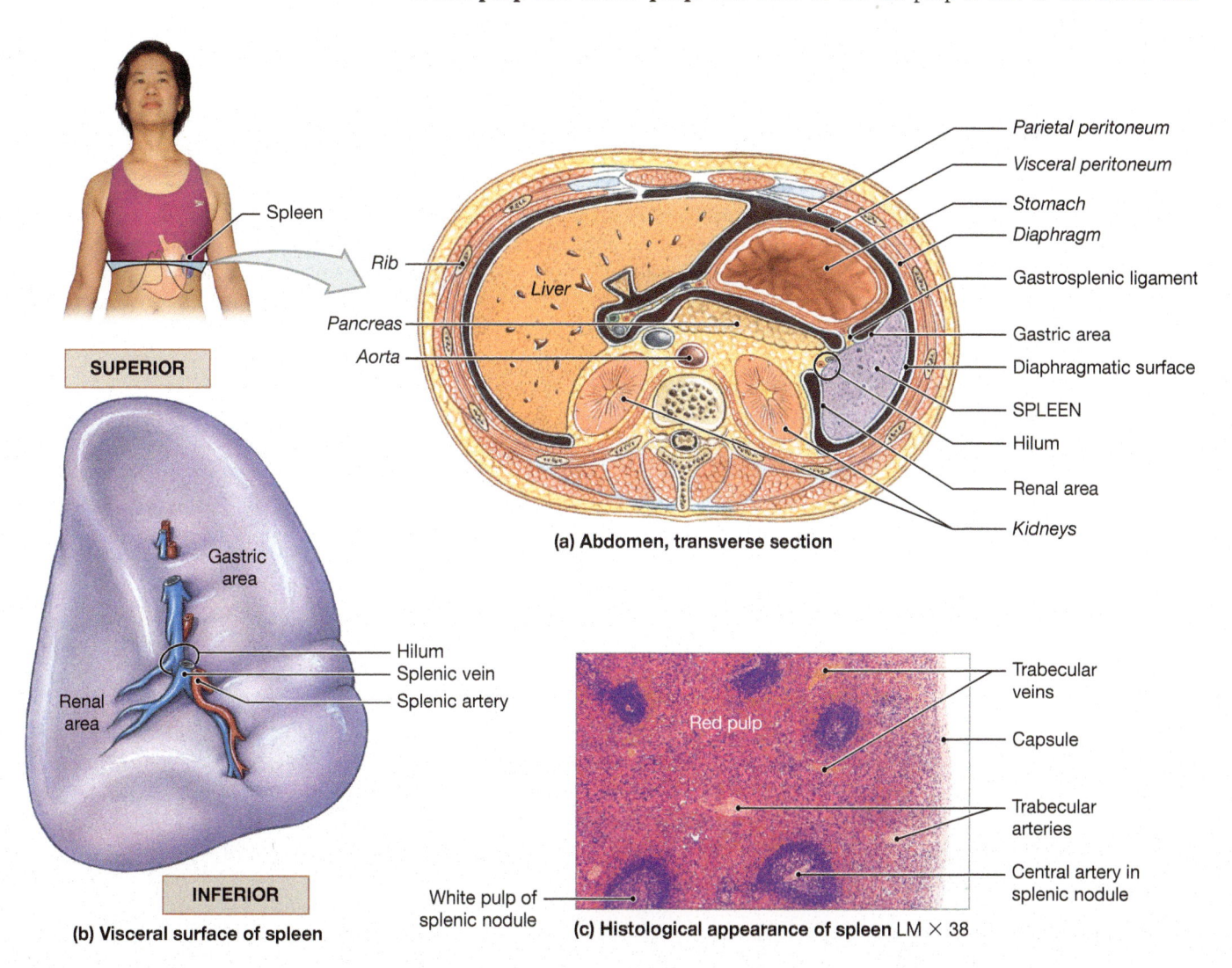

Figure 24.6   The Spleen
(a) Transverse section through the trunk that shows the typical position of the spleen in the abdominopelvic cavity. The shape of the spleen roughly conforms to the shapes of adjacent organs. (b) External appearance of the intact spleen that shows major anatomical landmarks. (c) Histological appearance of the spleen. White pulp is dominated by lymphocytes; it appears purple because the nuclei of lymphocytes stain very dark. Red pulp contains a preponderance of red blood cells. (LM × 38)

filters through; white pulp appears blue because the lymphocyte nuclei stains. Blood vessels and lymphatic vessels pass in and out of the spleen at the hilus. Branches of the splenic artery, called **trabecular arteries**, are distributed in the red pulp. **Central arteries** occur in the middle of white pulp. Capillaries of the trabecular arteries open into the red pulp. As blood flows through the red pulp, free and fixed phagocytes in the pulp remove abnormal red blood cells and other antigens from the blood. Upon exposure to the antigens, the lymphocytes of the red pulp become sensitized to them and produce antibodies to counteract them. Blood drains from the sinuses of the red pulp into trabecular veins that eventually empty into the splenic vein.

QuickCheck Questions

3.1 Where is the spleen located?

3.2 What tissues are in the white pulp?

3.3 Which vessels open into the red pulp?

3 Materials

- ☐ Torso model or chart that shows the spleen
- ☐ Compound microscope
- ☐ Prepared slide of spleen

Procedures

1. Review the anatomy of the spleen in Figure 24.6.
2. Locate the spleen on the torso model or chart. Identify the hilus, splenic artery, and splenic vein. On the visceral surface, locate the gastric area of the spleen, which is in contact with the stomach, and the renal area, which is in contact with the kidneys.
3. Examine the spleen slide at low magnification and identify the dark-stained regions of white pulp and the lighter regions of red pulp. Examine several white pulp trabeular artery masses for the presence of an artery. ■

female breast — A Regional Look

The female breast has milk-producing mammary glands embedded in a pectoral fat pad that lies against the pectoralis major muscle. An extensive network of lymphatic vessels and lymph nodes collects and filters lymph from the breast (Figure 24.7). The lymph nodes are of clinical importance in cases of breast and other cancers because cancer cells can enter the lymphatic vessels and spread to other parts of the body (*metastasize*) via the lymph. Breast cancer is classified according to the extent to which metastasizing cancer cells have invaded the lymph nodes. Treatment typically begins with the removal of the tumor and a biopsy of the axillary lymph nodes. A *mastectomy* is removal of the breast. A *radical mastectomy* involves removal of the breast plus the regional lymphatic vessels, including the axillary lymph nodes.

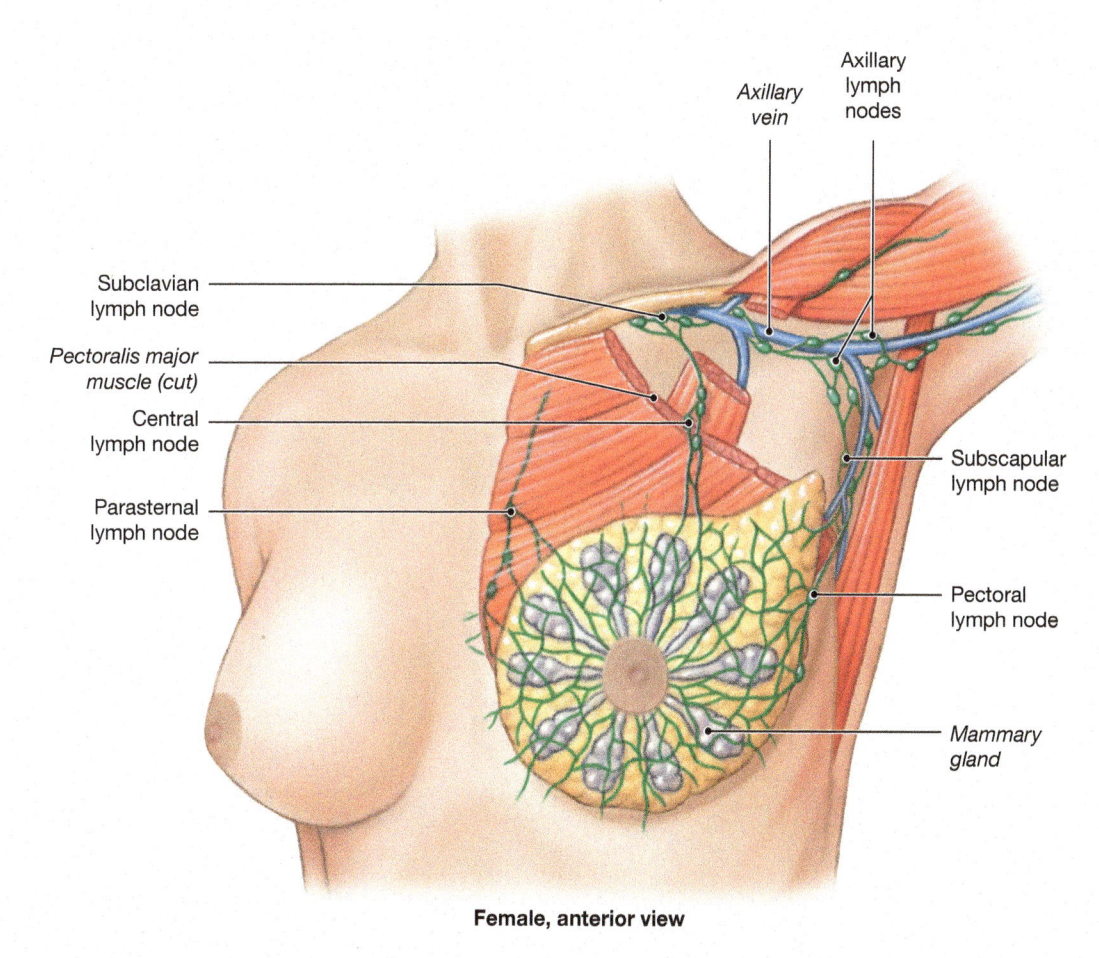

Female, anterior view

Figure 24.7 Lymphatic Vessels of the Female Breast
Superficial and deep lymphatic vessels and nodes in the female breast and chest.

LAB REPORT

EXERCISE 24: Lymphatic System

Name _____

Date _____

Section _____

A. Matching

Match each structure in the left column with its correct description from the right column.

_____ 1. efferent lymphatic vessel
_____ 2. medullary cords
_____ 3. cisterna chyli
_____ 4. right lymphatic duct
_____ 5. red pulp
_____ 6. lymph node
_____ 7. thoracic duct
_____ 8. white pulp
_____ 9. lymph
_____ 10. afferent lymphatic vessel

A. empties into circulation where right internal jugular and right subclavian veins meet
B. empties into lymph node
C. splenic tissue through which blood flows
D. liquid in lymphatic vessels
E. lymphocytes deep in lymph node
F. empties into left subclavian vein
G. full of macrophages and lymphocytes
H. drains lymph node
I. spleen tissue that surrounds trabecular artery
J. sac-like region of thoracic duct

B. Short-Answer Questions

1. Describe the organization inside a lymph node.

2. Discuss the major functions of the lymphatic system.

3. Explain how lymph is returned to the blood.

4. Describe the anatomy of the spleen.

EXERCISE 24 — LAB REPORT

C. **Analysis and Application**

1. How are blood plasma, extracellular fluid, and lymph interrelated?

2. How does the way lymph drains from the right lymphatic duct differ from the way it drains from the thoracic duct?

3. How can cancer cells spread by entering the lymphatic system?

EXERCISE 25

The Respiratory System

OBJECTIVES

On completion of this exercise, you should be able to:

- Identify and describe the structures of the nasal cavity.
- Distinguish among the three regions of the pharynx.
- Identify and describe the cartilages and ligaments of the larynx.
- Identify the gross and microscopic structure of the trachea.
- Name the progressively smaller and smaller branches of the bronchial tree.
- Identify and describe the gross and microscopic structure of the lungs.

LAB ACTIVITIES

1 Nose and Pharynx 399
2 Larynx 401
3 Trachea and Bronchial Tree 403
4 Lungs 406
  A Regional Look: Airway 409

All cells require a constant supply of oxygen (O_2) for the oxidative reactions of mitochondrial ATP production. A major byproduct of these reactions is carbon dioxide (CO_2). The respiratory system exchanges these two gases between the atmosphere and the blood. Specialized organs of the airway filter, warm, and moisten the inhaled air before it enters the lungs. Once the air is in the lungs, the O_2 gas in the air diffuses into the surrounding capillaries to oxygenate the blood. As the blood takes up this oxygen, CO_2 gas in the blood diffuses into the lungs and is exhaled. Pulmonary veins return the oxygenated blood to the heart, where it is pumped into arteries of the systemic circulation.

The respiratory system, shown in Figure 25.1, consists of the nose, nasal cavity, sinuses, pharynx, larynx, trachea, bronchi, and lungs. The **upper respiratory system** includes the nose, nasal cavity, sinuses, and pharynx. These structures filter, warm, and moisten air before it enters the **lower respiratory system**, which consists of the larynx, trachea, bronchi, and lungs. The larynx regulates the opening into the lower respiratory system and produces speech sounds. The trachea and bronchi maintain an open airway to the lungs, where gas exchange occurs.

LAB ACTIVITY 1 Nose and Pharynx

The primary route for air that enters the respiratory system is through two openings, the **external nares** (NAR-ēz), or nostrils (Figure 25.2). Just inside each external naris is an expanded **nasal vestibule** (VES-ti-byool) that contains coarse hairs. The hairs help to prevent large airborne materials such as dirt particles and insects from entering the respiratory system. The external portion of the nose is composed of **nasal cartilages** that form the bridge and tip of the nose.

The **nasal cavity** is the airway from the external nares to the superior part of the pharynx. The perpendicular plate of the ethmoid and the vomer create the

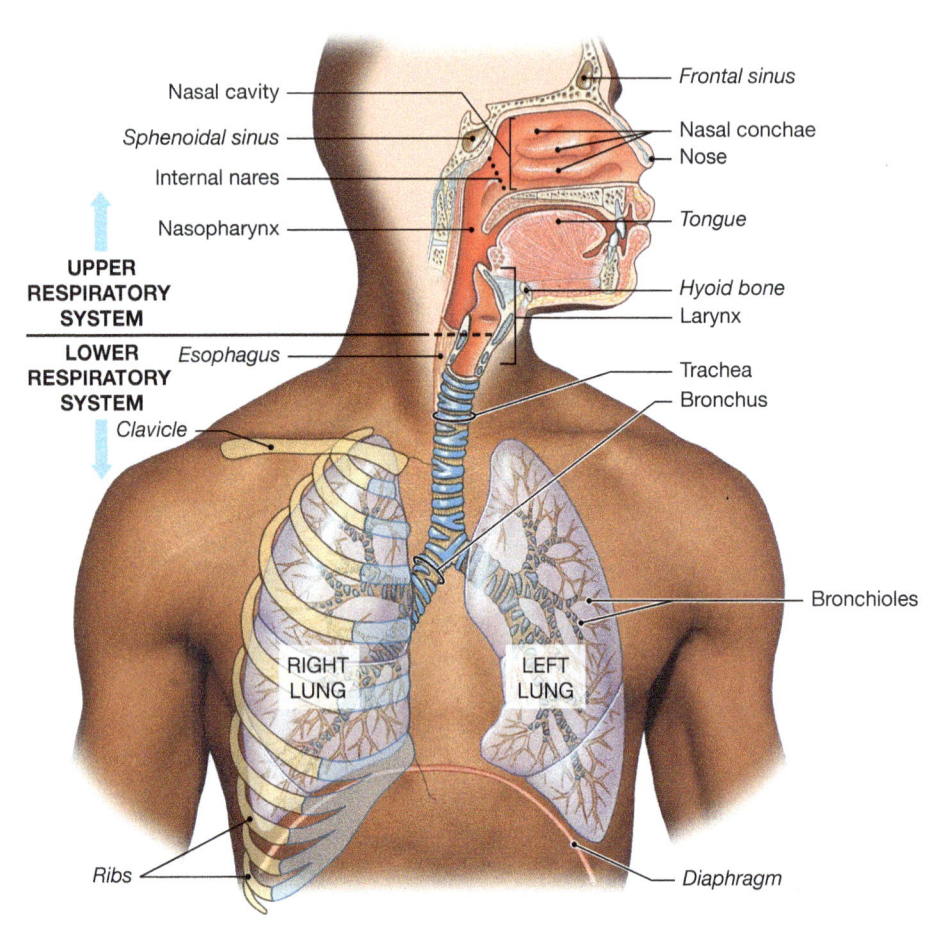

Figure 25.1 Structures of the Respiratory System

nasal septum which divides the nasal cavity into right and left sides. The **superior**, **middle**, and **inferior nasal conchae** are bony shelves that project from the lateral walls of the nasal cavity. The distal edge of each nasal concha curls inferiorly and forms a tube, or **meatus**, that causes inhaled air to swirl in the nasal cavity. This turbulence moves the air across the sticky epithelial lining, where dust and debris are removed. The floor of the nasal cavity is the superior portion of the **hard palate**, formed by the maxillae, palatine bones, and muscular **soft palate**. Hanging off the posterior edge of the soft palate is the conical **uvula** (Ū-vū-luh). The **internal nares** are the two posterior openings of the nasal cavity that connect with the superior portion of the pharynx.

The throat, or **pharynx** (FAR-inks), is divided into three regions: nasopharynx, oropharynx, and laryngopharynx. The **nasopharynx** (nā-zō-FAR-inks) is superior to the soft palate and serves as a passageway for airflow from the nasal cavity. Located on the posterior wall of the nasopharynx is the pharyngeal tonsil (Exercise 24). On the lateral walls of the nasopharynx are the openings of the auditory (pharyngotympanic) tubes (Exercise 19). The nasopharynx is lined with a pseudostratified ciliated columnar epithelium that functions to warm, moisten, and clean inhaled air. When a person eats, food pushes past the uvula, and the soft palate raises to prevent the food from entering the nasopharynx.

The **oropharynx**, which extends inferiorly from the soft palate, is connected to the oral cavity at an opening called the **fauces** (FAW-sēz). The oropharynx contains the palatine and lingual tonsils (Exercise 24).

The **laryngopharynx** (la-rin-gō-FAR-inks) is located between the hyoid bone and the entrance to the esophagus, which is the muscular tube that connects the

EXERCISE 25 The Respiratory System   401

(a) Anterior view

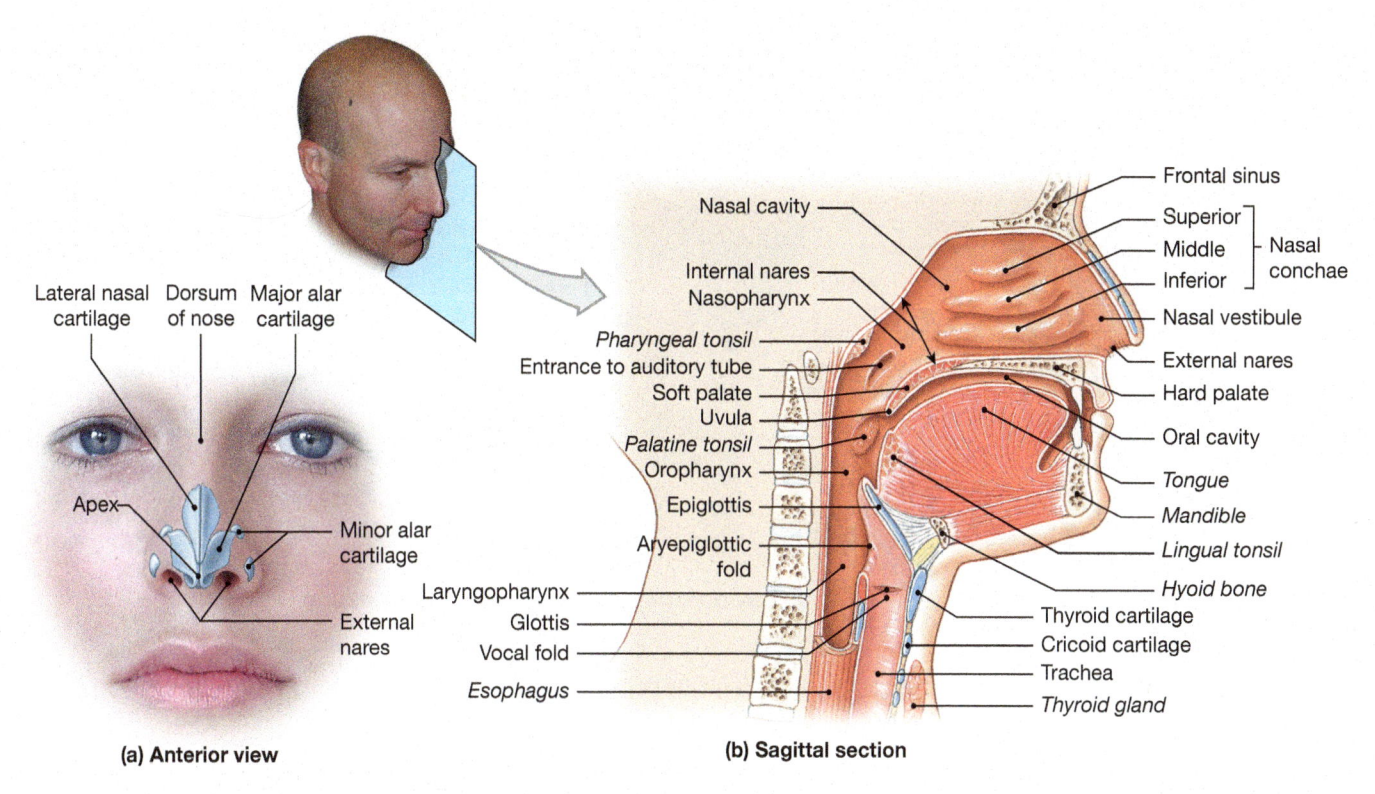

(b) Sagittal section

Figure 25.2 The Nose, Nasal Cavity, and Pharynx
(a) The nasal cartilages and external landmarks on the nose. (b) Diagrammatic view of the head and neck in sagittal section.

oral cavity with the stomach. (The esophagus is studied as part of the digestive system in Exercise 26.) The oropharynx and laryngopharynx have a stratified squamous epithelium as abrasion protection from swallowed food that passes through to the esophagus.

QuickCheck Questions

1.1 What are the components of the upper respiratory system?
1.2 What are the components of the lower respiratory system?
1.3 What are the passageways into and out of the nasal cavity?
1.4 What are the three regions of the pharynx?

1 Materials

☐ Head model
☐ Respiratory system chart
☐ Hand mirror

Procedures

1. Review the gross anatomy of the nose and pharynx in Figure 25.2. Locate these structures on the head model and respiratory system chart.
2. Using the hand mirror, examine the inside of your mouth. Locate your hard and soft palates, uvula, fauces, palatine tonsils, and oropharynx. ■

LAB ACTIVITY 2 Larynx

The **larynx** (LAR-inks), or voice box, lies inferior to the laryngopharynx and anterior to cervical vertebral C_4–C_7. It consists of nine cartilages held together by **laryngeal ligaments**. The airway through the larynx is the **glottis** (See Figure 25.2b).

Three large unpaired cartilages form the body of the larynx (Figure 25.3). The first cartilage, the **epiglottis** (ep-i-GLOT-is), is the flap of elastic cartilage that lowers to cover the glottis during swallowing and helps direct the food to the esophagus.

402 EXERCISE 25 The Respiratory System

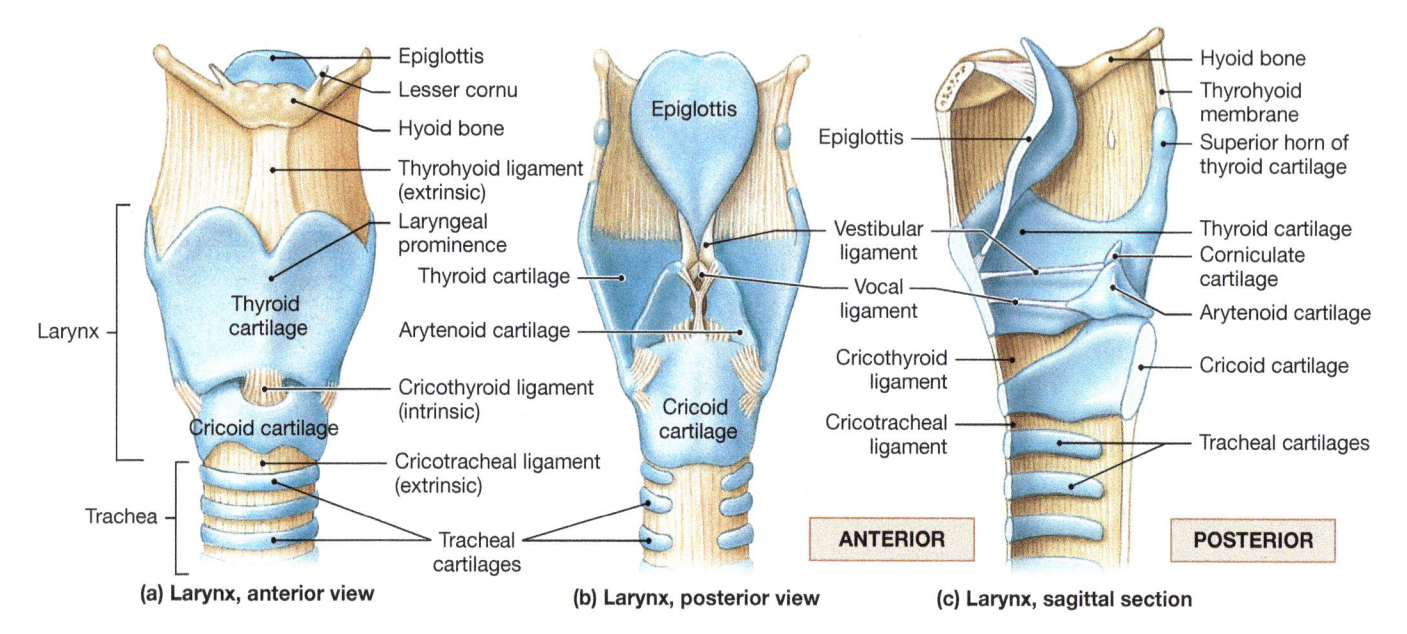

Figure 25.3 The Larynx
(a) Anterior view. (b) Posterior view. (c) A sagittal section through the larynx.

The **thyroid cartilage**, or Adam's apple, is composed of hyaline cartilage. It is visible under the skin on the anterior neck, especially in males. The **cricoid** (KRĪ-koyd) **cartilage** is a ring of hyaline cartilage that forms the base of the larynx.

The larynx also has three pairs of smaller cartilages. The **arytenoid** (ar-i-TĒ-noyd) **cartilages** articulate with the superior border of the cricoid cartilage. **Corniculate** (kor-NIK-ū-lāt) **cartilages** articulate with the arytenoid cartilages and are involved in the opening and closing of the glottis and in the production of sound. The **cuneiform** (kū-NĒ-i-form) **cartilages** are club-shaped cartilages that are anterior to the corniculate cartilages.

Two pairs of folds span the glottis from the thyroid cartilage (Figure 25.4). The superior folds are the **vestibular ligaments**, or *false vocal cords*. They prevent foreign materials from entering the glottis and close the glottis during coughing and sneezing. The inferior folds are the **vocal ligaments**, also called the *true vocal cords*, which vibrate to produce speech and other sounds.

QuickCheck Questions

2.1 How many pieces of cartilage are in the larynx?
2.2 What are the glottis and the epiglottis?
2.3 What are the structures that produce speech?

2 Materials

- ☐ Larynx model
- ☐ Torso model
- ☐ Respiratory system chart

Procedures

1. Review the gross anatomy of the larynx in Figure 25.3.
2. On the larynx model, torso model, or respiratory system chart, do the following:
   - Locate the thyroid cartilage. Is it continuous around the larynx?
   - Locate the cricoid cartilage. Is it continuous around the larynx?
   - Study the position of the epiglottis. How does it act like a chute to direct food into the esophagus?
   - Open the larynx model, and identify the arytenoid, corniculate, and cuneiform cartilages.
   - Locate the vestibular ligaments and the vocal ligaments.

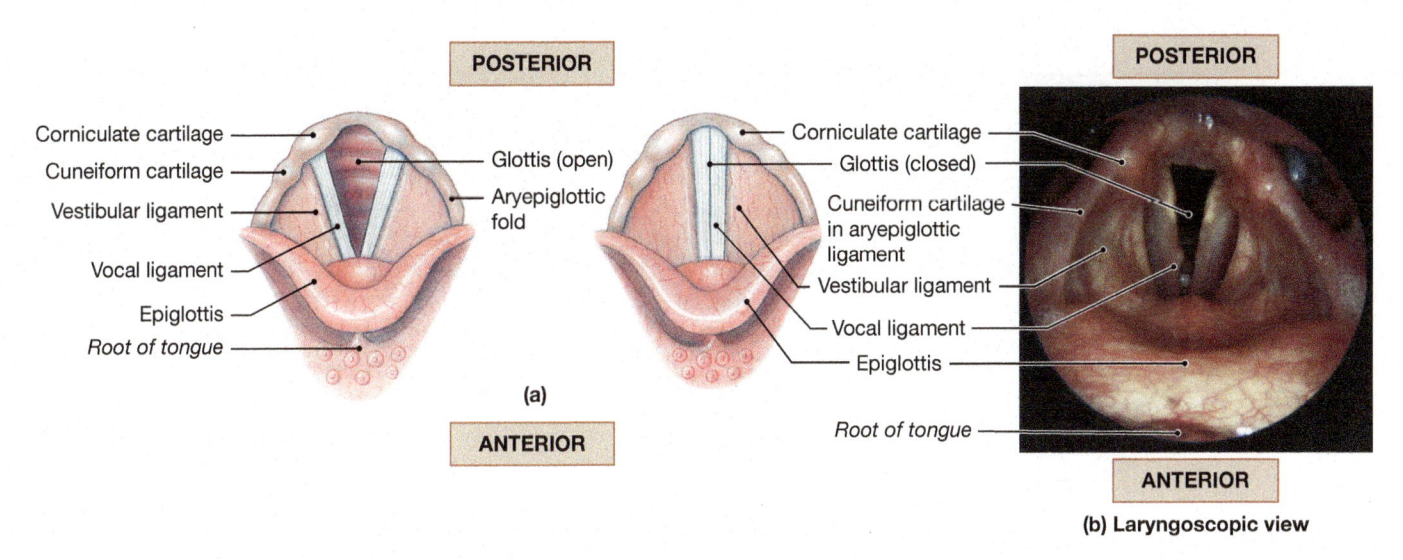

Figure 25.4 The Glottis
(a) Diagrammatic superior view of the entrance to the larynx, with the glottis open and closed. (b) Fiber-optic view of the entrance to the larynx, which corresponds to the image on the right in (a). Note that the glottis is almost completely closed by the vocal ligaments.

3. Put your finger on your thyroid cartilage and swallow. How does the cartilage move when you swallow? Is it possible to swallow and make a sound simultaneously?
4. While holding your thyroid cartilage, first make a high-pitched sound and then make a low-pitched sound. Describe the tension in your throat muscles for each sound, and relate the muscle tension to the tension in the vocal ligaments. ■

LAB ACTIVITY 3 — Trachea and Bronchial Tree

The **trachea** (TRĀ-kē-uh), or windpipe, is a tubular structure that is approximately 11 cm (4.25 in.) long and 2.5 cm (1 in.) in diameter (Figure 25.5). It lies anterior to the esophagus and can be felt on the front of the neck inferior to the thyroid cartilage of the larynx. Along the length of the trachea are 15 to 20 C-shaped **tracheal cartilages** of hyaline cartilage that keep the airway open. The **trachealis muscle** holds the two tips of each C-shaped tracheal cartilage together posteriorly. This muscle allows the esophagus diameter to increase during swallowing so that the esophagus wall presses against the adjacent trachea wall and decreases the trachea diameter momentarily.

The trachea is lined with a pseudostratified ciliated columnar epithelium that constantly sweeps the airway clean. Interspersed in the epithelium are goblet cells that secrete mucus to trap particles present in the inhaled air.

The trachea divides, at a ridge called the **carina**, into the left and right **primary bronchi** (BRONG-kī; singular *bronchus*). The right primary bronchus is wider and more vertical than the left primary bronchus. (For this reason, objects that are accidentally inhaled often enter the right primary bronchus.) Each bronchus branches into increasingly smaller passageways to conduct air into the lungs (Figure 25.6). This branching pattern formed by the divisions of the bronchial structures is called the **bronchial tree**. At the superior terminus of the tree, the primary bronchi branch into as many **secondary bronchi** as there are lobes in each lung. The right lung has three lobes, and each lobe receives a secondary bronchus to supply it with air. The left lung has two lobes, and thus two secondary bronchi branch off the left primary bronchus.

The secondary bronchi divide into **tertiary bronchi**, also called *segmental bronchi* (See Figure 25.6). Smaller divisions called **bronchioles** branch into

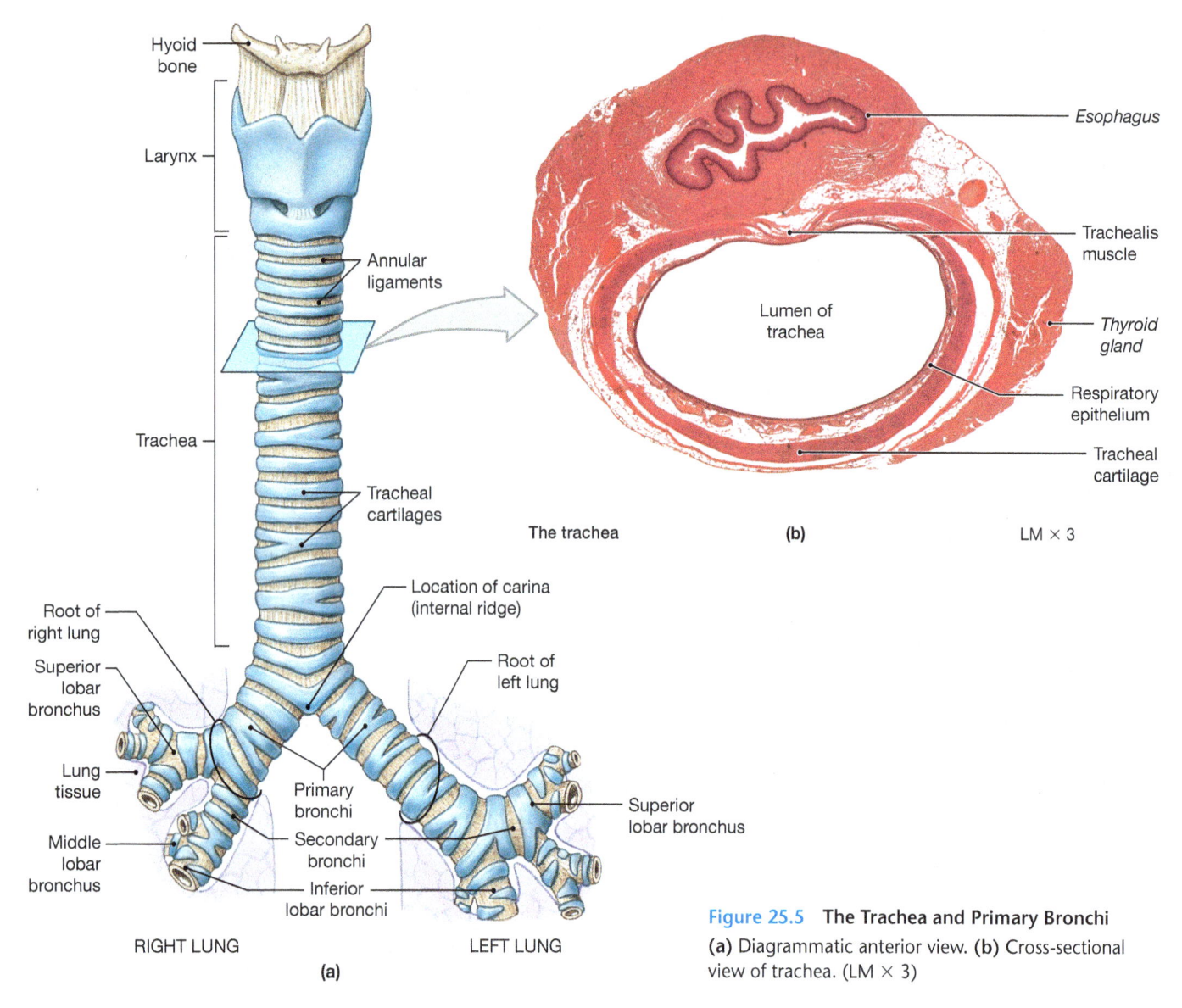

404 EXERCISE 25 The Respiratory System

Figure 25.5 The Trachea and Primary Bronchi
(a) Diagrammatic anterior view. (b) Cross-sectional view of trachea. (LM × 3)

terminal **bronchioles**. The terminal bronchioles branch into **respiratory bronchioles**, which further divide into the narrowest passageways, the **alveolar ducts**.

As the bronchial tree branches from the primary bronchi to the respiratory bronchioles, cartilage is gradually replaced with smooth muscle tissue. The epithelial lining of the bronchial tree also changes from pseudostratified ciliated columnar epithelium at the superior end of the tree to simple squamous epithelium at the inferior end.

Clinical Application   Asthma

Asthma (AZ-muh) is a condition that occurs when the smooth muscle that encircles the delicate bronchioles contracts and reduces the diameter of the airway. The airway is further compromised by increased mucus production and inflammation of the epithelial lining. The individual has difficulty breathing, especially during exhalation, as the narrowed passageways collapse under normal respiratory pressures. An asthma attack can be triggered by a number of factors, including allergies, chemical sensitivities, air pollution, stress, and emotion. **Bronchodilator** drugs are used to relax the smooth muscle and open the airway; other drugs reduce inflammation of the mucosa. *Albuterol* is an important bronchodilator, and is usually administered as an inhalant sprayed from a nebulizer.

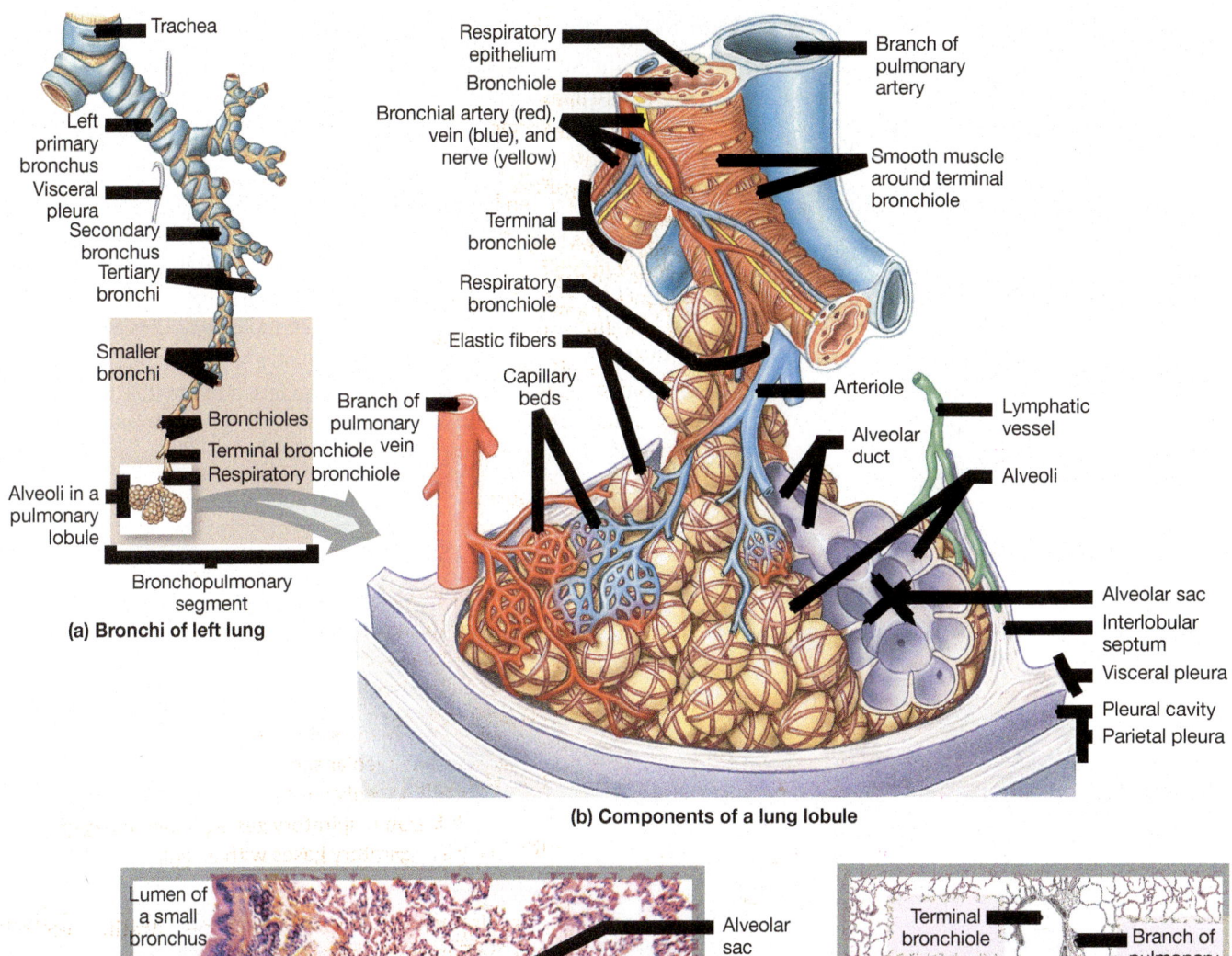

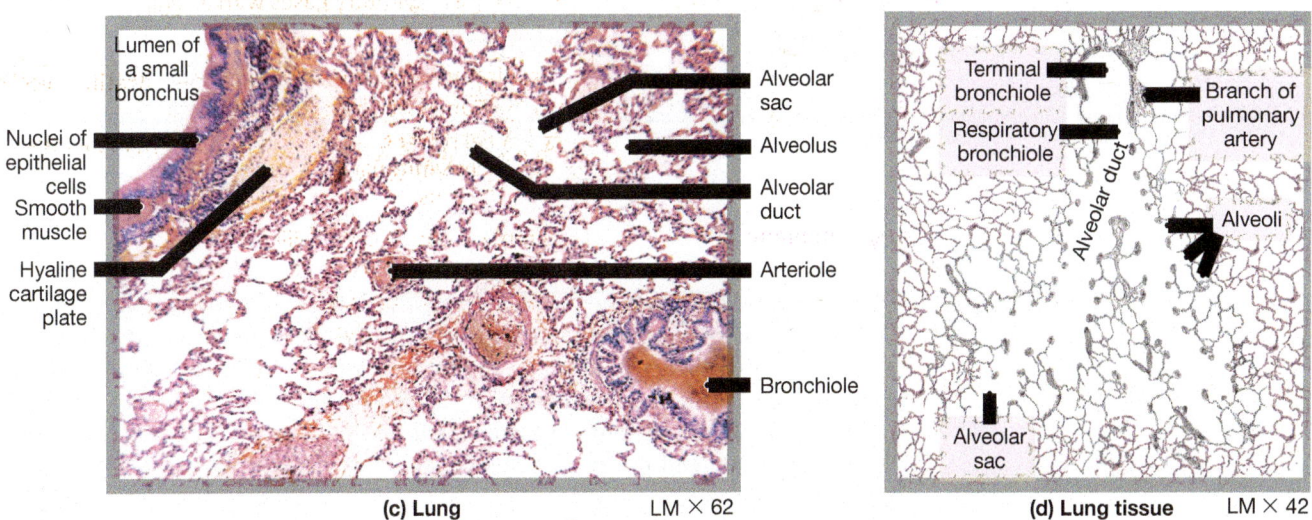

Figure 25.6 **Bronchi, Lobules, and Bronchioles of the Lung**
(a) Branching pattern of bronchi in the left lung, simplified. (b) Structure of a single pulmonary lobule, part of a bronchopulmonary segment. (c) Transverse section of lung that shows a hyaline cartilage plate next to a small bronchus. (LM × 62) (d) Distribution of a respiratory bronchiole that supplies a portion of a lobule. (LM × 42)

405

QuickCheck Questions

3.1 What is the epithelium that lines the trachea?

3.2 What is the bronchial tree?

3 Materials

- ☐ Compound microscope
- ☐ Prepared slide of trachea
- ☐ Head model
- ☐ Torso model
- ☐ Lung model
- ☐ Respiratory system chart

Procedures

1. Review the gross anatomy of the trachea in Figure 25.5. Locate these structures on the head, torso, and lung models and the respiratory system chart. Palpate your trachea for the tracheal cartilages.

2. On the trachea slide, locate the structures labeled in Figure 25.5b. Sketch a section of the trachea in the space provided.

3. Study the bronchial tree on the torso or lung models, and identify the primary bronchi, secondary bronchi, tertiary bronchi, bronchioles, terminal bronchioles, and respiratory bronchioles. ■

Trachea section

LAB ACTIVITY 4 Lungs

The lungs are a pair of cone-shaped organs that lie in the thoracic cavity (Figure 25.7). Each lung sits inside a pleural cavity located between the two layers of the pleura (Exercise 1). The **apex** is the conical top of each lung, and the broad inferior portion is the **base**. The anterior, lateral, and posterior surfaces of each lung face the thoracic cage, and the medial surface faces the mediastinum. The heart lies on a medial concavity of the left lung called the **cardiac notch**. Each lung has a **hilus**, which is a slit on the medial surface where the bronchi, blood vessels, lymphatic vessels, and nerves reach the lung.

As noted earlier, each lung is divided into lobes, two in the left lung and three in the right lung. Both lungs have an **oblique fissure** that forms the lobes, and the right lung also has a **horizontal fissure**. The oblique fissure of the left lung separates the lung into its **superior** and **inferior lobes**. The oblique fissure of the right lung separates the **middle lobe** from the **inferior lobe**, and the horizontal fissure separates the middle lobe from the superior lobe.

Air enters the lobes of the lungs via the tertiary bronchi, with ten tertiary bronchi that conduct air to the right lung and nine that run to the left lung. Inside a lobe, the region supplied by each tertiary bronchi is called a **bronchopulmonary**

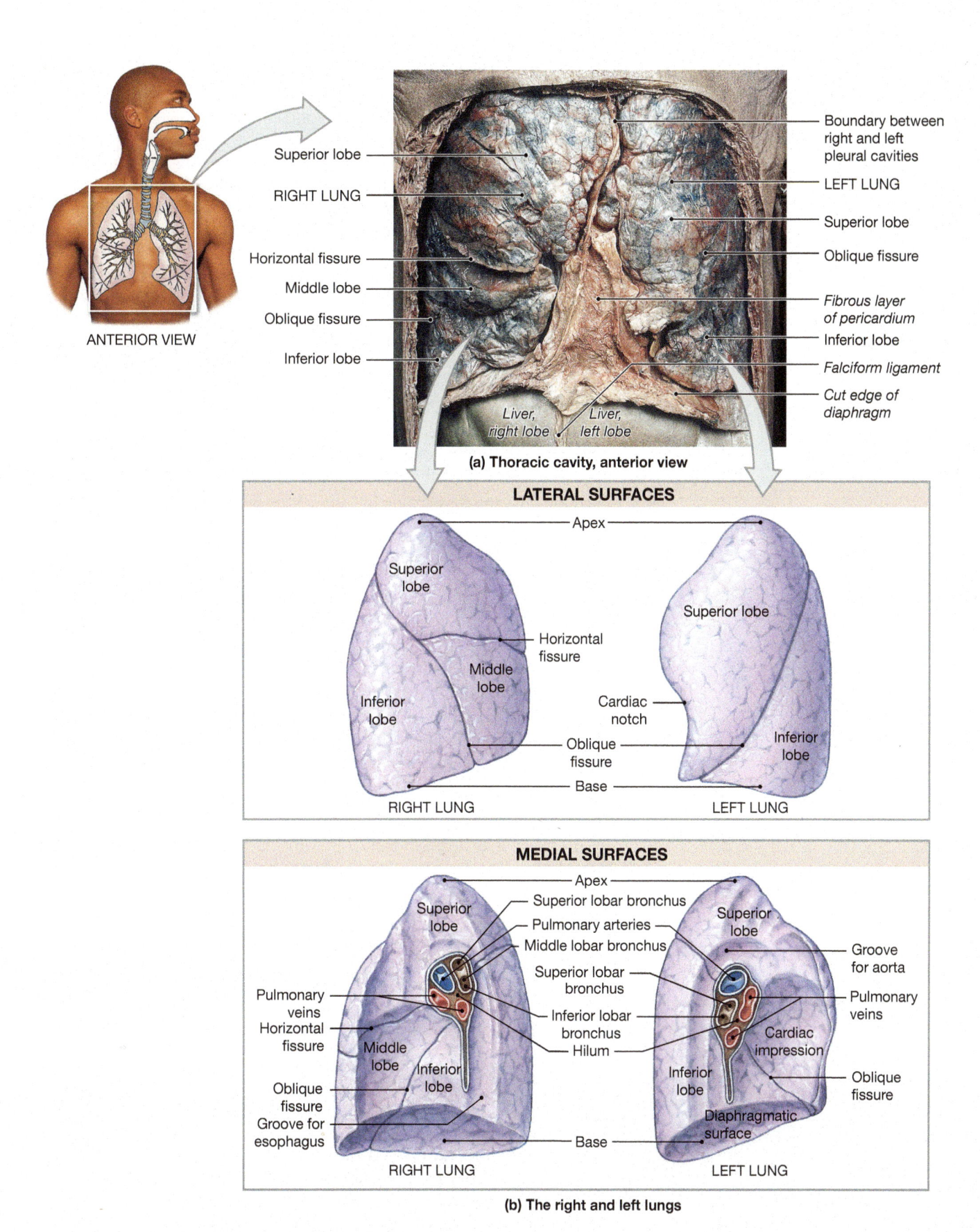

Figure 25.7 Superficial Anatomy of the Lungs

(a) Anterior view of the opened chest that shows the positions of the right and left lungs. (b) Diagrammatic views of the lateral and medial surfaces of the isolated right and left lungs.

407

408 EXERCISE 25 The Respiratory System

segment (See the shaded region in Figure 25.6a). Subregions within each bronchopulmonary segment are called **lobules**, and each lobule is made up of numerous tiny air pockets called **alveoli** (al-VĒ-ō-lī; singular *alveolus*). Groups of alveoli clustered together are called **alveolar sacs** (See Figure 25.6c and d). Each lobule is served by a single terminal bronchiole. Inside a lobule, at the finest level of the bronchial tree, each alveolar duct serves a number of alveolar sacs.

The walls of the alveoli are constructed of simple squamous epithelium. Scattered throughout the simple squamous epithelium are **surfactant** (sur-FAK-tant) **cells** that secrete an oily coating to prevent the alveoli from sticking together after exhalation. Also in the alveolar wall are macrophages that phagocytize debris. Pulmonary capillaries cover the exterior of the alveoli, and gas exchange occurs across the thin alveolar walls. Oxygen from inhaled air diffuses through the simple squamous epithelium of the alveolar wall, moves across the basal lamina membrane and the endothelium of the capillary, and enters the blood. The thickness of the combined alveolar wall and capillary wall is only about 0.5 mm, a size that permits rapid gas exchange between the alveoli and blood.

QuickCheck Questions

4.1 How many lobes does each lung have?

4.2 Which lung has the cardiac notch?

4 Materials

- ☐ Compound microscope
- ☐ Prepared slide of lung
- ☐ Torso model
- ☐ Respiratory system chart

Procedures

1. Review the gross anatomy of the lungs in Figure 25.7. Locate these structures on the torso models and on the respiratory system chart.

2. On the model, examine the right lung, and observe how the horizontal and oblique fissures divide it into three lobes. Note how the oblique fissure separates the left lung into two lobes.

3. Examine the model for the parietal pleura that lines the thoracic wall. Where is the pleural cavity relative to the parietal pleura?

4. On the prepared slide:

   a. Identify the alveoli, using Figure 25.6 as a reference.

   b. Locate an area where the alveoli appear to have been scooped out. This passageway is an alveolar duct. Follow the duct to its end, and observe the many alveolar sacs serviced by the duct.

   c. At the opposite end of the duct, look for the thicker wall of the respiratory bronchiole and blood vessels.

   d. In the space provided, sketch an alveolar duct and several alveolar sacs. ■

Alveolar duct and sacs

airway
A Regional Look

The anatomy of the neck is very complex because the neck must flex, extend, pivot, and rotate to reposition the head while at the same time always keep the airway open. Notice in the cadaver photograph in Figure 25.8 how the hard and soft palates separate the mouth from the nasal cavity. This division between the mouth of the digestive system and the nose of the respiratory system permits breathing while chewing and processing food in the mouth.

Movement in the epiglottis is another facet of the neck's complexity. When swallowed food contacts it, the epiglottis depresses and helps direct the food into the esophagus, which is the muscular food tube that lies posterior to the larynx and trachea, and thus protects the delicate vestibular folds and vocal folds of the larynx. Notice the laryngeal and tracheal cartilages, which support the walls of the airway and prevent them from collapsing during breathing. The esophagus is muscular and is not kept open the way the respiratory airway is. Muscle contractions during swallowing force food down the esophagus and into the stomach.

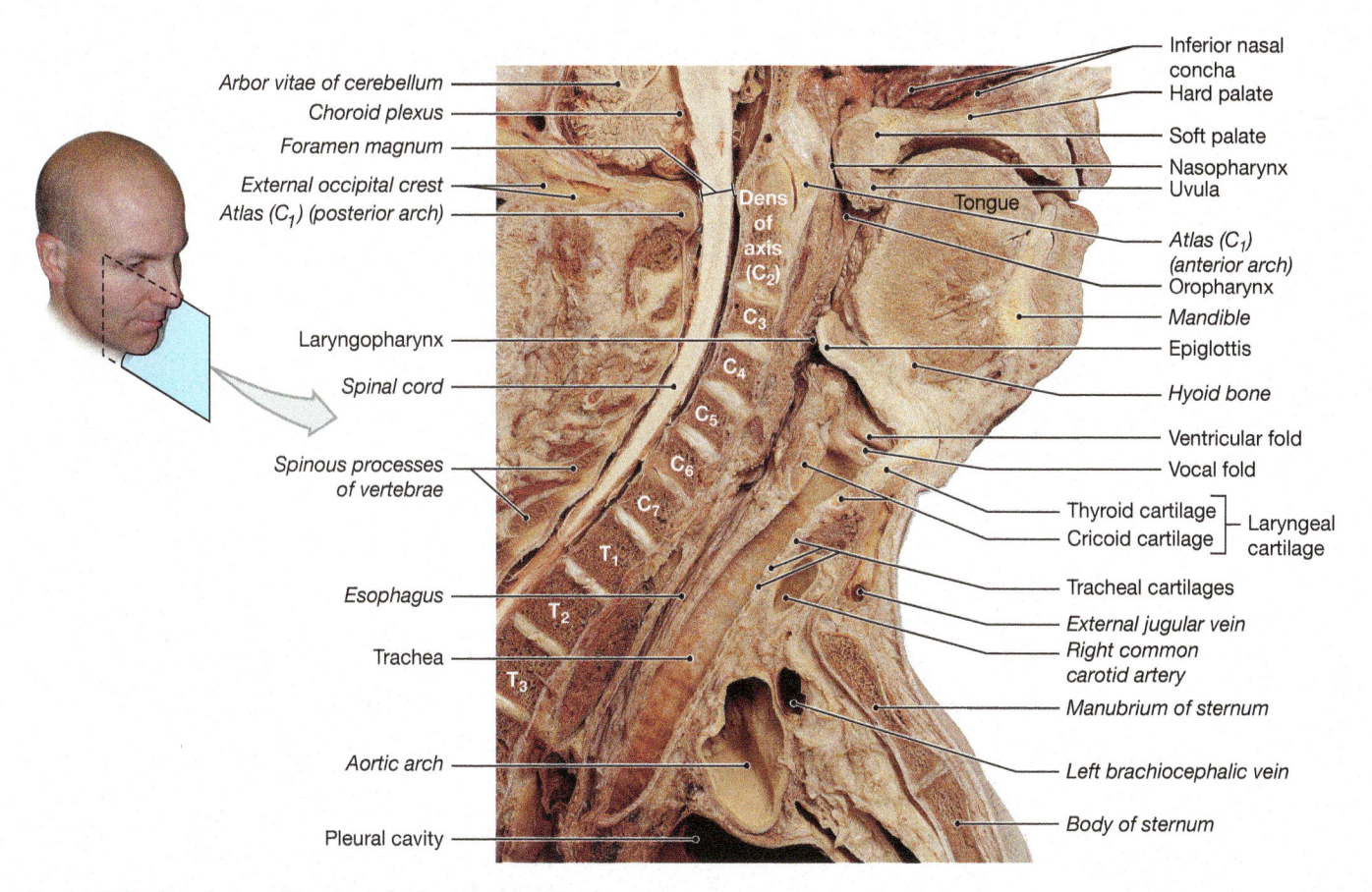

Figure 25.8 Respiratory Structures in the Head and Neck
The nasal cavity and pharynx as seen in a sagittal section of the head and neck.

409

EXERCISE 25

LAB REPORT

The Respiratory System

A. Matching

Match each structure in the left column with its correct description from the right column.

___	1. C-shaped rings	A.	voice box
___	2. internal nares	B.	elastic cartilage flap of larynx
___	3. cricoid cartilage	C.	tubular passageway of nasal concha
___	4. meatus	D.	found in left lung
___	5. epiglottis	E.	connect nasal cavity with throat
___	6. larynx	F.	tracheal cartilage
___	7. vocal ligament	G.	true vocal cord
___	8. cardiac notch	H.	false vocal cord
___	9. external nares	I.	nostrils
___	10. three lobes	J.	base of larynx
___	11. thyroid cartilage	K.	found in right lung
___	12. vestibular ligament	L.	Adam's apple

B. Labeling

Label the ten features of the trachea and bronchial tree numbered in Figure 25.9.

C. Short-Answer Questions

1. What are the functions of the superior, middle, and inferior nasal conchae?

2. Where is the pharyngeal tonsil located?

3. Trace a breath of air from the external nares through the respiratory system to the alveolar sacs.

EXERCISE 25 — LAB REPORT

Figure 25.9 **The Anatomy of the Trachea**
A diagrammatic anterior view.

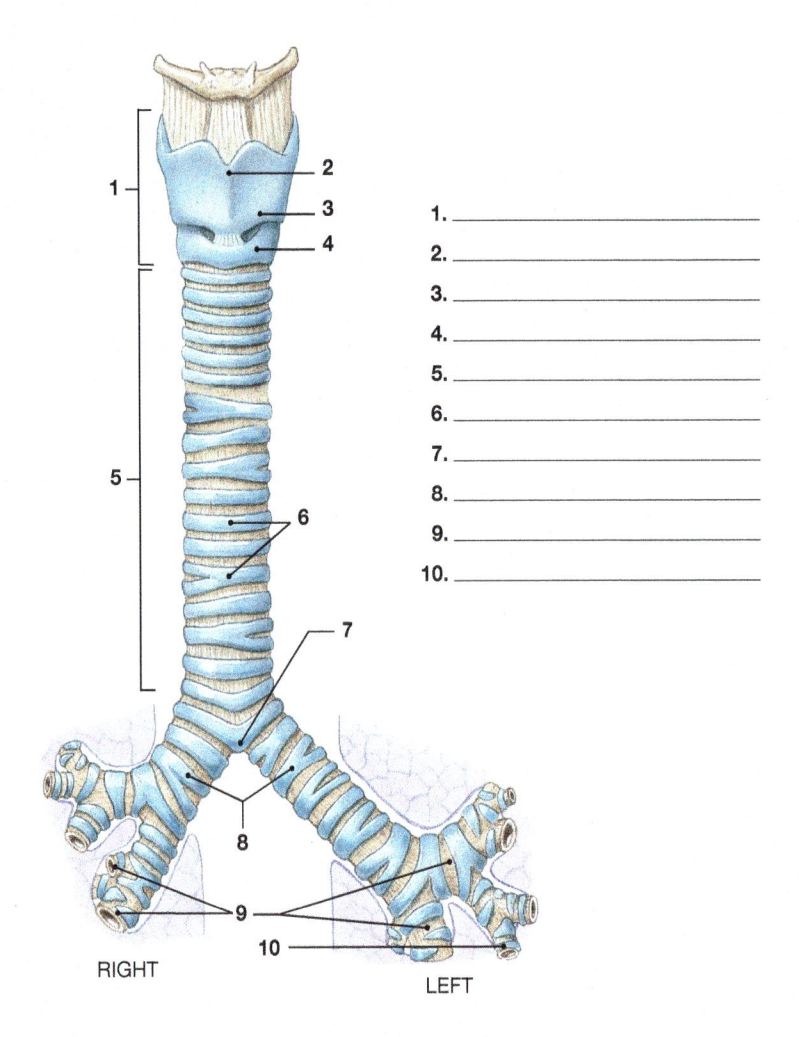

1. _____
2. _____
3. _____
4. _____
5. _____
6. _____
7. _____
8. _____
9. _____
10. _____

RIGHT   LEFT

D. Analysis and Application

1. Where do goblet cells occur in the respiratory system, and what function do they serve?

2. What tissue lines the pharynx where it serves as a common passageway for air and food?

3. How are speech sounds produced in the larynx?

4. How does an asthma attack cause difficulty in breathing?

EXERCISE 26

The Digestive System

OBJECTIVES

On completion of this exercise, you should be able to:

- Identify the major layers and tissues of the digestive tract.
- Identify the anatomy of the digestive system on laboratory models and charts.
- Describe the histological structure of the various digestive organs.
- Trace the secretion of bile from the liver to the duodenum.
- List the organs of the digestive tract and the accessory organs that empty into them.

LAB ACTIVITIES

1. Mouth   415
2. Pharynx and Esophagus   418
3. Stomach   419
4. Small Intestine   422
5. Large Intestine   425
6. Liver and Gallbladder   428
7. Pancreas   430

A Regional Look: Upper Abdomen   432

The five major processes of digestion are: (1) ingestion of food into the mouth; (2) movement of food through the digestive tract; (3) mechanical and enzymatic digestion of food; (4) absorption of nutrients into the blood; and (5) formation and elimination of indigestible material and waste.

The **digestive tract** is a muscular tube that extends from the mouth to the anus, a tube formed by the various hollow organs of the digestive system. Accessory organs outside the digestive tract plus the tract organs make up the **digestive system**. The accessory organs—salivary glands, teeth, liver, gallbladder, and pancreas—manufacture enzymes, hormones, and other compounds and secrete these substances onto the inner lining of the digestive tract. Food does not pass through the accessory organs.

The wet mucosal layer that lines the mouth and the rest of the digestive tract is a mucous membrane. Glands drench the tissue surface with enzymes, mucus, hormones, pH buffers, and other compounds to orchestrate the step-by-step breakdown of food as it passes through the digestive tract.

The histological organization of the digestive tract is similar throughout the length of the tract, and most of the tract consists of four major tissue layers: mucosa, submucosa, muscularis externa, and serosa (Figure 26.1). Each region of the digestive tract has anatomical specializations reflecting that region's role in digestion. The **mucosa** is the superficial layer exposed at the lumen of the tract. Three distinct layers in the mucosa can be identified: the mucosal epithelium, the lamina propria, and a thin layer of smooth muscle called the **muscularis** (mus-kū-LAR-is) **mucosae**. The mucosal epithelium is the superficial layer exposed

414　EXERCISE 26　The Digestive System

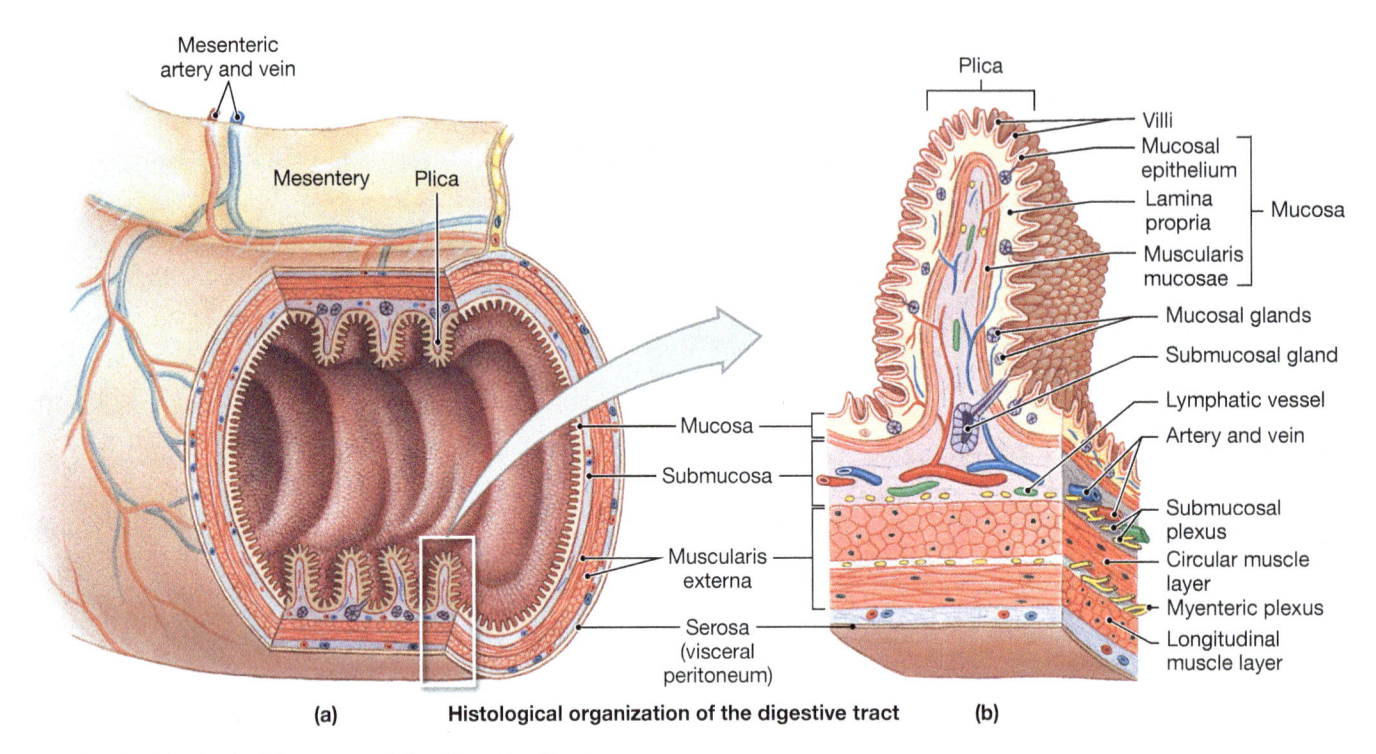

Figure 26.1　Histological Structure of the Digestive Tract
(a) Three-dimensional view of the organization of the digestive tract. (b) Detailed view of the histological features of the digestive tract.

to the lumen of the tract. From the mouth to the esophagus, the mucosal epithelium is stratified squamous epithelium that protects the mucosa from abrasion during swallowing. The mucosal epithelium in the stomach, small intestine, and large intestine is simple columnar epithelium, because food in these parts of the tract is liquid and less abrasive. Deep to the mucosal epithelium is the lamina propria, which is a layer of connective tissue that attaches the epithelium and contains blood vessels, lymphatic vessels, and nerves. The muscularis mucosae is the deepest layer of the mucosa and in most organs has two layers, an inner circular layer that wraps around the tract and an outer longitudinal layer that extends along the length of the tract.

Deep to the mucosa is the **submucosa**, a loose connective-tissue layer that contains blood vessels, lymphatic vessels, and nerves. Deep to the submucosa is a network of sensory and autonomic nerves, the **submucosal plexus**, that controls the tone of the muscularis mucosae.

Deep to the submucosal plexus is the **muscularis externa** layer, which is made up of two layers of smooth muscle tissue. Near the submucosa is a superficial circular muscle layer that wraps around the digestive tract; when this muscle contracts, the tract gets narrower. The deep layer is longitudinal muscle, with cells oriented parallel to the length of the tract. Contraction of this muscle layer shortens the tract. The layers of the muscularis externa produce waves of contraction called **peristalsis** (per-i-STAL-sis), which move materials along the digestive tract. Between the circular and longitudinal muscle layers is the **myenteric** (mī-en-TER-ik) **plexus**, which is comprised of nerves that control the activity of the muscularis externa.

The deepest layer of the digestive tract is called the **adventitia** in the mouth, pharynx, esophagus, and inferior part of the large intestine and either the **serosa** or **visceral peritoneum** (Exercise 1) in the rest of the digestive tract. The adventitia is a network of collagen fibers, and the serosa is a serous membrane of loose connective tissue that attaches the digestive tract to the abdominal wall.

EXERCISE 26 The Digestive System   415

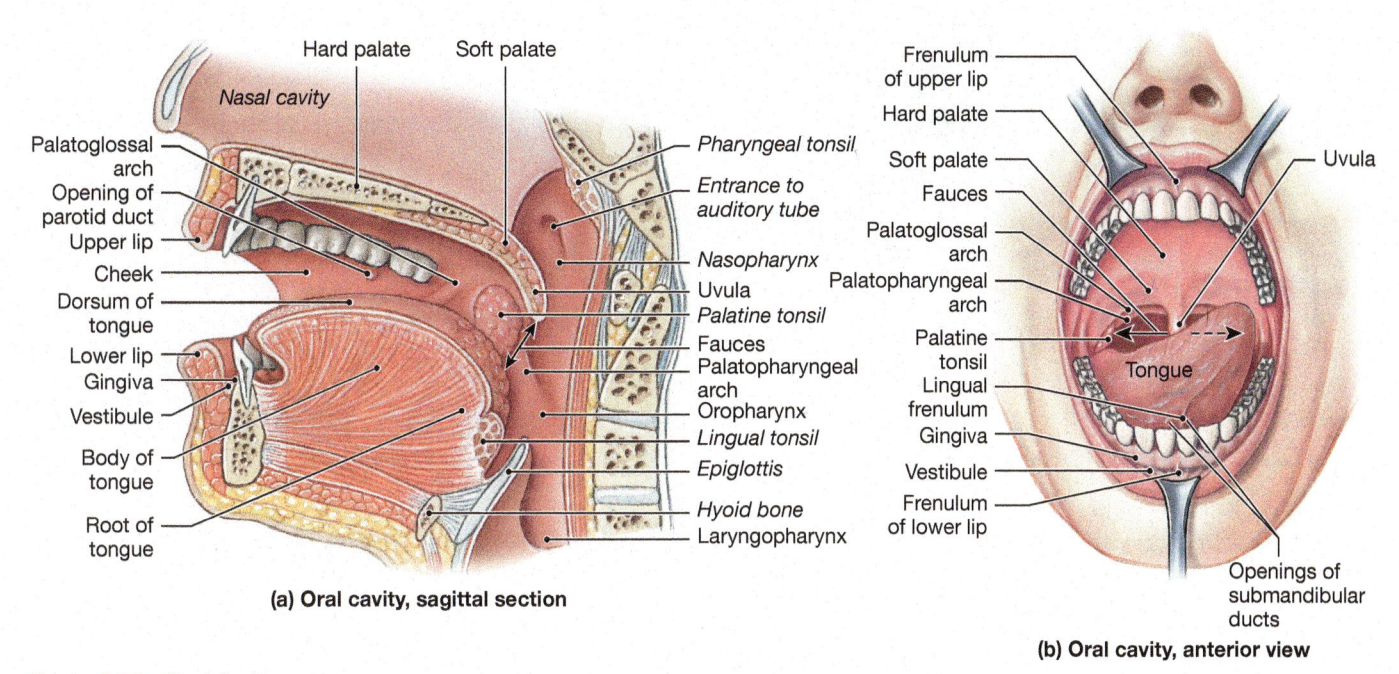

Figure 26.2 Oral Cavity
(a) The oral cavity in sagittal section. (b) Anterior view of the oral cavity, as seen through the open mouth.

LAB ACTIVITY 1   Mouth

The mouth (Figure 26.2) is formally called either the **oral cavity** or the **buccal** (BUK-al) **cavity** and is defined by the space from the lips, or **labia**, posterior to the fauces (Exercise 25). The cone-shaped uvula is suspended from the cavity roof just anterior to the fauces. The lateral walls of the cavity are composed of the **cheeks**, and the roof is the hard palate and the soft palate. The **vestibule** is the region between the teeth and the interior surface of the mouth; thus the vestibule is bounded by the teeth and the cheeks laterally and by the teeth and the upper and lower lips anteriorly. The floor of the mouth is muscular, mostly because of the muscles of the **tongue**. A fold of tissue, the **lingual frenulum** (FREN-ū-lum), anchors the tongue yet allows free movement for food processing and speech. Between the posterior base of the tongue and the roof of the mouth is the **palatoglossal** (pal-a-tō-GLOS-al) **arch**. At the fauces is the **palatopharyngeal arch**.

The mouth contains two structures that act as digestive-system accessory organs: the salivary glands and the teeth. Three pairs of major salivary glands, illustrated in Figure 26.3, produce the majority of the saliva, enzymes, and mucus of the oral cavity. The largest, the **parotid** (pa-ROT-id) **glands**, are anterior to each ear between the skin and the masseter muscle. The **parotid ducts** (*Stensen's ducts*) pierce the masseter and enter the oral cavity to secrete saliva at the upper second molar.

The **submandibular glands** are medial to the mandible and extend from the mandibular arch posterior to the ramus. The **submandibular ducts** (*Wharton's ducts*) pass through the lingual frenulum and open at the swelling on the central margin of this tissue. Submandibular secretions are thicker than that of the parotid glands because of the presence of **mucin**, a thick mucus that helps to keep food in a **bolus**, or ball, for swallowing. Notice the mucous cells in Figure 26.3b.

The **sublingual** (sub-LING-gwal) **glands** are located deep to the base of the tongue. These glands secrete mucus-rich saliva into numerous **sublingual ducts** (*ducts of Rivinus*) that open along the base of the tongue.

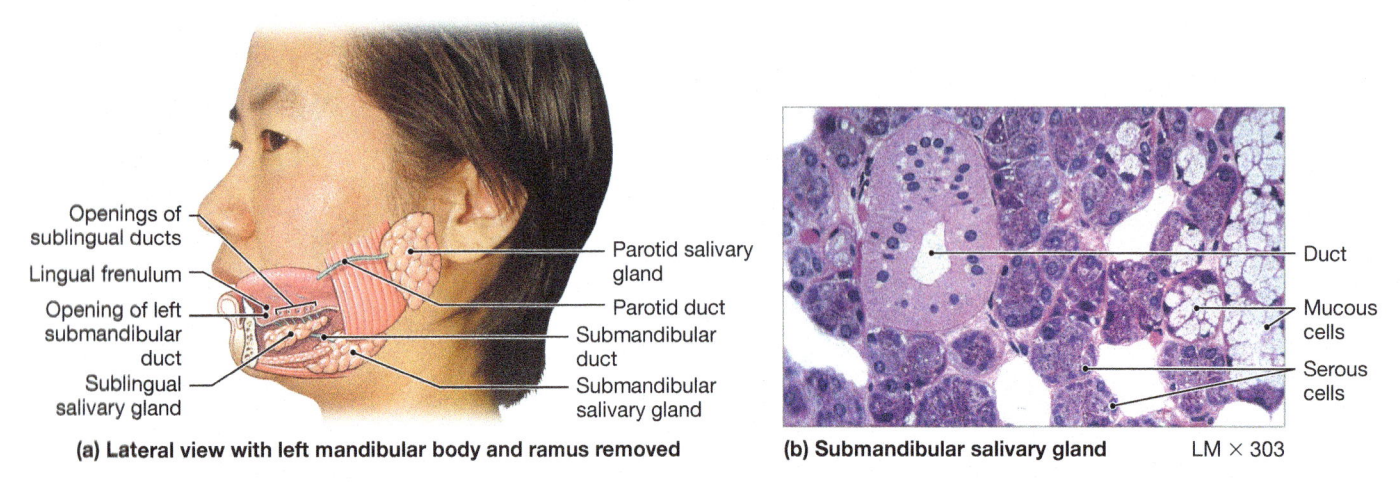

Figure 26.3 Salivary Glands
(a) Lateral view with left mandibular body and ramus removed
(b) Submandibular salivary gland   LM × 303

(a) Lateral view that shows the relative positions of the salivary glands and ducts on the left side of the head. For clarity, the left ramus and body of the mandible have been removed. (b) The submandibular gland secretes a mixture of mucins, which are produced by mucous cells, and enzymes, which are produced by serous cells. (LM × 303)

The teeth are accessory digestive structures for chewing, or **mastication** (mas-ti-KĀ-shun). The **occlusal surface** is the superior area where food is ground, snipped, and torn by the tooth. Figure 26.4a details the anatomy of a typical adult tooth. The tooth is anchored in the alveolar bone of the jaw by a strong **periodontal ligament** that lines the embedded part of the tooth, the **root**. The **crown** is the portion of the tooth above the **gingiva** (JIN-ji-va), or gum. The crown and root meet at the **neck**, where the gingiva forms the **gingival sulcus**, a tight seal around the tooth. Although a tooth has many distinct layers, only the inner **pulp cavity** is filled with living tissue, the **pulp**. Supplying the pulp are blood vessels, lymphatic vessels, and nerves, all of which enter the pulp cavity through the **apical foramen** at the inferior tip of the narrow U-shaped tunnel in the tooth root called the **root canal**. Surrounding the pulp cavity is **dentin** (DEN-tin), a hard, non-living solid similar to bone matrix. Dentin makes up most of the structural mass of a tooth. In the root portion of the tooth, the dentin is covered by **cementum**, a material that provides attachment for the periodontal ligament. The crown is covered with **enamel**, the hardest substance produced by living organisms. Because of this hard enamel, which does not decompose, teeth are often used to identify accident victims and skeletal remains that have no other identifying features.

Humans have two sets of teeth during their lifetime. The first set, the **deciduous** (dē-SID-ū-us; *decidua,* to shed) **dentition**, starts to appear at about the age of six months and is replaced by the **secondary dentition** (*permanent dentition*) starting at around the age of six years. The deciduous teeth are commonly called the *primary teeth, milk teeth,* or *baby teeth.* There are 20 of them, 5 in each jaw quadrant. (The mouth is divided into four quadrants: upper right, upper left, lower right, lower left.) Moving laterally from the midline of either jaw, the deciduous teeth are the **central incisor**, **lateral incisor**, **cuspid** (*canine*), **first molar**, and **second molar**. The secondary dentition consists of 32 teeth; each quadrant contains a central incisor; lateral incisor; cuspid; **first** and **second bicuspids** (*premolars*); and first, second, and third molars (Figure 26.4b). The third molar is also called the *wisdom tooth.*

Each tooth is specialized for processing food. The incisors are used for snipping and biting off pieces of food. The cuspid is like a fang and is used to pierce and tear food. Molars are for grinding and processing the food for swallowing.

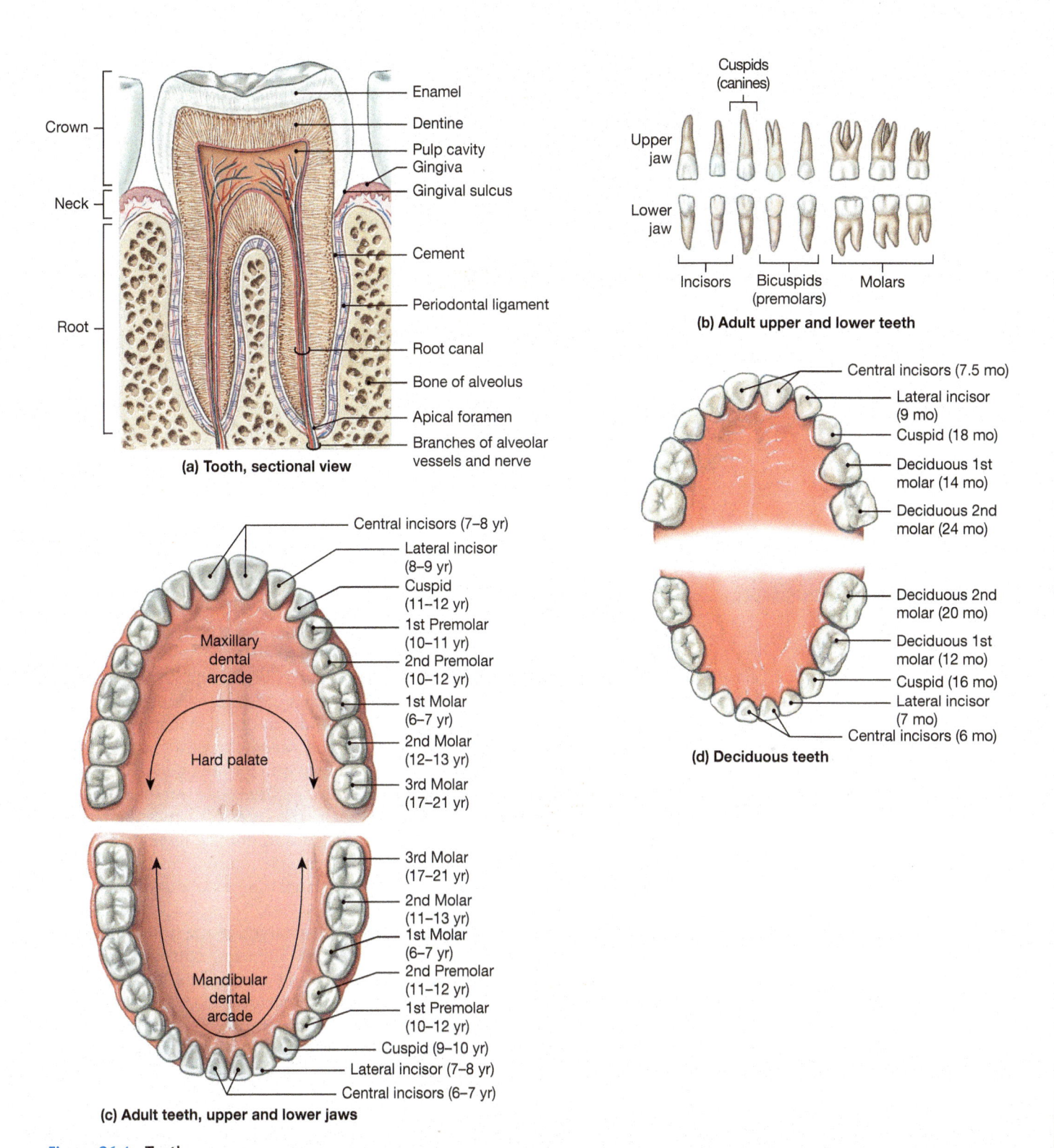

Figure 26.4 Teeth
(a) Diagrammatic section through a typical adult tooth. (b) The secondary teeth. (c) The secondary teeth, with the age at eruption given in years. (d) The deciduous teeth, with the age at eruption given in months.

417

QuickCheck Questions

1.1 Which two mouth structures are digestive system accessory organs?

1.2 Where is each salivary gland located?

1.3 What are the main layers of a tooth?

1 Materials

- ☐ Head model
- ☐ Digestive system chart
- ☐ Tooth model
- ☐ Hand mirror

Procedures

1. Review the mouth anatomy presented in Figures 26.2 and 26.3.
2. Identify the anatomy of the mouth on the head model and digestive system chart.
3. Identify each salivary gland and duct on the head model or digestive system chart.
4. Use the hand mirror to locate your uvula, fauces, and palatoglossal arch. Lift your tongue and examine your submandibular duct.
5. Review the tooth anatomy in Figure 26.4.
6. Use the mirror to examine your teeth. Locate your incisors, cuspids, bicuspids, and molars. How many teeth do you have? Are you missing any because of extractions? Do you have any wisdom teeth? ■

LAB ACTIVITY 2 — Pharynx and Esophagus

As noted in Exercises 25, the pharynx is a passageway for both nutrients and air, and is divided into three anatomical regions—nasopharynx, oropharynx, and laryngopharynx (see Figure 26.3). The nasopharynx is superior to the oropharynx, which is located directly posterior to the oral cavity. Muscles of the soft palate contract during swallowing and close the passageway to the nasopharynx to prevent food from entering the nasal cavity. When you swallow a bolus of food, it passes through the fauces into the oropharynx and then into the laryngopharynx. Toward the base of this area, the pharynx branches into the larynx of the respiratory system and the esophagus that leads to the stomach. The epiglottis closes the larynx so that swallowed food enters only the esophagus and not the respiratory passageways. The lumen of the oropharynx and laryngopharynx is lined with stratified squamous epithelium to protect the walls from abrasion as swallowed food passes through this region of the digestive tract.

The food tube, or **esophagus**, connects the pharynx to the stomach. It is inferior to the pharynx and posterior to the trachea. The esophagus is approximately 25 cm (10 in.) long. It pierces the diaphragm at the **esophageal hiatus** (hī-Ā-tus) and connects with the stomach in the abdominal cavity. At the stomach, the esophagus terminates in a **lower esophageal sphincter**, which is a muscular valve that prevents stomach contents from backwashing into the esophagus. The four layers of the esophagus are shown in Figure 26.5, along with the three regions of the mucosa.

Clinical Application — Acid Reflux

Acid reflux, also commonly called *heartburn*, occurs when stomach acid backflows into the esophagus and irritates the mucosal lining. The term *reflux* refers to a backflow, or regurgitation, of liquid; in this case, gastric juice (which is acidic). Some individuals have a weakened lower esophageal sphincter that allows the gastric juices to reflux during gastric mixing. Recent studies indicate that acid reflux is a major cause of esophageal and pharyngeal cancer. ▶

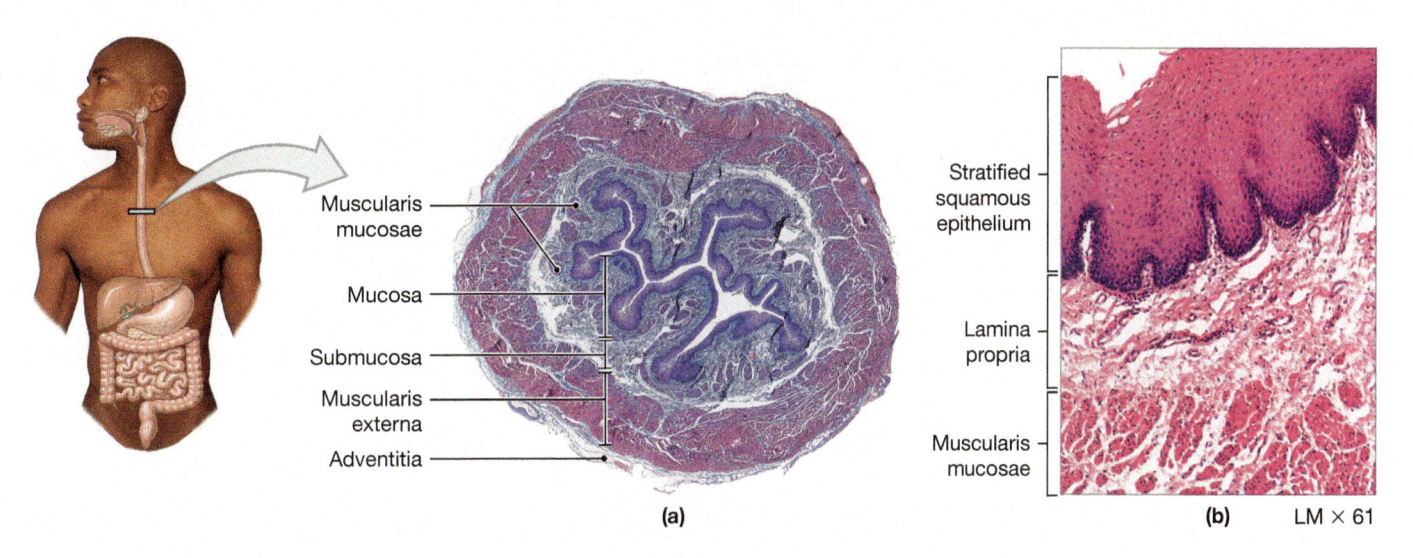

Figure 26.5 Esophagus
(a) Low-power view of a transverse section through the esophagus. (b) The esophageal mucosa. (LM × 61)

QuickCheck Questions
2.1 What are the three regions of the pharynx?
2.2 Which parts of the digestive tract does the esophagus connect?
2.3 Where is the esophageal hiatus?

2 Materials

- Head model
- Digestive system chart
- Torso model
- Hand mirror
- Compound microscope
- Prepared slide of esophagus

Procedures

1. Identify the anatomy of the pharynx and esophagus on the head model and digestive system chart.
2. Put your finger on your Adam's apple (thyroid cartilage of the larynx) and swallow. How does your larynx move, and what is the purpose of this movement?
3. Identify the anatomy of the esophagus on the torso model and digestive system chart.
4. Place the esophagus slide on the microscope and focus on the specimen at low magnification. Observe the organization of the esophageal wall and identify the mucosa, submucosa, muscularis externa with its inner circular and outer longitudinal layers, and adventitia. Use Figure 26.5 for reference during your observations.
5. Increase the magnification and study the mucosa. Distinguish among the mucosal epithelium, which is stratified squamous epithelium, the lamina propria, and the muscularis mucosae. ∎

LAB ACTIVITY 3 Stomach

The stomach is the J-shaped organ just inferior to the diaphragm (Figure 26.6). The four major regions of the stomach are the **cardia** (KAR-dē-uh), where the stomach connects with the esophagus; the **fundus** (FUN-dus), the superior rounded area; the **body** the middle region; and the **pylorus** (pī-LOR-us), which is the narrowed distal end connected to the small intestine. The **pyloric sphincter** (also called the *pyloric valve*) controls movement of material from the stomach into the small intestine. The lateral, convex border of the stomach is the **greater curvature** and the medial concave stomach margin is the **lesser curvature**. Extending from the

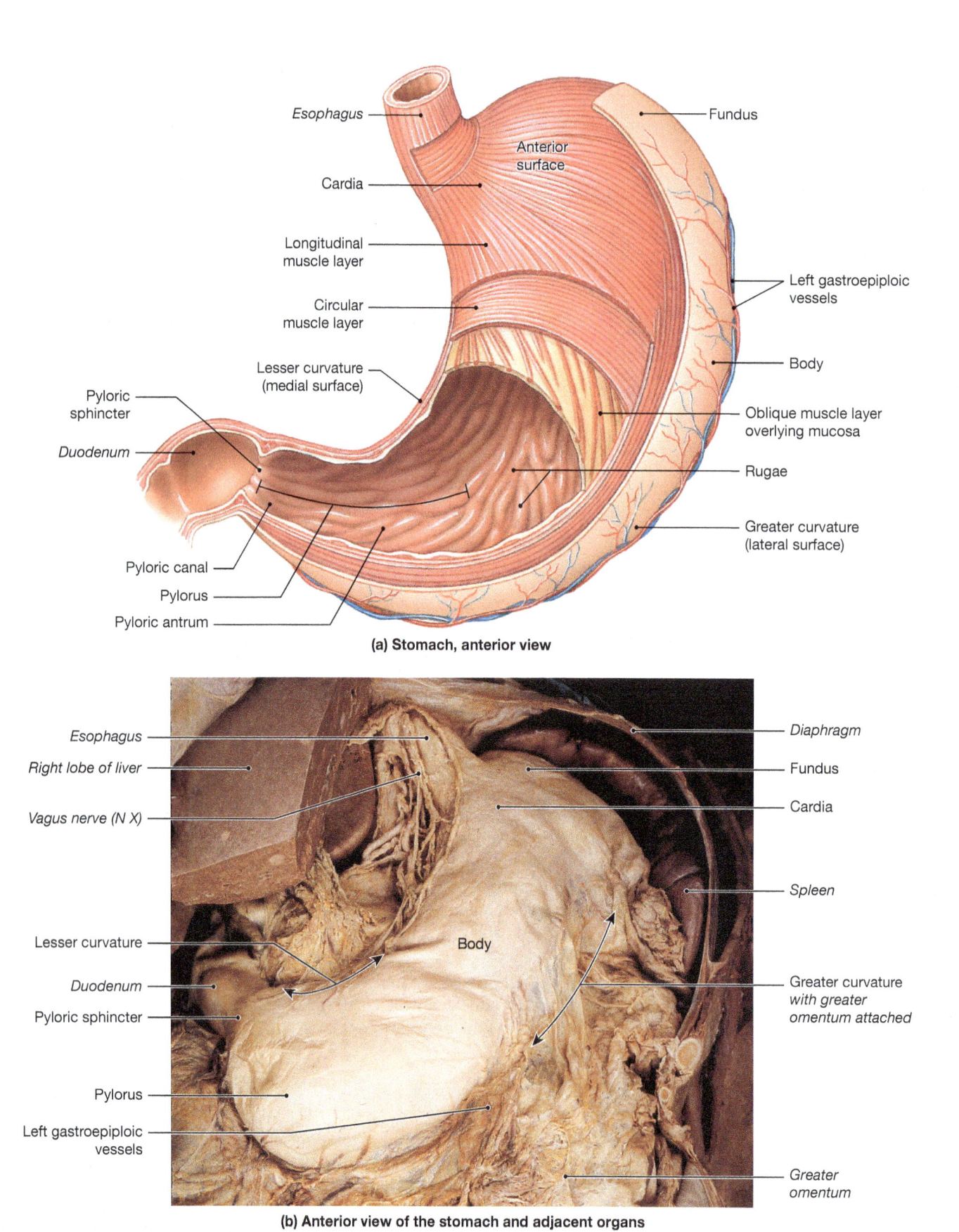

Figure 26.6 Gross Anatomy of the Stomach
(a) External and internal anatomy of the stomach. (b) Anterior view of the superior portion of the abdominal cavity after removal of the left lobe of the liver and part of the lesser omentum.

420

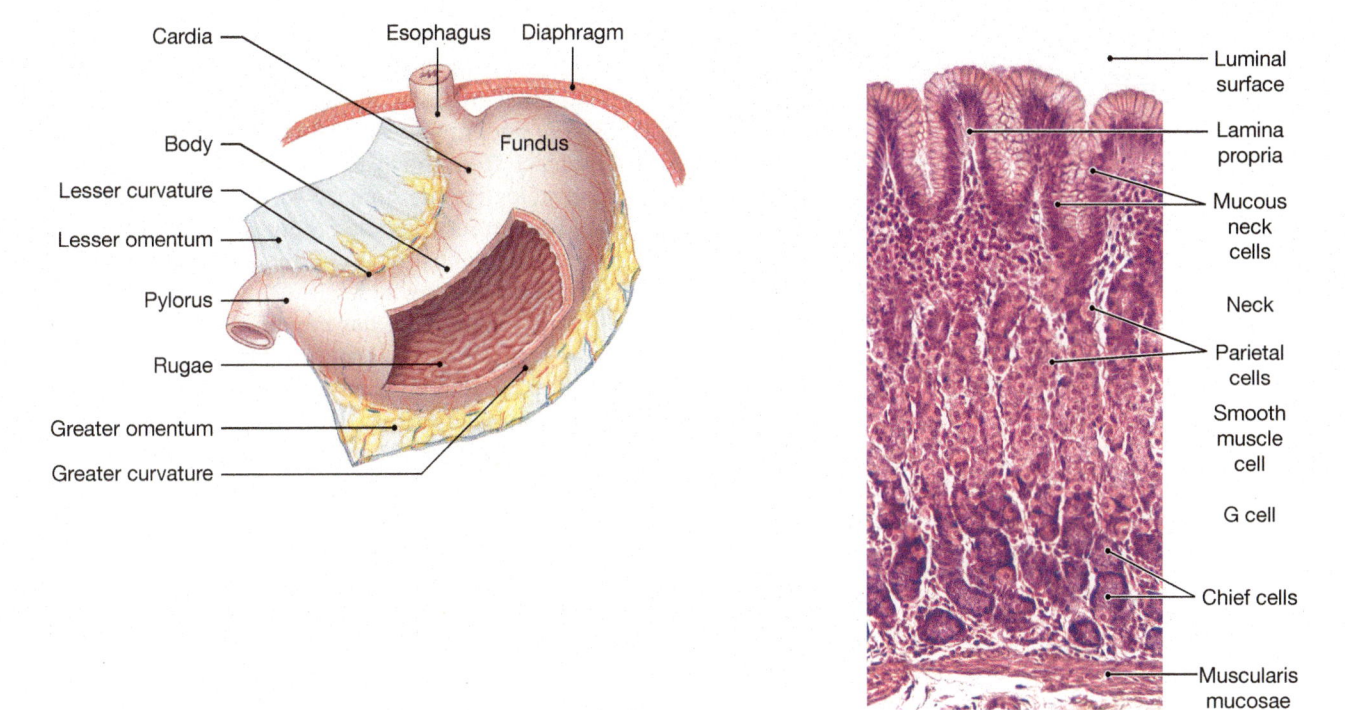

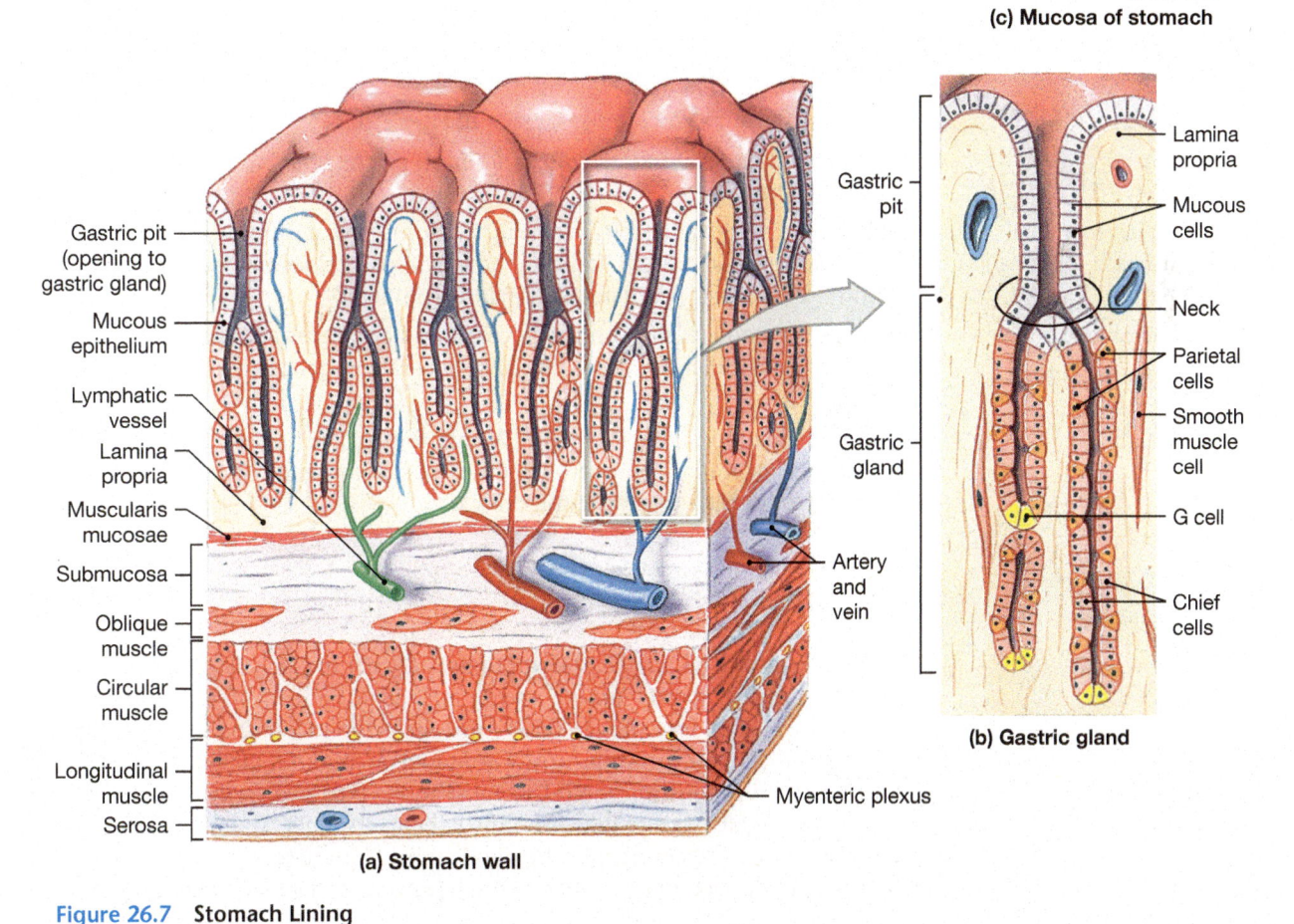

Figure 26.7 Stomach Lining
(a) Diagrammatic view of the organization of the stomach wall. (b) Detail of a gastric gland. (c) Section through gastric pits and gastric glands. (LM × 200)

421

greater curvature is the **greater omentum** (ō-MEN-tum), commonly referred to as the *fatty apron*. This fatty layer is part of the serosa of the stomach wall. Its functions are to protect the abdominal organs and to attach the stomach and part of the large intestine to the posterior abdominal wall. The **lesser omentum**, also part of the serosa, suspends the stomach from the liver.

Figure 26.7 shows the histology of the stomach wall. The mucosal epithelium is simple columnar and folds deep into the lamina propria as **gastric pits** that extend to the base of **gastric glands**. The glands consist of numerous **parietal cells**, which secrete hydrochloric acid, and **chief cells** which release an inactive protein-digesting enzyme called pepsinogen. The submucosa has **rugae** (ROO-gē), which are folds that enable the stomach to expand as it fills with food. Unlike what is found in other regions of the digestive tract, the muscularis externa of the stomach contains three layers of smooth muscle instead of two. The superficial layer (closest to the stomach lumen) is an oblique layer, which is surrounded by a circular layer and then a deep longitudinal layer. The three muscle layers contract and churn stomach contents, then mix gastric juice and liquefy the food into **chyme**. As mentioned previously, the serosa is expanded into the greater and lesser omenta.

QuickCheck Questions

3.1 What are the four major regions of the stomach?

3.2 How is the muscularis externa of the stomach unique?

3.3 Which structure of the stomach allows the organ to distend?

3 Materials

- ☐ Torso model
- ☐ Digestive system chart
- ☐ Preserved animal stomach (optional)
- ☐ Compound microscope
- ☐ Prepared slide of stomach

Procedures

1. Review the anatomy of the stomach in Figure 26.6.
2. Identify the gross anatomy of the stomach on the torso model and digestive system chart.
3. If specimens are available, examine the stomach of a cat or other animal. Locate the rugae, cardia, fundus, body, pylorus, greater and lesser omenta, lower esophageal sphincter, and pyloric sphincter.
4. Place the stomach slide on the microscope stage, focus at low magnification, and observe the rugae. Note how the mucosal epithelium is simple columnar.
5. Increase the magnification and locate the numerous gastric pits, which appear as invaginations along the rugae. Within the pits, distinguish between parietal cells, which are more numerous in the upper areas, and chief cells, which have nuclei at the basal region of the cells. ■

LAB ACTIVITY 4 Small Intestine

The small intestine (Figure 26.8) is approximately 6.4 m (21 ft) long and composed of three segments: duodenum, jejunum, and ileum. Sheets of serous membrane called the **mesenteries** (MEZ-en-ter-ēz) **proper** extend from the serosa to support and attach the small intestine to the posterior abdominal wall. The first 25 cm (10 in.) is the **duodenum** (doo-ō-DĒ-num) and is attached to the distal region of the pylorus. Digestive secretions from the liver, gallbladder, and pancreas flow into ducts that merge and empty into the duodenum. This anatomy is described further in the upcoming section on the liver. The **jejunum** (je-JOO-num) is approximately 3.6 m (12 ft) long and is the site of most nutrient absorption. The last 2.6 m (8 ft) is the **ileum** (IL-ē-um), which terminates at the **ileocecal** (il-ē-ō-SĒ-kal) **valve** and empties into the large intestine.

EXERCISE 26 The Digestive System   423

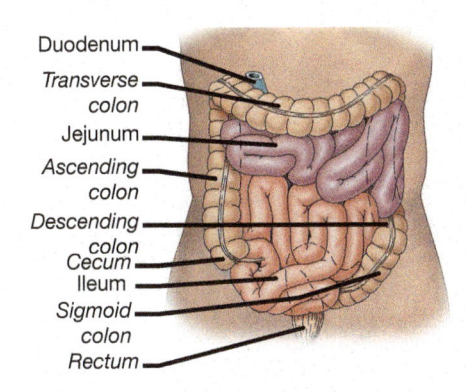

Figure 26.8 Regions of the Small Intestine
Position of the duodenum, jejunum, and ileum in the abdominopelvic cavity.

The small intestine is the site of most digestive and absorptive activities and has specialized folds to increase the surface area for these functions (Figures 26.8 and 26.9). The submucosa and mucosa are creased together into large folds called **plicae circulares** (PLĪ-sē sir-kyoo-LAR-ēz). Along the plicae, the lamina propria is pleated into small, finger-like **villi** that are lined with simple columnar epithelium. The epithelial cells have a **brush border** of minute cell-membrane extensions or folds called **microvilli**.

The epithelium at the base of the villi forms a pocket of cells called **intestinal glands** (*crypts of Lieberkuhn*) that secrete intestinal juice rich in enzymes and pH buffers to neutralize stomach acid. Interspersed among the columnar cells are oval mucus-producing goblet cells. In the middle of each villus is a **lacteal** (LAK-tē-al), which is a lymphatic vessel that absorbs fatty acids and monoglycerides from lipid digestion. In the submucosa are scattered mucus-producing **submucosal** (Brunner's) **glands**. The submucosa of the ileum has **aggregate lymphoid nodules**, also called **Peyer's patches**, which are large lymphatic nodules that prevent bacteria from entering the blood.

QuickCheck Questions

4.1 What are the three major regions of the small intestine?
4.2 What is a plica?
4.3 Where are the intestinal glands located?

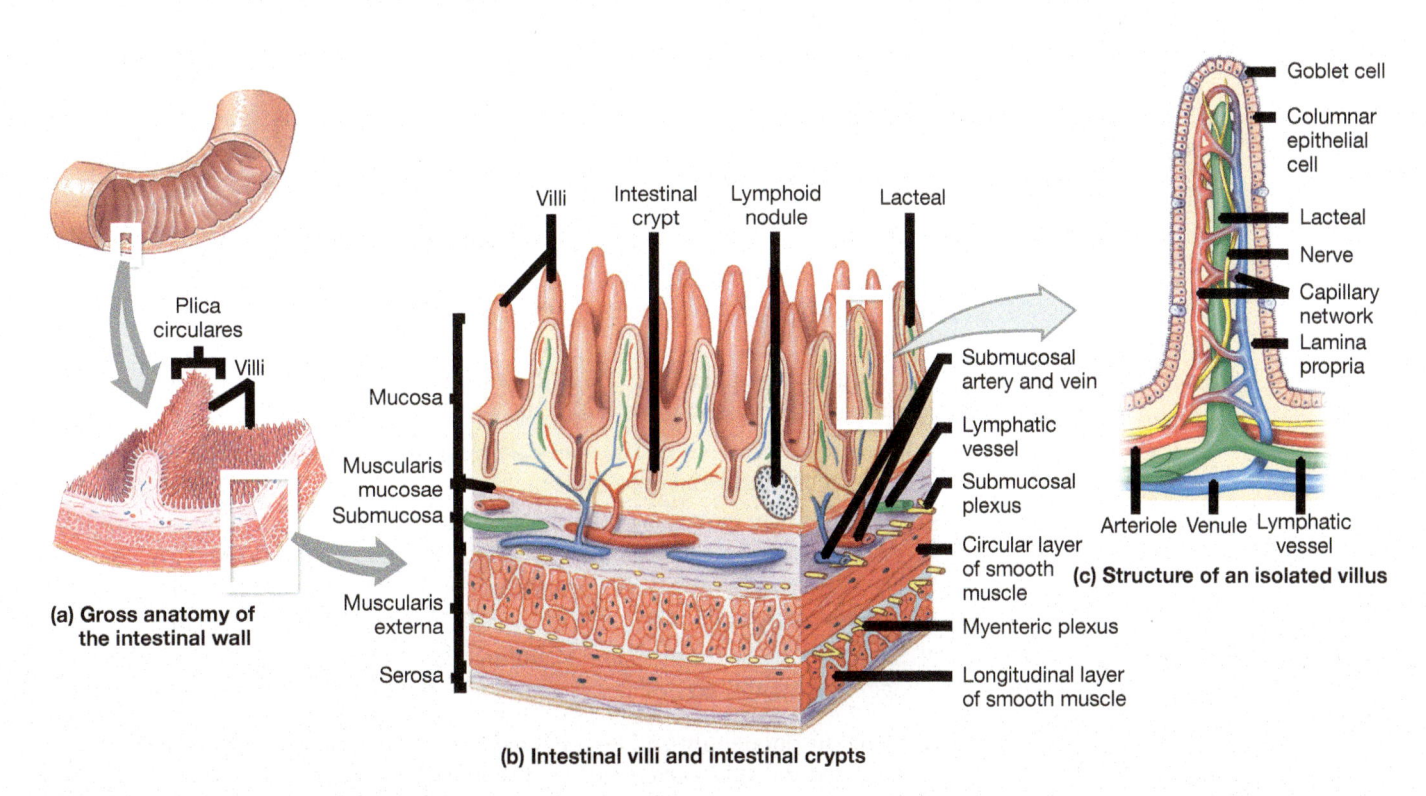

Figure 26.9 The Intestinal Wall
(a) Characteristic features of the intestinal lining. (b) Organization of the intestinal wall. (c) Diagrammatic view of a single villus that shows the capillary and lymphatic supply.

424 EXERCISE 26 The Digestive System

(d) Transverse section of duodenum LM × 65

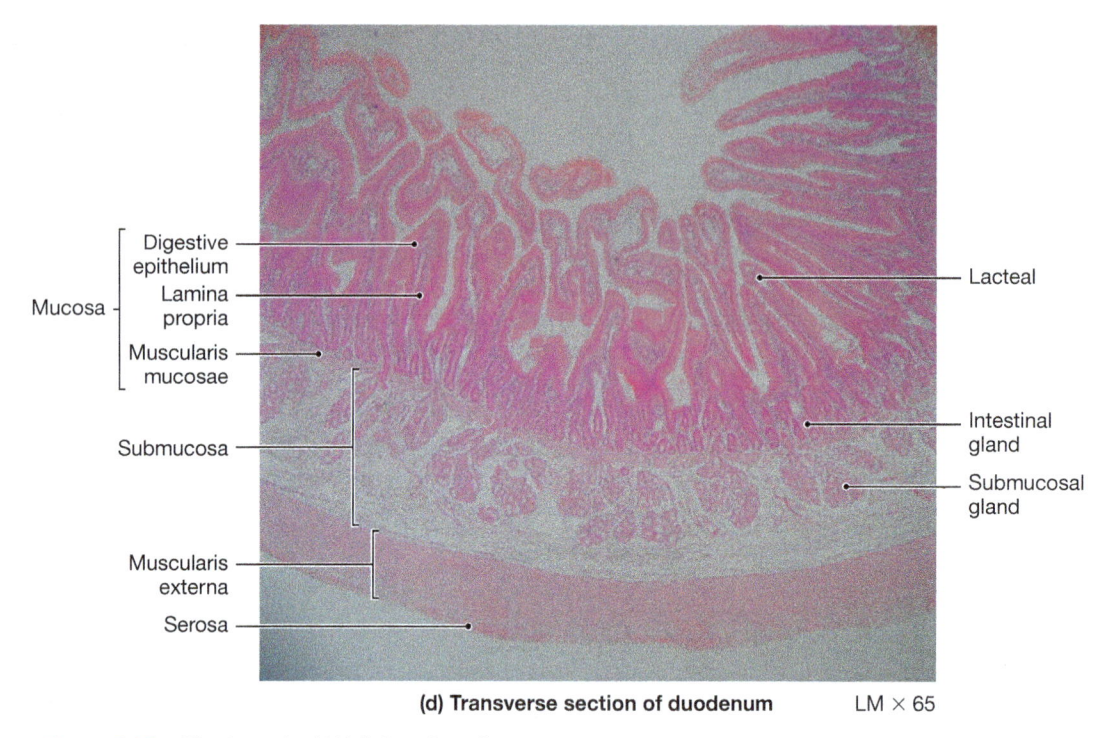

Figure 26.9 The Intestinal Wall *(continued)*
(d) Photomicrograph of the duodenal wall. (LM × 65)

4 Materials

- ☐ Torso model
- ☐ Digestive system chart
- ☐ Preserved animal intestines (optional)
- ☐ Compound microscope
- ☐ Prepared slides of duodenum and ileum

Procedures

1. Review the regions and organization of the small intestine in Figures 26.8 and 26.9.
2. Identify the anatomy of the small intestine on the torso model and the digestive system chart.
3. If a specimen is available, examine a segment of the small intestine of a cat or other animal.
4. Examine the duodenum slide at low magnification, and identify the mucosa, submucosa, muscularis externa, and serosa.
5. At medium power, note the many glands in the submucosa. Trace the ducts of the glands to the mucosal surface.
6. Increase the magnification to high and identify the simple columnar epithelium, goblet cells, lamina propria, and muscularis mucosae, using Figure 26.10 as a guide. The lacteals appear as empty ducts in the lamina propria of the villi. At the base of the villi, locate the intestinal glands.
7. Switch to the ileum slide, and examine it at each magnification. Locate the aggregate lymphoid nodules in the submucosa.
8. Draw the duodenum and ileum at medium magnification in the spaces provided. Label the layers and glands on each sketch. ■

Duodenum at medium magnification

Ileum at medium magnification

LAB ACTIVITY 5   Large Intestine

The large intestine is the site of electrolyte and water absorption and waste compaction. It is approximately 1.5 m (5 ft) long and is divided into two regions: the colon (KŌ-lin), which makes up most of the intestine, and the rectum. Figure 26.10 details the gross anatomy. The ileocecal valve regulates what enters the colon from the ileum. The first part of the colon, which is a pouch-like **cecum** (SĒ-kum), is located in the right lumbar region. At the medial floor of the cecum is the worm-like **appendix**. Distal to the cecum, the **ascending colon** travels up the right side of the abdomen, bends left at the **right colic (hepatic) flexure**, and crosses the abdomen inferior to the stomach as the **transverse colon**. The **left colic (splenic) flexure** turns the colon inferiorly to become the **descending colon**.

The S-shaped **sigmoid** (SIG-moyd) **colon** passes through the pelvic cavity to join the **rectum**, which is the last 15 cm (6 in.) of the large intestine and the end of digestive tract. The opening of the rectum, the **anus** is controlled by an **internal anal sphincter** of smooth muscle and an **external anal sphincter** of skeletal muscle. Longitudinal folds called **anal columns** occur in the rectum where the digestive epithelium changes from simple columnar to stratified squamous.

In the colon, the longitudinal layer of the muscularis externa is modified into three bands of muscle that is collectively called the **taenia coli** (TĒ-nē-a KŌ-lī). The muscle tone of the taenia coli constricts the colon wall into pouches called **haustra** (HAWS-truh, singular *haustrum*), which permit the colon wall to expand and stretch.

The wall of the colon lacks plicae and villi (Figure 26.11). It is thinner than the wall of the small intestine and contains more glands. The mucosal epithelium is simple columnar epithelium that folds into intestinal glands lined by goblet cells.

QuickCheck Questions

5.1 What are the major regions of the colon?

5.2 Where is the appendix located?

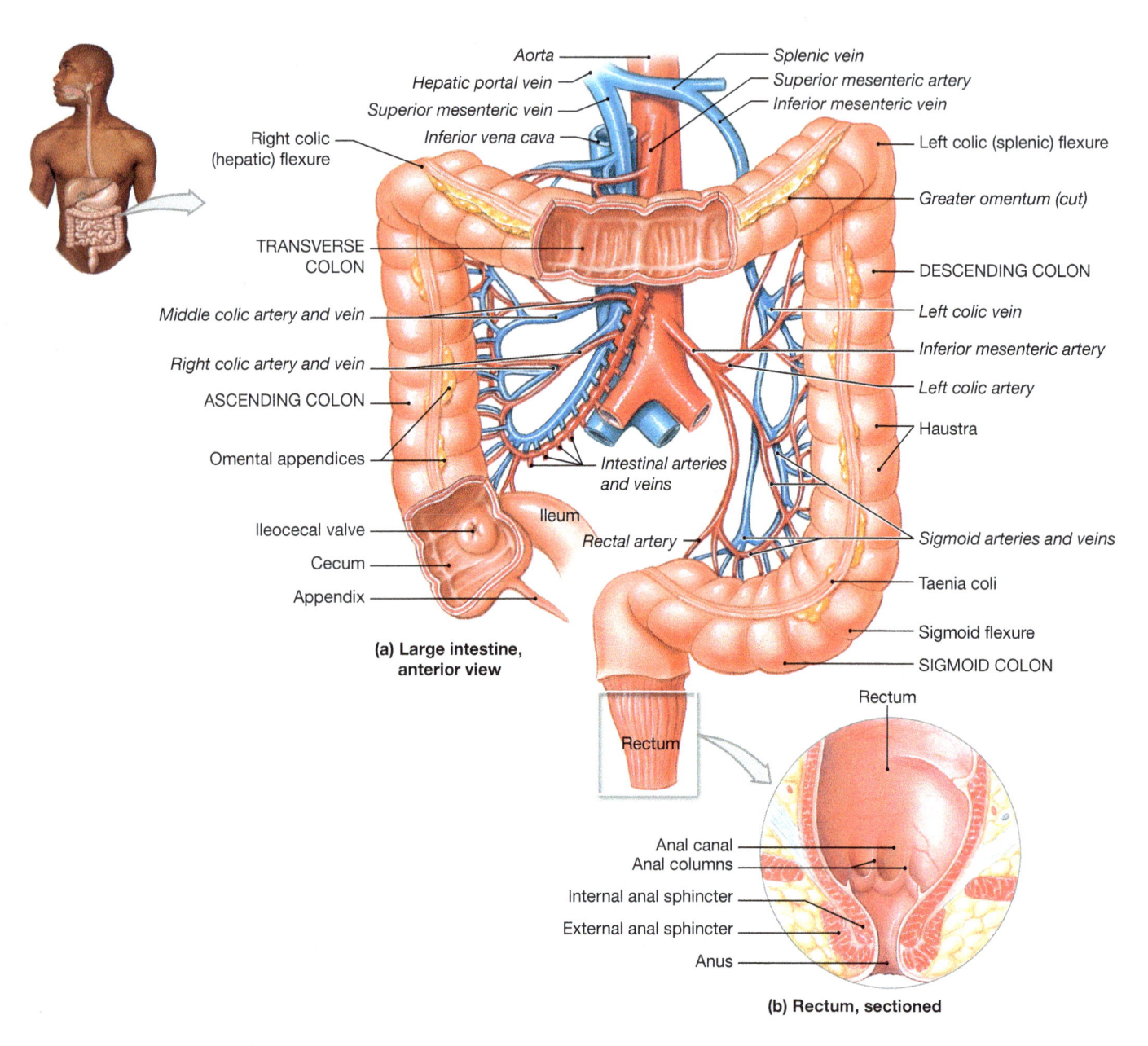

Figure 26.10 Large Intestine
(a) Gross anatomy and regions of the large intestine. (b) Rectum and anus.

EXERCISE 26 The Digestive System 427

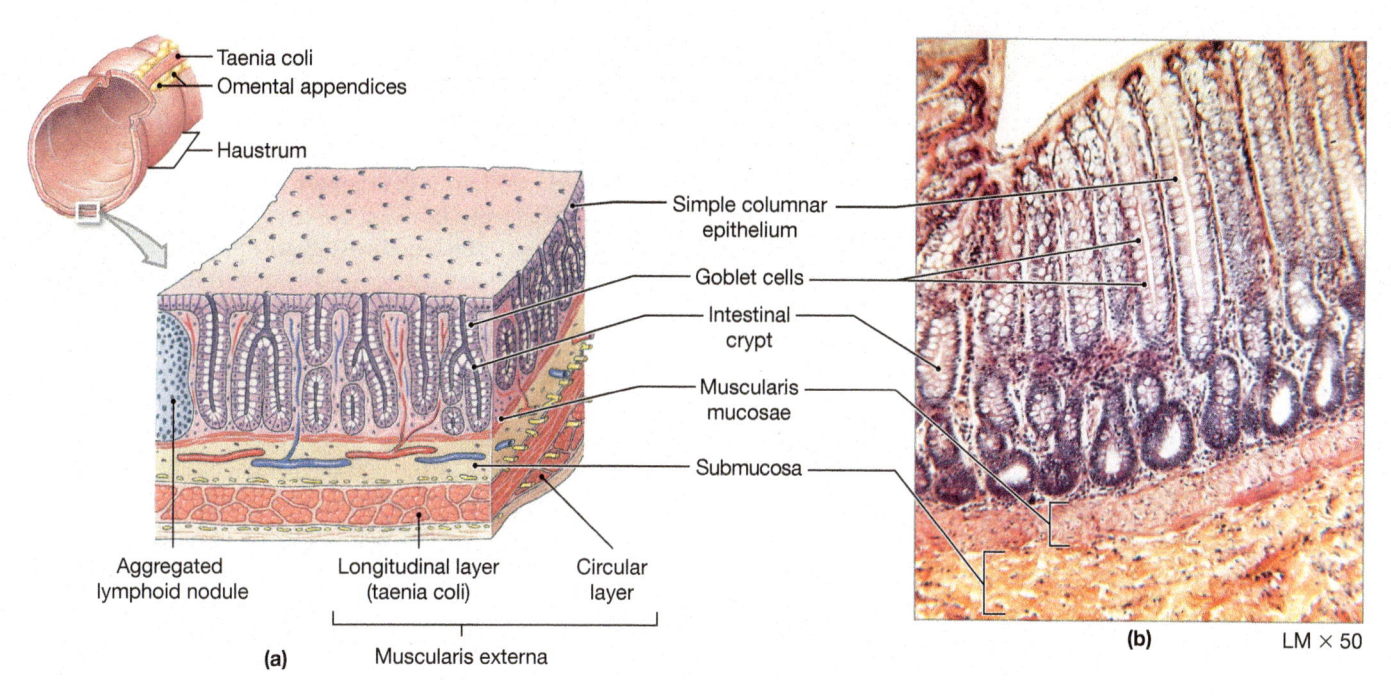

Figure 26.11 Wall of the Colon
(a) Three-dimensional view. (b) Photomicrograph that shows detail of mucosal and submucosal layers. (LM × 50)

5 Materials

- ☐ Torso model
- ☐ Digestive system chart
- ☐ Preserved animal intestines (optional)
- ☐ Compound microscope
- ☐ Prepared slide of large intestine

Procedures

1. Review the anatomy of the large intestine in Figures 26.10 and 26.11.
2. Identify the gross anatomy of the large intestine on the torso model and the digestive system chart.
3. If a specimen is available, examine the colon of a cat or other animal. Locate each region of the colon, the left and right colic flexures, the taenia coli, and the haustrae.
4. View the microscope slide of the large intestine and locate the intestinal glands. Distinguish between the simple columnar cells and goblet cells. ■

428   EXERCISE 26   The Digestive System

LAB ACTIVITY 6 — Liver and Gallbladder

The liver is located in the right upper quadrant of the abdomen and is suspended from the inferior of the diaphragm by the **coronary ligament** (Figure 26.12). Historically, the liver has been divided into four lobes that are visible in gross observation. Current medical and surgical classification of the liver is based on vascular supply to individual segments; however, the blood vessels are apparent only in dissection. For gross observations, we shall use the four-lobes description. The **right** and **left lobes** are separated by the **falciform ligament**, which attaches the lobes to the abdominal wall. Within the falciform ligament is the **round ligament**, where the fetal umbilical vein passed. The square **quadrate lobe** is located on the inferior surface of the right lobe, and the **caudate lobe** is posterior, near the site of the inferior vena cava.

Each lobe is organized into approximately 100,000 smaller lobules (Figure 26.13). In the lobules, cells called **hepatocytes** (he-PAT-ō-sīts) secrete **bile**, a watery substance that acts like dish soap and breaks down the fat in ingested food. The bile is released into small ducts called **bile canaliculi**, which empty into **bile ductules** (DUK-tūlz) that surround each lobule. Progressively larger ducts drain bile into the **right** and **left hepatic ducts**, which then join a **common hepatic duct**. Blood flows through spaces called **sinusoids**; each sinusoid receives blood from a branch of either the hepatic artery or the hepatic portal vein. The sinusoids empty into a **central vein** in the middle of each lobule. Hepatocytes that line the sinusoids phagocytize worn-out blood cells and reprocess the hemoglobin pigments for new blood cells.

The **gallbladder** is a small, muscular sac that stores and concentrates bile salts used in the digestion of lipids. It is located inferior to the right lobe of the liver.

Figure 26.14 details the ducts of the liver and gallbladder. The common hepatic duct from the liver meets the **cystic duct** of the gallbladder to form the **common bile duct**. This duct passes through the lesser omentum and continues on to a junction called the **duodenal ampulla** (am-PUL-uh). The ampulla projects into the lumen of the duodenum at the **duodenal papilla**. A band of muscle called the **hepatopancreatic sphincter** (*sphincter of Oddi*) regulates the flow of bile and other secretions into the duodenum.

QuickCheck Questions

6.1 What are the four visible lobes of the liver?

6.2 How does bile enter the small intestine?

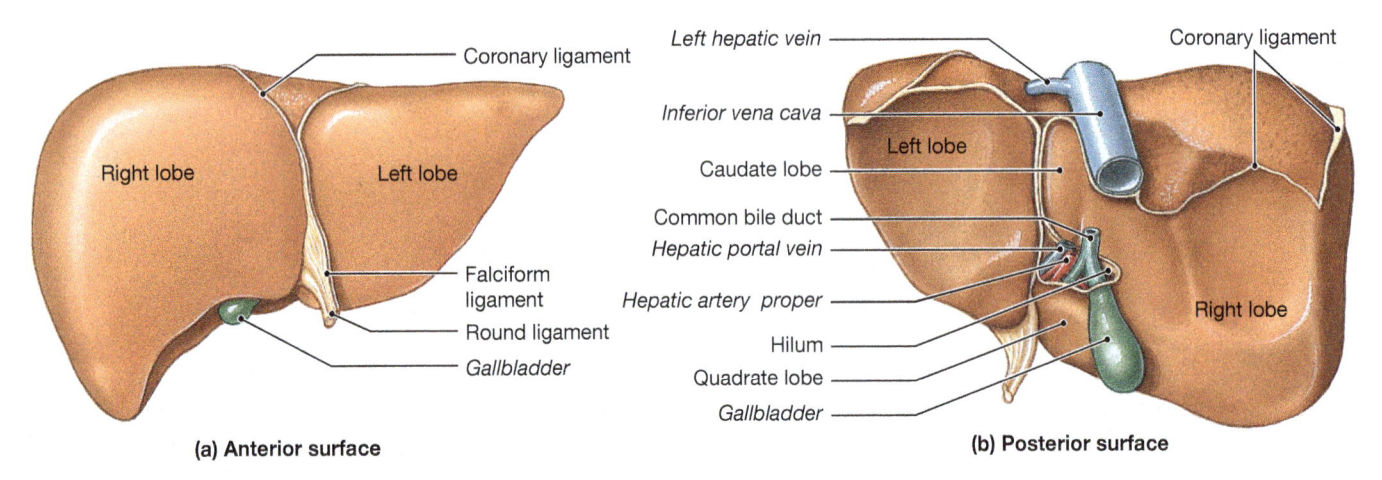

Figure 26.12 Anatomy of the Liver
(a) Anterior surface of the liver. (b) Posterior surface of the liver.

(a) Lobular organization

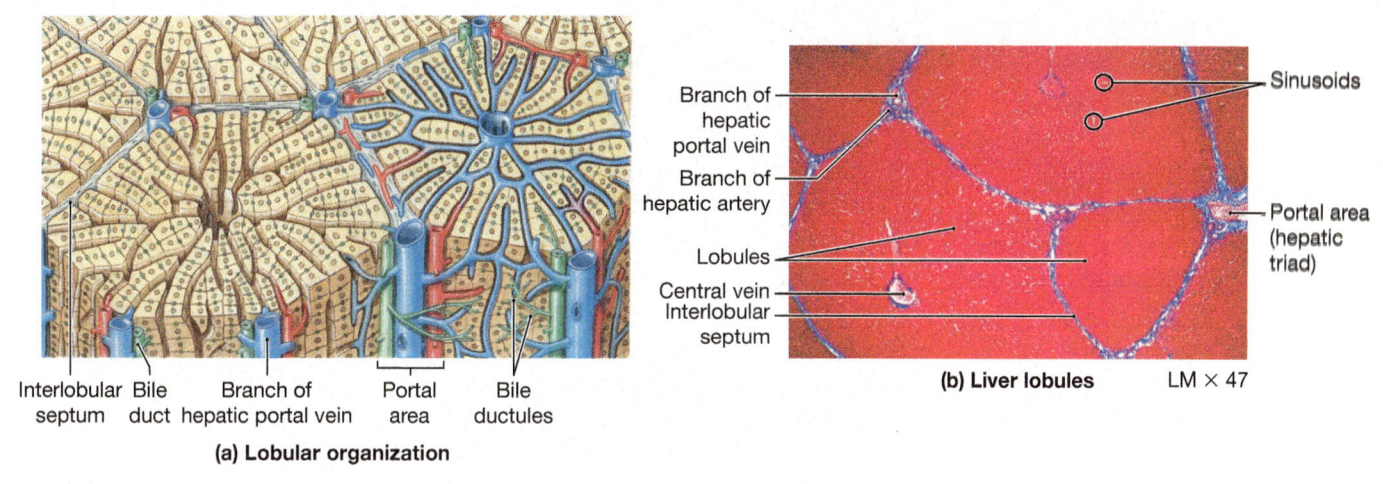

(b) Liver lobules LM × 47

Figure 26.13 **Liver Histology**
(a) Diagrammatic view of lobular organization. (b) Light micrograph that shows a section through a liver lobule. (LM × 47)

(a) Gallbladder and associated ducts

(b) Hepatopancreatic sphincter

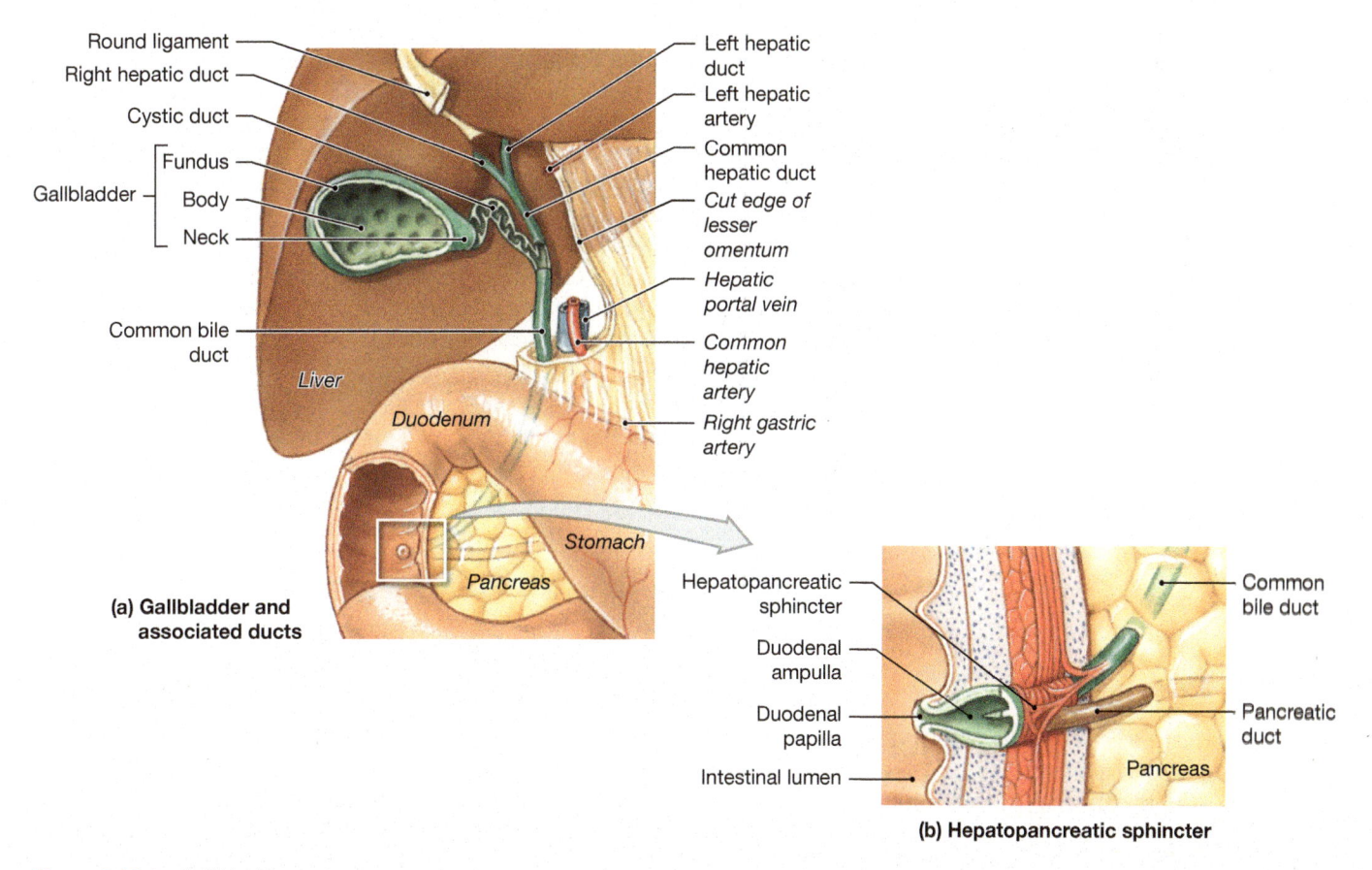

Figure 26.14 **Gallbladder**
(a) View of the inferior surface of the liver that shows the position of the gallbladder and ducts that transport bile from the liver to the gallbladder and duodenum. A portion of the lesser omentum has been cut away. (b) Interior view of the duodenum that shows the duodenal ampulla and related structures.

6 Materials

- ☐ Torso model
- ☐ Digestive system chart
- ☐ Liver model
- ☐ Preserved animal liver and gallbladder (optional)
- ☐ Compound microscope
- ☐ Prepared slide of liver

Procedures

1. Review the anatomy of the liver in Figures 26.12 and 26.13.
2. Review the anatomy of the gallbladder in Figure 26.14.
3. Identify the gross anatomy of the liver and gallbladder on the torso model, liver model, and digestive system chart. Trace the ducts that transport bile from the liver and gallbladder into the small intestine.
4. If specimens are available, examine the liver and gallbladder of a cat or other animal. Locate each liver lobe; the falciform and round ligaments; and the hepatic, cystic, and common bile ducts.
5. Examine the liver slide at low magnification and identify the many lobules. Notice hepatocytes that line the sinusoids and the central vein of each lobule. In humans, the lobules are not well-defined. Pigs and other animals have a connective tissue septum around each lobule; this septum can be seen in Figure 26.14b. ■

LAB ACTIVITY 7 Pancreas

The **pancreas** is a gland (Exercise 20) that is located posterior to the stomach. The pancreas **head** is adjacent to the duodenum, the **body** is the central region, and the **tail** tapers to the distal end of the gland (Figure 26.15). The pancreas is characterized as a *double gland,* which means that it has both endocrine and exocrine functions. The endocrine cells occur in **pancreatic islets** and secrete hormones for sugar metabolism. Most of the glandular epithelium of the pancreas has an exocrine function. These exocrine cells, called **acini** (AS-i-nī) **cells**, secrete pancreatic juice into small ducts called **acini** located in the pancreatic glands. The acini drain into progressively larger ducts that merge as the **pancreatic duct** and, in some individuals, an **accessory pancreatic duct**. The pancreatic duct joins the common bile duct at the duodenal ampulla (see Figure 26.14b).

QuickCheck Questions

7.1 What are the exocrine and endocrine functions of the pancreas?
7.2 Where does the pancreatic duct connect to the duodenum?

7 Materials

- ☐ Torso model
- ☐ Digestive system chart
- ☐ Preserved animal pancreas (optional)
- ☐ Compound microscope

Procedures

1. Review the anatomy of the pancreas in Figure 26.16.
2. Identify the anatomy of the pancreas on the torso model and the digestive system chart.
3. If a specimen is available, examine the pancreas of a cat or other animal. Locate the head, body, and tail of the organ and the pancreatic duct.
4. On the pancreas slide, observe the numerous oval pancreatic ducts at low and medium magnifications. The exocrine cells are the dark-stained acini cells that surround groups of endocrine cells, which are the light-stained pancreatic islets. ■

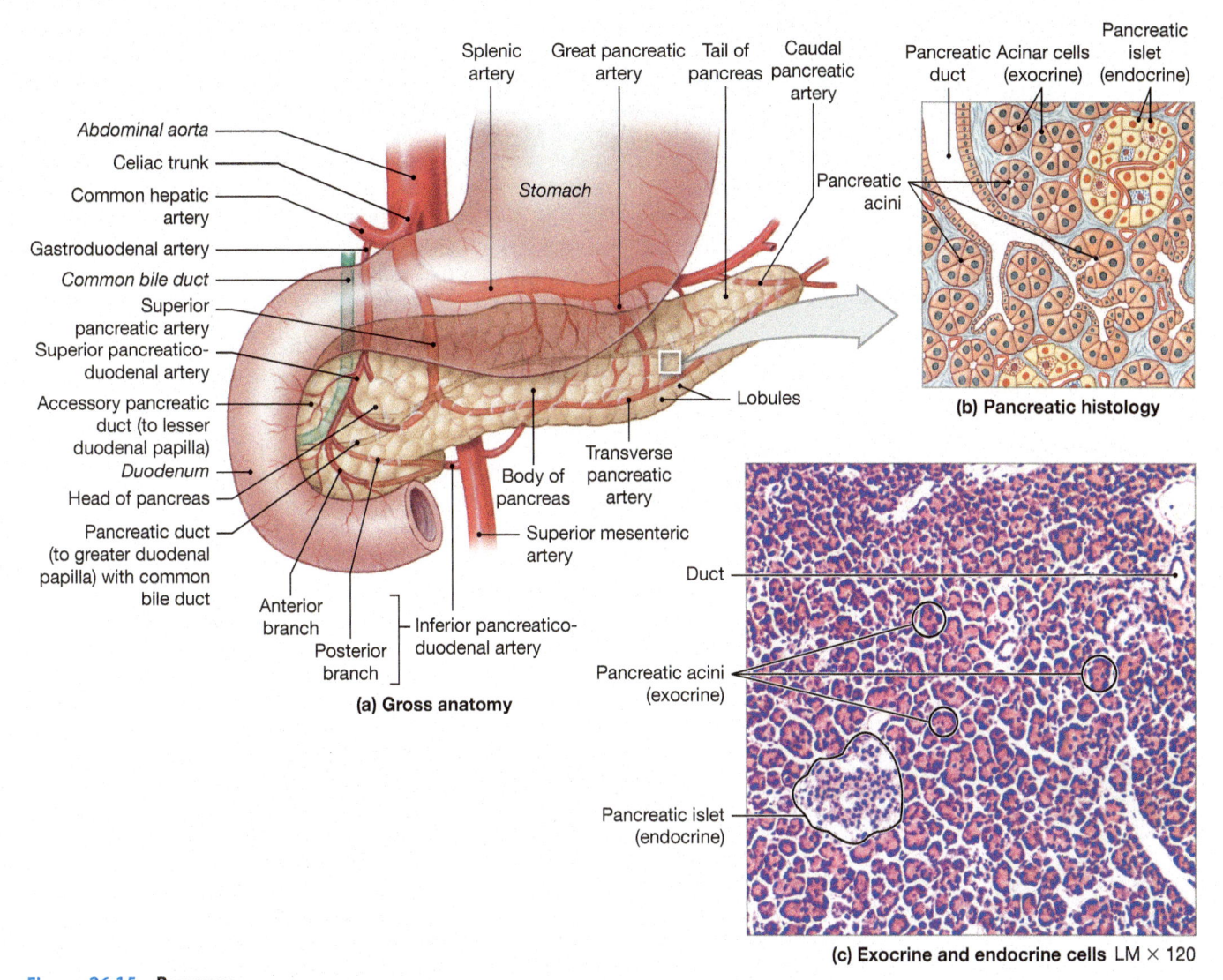

Figure 26.15 Pancreas
(a) Gross anatomy of the pancreas. The head is tucked into a C-shaped curve of the duodenum that begins at the pylorus of the stomach. The cellular organization of the pancreas is shown (b) diagrammatically and (c) in a micrograph. (LM × 120)

upper abdomen
A Regional Look

Figure 26.16 shows a cadaver photograph of the liver in horizontal section. The fact that the pleural cavity and cut edge of the diaphragm are visible tells you that the image is of the inferior side of the section. Locate the falciform ligament and note how it attaches the liver to the abdominal wall. Note the close association between the liver and the inferior vena cava. Recall that the liver processes the blood from the digestive tract and then this blood drains into the inferior vena cava.

The section at this level of the abdomen reveals how the liver occupies the right upper quadrant of the abdomen and extends into the left upper quadrant. On the left side, the stomach with its folded interior is sectioned. The spleen and left kidney are visible in the photograph; the right kidney is positioned lower than the left kidney because of the size of the liver and is not visible in the photograph. The gallbladder, small intestine, and large intestine are inferior to this section and therefore are not visible in the photograph.

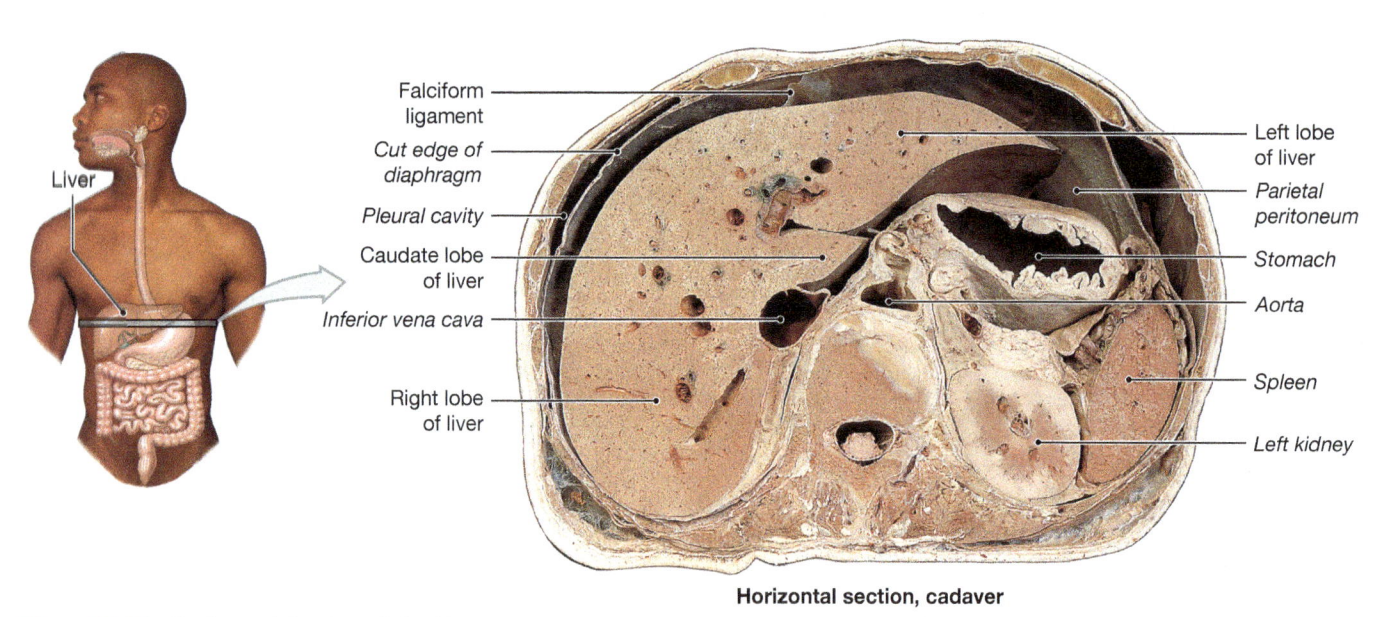

Figure 26.16 Horizontal Section of the Liver
Horizontal sectional view through the upper abdomen that shows the position of the liver relative to other visceral organs.

EXERCISE 26

LAB REPORT

The Digestive System

Name _____

Date _____

Section _____

A. Matching

Match each structure in the left column with its correct description from the right column.

_____ 1. pyloric sphincter
_____ 2. incisor
_____ 3. esophageal hiatus
_____ 4. taenia coli
_____ 5. haustra
_____ 6. muscularis mucosae
_____ 7. muscularis externa
_____ 8. gingiva
_____ 9. rugae
_____ 10. plicae
_____ 11. molar
_____ 12. duodenal ampulla

A. muscle layer of mucosa
B. submucosal folds of stomach wall
C. pouches in colon wall
D. folds of intestinal wall
E. tooth used for snipping food
F. longitudinal muscle of colon
G. gum that surrounds a tooth
H. tooth used for grinding
I. valve between stomach and duodenum
J. duct that transports both bile and pancreatic juice
K. passageway for esophagus in diaphragm
L. major muscle layer deep to submucosa

B. Short-Answer Questions

1. What are the four major layers of the digestive tract wall?

2. What are the three sublayers of the mucosa?

3. Name the accessory organs of the digestive system.

4. Describe the gross anatomy of the large intestine.

EXERCISE 26 — LAB REPORT

C. Drawing

Draw and label a cross-section of the wall of the small intestine, and show the structures that increase the wall surface area for maximum digestion and absorption of food.

D. Analysis and Application

1. Trace a drop of bile from the point where it is produced to the point where it is released into the intestinal lumen.

2. If the duodenal ampulla is blocked, what materials cannot be released into the lumen of the small intestine?

3. List the modifications of the small intestine wall that increase surface area.

EXERCISE 27

The Urinary System

OBJECTIVES

On completion of this exercise, you should be able to:

- Identify and describe the basic anatomy of the urinary system.
- Explain the function of the kidney.
- Identify the basic components of the nephron.
- Trace the flow of blood through the kidney.
- Describe the differences between the male and female urinary tracts.

LAB ACTIVITIES

1. Kidney   435
2. Nephron   436
3. Blood Supply to the Kidney   440
4. Ureters, Urinary Bladder, and Urethra   441
5. Sheep Kidney Dissection   444

    A Regional Look: Urinary Bladder   446

The primary function of the urinary system is to control the composition, volume, and pressure of the blood. The system exerts this control by adjusting both the volume of the liquid portion of the blood (the *plasma*, Exercise 4) and the concentration of solutes in the blood as it passes through the kidneys. Any excess water and solutes that accumulate in the blood and waste products and are eliminated from the body via the urinary system. These eliminated products are collectively called *urine*. The urinary system is comprised of a pair of kidneys, a pair of ureters, a urinary bladder, and a urethra.

LAB ACTIVITY 1 Kidney

The kidneys lie on the posterior surface of the abdomen on either side of the vertebral column between vertebrae T_{12} and L_3. The right kidney is typically lower than the left kidney because of the position of the liver. The kidneys are *retroperitoneal*, which means that they are located outside of the peritoneal cavity, behind the parietal peritoneum. Each kidney is secured in the abdominal cavity by three layers of tissue: renal fascia, adipose capsule, and renal capsule. Superficially, the **renal fascia** anchors the kidney to the abdominal wall. The **adipose capsule** is a mass of adipose tissue that envelopes the kidney and protects it from trauma as well as helping to anchor it to the abdominal wall. Deep to the adipose capsule, on the surface of the kidney, the fibrous tissue of the **renal capsule** protects the kidney from trauma and infection.

A kidney is about 13 cm (5 in.) long and 2.5 cm (1 in.) thick. The medial aspect contains a hilus through which blood vessels, nerves, and other structures enter and exit the kidney (Figure 27.1). The hilus also leads to a cavity in the kidney called the

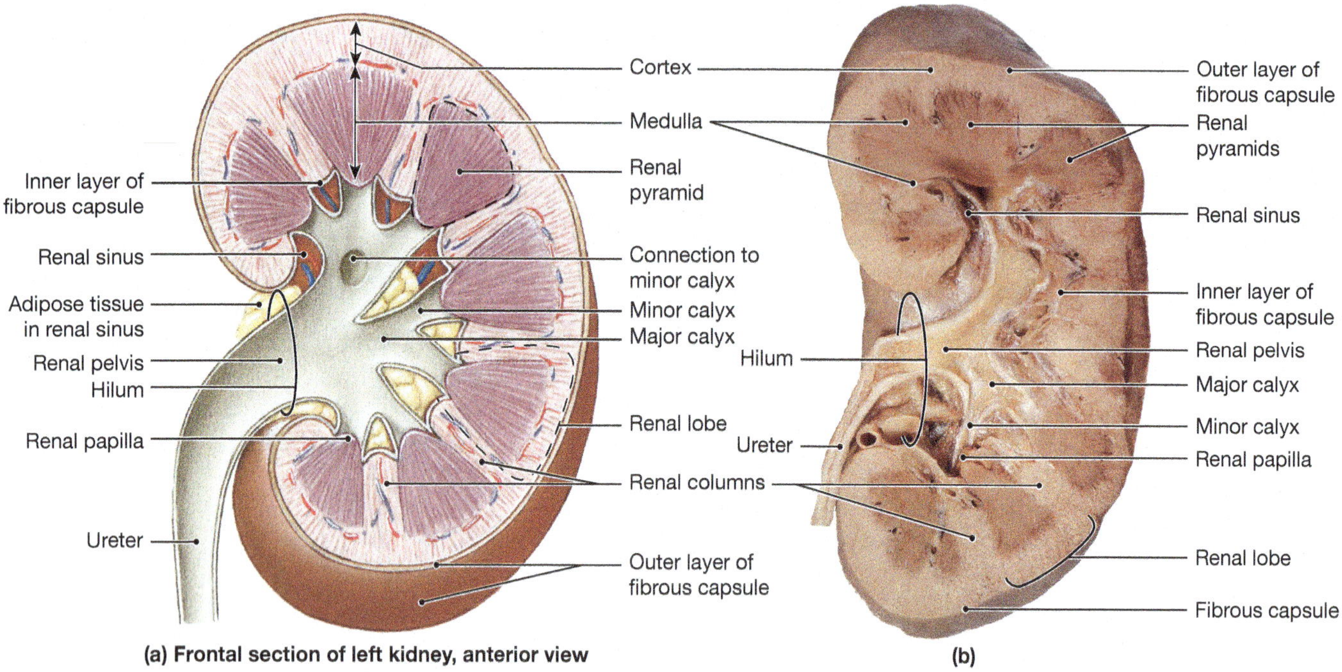

436 EXERCISE 27 The Urinary System

(a) Frontal section of left kidney, anterior view

(b)

Figure 27.1 Structure of the Kidney
(a) Diagrammatic view of a frontal section through the left kidney. The renal fascia and adipose capsule have been removed in this rendering. **(b)** Frontal section of the left kidney.

renal sinus. The **cortex** is the outer, light-red layer of the kidney, located just deep to the renal capsule. Deep to the cortex is a region called the **medulla**, which consists of triangular **renal pyramids** that project toward the kidney center. Areas of the cortex that extend between the renal pyramids are **renal columns**. At the apex of each renal pyramid is a **renal papilla** that empties urine into a small cup-like space called the **minor calyx** (KĀ-liks). Several minor calyces (KĀL-i-sēz) empty into a common space, the **major calyx**. These larger calyces merge to form the **renal pelvis**.

QuickCheck Questions

1.1 What is the hilus?
1.2 Where are the renal pyramids located?
1.3 Where are the renal columns located?

1 Materials

- ☐ Kidney model
- ☐ Kidney chart

Procedures

1. Review the anatomy of the kidney in Figure 27.1.
2. Locate each structure shown in Figure 27.1 on the kidney model or chart. ■

LAB ACTIVITY 2 Nephron

Each kidney contains more than 1 million microscopic tubules called **nephrons** (NEF-rons) that produce urine. As blood circulates through the blood vessels of the kidney, blood pressure forces materials such as water, excess ions, and waste products out of the blood and into the nephrons. This aqueous solution, called **filtrate**, circulates through the nephrons, and as this circulation takes place, any substances in the filtrate still needed by the body move back into the blood. The remaining filtrate is excreted as urine.

Approximately 85 percent of the nephrons are **cortical nephrons,** which are found in the cortex and barely penetrate into the medulla (Figure 27.2). The remaining

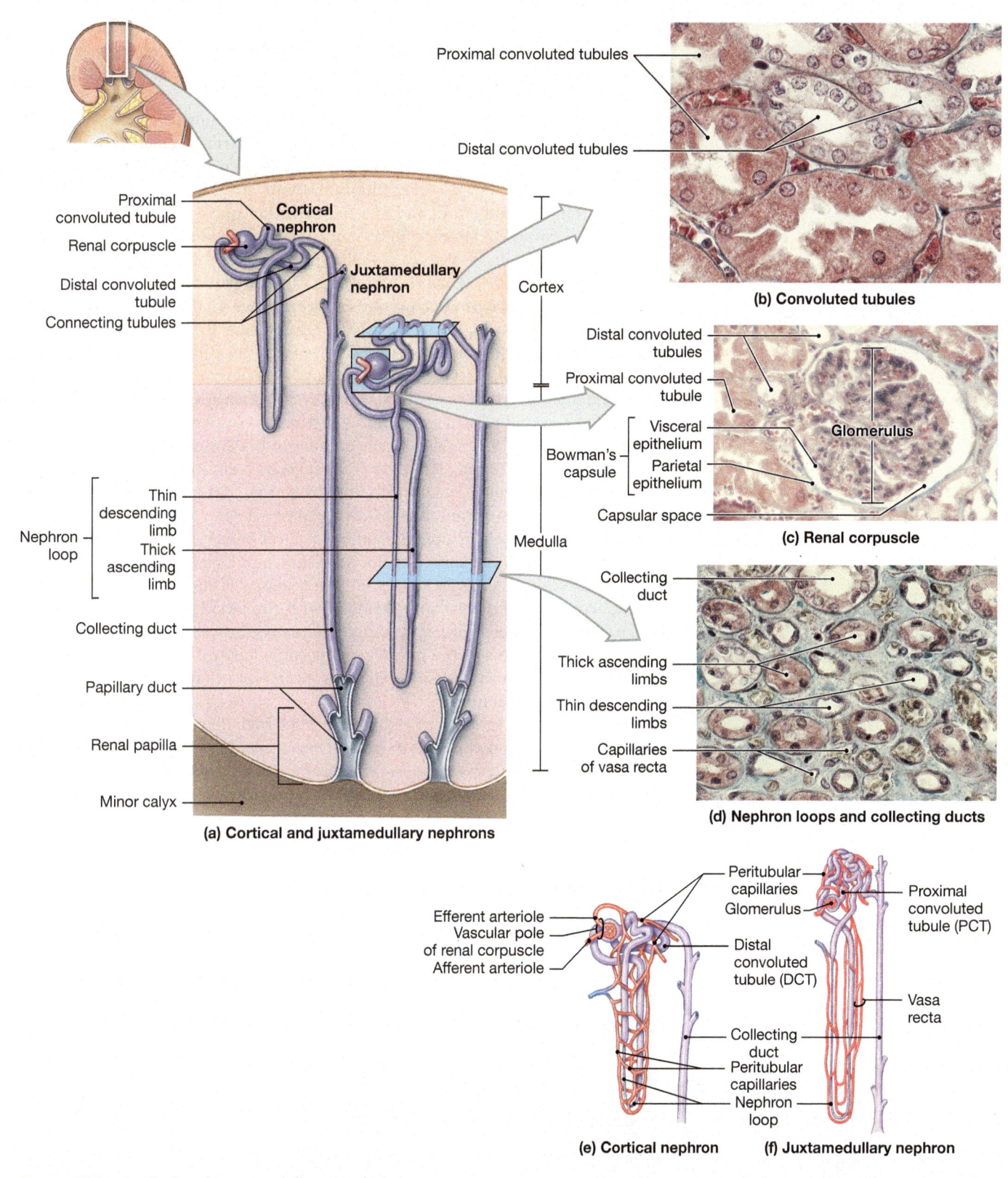

Figure 27.2 Cortical and Juxtamedullary Nephrons

(a) In a cortical nephron, the loop of Henle extends only a short distance into the medulla. In a juxtamedullary nephron, the loop extends far into the medulla. In both types, the filtrate moves from renal capsule to renal tubule to connecting tubule. (b) Proximal and distal convoluted tubules. (c) The renal corpuscle includes the glomerulus and Bowman's capsule. (d) Loops of Henle, collecting ducts and vasa recta capillaries. (e) The circulation to a cortical nephron. (f) The circulation to a juxtamedullary nephron.

15 percent are **juxtamedullary** (juks-ta-MED-ū-lar-ē) **nephrons,** located primarily at the junction of the cortex and the medulla and extend deep into the medulla before turning back toward the cortex. These longer nephrons produce a urine that is more concentrated than that produced by the cortical nephrons.

Each nephron, whether cortical or juxtamedullary, consists of two regions: a renal corpuscle and a renal tubule (Figure 27.2). The **renal corpuscle** is where blood is filtered. It consists of a **Bowman's capsule** that houses a capillary called the **glomerulus** (glō-MER-ū-lus). As filtration occurs, materials are forced out of the blood that is in the glomerulus and into the **capsular space** in Bowman's capsule.

The renal corpuscle empties filtrate into the **renal tubule,** which consists of twisted and straight ducts composed mainly of cuboidal epithelium. The first segment of the renal tubule, which is found right after Bowman's capsule, is a twisted segment called the **proximal convoluted tubule** (see Figure 27.3). The **loop of Henle** (HEN-lē) is a straight portion that begins where the proximal convoluted tubule turns toward the medulla. The loop has both thick portions near the cortex and thin portions that extend into the medulla. The **descending limb** is mostly a thin tubule that turns back toward the cortex as the **ascending limb**. The ascending limb leads to a second twisted segment, the **distal convoluted tubule**. The nephron ends where the distal convoluted tubule empties into a **connecting tubule** that in turn drains into a **collecting duct**. Several nephrons drain into the same collecting duct. These ducts merge with the collecting ducts from still other nephrons, and all of them empty into a common **papillary duct** that empties urine into a minor calyx. There are between 25 and 35 papillary ducts per renal pyramid. Cross-sections of convoluted tubules and collecting ducts are shown in Figure 27.2b and d.

At its superior end, the ascending limb of the loop of Henle twists back toward the renal corpuscle and comes into contact with the blood vessel that supplies its glomerulus. This point of contact is called the **juxtaglomerular complex** (see Figure 27.3). Here the cells of the renal tubule become tall and crowded together and form the **macula densa** (MAK-ū-la DEN-sa), which monitors NaCl concentrations in this area of the renal tubule.

The renal corpuscle is specialized for filtering blood. Bowman's capsule has a superficial layer called the **parietal epithelium** (*capsular epithelium*) and a deep **visceral epithelium** (*glomerular epithelium*), the latter of which wraps around the surface of the glomerulus (see Figure 27.3b). Between these two layers is the capsular space. The visceral epithelium consists of specialized cells called **podocytes**. These cells wrap extensions called **pedicels** (PED-i-selz) around the endothelium of the glomerulus. Small gaps between the pedicels are pores called **filtration slits**. In order to be filtered out of the blood that passes through the glomerulus, a substance must be small enough to pass through the capillary endothelium and its basement membrane and squeeze through the filtration slits to enter the capsular space. Any substance that can pass through these layers is removed from the blood as part of the filtrate. The filtrate, therefore, contains both essential materials and wastes.

QuickCheck Questions

2.1 What are the two main regions of a nephron?

2.2 What are the two kinds of nephrons?

2 Materials

- ☐ Kidney model
- ☐ Nephron model
- ☐ Compound microscope
- ☐ Prepared slide of kidney

Procedures

1. Review the nephron anatomy in Figures 27.2 and 27.3.
2. Identify each structure of a kidney on the kidney model.
3. Identify each structure of a nephron on the nephron model.
4. Observe the kidney slide at low and medium magnification and determine whether or not the renal capsule is present on the specimen. Increase the magnification to high and identify the renal cortex and, if it is present, the renal medulla. (The medulla is usually not present on most kidney slides.)

EXERCISE 27 The Urinary System   439

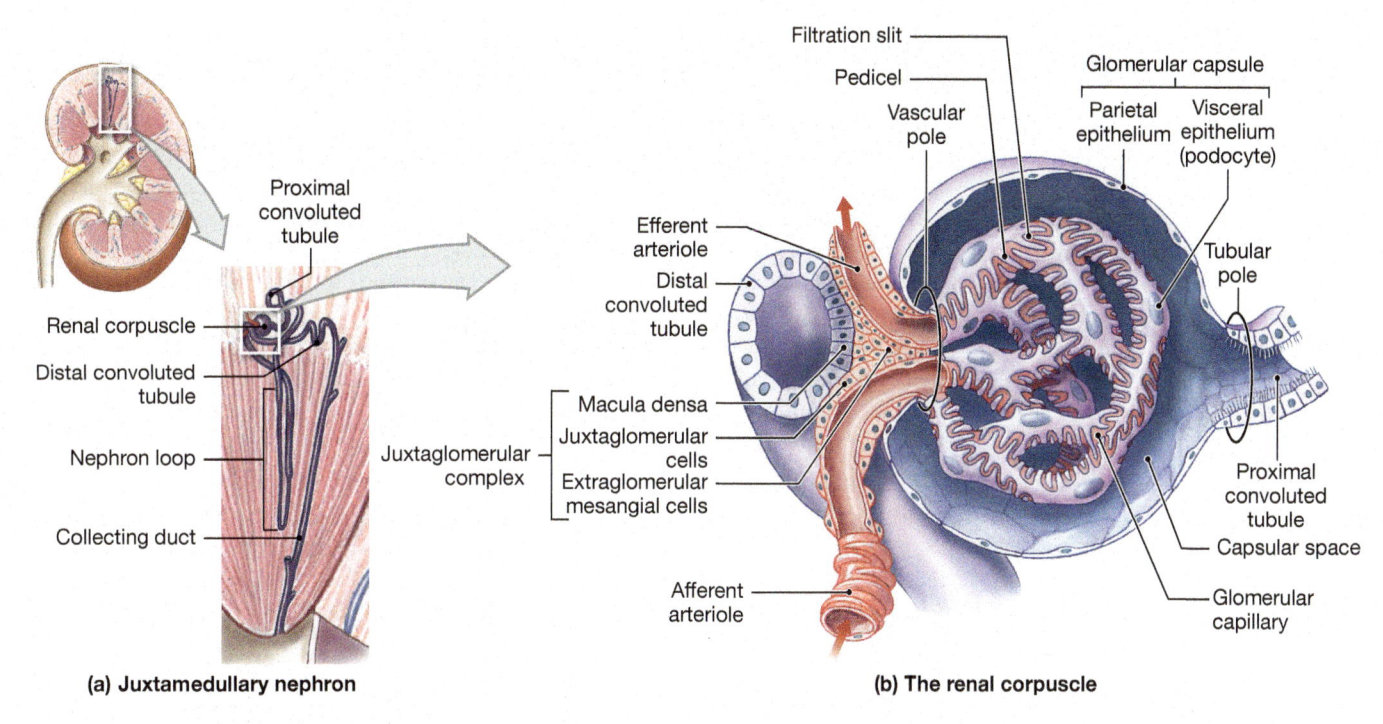

Figure 27.3 The Renal Corpuscle
(a) The renal corpuscle of a juxtamedullary nephron. (b) Structural features of a renal corpuscle.

5. Examine several renal tubules, which are visible as ovals on the slide. Use the micrographs in Figure 27.2 as a guide.

6. Locate a renal corpuscle, which appears as a small knot in the cortex. Distinguish among the parietal and visceral epithelia of Bowman's capsule, the capsular space, and the glomerulus. The visceral epithelium is visible as the cells that cover the glomerulus.

7. Draw a section of the kidney slide in the space provided. Label the cortex, a renal corpuscle, and a renal tubule. ∎

Cross section of kidney

LAB ACTIVITY 3 — Blood Supply to the Kidney

Each minute, approximately 25 percent of the total blood volume travels through the kidneys. This blood is delivered to a kidney by the **renal artery**, which branches off the abdominal aorta. Once it enters the hilus, the renal artery divides into five **segmental arteries**, which then branch into **interlobar arteries**, which pass through the renal columns (Figure 27.4). The interlobar arteries divide into **arcuate** (AR-kū-āt) **arteries**, which cross the bases of the renal pyramids and enter the renal cortex as **interlobular arteries**.

In the nephron, an **afferent arteriole** branches off from one of the interlobular arteries that serve the nephron, passes into Bowman's capsule, and supplies blood to the glomerulus. An **efferent arteriole** drains the blood from the glomerulus and branches into capillaries that surround the nephrons and reabsorb water, nutrients,

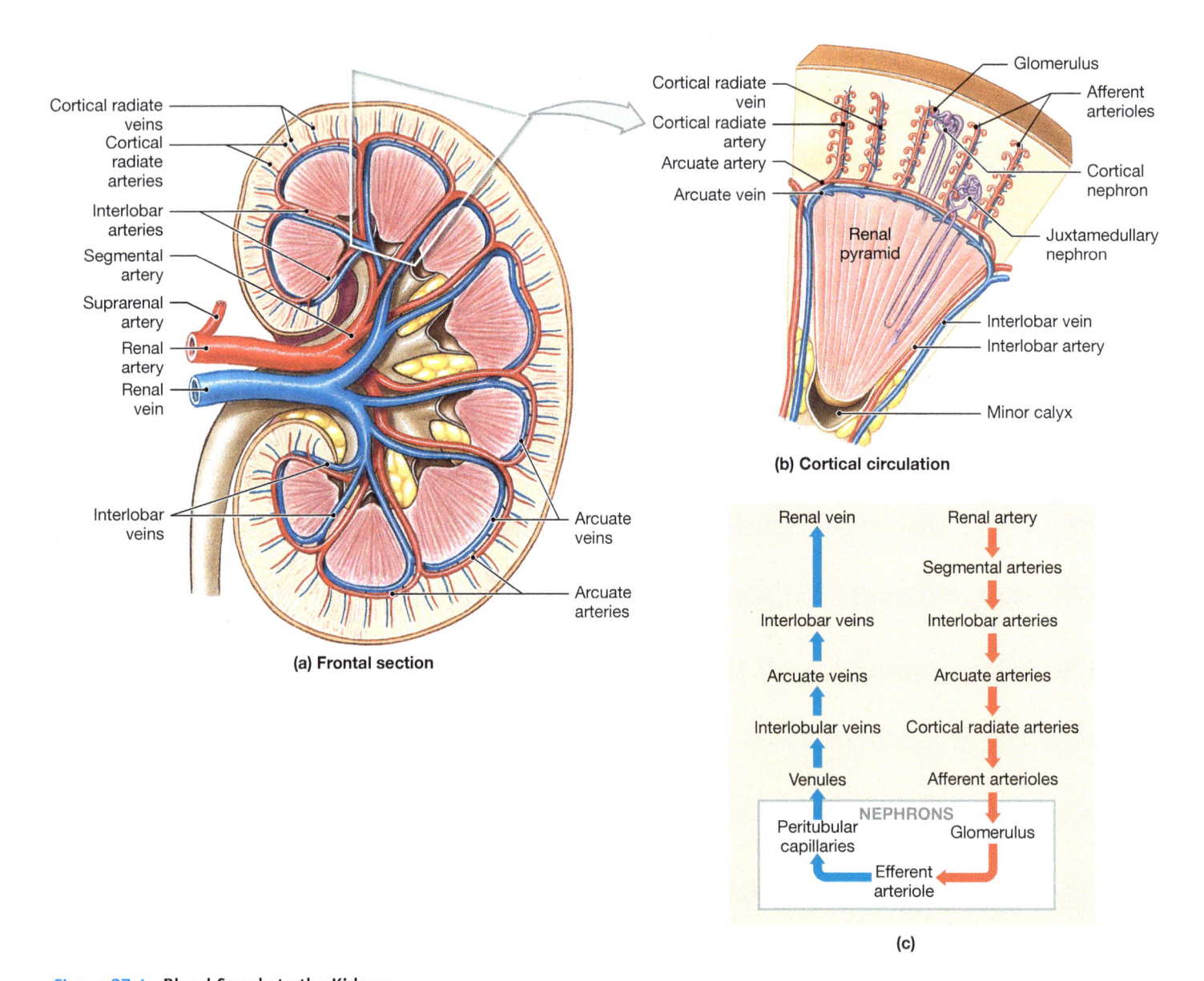

Figure 27.4 Blood Supply to the Kidneys
(a) Sectional view that shows major arteries and veins. (b) Detailed view of cortical arteries, veins, and arterioles. Note the interface between the arteriole and the nephron. (c) Circulation pathway through a cortical nephron.

and ions from the filtrate in the renal tubule (see Figure 27.2e and f). **Peritubular capillaries** in the cortex surround cortical nephrons and parts of juxtamedullary nephrons. The loops of Henle of juxtamedullary nephrons have thin vessels that are collectively called the **vasa recta**. Both the peritubular capillaries and the vasa recta are involved in reabsorbing materials from the filtrate of the renal tubules back into the blood. Both networks drain into **interlobular veins**, which then drain into **arcuate veins** along the base of the renal pyramids. **Interlobar veins** pass through the renal columns and join the **renal vein**, which drains into the inferior vena cava. Although there are segmental arteries, there are no segmental veins.

QuickCheck Questions

3.1 Which vessel branches from the abdominal aorta to supply blood to the kidney?

3.2 Where are the interlobar and interlobular arteries located?

3 Materials

- Kidney model
- Nephron model

Procedures

1. Review the blood vessels depicted in Figure 27.4.
2. On the kidney and nephron models, trace the blood vessels that supply and drain the kidneys. ■

LAB ACTIVITY 4 Ureters, Urinary Bladder, and Urethra

Each kidney has a single **ureter**, which is a muscular tube that transports urine from the renal pelvis to the **urinary bladder**, which is a hollow, muscular organ that stores urine temporarily. The two ureters conduct urine from kidney to bladder by means of gravity and peristalsis. The ureter, which is detailed in Figure 27.5, is lined with a mucosa that consists of a layer of transitional epithelium that covers a lamina propria. The function of this mucus-producing covering is to protect the ureteral walls from the acidic urine. The ureters enter the bladder low on the posterior bladder surface.

In males, the urinary bladder lies between the pubic symphysis and the rectum (Figure 27.6a). In females, it is posterior to the pubic symphysis, inferior to the uterus, and superior to the vagina (Figure 27.6b).

Histological details of the bladder are shown in Figure 27.6c. Because the bladder is a passageway to the external environment, the mucosal lining that faces the lumen is the superficial layer. This mucosa is made up of transitional epithelium that overlies a lamina propria. The transitional epithelium consists of many different cell shapes to facilitate the stretching and recoiling of the bladder wall. Folds in the mucosa called **rugae** allow the bladder wall to expand and shrink as it fills with urine and then empties. The submucosa is deep to the mucosa. Deep to the submucosa, the muscular wall of the bladder is known as the **detrusor** (de-TROO-sor) **muscle**.

A single duct, the **urethra**, drains urine from the bladder out of the body. Around the opening to the urethra are two sphincter muscles that control the voiding of urine from the bladder, the **internal urethral sphincter** and the **external urethral sphincter** (Figure 27.6c). In males, the **prostatic urethra** is the portion of the urethra that passes through the prostate gland, which is located inferior to the bladder. In males, the urethra passes through the penis and opens at the distal tip of the penis. In females, the urethra is separate from the reproductive organs and opens anteriosuperior to the vaginal opening. A transverse section through the female urethra is shown in Figure 27.6b.

The point where the urethra exits the bladder plus the two points where the ureters enter the bladder on its posterior surface define a triangular area of the bladder wall called the **trigone** (TRĪ-gōn). In this region, the lumenal bladder wall is smooth rather than folded into rugae.

442   EXERCISE 27   The Urinary System

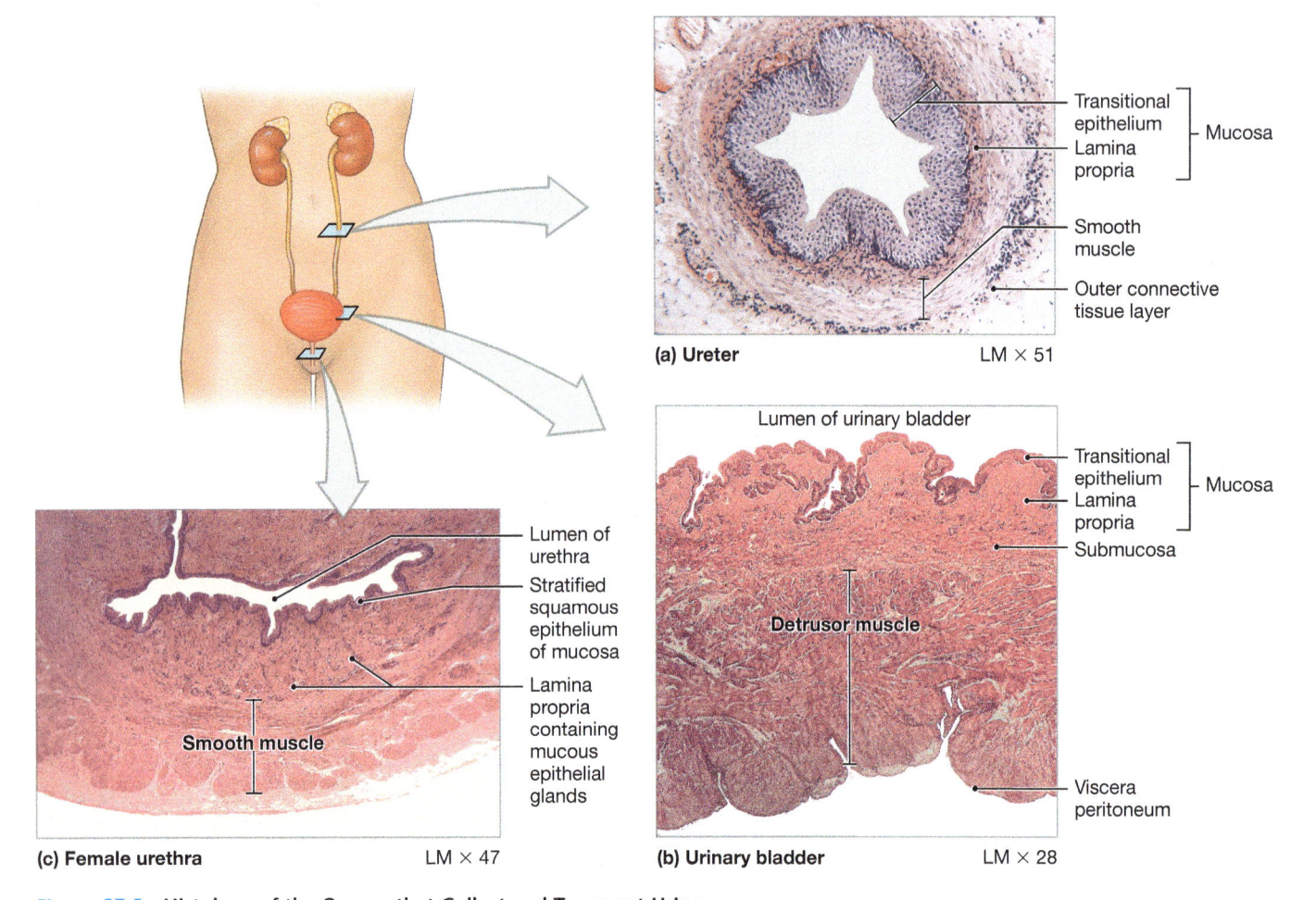

Figure 27.5 Histology of the Organs that Collect and Transport Urine
(a) Transverse section through the ureter. A thick layer of smooth muscle surrounds the lumen. (LM × 51) (b) Wall of the urinary bladder. (LM × 28) (c) Transverse section through the female urethra. (LM × 47)

Clinical Application

Floating Kidneys

Nephroptosis, or "floating kidney" is the condition that results when the integrity of either the adipose capsule or the renal fascia is jeopardized, often because of excessive weight loss. There is less adipose tissue available to secure the kidneys around the renal fascia. This lack of support can result in the pinching or kinking of one or both ureters, which prevents the normal flow of urine to the urinary bladder.

QuickCheck Questions

4.1 Where do the ureters join the urinary bladder?
4.2 What is the trigone?

4 Materials

☐ Urinary system model
☐ Urinary system chart

Procedures

1. Review the anatomy of the lower urinary tract in Figure 27.6.
2. Locate the ureters on the urinary system model or chart. Trace the path that urine follows from the renal papilla to the ureter.
3. On the model, examine the wall of the urinary bladder. Identify the trigone and the rugae. Which structures control emptying of the bladder?
4. On the model, examine the urethra. How does the male urethra differ from the female urethra?

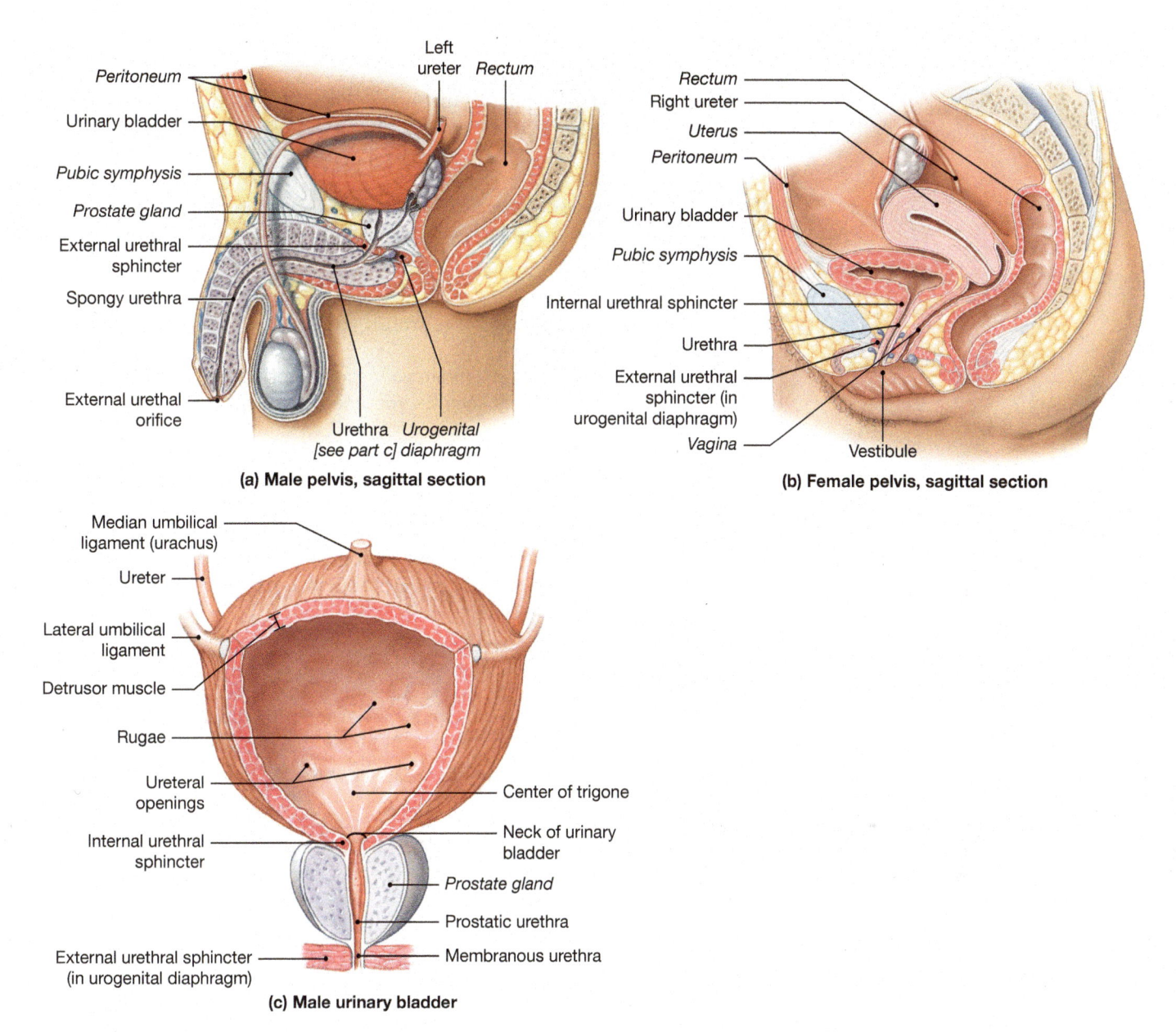

Figure 27.6 Organs for Conducting and Storing Urine

The ureter, urinary bladder, and urethra (a) in the male and (b) in the female. (c) The urinary bladder, prostate gland, and urethra in the male.

443

LAB ACTIVITY 5 | Sheep Kidney Dissection

The sheep kidney is very similar to the human kidney in both size and anatomy. Dissection of a sheep kidney reinforces your observations of kidney models in the laboratory.

> ⚠️ *Safety Alert:* **Dissecting a Kidney**
>
> You must—repeat, *must*—practice the highest level of laboratory safety while handling and dissecting the kidney. Keep the following guidelines in mind during the dissection:
>
> 1. Wear gloves and safety glasses to protect yourself from the fixatives used to preserve the specimen.
> 2. Do not dispose of the fixative from your specimen. You will later store the specimen in the fixative to keep the specimen moist and to keep it from decaying.
> 3. Be extremely careful when using a scalpel or other sharp instrument. Always direct cutting and scissor motions away from you to prevent an accident if the instrument slips on moist tissue.
> 4. Before cutting a given tissue, make sure it is free from underlying or adjacent tissues so that they will not be accidentally severed.
> 5. Never discard tissue in the sink or trash. Your instructor will inform you of the proper disposal procedure. ▲

QuickCheck Questions

5.1 What type of safety equipment should you wear during the sheep kidney dissection?

5.2 How should you dispose of the sheep kidney and scrap tissue?

5 Materials

- ☐ Gloves
- ☐ Safety glasses
- ☐ Dissecting tools
- ☐ Dissecting pan
- ☐ Preserved sheep kidney

Procedures

1. Put on gloves and safety glasses, and clear your workspace before obtaining your dissection specimen.
2. Rinse the kidney with water to remove excess preservative. Minimize your skin and mucous membrane exposure to the preservatives.
3. Examine the external features of the kidney. Using Figure 27.7 as a guide, locate the hilus. Locate the renal capsule and gently lift it by teasing with a needle. Below this capsule is the light pink cortex.
4. With a scalpel, make a longitudinal cut to divide the kidney into anterior and posterior portions. A single long, smooth cut is less damaging to the internal anatomy than a sawing motion.
5. Distinguish between the cortex and the darker medulla, which is organized into many triangular renal pyramids. The base of each pyramid faces the cortex, and the tip narrows into a renal papilla.
6. The renal pelvis is the large, expanded end of the ureter. Extending from this area are the major calyces and then the smaller minor calyces into which the renal papillae project.
7. Upon completion of the dissection, dispose of the sheep kidney as directed by your instructor; then wash your hands and dissecting instruments. ■

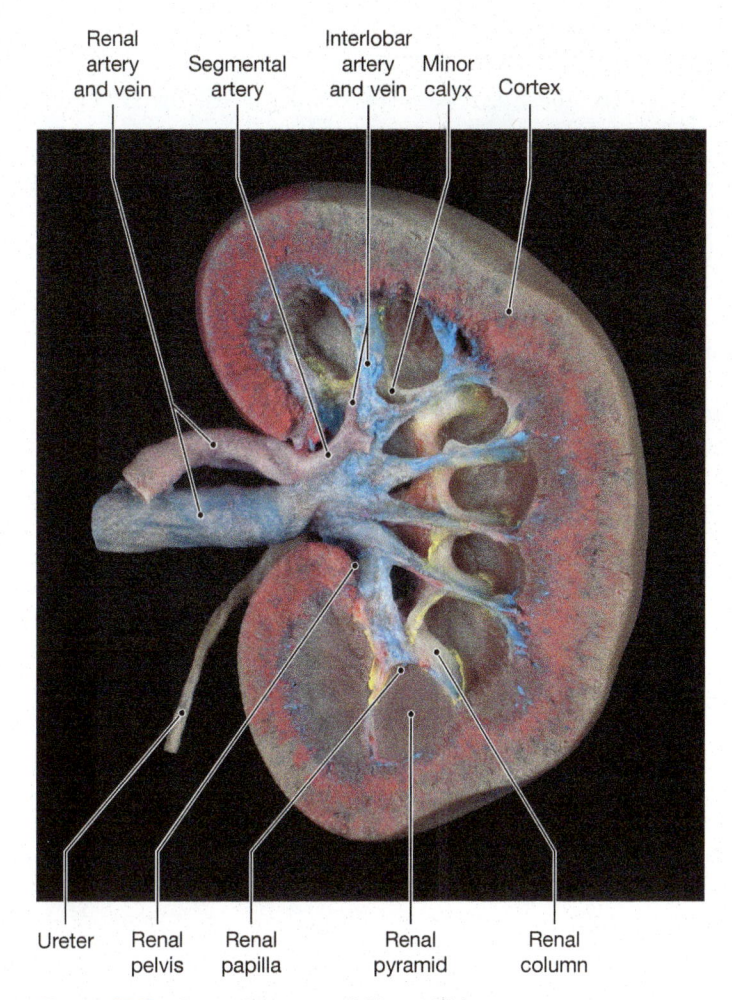

Figure 27.7 Gross Anatomy of Sheep Kidney
A frontal section of a sheep kidney that has been injected with latex dye to highlight arteries (red), veins (blue), and the urinary passageways (yellow).

urinary bladder

A Regional Look

The urinary bladder in females is positioned anterioinferior to the uterus (Figure 27.8). In pregnancy, the uterus serves as the womb for the gestating fetus. During the first months of pregnancy, embryonic and fetal growth involves development and maturation of organ systems rather than growth in the size of the fetus. In the later stage of pregnancy, however, the fetus grows rapidly and more than doubles in size. As the uterus expands, it compresses the urinary bladder and reduces the volume of urine that can be stored. The filtration rate in the woman's kidneys increases significantly to remove fetal wastes that have diffused across the placental membranes and entered the maternal blood. During the final months of the pregnancy, the woman urinates frequently because of her increased filtration rate and the increasing pressure exerted on the bladder by the fetus.

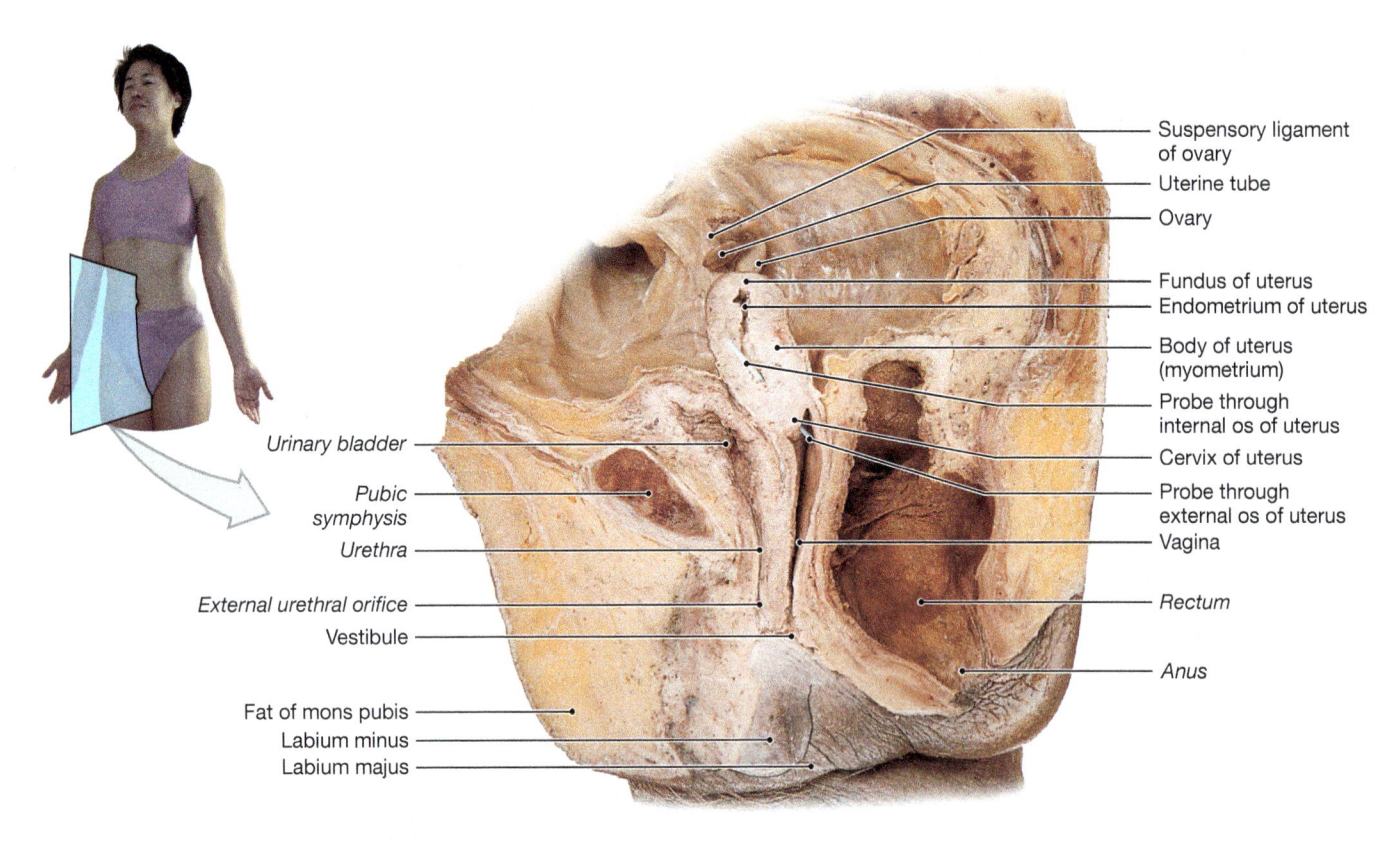

Figure 27.8 Pelvic Region of the Female
A sagittal section of the female pelvis that shows the relationship of the urinary bladder to the uterus and other reproductive organs.

EXERCISE 27

LAB REPORT

The Urinary System

A. Matching

Match each term in the left column with its correct description from the right column.

_____	1. renal papilla	A.	drains into collecting duct
_____	2. cortex	B.	located at base of pyramid
_____	3. Bowman's capsule	C.	covers surface of kidney
_____	4. loop of Henle	D.	surrounds glomerulus
_____	5. renal pelvis	E.	surrounds renal pelvis
_____	6. connecting tubule	F.	entrance for blood vessels
_____	7. efferent arteriole	G.	extends into minor calyx
_____	8. renal sinus	H.	tissue between pyramids
_____	9. renal pyramid	I.	U-shaped tubule
_____	10. arcuate artery	J.	transports urine to bladder
_____	11. hilus	K.	functional unit of kidney
_____	12. renal columns	L.	drains glomerulus
_____	13. renal capsule	M.	outer layer of kidney
_____	14. nephron	N.	medulla component
_____	15. ureter	O.	drains major calyx

B. Drawing

Draw and label the features of a nephron.

EXERCISE 27 — LAB REPORT

C. Short-Answer Questions

1. Describe the components of the renal corpuscle.

2. What are two differences between cortical and juxtamedullary nephrons?

3. How does the urethra in males differ from that in females?

4. Where are the internal and external urethral sphincters located?

D. Analysis and Application

1. List the layers in the renal corpuscle through which filtrate must pass to enter the capsular space.

2. Trace a drop of blood from the abdominal aorta, through a kidney, and into the inferior vena cava.

3. Trace a drop of filtrate from the glomerulus to the renal papillae.

4. Trace a drop of urine from a minor calyx to the urinary bladder.

EXERCISE 28

The Reproductive System

OBJECTIVES

On completion of this exercise, you should be able to:

- Describe the location of the male gonad.
- Identify the male testes, ducts, and accessory glands.
- Describe the composition of semen.
- Identify the three regions of the male urethra.
- Identify the structures of the penis.
- Identify the female ovaries, ligaments, uterine tubes, and uterus.
- Describe and recognize the three main layers of the uterine wall.
- Identify the vagina and the features of the vulva.
- Identify the structures of the mammary glands.
- Compare gamete formation in males and females.

LAB ACTIVITIES

1. Male: Testes and Spermatogenesis 449
2. Male: Epididymis and Ductus Deferens 453
3. Male: Accessory Glands 454
4. Male: Penis 456
5. Female: Ovaries and Oogenesis 456
6. Female: Uterine Tubes and Uterus 460
7. Female: Vagina and Vulva 463
8. Female: Mammary Glands 464
   A Regional Look: Pelvis 466

Whereas all the other systems of the body function to support the continued life of the organism, the reproductive system functions to ensure continuation of the species. The primary sex organs, or **gonads** (GŌ-nads), of the male and female are the **testes** (TES-tēz; singular *testis*) and **ovaries**, respectively. The testes produce the male sex cells, **spermatozoa** (sper-ma-tō-ZŌ-uh; singular **spermatozoon**; also called *sperm cell*), and the ovaries produce the female sex cells, **ova** (singular **ovum**). These reproductive cells, collectively called **gametes** (GAM-ēts), are the parental cells that combine and become a new life. The gonads have important endocrine functions and secrete hormones that support maintenance of the male and female sex characteristics. The gametes are stored and transported in ducts, and several accessory glands in the reproductive system secrete products to protect and support the gametes.

LAB ACTIVITY 1   Male: Testes and Spermatogenesis

In addition to the pair of testes, the male reproductive system consists of ducts, glands, and the penis (Figure 28.1). The testes are located outside the pelvic cavity, and the ducts transport the spermatozoa produced in the testes to inside the pelvic cavity, where glands add secretions to form a mixture called **semen** (SĒ-men), the liquid that is ejaculated into the female during intercourse. The testes are located in the **scrotum** (SKRŌ-tum), a pouch of skin that hangs from the pubis region. The pouch is divided into two compartments, each containing one testis, also called a

450 EXERCISE 28 The Reproductive System

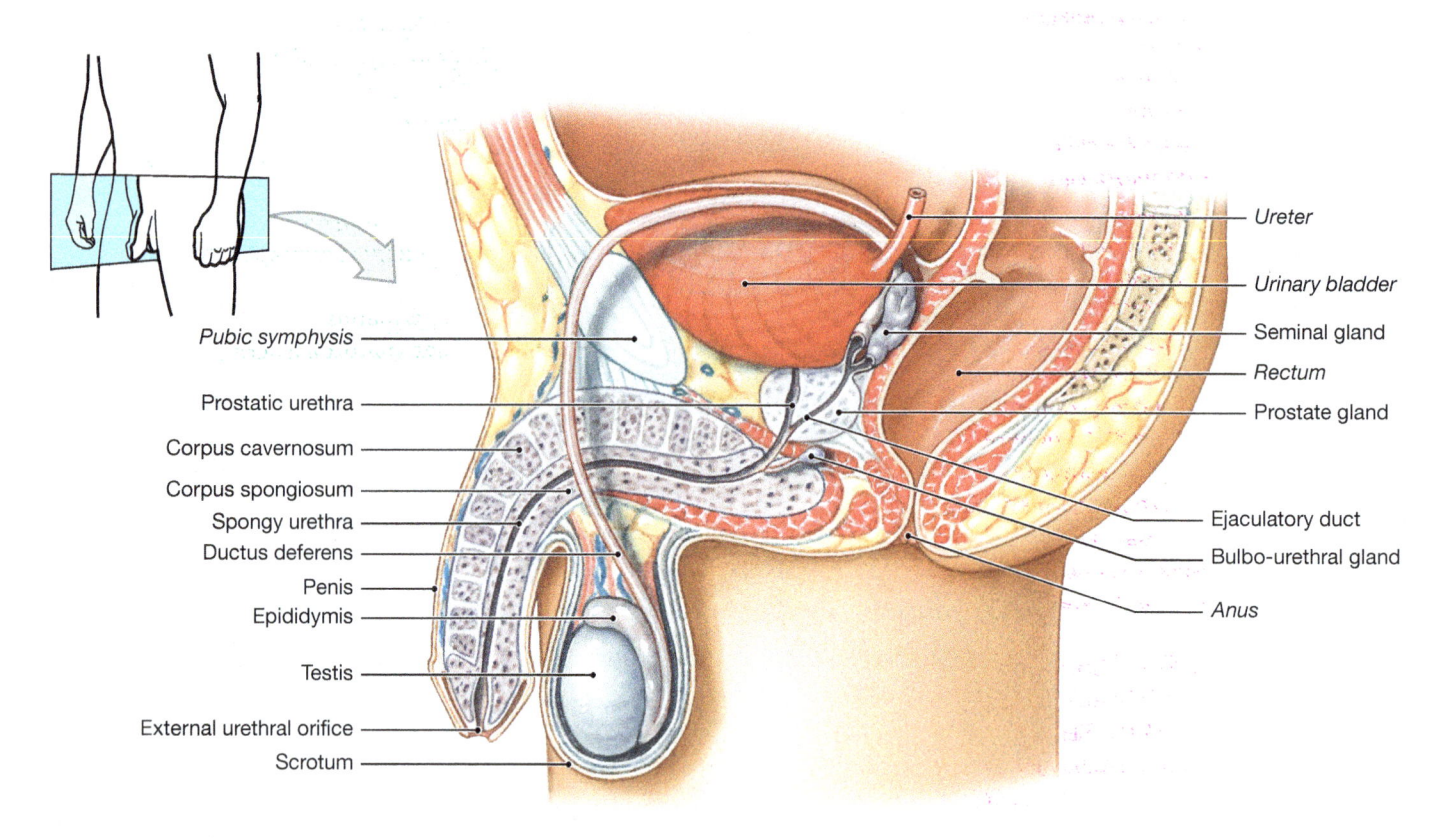

Labels (figure): Pubic symphysis, Prostatic urethra, Corpus cavernosum, Corpus spongiosum, Spongy urethra, Ductus deferens, Penis, Epididymis, Testis, External urethral orifice, Scrotum, Ureter, Urinary bladder, Seminal gland, Rectum, Prostate gland, Ejaculatory duct, Bulbo-urethral gland, Anus

Figure 28.1 Male Reproductive System
Sagittal section of the male reproductive organs.

testicle. The **dartos** (DAR-tōs) **muscle** forms part of the septum that separates the testes and is responsible for the wrinkling of the scrotum skin.

Each testis is about 5 cm (2 in.) long and 2.5 cm (1 in.) in diameter. Compartments in the testis called **lobules** contain highly coiled **seminiferous** (se-mi-NIF-er-us) **tubules** (Figure 28.2). Between the seminiferous tubules are small clusters of cells called **interstitial cells** that secrete **testosterone**, the male sex hormone. Testosterone is responsible for the male sex drive and for development and maintenance of the male secondary sex characteristics, such as facial hair and increased muscle and bone development.

Millions of spermatozoa are produced each day by the seminiferous tubules, in a process called **spermatogenesis** (sper-ma-tō-JEN-e-sis), which is shown in Figure 28.2b. During this process, cells go through a series of cell divisions, called **meiosis** (mī-Ō-sis), that ultimately reduce the number of chromosomes in each cell to one-half the initial number. Cells that contain this lower number of chromosomes are called **haploid** (HAP-loyd) cells. In females, a similar process (called *oogenesis* and discussed in Lab Activity 5) occurs in an ovary to produce a haploid ovum. When a haploid spermatozoon with its 23 chromosomes joins a haploid ovum with its 23 chromosomes, the resulting fertilized ovum has all 46 chromosome and is **diploid** (DIP-loyd). From this first new diploid cell, called the **zygote**, an incomprehensible number of divisions ultimately shape a new human.

The term **somatic cells** refers to all the cells in the body except the cells that produce gametes. **Mitosis** (mī-TŌ-sis) is cell division in somatic cells, where one parent cell divides to produce two identical diploid daughter cells. Meiosis, as just noted above, is cell division of cells in the testes and ovaries that produces haploid gametes. Meiosis occurs in two cycles, meiosis I and II, and in many ways is similar to mitosis. For simplicity, Figure 28.2 illustrates meiosis in a diploid cell that contains three chromosome pairs (six individual chromosomes) instead of the 23 pairs found in humans.

When a male reaches puberty, hormones stimulate the testes to begin spermatogenesis. Cells called **spermatogonia** (sper-ma-tō-GŌ-nē-uh) that are located

EXERCISE 28 The Reproductive System   451

SPERMATOGENESIS

MITOSIS of spermatogonium (diploid)

Primary spermatocyte (diploid)

DNA replication Synapsis and tetrad formation

Primary spermatocyte — Tetrad

MEIOSIS I

Secondary spermatocytes

Spermatids (haploid)

SPERMIOGENESIS (physical maturation)

Spermatozoa (haploid)

(a) Seminiferous tubules

- Seminiferous tubules that contain early spermatids
- Seminiferous tubules that contain late spermatids
- Seminiferous tubules that contain spermatozoa

(b) Sperm production

(c) Seminiferous tubule LM × 983

Labels on (c):
- Interstitial cells
- Spermatogonia
- Nurse cells
- Dividing spermatocytes
- Heads of maturing spermatozoa
- Connective tissue capsule
- Spermatids
- Lumen

(d) Wall of seminiferous tubule

Labels on (d):
- Spermatids completing spermiogenesis
- Initial spermiogenesis
- Luminal compartment
- Secondary spermatocyte in Meiosis II
- Level of blood-testis barrier
- Fibrocyte
- Connective tissue capsule
- Interstitial cells
- LUMEN
- Spermatids beginning spermiogenesis
- Secondary spermatocyte
- Primary spermatocyte preparing for Meiosis I
- Nurse cells
- Capillary
- Spermatogonium
- Basal compartment

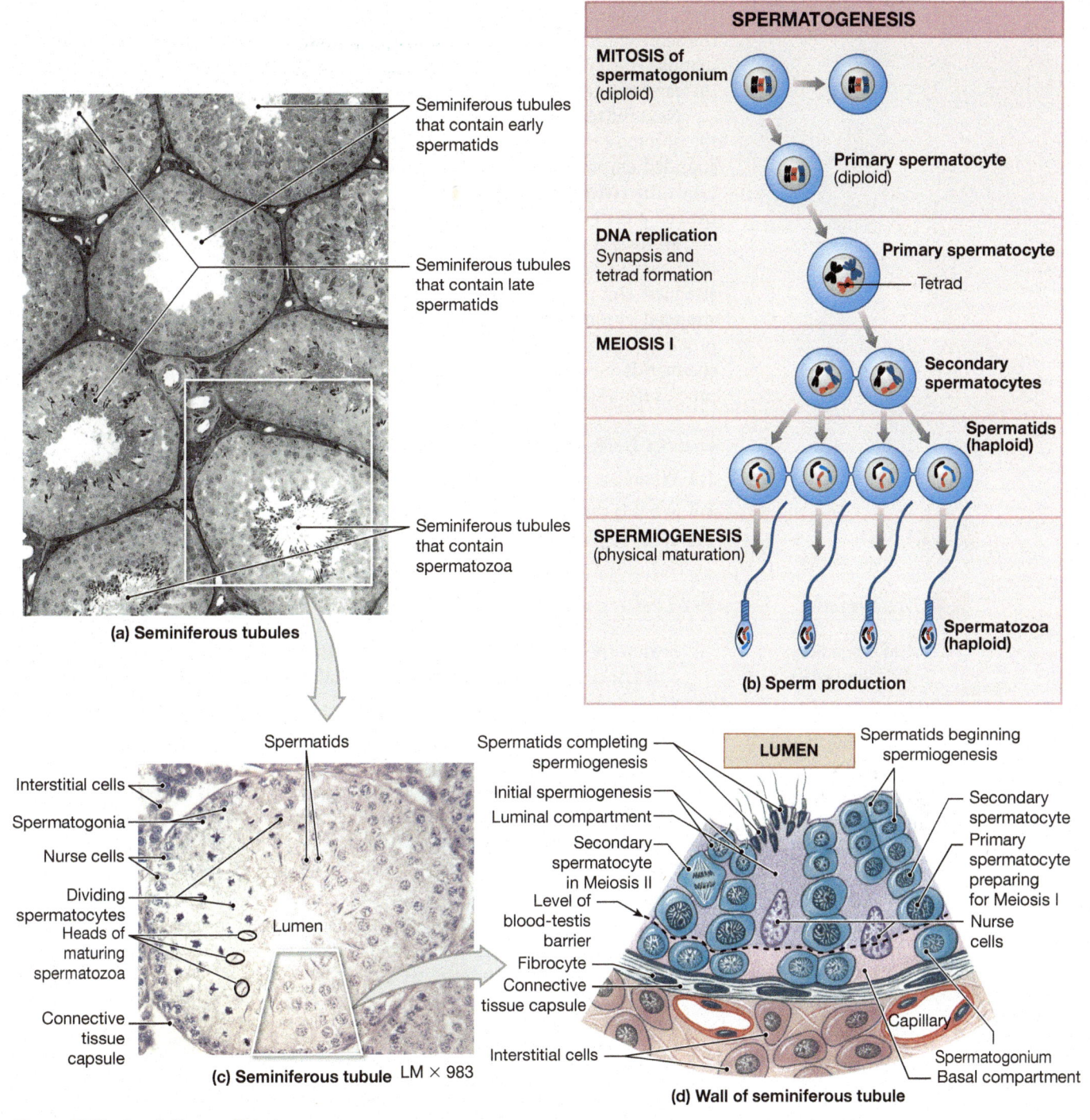

Figure 28.2 Seminiferous Tubules and Meiosis

(a) Sectional view of a number of seminiferous tubules. (b) Steps in meiosis. Human gametes contain 46 chromosomes in diploid stages, but for clarity only three chromosomes are shown here. (c) Sectional view of one seminiferous tubule. (LM × 983) (d) Detail of seminiferous tubule wall.

in the outer wall of the seminiferous tubules divide by mitosis and produce, in addition to new (haploid) spermatogonia, some diploid **primary spermatocytes** (sper-MA-tō-sīts). A primary spermatocyte prepares for meiosis by duplicating its genetic material. After replication, each chromosome is double-stranded and consists of two **chromatids**. Thus each original pair of chromosomes, which are called **homologous chromosomes**, now consists of four chromatids. The primary spermatocyte is now ready to proceed into meiosis.

Meiosis I begins as the nuclear membrane of the primary spermatocyte dissolves and the chromatids condense into chromosomes. The homologous chromosomes match into pairs in a process called **synapsis**, and the four chromatids of the pair are collectively called a **tetrad**. Because each chromatid in a tetrad belongs to the same chromosome pair, genetic information may be exchanged between chromatids. This **crossing over,** or mixing, of the genes contained in the chromatids increases the genetic variation within the population.

Next the tetrads line up in the middle of the cell, and the critical step of reducing the chromosome number to haploid occurs. The tetrads separate, and the double-stranded chromosomes move to opposite sides of the cell. This separation step is called the **reduction division** of meiosis because haploid cells are produced. Next, the cell pinches apart into two haploid **secondary spermatocytes**.

Meiosis II is necessary because, although the secondary spermatocytes are haploid, they have double-stranded chromosomes that must be reduced to single-stranded chromosomes. The process is similar to mitosis, in which the double-stranded chromosomes line up and separate. The two secondary spermatocytes produce four haploid **spermatids** that contain single-stranded chromosomes. The spermatids are what develop into spermatozoa, as Figure 28.3b indicates. The entire process from spermatogonia to mature spermatozoa takes approximately 9 weeks.

QuickCheck Questions

1.1 Where are spermatozoa produced in the male?

1.2 What is the name of the cell that divides to produce a primary spermatocyte?

1.3 What is a tetrad?

1 Materials

- ☐ Male urogenital model and chart
- ☐ Meiosis models
- ☐ Compound microscope
- ☐ Prepared slide of testis

Procedures

1. Review the male anatomy in Figures 28.1 and 28.2. Locate the scrotum, testes, and associated anatomy on the urogenital model and chart.
2. Identify the different cell types shown on the meiosis models.
3. Examine the testis slide, using the micrographs in Figure 28.2 for reference. Scan the slide at low magnification and observe the many seminiferous tubules. Increase the magnification and locate the interstitial cells between the tubules. At high power, pick a seminiferous tubule that has distinct cells within the walls. Identify the spermatogonia, primary and secondary spermatocytes, and spermatids. Spermatozoa are visible in the lumen of the tubule.
4. In the space provided, draw a section of a seminiferous tubule, and label the spermatogonia, primary spermatocytes, secondary spermatocytes, spermatids, and spermatozoa. ■

Seminiferous tubule

EXERCISE 28 The Reproductive System 453

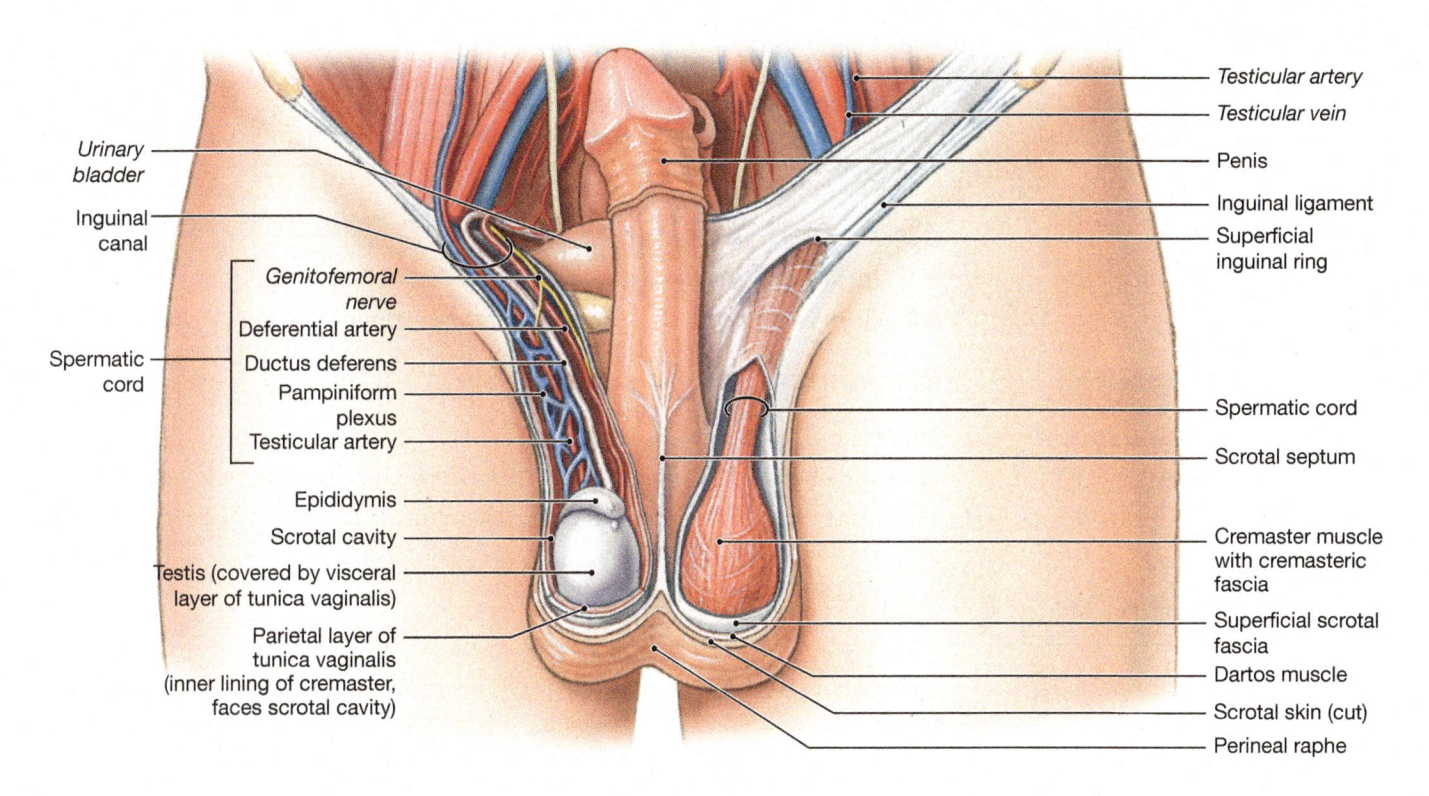

Figure 28.3 Male Reproductive System in Anterior View

LAB ACTIVITY 2 — Male: Epididymis and Ductus Deferens

After spermatozoa are produced in the seminiferous tubules, they move into the **epididymis** (ep-i-DID-i-mus), a highly-coiled tubule located on the posterior of the testis and visible in Figure 28.3. The spermatozoa mature in the epididymis and are stored until ejaculation out of the male reproductive system. Peristalsis of the smooth muscle of the epididymis propels the spermatozoa into the **ductus deferens** (DUK-tus DEF-e-renz), or *vas deferens*, the duct that empties into the urethra. The ductus deferens is 46 to 50 cm (18 to 20 in.) long and is lined with pseudostratified columnar epithelium. Peristaltic waves propel spermatozoa toward the urethra.

Within the scrotum, the ductus deferens ascends into the pelvic cavity as part of the **spermatic cord**. Other structures in the spermatic cord include blood and lymphatic vessels, nerves, and the **cremaster** (krē-MAS-ter) **muscle**, which encases the testes and raises or lowers them to maintain an optimum temperature for spermatozoa production. The ductus deferens passes through the **inguinal** (ing-gwi-nal) **canal** in the lower abdominal wall to enter the body cavity. This canal is a weak area and is frequently injured. An **inguinal hernia** occurs when portions of intestine protrude through the canal and slide into the scrotum. The ductus deferens continues around the posterior of the urinary bladder and widens into the **ampulla** (am-PŪL-la) before joining the seminal vesicle at the ejaculatory duct.

Clinical Application

Vasectomy

A common method of birth control for men is a procedure called **vasectomy** (vaz-EK-to-mē). Two small incisions are made in the scrotum, and a small segment of the ductus deferens on each side is removed. A vasectomized man still produces spermatozoa, but because the duct that transports them from the epididymis to the urethra is removed, the semen that is ejaculated contains no spermatozoa. As a result, no female ovum can be fertilized. Men who have had a vasectomy still produce testosterone and have a normal sex drive. They have orgasms, and the ejaculate is approximately the same volume as in men who have not been vasectomized.

QuickCheck Questions

2.1 Where are spermatozoa stored?

2.2 Where does the ductus deferens enter the abdominal cavity?

2 Materials

- Male urogenital model and chart

Procedures

1. Review the anatomy in Figure 28.3.
2. Locate the epididymis and ductus deferens on the laboratory model and chart.
3. Locate the spermatic cord, the cremaster muscle, and the inguinal canal on the model and chart.

LAB ACTIVITY 3 — Male: Accessory Glands

Three accessory glands—seminal vesicles, prostate gland, and bulbo-urethral glands—produce fluids that nourish, protect, and support the spermatozoa (Figure 28.4). The spermatozoa and fluids from these glands mix together as semen. The average number of spermatozoa per milliliter of semen is between 50 and 150 million, and the average volume of ejaculate is between 2 and 5 mL.

The **seminal** (SEM-i-nal) **vesicles** are a pair of glands posterior and lateral to the urinary bladder. Each gland is approximately 15 cm (6 in.) long and merges with the ductus deferens into an **ejaculatory duct**. The seminal vesicles contribute about 60 percent of the total volume of semen. They secrete a viscous, alkaline **seminal fluid** that contain the sugar fructose. The alkaline nature of this liquid neutralizes the acidity of the male urethra and the female vagina. The fructose provides the energy needed by each spermatozoon for beating its flagellum tail to propel the cell on its way to an ovum. Seminal fluid also contains fibrinogen, which causes the semen to temporarily clot after ejaculation.

The **prostate** (PROS-tāt) **gland** is a single gland just inferior to the urinary bladder. The ejaculatory duct passes into the prostate gland and empties into the first segment of the urethra, the **prostatic urethra**. The prostate gland secretes a milky-white, slightly acidic liquid that contains clotting enzymes to coagulate the semen. These secretions contribute about 20 to 30 percent of the semen volume.

The prostatic urethra exits the prostate gland and passes through the floor of the pelvis as the **membranous urethra**. A pair of **bulbo-urethral** (bul-bō-ū-RĒ-thral) **glands**, also called *Cowper's glands,* occur on either side of the membranous urethra and add an alkaline mucus to the semen. Before ejaculation, the bulbo-urethral secretions neutralize the acidity of the urethra and lubricate the end of the penis for sexual intercourse. These glands contribute about 5 percent of the volume of semen.

QuickCheck Questions

3.1 What are the three accessory glands that contribute to the formation of semen?

3.2 Where is the membranous urethra located?

3 Materials

- Male urogenital model and chart

Procedures

1. Review the anatomy in Figure 28.4.
2. On the model or chart, trace both ductus deferens through the inguinal canal, behind the urinary bladder, to where each unites with a seminal vesicle. Identify the swollen ampulla of the ductus deferens.

EXERCISE 28 The Reproductive System 455

(b) Ductus deferens LM × 111

(c) Seminal gland LM × 44

(d) Bulbo-urethral gland LM × 148

(e) Prostate gland LM × 47

(a) Posterior view of bladder and prostate gland

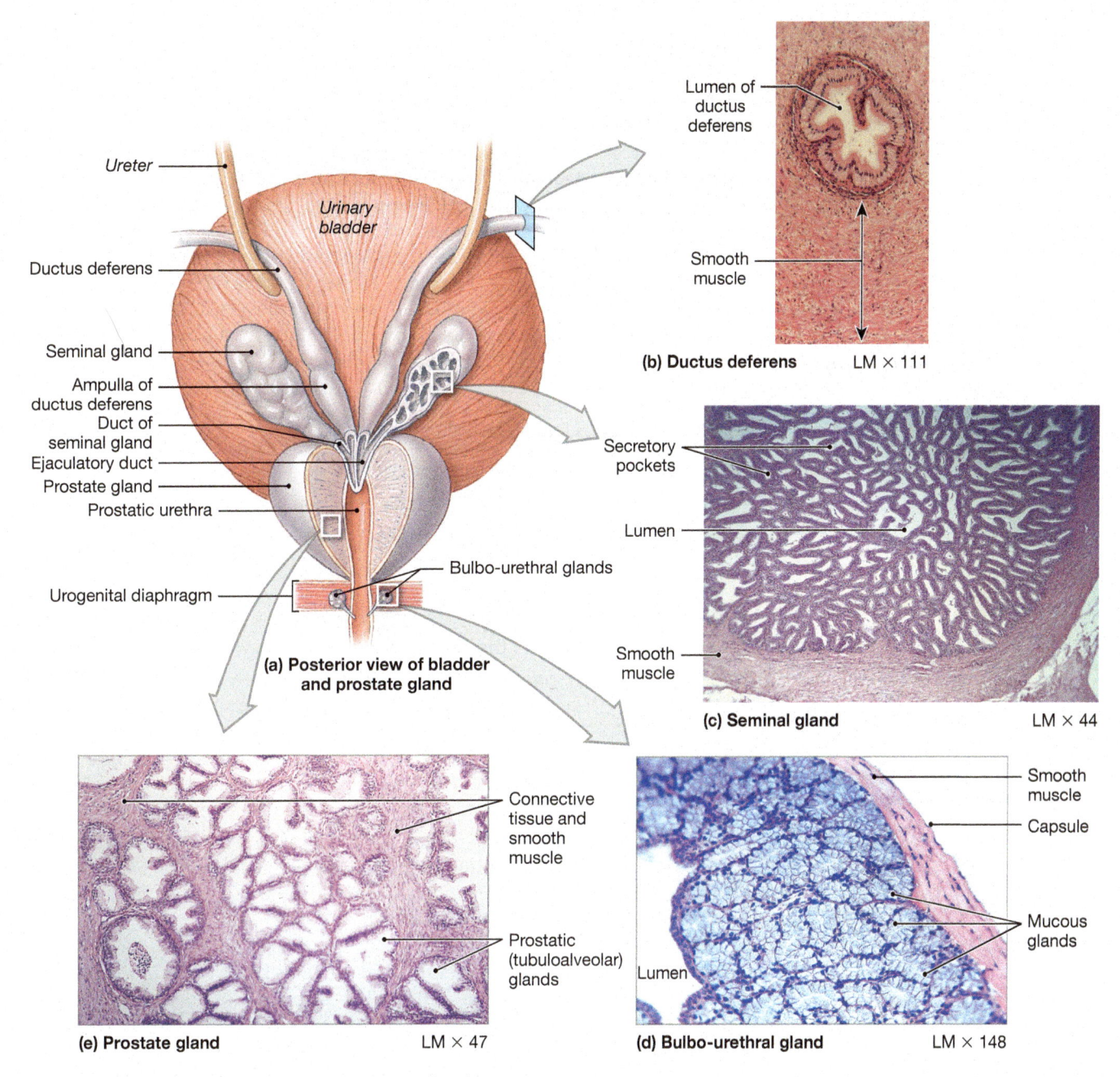

Figure 28.4 **Ductus Deferens and Accessory Glands**
(a) Posterior view of the prostate gland that shows subdivisions of the ductus deferens in relation to surrounding structures. (b) Ductus deferens that shows the smooth muscle around the lumen. (SEM × 42, LM × 111) (© R. G. Kessel and R. H. Kardon, *Tissues and Organs: A Text-Atlas of Scanning Electron Microscopy*, W. H. Freeman & Co., 1979. All rights reserved.) Sections of (c) the seminal vesicle (LM × 44), (d) a bulbo-urethral gland (LM × 148), and (e) the prostate gland. (LM × 47)

3. Identify the prostate gland, and note the ejaculatory duct that drains the ductus deferens and the seminal vesicle on each side of body. Identify the prostatic urethra that passes from the urinary bladder through the prostate gland.
4. Find the membranous urethra in the muscular pelvic floor. Identify the small bulbo-urethral glands on either side of the urethra. ■

LAB ACTIVITY 4 — Male: Penis

The **penis** (PĒ-nis), which is detailed in Figure 28.5, is the male copulatory organ that delivers semen into the vagina of the female. The penis is cylindrical and has an enlarged, acorn-shaped head called the **glans**. Around the base of the glans is a margin called the **corona** (crown). On an uncircumcised penis, the glans is covered with a loose-fitting skin called the **prepuce** (PRĒ-pūs) or *foreskin*. **Circumcision** (ser-kum-SIZH-un) is surgical removal of the prepuce. The **spongy urethra** transports both semen and urine through the penis and ends at the **external urethral orifice** in the tip of the glans. The **root** of the penis anchors the penis to the pelvis. The **body** consists of three cylinders of erectile tissue: a pair of dorsal **corpora cavernosa** (KOR-pōr-a ka-ver-NŌ-sa), and a single ventral **corpus spongiosum** (spon-jē-Ō-sum). During sexual arousal, the three erectile tissues become engorged with blood and cause the penis to stiffen into an erection.

QuickCheck Questions

4.1 What is the enlarged structure at the tip of the penis?

4.2 Which structures fill with blood during erection?

4.3 What duct transports urine and semen in the penis?

4 Materials

- Male urogenital model and chart

Procedures

1. Review the anatomy of the penis in Figure 28.5.
2. Identify the glans, corona, body, and root of the penis on the model or chart.
3. On the model, identify the corpora cavernosa and the corpus spongiosum. ∎

LAB ACTIVITY 5 — Female: Ovaries and Oogenesis

The female reproductive system, which is highlighted in Figure 28.6, includes two ovaries, two uterine tubes, the uterus, the vagina, external genitalia, and two mammary glands. **Gynecology** is the branch of medicine that deals with the care and treatment of the female reproductive system.

Formation of the female gamete, the ovum (or *egg*), is called **oogenesis** (ō-ō-JEN-e-sis) and occurs in the ovaries. In a female fetus being carried in its mother's uterus, meiosis I begins when cells called **oogonia** (ō-ō-GŌ-nē-uh, singular *oogonium*) divide by mitosis and produce **primary oocytes** (Ō-ō-sīts), which remain suspended in this stage until the child reaches puberty. At puberty, each month, a primary oocyte divides into two **secondary oocytes**. One of the secondary oocytes is much smaller than its sister cell and is a nonfunctional cell called the **first polar body** (Figure 28.7). The other secondary oocyte remains suspended in meiosis II until it is ovulated. If fertilization occurs, the secondary oocyte completes meiosis II and divides into another polar body, called the **second polar body**, and an ovum. The haploid ovum and haploid spermatozoon combine their haploid chromosomes and become the first cell of the offspring, the diploid zygote.

Note from Figure 28.8 that females produce only a single ovum by oogenesis, whereas in males spermatogenesis results in four spermatozoa (see Figure 28.3b).

Each almond-sized ovary contains from 100,000 to 200,000 oocytes clustered in groupings called **egg nests**. Within the nests are **primordial follicles**, which are primary oocytes surrounded by follicular cells. Figure 28.8 details the monthly ovarian cycle, during which hormones stimulate the follicular cells of the primordial follicles to proliferate and produce several **primary follicles**, each one a primary oocyte that is surrounded by follicular cells. These follicles increase in size, and a few become **secondary follicles** that contain primary oocytes. Eventually, one secondary follicle

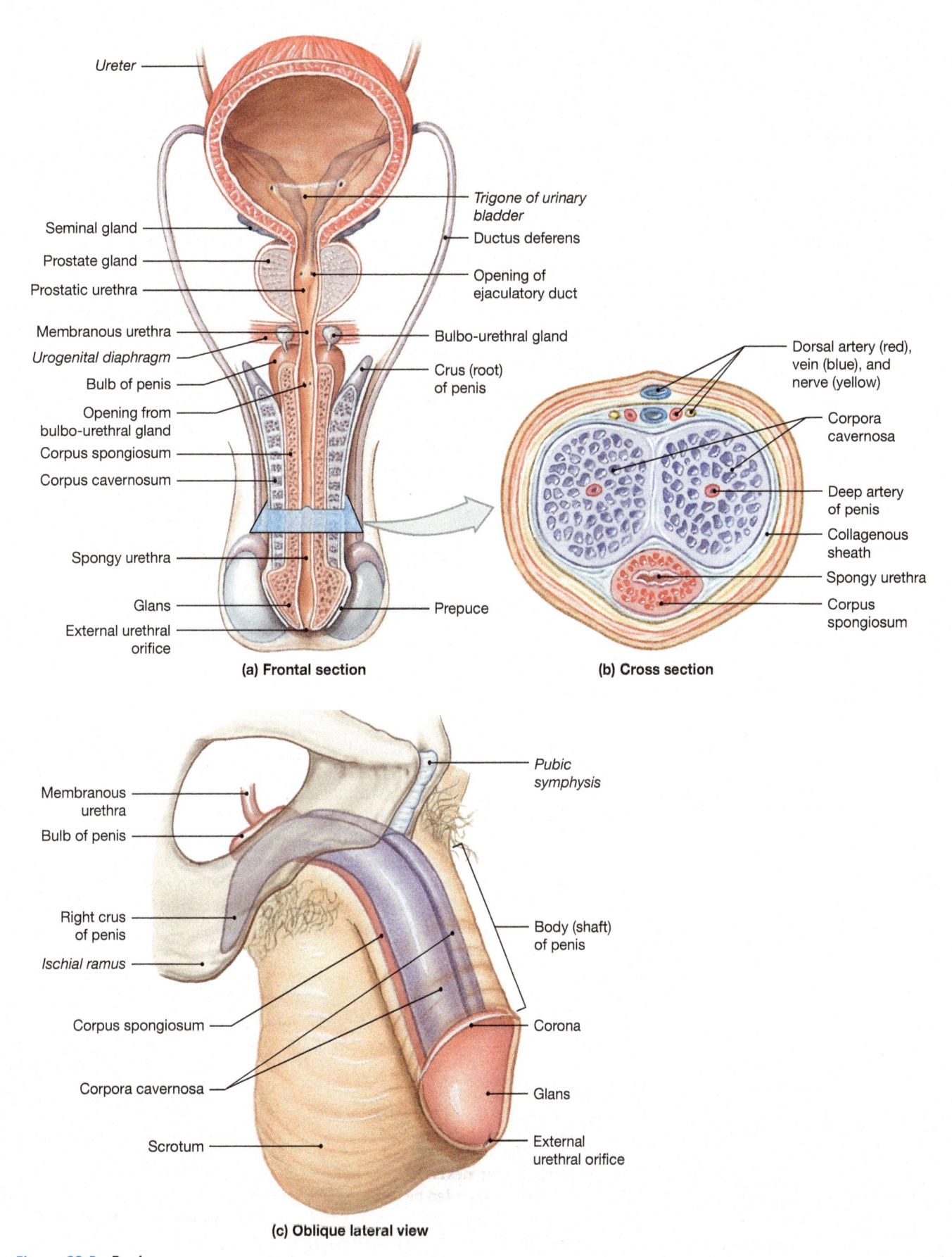

Figure 28.5 Penis

(a) Frontal section through the penis and associated organs. (b) Sectional view through the penis. (c) Anterior and lateral view of the penis that shows positions of the erectile tissues.

457

458 EXERCISE 28 The Reproductive System

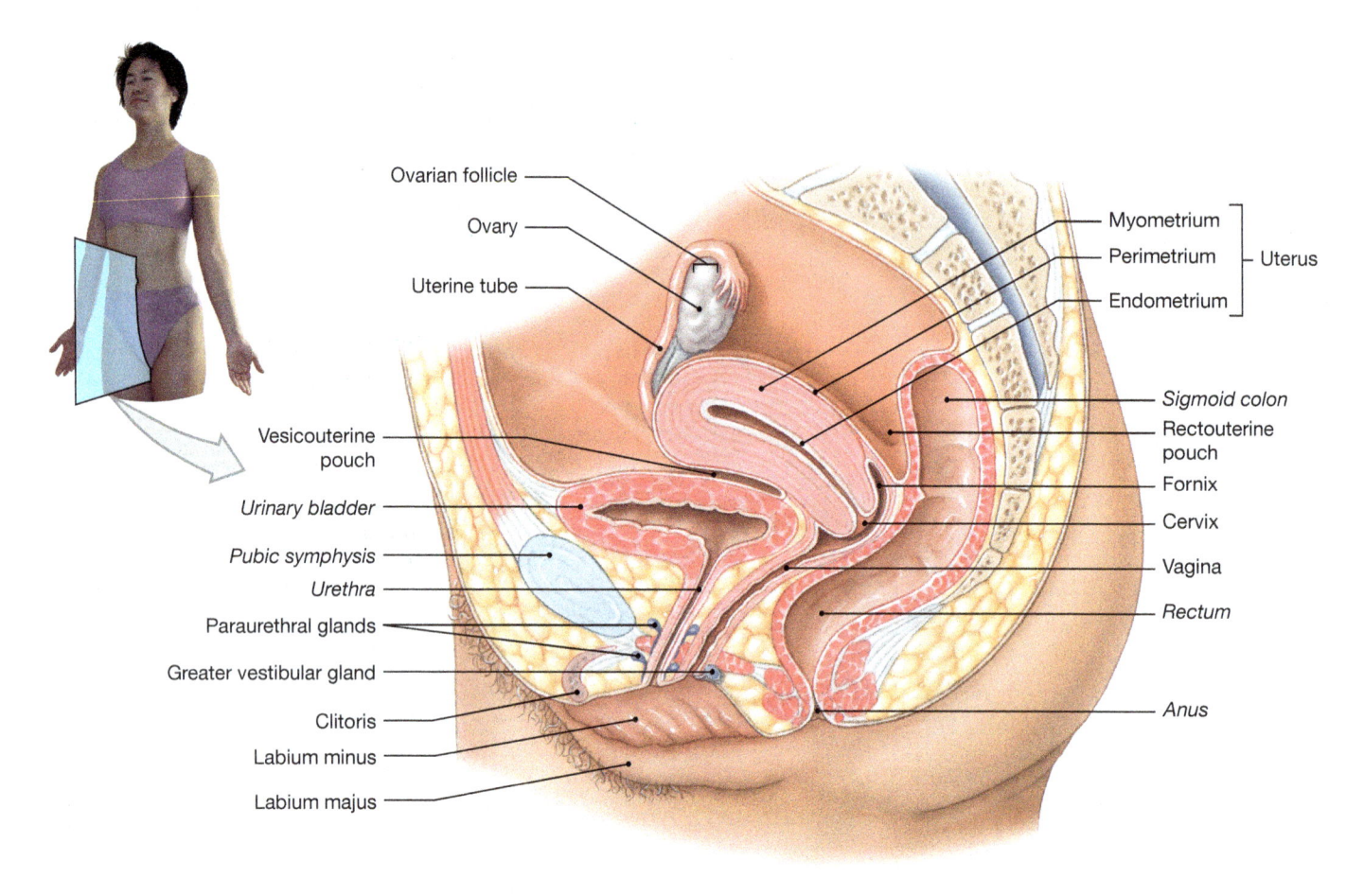

Figure 28.6 Sagittal Section of the Female Reproductive System

develops into a **tertiary follicle**, also called a *mature Graafian* (GRAF-ē-an) *follicle*. By now the oocyte has completed meiosis I and is a secondary oocyte that is starting meiosis II. The tertiary follicle fills with liquid and ruptures, which then casts out the secondary oocyte during ovulation. This follicle secretes **estrogen**, the hormone that stimulates rebuilding of the spongy lining of the uterus. After ovulation, the follicular cells of the tertiary follicle become the **corpus luteum** (LOO-tē-um) and secrete primarily the hormone **progesterone** (prō-JES-ter-ōn), which prepares the uterus for pregnancy. If the secondary oocyte is not fertilized, the corpus luteum degenerates into the **corpus albicans** (AL-bi-kanz), and most of the rebuilt lining of the uterus is shed as the menstrual flow.

QuickCheck Questions

5.1 Where are ova produced in the female?

5.2 Which structure ruptures during ovulation to release an ovum?

5.3 What are polar bodies?

5 Materials

☐ Female reproductive system model and chart

☐ Compound microscope

☐ Prepared slide of ovary

Procedures

1. Review the female anatomy in Figure 28.6.

2. Locate the ovaries, uterine tubes, uterus, and vagina on the laboratory model or chart.

3. Using Figure 28.8 as a reference, scan the ovary slide at low magnification, and locate an egg nest along the periphery of the ovary.

EXERCISE 28   The Reproductive System   459

Figure 28.7 Oogenesis
In oogenesis, a single primary oocyte produces an ovum and two nonfunctional polar bodies.

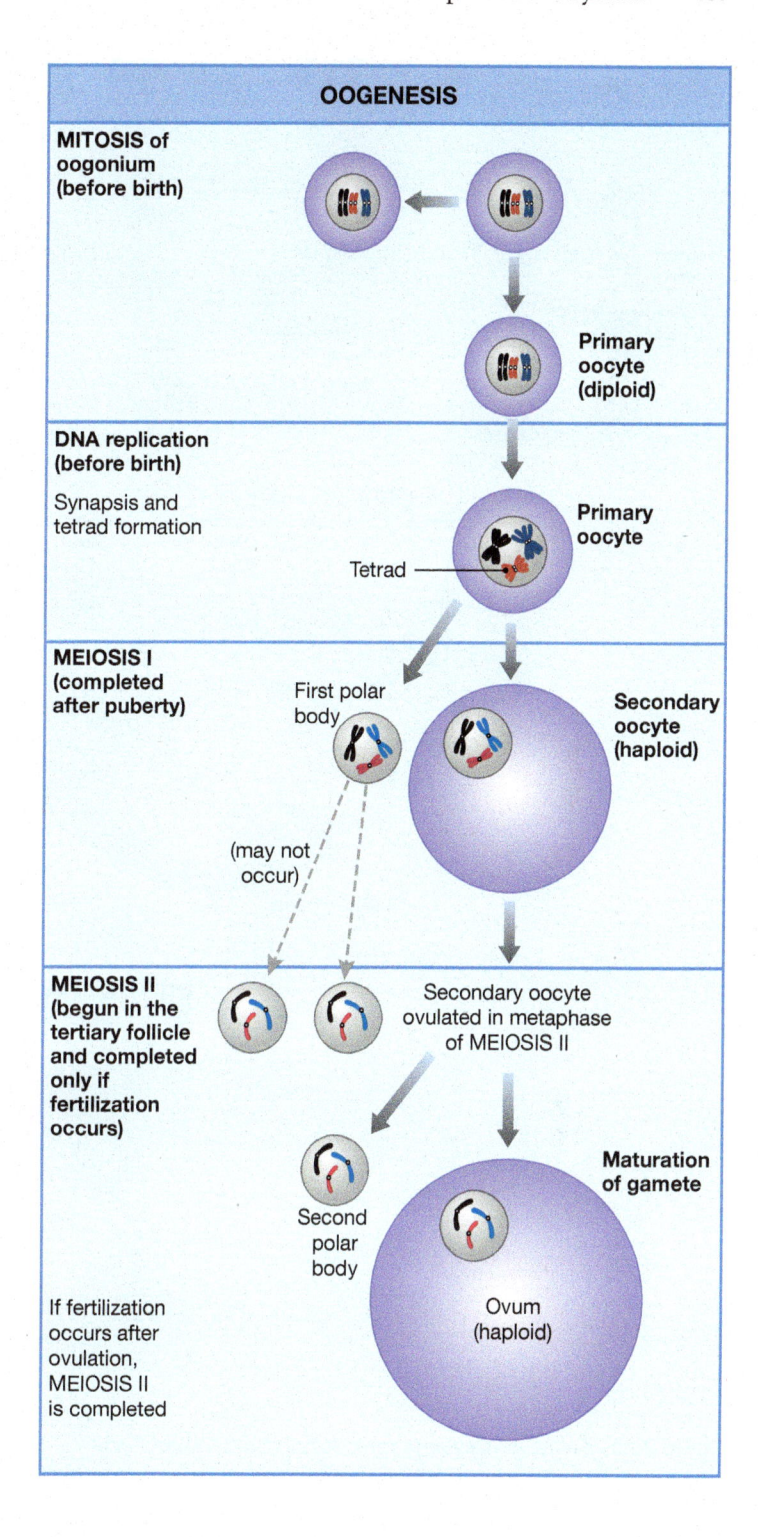

OOGENESIS

MITOSIS of oogonium (before birth) → Primary oocyte (diploid)

DNA replication (before birth)
Synapsis and tetrad formation → Primary oocyte (Tetrad)

MEIOSIS I (completed after puberty) → First polar body (may not occur) / Secondary oocyte (haploid)

MEIOSIS II (begun in the tertiary follicle and completed only if fertilization occurs) → Secondary oocyte ovulated in metaphase of MEIOSIS II → Second polar body / Ovum (haploid) — Maturation of gamete

If fertilization occurs after ovulation, MEIOSIS II is completed

4. Identify the primary follicles, which are larger than the primordial follicles in the nests. In the primary-follicle stage, the oocyte has increased in size and is surrounded by follicular cells.

5. Identify some secondary follicles, which are larger than primary follicles and have a separation between the outer and inner follicular cells.

6. Identify some tertiary follicles, which are easily distinguished by the large, liquid-filled space they contain. ■

460 EXERCISE 28 The Reproductive System

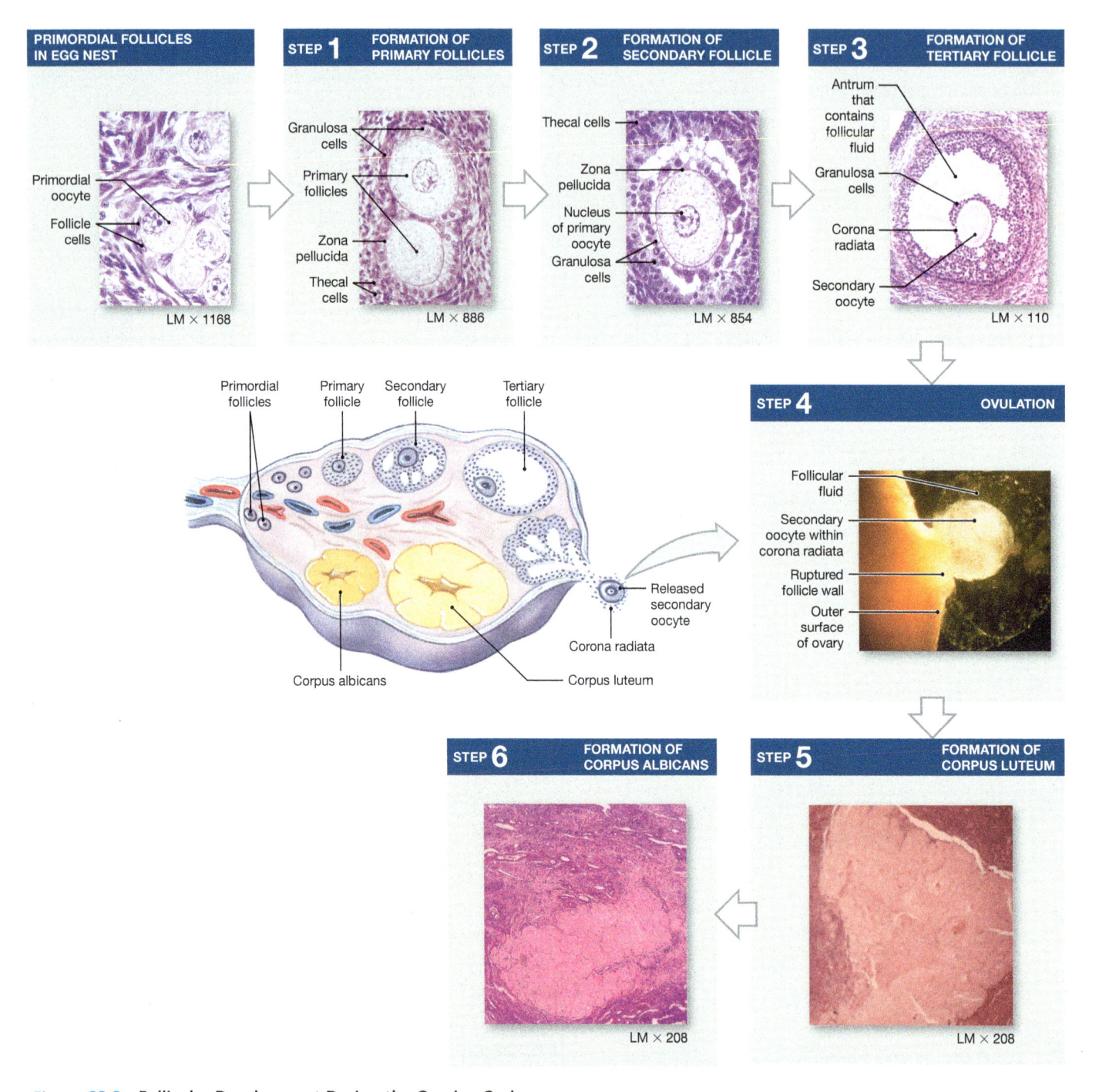

Figure 28.8 Follicular Development During the Ovarian Cycle

LAB ACTIVITY 6 — Female: Uterine Tubes and Uterus

The two **uterine tubes**, commonly called *fallopian tubes*, end in finger-like projections called **fimbriae** (FIM-brē-ē). These projections sweep over the surface of the ovary to capture an ovum released during ovulation and draw it into the expanded **infundibulum** region of the uterine tube (Figure 28.9). Once the ovum is inside the uterine tube, ciliated epithelium transports it toward the uterus. The tube widens midway along its length in the **ampulla** and then narrows at the **isthmus** (IS-mus) to enter the uterus. Fertilization of the ovum usually occurs between the infundibulum and the ampulla of the uterine tube.

EXERCISE 28 The Reproductive System 461

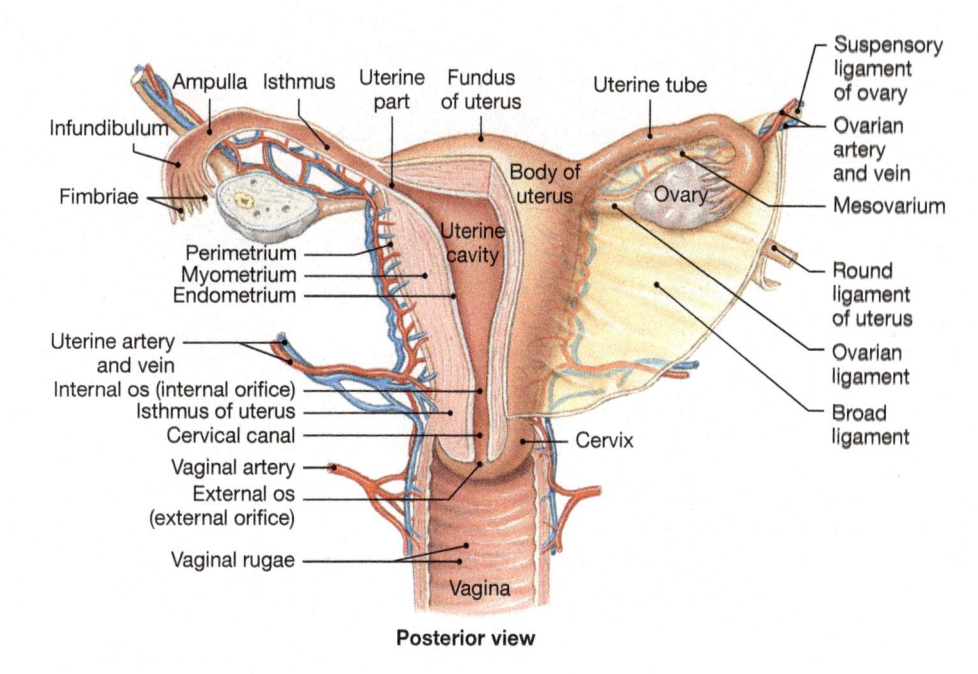

Figure 28.9 Uterus
Posterior view with the left side of the uterus, left uterine tube, and left ovary shown in section.

Clinical Application

Tubal Ligation

Permanent birth control for females involves removing a small segment of the uterine tubes in a process called **tubal ligation**. The female still ovulates, but the spermatozoa cannot reach the ova to fertilize them. The female still has a monthly menstrual period.

The **uterus,** the pear-shaped muscular organ located between the urinary bladder and the rectum, is the site where a fertilized ovum is implanted and where the fetus develops during pregnancy. The uterus consists of three major regions: fundus, body, and cervix. The superior, dome-shaped portion of the uterus is the **fundus,** and the inferior, narrow portion is the **cervix** (SER-viks). The rest of the uterus is called the **body.** Within the uterus is a space called the **uterine cavity** that narrows at the cervix as the **cervical canal.**

A double-layered fold of peritoneum called the **mesovarium** (mez-ō-VA-rē-um) holds the ovaries to the **broad ligament** of the uterus (see Figure 28.9). The **suspensory ligaments** hold the ovaries to the wall of the pelvis, and the **ovarian ligaments** hold the ovaries to the uterus. The **round ligaments** extend laterally from the ovaries and provide posterior support.

The uterine wall consists of three main layers: perimetrium, myometrium, and endometrium. The **perimetrium** is the outer covering of the uterus. It is an extension of the visceral peritoneum and is therefore also called the *serosa*. The thick middle layer, the **myometrium** (mī-ō-MĒ-trē-um), is composed of three layers of smooth muscle and is responsible for the powerful contractions during labor. The inner layer, the **endometrium** (en-dō-MĒ-trē-um), consists of two layers, a basilar zone and a functional zone (Figure 28.10). The **basilar zone** covers the myometrium and produces a new functional zone each month. The **functional zone** is very glandular and is highly vascularized to support an implanted embryo. This is the layer that is shed each cycle during menstruation.

462   EXERCISE 28   The Reproductive System

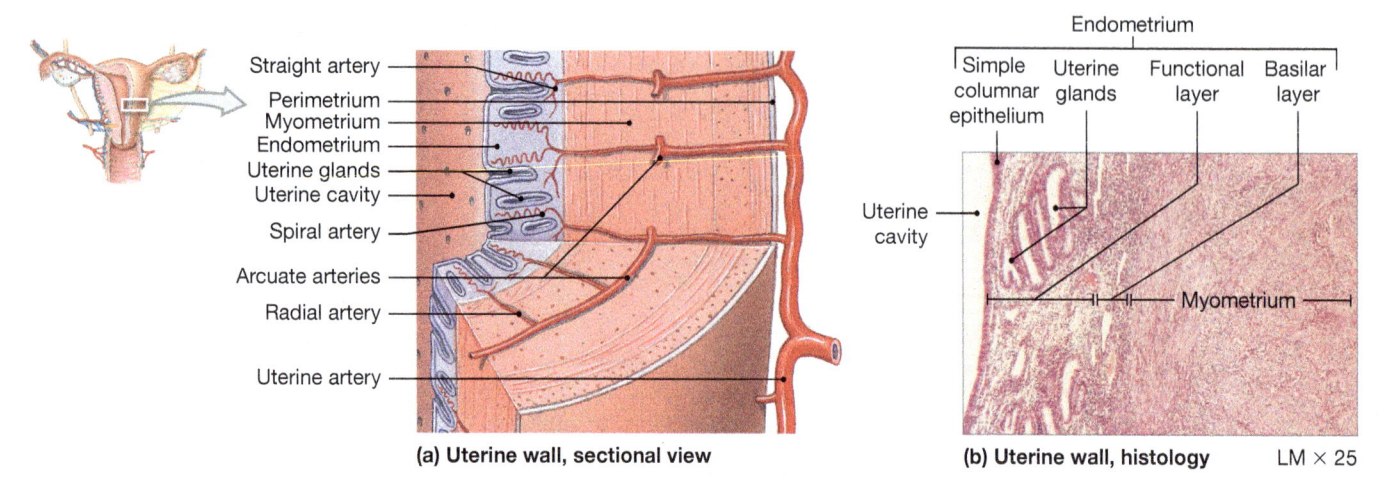

(a) Uterine wall, sectional view

Labels: Straight artery, Perimetrium, Myometrium, Endometrium, Uterine glands, Uterine cavity, Spiral artery, Arcuate arteries, Radial artery, Uterine artery

(b) Uterine wall, histology   LM × 25

Labels: Endometrium (Simple columnar epithelium, Uterine glands, Functional layer, Basilar layer), Uterine cavity, Myometrium

Figure 28.10   Uterine Wall
(a) Sectional view of the uterine wall that show's the myometrial and endometrial regions and the circulatory supply to the endometrium.
(b) Structure of the endometrium. (LM × 25)

QuickCheck Questions

6.1 What structure transports an ovum from the ovary to the uterus?

6.2 What are the three layers of the uterine wall?

6.3 Which layer of the uterine wall is shed during menses?

6   Materials

- Female reproductive system model and chart
- Compound microscope
- Prepared slide of uterine tissue

Procedures

1. Review the uterus and associated anatomy presented in Figure 28.9.
2. Identify the uterine tubes, the ampulla, and the isthmus on the laboratory model chart.
3. On the model, identify the fundus, body, and cervix.
4. Using Figure 28.10 for reference, scan the uterine-tissue slide at low and medium magnifications, and locate the thick myometrium composed of smooth muscle tissue. Identify the endometrium.
5. Draw a section of the uterine wall in the space provided. ■

Layers of the uterine wall

LAB ACTIVITY 7 — Female: Vagina and Vulva

The **vagina** (va-JĪ-na) is a muscular tube approximately 10 cm (4 in.) long. It is lined with stratified squamous epithelium and is the female copulatory organ, the pathway for menstrual flow, and the lower birth canal. The **fornix** is the pouch formed where the uterus protrudes into the vagina. The **vaginal orifice** is the external opening of the vagina. This opening may be partially or totally occluded by a thin fold of vascularized mucus membrane called the **hymen** (HĪ-men). On either side of the vaginal orifice are openings of the **greater vestibular glands,** glands that produce a mucous secretion that lubricates the vaginal entrance for sexual intercourse. These glands are similar to the bulbo-urethral glands of the male.

The **vulva** (VUL-vuh), which is the collective name for the female **external genitalia** (jen-i-TĀ-lē-uh), includes the following structures (Figure 28.11):

- The **mons pubis** is a pad of adipose tissue over the pubic symphysis. The mons is covered with skin and pubic hair and serves as a cushion for the pubic symphysis.
- The **labia** (LĀ-bē-uh) **majora** are two fatty folds of skin that extend from the mons pubis and continue posteriorly. They are homologous to the scrotum of the male. They usually have pubic hair and contain many sudoriferous (sweat) and sebaceous (oil) glands.
- The **labia minora** (mi-NOR-uh) are two smaller parallel folds of skin that contain many sebaceous glands. This pair of labia lacks hair.

Inferior view

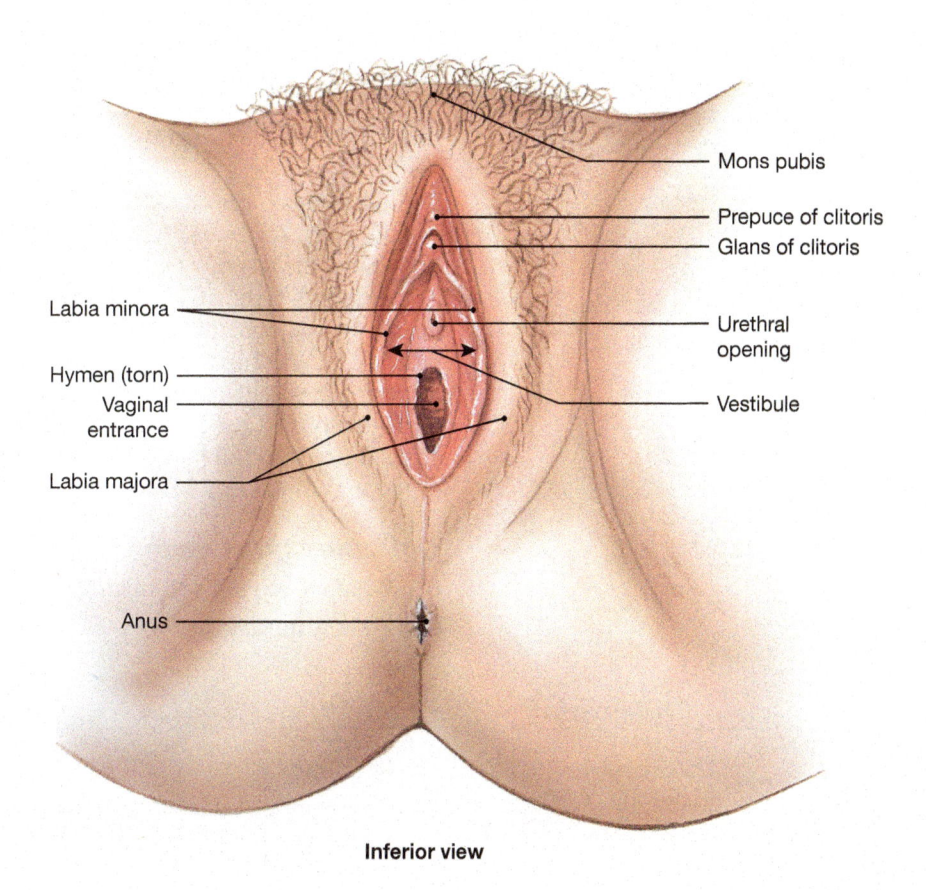

Figure 28.11 Female External Genitalia
Inferior view.

- The **clitoris** (KLIT-ō-ris) is a small, cylindrical mass of erectile tissue analogous to the penis. Like the penis, the clitoris contains a small fold of covering skin called the prepuce. The exposed portion of the clitoris is called the **glans**.
- The **vestibule** is the area between the labia minora that contains the vaginal orifice, hymen, and external urethral orifice.
- **Paraurethral glands** (Skene's glands) surround the urethra.
- The **perineum** is the area between the legs from the clitoris to the anus. This area is of clinical significance because of the tremendous pressure exerted on it during childbirth. If the vagina is too narrow during childbirth, an **episiotomy** (e-pē-zē-OT-uh-mē) is performed by making a small incision at the base of the vaginal opening toward the anus to expand the vaginal opening.

QuickCheck Questions

7.1 Where is the mons pubis located?

7.2 The vestibule is between what two sets of folds?

7.3 Which female organ has a glans?

7 Materials

☐ Female reproductive system model and chart

Procedures

1. Review the vulvar anatomy in Figure 28.11.
2. Locate the vagina and vaginal orifice on the laboratory model or chart. Examine the fornix, which is the point where the cervix and vagina connect.
3. Locate each component of the vulva. Note the positions of the clitoris, urethra, and vagina. ■

LAB ACTIVITY 8 Female: Mammary Glands

The **mammary glands** (Figure 28.12) are modified sweat glands that, in the process called **lactation** (lak-TĀ-shun), produce milk to nourish a newborn infant. At puberty, the release of estrogens stimulates an increase in the size of these glands. Fat deposition is the major contributor to the size of the breast, and size does not influence the amount of milk produced. Each gland consists of 15 to 20 lobes separated by fat and connective tissue. Each lobe contains smaller lobules that contain milk-secreting cells called **alveoli**. **Lactiferous** (lak-TIF-e-rus) **ducts** drain milk from the lobules toward the **lactiferous sinuses**. These sinuses empty the milk at the raised portion of the breast called the *nipple*. A circular pigmented area called the **areola** (a-RĒ-ō-luh) surrounds the nipple.

QuickCheck Questions

8.1 What are the milk-producing cells of the breast called?

8.2 What is the areola?

8 Materials

☐ Breast model

Procedures

1. Review the anatomy of the breast presented in Figure 28.12.
2. On the model, trace the pathway of milk from a lobule to the surface of the nipple. ■

(a) Left breast

(b) Active mammary gland LM × 131

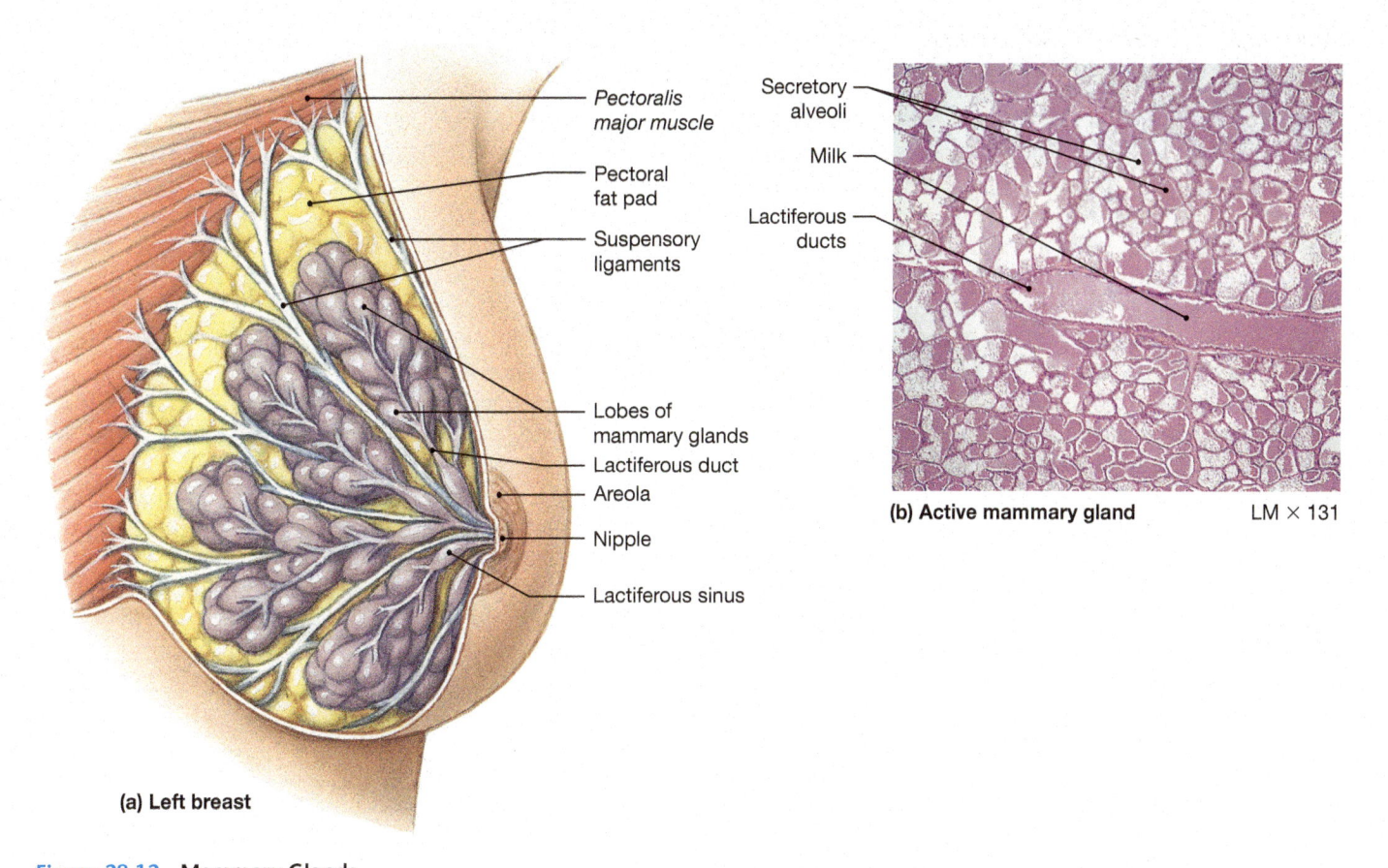

Figure 28.12 Mammary Glands
(a) Mammary gland of the left breast. **(b)** Active mammary gland of a lactating woman. (LM × 131)

(a)

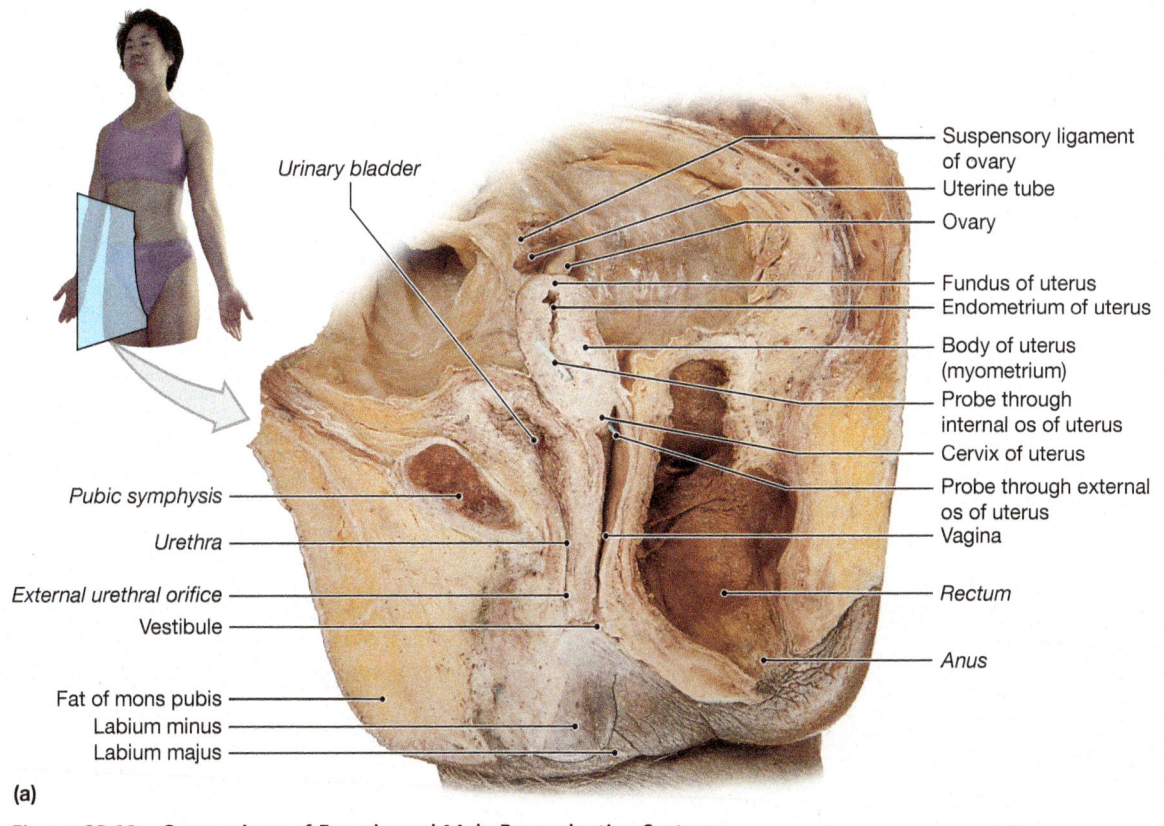

Figure 28.13 Comparison of Female and Male Reproductive Systems
(a) Female. **(b)** Male (next page).

465

pelvis   A Regional Look

The cadaver dissections in Figure 28.13 compare the female and male reproductive systems in midsagittal section. Most obvious are the differences in the external genitalia. Females have folds of skin, the labia, that protect and cover the vaginal entrance. In males, the external genitalia include the penis and scrotum. The penis extends from a base between the legs where the bulbospongiosus muscle of the pelvic floor is located. This muscle, along with others, contracts during orgasm to propel spermatozoa out of the urethra. In females, corresponding muscles contract during orgasm.

In males, the gametes are produced in the testes, which are located in the scrotum and outside of the pelvic cavity. Muscles associated with the testes control the temperature within the testes by moving the testes closer to or farther away from the warm pelvic floor. In females, the ovaries are internal and are coupled with the fimbrae of the uterine tubes so that an ovum can be transferred to the uterus.

In males, the urethra is a dual-function passageway for voiding urine from the urinary bladder and transporting semen during ejaculation; in females, the urethra is exclusively a urinary system structure and serves no reproductive function.

(b)

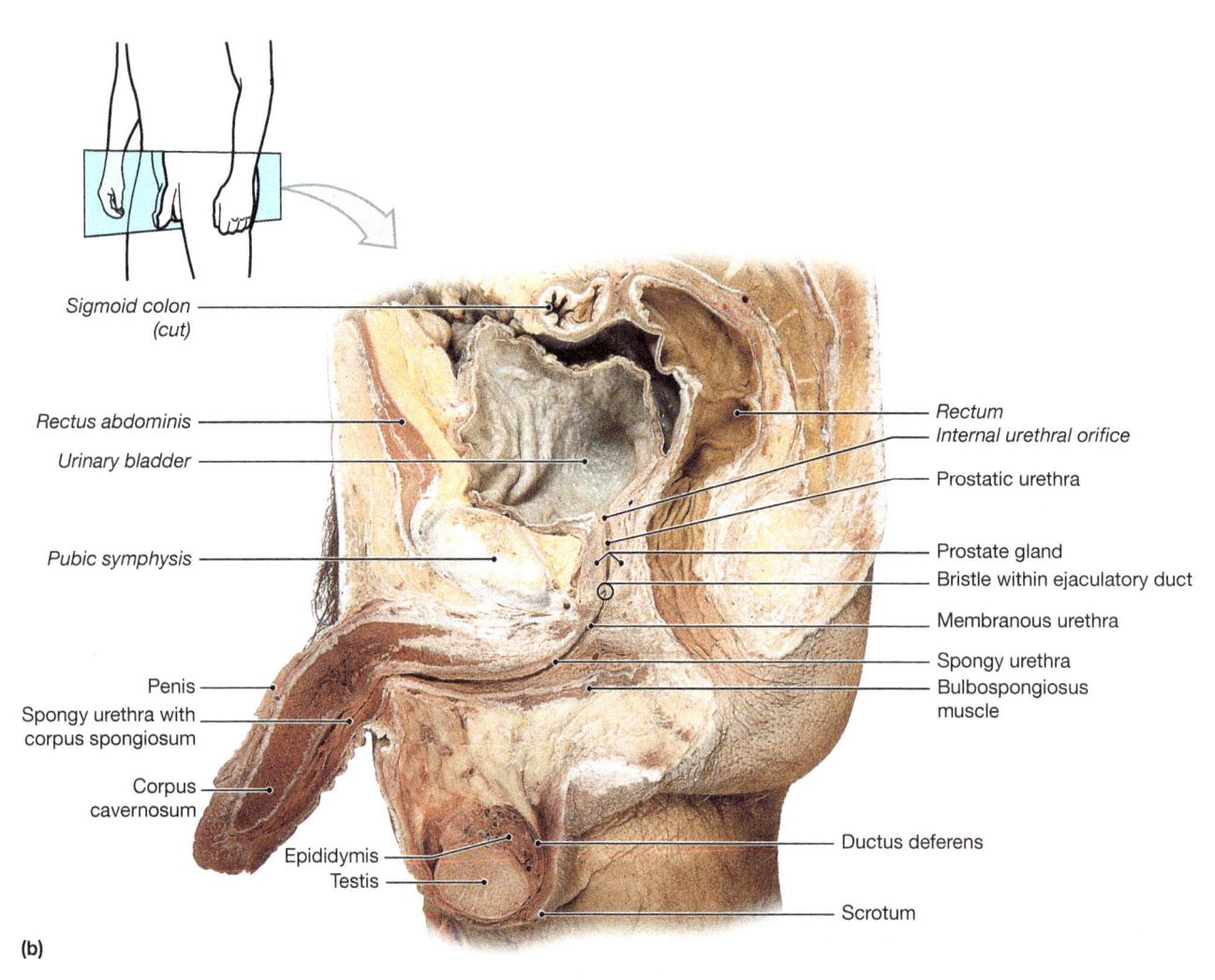

Figure 28.13 Comparison of Female and Male Reproductive Systems *(continued)*

EXERCISE 28

LAB REPORT

The Reproductive System

Name _____

Date _____

Section _____

A. Matching

Match each structure in the left column with its correct description from the right column.

_____ 1. epididymis   A. site of spermatozoa storage
_____ 2. infundibulum   B. site of spermatozoa production
_____ 3. vestibule   C. space between labia minora
_____ 4. ductus deferens   D. paired erectile cylinders
_____ 5. bulbo-urethral glands E. small glands in pelvic floor
_____ 6. corpora cavernosa   F. flared end of uterine tube
_____ 7. prepuce   G. female external genitalia
_____ 8. labia minora   H. first segment of urethra in males
_____ 9. myometrium   I. also called foreskin
_____ 10. prostatic urethra J. small fold that lacks pubic hair
_____ 11. membranous urethra K. muscular layer of uterine wall
_____ 12. labia majora   L. uterine protrusion into vagina
_____ 13. vulva   M. large fold with pubic hair
_____ 14. seminiferous tubules N. transports spermatozoa to urethra
_____ 15. cervix   O. portion of urethra in pelvic floor

B. Short-Answer Questions

1. List the three layers of the uterus, from superficial to deep.

2. Describe the gross anatomy of the female breast.

3. List the components of the vulva.

EXERCISE 28

LAB REPORT

4. How is temperature regulated in the testes for maximal spermatozoa production?

5. Name the three regions of the male urethra.

6. What are the three accessory glands of the male reproductive system?

C. Analysis and Application

1. How would removal of the testes affect endocrine function and reproductive function?

2. Explain the division sequence that leads to four spermatids in male meiosis but only one ovum in female meiosis.

3. How does a vasectomy or a tubal ligation sterilize an individual?

4. How are the clitoris and the penis similar to each other?

EXERCISE 29

Development

OBJECTIVES

On completion of this exercise, you should be able to:

- Describe the process of fertilization and early cleavage to the blastocyst stage.
- Describe the process of implantation and placenta formation.
- List the three germ layers and the embryonic fate of each.
- List the four extraembryonic membranes and the function of each.
- Describe the general developmental events of the first, second, and third trimesters.
- List the three stages of labor.

LAB ACTIVITIES

1. First Trimester: Fertilization, Cleavage, and Blastocyst Formation 469
2. First Trimester: Implantation and Gastrulation 472
3. First Trimester: Extraembryonic Membranes and the Placenta 476
4. Second and Third Trimesters and Birth 478

The cell theory of biology states that cells come from preexisting cells. In animals, gametes from the parents unite and form a new cell, the zygote, that has inherited the parental genetic material. The zygote quickly develops into an **embryo**, the name given to the organism for approximately the first two months after fertilization. By the end of the second month, most organ systems have started to form, and the embryo is then called a **fetus**.

In humans, the prenatal period of development occurs over a nine-month **gestation** (jes-TĀ-shun) that is divided into three three-month trimesters. During the first trimester, the embryo develops cell layers that are precursors to organ systems. The second trimester is characterized by growth in length, mass gain, and the appearance of functional organ systems. In the third trimester, increases in length and mass occur, and all organ systems either become functional or are prepared to become functional at birth. After 38 weeks of gestation, the uterus begins to rhythmically contract to deliver the fetus into the world. Although maternal changes occur during the gestation period, this exercise focuses on the development of the fetus.

Morphogenesis (mor-fō-JEN-uh-sis) is the general term for all the processes involved in the specialization of cells in the developing fetus and the migration of those cells to produce anatomical form and function.

LAB ACTIVITY 1 First Trimester: Fertilization, Cleavage, and Blastocyst Formation

Fertilization occurs when the spermatozoon and ovum join their haploid nuclei to produce a diploid zygote, the genetically unique cell that develops into an individual.

469

470 EXERCISE 29 Development

The male ejaculates approximately 300 million spermatozoa into the female's reproductive tract during intercourse. Once exposed to the female's reproductive tract, the spermatozoa complete a process called **capacitation** (ka-pas-i-TĀ-shun), during which they increase their motility and become capable of fertilizing an ovum. Most spermatozoa do not survive the journey through the vagina and uterus, and only an estimated 100 of them reach the ampulla. Normally, only a single ovum is released from a single ovary during one ovulation cycle.

Figure 29.1 illustrates fertilization. Ovulation releases a secondary oocyte from the ovary, and the oocyte begins moving along the uterine tube. Layers of follicular cells still encase the ovulated oocyte and now constitute a layer called the **corona radiata** (kō-RŌ-nuh rā-dē-A-tuh). Spermatozoa that reach the oocyte in the uterine tube must pass through the corona radiata to reach the cell membrane of the oocyte. Spermatozoa that swarm around the oocyte release an enzyme called *hyaluronidase*. The combined action of the hyaluronidase contributed by all the spermatozoa eventually creates a gap between some coronal cells, and a single spermatozoon slips into the oocyte.

The membrane of the oocyte instantly undergoes chemical and electrical changes that prevent additional spermatozoa from entering the cell. The oocyte, which is suspended in meiosis II since ovulation, now completes meiosis, while the spermatozoon prepares the paternal chromosomes for the union with the maternal chromosomes. Each set of nuclear material is called a **pronucleus**. Within 30 hours of fertilization, the male and female pronuclei come together in **amphimixis** (am-fi-MIK-sis) and undergo the first **cleavage**, which is a mitotic division that results in two cells, each called a **blastomere** (BLAS-tō-mēr). During cleavage, the existing cell mass of the ovum is subdivided by each cell division. (In other words, there is no increase in the mass of the zygote at this time.)

As the zygote slowly descends in the uterine tube toward the uterus, cleavages occur approximately every 12 hours. By the third day, the blastomeres are organized into a solid ball of nearly identical cells called a **morula** (MOR-ū-la), which is shown in Figure 29.2. Around day six, the morula has entered the uterus and changed into a **blastocyst** (BLAS-to-sist), a hollow ball of cells with an internal cavity called the **blastocoele** (BLAS-tō-sel). Now the process of **differentiation**, or specialization, begins. The blastomeres that make up the blastocyst are now of various sizes and have migrated into two regions. Cells on the outside compose the **trophoblast** (TRŌ-fō-blast), which will burrow into the uterine lining and eventually form part of the placenta. Cells clustered inside the blastocoele form the **inner cell mass**, which will develop into the embryo.

QuickCheck Questions

1.1 Where does fertilization normally occur?

1.2 What is a morula?

1.3 What is a blastocyst?

1 Materials

- ☐ Fertilization model
- ☐ 6, 10, and 12-day embryo models

Procedures

1. Review the steps of fertilization in Figure 29.1 and those of cleavage in Figure 29.2.
2. On the fertilization model, note how the male and female pronuclei join to create the diploid zygote.
3. On the embryo models, identify some blastomere cells and the morula.
4. On the embryology models, identify the blastocoele, and the trophoblast. ■

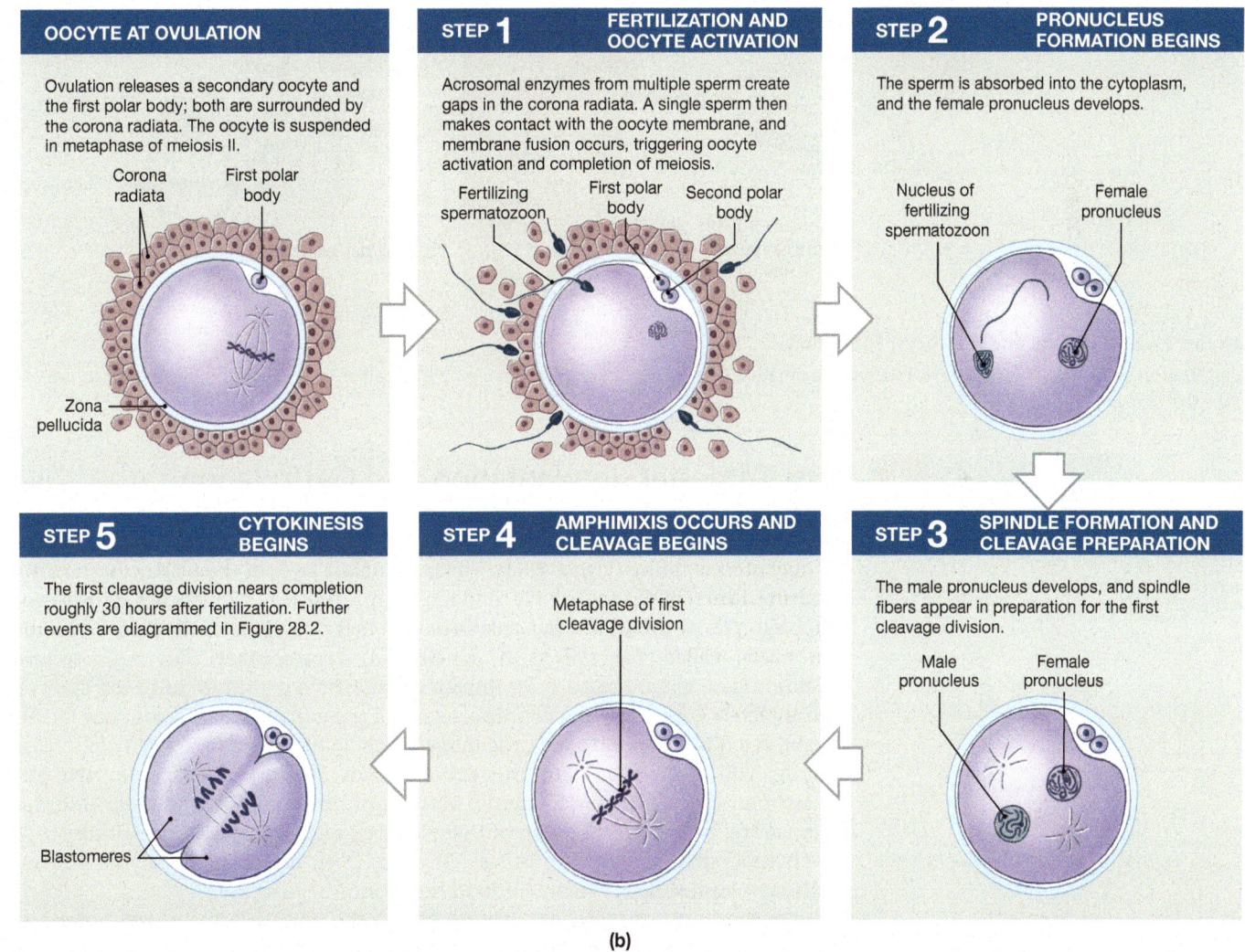

Figure 29.1 Fertilization
(a) A secondary oocyte surrounded by spermatozoa. Notice the difference in size between the two types of gametes. (b) Fertilization and the preparations for first cleavage.

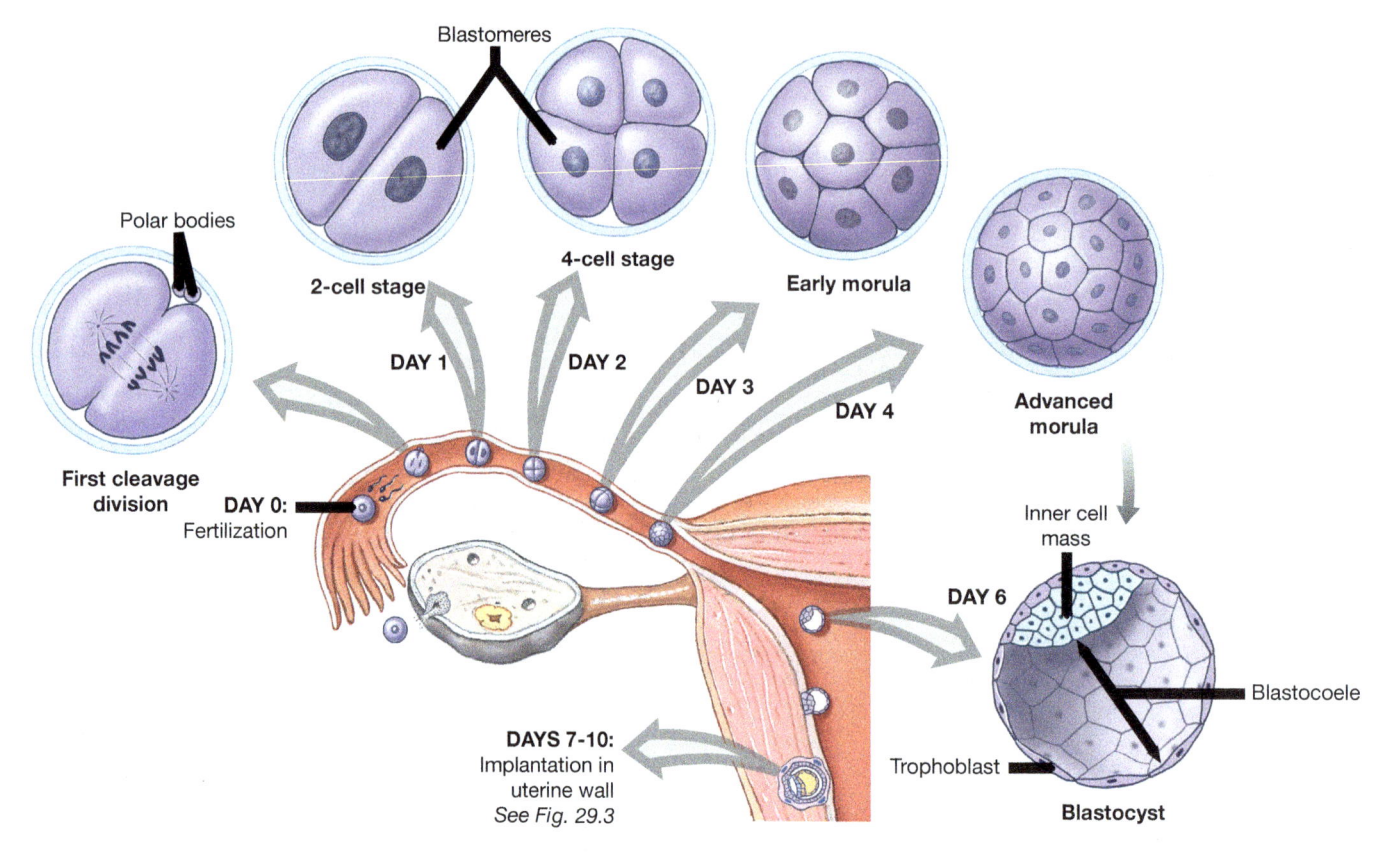

Figure 29.2 Cleavage and Blastocyst Formation

Fertilization occurs in the ampulla. It takes approximately 6 days for the embryo, now a hollow ball of cells called the blastocyst, to pass into the uterus.

LAB ACTIVITY 2 — First Trimester: Implantation and Gastrulation

Implantation begins on day seven or eight, when the blastocyst touches the spongy uterine lining (Figure 29.3). The trophoblast layer of the blastocyst burrows into the functional zone of the endometrium. The cell membranes of the trophoblast cells dissolve, and the cells mass together as a cytoplasmic layer of multiple nuclei called the **syncytial** (sin-SISH-al) **trophoblast**. The cells secrete hyaluronidase to erode a path for implantation, which continues until the embryo is completely covered by the functional zone of the endometrium, about day 14. To establish a diffusional link with the maternal circulation, the syncytial trophoblast sprouts villi that erode into the endometrium and create spaces in the endometrium called **lacunae**. Maternal blood from the endometrium seeps into the lacunae and bathes the villi with nutrients and oxygen. These materials diffuse into the blastocyst to support the inner cell mass. Deep to the syncytial layer is the **cellular trophoblast**, which will soon help form the placenta.

On the ninth or tenth day (Figure 29.4a), the middle layers of the inner cell mass drop away from the layer next to the cellular trophoblast. This movement of cells forms the **amniotic** (am-nē-OT-ik) **cavity**. The inner cell mass organizes into a **blastodisc** (BLAS-tō-disk) made up of two cell layers: the **epiblast** (EP-i-blast), or *superficial layer*, on top and the **hypoblast** (HĪ-pō-blast), or *deep layer*, that faces the blastocoele.

Within the next few days, cells begin to migrate in the process called **gastrulation** (gas-troo-LĀ-shun) (Figure 29.4b). Cells of the epiblast move toward the medial plane of the blastodisc to a region known as the **primitive streak**. As cells

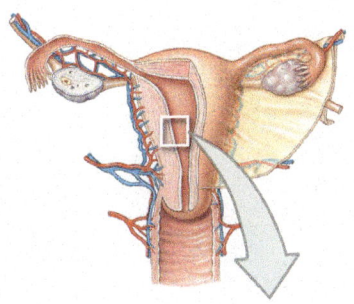

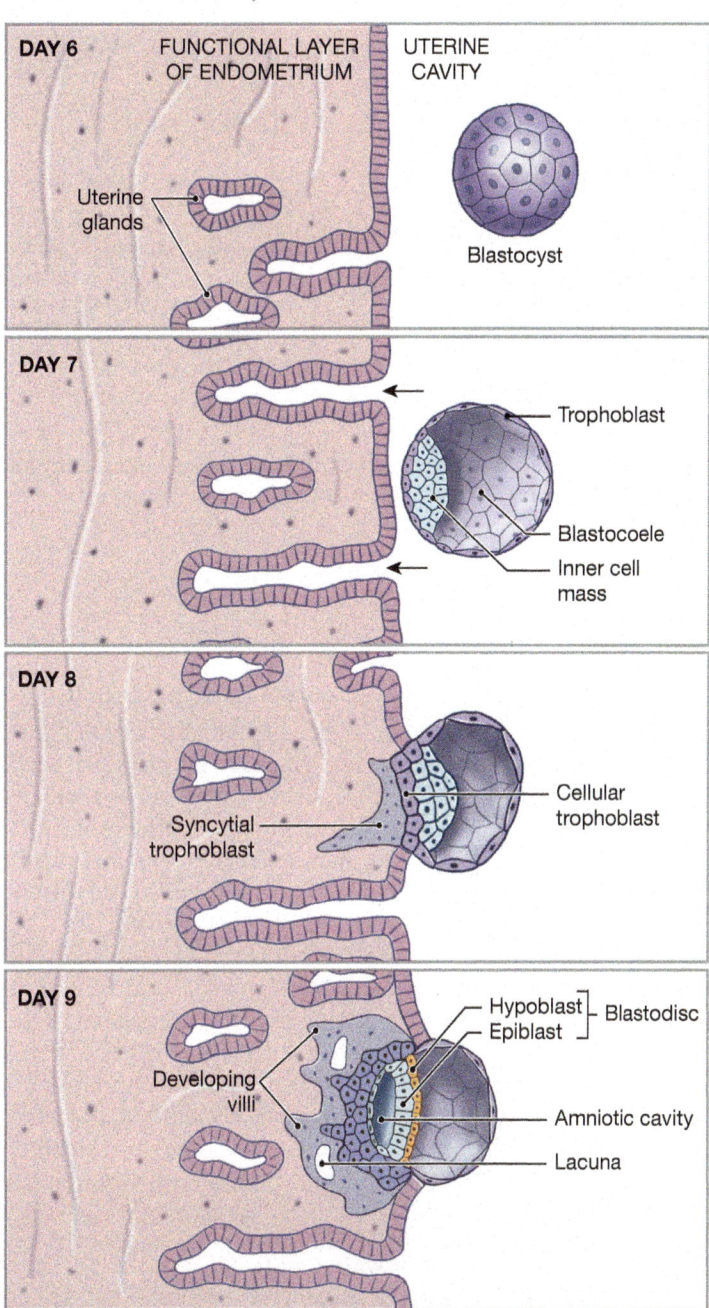

Figure 29.3 Implantation
The syncytial trophoblast of the embryo secretes enzymes that allow the trophoblast to implant into the wall for gestation.

474 EXERCISE 29 Development

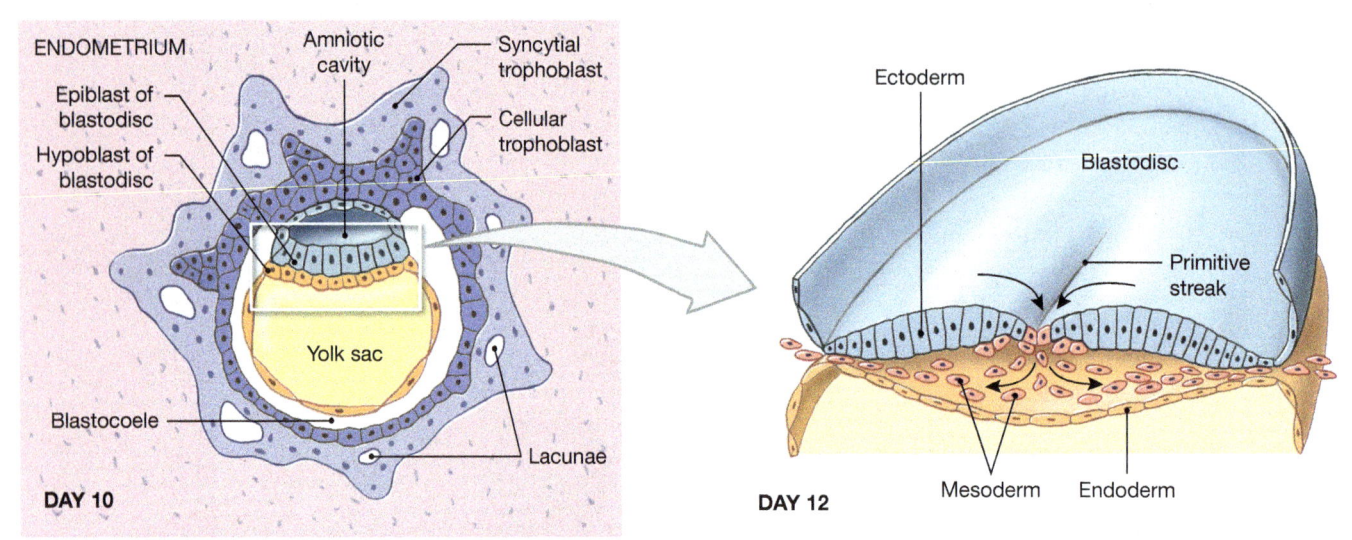

(a) The blastodisc begins as two layers: the *epiblast*, facing the amniotic cavity, and the *hypoblast*, exposed to the blastocoele. Migration of epiblast cells around the amniotic cavity is the first step in the formation of the amnion. Migration of hypoblast cells creates a sac that hangs below the blastodisc. This is the first step in yolk sac formation.

(b) Migration of epiblast cells into the region between epiblast and hypoblast gives the blastodisc a third layer. From the time this process (gastrulation) begins, the epiblast is called *ectoderm*, the hypoblast *endoderm*, and the migrating cells *mesoderm*.

Figure 29.4 Blastodisc Organization and Gastrulation
(a) On about day 10, the two cell layers of the blastodisc—the epiblast and the hypoblast—move away from the trophoblast to form the amniotic cavity. **(b)** By day 12 the epiblast and hypoblast separate from each other, and the three tissue-producing germ layers—ectoderm, mesoderm, and endoderm—are established.

arrive at the primitive streak, infolding, or **invagination**, occurs, and cells are liberated into the region between the epiblast and the hypoblast, and produce three cell layers in the embryo. The epiblast becomes the **ectoderm**, the hypoblast is now the **endoderm**, and the cells proliferating between the two layers form the **mesoderm**. These three layers, called **germ layers**, each produce specialized tissues that contribute to the formation of the organ systems. The ectoderm forms the nervous system, skin, hair, and nails. The mesoderm contributes to the development of the skeletal and muscular systems, and the endoderm forms part of the lining of the respiratory and digestive systems.

By the end of the fourth week of development, the embryo is distinct and has a **tail fold** and a **head fold**. The dorsal and ventral surfaces and the right and left sides are well-defined. The process of **organogenesis** begins as organ systems develop from the germ layers. In Figure 29.5b, the heart is clearly visible and has beat since the third week of growth. **Somites** (so-MĪ-tis), which are embryonic precursors of skeletal muscles, appear. Elements of the nervous system are also developing. Buds for the upper and lower limbs and small discs for the eyes and ears are also present. By week eight, fingers and toes are present, and the embryo is now usually called the fetus, as noted earlier. At the end of the third month, the first trimester is completed, and every organ system has appeared in the fetus.

QuickCheck Questions

2.1 Where does implantation normally occur?

2.2 What is the syncytial trophoblast?

2.3 What are the two cellular layers of the blastodisc?

EXERCISE 29 Development 475

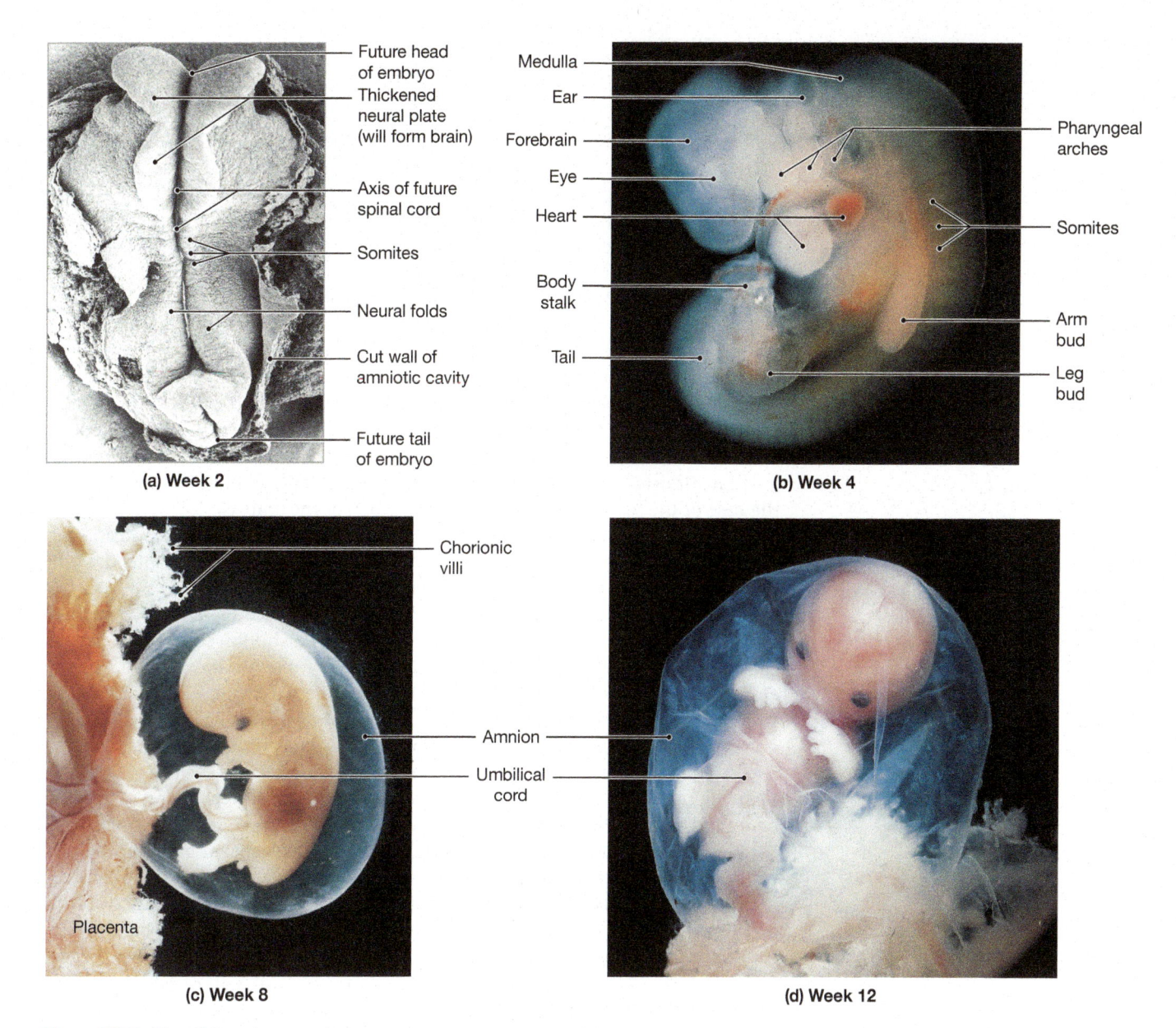

(a) Week 2

(b) Week 4

(c) Week 8

(d) Week 12

Figure 29.5 First Trimester
(a) Scanning electron micrograph of the superior surface of a monkey embryo after two weeks of development. A human embryo at this stage looks essentially the same. (b–d) Fiber-optic views of human embryos at four, eight, and twelve weeks.

2 Materials

- 6-, 10-, and 12-day embryo models and charts

Procedures

1. Review the anatomy of the blastocyst during implantation in Figure 29.3.
2. On the 6-day model or chart, locate the cellular and syncytial trophoblasts. The model may show the development of villi where the syncytial trophoblast has dissolved the endometrium for implantation.
3. Review the anatomy of the blastodisc in Figure 29.4a.
4. On the 10-day model or chart, examine the blastodisc and identify the epiblast and the hypoblast.
5. On the 12-day model or chart, identify the ectoderm, mesoderm, and endoderm. ■

LAB ACTIVITY 3

First Trimester: Extraembryonic Membranes and the Placenta

Four extraembryonic membranes develop from the germ layers: the yolk sac, amnion, chorion, and allantois (Figure 29.6). These membranes lie outside the blastodisc and provide protection and nourishment for the embryo/fetus. The **yolk sac** is the first membrane to appear, around the 10th day. Initially, cells from the hypoblast form a pouch under the blastodisc. Mesoderm reinforces the yolk sac, and blood vessels appear. As the syncytial trophoblast develops more villi, the yolk sac's importance in providing nourishment for the embryo diminishes.

While the yolk sac is forming, cells in the epiblast portion of the blastodisc migrate to line the inner surface of the amniotic cavity with a membrane called the **amnion** (AM-nē-on), the "water bag." As with the yolk sac, the amnion is soon reinforced with mesoderm. Embryonic growth continues, and by week 10 the amnion has mushroomed and envelops the embryo in a protective environment of **amniotic fluid** (see Figure 29.6).

The **allantois** (a-LAN-tō-is) develops from the endoderm and mesoderm near the base of the yolk sac. The allantois forms part of the embryonic urinary bladder and contributes to the **body stalk**, which is the tissue between the embryo and the developing chorion. Blood vessels pass through the body stalk and into the villi that protrude into the lacunae of the endometrium.

The outer extraembryonic membrane is the **chorion** (KOR-ē-on), which is formed by the cellular trophoblast and mesoderm. The chorion completely encases the embryo and the blastocoele. In the third week of growth, the chorion extends **chorionic villi** and blood vessels into the endometrial lacunae to establish the structural framework for the development of the **placenta** (pla-SENT-uh), the temporary organ through which nutrients, blood gases, and wastes are exchanged between the mother and the embryo. The embryo is connected to the placenta by the body stalk. The **yolk stalk**, where the yolk sac attaches to the endoderm of the embryo, and the body stalk together form the **umbilical cord**. Inside the umbilical cord are two **umbilical arteries**, which transport deoxygenated blood to the placenta, and a single **umbilical vein**, which returns oxygenated blood to the embryo.

By the fifth week of development, the chorionic villi have enlarged only where they face the uterine wall, and villi that face the uterine cavity become insignificant (see Figure 29.6). Only the part of the chorion where the villi develop becomes the placenta. The rest of this membrane remains chorion, as the week 10 part of Figure 29.6 indicates. Thus, the placenta does not completely surround the embryo. The placenta is in contact with the area of the endometrium called the **decidua basalis** (dē-SID-ū-a ba-SA-lis). The rest of the endometrium, where villi are absent, isolates the embryo from the uterine cavity and is called the **decidua capsularis** (kap-sū-LA-ris). The endometrium on the wall opposite the embryo is called the **decidua parietalis**.

QuickCheck Questions

3.1 What are the four extraembryonic membranes?

3.2 Which membrane gives rise to the placenta?

3 Materials

- 3-week, 5-week, and 10-week embryo models and charts
- Placenta model or biomount

Procedures

1. Review the extraembryonic membranes in Figure 29.6.
2. On the embryology models or charts, locate the yolk sac and the amnion. How does each of these membranes form, and what is the function of each?
3. On the embryology models or charts, locate the allantois and the chorion. How does each of these membranes form? Describe the chorionic villi and their significance to the embryo.

(a) WEEK 2

Migration of mesoderm around the inner surface of the trophoblast creates the chorion. Mesodermal migration around the outside of the amniotic cavity, between the ectodermal cells and the trophoblast, forms the amnion. Mesodermal migration around the endodermal pouch creates the yolk sac.

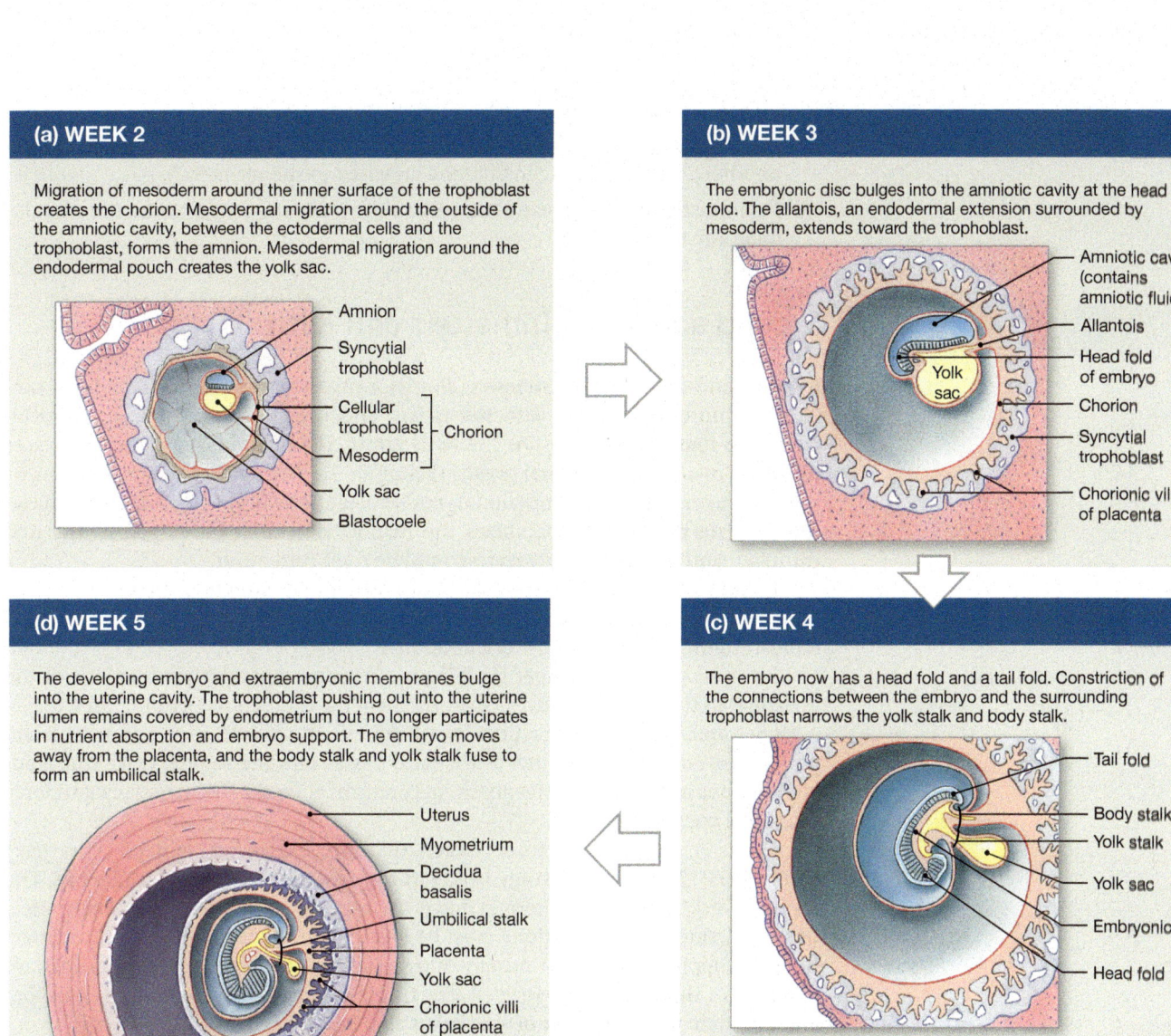

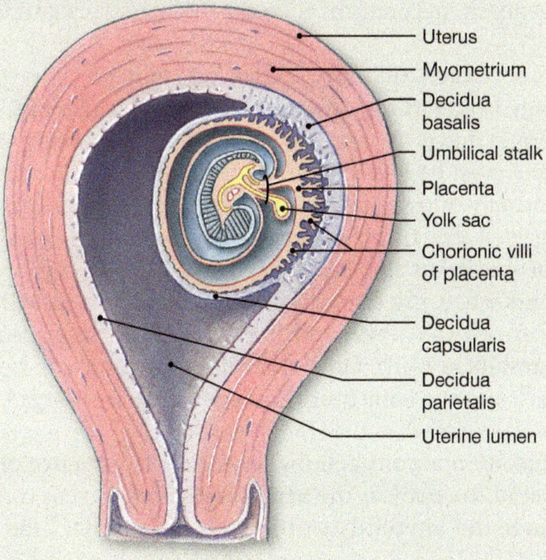

- Amnion
- Syncytial trophoblast
- Cellular trophoblast ⎤
- Mesoderm ⎦ Chorion
- Yolk sac
- Blastocoele

(b) WEEK 3

The embryonic disc bulges into the amniotic cavity at the head fold. The allantois, an endodermal extension surrounded by mesoderm, extends toward the trophoblast.

- Amniotic cavity (contains amniotic fluid)
- Allantois
- Head fold of embryo
- Chorion
- Syncytial trophoblast
- Chorionic villi of placenta

(c) WEEK 4

The embryo now has a head fold and a tail fold. Constriction of the connections between the embryo and the surrounding trophoblast narrows the yolk stalk and body stalk.

- Tail fold
- Body stalk
- Yolk stalk
- Yolk sac
- Embryonic gut
- Head fold

(d) WEEK 5

The developing embryo and extraembryonic membranes bulge into the uterine cavity. The trophoblast pushing out into the uterine lumen remains covered by endometrium but no longer participates in nutrient absorption and embryo support. The embryo moves away from the placenta, and the body stalk and yolk stalk fuse to form an umbilical stalk.

- Uterus
- Myometrium
- Decidua basalis
- Umbilical stalk
- Placenta
- Yolk sac
- Chorionic villi of placenta
- Decidua capsularis
- Decidua parietalis
- Uterine lumen

(e) WEEK 10

The amnion has expanded greatly, and fills the uterine cavity. The fetus is connected to the placenta by an elongated umbilical cord that contains a portion of the allantois, blood vessels, and the remnants of the yolk stalk.

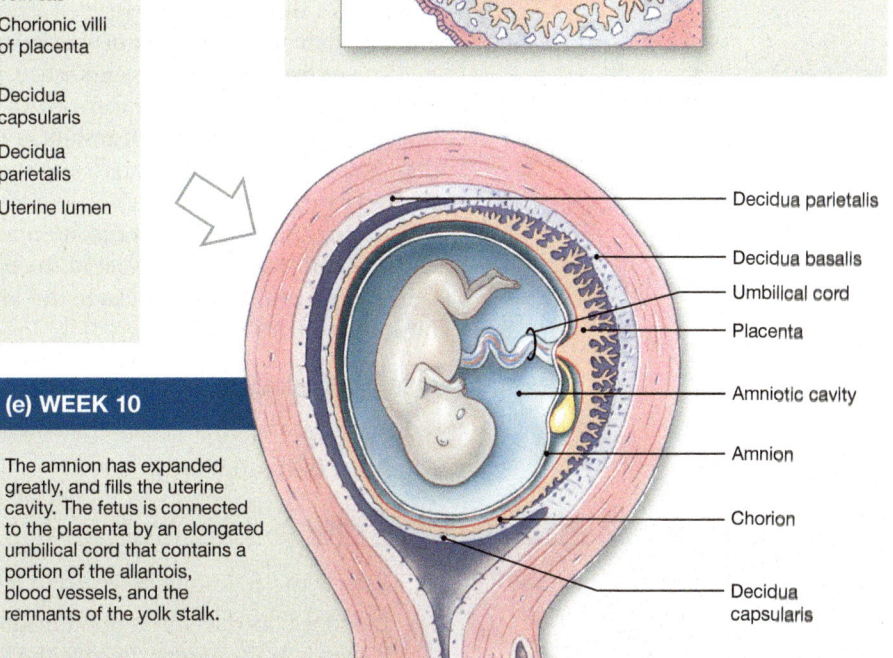

- Decidua parietalis
- Decidua basalis
- Umbilical cord
- Placenta
- Amniotic cavity
- Amnion
- Chorion
- Decidua capsularis

Figure 29.6 **Extraembryonic Membranes and Placenta Formation**

478 EXERCISE 29 Development

4. On the placenta model or biomount, note the appearance of the various placental surfaces. Are there any differences in appearance from one surface to another? Is the amniotic membrane attached to the placenta?

5. Examine the umbilical cord attached to the placenta, and describe the vascular anatomy in the cord. ■

LAB ACTIVITY 4 Second and Third Trimesters and Birth

By the start of the second trimester, all major organ systems have started to form. Growth during the second trimester is fast, and the fetus doubles in size and increases its mass by 50 times. As the fetus grows, the uterus expands and displaces the other maternal abdominal organs (Figure 29.7). The fetus begins to move as its muscular system becomes functional, and articulations begin to form in the skeleton. The nervous system organizes the neural tissue that developed in the first trimester, and many sensory organs complete their formation.

During the third trimester, all organ systems complete their development and become functional, and the fetus responds to sensory stimuli such as a hand-rubbing across the mother's abdomen.

Birth, or **parturition** (par-tūr-ISH-un), involves muscular contractions of the uterine wall to expel the fetus. Delivering the fetus is much like pulling on a turtleneck sweater. Muscle contractions must stretch the cervix over the fetal head, pulling the uterine wall thinner as the fetus passes into the vagina. Once true labor contractions begin, positive feedback mechanisms increase the frequency and force of uterine contractions.

Labor is divided into three stages: dilation, expulsion, and placental (see Figure 29.8). The **dilation stage** begins at the onset of true labor contractions. The cervix dilates, and the fetus moves down the cervical canal. To be maximally effective at dilation, the contractions must be less than 10 minutes apart. Each contraction lasts approximately one minute and spreads from the upper cervix downward to *efface*, or thin, the cervix for delivery. Contractions usually rupture the amnion, and amniotic fluid flows out of the uterus and the vagina.

The **expulsion stage** occurs when the cervix is dilated completely, usually to 10 cm, and the fetus passes through the cervix and the vagina. This stage usually lasts less than two hours and results in birth. Once the baby is breathing independently, the umbilical cord is cut, and the baby must now rely on its own organ systems to survive.

During the **placental stage**, uterine contractions break the placenta free of the endometrium and deliver it out of the body as the **afterbirth**. This stage is usually short, and many women deliver the afterbirth within 5 to 10 minutes after the birth of the fetus.

QuickCheck Questions

4.1 What are the three stages of labor?

4.2 What is the afterbirth?

4 Materials

- ☐ Second-trimester model
- ☐ Third-trimester model
- ☐ Parturition model

Procedures

1. Review the anatomy presented in Figure 29.7 and 29.8.

2. Describe how the fetus is positioned in the uterus in the second-trimester model. If shown in the model, describe the location of the amnion and the placenta.

3. Describe how the fetus is positioned in the uterus in the third-trimester model. If shown in the model, describe the location of the amnion and the placenta.

4. Using the parturition model as an aid, describe the contractions that force the fetus out of the uterus. ■

(a) Pregnancy at four months

Labels: Placenta; Umbilical cord; Fetus at 16 weeks; Uterus; Amniotic fluid; Cervix; Vagina

(b) Pregnancy at three to nine months

Labels: 9 months, 8 months, 7 months, 6 months, 5 months, 4 months, 3 months; After dropping, in preparation to delivery

(c) Pregnant female (full-term infant)

Labels: Liver; Small intestine; Stomach; Transverse colon; Fundus of uterus; Umbilical cord; Placenta; Urinary bladder; Pubic symphysis; Vagina; Urethra; Pancreas; Aorta; Common iliac vein; Cervical (mucus) plug in cervical canal; External os; Rectum

(d) Nonpregnant female

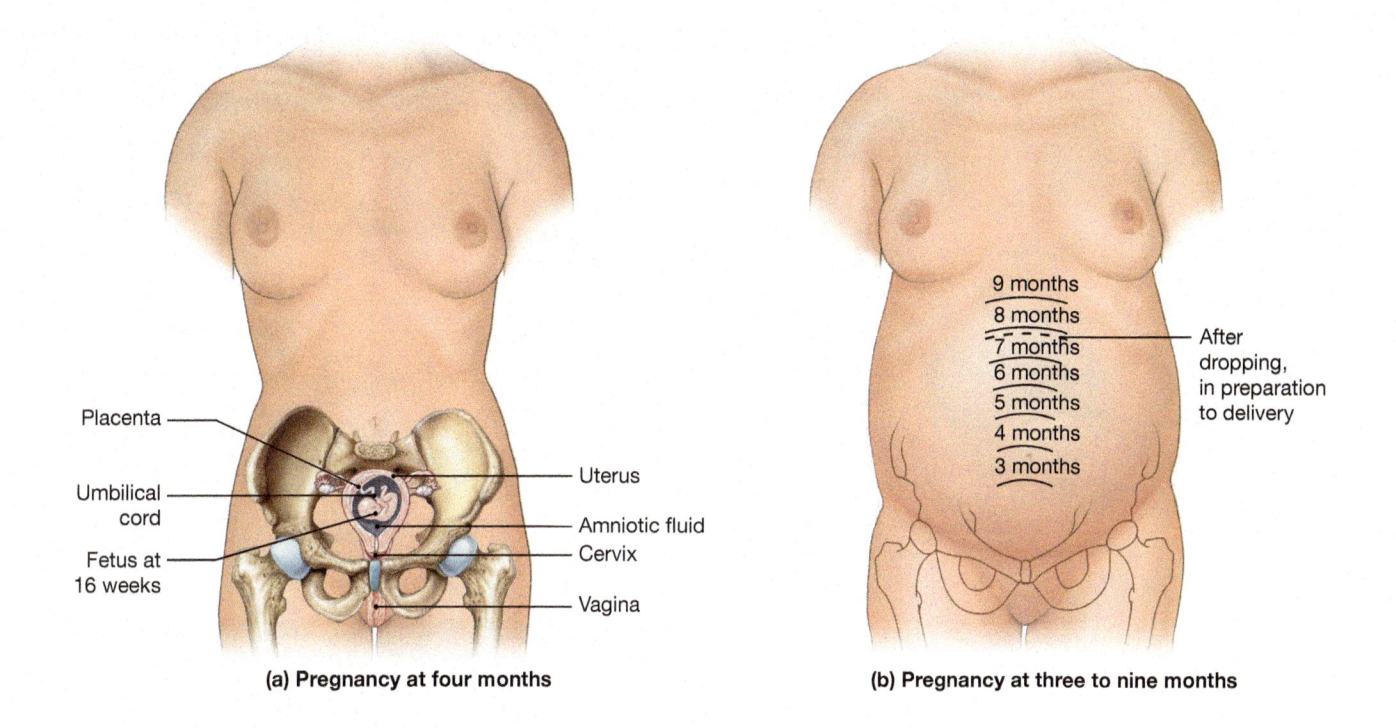

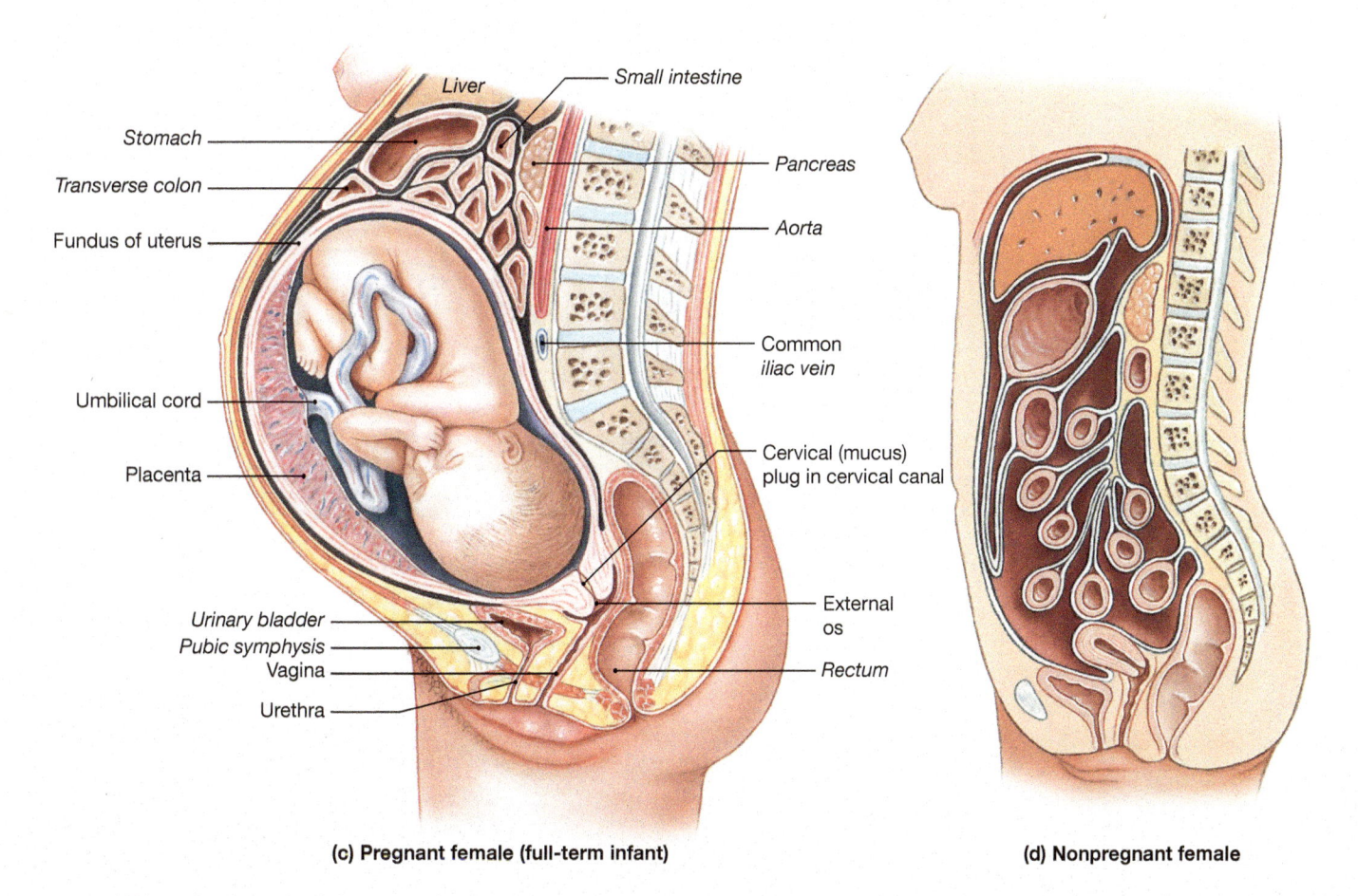

Figure 29.7 The Growth of the Uterus and Fetus
(a) Pregnancy at four months (16 weeks), showing the position of the uterus, fetus, and placenta. (b) Changes in the size of the uterus during the second and third trimesters. (c) Pregnancy at full term. Note the position of the uterus and fetus and the displacement of abdominal organs relative to part (d). (d) Organ position and orientation in a nonpregnant female.

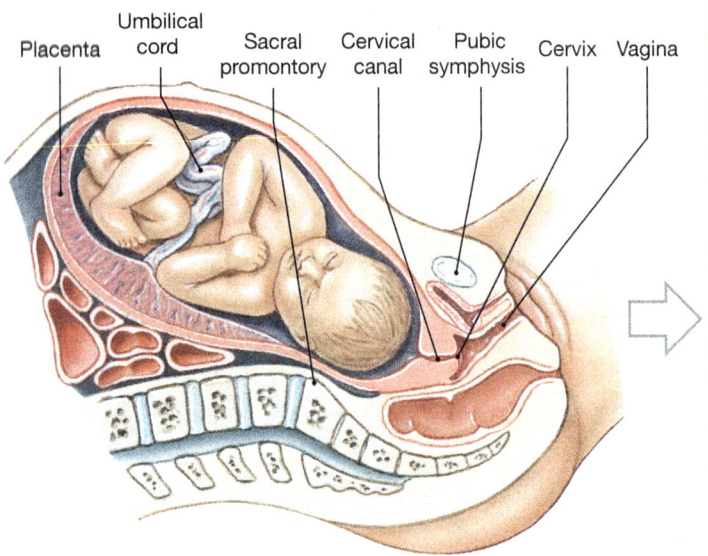

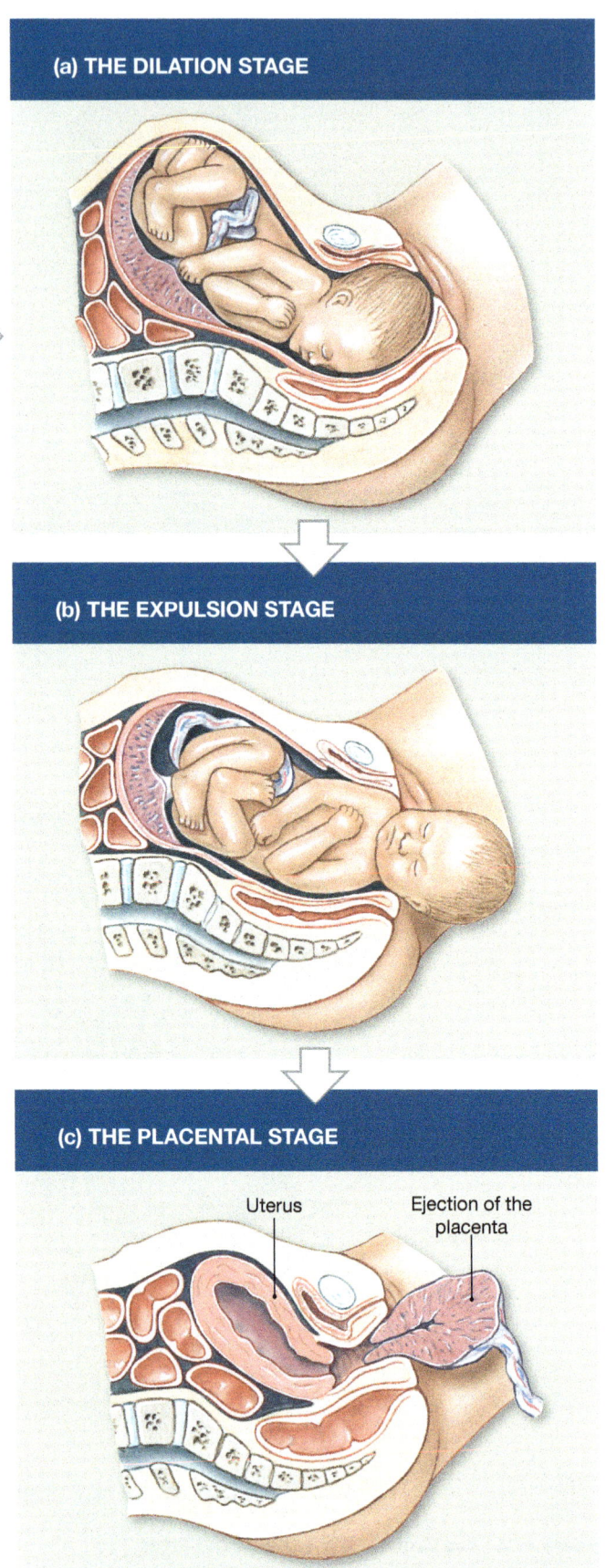

Figure 29.8 The Stages of Labor
At birth, the cervix dilates and the myometrium contracts to deliver the fetus. After the baby is born, the placenta is expelled.

EXERCISE 29

LAB REPORT

Development

A. Matching

Match each structure in the left column with its correct description from the right column.

_____	1. blastomere	A.	has villi
_____	2. amnion	B.	forms at primitive streak
_____	3. allantois	C.	cavity of blastocyst
_____	4. morula	D.	forms part of urinary bladder
_____	5. blastocoele	E.	becomes ectoderm
_____	6. syncytial trophoblast	F.	migration of cells
_____	7. chorion	G.	water bag
_____	8. mesoderm	H.	birth
_____	9. endoderm	I.	cell produced by early cleavage
_____	10. epiblast	J.	produces lining of respiratory tract
_____	11. parturition	K.	solid ball of cells
_____	12. gastrulation	L.	erodes endometrium

B. Labeling

Label the anatomy of the embryo and fetus in Figure 29.9.

C. Short-Answer Questions

1. Describe the process of fertilization.

2. Why are so many spermatozoa required for successful fertilization?

3. Discuss the formation of the three germ layers.

481

EXERCISE 29   LAB REPORT

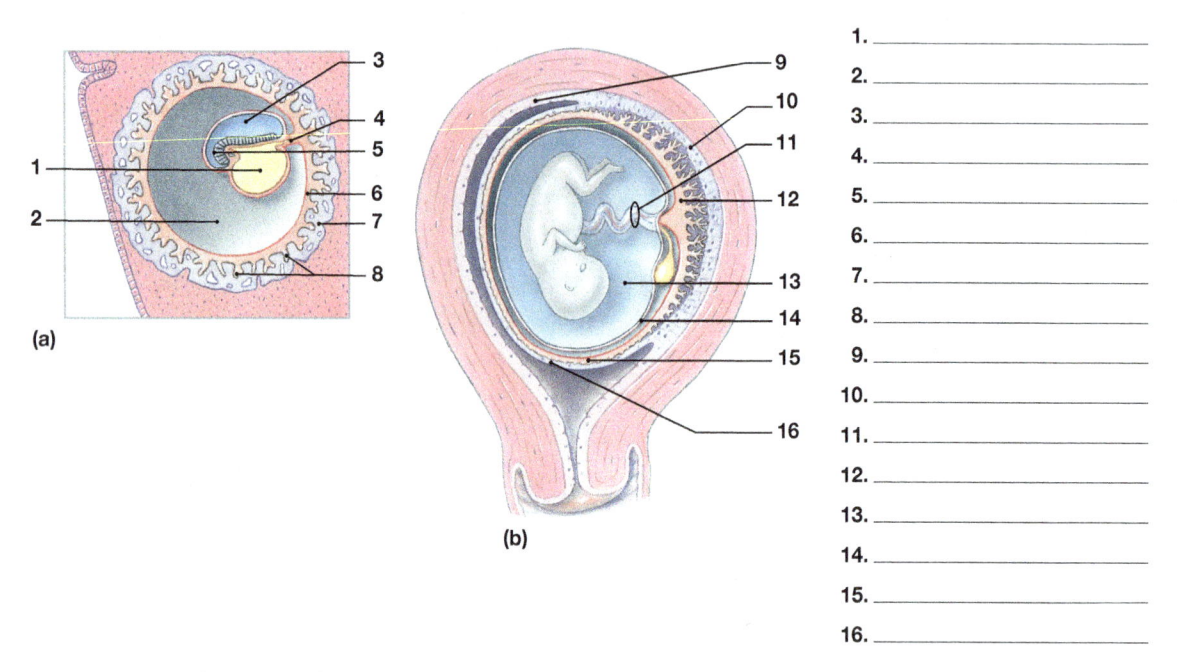

Figure 29.9 Embryonic and Fetal Development
(a) Embryo at 3 weeks. (b) Fetus at 10 weeks.

1. _____
2. _____
3. _____
4. _____
5. _____
6. _____
7. _____
8. _____
9. _____
10. _____
11. _____
12. _____
13. _____
14. _____
15. _____
16. _____

4. Describe the structure of the blastocyst.

5. List the three stages of labor and explain what takes place in each stage.

D. Analysis and Application

1. How does the amniotic cavity form?

2. What is the function of each of the four extraembryonic membranes?

3. How does a fetus obtain nutrients and gases from the maternal blood?

4. From what structures do the four major tissue groups of the body arise?

EXERCISE 30

Surface Anatomy

OBJECTIVES

On completion of this exercise, you should be able to:

- Describe the surface anatomy of the head, neck, and trunk.
- Describe the surface anatomy of the shoulder and upper limb.
- Describe the surface anatomy of the pelvis and lower limb.
- Identify the major surface features on your body or on the body of a partner.

LAB ACTIVITIES

1 Head, Neck, and Trunk 483
2 Shoulder and Upper Limb 485
3 Pelvis and Lower Limb 489

Surface anatomy is the study of anatomical landmarks that can be identified on the body surface. Most of the features are either skeletal structures or muscles and tendons. A regional approach to surface anatomy is presented in this exercise. Because the models in the photographs are muscular and have little body fat, all the anatomical landmarks discussed in this exercise are easily seen. Depending on your body type, it may be difficult to precisely identify a structure on yourself.

LAB ACTIVITY 1 — Head, Neck, and Trunk

The head is a complex region where many body systems are integrated for such vital functions as breathing, eating, and speech production. Main surface features include the eyes and eyebrows, the nose, mouth, and ears. The zygomatic bone is the prominent cheek bone (Figure 30.1a). The surface anatomy of the head is divided into regions that correspond to the underlying bones.

The sternocleidomastoid muscle divides the neck into an **anterior cervical triangle** and a **posterior cervical triangle** (Figure 30.1b). The anterior cervical triangle lies inferior to the mandible and anterior to the sternocleidomastoid muscle. It is subdivided into four smaller triangles, as shown in Figure 30.1c. The **suprahyoid triangle** is the superior region of the anterior neck. Inferior is the **submandibular triangle**. The **superior carotid triangle** is at the midpoint of the neck and surrounds the thyroid cartilage of the larynx. The pulse of the carotid artery is often palpated within this region. The base of the neck is the **inferior carotid triangle**.

The trunk is comprised of the **thorax**, or chest, the **abdominopelvic region**, and the **back** (Figures 30.2 and 30.3). The jugular notch marks the boundary between the neck and the thorax. The pectoralis major muscles, nipples, and umbilicus are prominent on the thorax. The jugular notch of the sternum can be palpated at the base of the neck. The inferior sternum is the xiphoid process. When CPR is being performed, it is critical that the xiphoid process is not pushed on during chest compressions. Muscles of the abdomen are difficult to palpate on most individuals because of the presence of a layer of adipose tissue along the waistline. The commonly-called "six pack" is the rectus abdominis muscle and the linea alba.

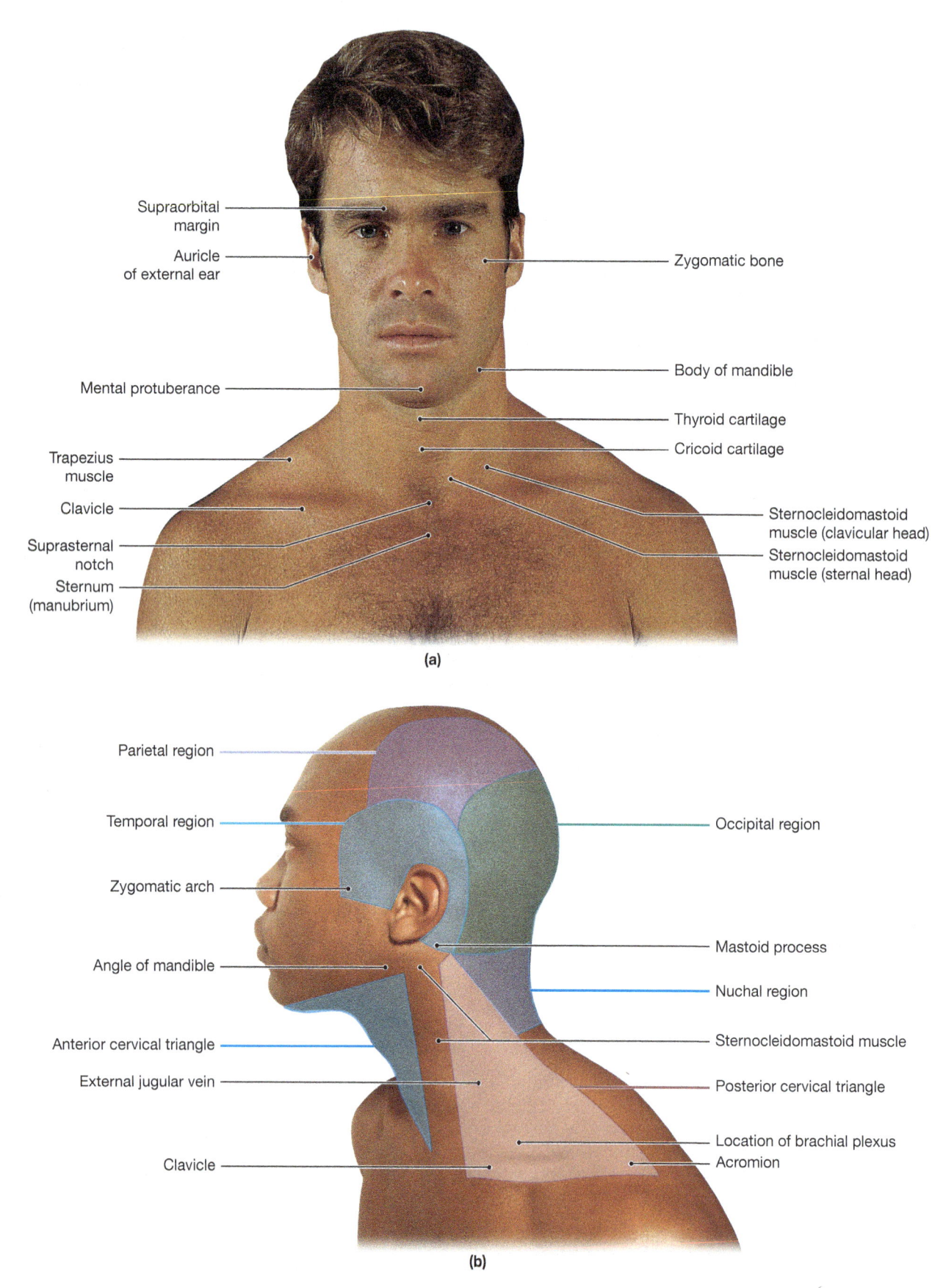

Figure 30.1 Surface Anatomy of Head and Neck

(a) Anterior view of head, neck, and upper trunk. (b) The neck is divided into two anatomical regions: an anterior cervical triangle and a posterior cervical triangle. The anterior triangle extends from the anterior midline to the anterior border of the sternocleidomastoid muscle. The posterior triangle extends between the posterior border of the sternocleidomastoid muscle and the anterior border of the trapezius muscle.

484

EXERCISE 30 Surface Anatomy   485

KEY TO DIVISIONS OF THE ANTERIOR CERVICAL TRIANGLE

- SHT Suprahyoid triangle
- SMT Submandibular triangle
- SCT Superior carotid triangle
- ICT Inferior carotid triangle

Labels on figure:
- Angle of mandible
- Site for palpation of submandibular gland and submandibular lymph nodes
- Hyoid bone
- Thyroid cartilage
- Trapezius muscle
- Supraclavicular fossa
- Omohyoid muscle
- ANTERIOR CERVICAL TRIANGLE
- Jugular notch
- Mastoid process
- Sternocleidomastoid region
- External jugular vein beneath platysma
- Site for palpation of carotid pulse
- POSTERIOR CERVICAL TRIANGLE
- Location of brachial plexus
- Acromion
- Clavicle
- Sternocleidomastoid muscle (clavicular head [lateral] and sternal head [medial])

(c)

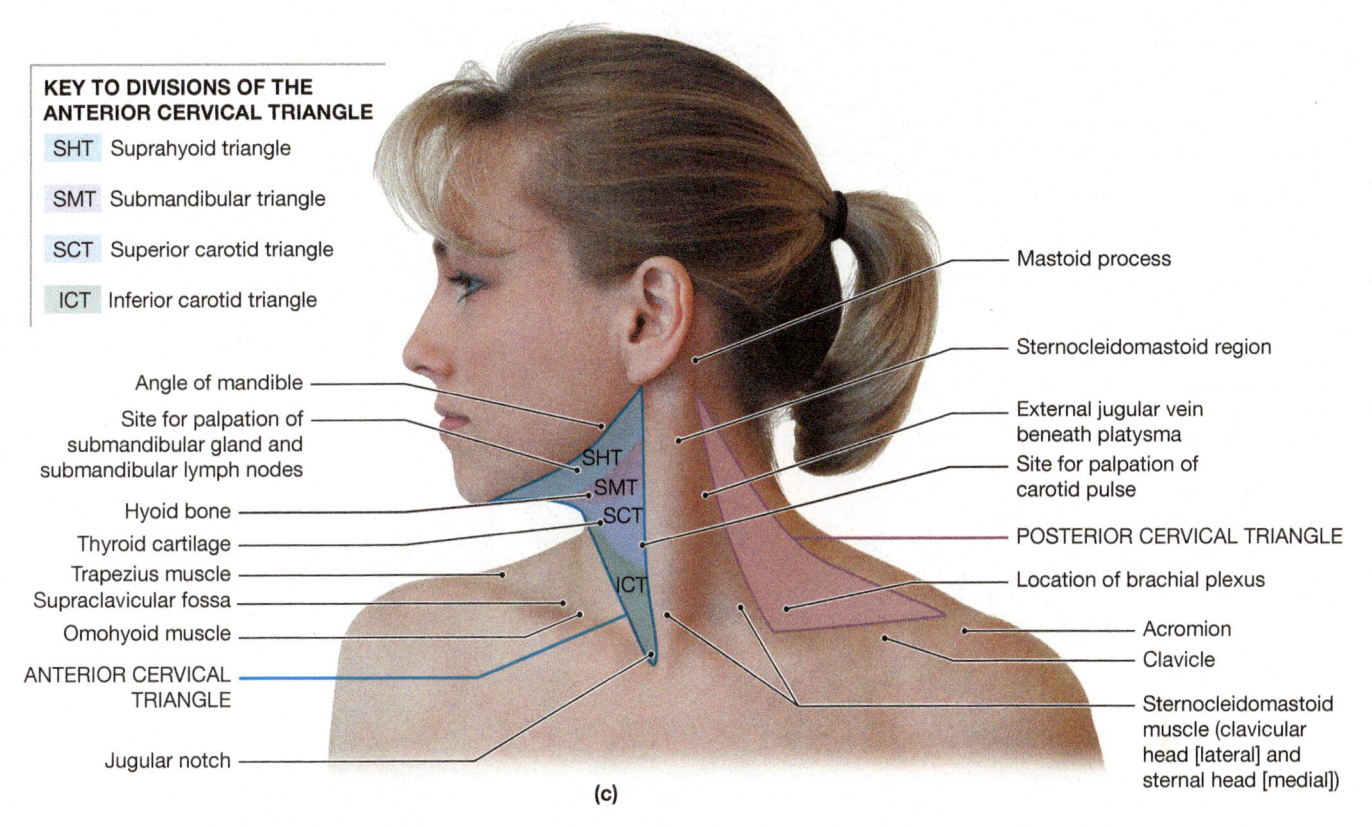

Figure 30.1 Surface Anatomy of Head and Neck *(continued)*
(c) The anterior cervical triangle is subdivided into four smaller triangles.

QuickCheck Questions

1.1 What are the two main regional divisions of the neck?
1.2 In what region of the neck is the pulse easily felt?
1.3 What are the main surface features of the thorax?

1 Materials

☐ Lab partner
☐ Mirror

Procedures

1. Review the surface anatomy of the head, neck, and trunk in Figures 30.1, 30.2, and 30.3.
2. On your lab partner or yourself in a mirror, identify the regions of the head and neck identified in Figures 30.1 and 30.2.
3. Palpate around the superior part of the neck and attempt to detect the submandibular salivary gland or a cervical lymph gland. Palpate midway up your neck and find the pulse in the carotid artery. This region of the neck is the superior carotid triangle.
4. Locate the thyroid cartilage and use it as reference in identifying the divisions of the anterior cervical triangle of the neck.
5. On your partner or yourself, locate the rectus abdominis muscle and the linea alba. ■

LAB ACTIVITY 2 Shoulder and Upper Limb

The surface anatomy of the shoulder includes the deltoid muscle and bony features of the clavicle and scapula. The scapular spine and acromion are easy to palpate, as is the sternal end and body of the clavicle.

486   EXERCISE 30   Surface Anatomy

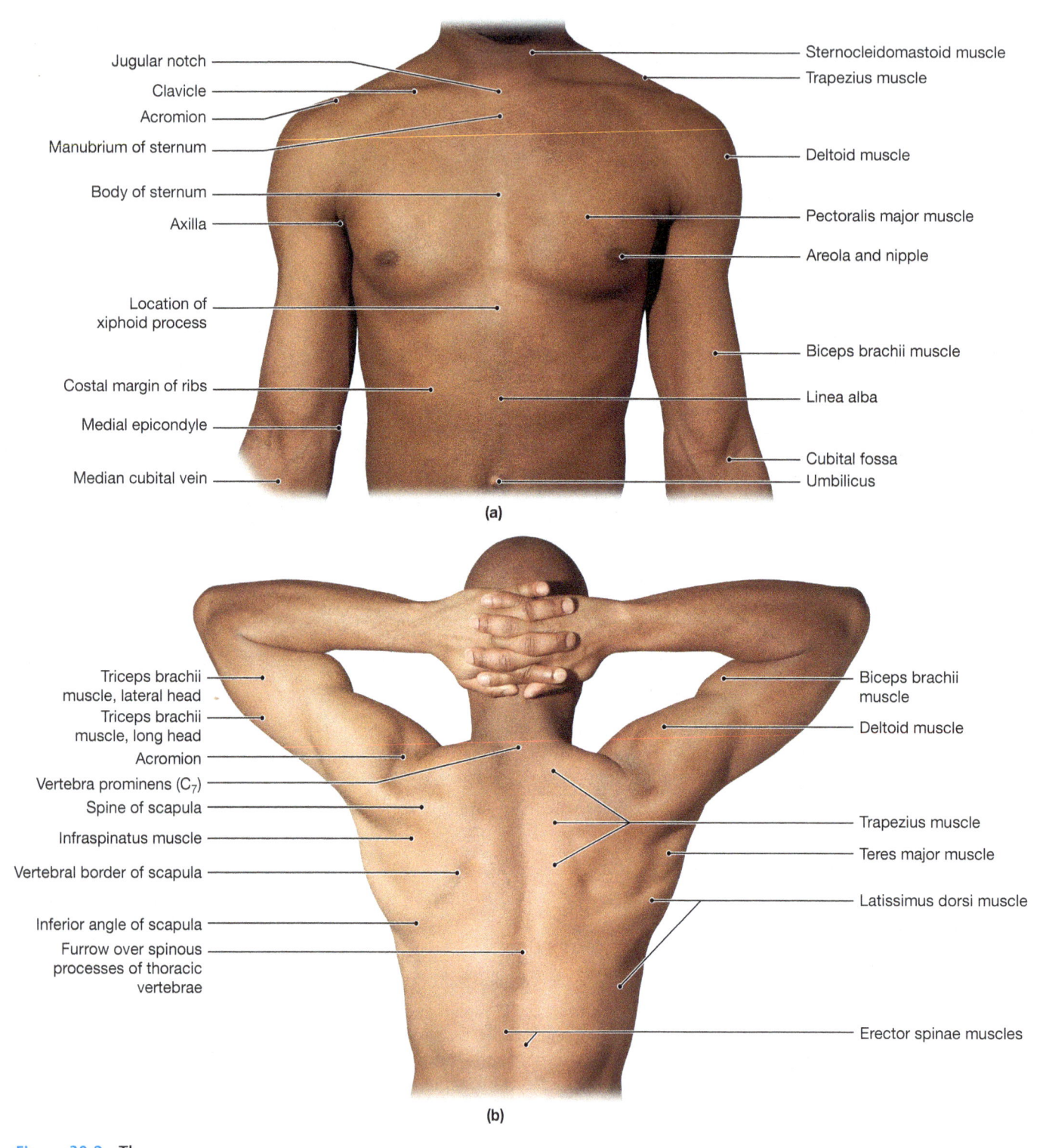

(a) Anterior view labels: Jugular notch, Clavicle, Acromion, Manubrium of sternum, Body of sternum, Axilla, Location of xiphoid process, Costal margin of ribs, Medial epicondyle, Median cubital vein, Sternocleidomastoid muscle, Trapezius muscle, Deltoid muscle, Pectoralis major muscle, Areola and nipple, Biceps brachii muscle, Linea alba, Cubital fossa, Umbilicus

(b) Posterior view labels: Triceps brachii muscle, lateral head; Triceps brachii muscle, long head; Acromion; Vertebra prominens (C₇); Spine of scapula; Infraspinatus muscle; Vertebral border of scapula; Inferior angle of scapula; Furrow over spinous processes of thoracic vertebrae; Biceps brachii muscle; Deltoid muscle; Trapezius muscle; Teres major muscle; Latissimus dorsi muscle; Erector spinae muscles

Figure 30.2   Thorax
(a) Anterior view. (b) Posterior view.

Anatomy visible on the surface of the arm includes the biceps brachii, brachialis, and triceps brachii muscles (Figure 30.4). In some individuals, the median cubital vein is clearly visible in the anterior of the elbow, the *antecubitis* (Figure 30.5). The olecranon of the ulna, which forms the point of the elbow, can be felt on the posterior surface of the elbow (see Figure 30.4).

Muscles that flex the wrist and hand are positioned on the anterior forearm (see Figure 30.5). The extensors are located posteriorly (Figure 30.4b). The tendons to the extensor muscles are clearly visible on the posterior surface of the hand.

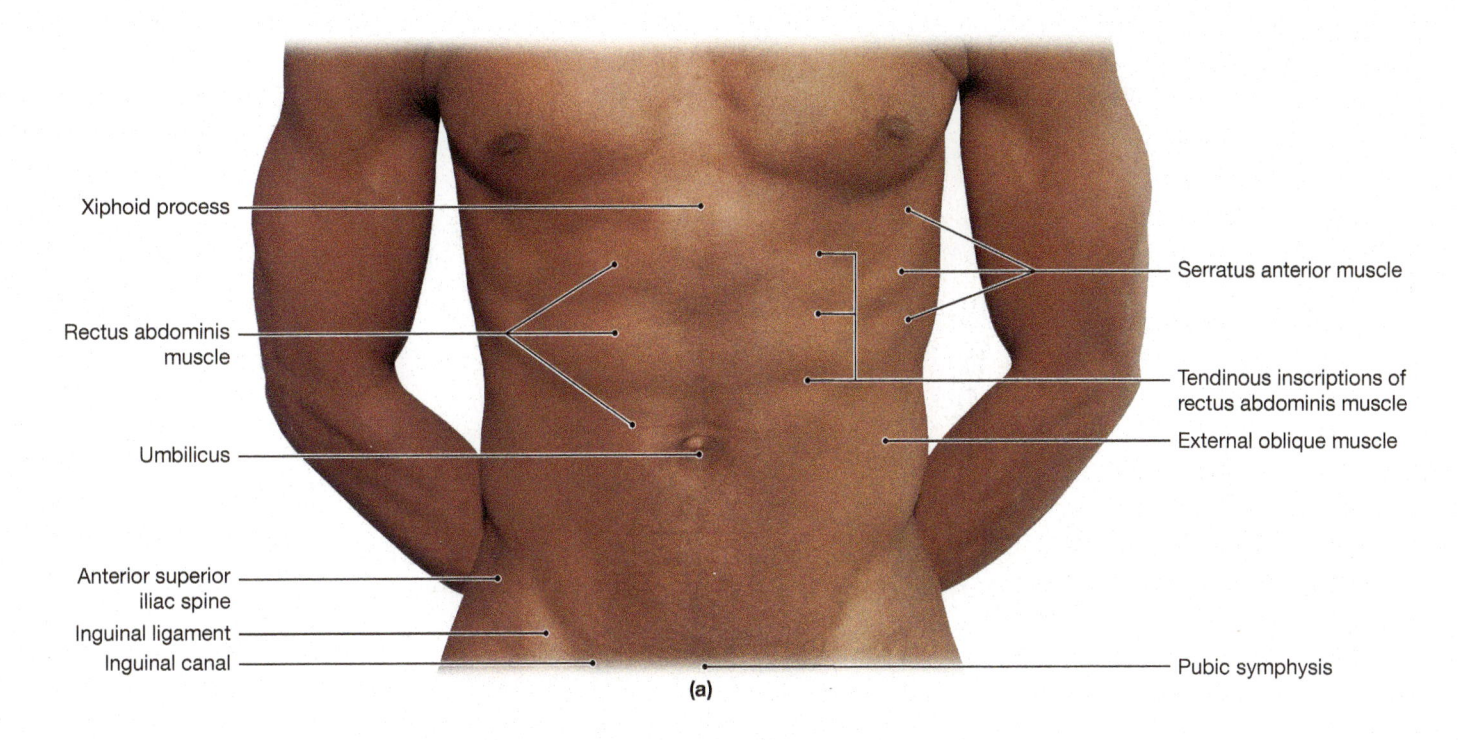

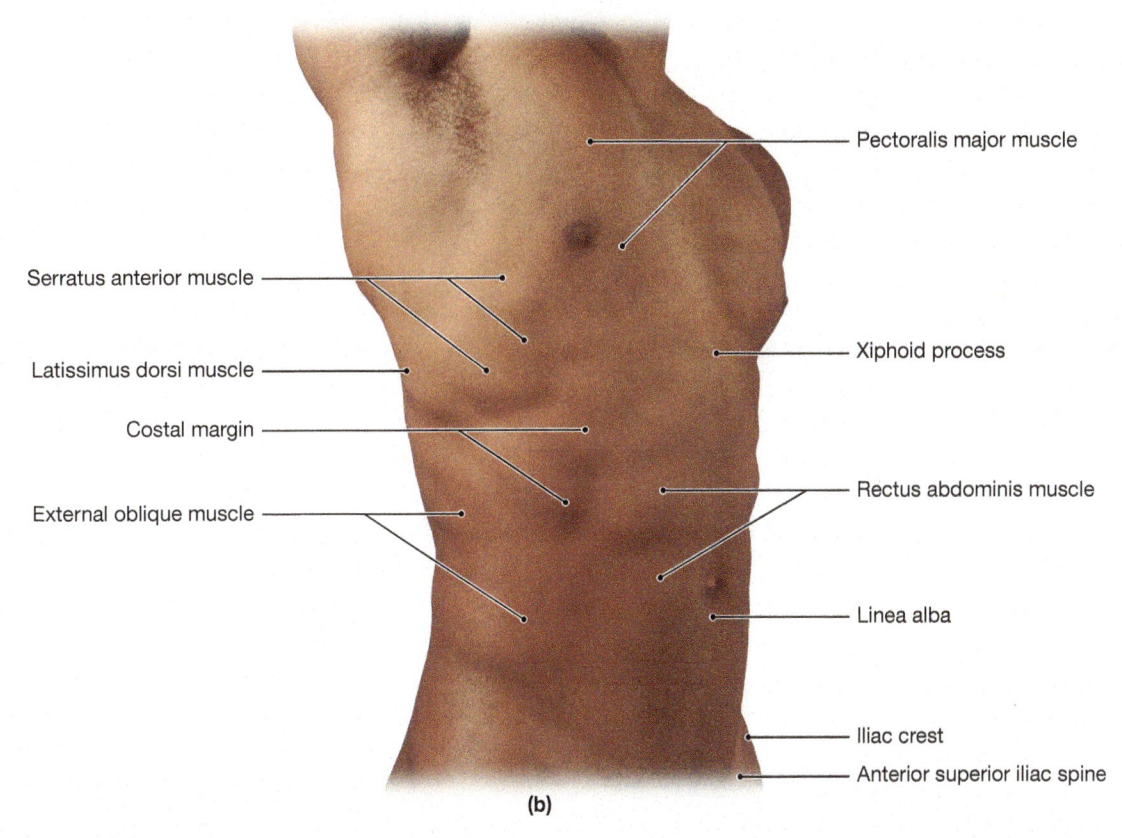

Figure 30.3 Abdominal Wall
(a) Anterior view. (b) Anterolateral view.

487

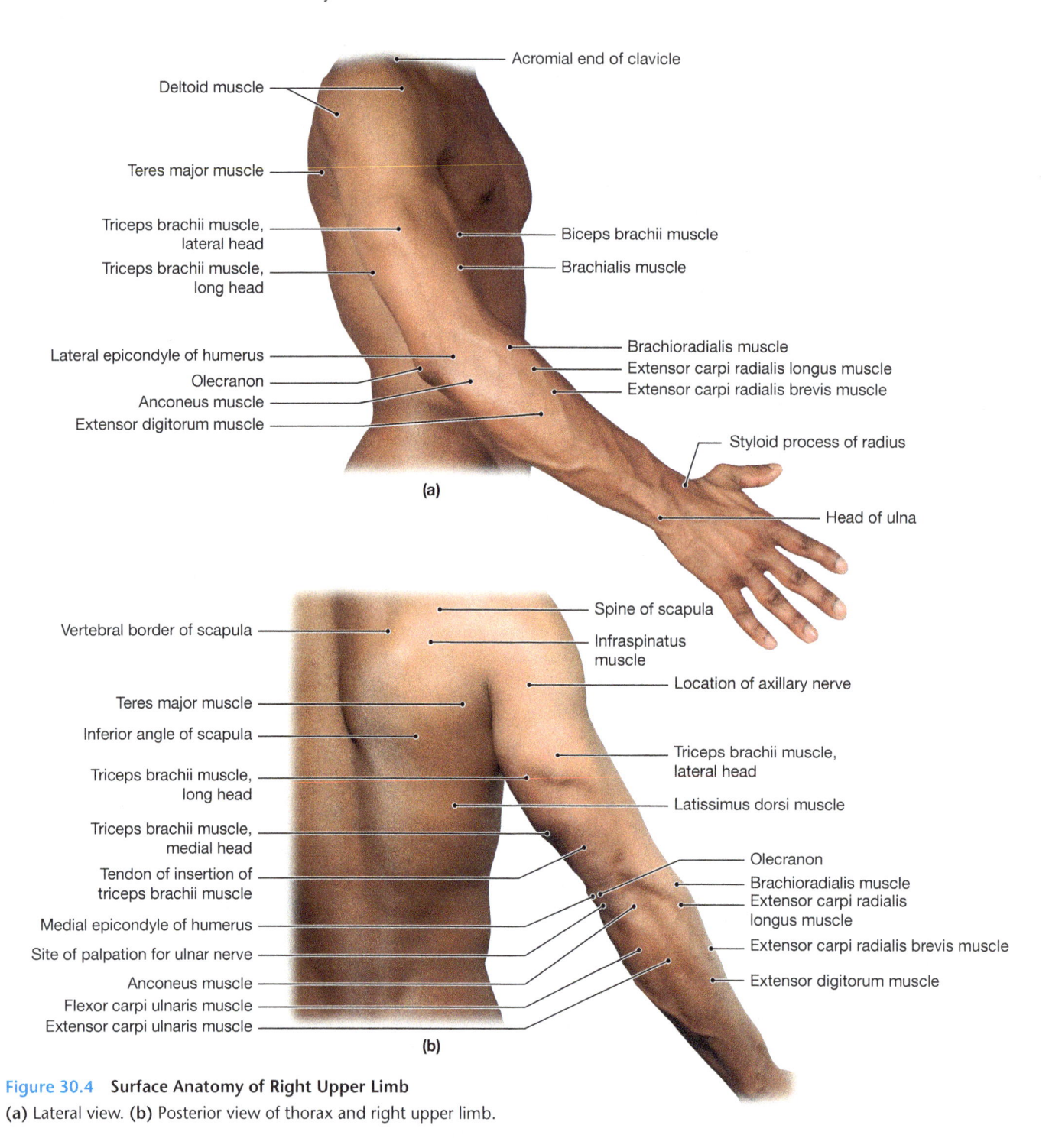

488 EXERCISE 30 Surface Anatomy

Figure 30.4 Surface Anatomy of Right Upper Limb
(a) Lateral view. (b) Posterior view of thorax and right upper limb.

QuickCheck Questions

2.1 Where are the flexor muscles of the wrist located on the forearm?
2.2 What are the three muscles of the arm?

2 Materials

- ☐ Upper limb model
- ☐ Lab partner

Procedures

1. Review the surface anatomy of the shoulder and upper limb in Figures 30.4 and 30.5. Identify the superficial muscles on the upper limb model.
2. On your lab partner or on yourself, identify the muscles of the arm. Try to distinguish between the biceps brachii and the deeper brachialis muscles. Have

EXERCISE 30 Surface Anatomy   489

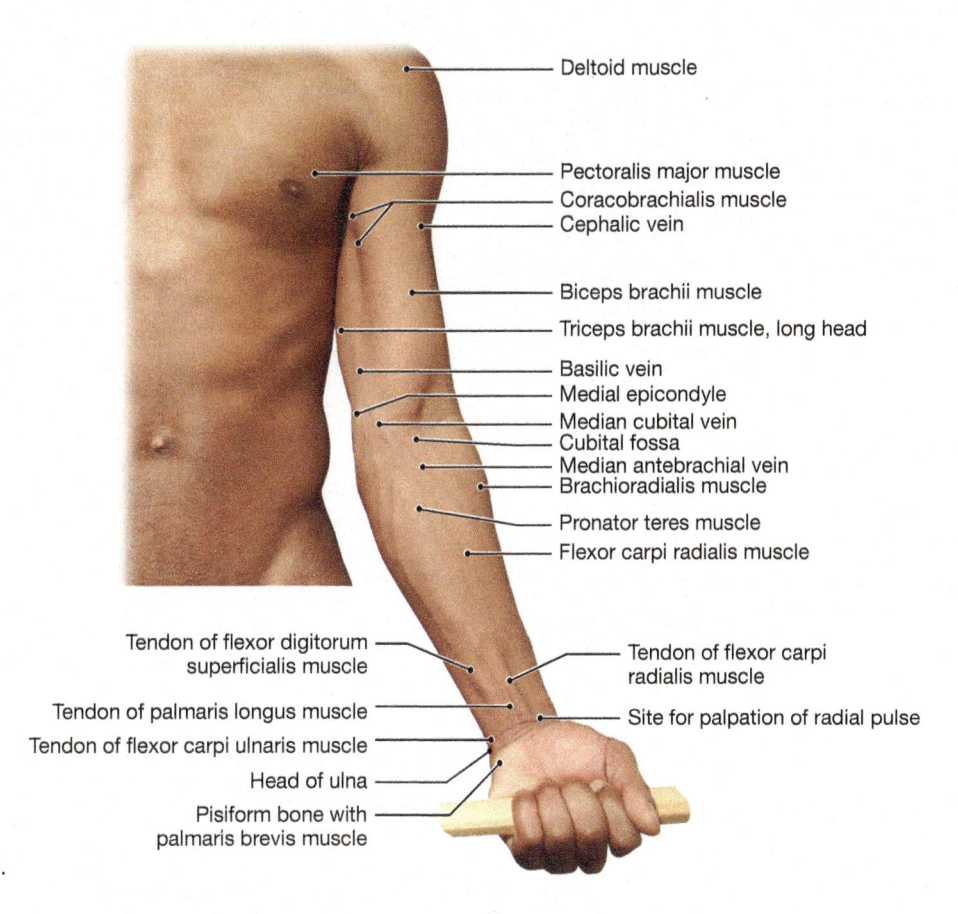

Figure 30.5 Arm, Forearm, and Wrist
Anterior view of left arm, forearm, and wrist.

Labels: Deltoid muscle; Pectoralis major muscle; Coracobrachialis muscle; Cephalic vein; Biceps brachii muscle; Triceps brachii muscle, long head; Basilic vein; Medial epicondyle; Median cubital vein; Cubital fossa; Median antebrachial vein; Brachioradialis muscle; Pronator teres muscle; Flexor carpi radialis muscle; Tendon of flexor digitorum superficialis muscle; Tendon of palmaris longus muscle; Tendon of flexor carpi ulnaris muscle; Head of ulna; Pisiform bone with palmaris brevis muscle; Tendon of flexor carpi radialis muscle; Site for palpation of radial pulse

your partner repeatedly clench her or his hand into a fist and then relax it while you palpate the tendons of the biceps brachii and brachialis muscles.

3. Examine the tendons at your wrist and on the posterior surface of your hand. Determine the action of the muscles of each group of tendons. ■

LAB ACTIVITY 3 Pelvis and Lower Limb

The surface anatomy of the pelvis and thigh is shown in Figure 30.6. The iliac crests mark the superior border of the hips. The gluteus maximus is easily located on the posterior of the pelvis. The rectus femoris, sartorius, vastus lateralis, and vastus medialis muscles are visible on the anterior thigh, and the hamstrings are seen in the posterior view. Adductor muscles are positioned on the medial thigh; the tensor fasciae latae muscle and iliotibial tract are on the lateral thigh.

At the posterior knee, the popliteal fossa is a palpation site for the popliteal artery. The gastrocnemius and soleus muscles of the leg are well-defined on many individuals. The tibialis anterior muscle is along the lateral edge of the tibial diaphysis. The tendons of the fibularis muscles pass immediately posterior to the lateral malleolus of the fibula.

The surface anatomy of the leg and foot is shown in Figure 30.7. The calcaneal tendon inserts on the calcaneus of the foot. Visible on the superior (dorsal) surface of the foot is the tendon of the extensor hallucis longus, which inserts on the big toe, and the tendons of the extensor digitorum longus muscle that insert on toes 2–5.

QuickCheck Questions

3.1 What is the general action of the muscles on the medial thigh?

3.2 What muscles insert on the calcaneal tendon?

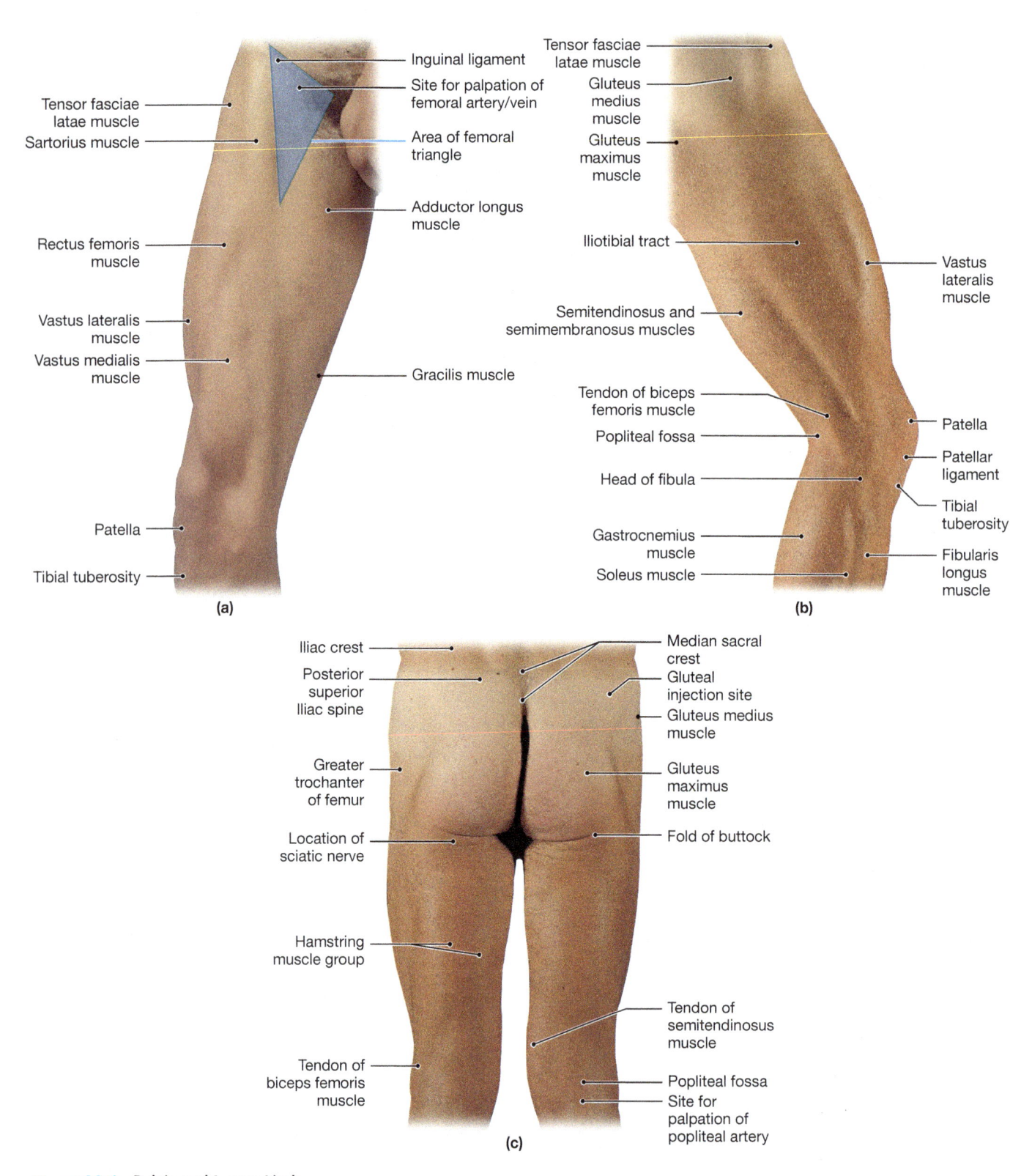

Figure 30.6 Pelvis and Lower Limb
(a) Anterior surface of right thigh. The boundaries of the femoral triangle are the inguinal ligament, medial border of the sartorius, and lateral border of the adductor longus muscle. (b) Lateral surface of right thigh and gluteal region. (c) Posterior surface of right thigh and gluteal region.

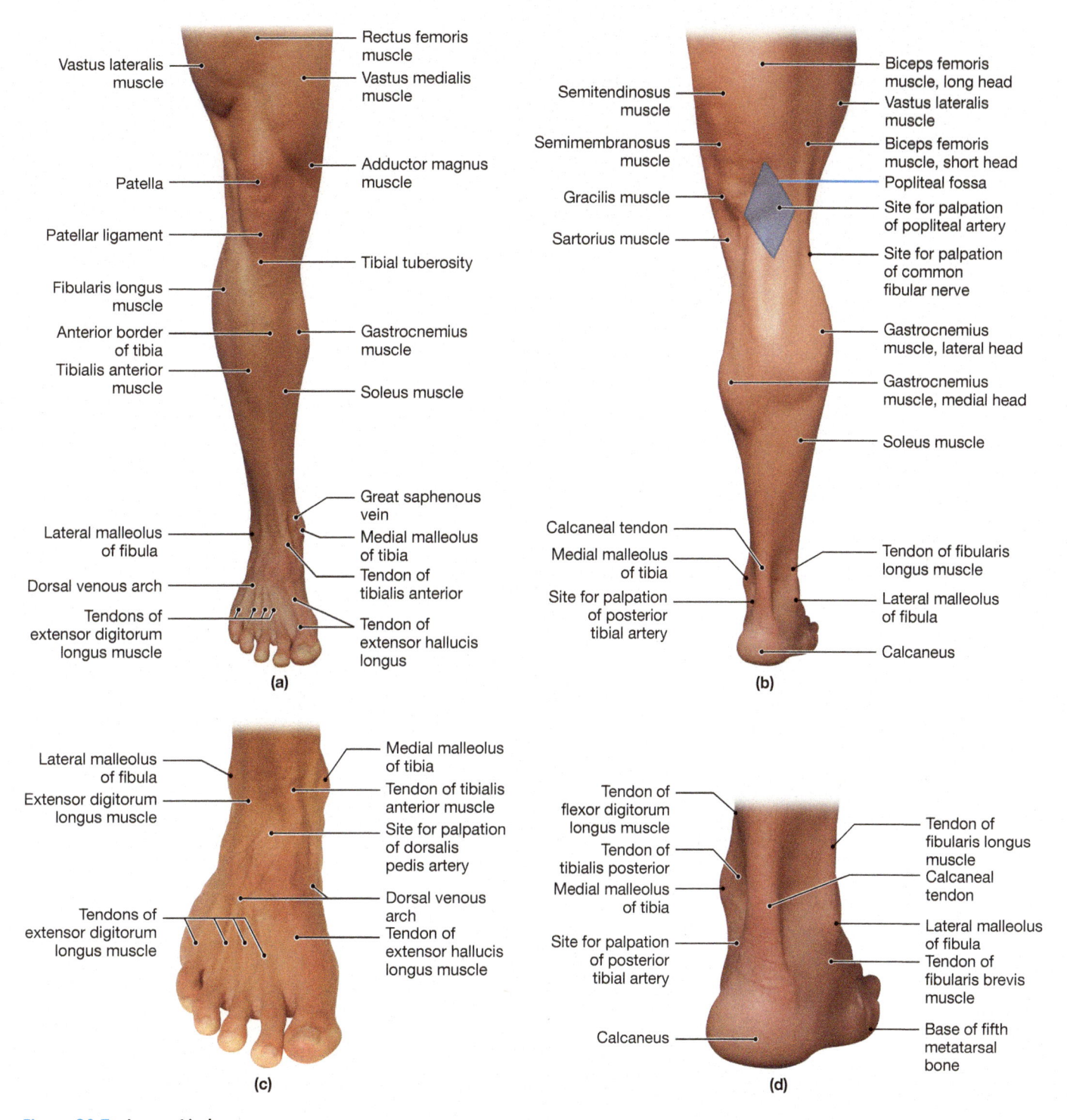

Figure 30.7 Lower Limb
(a) Anterior view of right lower limb. (b) Posterior view of right lower limb. (c) Anterior view of right ankle and foot. (d) Posterior view of right ankle and foot.

3 Materials

- Lower limb model
- Lab partner

Procedures

1. Review the surface anatomy of the pelvis and lower limb in Figures 30.6 and 30.7. Identify the tibialis anterior, gastrocnemius, and soleus muscles on the lower limb model.
2. On your lab partner or on yourself, identify the muscles of the leg. Try to distinguish between the gastrocnemius and the soleus muscles.
3. Examine the tendons at your heel and on the anterior surface of your foot. Determine the action of the muscles of each group of tendons. ■

EXERCISE 30

LAB REPORT

Name _____

Date _____

Section _____

Surface Anatomy

A. Matching

Match each structure in the left column with its correct anatomical region from the right column.

_____ 1. muscle that divides the neck into anterior and posterior cervical triangles

_____ 2. neck region immediately inferior to body of mandible

_____ 3. neck region immediately inferior to suprahyoid triangle

_____ 4. base of neck

_____ 5. neck region that contains thyroid cartilage

_____ 6. olecranon process

_____ 7. xiphoid process

_____ 8. sartorius muscle

_____ 9. gastrocnemius muscle

_____ 10. brachialis

A. elbow
B. thigh
C. superior carotid triangle
D. sternum
E. arm
F. leg
G. suprahyoid triangle
H. submandibular triangle
I. inferior carotid triangle
J. sternocleidomastoid muscle

B. Short-Answer Questions

1. Name at least five surface features of the upper limb.

2. What tendon is visible at the posterior ankle, and which muscles insert on this tendon?

493

EXERCISE 30 — LAB REPORT

C. Analysis and Application

1. Sam is injured on the anterior surface of his arm. What muscles may be involved in his injury?

2. An accident victim is not breathing and needs CPR. What structure of the thorax should you first palpate to properly position your hands for chest compressions? How should you then position your hands in relation to this structure?

DISSECTION EXERCISE 1

Muscles of the Cat

OBJECTIVES

On completion of this exercise, you should be able to:

- Compare and contrast the anatomy of human muscles and cat muscles.
- Identify the muscles of the cat back and shoulder.
- Identify the muscles of the cat neck.
- Identify the muscles of the cat chest and abdomen.
- Identify the muscles of the cat forelimb and hindlimb.

LAB ACTIVITIES

1. Preparing the Cat for Dissection 496
2. Superficial Muscles of the Back and Shoulder 498
3. Deep Muscles of the Back and Shoulder 500
4. Superficial Muscles of the Neck, Abdomen, and Chest 500
5. Deep Muscles of the Chest and Abdomen 505
6. Muscles of the Forelimb 505
7. Muscles of the Thigh 509
8. Muscles of the Lower Hindlimb 512

Some muscles found in four-legged animals are lacking in humans, and some muscles that are fused in humans occur as separate muscles in four-legged animals. Despite these small differences, studying muscles in four-legged animals is an excellent way to learn about human muscles. This exercise on cat muscles is intended to complement your study of human muscles.

The terminology used to describe the position and location of body parts in four-legged animals differs slightly from that used for the human body because four-legged animals move forward head first, with the abdominal surface parallel to the ground. Anatomical position for a four-legged animal is all four limbs on the ground. *Superior* refers to the back (dorsal) surface, and *inferior* relates to the belly (ventral) surface. *Cephalic* means toward the front or anterior, and *caudal* refers to posterior structures.

⚠ **Safety Tip: Cat Dissection Basics**

You *must* practice the highest level of laboratory safety while handling and dissecting the cat. Keep the following guidelines in mind during the dissection:

1. Wear gloves and safety glasses to protect yourself from the fixatives used to preserve the specimen.

DISSECTION EXERCISE 1 Muscles of the Cat

2. Do not dispose of the fixative from your specimen. You will later store the specimen in the fixative to keep the specimen moist and to keep it from decaying.
3. Be extremely careful when using a scalpel or other sharp instrument. Always direct cutting and scissor motion away from yourself to prevent an accident if the instrument slips on moist tissue.
4. Before cutting a given tissue, make sure it is free from underlying or adjacent tissues so that they will not be accidentally severed.
5. Never discard tissue in the sink or trash. Your instructor will inform you of the proper disposal procedure. ▲

LAB ACTIVITY 1 Preparing the Cat for Dissection

Read this entire section and familiarize yourself completely with the exercise before proceeding. You must exercise care to prevent muscle damage as you remove the cat's skin. The degree to which the skin is attached to the underlying connective tissue varies from one part of the body to another. For instance, in the abdominopelvic region the skin is loosely attached to the body, with a layer of subcutaneous fat between the skin and the underlying muscle; whereas in the thigh, the skin is held tightly to underlying muscle by tough fascia.

When it is time to remove the skin, remove it as a single intact piece and save it to wrap the body for storage. The skin wrapping keeps the body moist and keeps the muscles from drying out. If the skin does not cover the body completely, place paper towels dampened with fixative on any uncovered sections. Never rinse the cat with water; doing so will remove the preservative and promote the growth of mold. Always remoisten the body, skin, and other wrappings with fixative prior to storage. Place the cat in the plastic storage bag provided and seal the bag. Fill out a name tag, attach it to the bag, and place the bag in the assigned storage area.

1 Materials

☐ Gloves
☐ Safety glasses
☐ Dissecting tools
☐ Dissecting tray
☐ Preserved cat

Procedures

1. Place the cat dorsal side down on a dissecting tray.
2. With a scalpel, make a short, *shallow* incision on the midline of the ventral surface just anterior to the tail (line 2a in Figure D1.1a). Caution: Be sure to make your incision shallow, for the skin is very thin and too deep a cut will damage the underlying muscles. Using Figure D1.2 as a guide, insert a blunt probe or your finger to separate the skin from the underlying muscle and connective fibers. Place the *blunt-end blade* of a pair of scissors under the skin, and extend your incision anteriorly by cutting the separated skin. Continue to separate with the probe and cut with the scissors until you have an incision that extends up to the neck (line 2b in Figure D1.1a).
3. On either side of the midventral incision, use either your fingers or a blunt probe to gently separate as much skin as possible from the underlying muscle and connective fibers. If you use a scalpel, keep the blade facing the skin, and take care not to damage the musculature.
4. From the ventral surface, make several incision lines. Always, before cutting any skin, use either a blunt probe or your finger to separate the skin from the underlying tissue. The incisions to be made are:
    a. complete encirclement of the neck (line 4a in Figures D1.1a and b)
    b. on the lateral side of each forelimb from the ventral incision to the wrist (4b), completely cutting the skin around the wrists
    c. complete encirclement of the pelvis (4c)
    d. on the lateral side of each hindlimb from the ventral incision to the ankle (4d), completely cutting the skin around the ankles

DISSECTION EXERCISE 1   Muscles of the Cat   497

Figure D1.1 Cat Skinning Incision Lines
(a) Ventral view. (b) Dorsal view.

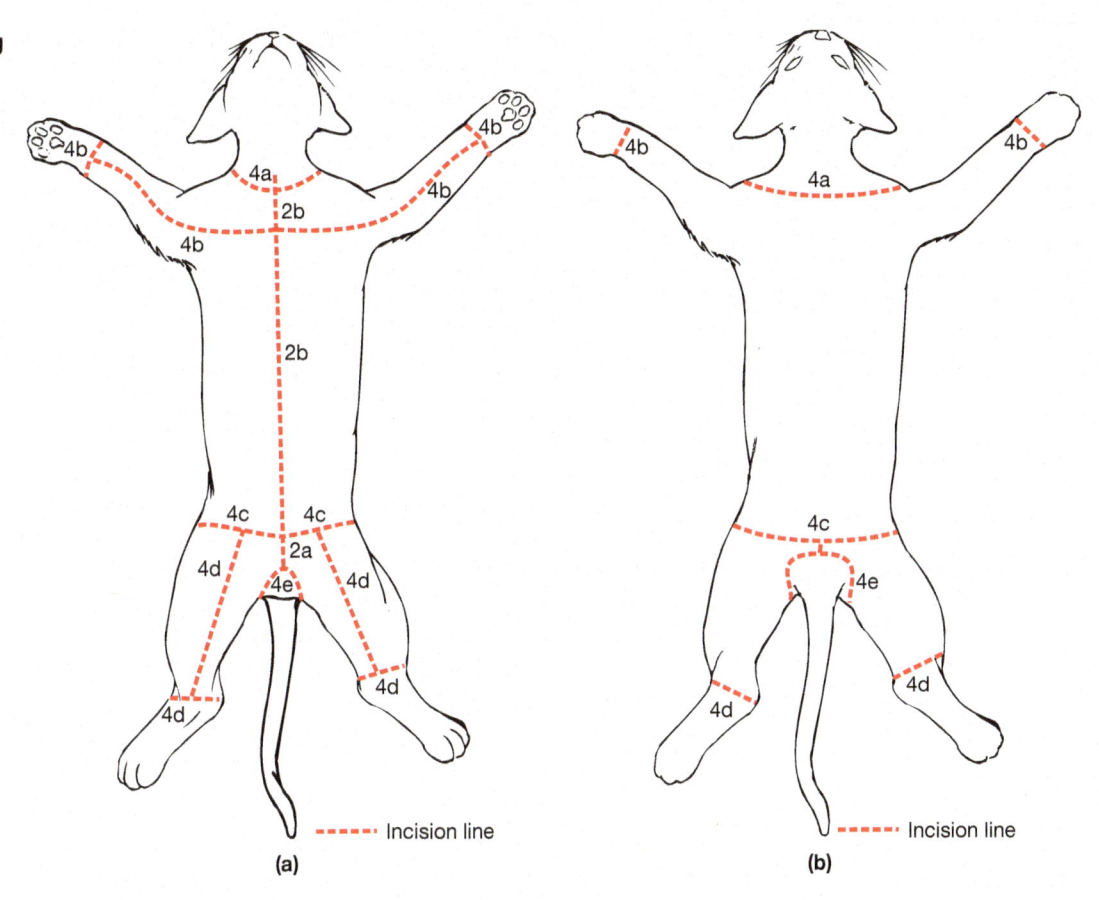

(a)
(b)
------ Incision line

e. encircle the anus, the genital organs, and the tail at its base (4e). Use your fingers to loosen the anal/genital skin as much as possible. If the skin does not come off easily, free it by cutting with the scalpel.

5. Use your fingers to free the entire skin from the underlying connective tissue. Work from the ventral surface toward the dorsal surface at the posterior of the body, and then work on the ventral surface from the posterior end of the body toward the neck. Continue working from the ventral to the dorsal surface, freeing the skin from the underlying connective tissue and other attached structures. The only skin remaining on the body once you have completed this step should be the skin on the head, on the tail, and on the paws below the wrist and ankle joints.

6. Remove all remaining skin from the side of the neck up to the ear. Be careful not to sever the external jugular vein, which is the large blue vein lying on the ventral surface of the neck. Free this vein from the underlying muscles, and clean the connective tissue from the back of the shoulder and the ventral and lateral surfaces of the neck. Do not remove the connective tissue from the midline of the back because this is the origin of the trapezius muscle group.

7. Depending on your specimen, you may observe some or all of the following: thin red or blue latex-injected blood vessels that resemble rubber bands and project between muscles and skin; in female cats, mammary glands between skin and underlying muscle; or cutaneous nerves, which are small, white, cord-like structures that extend from muscles to skin. In male cats, leave undisturbed the skin associated with the external genitalia; you will remove it later when you dissect the reproductive system.

8. Before you begin dissecting muscles, remove as much extraneous fat, hair, platysma, and cutaneous maximus muscle and loose fascia as possible. If at

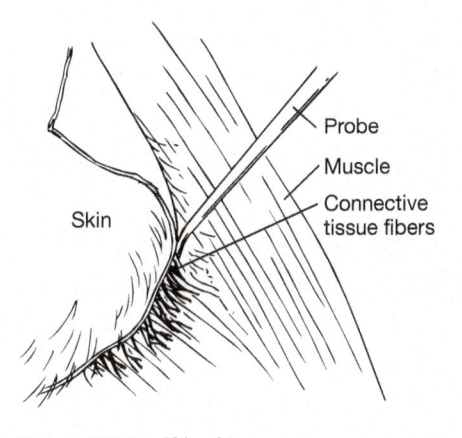

Figure D1.2 Skinning

DISSECTION EXERCISE 1 Muscles of the Cat

any time you are confused about the nature of any of the material, check with your instructor before removing it.

9. To store your specimen, wrap it in the skin and moisten it with fixative. Use paper towels if necessary to cover the entire specimen. Return it to the storage bag and seal the bag securely. Label the bag with your name and place it in the storage area as indicated by your instructor. ■

LAB ACTIVITY 2 Superficial Muscles of the Back and Shoulder

Begin your dissection with the superficial muscles in the upper back region between the two forelimbs (Figure D1.3).

2 Materials

- Gloves
- Safety glasses
- Dissecting tools
- Dissecting tray
- Preserved cat, skin removed

Procedures

1. Locate the **trapezius group** of muscles that cover the dorsal surface of the neck and the scapula. The single trapezius in humans occurs as three distinct muscles in cats. A prefix describes the insertion of each muscle:
    a. The **spinotrapezius** is the most posterior trapezius muscle. It originates on the spinous processes of posterior thoracic vertebrae and inserts on the scapular spine. It pulls the scapula dorsocaudad.
    b. The **acromiotrapezius** is a large muscle anterior to the spinotrapezius. The almost-square acromiotrapezius originates on the spinous processes of cervical and anterior thoracic vertebrae and inserts on the scapular spine. It holds the scapula in place.
    c. The **clavotrapezius** is a broad muscle anterior to the acromiotrapezius. It originates on the lambdoidal crest and axis and inserts on the clavicle. It draws the scapula cranially and dorsally.

2. The large, flat muscle posterior to the trapezius group is the **latissimus dorsi**. Its origin is on the spines of thoracic and lumbar vertebrae, and it inserts on the medial side of the humerus. The latissimus dorsi acts to pull the forelimb posteriorly and dorsally.

3. The **levator scapulae ventralis** is a flat, strap-like muscle that lies on the side of the neck between the clavotrapezius and the acromiotrapezius. The occipital bone and atlas are its origin, and the vertebral border of the scapula is its insertion. This muscle, which does not have a counterpart in humans, pulls the scapula toward the head.

4. The **deltoid group** comprises the shoulder muscles lateral to the trapezius group. The following three cat muscles are equivalent to the single deltoid muscle in the human:
    a. The **spinodeltoid** is the most posterior of the deltoid group. It originates on the scapula ventral to the insertion of the acromiotrapezius and inserts on the proximal humerus. The action of this muscle is to flex the humerus and rotate it laterally.
    b. The **acromiodeltoid** is the middle muscle of the deltoid group. It originates on the acromion process of the scapula deep to the levator scapulae ventralis and inserts on the proximal end of the humerus. The acromiodeltoid flexes the humerus and rotates it laterally.
    c. The **clavodeltoid**, also called the clavobrachialis, originates on the clavicle and inserts on the ulna. It is a continuation of the clavotrapezius below the clavicle and extends down the forelimb from the clavicle. It functions to flex the humerus. ■

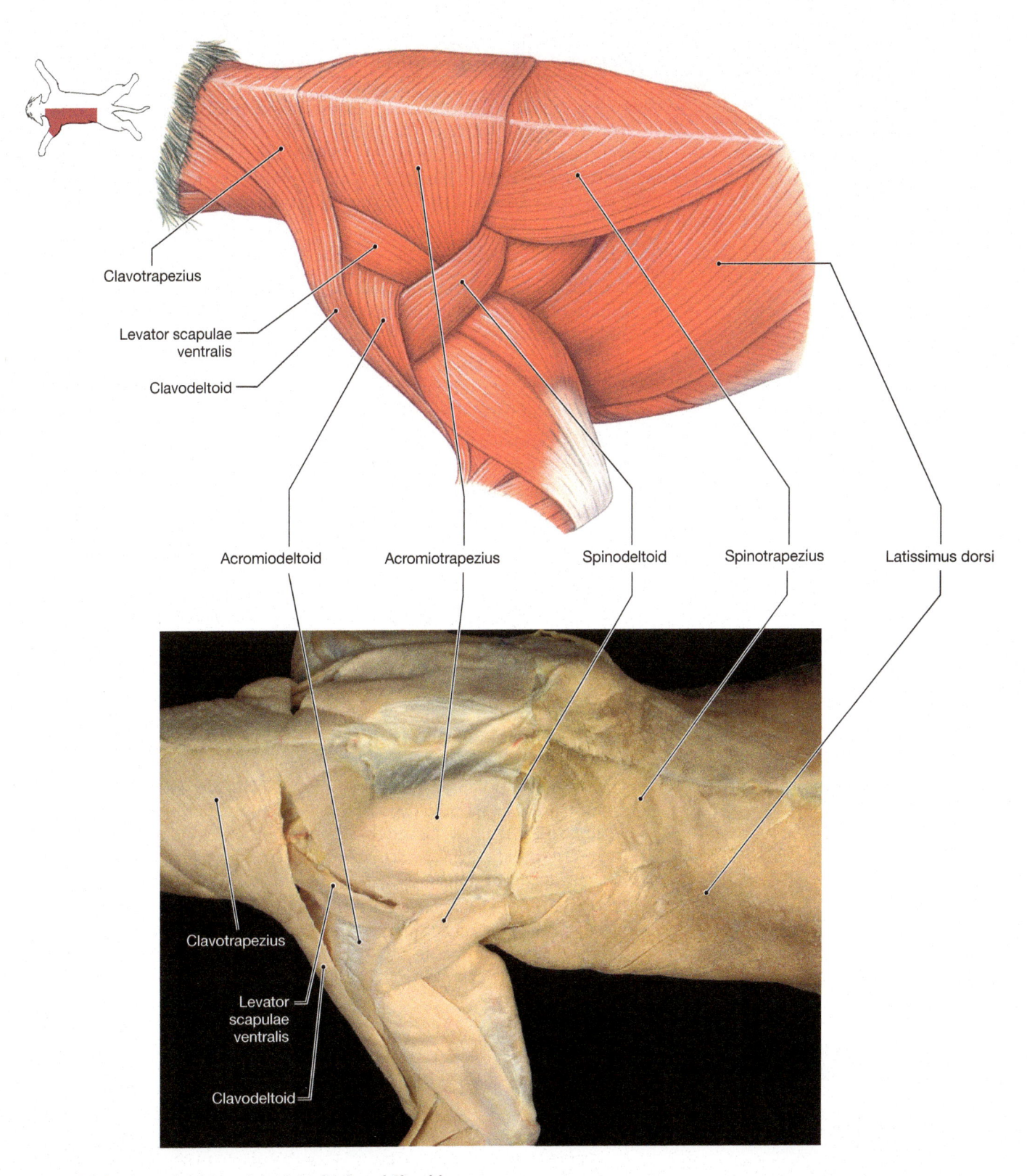

Figure D1.3 Superficial Muscles of the Back and Shoulder
Illustration and photo.

499

LAB ACTIVITY 3 — Deep Muscles of the Back and Shoulder

Working on the left side of the cat, expose deeper muscles of the shoulder and back by transecting (cutting) the three muscles of the trapezius group and the latissimus dorsi muscles in the middle. Reflect these overlying muscles to expose the muscles underneath. (To *reflect* a muscle means to fold it back out of the way.) Use Figure D1.4 as a guide as you look at the following deep muscles.

Materials

- Gloves
- Safety glasses
- Dissecting tools
- Dissecting tray
- Preserved cat, skin removed

Procedures

1. Deep to the acromiotrapezius, the **supraspinatus** occupies the lateral surface of the scapula in the supraspinous fossa. It originates on the scapula and inserts on the humerus. It functions to extend the humerus.

2. On the lateral surface of the scapula, the **infraspinatus** occupies the infraspinous fossa. It originates on the scapula, inserts on the humerus, and causes the humerus to rotate laterally.

3. The **teres major** occupies the axillary border of the scapula, where it has its origin. It inserts on the proximal end of the humerus, and acts to rotate the humerus and draw this bone posteriorly.

4. The **rhomboideus group** connects the spinous processes of cervical and thoracic vertebrae with the vertebral border of the scapula. The muscles of this group hold the dorsal part of the scapula to the body.
   a. The posterior muscle of this group is the **rhomboideus major**. This fan-shaped muscle originates on the spinous processes and ligaments of posterior cervical and anterior thoracic vertebrae and inserts on the dorsal posterior angle of the scapula. It draws the scapula dorsally and anteriorly.
   b. The **rhomboideus minor** is just anterior to the rhomboideus major. Its origin is on the spines of posterior cervical and anterior thoracic vertebrae. This muscle inserts along the vertebral border of the scapula and draws the scapula forward and dorsally.
   c. The narrow, ribbon-like **rhomboideus capitis** is the anterolateral muscle of the rhomboideus group. It originates on the spinous nuchal line and inserts on the vertebral border of the scapula. It elevates and rotates the scapula. The rhomboideus capitis muscle does not have a counterpart in humans.

5. The **splenius** is a broad, flat, thin muscle that covers most of the lateral surface of the cervical and thoracic vertebrae. It is deep to the rhomboideus capitis. The splenius has its origin on the spines of thoracic vertebrae and its insertion on the superior nuchal line of the skull. It acts to both turn and raise the head. ■

LAB ACTIVITY 4 — Superficial Muscles of the Neck, Abdomen, and Chest

Materials

- Gloves
- Safety glasses
- Dissecting tools
- Dissecting tray
- Preserved cat, skin removed

Procedures

1. Locate the following muscles of the neck:
   a. The **sternomastoid** is a large, V-shaped muscle between the sternum and the head (Figure D1.5). This muscle originates on the manubrium of the sternum and passes obliquely around the neck to insert on the superior nuchal line and on the mastoid process. It turns and depresses the head. This muscle is the counterpart of the sternocleidomastoid in humans.

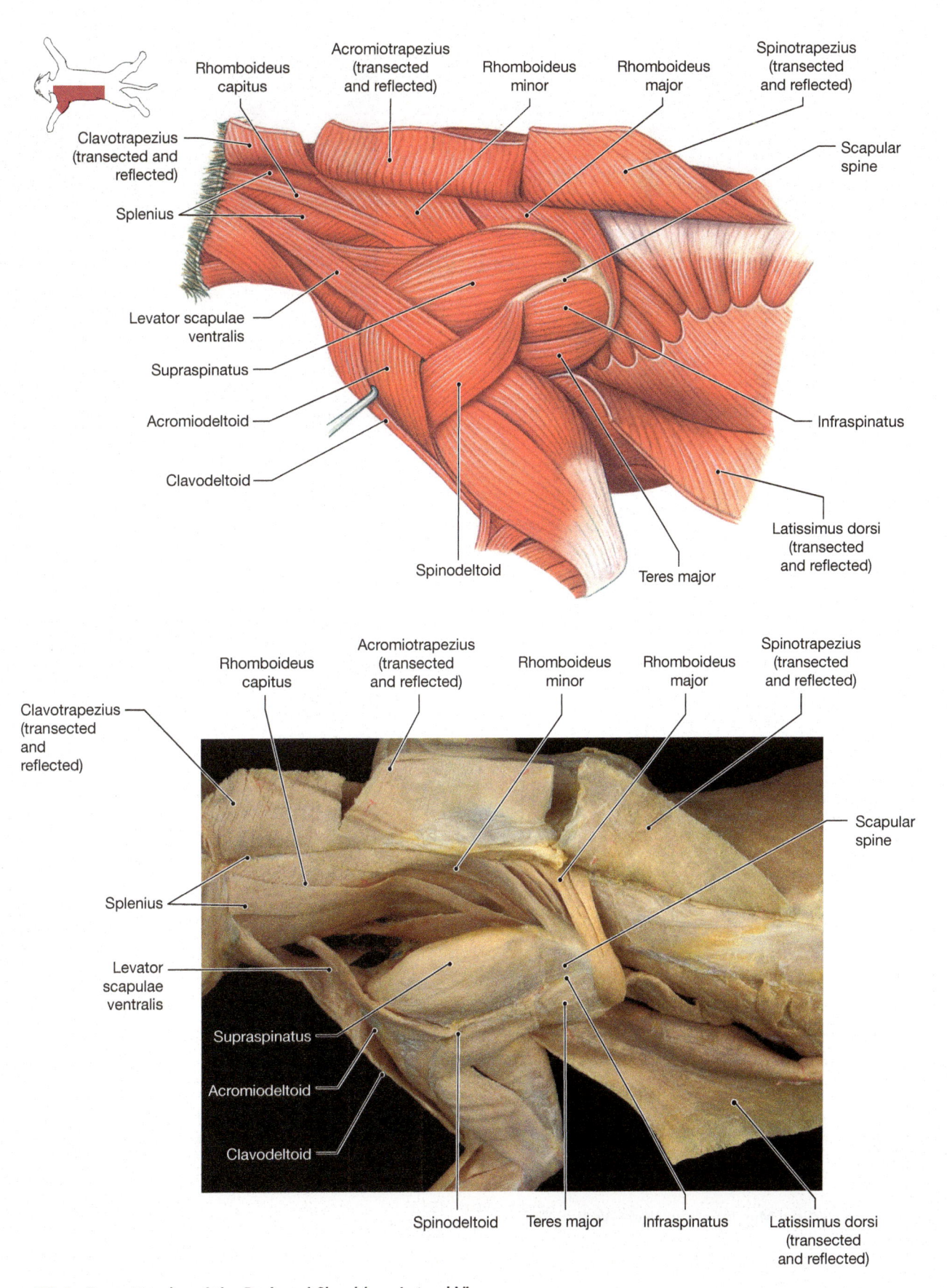

Figure D1.4 Deep Muscles of the Back and Shoulder—Lateral View
Illustration and photo.

501

(a) Superficial neck muscles

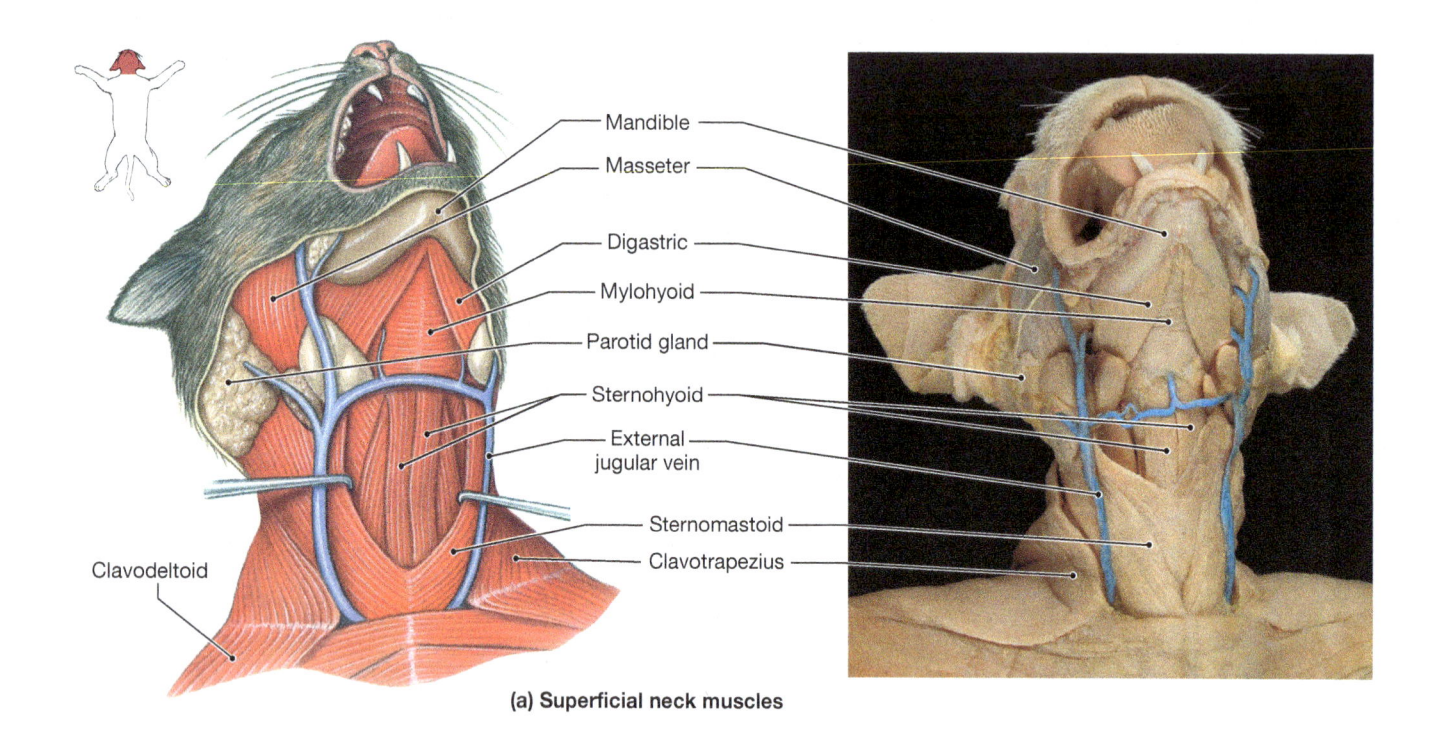

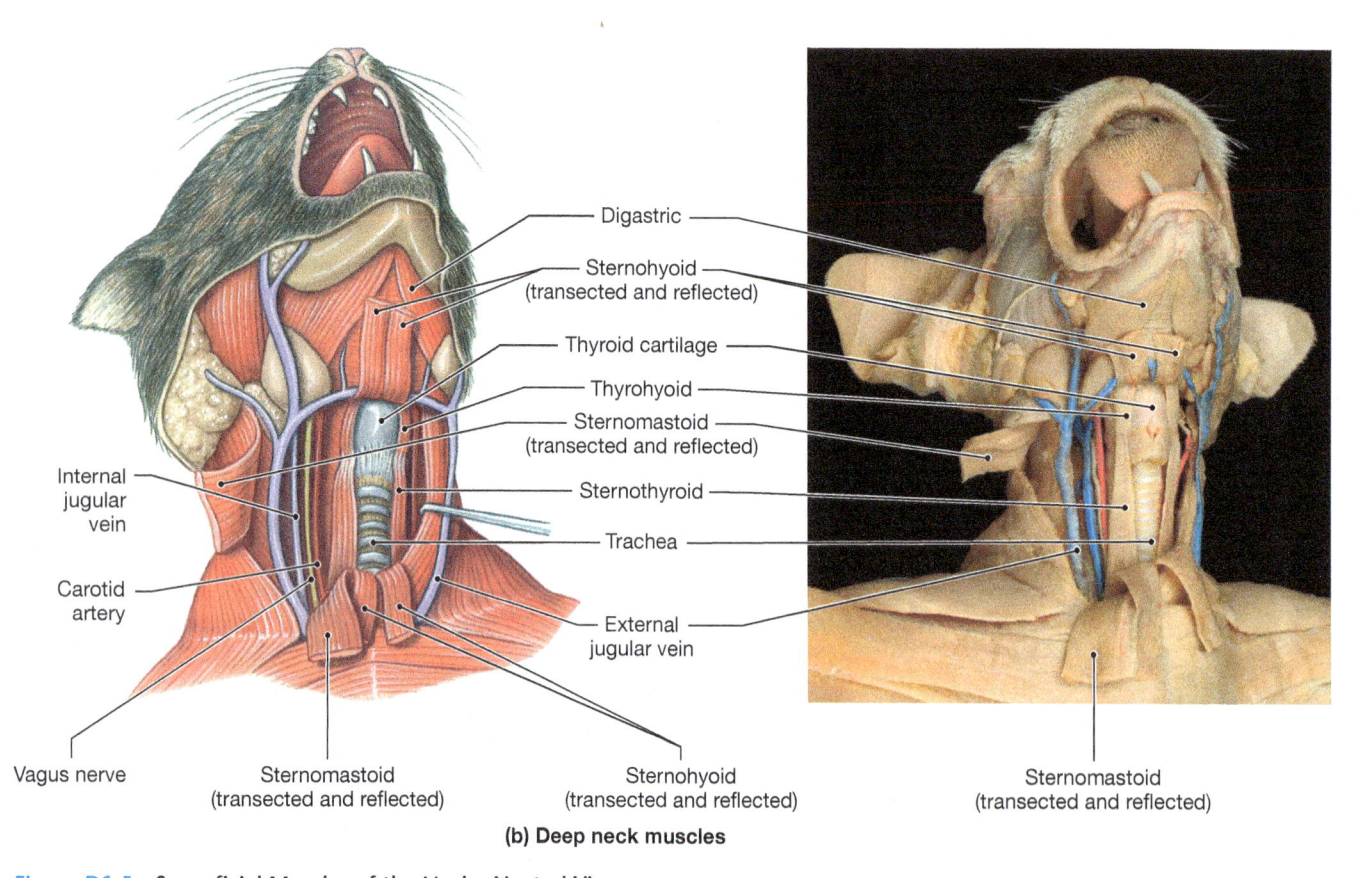

(b) Deep neck muscles

Figure D1.5 Superficial Muscles of the Neck—Ventral View
Illustrations and photos.

b. The **sternohyoid** is a narrow muscle that lies over the larynx, along the midventral line of the neck. Its origin is the costal cartilage of the first rib, and it inserts on the hyoid bone. It acts to depress the hyoid bone.

c. The **digastric** is a superficial muscle that extends along the inner surface of the mandible. It originates on the occipital bone and mastoid process and functions as a depressor of the mandible.

d. The **mylohyoid** is a superficial muscle that runs transversely along the midline and passes deep to the digastrics. It originates on the mandible and inserts on the hyoid bone, and its function is to raise the floor of the mouth.

e. The **masseter** is the large muscle mass anteroventral to the parotid gland at the angle of the jaw. This cheek muscle originates on the zygomatic bone. It inserts on the posterolateral surface of the dentary bone and elevates the mandible.

2. Three layers of muscle form the lateral abdominal wall and insert on a fourth muscle along the ventral midline. Follow the steps below to locate these muscles. Because the three layers are very thin, be careful as you separate them. Collectively, these muscles act to compress the abdomen. They are all shown in Figure D1.6.

    a. The **external oblique** is the most superficial of the lateral abdominal muscles. It originates on posterior ribs and lumbodorsal fascia and *inserts on the linea alba* from the sternum to the pubis. Its fibers run from anterodorsal to posteroventral, and it acts to compress the abdomen.

    b. Cut and reflect the external oblique to view the **internal oblique**, which lies deep to the external oblique and is the second of the three lateral layers of the abdominal wall. The fibers of the internal oblique run perpendicular to those of the external oblique in a posterodorsal-to-anteroventral orientation. The internal oblique originates on the pelvis and lumbodorsal fascia and *inserts on the linea alba,* where it functions to compress the abdomen.

    c. Cut and reflect the internal oblique to see the third muscle layer of the abdominal wall, the **transverse abdominis**. This layer is deep to the internal oblique, has fibers that run transversely across the abdomen, and forms the deepest layer of the abdominal wall. It originates on the posterior ribs, lumbar vertebrae, and ilium and *inserts on the linea alba.* It acts to compress the abdomen.

    d. The **rectus abdominis** is the long, ribbon-like muscle alongside the midline of the ventral surface of the abdomen. It originates on the pubic symphysis and inserts on the costal cartilage. It compresses the internal organs of the abdomen.

3. The **pectoralis group** consists of the large muscles that cover the ventral surface of the chest (Figure D1.6). They arise from the sternum and mostly attach to the humerus. There are four subdivisions in the cat but only two in humans. In the cat, the relatively large degree of fusion gives the pectoral muscles the appearance of a single muscle. This fusion makes the chest rather difficult to dissect, because the muscles do not separate from one another easily.

    a. The **pectoantebrachialis** is the most superficial of the pectoral muscles. It originates on the manubrium, inserts on fascia of the forearm, and adducts the humerus. It has no counterpart in humans.

    b. The broad, triangular **pectoralis major** is posterior to the pectoantebrachialis. The origin of the pectoralis major is on the sternum, and its

504 DISSECTION EXERCISE 1 Muscles of the Cat

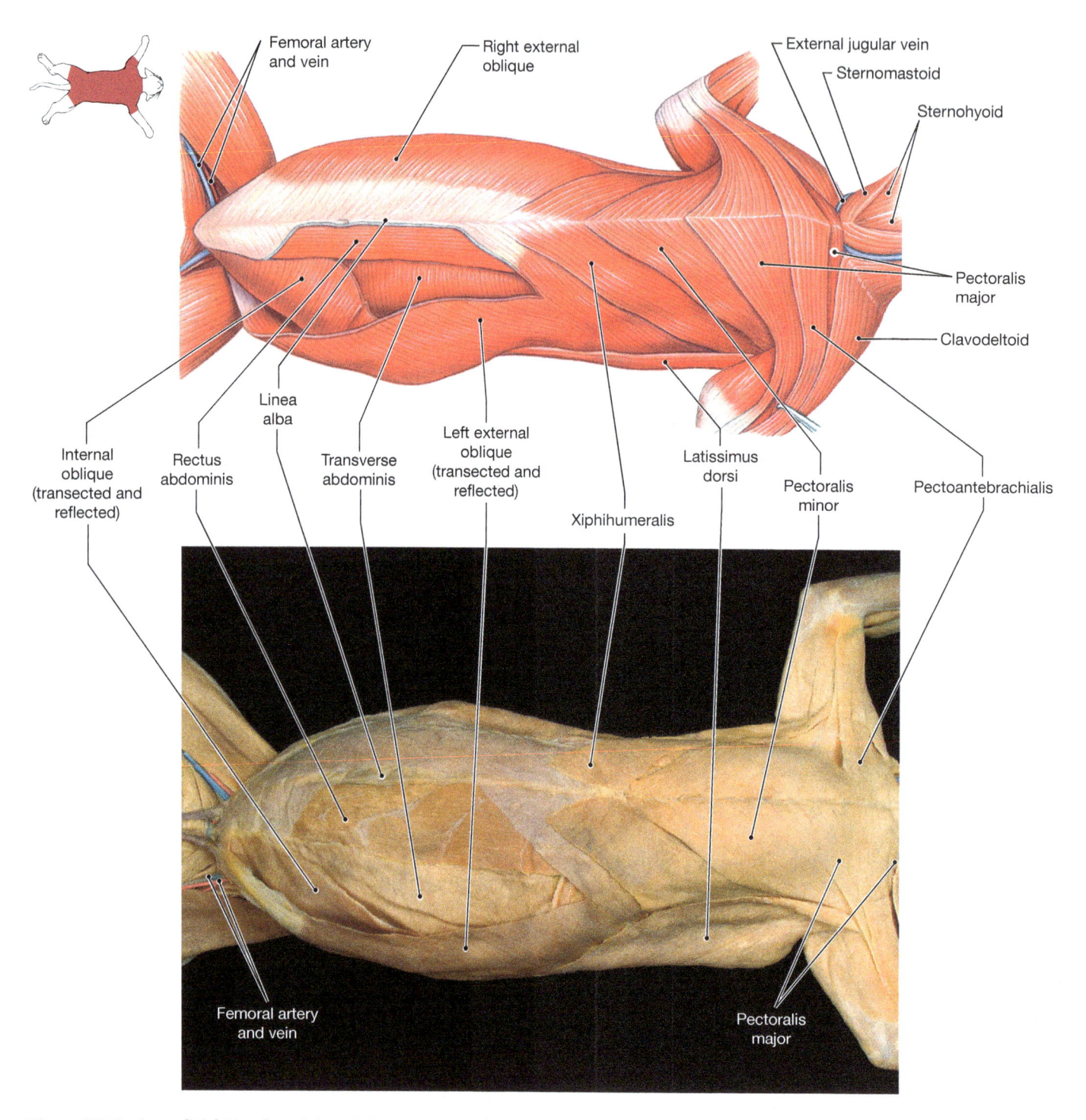

Figure D1.6 Superficial Muscles of the Abdomen—Ventral View Illustration and photo.

insertion is on the posterior humerus. It functions to adduct the humerus.

c. Posterior to the pectoralis major is the **pectoralis minor**. This is the broadest and thickest muscle of the group. It extends posteriorly to the pectoralis major. It originates on the sternum, inserts near the proximal end of the humerus, and acts to adduct the humerus.

d. The thin **xiphihumeralis** is posterior to the posterior edge of the pectoralis minor. The xiphihumeralis originates on the xiphoid process of the sternum and inserts by a narrow tendon on the humerus. It adducts and helps rotate the humerus. Humans do not have a counterpart to the xiphihumeralis muscle. ∎

LAB ACTIVITY 5 — Deep Muscles of the Chest and Abdomen

Materials

- Gloves
- Safety glasses
- Dissecting tools
- Dissecting tray
- Preserved cat, skin removed

Procedures

1. On the lateral thoracic wall, just ventral to the scapula, a large, fan-shaped muscle, the **serratus ventralis**, originates by separate bands from the ribs (Figure D1.7). It passes ventrally to the scapula and inserts on the vertebral border of the scapula. It is homologous to the serratus anterior and the levator scapulae muscles of humans and functions to pull the scapula toward the head and ventrally.

2. The **serratus dorsalis** is a serrated muscle that lies medial to the serratus ventralis. Its origin is along the mid-dorsal cervical, thoracic, and lumbar regions, and it inserts on the ribs. It acts to pull the ribs craniad and outward.

3. The **scalenus medius** is a three-part muscle on the lateral surface of the trunk. The bands of muscle unite anteriorly. The scalenus medius originates on the ribs and inserts on the transverse processes of the cervical vertebrae. It acts to flex the neck and draw the ribs anteriorly.

4. The **intercostal group** consists of two layers of muscles between the ribs that move the ribs during respiration.

    a. The **external intercostals** are deep to the external oblique. The muscle fibers of the external intercostals run obliquely from the posterior border of one rib to the anterior border of the next rib. Their origin is on the caudal border of one rib, and their insertion is on the cranial border of the next rib. They lift the ribs during inspiration.

    b. The **internal intercostals** are toward the midline and deep to the external intercostals. (They are not visible in Figure D1.7.) Bisect one external intercostal muscle to expose an internal intercostal. The fibers of the intercostals run at oblique angles: the internal from medial to lateral and the external from lateral to medial. The origin of the internal intercostals is on the superior border of the rib below, and they insert on the inferior border of the rib above. They draw adjacent ribs together and depress ribs during active expiration. ∎

LAB ACTIVITY 6 — Muscles of the Forelimb

Materials

- Gloves
- Safety glasses
- Dissecting tools
- Dissecting tray
- Preserved cat, skin removed

Procedures

1. Observe the **epitrochlearis**, which is a broad, flat muscle covering the medial surface of the upper forelimb (Figure D1.8). This muscle appears to be an extension of the latissimus dorsi and originates on the fascia of the latissimus dorsi. It inserts on the olecranon process of the ulna, where it acts to extend the forelimb. It is not found in humans.

2. The **biceps brachii** is a convex muscle interior to the pectoralis major and pectoralis minor on the ventromedial surface of the humerus. It originates on

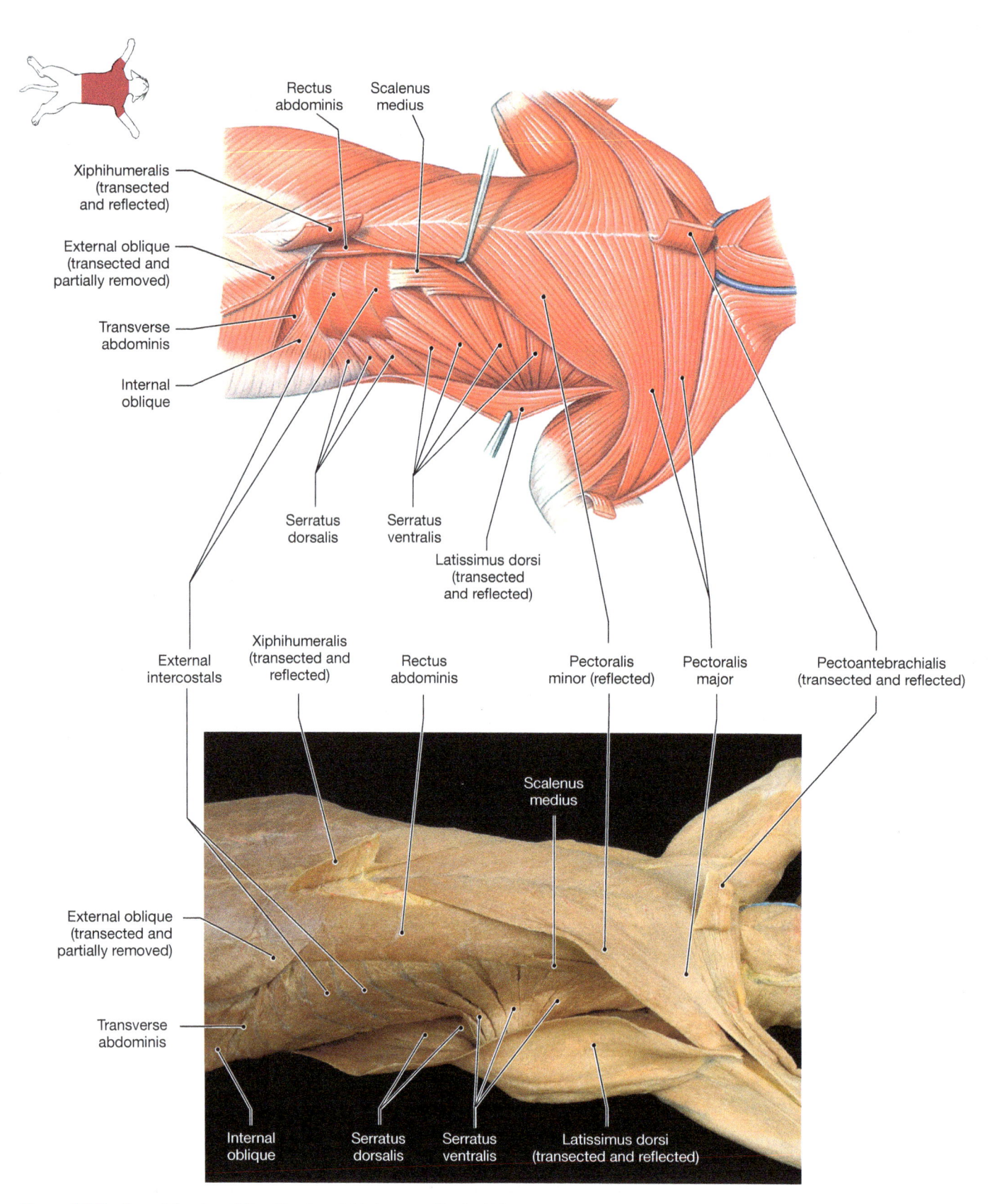

Figure D1.7 Deep Muscles of the Chest and Abdomen—Ventral View
Illustration and photo.

DISSECTION EXERCISE 1   Muscles of the Cat   507

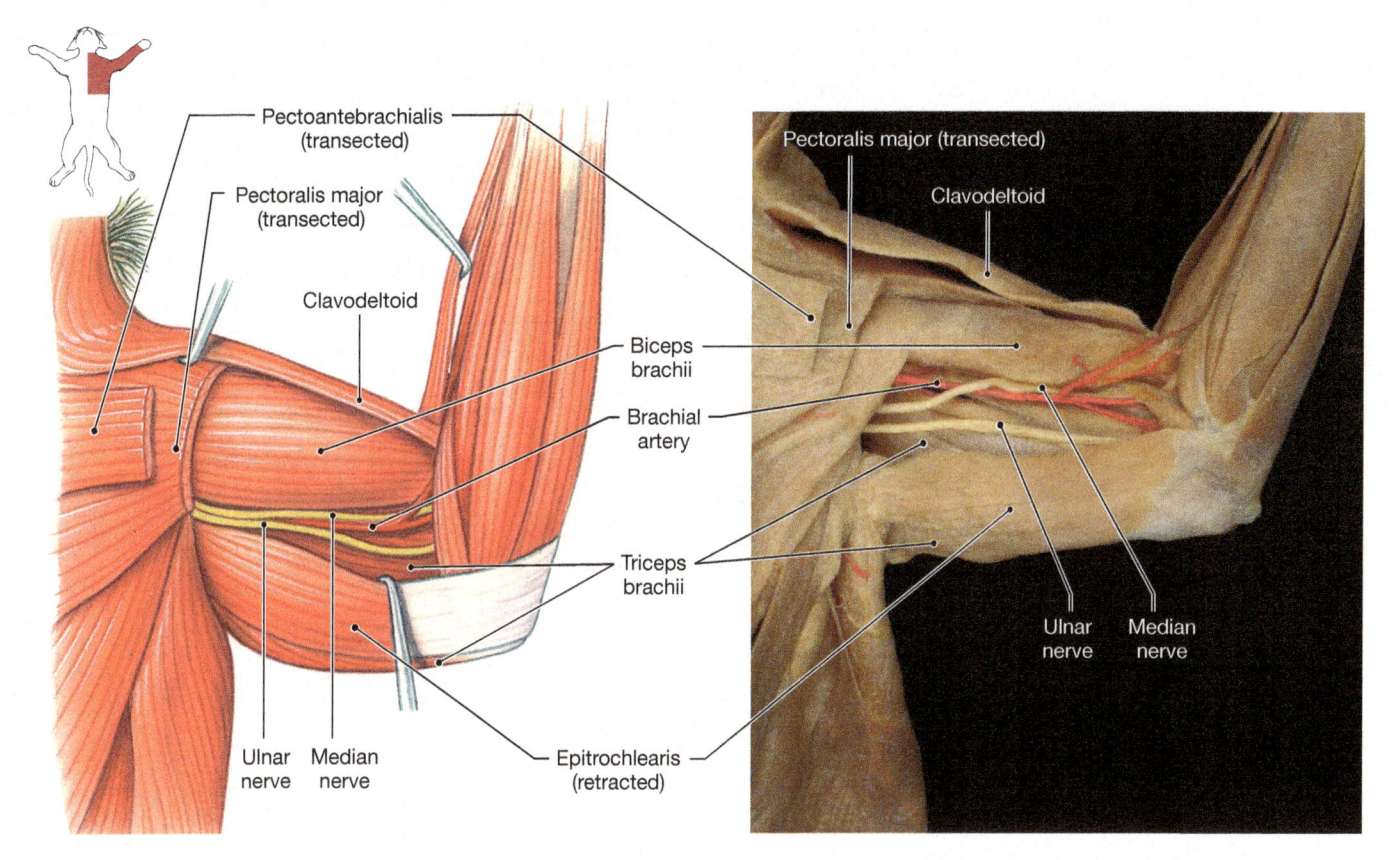

Figure D1.8   Muscles of the Upper Forelimb—Medial View
Illustration and photo.

the scapula and inserts on the radial tuberosity near the proximal end of the radius. It functions as a flexor of the forelimb.

3. Observe the ribbon-like **brachioradialis** muscle on the lateral surface of the humerus (Figure D1.9). Its origin is on the mid-dorsal border of the humerus, and its insertion is on the distal end of the radius on the styloid process. It supinates the front paw (manus).

4. The **extensor carpi radialis** is deep to the brachioradialis. There are two parts to this extensor muscle: the shorter, triangular **extensor carpi radialis brevis** and the superficial **extensor carpi radialis longus**. Both originate on the lateral surface of the humerus above the lateral epicondyle and insert on the bases of the second and third metacarpals. Both extensor carpi radialis muscles cause extension of the paw.

5. The **pronator teres**, a narrow muscle next to the extensor carpi radialis, runs from its point of origin on the medial epicondyle of the humerus and gets smaller as it approaches insertion on the radius. It rotates the radius for pronation.

6. The **flexor carpi radialis**, found adjacent to the pronator teres, originates on the medial epicondyle of the humerus and inserts on the second and third metacarpals. It acts to flex the wrist.

7. The large, flat muscle in the center of the medial surface of the forelimb is the **palmaris longus**. Its origin is on the medial epicondyle of the humerus, and it inserts on all digits. The palmaris longus flexes the digits.

8. The flat muscle on the posterior edge of the forelimb is the **flexor carpi ulnaris**. It arises from a two-headed origin—on the medial epicondyle of the humerus and on the olecranon process on the ulna—and inserts by a single tendon on the ulnar side of the carpals. It is a flexor of the wrist.

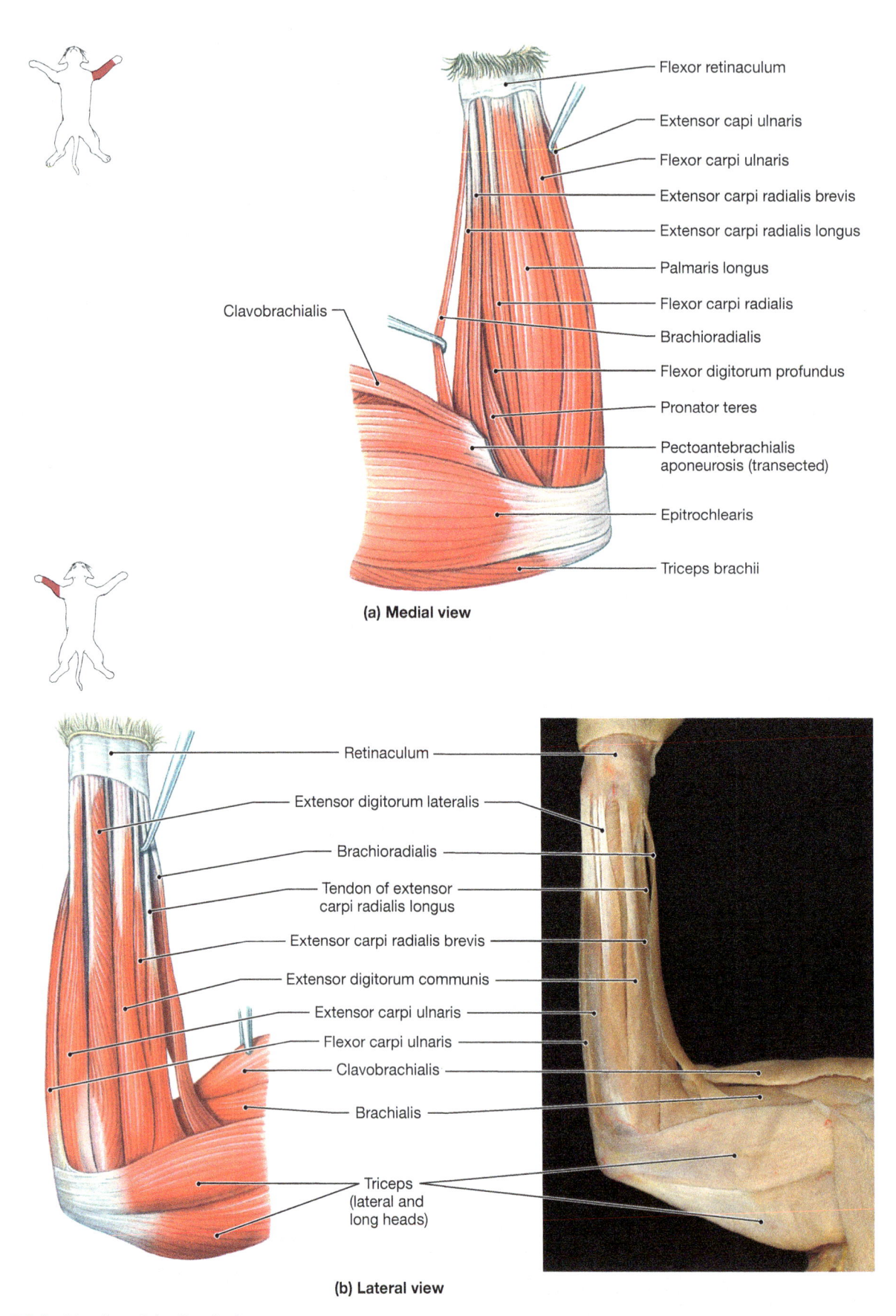

Figure D1.9 Muscles of the Forelimb
Illustrations and photos.

508

9. The **brachialis** is on the ventrolateral surface of the humerus (Figure D1.9b). This muscle originates on the lateral side of the humerus and inserts on the proximal end of the ulna. It functions to flex the forelimb.

10. The **triceps brachii** is the largest superficial muscle of the upper forelimb. It is located on the lateral and posterior surfaces of the forelimb. As the name implies, the triceps brachii has its origins on three heads. The **long head** is the large muscle mass on the posterior surface and originates on the lateral border of the scapula. The **lateral head**, which lies next to the long head on the lateral surface, originates on the deltoid ridge of the humerus. The **small medial** head lies deep to the lateral head and originates on the shaft of the humerus. All three heads have a single insertion on the olecranon process of the ulna. The function of the triceps brachii is to extend the forelimb. ■

LAB ACTIVITY 7 Muscles of the Thigh

The hindlimb of cats and other four-legged animals looks different from the human leg. Cats have long metatarsals in the arch of the foot. This makes the ankle very high, and some people might even mistake the ankle for the knee joint. Place your hand on the cat's pelvis and then slide your hand onto the thigh. Locate the distal joint, the knee, and then the ankle.

7 Materials

- ☐ Gloves
- ☐ Safety glasses
- ☐ Dissecting tools
- ☐ Dissecting tray
- ☐ Preserved cat, skin removed

Procedures

1. The **sartorius** is a wide, superficial muscle covering the anterior half of the medial aspect of the thigh (Figure D1.10). It originates on the ilium and inserts on the tibia. The sartorius adducts and rotates the femur and extends the tibia.

2. The **gracilis** is a broad muscle that covers the posterior portion of the medial aspect of the thigh. The gracilis originates on the ischium and pubic symphysis and inserts on the medial proximal surface of the tibia. It adducts the thigh and draws it posteriorly.

3. The **tensor fasciae latae** is a triangular muscle located posterior to the sartorius (Figures D1.10b and D1.11). This muscle originates on the crest of the ilium and inserts into the fascia lata. The tensor fasciae latae extends the thigh.

4. Posterior to the tensor fasciae latae lies the **gluteus medius** (Figure D1.11b). This is the largest of the gluteus muscles, and it originates on both the ilium and the transverse processes of the last sacral and first caudal vertebrae. It inserts on the greater trochanter of the femur and acts to abduct the thigh.

    Just posterior to the gluteus medius, locate the **gluteus maximus,** a small, triangular hip muscle. It originates on the transverse processes of the last sacral and first caudal vertebrae and inserts on the proximal femur. It abducts the thigh.

5. The **adductor femoris,** deep to the gracilis, is a large muscle with an origin on the ischium and pubis. It inserts on the femur and acts to adduct the thigh.

    Anterior to the adductor femoris is the **adductor longus** muscle, visible in Figure D1.10. It is a narrow muscle that originates on the ischium and pubis and inserts on the proximal surface of the femur. It adducts the thigh.

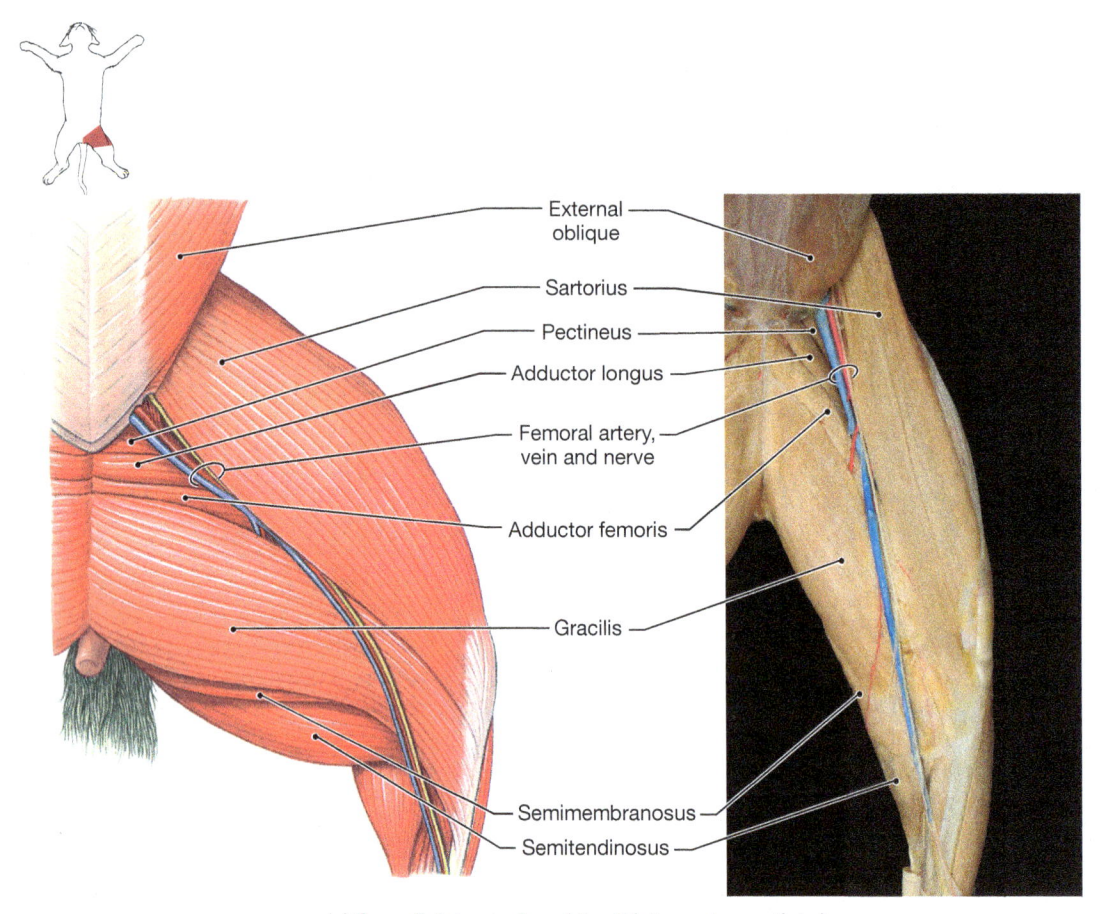

(a) Superficial muscles of the thigh, ventromedial view

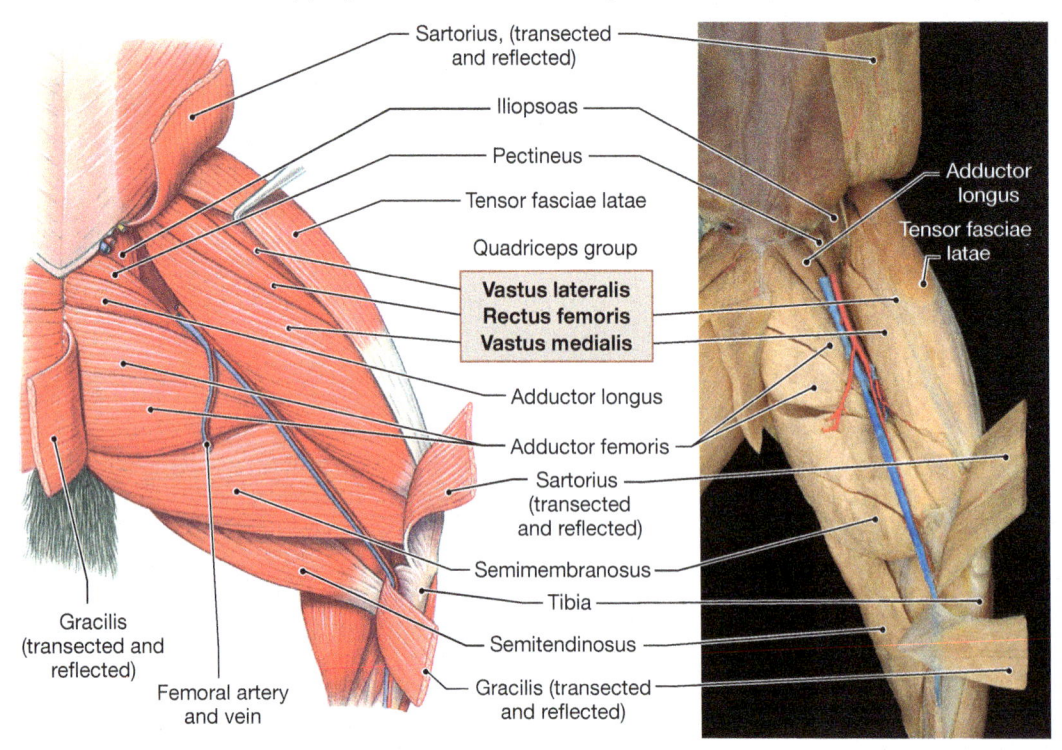

(b) Deep muscles of the thigh, ventromedial view

Figure D1.10 Muscles of the Thigh—Ventromedial View
Illustrations and photos.

DISSECTION EXERCISE 1 Muscles of the Cat 511

(a) Superficial muscles of the thigh, lateral view

(b) Deep muscles of the thigh, lateral view

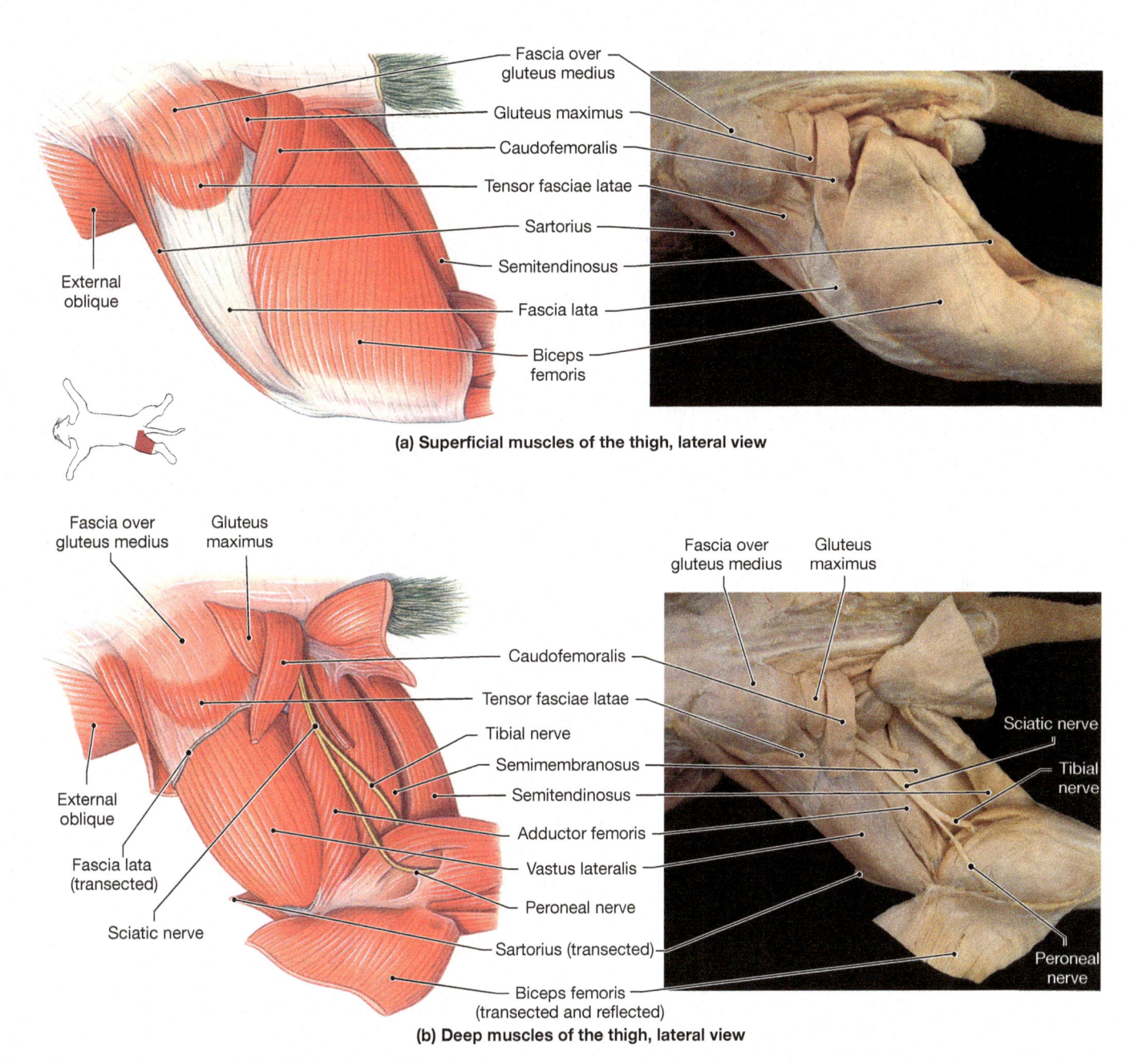

Figure D1.11 Muscles of the Thigh—Lateral View
Illustrations and photos.

6. The **pectineus** is anterior to the adductor longus (see Figure D1.10). The pectineus is a deep, small muscle posterior to both the femoral artery and the femoral vein. It originates on the anterior border of the pubis and inserts on the proximal end of the femur. It functions to adduct the thigh.

7. Locate the four large muscles that constitute the **quadriceps femoris group**. These muscles cover about one-half of the surface of the thigh. Collectively they insert into the patellar ligament and act as powerful extensors of the leg. Bisect the sartorius, and free both borders of the tensor fasciae latae. Reflect these muscles and observe that the muscles of the quadriceps femoris converge and insert on the patella.

   a. The **vastus lateralis** is the large, fleshy muscle on the anterolateral surface of the thigh deep to the tensor fasciae latae. It originates along the entire length of the lateral surface of the femur.

b. The **vastus medialis** is on the medial surface of the femur just under the sartorius. It originates on the shaft of the femur and inserts on the patellar ligament.

c. The **rectus femoris** is a small, cylindrical muscle between the vastus medialis and the vastus lateralis muscles. In humans the rectus femoris originates on the ilium, but in cats it originates on the femur.

d. The **vastus intermedius** (not visible in Figure D1.10) lies deep to the rectus femoris. Transect and reflect the rectus femoris to expose the vastus intermedius. It originates on the shaft of the femur.

8. On the posterior thigh are three muscles of the **hamstring group**. These muscles span the knee joint and act to flex the leg. They are all visible in Figure D1.11, and the latter two are also visible in Figure D1.10.

    a. The **biceps femoris** is a large, broad muscle that covers most of the lateral region of the thigh. It originates on the ischial tuberosity and inserts on the tibia. The biceps femoris flexes the leg and also abducts the thigh.

    b. The **semitendinosus** is visible under the posterior portion of the biceps femoris. The belly of the muscle is a uniform strap from the origin to the tendon at the insertion. The semitendinosus originates on the ischial tuberosity and inserts on the medial side of the tibia. It flexes the leg.

    c. The **semimembranosus** is a large muscle medial to the semitendinosus. It is seen best in the anteromedial view of the thigh. Transect the semitendinosus to view the semimembranosus in the posterior aspect. The semimembranosus originates on the ischium and inserts on the medial epicondyle of the femur and medial surface of the tibia. It extends the thigh. ∎

LAB ACTIVITY 8 — Muscles of the Lower Hindlimb

8 Materials

- ☐ Gloves
- ☐ Safety glasses
- ☐ Dissecting tools
- ☐ Dissecting tray
- ☐ Preserved cat, skin removed

Procedures

1. The **gastrocnemius**, or calf muscle, is on the posterior of the hindlimb (Figure D1.12). It has two heads of origin, medial on the knee's fascia and lateral on the distal end of the femur. The two bellies of the muscle unite at the calcaneal tendon and insert on the calcaneus. The gastrocnemius extends the foot.

2. The **flexor digitorum longus** is found between the gastrocnemius and the tibia. This muscle has two heads of origin: one on the distal end of the tibia and another on the head and shaft of the fibula. It inserts by four tendons onto the bases of the terminal phalanges. It acts as a flexor of the digits.

3. The **tibialis anterior** is on the anterior surface of the tibia. This muscle originates on the proximal ends of the tibia and fibula and inserts on the first metatarsal. It acts to flex the foot.

4. Deep to the gastrocnemius but visible on the lateral surface of the calf is the **soleus** (Figure D1.12b). The soleus originates on the fibula and inserts on the calcaneus. It extends the foot.

5. The **fibularis group**, also called the *peroneus muscles,* consists of three muscles deep to the soleus on the posterior and lateral surfaces.

    a. The **fibularis brevis** lies deep to the tendon of the soleus; it originates from the distal portion of the fibula and inserts on the base of the fifth metatarsal. The fibularis brevis extends the foot.

(a) Medial view

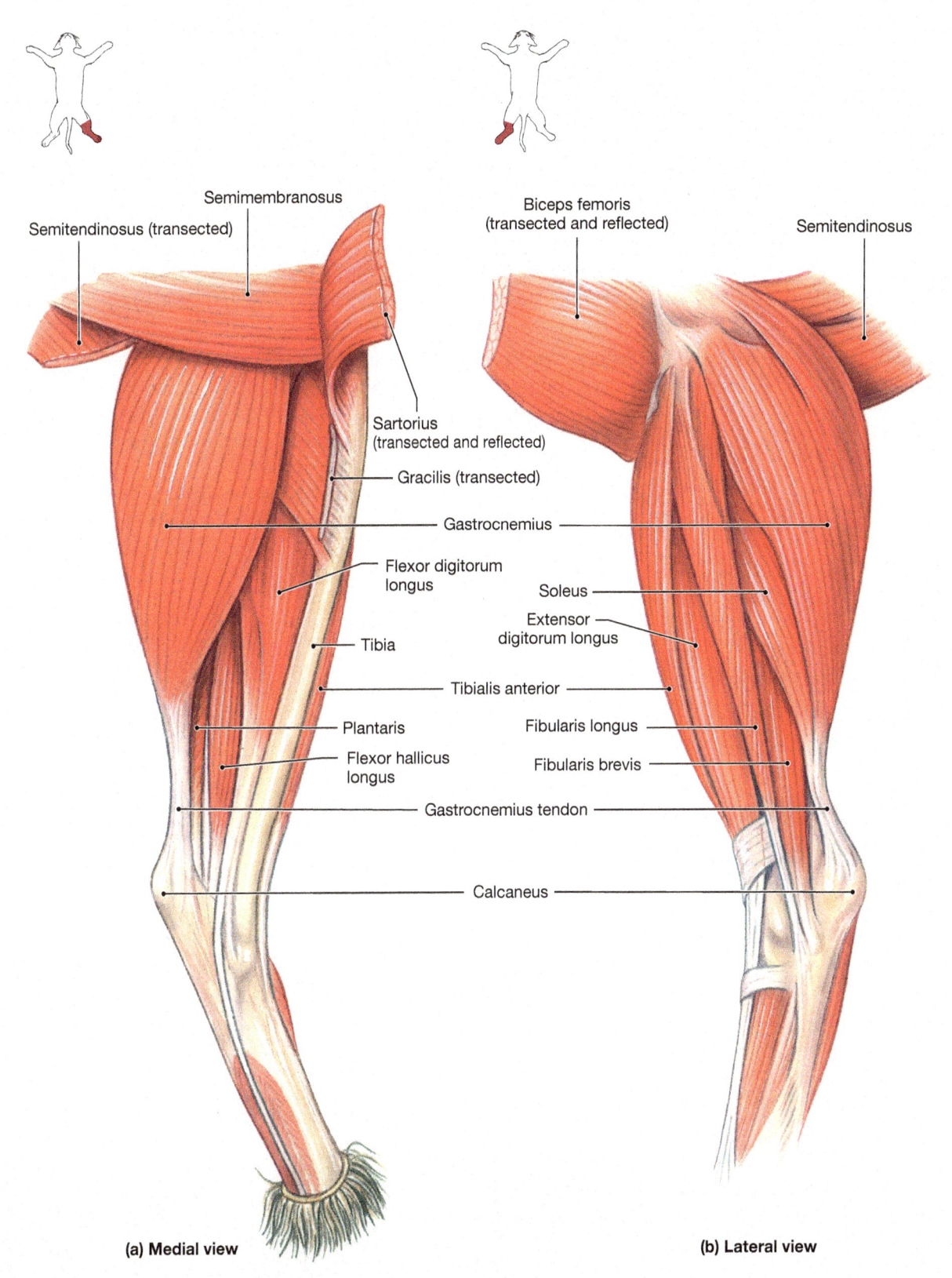

(b) Lateral view

Figure D1.12 Muscles of the Left Hindlimb

b. The **fibularis longus** is a long, thin muscle that lies on the lateral surface of the hindlimb; it originates on the proximal portion of the fibula and inserts by a tendon that passes through a groove on the lateral malleolus and turns medially to attach to the bases of the metatarsals. This muscle acts to flex the foot.

c. The **fibularis tertius** lies along the tendon of the fibularis longus; it originates on the fibula and inserts on the base of the fifth metatarsal. This muscle extends the fifth digit and flexes the foot.

6. On the anterolateral border of the tibia, the **extensor digitorum longus** originates on the lateral epicondyle of the femur. It inserts by long tendons on each of the five digits and functions to extend the digits. ■

Safety Tip: Storing the Cat and Cleaning Up

To store your specimen, wrap it in the skin and moisten it with fixative. Use paper towels if necessary to cover the entire specimen. Return it to the storage bag and seal the bag securely. Label the bag with your name, and place it in the storage area as indicated by your instructor.

Wash all dissection tools and the tray, and set them aside to dry.

Dispose of your gloves and any tissues from the dissection as indicated by your laboratory instructor. ▲

DISSECTION
LAB REPORT
EXERCISE 1
Muscles of the Cat

Name _____

Date _____

Section _____

A. **Matching**

Match each structure in the left column with its correct description from the right column.

_____ 1. acromiotrapezius
_____ 2. levator scapulae ventralis
_____ 3. spinodeltoid
_____ 4. sternomastoid
_____ 5. xiphihumeralis
_____ 6. epitrochlearis
_____ 7. pectineus
_____ 8. vastus intermedius
_____ 9. fibularis tertius
_____ 10. fibularis brevis

A. adducts thigh
B. called sternocleidomastoid in humans
C. deep to rectus femoris
D. extends foot
E. flat muscle of medial upper forelimb, extends forelimb
F. flexes and laterally rotates forelimb
G. flexes foot
H. originates on xiphoid process, adducts humerus
I. pulls scapula toward the head
J. square muscle, holds scapula in place

B. **Short-Answer Questions**

1. Describe the neck muscles of the cat.

2. Describe the flexor and extensor muscles of the cat forelimb.

3. Describe the abdominal muscles of the cat.

EXERCISE 1 — LAB REPORT

DISSECTION

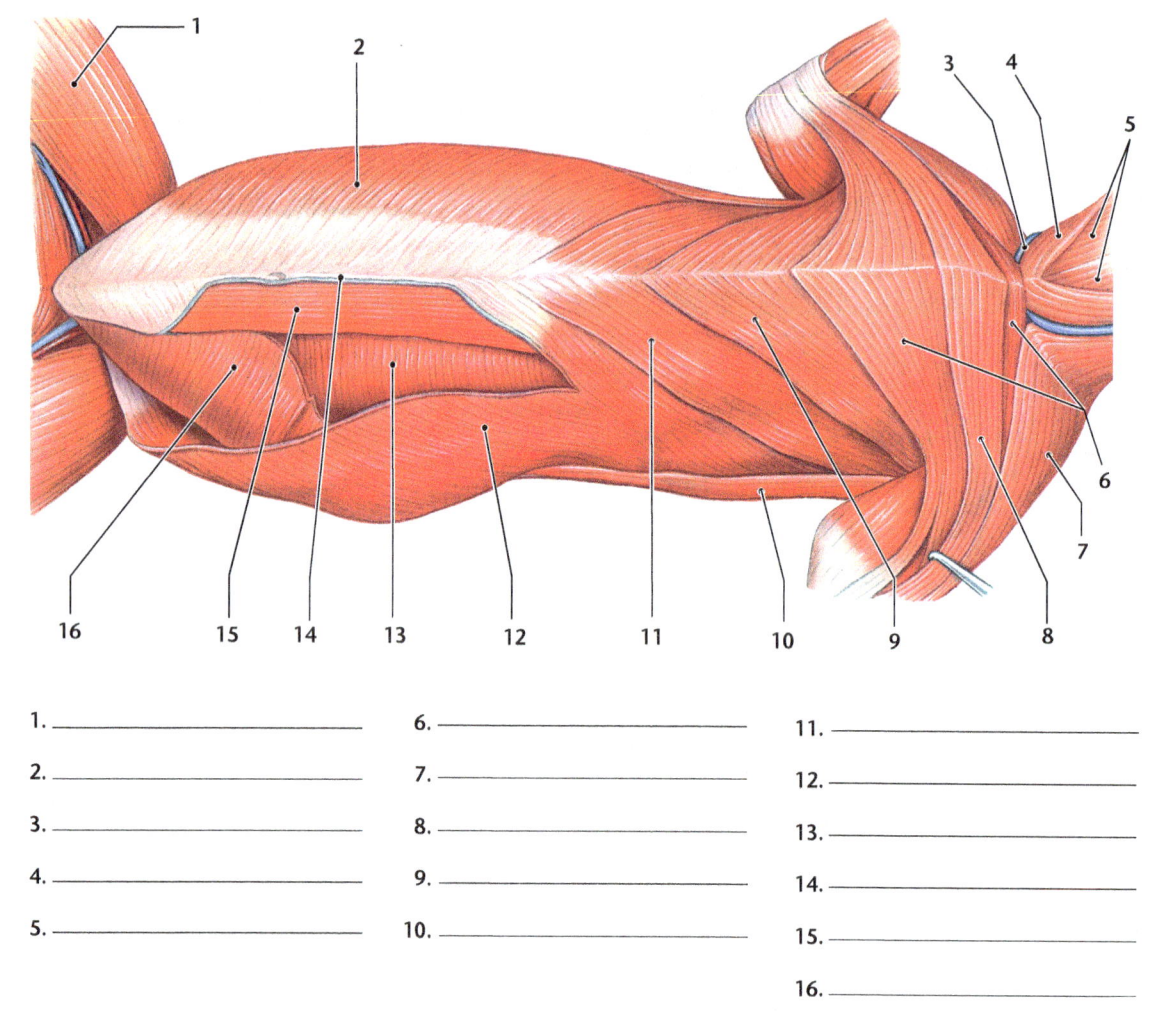

Figure D1.13   Major Muscles of the Cat

1. _____
2. _____
3. _____
4. _____
5. _____
6. _____
7. _____
8. _____
9. _____
10. _____
11. _____
12. _____
13. _____
14. _____
15. _____
16. _____

C. Labeling

Label the muscles in Figure D1.13.

D. Analysis and Application

1. Explain the differences between the cat and the human deltoid and trapezius muscles.

2. How are the superficial chest muscles of the cat different from chest muscles in humans?

3. Compare cat and human muscles of the thigh and leg/hindleg.

DISSECTION EXERCISE 2

Cat Nervous System

LAB ACTIVITIES

1. Preparing the Cat for Dissection   517
2. The Brachial Plexus   518
3. The Sacral Plexus   518
4. The Spinal Cord   521

OBJECTIVES

On completion of this exercise, you should be able to:

- Identify the major nerves of the feline brachial plexus.
- Identify the major nerves of the feline sacral plexus.
- Identify the feline spinal meninges and the dorsal and ventral roots of the spinal cord.

Cats, like humans, have pairs of spinal nerves that extend laterally from the various segments of the spinal cord. Humans have 31 pairs of spinal nerves; cats have 38 to 40 pairs, depending on whether some of the caudal nerves have fused (they are difficult to distinguish individually). In this exercise, you will identify the major nerves of the brachial and sacral plexuses of the feline nervous system. You will also dissect the sacral plexus and then examine the exposed spinal cord. This exercise complements the study of the human nervous system.

⚠️ **Safety Tip: Cat Dissection Basics**

Before beginning this dissection, review the Safety Tip presented at the beginning of Dissection Exercise 1. ▲

LAB ACTIVITY 1 Preparing the Cat for Dissection

If the cat was not skinned for muscle studies, refer to Dissection Exercise 1, Lab Activity 1: "Preparing the Cat for Dissection."

1 Materials

- ☐ Gloves
- ☐ Safety glasses
- ☐ Dissecting tools
- ☐ Dissecting tray
- ☐ String
- ☐ Preserved cat, skin removed

Procedures

1. Put on gloves and safety glasses and clear your workspace before obtaining your dissection specimen.
2. Secure the specimen ventral side up on the dissecting tray by spreading the limbs and tying them flat with lengths of string passing under the tray, one string for the two forelimbs and one string for the two hindlimbs.
3. Be sure to keep the specimen moist with fixative during the dissection. Keep the skin draped on areas not undergoing dissection.
4. Proceed to the next Lab Activity to dissect the cat. ■

LAB ACTIVITY 2 The Brachial Plexus

Dissection of the chest and forelimb reveals the brachial plexus, a network formed by the intertwining of cervical nerves C_6, C_7, and C_8 and thoracic nerve T_1. This plexus innervates muscles and other structures of the shoulder, forelimb, and thoracic wall. The nerves are delicate, so you must be careful not to damage or remove them during dissection. Use Figure D2.1 as a reference in identifying the nerves of the brachial plexus.

2 Materials

- ☐ Gloves
- ☐ Safety glasses
- ☐ Dissecting tools
- ☐ Dissecting tray
- ☐ String
- ☐ Preserved cat, skin removed

Procedures

1. Put on gloves and safety glasses, and clear your workspace before obtaining your dissection specimen.
2. Secure the specimen ventral side up on the dissecting tray by spreading the limbs and tying them flat with lengths of string passing under the tray. Use one string for the two forelimbs and one string for the two hindlimbs.
3. Reflect the left pectoralis major muscle, and observe the underlying blood vessels in the axilla. If your cat has been injected with colored latex the arteries are injected with red latex and the veins with blue latex. If the cat was triple-injected, the hepatic portal system will have yellow latex.
4. Use a probe and forceps to carefully remove fat and other tissue from around the vessels and nerves.
5. The largest nerve of the brachial plexus is the **radial nerve.** It lies dorsal to the axillary artery, which is the red-injected blood vessel in the axilla. The radial nerve supplies the triceps brachii muscle and other dorsal muscles of the forelimb. Trace this nerve from close to its origin near the midline to where it passes into the triceps muscle.
6. The **musculocutaneous nerve,** which is superior to the radial nerve, supplies the coracobrachial and biceps brachii muscles of the ventral forelimb and the skin of the forelimb. Trace this nerve into the musculature, and notice its two divisions.
7. Next notice the **median nerve,** which follows the red-injected brachial artery into the ventral forelimb. This nerve supplies the muscles of the ventral antebrachium of the forelimb.
8. The most posterior of the brachial plexus nerves is the **ulnar nerve.** It is often isolated from the other nerves once the surrounding fat and tissues have been removed from the plexus. Trace this nerve down the brachium to the elbow to where it supplies the muscles of the antebrachium. ■

LAB ACTIVITY 3 The Sacral Plexus

Dissection of the hindlimb exposes the three major nerves of the sacral plexus, the neural network that supplies the muscles and structures of the hip and hindlimb. As you reflect muscles to expose the nerves, be careful not to remove or damage the nerves. Use Figure D2.2 as a reference in identifying the nerves of the sacral plexus.

3 Materials

- ☐ Gloves
- ☐ Safety glasses
- ☐ Dissecting tools
- ☐ Dissecting tray
- ☐ String
- ☐ Preserved cat, skin removed

Procedures

1. Put on gloves and safety glasses, and clear your workspace before obtaining your dissection specimen.
2. Secure the specimen dorsal side up on the dissecting tray as described in Lab Activity 1.

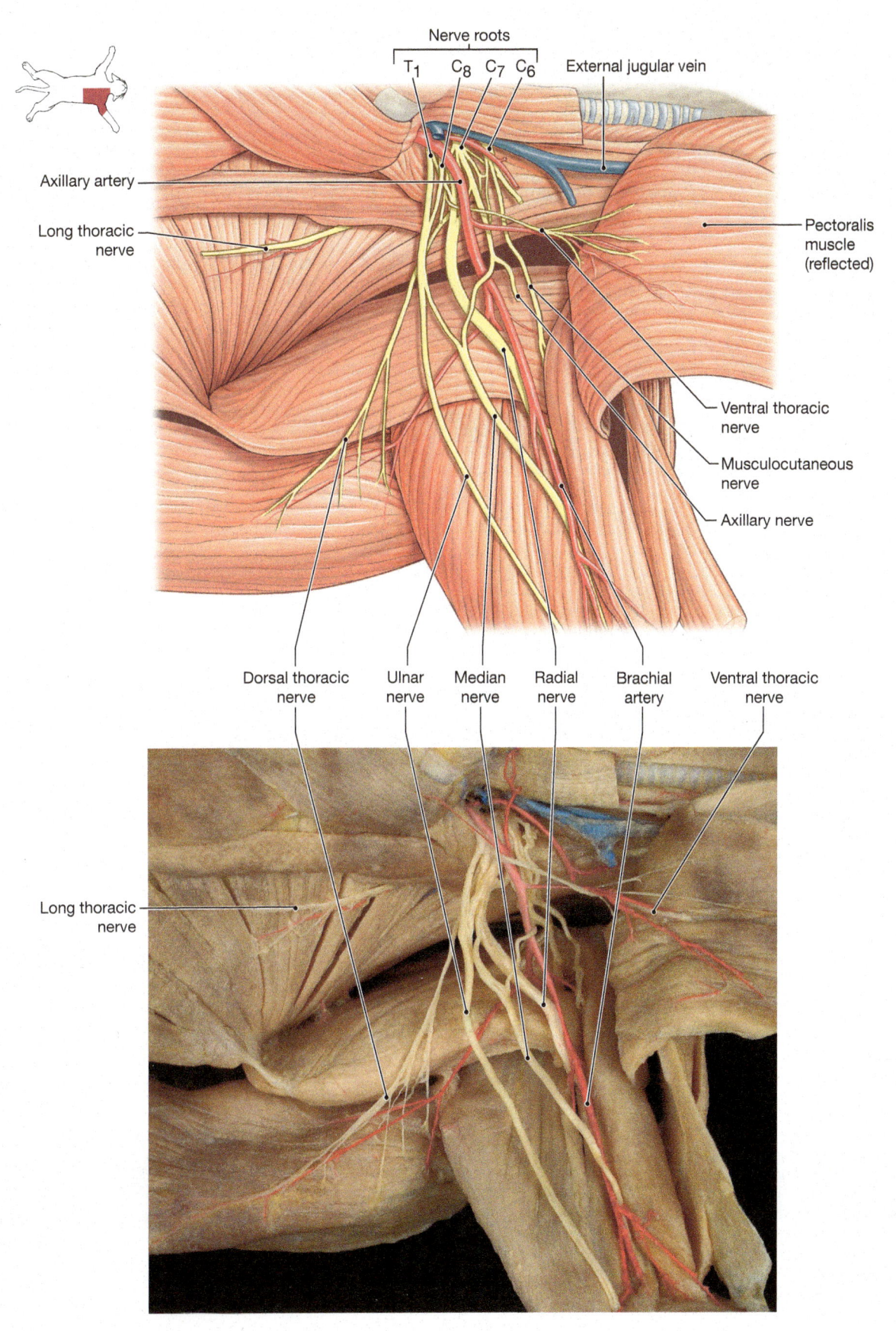

Figure D2.1 Brachial Plexus
Illustration and photo.

519

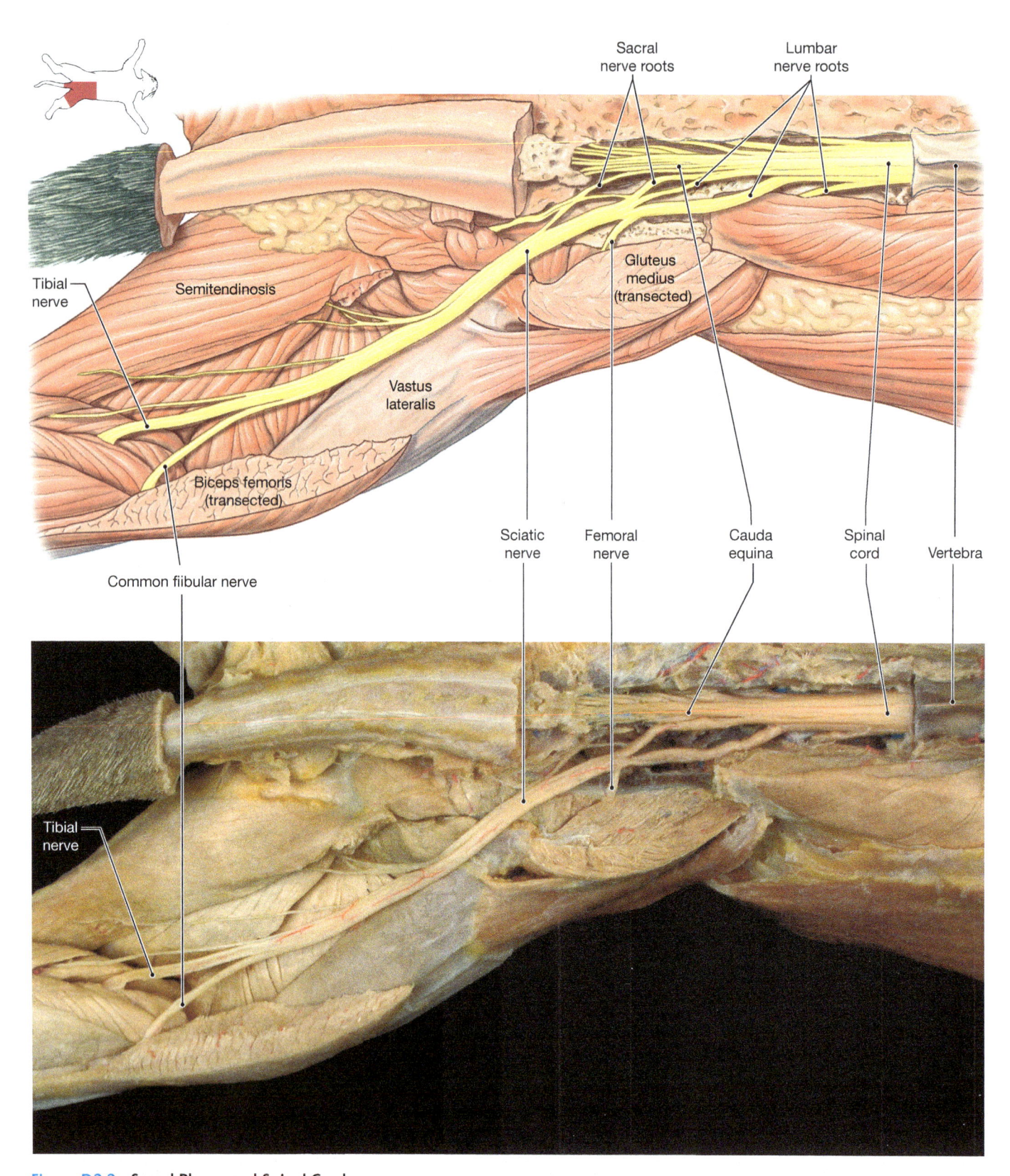

Figure D2.2 **Sacral Plexus and Spinal Cord**
Illustration and photo.

520

3. Reflect the cut ends of the biceps femoris muscle and note the large **sciatic nerve**. This nerve supplies the muscles of the hindlimb.

4. Follow the sciatic nerve down the hindlimb to the gastrocnemius muscle, where the nerve branches into two smaller nerves.

5. Identify the **tibial nerve** on the medial side of the hindlimb and the **common fibular nerve** (also called the *common peroneal nerve*) on the lateral side. The tibial and common fibular nerves supply the inferior part of the hindlimb. ■

LAB ACTIVITY 4 — The Spinal Cord

Dissection and reflection of the posterior muscles on the dorsal surface will expose the vertebral column. To save time, only a small section of the vertebral column will be dissected to study the spinal cord. Use care while using bone cutters to remove pieces of the vertebrae. Use Figure D2.2 as a reference in identifying the spinal cord.

4 Materials

- ☐ Gloves
- ☐ Safety glasses
- ☐ Dissecting tools
- ☐ Dissecting tray
- ☐ String
- ☐ Preserved cat, skin removed

Procedures

1. Put on gloves and safety glasses, and clear your workspace before obtaining your dissection specimen.

2. Secure the specimen ventral side down on the dissection tray, tying the limbs as described in Lab Activity 1.

3. Cut and reflect the large dorsal muscles that cover the vertebral column in the lumbar region of the back.

4. Use bone cutters to remove the vertebral arches of three vertebrae and thereby expose the **spinal cord**. Gently remove each piece of bone, using care not to tear or remove the membranes covering the spinal cord.

5. Examine the exposed spinal cord and identify the **dorsal roots** and **ventral roots** that form the spinal nerves.

6. Identify the outermost **dura mater** over the spinal cord.

7. Cut through the dura, and note the **arachnoid** membrane with its many fine extensions to the spinal cord.

8. Use a dissection pin to tease away a portion of the delicate **pia mater** lying on the surface of the spinal cord.

9. Note the **subarachnoid space** between the arachnoid and pia mater. Cerebrospinal fluid circulates in this space.

10. Remove a thin section of the spinal cord to view the internal organization. Identify the inner **gray horns** surrounded by the **white columns**. ■

⚠ **Safety Tip: Storing the Cat and Cleaning Up**

To store your specimen, wrap it in the skin and moisten it with fixative. Use paper towels if necessary to cover the entire specimen. Return it to the storage bag and seal the bag securely. Label the bag with your name, and place it in the storage area as indicated by your instructor.

Wash all dissection tools and the tray, and set them aside to dry.

Dispose of your gloves and any tissues from the dissection as indicated by your laboratory instructor. ▲

LAB REPORT

DISSECTION EXERCISE 2

Cat Nervous System

Name _____

Date _____

Section _____

A. Matching

Match each structure in the left column with its correct description from the right column.

_____ 1. musculocutaneous nerve
_____ 2. radial nerve
_____ 3. tibial nerve
_____ 4. median nerve
_____ 5. sciatic nerve
_____ 6. common fibular nerve
_____ 7. subarachnoid space
_____ 8. dura mater
_____ 9. ulnar nerve

A. serves muscles of antebrachium
B. serves triceps brachii muscle
C. medial nerve of lower hindlimb
D. serves biceps brachii
E. lateral nerve of lower hindlimb
F. large nerve of sacral plexus
G. posterior nerve of brachial plexus
H. site of cerebrospinal fluid circulation
I. outer membrane protecting spinal cord

B. Short-Answer Questions

1. Compare the brachial plexus nerves of cats and humans.

2. Describe the three major nerves of the sacral plexus.

3. What are the meningeal layers surrounding the spinal cord?

523

EXERCISE 2 — LAB REPORT

C. Analysis and Applications

1. Describe the major nerves of the forelimb.

2. How are white and gray matter organized in the spinal cord?

DISSECTION EXERCISE 3

Cat Endocrine System

OBJECTIVES

On completion of this exercise, you should be able to:

- Identify the main glands of the feline endocrine system.
- List the hormones produced by each gland and state the basic function of each.

LAB ACTIVITIES

1. Preparing the Cat for Dissection 525
2. Endocrine Glands of the Cat 527

In this exercise you will identify the major organs of the feline endocrine system. This exercise complements the study of the human endocrine system.

Safety Tip: Cat Dissection Basics

Before beginning this dissection, review the Safety Tip presented at the beginning of Dissection Exercise 1. ▲

LAB ACTIVITY 1 Preparing the Cat for Dissection

If the ventral body cavity has not been opened on your dissection specimen, complete the following procedures. Otherwise, skip to Lab Activity 2. Use Figure D3.1 as a reference as you dissect the ventral body cavity.

Materials

- ☐ Gloves
- ☐ Safety glasses
- ☐ Dissecting tools
- ☐ Dissecting tray
- ☐ String
- ☐ Preserved cat, skin removed

Procedures

1. Put on gloves and safety glasses, and clear your workspace before obtaining your dissection specimen.
2. Secure the specimen ventral side up on the dissecting tray by spreading the limbs and tying them flat with lengths of string passing under the tray. Use one string for the two forelimbs and one string for the two hindlimbs.
3. Use scissors to cut a midsagittal section through the muscles of the abdomen to the sternum.
4. To avoid cutting through the bony sternum, angle your incision laterally approximately 0.5 inch, and cut the costal cartilages. Continue the parasagittal section to the base of the neck.
5. Make a lateral incision on each side of the diaphragm. Use care not to damage the diaphragm or the internal organs. Spread the thoracic walls to observe the internal organs.
6. Make a lateral section across the pubic region and angled toward the hips. Spread the abdominal walls to expose the abdominal organs. ■

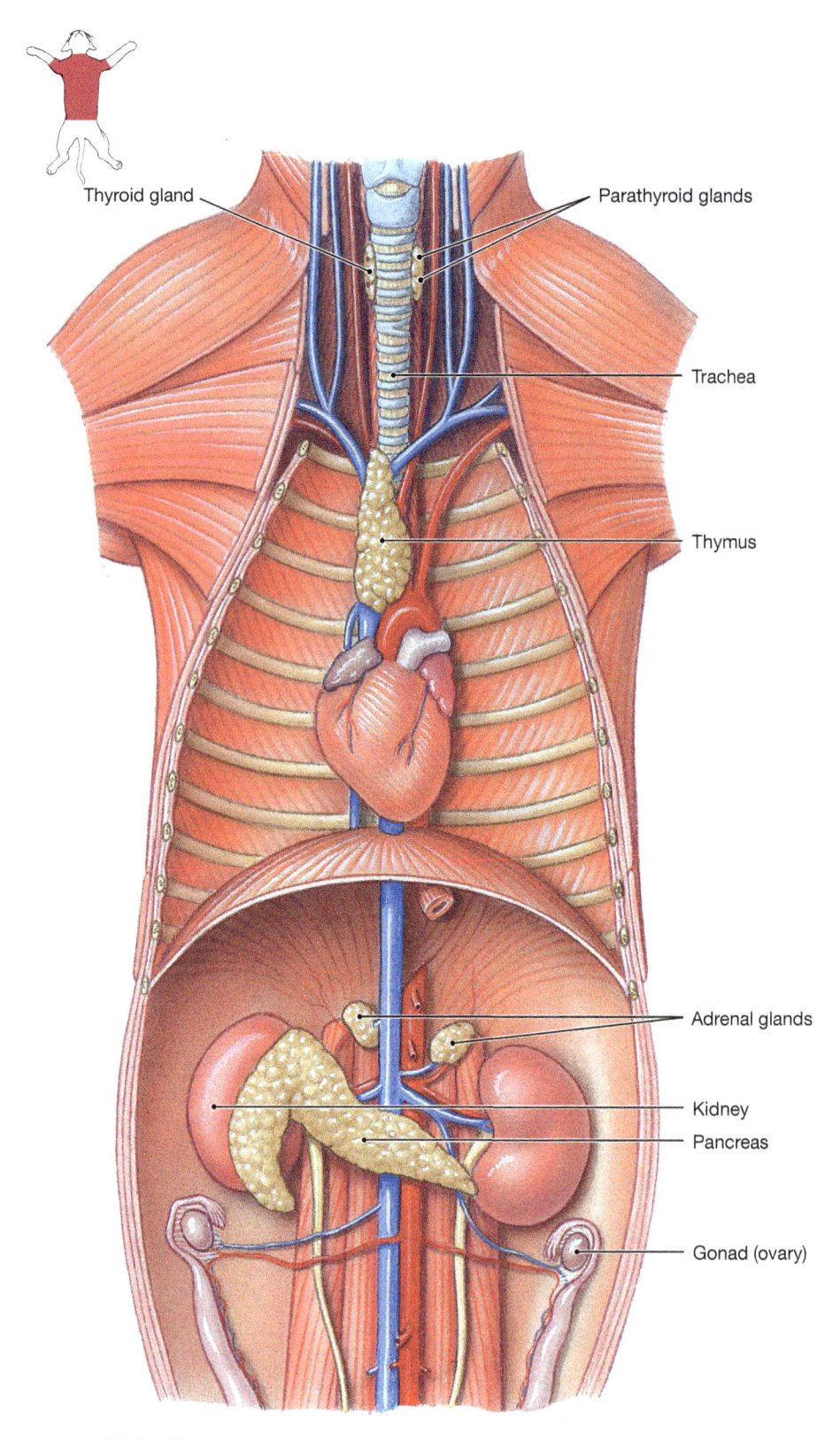

Figure D3.1 Cat Endocrine System

LAB ACTIVITY 2   **Endocrine Glands of the Cat**

When moving internal organs aside to locate the endocrine glands, take care not to rupture the digestive tract. Because hormones are transported by the bloodstream, take note of the vascularization of the endocrine glands. Many of these vessels will be identified in a later dissection of the cardiovascular system. Use Figure D3.1 as a reference in identifying the endocrine glands.

If you are continuing from Lab Activity 1, begin with step 3 of the following Procedures list.

2 Materials

- ☐ Gloves
- ☐ Safety glasses
- ☐ Dissecting tools
- ☐ Dissecting tray
- ☐ String
- ☐ Preserved cat, skin removed

Procedures

1. Put on gloves and safety glasses, and clear your workspace before obtaining your dissection specimen.

2. Secure the specimen ventral side up on the dissecting tray by spreading the limbs and tying them flat with lengths of string passing under the tray. Use one string for the two forelimbs and one string for the two hindlimbs.

3. Locate the windpipe, called the *trachea*, which passes through the midline of the neck. Spanning the trachea on both sides is the **thyroid gland**. Note that this gland is divided into two **lateral lobes** with a small band, the **isthmus**, and connects them. The thyroid gland produces triiodothyronine (T_3) and thyroxine (T_4), hormones that increase the metabolic rate of cells. Other cells in the thyroid gland secrete the hormone calcitonin (CT), which decreases blood plasma calcium ion concentration.

4. Locate the small masses of the **parathyroid gland** on the dorsal surface of each lateral lobe of the thyroid gland. The parathyroid gland secretes parathyroid hormone (PTH), which increases blood plasma calcium ion concentration.

5. Examine the superior surface of the heart and identify the **thymus**. This organ produces thymosin, a hormone that stimulates the development of the immune system. If your dissection specimen is an immature cat, the thymus should be large and conspicuous. In adult cats (and in adult humans, too), the thymus is gradually replaced by fat tissue.

6. Locate the **pancreas** between the stomach and small intestine. The pancreas is a "double gland" because it has both exocrine and endocrine functions. The exocrine part of the pancreas produces digestive enzymes. The endocrine part of the gland consists of islands of cells that secrete hormones that regulate the amount of sugar present in the blood. The hormone insulin decreases the amount of blood sugar by stimulating cellular intake of sugar. Glucagon has the opposite effect: it stimulates the liver to release sugar into the blood.

7. Gently move the abdominal organs to one side, and locate an **adrenal gland**. In humans the adrenal glands rest on the superior margin of the kidneys; in cats the adrenals are separate from the kidneys. The adrenal glands secrete a variety of hormones. The outer **adrenal cortex** secretes three types of hormones: mineralocorticoids to regulate sodium ions, glucocorticoids to fight stress, and androgen, which is converted to sex hormones. The inner **adrenal medulla** secretes epinephrine and norepinephrine commonly called adrenalin and noradrenaline into the blood during times of excitement, stress, and exercise.

8. Locate a **kidney**, positioned dorsal to the adrenal gland. The kidneys secrete renin, an enzyme that initiates the endocrine reflex for sodium regulation. The kidneys also secrete erythropoietin (EPO), the hormone that stimulates the bone marrow to produce red blood cells.

9. The **gonads**—testes in the male and ovaries in the female—produce sex hormones. If your dissection specimen is a male, locate the **testes** in the pouch-like scrotum positioned between the hindlimbs. The testes secrete the hormone testosterone, which stimulates maturation and maintenance of the male sex organs and promotes such male characteristics as muscle development and sex drive. Testes also produce the spermatozoa that fertilize eggs during reproduction.

10. If your specimen is a female, locate the small **ovaries** in the pelvic cavity. The ovaries secrete estrogen and progesterone, hormones that prepare the uterus for gestation of an embryo. Estrogen is also responsible for development of the female sex organs, breast development, and other adult female traits. ∎

Safety Tip: Storing the Cat and Cleaning Up

To store your specimen, wrap it in the skin and moisten it with fixative. Use paper towels if necessary to cover the entire specimen. Return it to the storage bag and seal the bag securely. Label the bag with your name, and place it in the storage area as indicated by your instructor.

Wash all dissection tools and the tray, and set them aside to dry.

Dispose of your gloves and any tissues from the dissection as indicated by your laboratory instructor. ▲

DISSECTION LAB REPORT

EXERCISE 3

Cat Endocrine System

Name _____
Date _____
Section _____

A. Matching

Match each structure in the left column with its correct description from the right column.

_____ 1. pancreas
_____ 2. adrenal medulla
_____ 3. thyroid gland
_____ 4. parathyroid gland
_____ 5. ovary
_____ 6. gonads

A. divided into two lobes called lateral lobes
B. secretes hormone that raises amount of calcium ions in blood
C. has both exocrine and endocrine functions
D. activated when body is under emotional stress
E. general term for sex organs
F. source of both estrogen and progesterone

B. Short-Answer Questions

1. How are hormones transported to the tissues of the body?

2. Describe the endocrine glands located in the neck.

3. Describe the location of the kidneys, and name one hormone and one enzyme produced by these organs.

529

EXERCISE 3

LAB REPORT

C. Analysis and Application

1. How would the body respond if the thyroid gland produced an insufficient amount of thyroid hormone?

2. When male cats are neutered, the testes are removed. What kind of changes are expected in a neutered animal?

3. Suppose a cat has not eaten all day. Describe the endocrine activity of the pancreas in this animal.

4. Which endocrine gland is activated in a cat that is being chased by a dog?

DISSECTION EXERCISE 4

Cat Circulatory System

OBJECTIVES

On completion of this exercise, you should be able to:

- Identify the major arteries and veins of the feline vascular system.
- Compare the circulatory vessels of the cat with those of the human.

LAB ACTIVITIES

1. Preparing the Cat for Dissection 532
2. Arteries That Supply the Head, Neck, and Thorax 532
3. Arteries That Supply the Shoulder and Forelimb (Medial View) 535
4. Arteries That Supply the Abdominal Cavity 536
5. Arteries That Supply the Hindlimb (Medial View) 538
6. Veins That Drain the Head, Neck, and Thorax 539
7. Veins That Drain the Forelimb (Medial View) 540
8. Veins That Drain the Abdominal Cavity 540
9. Veins That Drain the Hindlimb (Medial View) 542

In this exercise you will dissect the vascular system of the cat and identify the major arteries and veins. If your cat has been injected with colored latex the arteries are filled with red latex and the veins with blue latex. Because this exercise complements the study of the human circulatory system, be sure to note differences between the blood vessels of the cat and those of humans.

⚠ **Safety Tip: Cat Dissection Basics**
Before beginning this dissection, review the Safety Tip presented at the beginning of Dissection Exercise 1. ▲

LAB ACTIVITY 1 Preparing the Cat for Dissection

If the thoracic and abdominal cavities have not been opened on your dissection specimen, complete the following procedures. Otherwise, proceed to Lab Activity 2.

Materials

- ☐ Gloves
- ☐ Safety glasses
- ☐ Dissecting tools
- ☐ Dissecting tray
- ☐ Preserved cat, skin removed

Procedures

1. Put on gloves and safety glasses, and clear your workspace before obtaining your dissection specimen.
2. Lay the specimen ventral side up on the dissecting tray and expose the thoracic and abdominal organs by making a longitudinal midline incision through the muscles of the neck, thorax, and abdominal wall.
3. Avoid cutting through the muscular diaphragm so that you can identify the vessels and structures that pass between the thoracic and abdominal cavities. Your instructor will provide you with specific instructions for exposing these cavities and for isolating the blood vessels.
4. Occasionally, clotted blood fills the thoracic and abdominal cavities and must be removed. If you encounter clots, check with your instructor before proceeding. ∎

LAB ACTIVITY 2 Arteries That Supply the Head, Neck, and Thorax

Only those arteries typically injected with colored latex are listed for identification in this activity. Keep in mind that the cat has more arteries than listed here, and your instructor may assign additional vessels for you to identify. Use Figure D4.1 as a reference as you identify the vessels.

Materials

- ☐ Gloves
- ☐ Safety glasses
- ☐ Dissecting tools
- ☐ Dissecting tray
- ☐ Preserved cat, skin removed

Procedures

1. Locate the large arteries that leave the right and left ventricles of the heart:
   a. The artery connected to the right ventricle is the **pulmonary trunk**.
   b. The artery connected to the left ventricle is the **aorta**. The aorta is the main arterial blood vessel, and all major arteries except those to the lungs arise from it.
2. Trace these two large arteries, which direct blood away from the heart, to smaller arteries that direct blood to specific organs and tissues. The pulmonary trunk delivers deoxygenated blood to the lungs, and the aorta delivers oxygenated blood to the rest of the body.
   a. The aorta curves at the **aortic arch**—where vessels to the forelimbs, head, and neck arise—and then continues along the chest and abdomen on the left side of the vertebral column to the pelvis, where it divides into branches to supply the hindlimbs. The portion of the aorta anterior to the diaphragm is the **thoracic aorta**.
   b. The pulmonary trunk divides into the right and left branches of the **pulmonary artery.** The left branch of the pulmonary artery passes ventral to the aorta to reach the left lung; the right branch passes between the aortic arch and the heart to reach the right lung.
   c. The **ligamentum arteriosum** is a remnant of fetal circulation when the pulmonary trunk was connected to the aorta by a vessel called the ductus arteriosus. At birth, the ductus arteriosus closes and becomes the ligamentum arteriosum.
3. The feline aortic arch has two major branches: the **brachiocephalic artery** and the **left subclavian artery**. (In humans, there are three branches off this arch.) Via all its various branches, the brachiocephalic artery ultimately

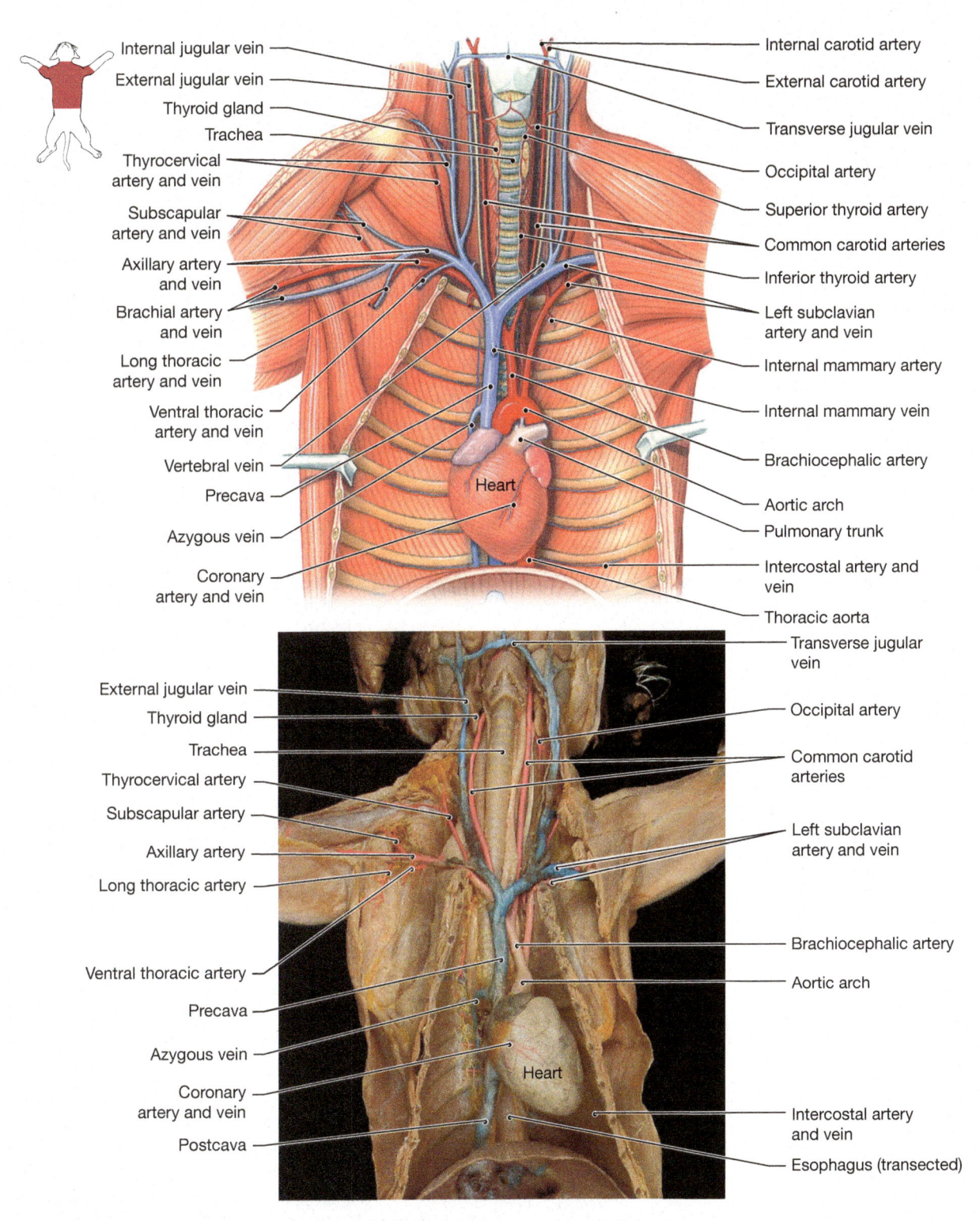

Figure D4.1 Arteries and Veins of the Chest and Neck
Illustration and photo.

533

supplies blood to the head and the right forelimb. The left subclavian artery supplies the left forelimb.

   a. Near the level of the second rib, the brachiocephalic artery branches into the **right subclavian artery** and the **left common carotid artery**. (Note: Although the right and left subclavians have different origins, once these vessels enter their respective axillae, all the smaller vessels that branch off from each subclavian are the same in the two forelimbs.)

   b. The **right common carotid artery** branches off of the right subclavian artery.

   c. Above the larynx, each common carotid artery divides into an **internal carotid artery** and an **external carotid artery**. The two internal carotid arteries enter the skull via the foramen lacerum and join the **posterior cerebral artery** (not shown in Figure D4.1).

   d. Each external carotid artery turns medially near the posterior margin of the masseter and continues as an **internal maxillary artery**. Each internal maxillary artery gives off several branches and then branches to form the carotid plexus surrounding the maxillary branch of the trigeminal nerve near the foramen rotundum. Through various branches, each internal maxillary and each carotid plexus carries blood to the brain, eye, and other deep structures of the head.

4. Locate the left and right **superior thyroid arteries,** one on either side of the thyroid gland. Branching from its respective common carotid artery at the cranial end of the thyroid gland, each superior thyroid artery supplies blood to the thyroid gland, to superficial laryngeal muscles, and to ventral neck muscles.

5. Next locate an **internal mammary artery,** which arise from the ventral surface of either subclavian at about the level of the vertebral artery (discussed next) and passing caudally to the ventral thoracic wall. Branches of both internal mammary arteries supply adjacent muscles, pericardium, mediastinum, and diaphragm.

6. Trace either **vertebral artery** as it arises from the dorsal surface of either subclavian and passes cranially through the transverse foramen of the cervical vertebrae. Each vertebral artery gives off branches to the deep neck muscles and to the spinal cord near the foramen magnum:

   a. Distal to the vertebral artery, a **costocervical artery** arises from the dorsal surface of either subclavian artery. Each costocervical artery sends branches to the deep muscles of the neck and shoulder and to the first two costal interspaces.

   b. A **thyrocervical artery** arises from the cranial aspect of either subclavian (distal to the costocervical artery) and passes cranially and laterally, and supplies the muscles of the neck and chest.

7. On the descending thoracic aorta, note where 10 pairs of **intercostal arteries** are given off to the interspaces between the last 11 ribs. Also note the paired **bronchial arteries,** which arise from the thoracic aorta and supply the bronchi. Next look for the **esophageal arteries,** several small vessels of varying origin along the thoracic aorta that supply the esophagus.

8. Follow either subclavian artery from its origin, which moves away from the heart. While still deep in the thoracic cavity, this vessel is called the **axillary artery** because it ultimately passes through the axilla. A right axillary artery is visible in Figure D4.1. Locate the vessels that branch from either axillary artery:

   a. From the ventral surface of the axillary artery, just lateral to the first rib, the **ventral thoracic artery** arises and passes caudally to supply the medial ends of the pectoral muscles.

DISSECTION EXERCISE 4 Cat Circulatory System 535

b. The **long thoracic artery** arises lateral to the ventral thoracic artery, passing caudally to the pectoral muscles and the latissimus dorsi.

c. A third artery branching off the subclavian artery is the **subscapular artery**, which passes laterally and dorsally between the long head of the triceps and the latissimus dorsi to supply the dorsal shoulder muscles. It gives off two branches, the **thoracodorsal artery** and the posterior **humeral circumflex artery**.

Arteries That Supply the Shoulder and Forelimb (Medial View)

LAB ACTIVITY 3

The axillary arteries supply blood to the brachium. Use Figure D4.2 as a reference as you identify the vessels.

Materials

- Gloves
- Safety glasses
- Dissecting tools
- Dissecting tray
- Preserved cat, skin removed

Procedures

1. Follow the axillary artery to where it enters the forelimb and is now called the **brachial artery**. Distal to the elbow, the brachial artery gives rise to the **radial artery** and the **ulnar artery**.

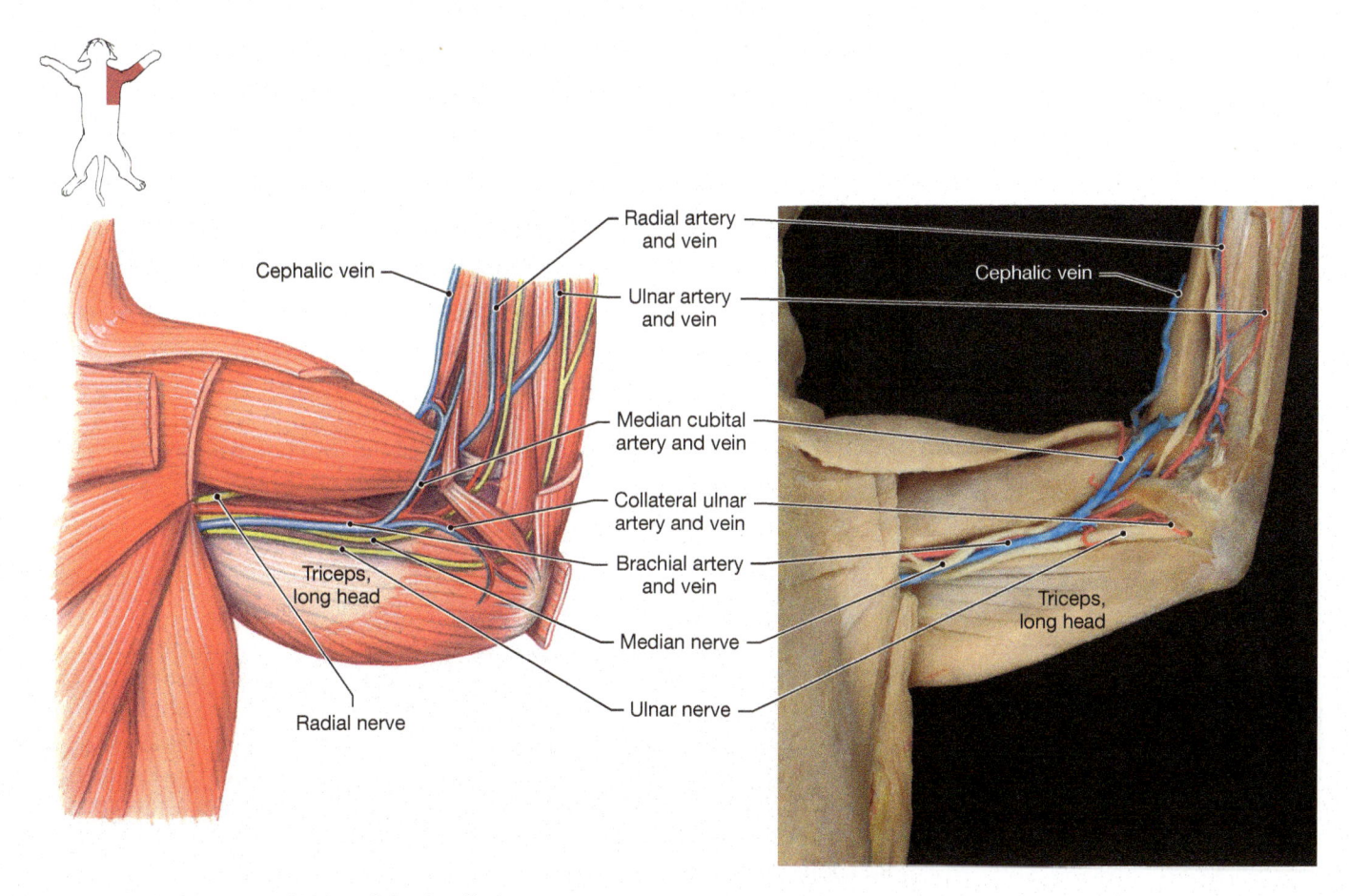

Figure D4.2 Arteries and Veins of the Forelimb
Illustration and photo.

LAB ACTIVITY 4 — Arteries That Supply the Abdominal Cavity

The aorta posterior to the diaphragm is called the **abdominal aorta**. As it passes through the abdomen, arteries to the digestive organs, spleen, urinary system, and reproductive system branch off of the abdominal aorta. As you identify the vessels, note the pattern of paired and unpaired vessels along the abdominal aorta. Use Figure D4.3 as a reference to identify these vessels in the cat.

4 Materials

- ☐ Gloves
- ☐ Safety glasses
- ☐ Dissecting tools
- ☐ Dissecting tray
- ☐ Preserved cat, skin removed

Procedures

1. A single vessel, the **celiac trunk**, is the first arterial branch off the abdominal aorta. Notice how it divides into three branches: the hepatic, left gastric, and splenic arteries.

    a. Along the cranial border of the gastrosplenic part of the pancreas lies the **hepatic artery**. It turns cranially near the pylorus, and lies in a fibrous sheath together with the portal vein and the common bile duct. Its branches include the **cystic artery** to the gallbladder and liver and the **gastroduodenal artery** near the pylorus.

    b. Along the lesser curvature of the stomach lies the **left gastric artery**, which supplies many branches to both dorsal and ventral stomach walls.

    c. The **splenic artery** is the largest branch of the celiac artery. It supplies at least two branches to the dorsal surface of the stomach and divides into anterior and posterior branches to supply these portions of the spleen.

2. Just posterior to the celiac trunk, find the unpaired **superior mesenteric artery**. It divides into numerous intestinal branches that supply the small and large intestines.

3. Next notice the paired **adrenolumbar arteries** just posterior to the superior mesenteric artery. These two arteries, which supply the adrenal glands, pass laterally along the dorsal body wall and give rise to **phrenic** and **adrenal arteries** and then supply the muscles of the dorsal body wall. The phrenic artery branches off the adrenolumbar artery and supplies the diaphragm.

4. Locate the paired **renal arteries** that emerge from the abdominal aorta to supply the kidneys. In some specimens, each renal artery gives rise to an adrenolumbar artery. Often double renal arteries supply each kidney.

5. Locate the paired **gonadal arteries** as they arise from the aorta near the caudal ends of the kidneys.

    a. In females, the gonadal arteries are called the **ovarian arteries**. They pass laterally in the broad ligament to supply the ovaries. Each artery gives a branch to the cranial end of the corresponding uterine horn.

    b. In the male, the gonadal arteries are called the **spermatic arteries**. They lie on the surface of the iliopsoas muscle, and pass caudally to the internal inguinal ring and through the inguinal canal to the testes.

6. Next find the **lumbar arteries,** which are seven pairs of arteries that arise from the dorsal surface of the aorta and supply the muscles of the dorsal abdominal wall.

7. The unpaired **inferior mesenteric artery** arises from the abdominal aorta near the last lumbar vertebra. Close to its origin, notice how this vessel divides into the **left colic artery**, which passes anteriorly to supply the descending colon, and the **superior hemorrhoidal artery**, which passes posteriorly to supply the rectum.

8. Locate the paired **iliolumbar arteries** as they arise near the inferior mesenteric artery and pass laterally across the iliopsoas muscle to supply the muscles of the dorsal abdominal wall.

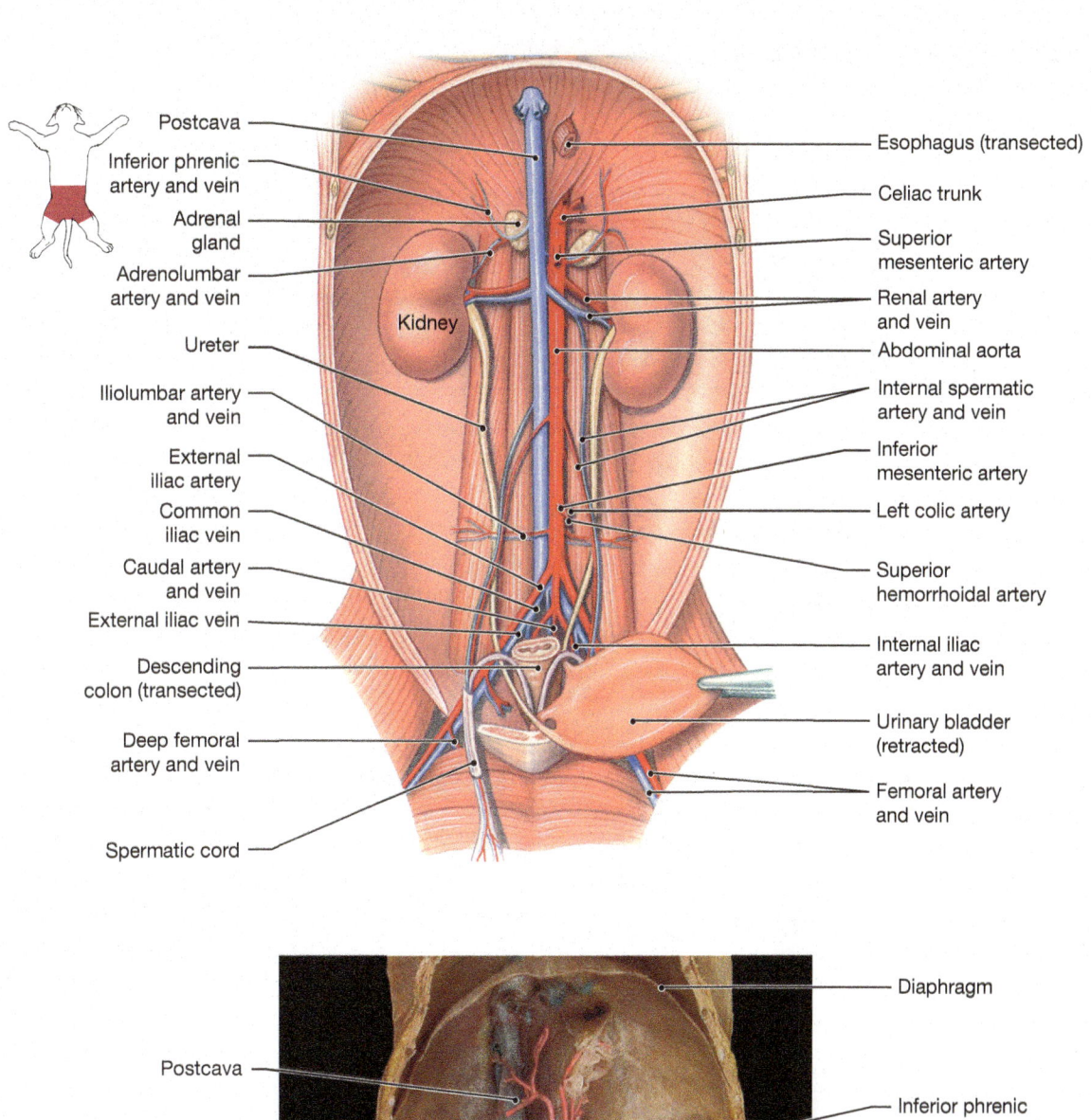

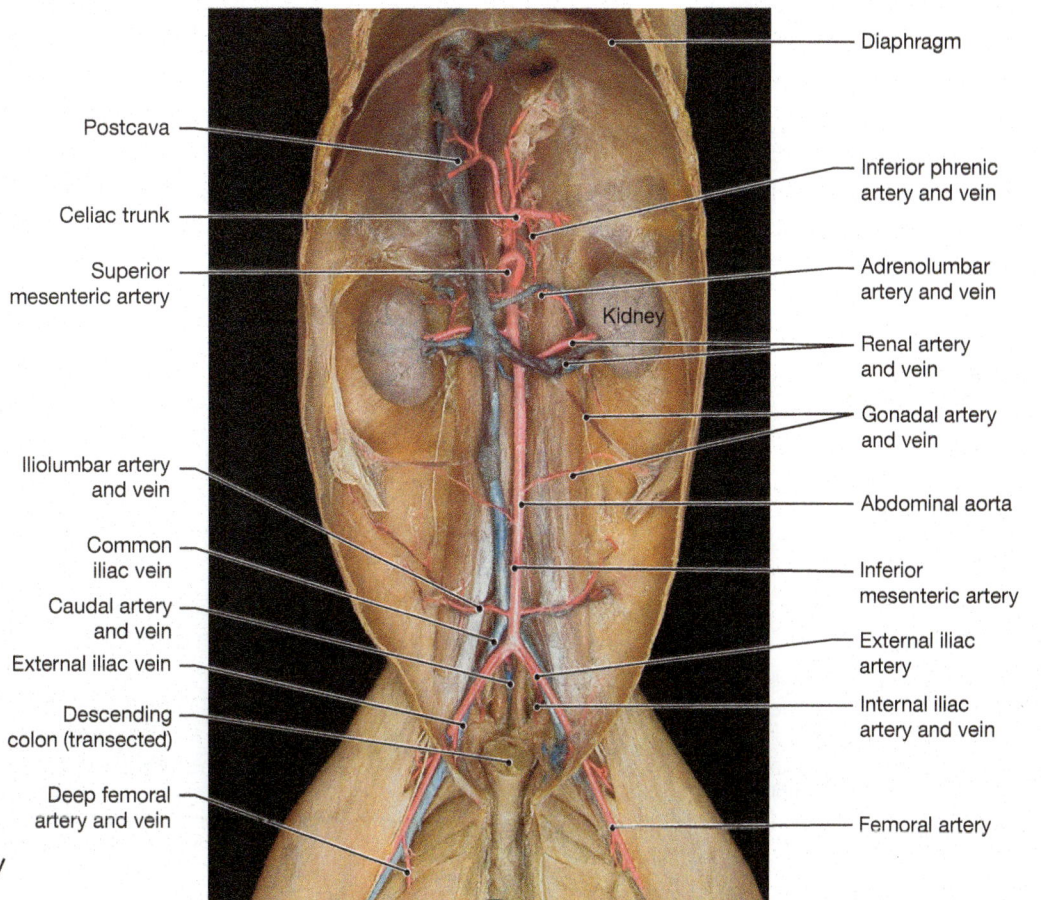

Figure D4.3 Arteries and Veins of the Abdominal Cavity
Illustration and photo.

538 DISSECTION EXERCISE 4 Cat Circulatory System

9. Near the sacrum, the abdominal aorta branches into three vessels that you should search for next. The **right** and **left external iliac arteries** lead toward the hindlimbs, and the single **internal iliac artery** serves the tail. Unlike humans, cats do not have common iliac arteries.

10. Last, look for the first branch of the internal iliac artery, called the **umbilical artery**, and the **caudal (medial sacral) artery**, which passes into the tail. ∎

LAB ACTIVITY 5 — Arteries That Supply the Hindlimb (Medial View)

The external iliac arteries enter the thigh to supply blood to the hindlimb. Use Figure D4.4 as a reference to locate these vessels in the cat.

5 Materials

- ☐ Gloves
- ☐ Safety glasses
- ☐ Dissecting tools
- ☐ Dissecting tray
- ☐ Preserved cat, skin removed

Procedures

1. Follow either one of the external iliac arteries, and note how, just prior to leaving the abdominal cavity, it gives off the **deep femoral artery**, which passes between the iliopsoas and the pectineus to supply the muscles of the thigh (see Figure D4.3).

2. Each external iliac artery continues outside the abdominal cavity as a **femoral artery** (Figure D4.4). Note a femoral artery that lies on the medial surface of the thigh. At the posterior knee, this artery is called the **popliteal artery** and divides into the **anterior** and **posterior tibial arteries**. ∎

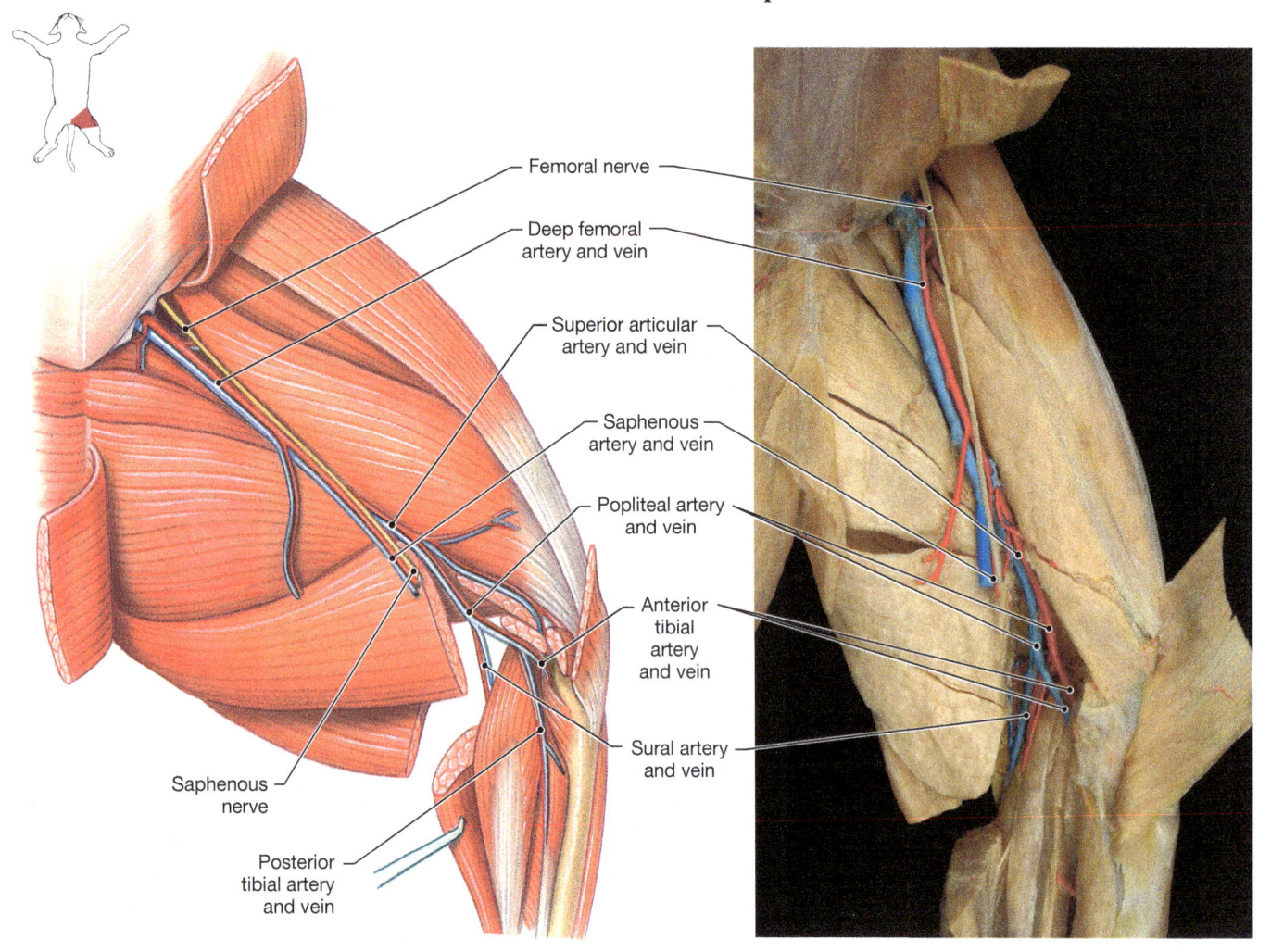

Figure D4.4 Arteries and Veins of the Hindlimb
Illustration and photo.

LAB ACTIVITY 6   **Veins That Drain the Head, Neck, and Thorax**

Veins usually follow the pattern of the corresponding arteries. Trace the veins in the direction of venous blood flow, which is away from the extremities toward the heart. Refer to Figure D4.1 for the veins of the head, neck, and thorax.

Materials

- Gloves
- Safety glasses
- Dissecting tools
- Dissecting tray
- Preserved cat, skin removed

Procedures

1. Gently pull the heart away from a lung and examine the exposed lung root. Each lobe of the lung has a **pulmonary vein** (which usually is not injected with colored latex) that passes oxygenated blood toward the dorsal side of the heart to enter the left atrium.

2. Find the two major veins that deliver deoxygenated blood to the right atrium of the heart: the precava and the postcava veins. (In humans, these veins are called the superior vena cava and the inferior vena cava.) The **precava** is the large vessel that drains blood from the head, neck, and forelimbs to the right atrium. Its principal tributaries include the internal and external jugular veins, the subscapular veins, and the axillary veins. The **postcava** is the large vessel that returns blood from the abdomen, hindlimbs, and pelvis to the right atrium. It drains blood from numerous vessels of the abdomen.

3. Observe the smaller vessels that feed into the precava:

   a. The paired **internal thoracic (mammary) veins** unite at a small stem that joins the precava. The internal thoracic veins lie on either side of the body midline and drain the ventral chest wall.

   b. Push the heart toward the left lung to see the **azygous vein** that arches over the root of the right lung and join the precava near the right atrium. The **intercostal veins** drain blood from the intercostal muscles into the azygous vein.

4. Locate the **axillary vein,** the major vessel that drains the forelimb in the axilla. The **subscapular vein** is the largest tributary to the axillary vein. A small vessel that empties into the axillary vein on its ventral surface is the **ventral thoracic vein**. It is located near the subscapular vein. The **long thoracic vein** is another small vessel that empties into the axillary vein. It is distal to the ventral thoracic vein.

5. Draining each side of the head is the **external jugular vein**, which merges with the subclavian vein. Each external jugular vein joins with a subclavian vein to form the brachiocephalic vein. The **internal jugular vein** drains the brain and spinal cord and joins the external jugular vein near its union with the brachiocephalic vein. Just superior to the hyoid bone, find the **transverse jugular vein** as it connects the left and right external jugular veins. At the shoulder, the external jugular vein receives the large **transverse scapular vein** that drains the dorsal surface of the scapula.

6. The subclavian veins drain into a pair of **brachiocephalic veins** that unite as the precava.

7. Notice the **costocervical** and **vertebral veins**, which form a common stem and dorsally connect with the brachiocephalic vein. The vertebral vein in some specimens empties into the precava vein. ∎

LAB ACTIVITY 7 — Veins That Drain the Forelimb (Medial View)

The veins of the forelimb drain into the axillary vein and the transverse scapular vein. Refer to Figure D4.2 for the veins of the forelimb.

7 Materials

- ☐ Gloves
- ☐ Safety glasses
- ☐ Dissecting tools
- ☐ Dissecting tray
- ☐ Preserved cat, skin removed

Procedures

1. In the forelimb, locate the **radial vein** on the lateral side and the **ulnar vein** medially.
2. Near the elbow, these veins drain into the **brachial vein**, which ascends the forelimb and becomes the axillary vein.
3. The **cephalic vein** ascends the forelimb and joins the transverse scapular vein, which drains into the external jugular vein. In humans, the cephalic vein joins the axillary vein. ■

LAB ACTIVITY 8 — Veins That Drain the Abdominal Cavity

The postcava receives blood from the major veins that drain the abdomen, pelvis, and hindlimbs. For ease of identification, the order in which we describe the veins that empty into the postcava is the order in which they appear as we move from the diaphragm to the tail and hindlimb. Refer to Figure D4.3 for the veins of the abdominal cavity.

8 Materials

- ☐ Gloves
- ☐ Safety glasses
- ☐ Dissecting tools
- ☐ Dissecting tray
- ☐ Preserved cat, skin removed

Procedures

1. Along the dorsal pelvis wall, a pair of common iliac veins join together as the postcava. Ventral to the postcava is the major artery of the abdomen, the abdominal aorta. The large **internal iliac veins** enter the common iliac vein in the pelvic cavity and drain the rectum, bladder, and internal reproductive organs. Distal to the joining of the internal iliac vein with the common iliac vein, the **external iliac vein** is a continuation of the femoral vein from the hindlimb. The **caudal vein** drains the tail and empties into the origin of the postcava.
2. Working toward the heart, identify in sequence the veins that drain into the postcava the iliolumbar veins and the **lumbar veins** that drain the abdominal muscles. Next are the gonadal veins that drain the reproductive organs. In males, these vessels drain the testes and are called the **internal spermatic veins**. Usually, the left internal spermatic vein empties into the renal vein. In females, the gonadal veins drain the ovaries and are called the **ovarian veins**. These vessels empty into the postcava vein. Note the paired **renal veins** as they drain the kidneys into the postcava. Occasionally, double renal veins drain each kidney. Also identify the **adrenolumbar veins**, which drain the adrenal glands.
3. The **hepatic veins** drain blood from the liver into the postcava. The postcava is in close contact with the liver, and as a result the hepatic veins are difficult to locate. Remove some liver tissue around the postcava, and try to expose the hepatic veins. In triple-injected specimens, the entire hepatic portal system is injected with yellow latex. Locate the **hepatic portal vein**, which carries blood from the intestines and other organs to the liver (Figure D4.5). Note that this vein has several veins that empty into it:

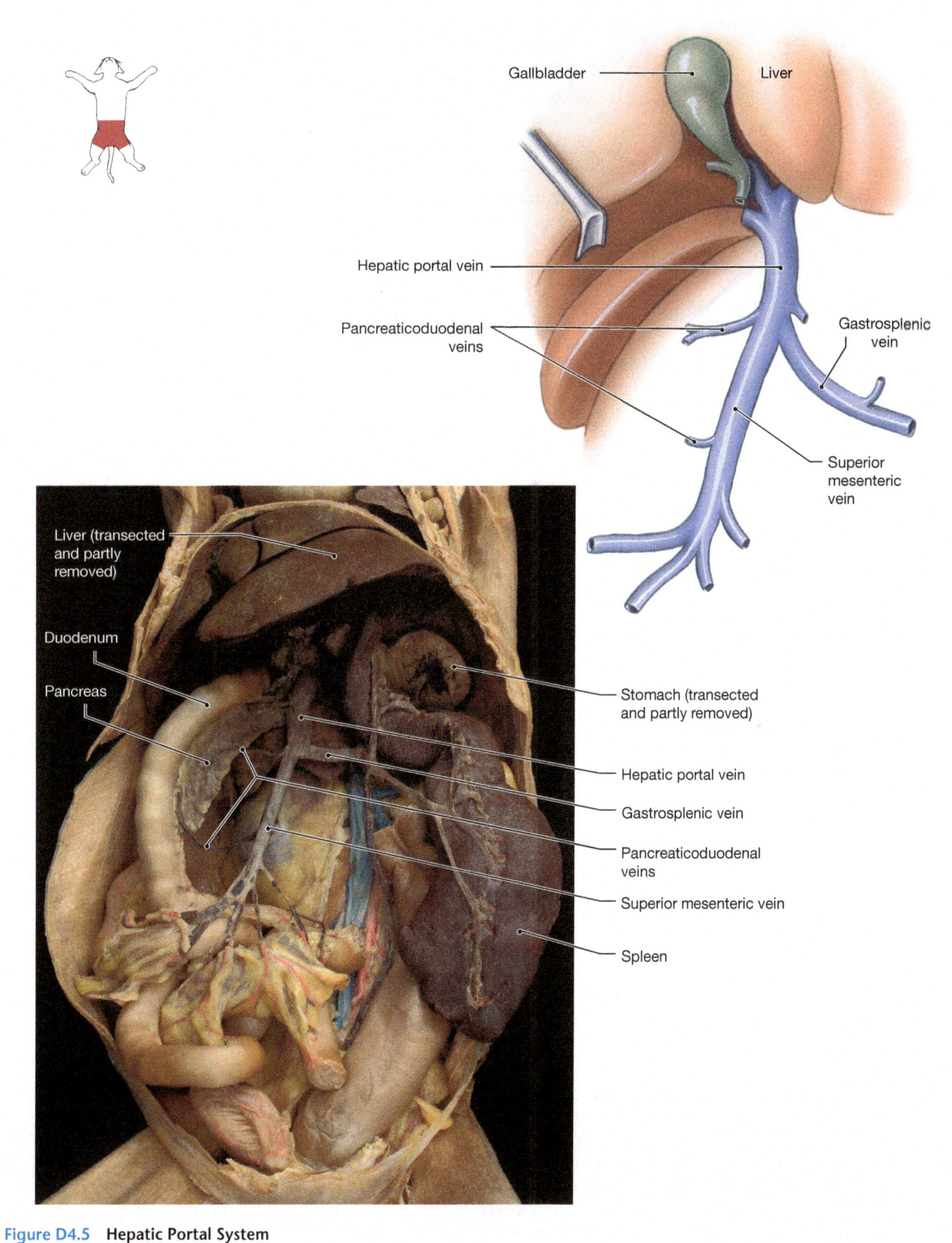

Figure D4.5 Hepatic Portal System
(a) Illustration that shows the hepatic portal system in isolation. (b) Photo that shows its location in the ventral body cavity.

a. The **superior mesenteric vein** is the large vein that drains the small and large intestines and the pancreas. It is the largest branch of the hepatic portal vein.

b. The **inferior mesenteric vein** follows along the inferior mesenteric artery. It drains part of the large intestine.

c. The **gastrosplenic vein**, which drains the stomach and spleen, lies on the dorsal side of the stomach.

4. Next find the **inferior phrenic veins** as they run from the diaphragm into the postcava just before the postcava pierces the diaphragm. ∎

LAB ACTIVITY 9: Veins That Drain the Hindlimb (Medial View)

The veins of the hindlimb basically correspond to the arteries in the region. Refer to Figure D4.4 for the veins of the hindlimb.

Materials

- Gloves
- Safety glasses
- Dissecting tools
- Dissecting tray
- Preserved cat, skin removed

Procedures

1. Find the medial and lateral branches of the **popliteal vein**, which drain the foot and calf region, and unite in the popliteal region to form the **saphenous vein**, which drains into the femoral vein. These vessels lie superficial to the muscles of the hindlimb.

2. Locate the superficial vein on the anterior surface of the thigh that is the **femoral vein**. As it enters the abdominal cavity, this vein becomes the external iliac vein. The **deep femoral vein** is a medial branch that empties into the femoral vein near the pelvis. ∎

> ⚠ **Safety Tip: Storing the Cat and Cleaning Up**
>
> To store your specimen, wrap it in the skin and moisten it with fixative. Use paper towels if necessary to cover the entire specimen. Return it to the storage bag and seal the bag securely. Label the bag with your name and place it in the storage area as indicated by your instructor.
>
> Wash all dissection tools and the tray, and set them aside to dry.
>
> Dispose of your gloves and any tissues from the dissection into a biohazard box or as indicated by your laboratory instructor. Wipe your work area clean and wash your hands. ▲

DISSECTION
LAB REPORT

EXERCISE 4

Cat Circulatory System

Name _____
Date _____
Section _____

A. **Matching**

Match each description in the left column with its correct definition from the right column.

_____ 1. superior vena cava in humans
_____ 2. major branch off right subclavian artery
_____ 3. drains blood from intercostals
_____ 4. inferior vena cava in humans
_____ 5. first major branch of aortic arch
_____ 6. vein supplying blood to liver
_____ 7. major branch off subscapular artery
_____ 8. major branch off brachiocephalic artery
_____ 9. vein that unites as small stem on precava
_____ 10. artery of the tail

A. right common carotid artery
B. brachiocephalic artery
C. internal thoracic vein
D. internal iliac artery
E. precava
F. azygous vein
G. hepatic portal vein
H. postcava
I. left common carotid artery
J. thoracodorsal artery

EXERCISE 4 — LAB REPORT

DISSECTION

B. Labeling

Identify the vessels in Figure D4.6.

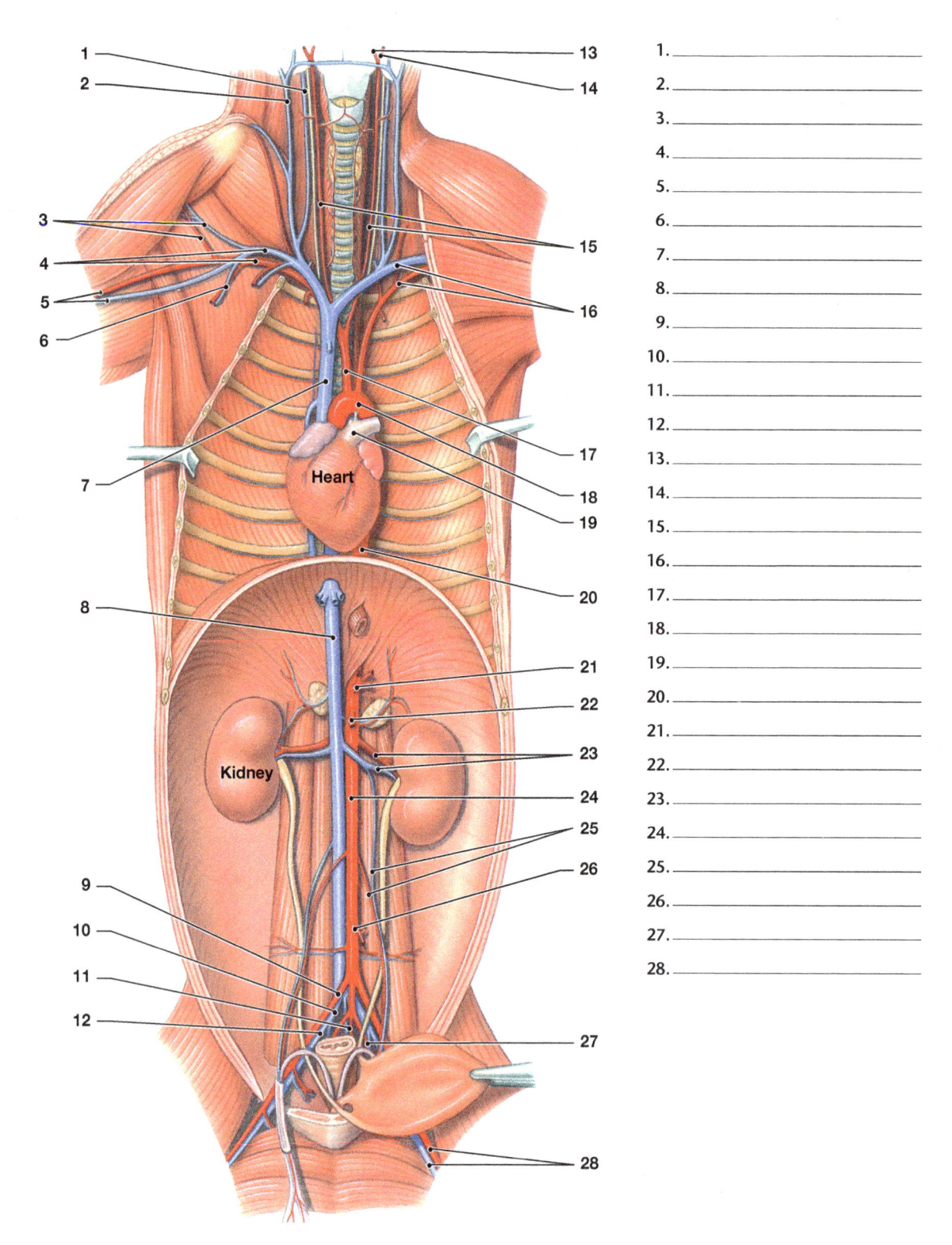

Figure D4.6 Major Arteries and Veins of the Cat

1. _____
2. _____
3. _____
4. _____
5. _____
6. _____
7. _____
8. _____
9. _____
10. _____
11. _____
12. _____
13. _____
14. _____
15. _____
16. _____
17. _____
18. _____
19. _____
20. _____
21. _____
22. _____
23. _____
24. _____
25. _____
26. _____
27. _____
28. _____

DISSECTION

LAB REPORT | **EXERCISE 4**

C. **Short-Answer Questions**

1. Trace the feline vascular route from the abdominal aorta to the intestines and into the postcava.

2. Give an example of a feline artery and its corresponding vein that are next to each other and have the same regional name.

3. Trace the return of blood to the heart starting in the saphenous vein.

4. List the vessels that transport blood to and from the lungs.

5. Why is the aorta called the major artery of systemic circulation?

6. Trace a drop of oxygenated blood from the heart to the right lower forelimb.

EXERCISE 4 — LAB REPORT

D. Analysis and Application

1. How is the branching off the aortic arch in cats different from this branching in humans?

2. Name three blood vessels found in cats but not in humans.

3. Describe the differences between the origins of the left and right common carotid arteries in cats and humans.

4. How is the branching off the abdominal aorta at the pelvis in cats different from this branching in humans?

DISSECTION EXERCISE 5

Cat Lymphatic System

LAB ACTIVITIES
1. Preparing the Cat for Dissection 547
2. The Cat Lymphatic System 549

OBJECTIVES

On completion of this exercise, you should be able to:

- Identify where the main feline lymphatic ducts empty into the vascular system.
- Identify lymph nodes in the feline jaw and tonsils in the mouth.
- Locate the feline spleen.

The lymphatic system protects the body against infection by producing lymphocyte blood cells, which manufacture antibodies to destroy microbes that have invaded the body. Another protective task of the lymphatic system is to collect liquid that has been forced out of blood capillaries and into extracellular spaces. This liquid, called **lymph**, is carried into the lymph nodes, where phagocytic cells remove debris and microbes from the lymph before the liquid passes out of the nodes and into the venous bloodstream. Because of this cleansing role, a node may occasionally itself become infected.

Major organs of the lymphatic system include the thymus gland, spleen, tonsils, lymph nodes, and lymphatic nodules. The nodules are similar to nodes but smaller and more scattered in the tissues of the digestive and other systems. Lymphatic vessels are long tubes similar to blood vessels except with a much lower fluid pressure. Movements of the body squeeze on the lymphatic vessels and push the contained lymph toward the heart, to the location near where the lymph is returned to the circulation.

This exercise complements the study of the human lymphatic system.

⚠️ **Safety Tip: Cat Dissection Basics**

Before beginning this dissection, review the Safety Tip presented at the beginning of Dissection Exercise 1. ▲

LAB ACTIVITY 1 Preparing the Cat for Dissection

If the ventral body cavity has not been opened on your dissection specimen, complete the following procedures. Otherwise, skip to Lab Activity 2. Use Figure D5.1 as a reference as you dissect the ventral body cavity.

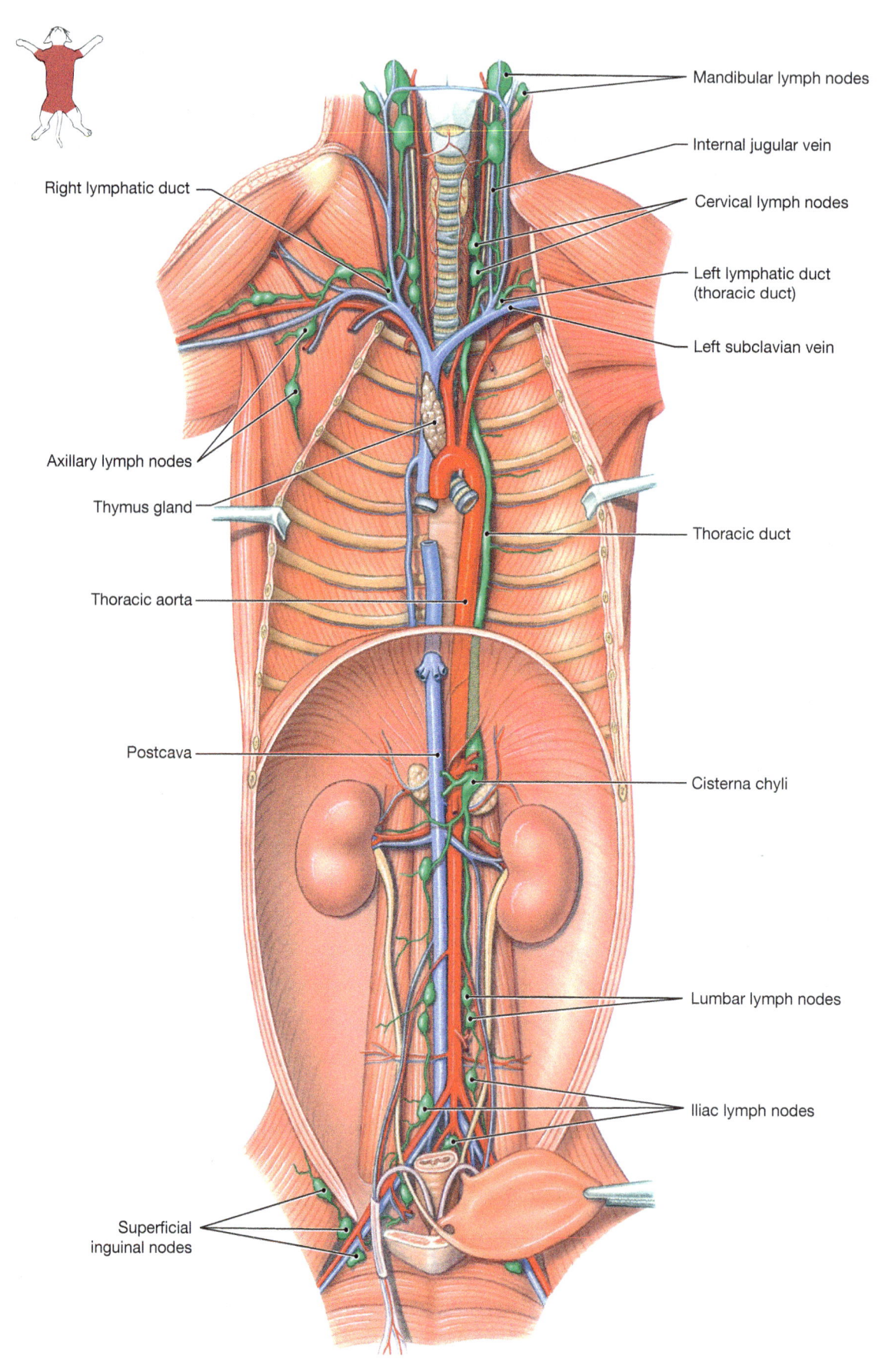

Figure D5.1 Major Lymphatic Ducts of the Cat

548

1 Materials

- ☐ Gloves
- ☐ Safety glasses
- ☐ Dissecting tools
- ☐ Dissecting tray
- ☐ String
- ☐ Preserved cat, skin removed

Procedures

1. Put on gloves and safety glasses, and clear your workspace before obtaining your dissection specimen.
2. Secure the specimen ventral side up on the dissecting tray by spreading the limbs and tying them flat with lengths of string passing under the tray. Use one string for the two forelimbs and one string for the two hindlimbs.
3. If the ventral body cavity has not been opened, use scissors to cut a midsagittal section through the muscles of the abdomen to the sternum.
4. To avoid cutting through the bony sternum, angle your incision laterally approximately 0.5 inch, and cut the costal cartilages. Continue the parasagittal section to the base of the neck.
5. Make a lateral incision on each side of the diaphragm. Use care not to damage the diaphragm or the internal organs. Spread the thoracic walls to observe the internal organs.
6. Make a lateral section across the pubic region and angled toward the hips. Spread the abdominal walls to expose the abdominal organs. ∎

LAB ACTIVITY 2 — The Cat Lymphatic System

Lymphatic vessels are thin and difficult to locate. The larger lymphatic ducts near the subclavian veins may be colored with some blue latex that leaked in when nearby veins were injected. Use Figure D5.1 as a reference during your dissection and observations.

2 Materials

- ☐ Gloves
- ☐ Safety glasses
- ☐ Dissecting tools
- ☐ Dissecting tray
- ☐ Preserved cat, skin removed

Procedures

1. Reflect the muscles and wall of the chest to expose the thoracic cavity, and move the organs to one side. Locate the thin, brown **thoracic duct** lying along the dorsal surface of the descending aorta. This duct receives lymph from both hindlimbs, the abdomen, the left forelimb, and the left side of the chest, neck, and head. Examine the duct, and notice how the internal valves cause the duct to expand over the valve and appear beaded. Trace the duct anteriorly to where it empties into the external jugular vein near the left subclavian vein as the **left lymphatic duct**, which is an alternative name for the thoracic duct. In humans, the left lymphatic (thoracic) duct empties not into the external jugular vein, but rather into the left subclavian vein.
2. Trace the thoracic duct from the thorax to the abdomen. Posterior to the diaphragm, the thoracic duct is dilated into a sac called the **cisterna chyli**, where other lymphatic ducts from the hindlimbs, pelvis, and abdomen drain.
3. Return to the thorax, and locate where the left lymphatic (thoracic) duct empties into the external jugular vein. Now move to the right side of the thorax, and examine this area closely. The **right lymphatic duct** drains into the external jugular vein where the vein empties into the right subclavian vein. The right lymphatic duct drains lymph from the right forelimb and from the right side of the chest, neck, and head. Remember from step 1 that the left lymphatic (thoracic) duct drains lymph from the hindlimbs, abdomen, left forelimb, and left side of the chest, head, and neck. Review Figure 24.3 to see more detail on this uneven drainage of the right and left lymphatic vessels in humans.
4. On either side of the jaw, between the mandible and ear, identify the brown kidney-bean-shaped **lymph nodes**. Phagocytes in the lymph nodes remove

debris and pathogens from the lymph. Although lymph nodes are distributed throughout the body, they are typically small and difficult to locate. The most prominent nodes are in the inguinal and axillary regions and in the neck and jaw.

5. Open the mouth and examine the roof, which is called the *palate*. Posteriorly, the palate stops where the mouth joins the pharynx (throat). Note the folds of tissue that form an arch between the mouth and pharynx. Along the lateral wall of the arches are a pair of small and round **palatine tonsils**. As in humans, the feline tonsils are lymphatic organs, and their enclosed lymphocytes help fight infection.

6. On the superior surface of the heart locate the **thymus gland**. It is important in the development of the immune system. It is larger in immature cats (and in humans, too) and gradually replaced by fat in adults.

7. Next locate the **spleen**, the red, flat organ on the left side just posterior to the stomach. This lymphatic organ removes worn-out red blood cells from circulation and assists in recycling the iron from hemoglobin. Antigens in the blood stimulate the spleen to activate the immune system. ■

Safety Tip: Storing the Cat and Cleaning Up

To store your specimen, wrap it in the skin and moisten it with fixative. Use paper towels if necessary to cover the entire specimen. Return it to the storage bag and seal the bag securely. Label the bag with your name, and place it in the storage area as indicated by your instructor.

Wash all dissection tools and the tray, and set them aside to dry.

Dispose of your gloves and any tissues from the dissection as indicated by your laboratory instructor. ▲

DISSECTION LAB REPORT

Name _____
Date _____
Section _____

EXERCISE 5

Cat Lymphatic System

A. Matching

Match each structure in the left column with its correct description from the right column.

_____ 1. spleen
_____ 2. thoracic duct
_____ 3. lymph node
_____ 4. right lymphatic duct
_____ 5. cisterna chili
_____ 6. thymus

A. drains lymph from part of one side of body only
B. small mass that filters lymph
C. expanded posterior of main lymphatic duct
D. gland over heart
E. drains lymph from hindlimbs and from part of one side of body
F. removes worn blood cells from circulation

B. Short-Answer Questions

1. Describe the location and function of the spleen.

2. What is the function of lymph nodes?

3. List the organs of the lymphatic system.

551

EXERCISE 5

LAB REPORT

C. Analysis and Application

1. Compare the way the feline thoracic duct drains into the venous system with how this duct drains into the venous system in humans.

2. Compare the return of lymph on the right and left sides of the body.

3. Describe how liquid circulates from the blood, into lymphatic vessels, and is returned to the blood.

DISSECTION EXERCISE 6

Cat Respiratory System

LAB ACTIVITIES
1. Preparing the Cat for Dissection 553
2. Nasal Cavity and Pharynx 554
3. Larynx and Trachea 555
4. Bronchi and Lungs 555

OBJECTIVES

On completion of this exercise, you should be able to:

- Identify the structures of the feline nasal cavity.
- Identify the three regions of the feline pharynx.
- Identify the cartilages and folds of the feline larynx.
- Identify the structures of the feline trachea and bronchi.
- Identify and describe the gross anatomy of the feline lung.
- Compare and contrast the respiratory anatomy of cats and humans.

The main function of the respiratory system is to oxygenate the blood and to remove carbon dioxide. In this exercise you will identify the major organs of the feline respiratory system. As this exercise is designed to accompany the exercise on the human respiratory system, be sure to note differences between the respiratory anatomy of cats and humans.

This dissection will cover the nose, pharynx, larynx, trachea, bronchi, and lungs.

⚠️ **Safety Tip: Cat Dissection Basics**

Before beginning this dissection, review the Safety Tip presented at the beginning of Dissection Exercise 1. ▲

LAB ACTIVITY 1   Preparing the Cat for Dissection

If the ventral body cavity has not been opened on your dissection specimen, complete the following procedures. Otherwise, skip to Lab Activity 2. Use Figure D6.2 as a reference as you dissect the ventral body cavity.

1 Materials

- ☐ Gloves
- ☐ Safety glasses
- ☐ Dissecting tools
- ☐ Dissecting tray
- ☐ String
- ☐ Preserved cat, skin removed

Procedures

1. Put on gloves and safety glasses, and clear your workspace before obtaining your dissection specimen.
2. Secure the specimen ventral side up on the dissecting tray by spreading the limbs and tying them flat with lengths of string passing under the tray. Use one string for the two forelimbs and one string for the two hindlimbs.
3. Use scissors to cut a midsagittal section through the muscles of the abdomen to the sternum.

554 DISSECTION EXERCISE 6 Cat Respiratory System

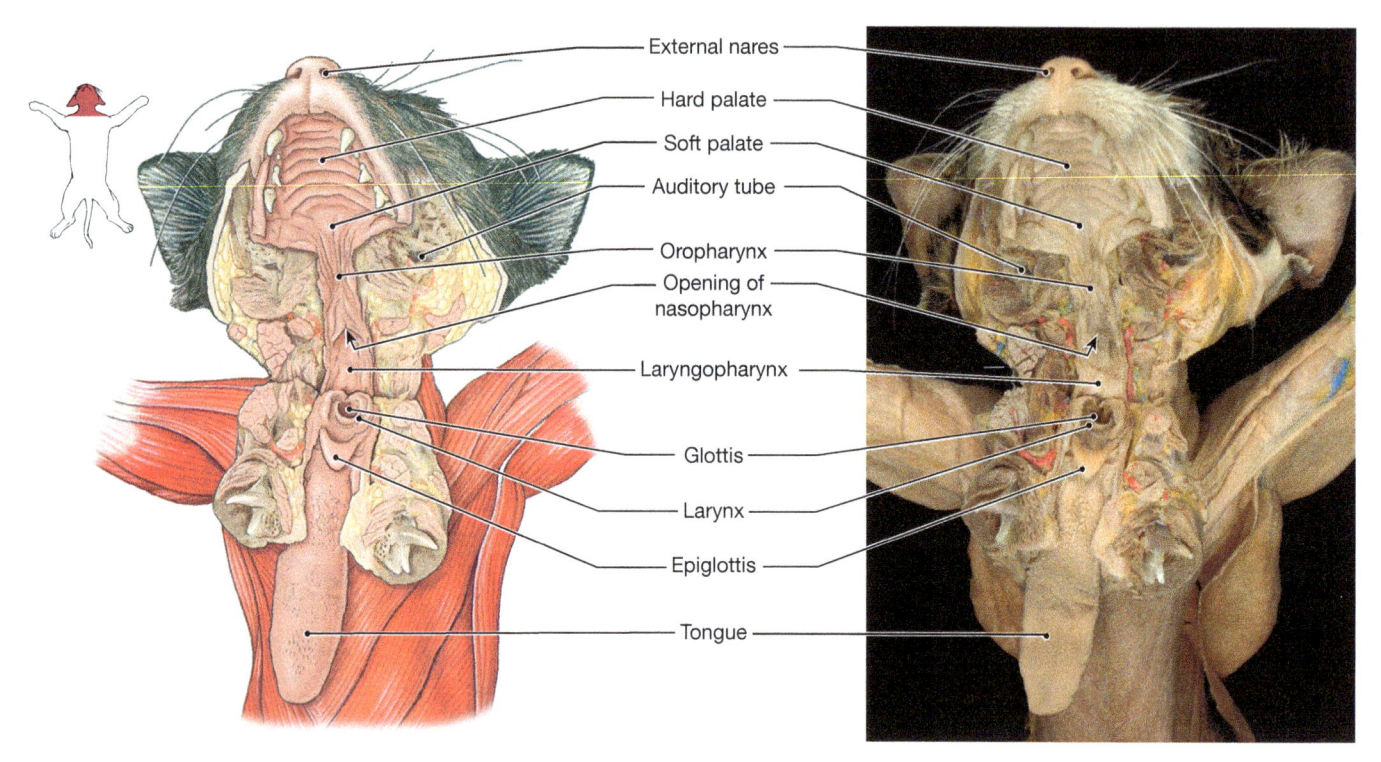

Figure D6.1 Oral Cavity of the Cat
Illustration and photo.

4. To avoid cutting through the bony sternum, angle your incision laterally approximately 0.5 inch, and cut the costal cartilages. Continue the parasagittal section to the base of the neck.

5. Make a lateral incision on each side of the diaphragm. Use care not to damage the diaphragm or the internal organs. Spread the thoracic walls to observe the internal organs. ∎

LAB ACTIVITY 2

Nasal Cavity and Pharynx

Air enters the nasal cavity through the two nostrils, which are called the **external nares** (Figure D6.1). Openings called the **internal nares** (also called **choanae**) connect the posterior of the nasal cavity with the **pharynx** (the throat), which is the cavity dorsal to the soft palate of the mouth. As in humans, the cat pharynx has three regions: the anterior **nasopharynx** behind the nose, the **oropharynx** behind the mouth, and the **laryngopharynx**, where the pharynx divides into the esophagus and larynx. Air must pass through an opening called the **glottis** to enter the larynx.

2 Materials

- ☐ Gloves
- ☐ Safety glasses
- ☐ Dissecting tools
- ☐ Dissecting tray
- ☐ Pipette or plastic tubing
- ☐ Preserved cat, skin removed

Procedures

1. Use bone cutters to cut through the angle of the mandible on one side only. Pull the lower jaw toward the uncut side, and leave the salivary glands intact on the uncut side. If necessary, secure the jaw with a pin so that it will not interfere with the rest of the dissection.

2. Carefully use a scalpel to cut through soft tissues, such as connective tissue and muscle, until you reach the pharynx.

3. Observe the internal nares and the three regions of the pharynx, and locate the glottis. ∎

LAB ACTIVITY 3 — Larynx and Trachea

The feline larynx has five cartilages, whereas the human larynx has nine cartilages (Figure D6.2). The **thyroid cartilage** is the large prominent ventral cartilage deep to the ventral neck muscles. Caudal to the thyroid cartilage is the **cricoid cartilage**, which is the only complete ring of cartilage in the respiratory tract. The paired **arytenoid cartilages** occupy the dorsal surface of the larynx anterior to the cricoid cartilage. A flap of cartilage called the **epiglottis** covers the glottis during swallowing and keeps food from entering the lower respiratory tract. Inside the larynx are vocal cords for production of sound. The anterior **vestibular ligaments**, commonly called the false vocal cords, protect the posterior **vocal ligaments** (true vocal cords), which vibrate and produce sounds.

Posteriorly, the larynx is continuous with the **trachea**, which is kept open by the C-shaped pieces of hyaline cartilage called the **tracheal rings**. On the dorsal side of the trachea is the food tube, the **esophagus**. Laterally, the common carotid arteries, internal jugular veins, and vagus nerve pass through the neck.

3 Materials

- Gloves
- Safety glasses
- Dissecting tools
- Dissecting tray
- Pipette or plastic tubing
- Preserved cat, skin removed

Procedures

1. Cut completely through the neck muscles to the body of a cervical vertebra. Carefully cut through any remaining connective tissue that may still be securing the larynx. Be careful not to cut the common carotid arteries or vagus nerves.
2. Locate the five cartilages of the larynx.
3. Identify the trachea and the tracheal rings. Also locate the esophagus, and on the lateral neck the common carotid arteries, internal jugular veins, and vagus nerve.
4. Expose and remove the larynx. Make a median incision on the dorsal surface of the larynx. Open the larynx to expose the elastic vocal cords between the thyroid and arytenoid cartilages. Identify the vestibular ligaments located anteriorly and the posterior pair of vocal ligaments. ■

LAB ACTIVITY 4 — Bronchi and Lungs

The trachea divides (bifurcates) into left and right **primary bronchi**. The bronchi penetrate the lungs at a slit-like **hilus** and branch repeatedly to supply the alveoli with air. The lungs of the cat have more lobes than the lungs of humans. The feline left lung has three lobes and the right lung has four: three main lobes and a fourth smaller mediastinal lobe (not shown in Figure D6.2). At the site of attachment, other structures such as the pulmonary artery and pulmonary veins enter and exit the lungs.

Each lung is enclosed in a serous membrane called the *pleura*. This membrane consists of the glistening **visceral pleura** on the lung surface and the **parietal pleura** that lines the thoracic wall. Between these layers is a small space called the **pleural cavity**, which contains **pleural fluid** secreted by the pleura.

4 Materials

- Gloves
- Safety glasses
- Dissecting tools
- Dissecting tray
- Pipette or plastic tubing
- Preserved cat, skin removed

Procedures

1. Trace the trachea to where it bifurcates into left and right primary bronchi. Note the hilus, where the bronchi enter the lungs. Examine this area for pulmonary arteries and pulmonary veins.
2. Using Figure D6.2 as a guide, identify the four lobes of the right lung and the three lobes of the left lung.
3. Next locate the pleura that surrounds each lung. Distinguish between the glossy visceral pleura on the lung surface and the parietal pleura on the thoracic wall.
4. The diaphragm is the sheet of muscle that divides the thoracic cavity from the abdominal cavity and is one of the major muscles involved in respiration.

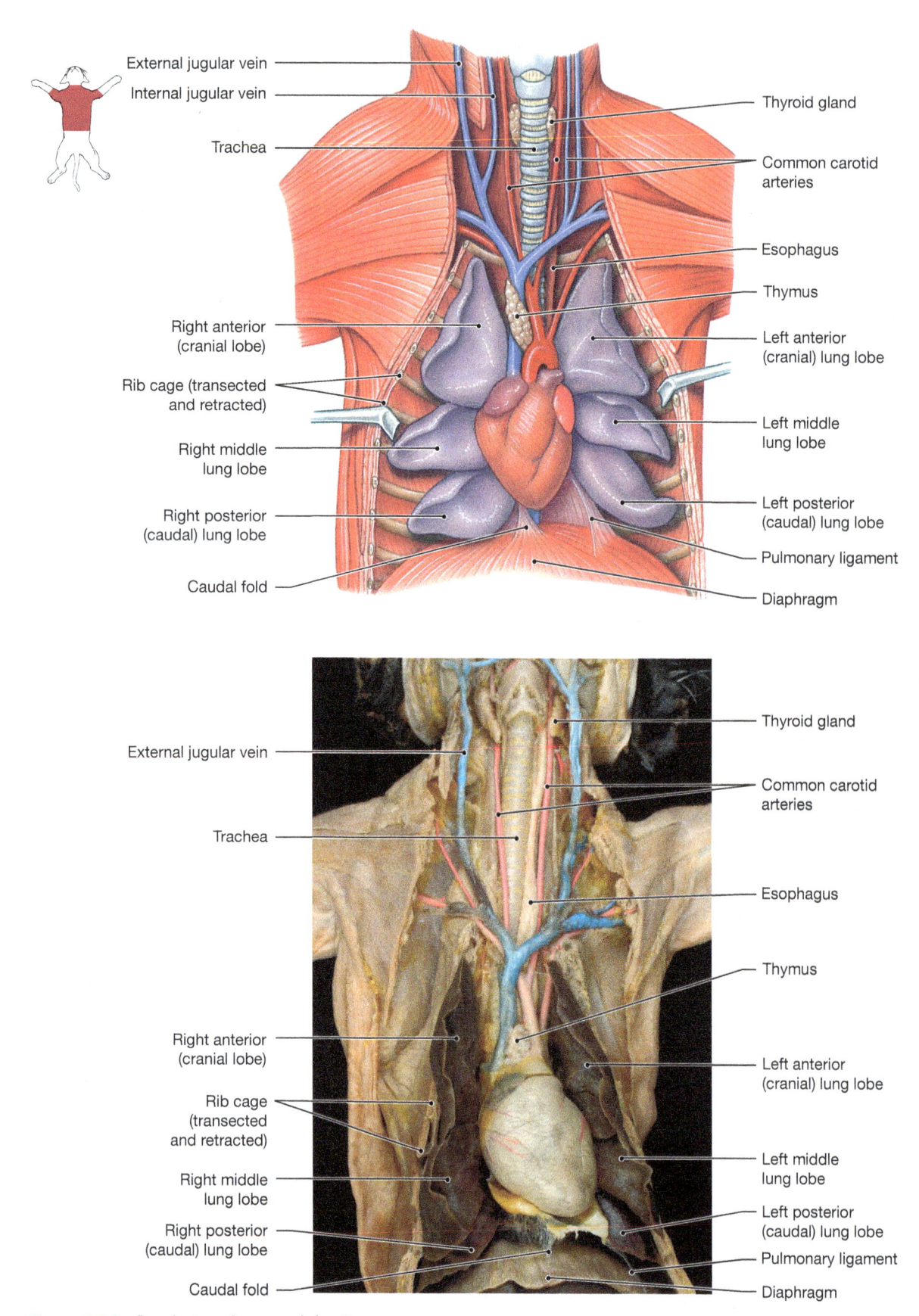

Figure D6.2 Respiratory System of the Cat
Illustration and photo. Note: Right lung's mediastinal lobe posterior to the heart not shown.

556

Locate the **phrenic nerve** that controls the diaphragm. This nerve should be clearly visible as a white "thread" along the heart.

5. Place a clean pipette or a piece of plastic tubing into the cat's mouth and push it into the laryngopharynx. Attempt to inflate the cat's lungs by gently exhaling into the tube. Observe the expansion of the lungs as they fill with air.

6. Remove a section of lung from a lobe. Observe the cut edge of the specimen and notice the spongy appearance. Are blood vessels visible? ■

⚠ **Safety Tip: Storing the Cat and Cleaning Up**

To store your specimen, wrap it in the skin and moisten it with fixative. Use paper towels if necessary to cover the entire specimen. Return it to the storage bag and seal the bag securely. Label the bag with your name, and place it in the storage area as indicated by your instructor.

Wash all dissection tools and the tray, and set them aside to dry.

Dispose of your gloves and any tissues from the dissection as indicated by your laboratory instructor. ▲

DISSECTION LAB REPORT

EXERCISE 6: Cat Respiratory System

Name _____
Date _____
Section _____

A. Matching

Match each short definition in the left column with its correct answer from the right column.

_____ 1. lung with three lobes
_____ 2. internal nares
_____ 3. external nares
_____ 4. lung with four lobes
_____ 5. complete ring of cartilage
_____ 6. largest cartilage of larynx
_____ 7. innervated by phrenic nerve
_____ 8. C-ring of cartilage
_____ 9. membrane on lung surface
_____ 10. membrane against thoracic wall
_____ 11. airway of larynx
_____ 12. passageway into lung

A. right lung
B. thyroid cartilage
C. visceral pleura
D. diaphragm
E. tracheal ring
F. choanae
G. glottis
H. left lung
I. cricoid cartilage
J. hilus
K. parietal pleura
L. nostrils

B. Short-Answer Questions

1. What are the three regions of the pharynx?

2. List the respiratory structures that air passes through from the external nares to the lungs.

559

EXERCISE 6 — LAB REPORT

3. Describe the similarities between the feline larynx and the human larynx.

4. Which structures produce vocal sounds?

C. Analysis and Application

1. Compare the gross anatomy of the lungs of cats and humans.

2. Knowing that the feline lungs have more lobes than the lungs in humans, speculate on how the feline secondary bronchi compare with human secondary bronchi.

3. Describe the cartilaginous structures of the respiratory system.

DISSECTION EXERCISE 7

Cat Digestive System

OBJECTIVES

On completion of this exercise, you should be able to:

- Identify the structures of the feline mouth, stomach, and small and large intestines.
- Identify the gross anatomy of the feline liver, gallbladder, and pancreas.
- Compare and contrast the digestive system of cats and humans.

LAB ACTIVITIES

1. Preparing the Cat for Dissection 561
2. The Oral Cavity, Salivary Glands, Pharynx, and Esophagus 562
3. The Abdominal Cavity, Stomach, and Spleen 564
4. The Small and Large Intestines 565
5. The Liver, Gallbladder, and Pancreas 566

In this exercise you will be looking at the major organs and structures of the feline digestive system. Because the feline digestive system is very similar to that of the human, this dissection complements the study of the human digestive system. There are some differences, however, and these are described in the lab activities.

> ⚠️ **Safety Tip: Cat Dissection Basics**
>
> Before beginning this dissection, review the Safety Tip presented at the beginning of Dissection Exercise 1. ▲

LAB ACTIVITY 1 — Preparing the Cat for Dissection

Materials

- ☐ Gloves
- ☐ Safety glasses
- ☐ Dissecting tools
- ☐ Dissecting tray
- ☐ String
- ☐ Preserved cat, skin removed

If the ventral body cavity has not been opened on your dissection specimen, complete the following procedures. Otherwise, skip to Lab Activity 2.

Procedures

1. Put on gloves and safety glasses, and clear your workspace before obtaining your dissection specimen.
2. Secure the specimen ventral side up on the dissecting tray by spreading the limbs and tying them flat with lengths of string passing under the tray. Use one string for the two forelimbs and one string for the two hindlimbs.

3. Make a longitudinal midline incision through the muscles of the neck, the bones of the sternum, and the muscles of the thorax and the abdominal wall.

4. Make a lateral incision on the cranial side of the diaphragm and another on the caudal side of the diaphragm. Use care not to damage the diaphragm or the internal organs. Spread the thoracic walls to observe the internal organs. ■

LAB ACTIVITY 2: The Oral Cavity, Salivary Glands, Pharynx, and Esophagus

The mouth is called the **oral cavity**. The **vestibule** is the space between the teeth and lips in the front section of the oral cavity and between the teeth and cheeks on the sides of the oral cavity. The roof of the oral cavity consists of the bony **hard palate** and the posterior **soft palate**. Salivary glands in the head secrete saliva into the oral cavity. The pharynx connects the mouth to the esophagus, which in turn delivers food to the stomach.

The dentition in cats is different from that in humans, in that cats have 30 teeth whereas humans have 32. The cat dental formula is:

$$\frac{3\text{-}1\text{-}3\text{-}1}{3\text{-}1\text{-}2\text{-}1}$$

The sequence of numbers from left to right represents the types and numbers of teeth: incisors–canines–premolars–molars. The numbers in the top row represent teeth on one side of the upper jaw, and those in the bottom row represent teeth on one side of the lower jaw. Thus cats have 3 incisors, 1 canine, 3 premolars, and 1 molar on each side of the upper jaw, for a total of 16 upper teeth. The lower jaw has a total of 14 teeth.

The dental formula for humans is

$$\frac{2\text{-}1\text{-}2\text{-}3}{2\text{-}1\text{-}2\text{-}3}$$

Materials

- ☐ Gloves
- ☐ Safety glasses
- ☐ Dissecting tools
- ☐ Dissecting tray
- ☐ Preserved cat, skin removed

Procedures

1. Observe the raised **papillae** on the surface of the **tongue**. Taste buds are located on the papillae. Note the spiny filiform papillae in the front and middle of the tongue. These function as combs when the cat grooms itself by licking its fur. Cats have more filiform papillae than humans. Lift the tongue, and identify the inferior **lingual frenulum**, which is the structure that attaches the tongue to the floor of the oral cavity (Figure D7.1).

2. Use the feline dental formula given earlier to identify the different teeth. Observe if any teeth are missing or damaged.

3. At the posterior of the oral cavity, locate the pharynx, the three regions of which were studied in Dissection Exercise 6. Recall that these regions are the nasopharynx dorsal to the soft palate, the oropharynx posterior to the oral cavity, and the laryngopharynx around the epiglottis and the opening to the esophagus.

4. Unless already done in a previous dissection exercise, remove the skin from one side of the head. Carefully remove the connective tissue between the jaw and the ear. Observe the small, dark, kidney-bean-shaped **lymph nodes** and the oatmeal-colored, textured **salivary glands**. Locate the large **parotid gland** inferior to the ear on the surface of the masseter muscle. The **parotid duct** passes over this muscle and enters the oral cavity. The **submandibular gland** is inferior to the parotid gland. The **sublingual gland** is anterior to the submandibular gland. Ducts of both glands open

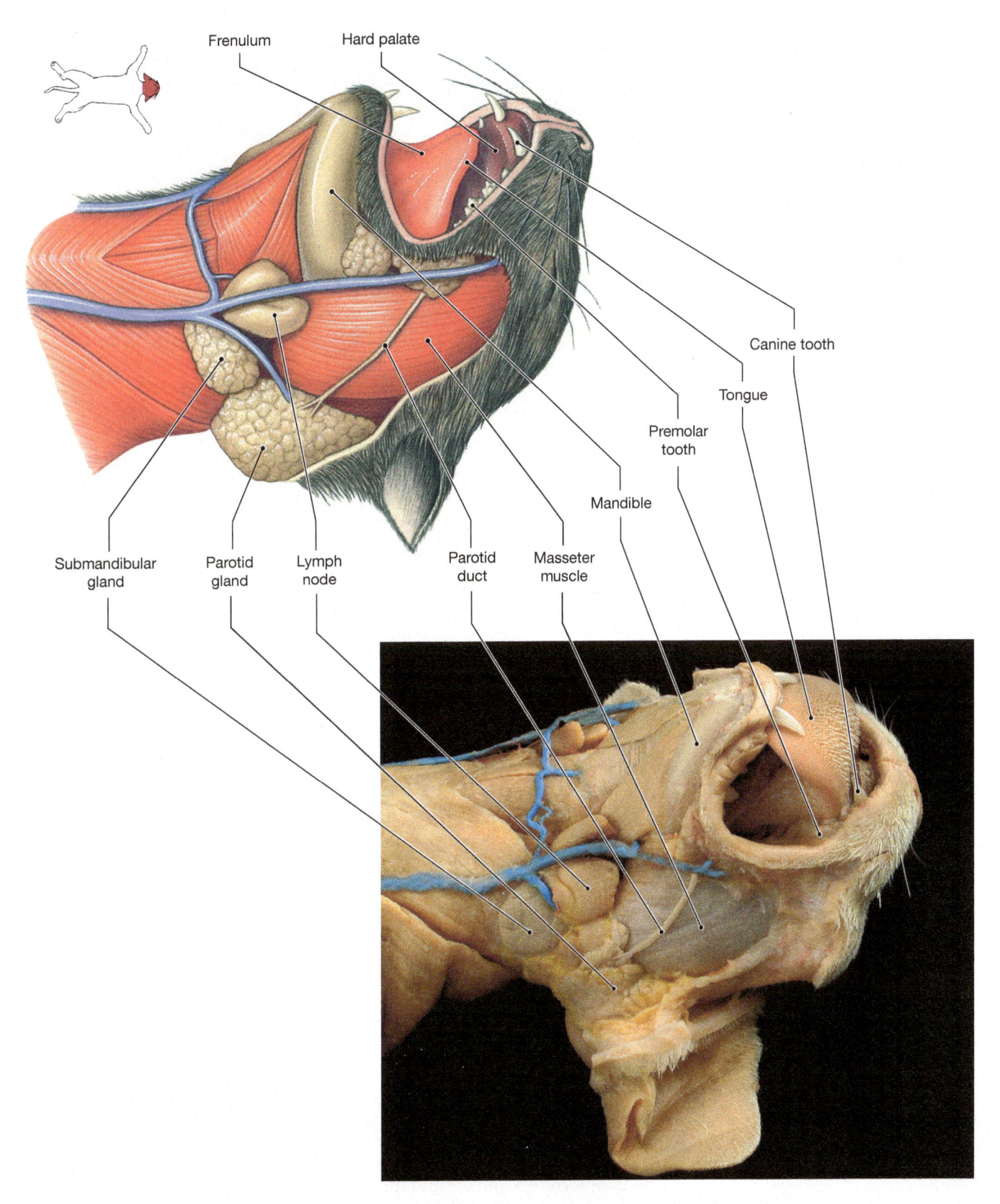

Figure D7.1 Facial Glands of the Cat
Illustration and photo.

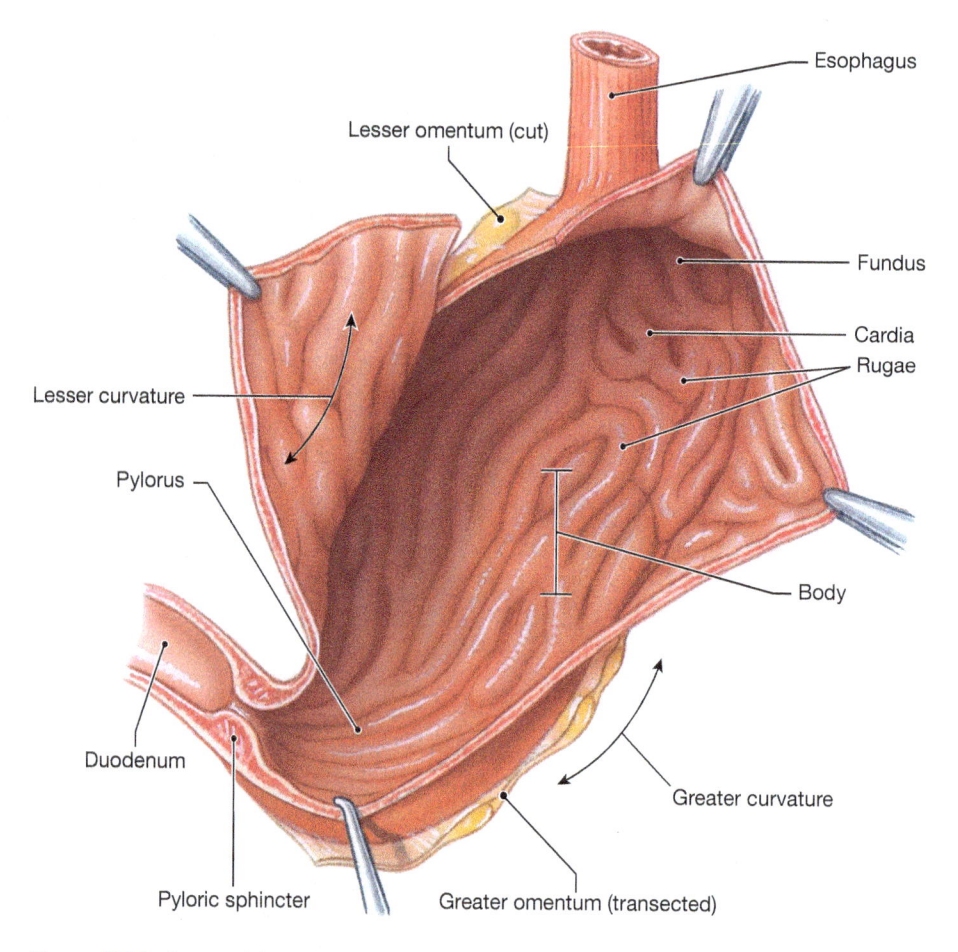

564 DISSECTION EXERCISE 7 Cat Digestive System

Figure D7.2 Internal Structure of the Stomach of the Cat

onto the floor of the mouth, but typically only the **submandibular duct** can be traced.

5. Return to the pharynx and identify the opening into the esophagus posterior to the epiglottis of the larynx. The esophagus connects the laryngopharynx to the stomach. Reflect the organs of the thoracic cavity and trace the esophagus through the diaphragm into the abdominal cavity, where it connects with the stomach. ■

LAB ACTIVITY 3 — The Abdominal Cavity, Stomach, and Spleen

The stomach (Figure D7.2) has four major regions: the **cardia**, at the entrance of the esophagus; the **fundus**, which is the dome-shaped pouch that rises above the esophagus; the **body**, the main portion of the stomach; and the **pylorus**, the posterior region of the stomach. The pylorus ends at the **pyloric sphincter**, the location where the digestive tube continues as the duodenum. The lateral margin of the stomach is convex and is called the **greater curvature**. The medial margin is concave and is called the **lesser curvature**.

The abdominal organs are protected by a fatty extension of the peritoneum from the greater curvature called the **greater omentum**. The **lesser omentum** is a peritoneal sheet of tissue on the lesser curvature that suspends the stomach from the liver.

3 Materials

- ☐ Gloves
- ☐ Safety glasses
- ☐ Dissecting tools
- ☐ Dissecting tray
- ☐ Preserved cat, skin removed

Procedures

1. Reflect the greater omentum to expose the abdominal organs. Note the attachment of the greater omentum to the stomach and the dorsal wall. Remove the greater omentum and discard the fatty tissue in the biohazard box, or as indicated by your instructor.
2. Locate the stomach and identify its four regions and the greater and lesser curvatures.
3. Make an incision through the stomach wall; run your scalpel along the greater curvature and continue about two inches past the pylorus and into the duodenum. Open the stomach and observe the pyloric sphincter. Large folds of the stomach mucosa, called **rugae**, are visible in the empty stomach.
4. Posterior to the stomach, in the abdominal cavity, observe a large, dark-brown organ, which is the **spleen**. ■

LAB ACTIVITY 4 The Small and Large Intestines

The small intestine has three regions (Figure D7.3). The first six inches is the C-shaped **duodenum**. It receives chyme from the stomach and secretions from the gallbladder, liver and pancreas. The **jejunum** comprises the bulk of the remaining length of the small intestine. The **ileum** is the last region of the small intestine and joins with the large intestine.

The large intestine is also divided into three regions. The first, following the terminus of the small intestine, is the **cecum**, which is wider than the rest of the large intestine and noticeably pouch-shaped. At this location is one difference between the feline and human digestive tracts: in humans, the appendix is attached to the cecum, but cats have no appendix. The greatest portion of the large intestine is the **colon**, which runs upward from the cecum, across the abdominal cavity, and then downward. The colon terminates in the third region of the large intestine, the **rectum**.

The intestines are surrounded by the peritoneum. Sheets of peritoneum, called **mesentery**, extend between the loops of intestines. The **mesocolon** is the mesentery of the large intestine.

4 Materials

- ☐ Gloves
- ☐ Safety glasses
- ☐ Dissecting tools
- ☐ Dissecting tray
- ☐ Hand lens
- ☐ Preserved cat, skin removed

Procedures

1. Identify the three portions of the small intestine: the duodenum, the jejunum, and the ileum. Rub your fingers around on the ileum at the point where it joins the large intestine to feel the **ileocecal sphincter**, the valve that controls the flow of chyme from the ileum into the cecum.
2. Extend the cut at the pylorus to several inches along the duodenum. Reflect the cut segment of the small intestine and secure it open with dissecting pins. Use a hand lens to observe the numerous **villi**, the duodenal **ampulla**, and the opening of the duct.
3. To view the large intestine, pull the loops of the small intestine to the cat's left and let them drape out of the body cavity.
4. Take note of the three parts of the colon. The **ascending colon** lies on the right side of the abdominal cavity and begins just superior to the cecum. The **transverse colon** extends across the abdominal cavity, and the **descending colon** runs on the left side of the posterior abdominal wall.
5. Next locate the rectum, which ends at the **anus**.
6. Examine the peritoneum that supports all three regions of the colon and attaches them to the posterior body wall. Here the peritoneum is called the **mesocolon**. ■

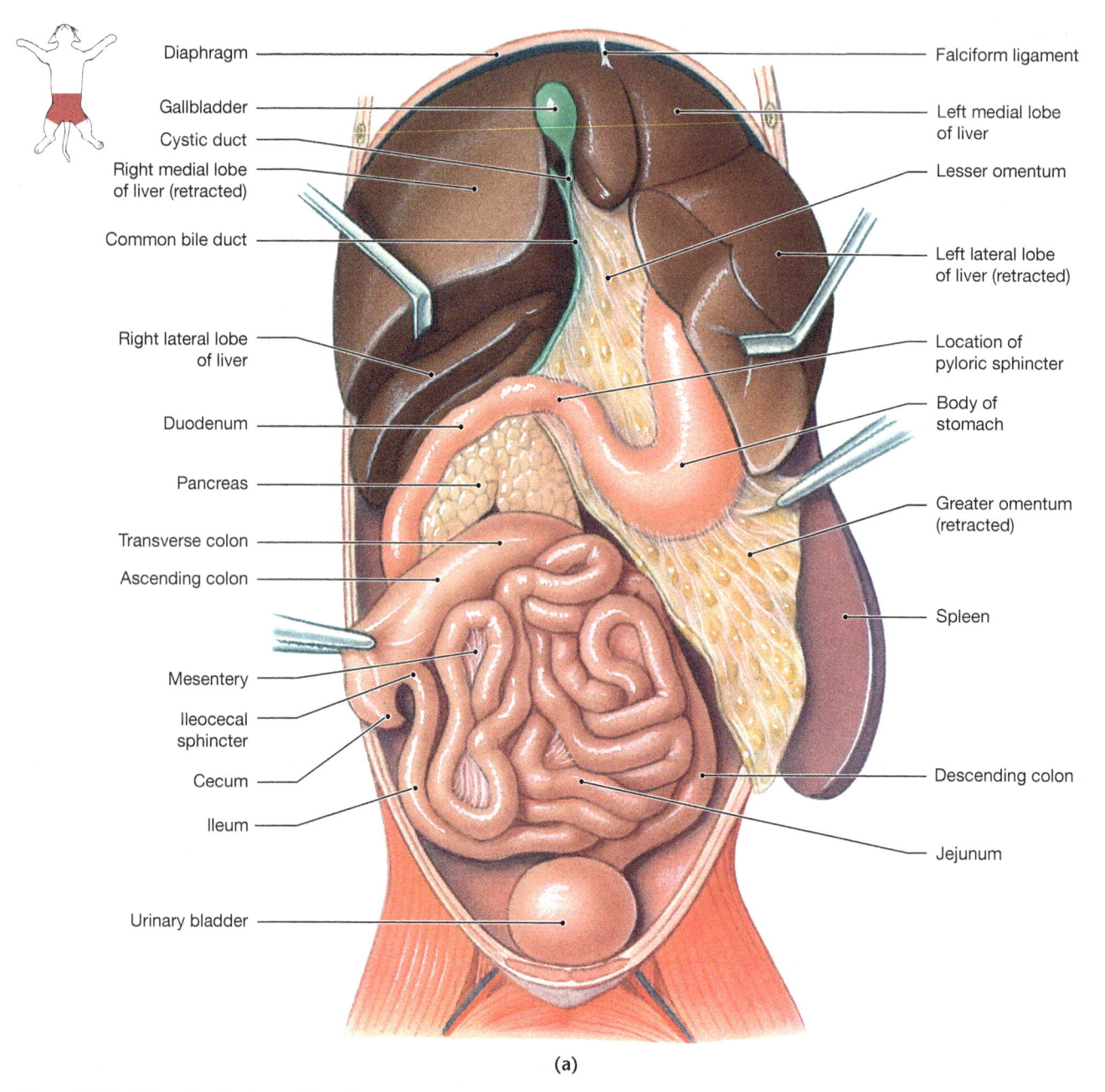

566 DISSECTION EXERCISE 7 Cat Digestive System

Figure D7.3 Digestive System of the Cat
(a) Illustration.

LAB ACTIVITY 5 — The Liver, Gallbladder, and Pancreas

The liver is the largest organ in the abdominal cavity and is located posterior to the diaphragm (see Figure D7.3). The liver is divided into five lobes: **right** and **left medial**, **right** and **left lateral**, and **caudate** (**posterior**). The liver in humans has only four lobes: right, left, caudate, and quadrate. The **falciform ligament** is a delicate membrane that attaches the liver superiorly to the diaphragm and abdominal wall. The gallbladder is a dark-green sac within a fossa in the right medial liver lobe. The liver produces bile, a substance that emulsifies lipids into small drops for digestion. The common hepatic duct transports bile from the liver. The cystic duct from

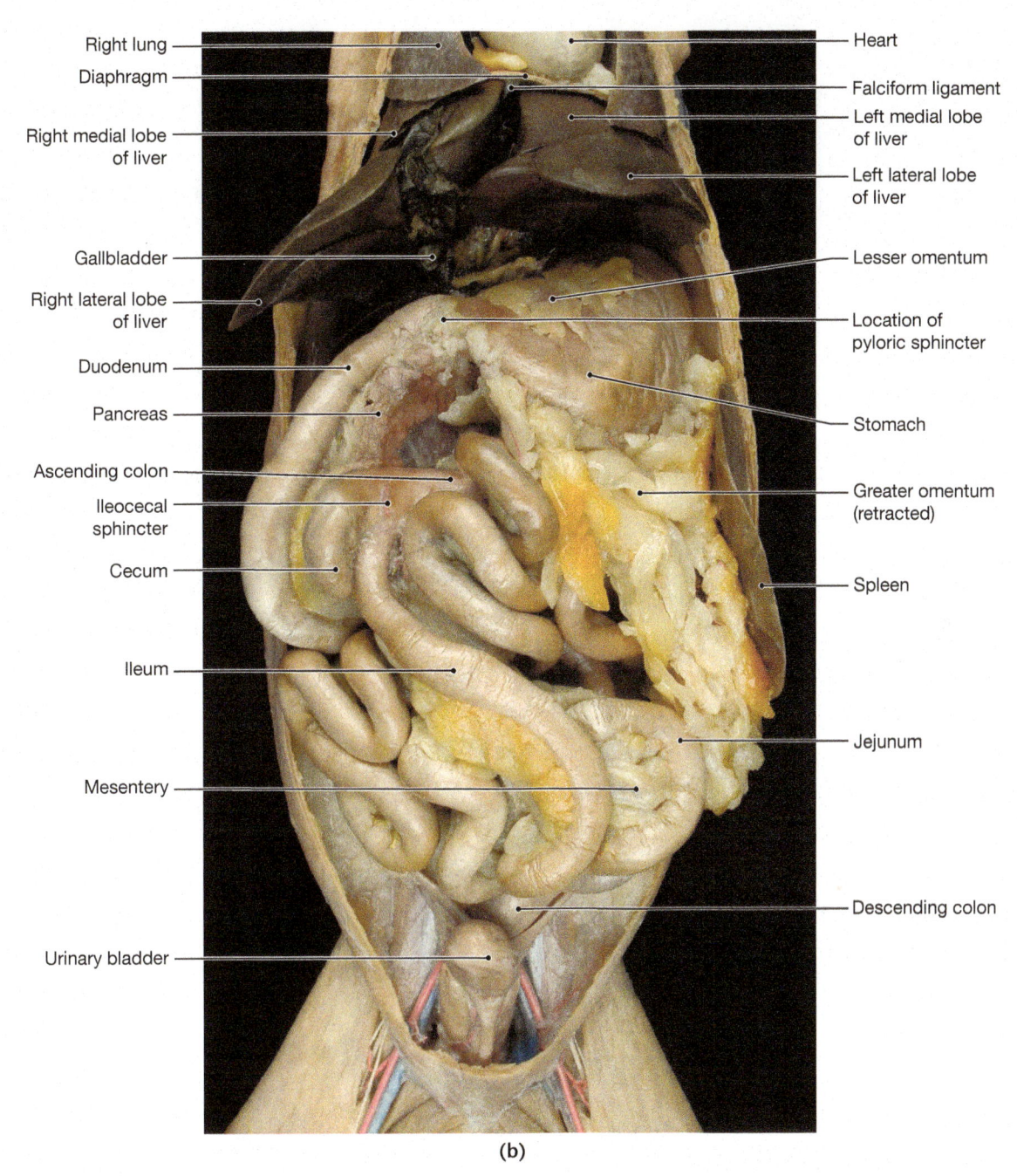

DISSECTION EXERCISE 7   Cat Digestive System   **567**

Figure D7.3 *(Continued)*
(b) Photo. Caudate lobe of liver not shown.

the gallbladder merges with the common hepatic duct as the common bile duct, which empties bile into the duodenum.

Posterior to the stomach and within the curvature of the duodenum lies the **pancreas**, the major glandular organ of the digestive system. In the cat, the pancreas has two regions, **head** and **tail**. The region that wraps around the duodenum is the head, and the portion that passes along the posterior surface of the stomach is the tail. In humans the pancreas has a broad middle portion called the *body*. The **pancreatic duct** (duct of Wirsung) transports pancreatic juice, which is rich in enzymes and buffers, to the duodenum. The common bile duct and the pancreatic duct join in the intestinal wall at the duodenal ampulla. Bile and pancreatic juice enter the duodenum from the ampulla.

5 Materials

- ☐ Gloves
- ☐ Safety glasses
- ☐ Dissecting tools
- ☐ Dissecting tray
- ☐ Preserved cat, skin removed

Procedures

1. Observe the large, brown liver posterior to the diaphragm and distinguish between the five lobes: right and left medial, right and left lateral, and caudate. Identify the gallbladder and the falciform ligament.

2. Tease the connective tissue away from the common hepatic duct, cystic duct, and common bile duct. Trace the common bile duct to its terminus at the duodenal wall.

3. Examine the head and tail of the pancreas. Expose the pancreatic duct and ampulla by using a teasing needle probe to scrape away the pancreatic tissue of the head portion. Trace the duct to the ampulla. The pancreatic and common bile ducts are adjacent to each other. ■

⚠ Safety Tip: Storing the Cat and Cleaning Up

To store your specimen, wrap it in the skin and moisten it with fixative. Use paper towels if necessary to cover the entire specimen. Return it to the storage bag and seal the bag securely. Label the bag with your name, and place it in the storage area as indicated by your instructor.

Wash all dissection tools and the tray, and set them aside to dry.

Dispose of your gloves and any tissues from the dissection as indicated by your laboratory instructor. ▲

DISSECTION

LAB REPORT

EXERCISE 7

Cat Digestive System

Name _____

Date _____

Section _____

A. Matching

Match each structure in the left column with its correct description from the right column.

_____ 1. greater omentum
_____ 2. lingual frenulum
_____ 3. pyloric sphincter
_____ 4. hard palate
_____ 5. cecum
_____ 6. lesser omentum
_____ 7. soft palate
_____ 8. liver
_____ 9. duodenal ampulla
_____ 10. pancreas
_____ 11. rugae
_____ 12. villi

A. site where bile empties into small intestine
B. muscular roof of mouth
C. valve of stomach
D. anchors tongue to floor of mouth
E. bony roof of mouth
F. fatty sheet that protects abdominal organs
G. pouch region of large intestine
H. folds of empty stomach
I. suspends stomach from liver
J. folds of small intestine
K. glandular organ near duodenum
L. soft organ with five lobes

B. Short-Answer Questions

1. Name the four parts of the peritoneum and describe their position in the feline abdominal cavity.

2. Trace a bite of food through the digestive tract from the mouth to the anus.

569

EXERCISE 7

LAB REPORT

DISSECTION

3. Describe the duodenal ampulla and the ducts that empty into it.

4. List the salivary glands in the feline.

C. Analysis and Application

1. Compare the dentition of cats and humans.

2. Compare the gross anatomy of the liver in cats and humans.

3. How does the cecum of the cat differ from that of humans?

# DISSECTION EXERCISE 8

Cat Urinary System

OBJECTIVES

On completion of this exercise, you should be able to:

- Locate and describe the gross anatomy of the feline urinary system.
- Identify the structures of the feline kidney.
- Identify the feline ureter, urinary bladder, and urethra.

LAB ACTIVITIES

1. Preparing the Cat for Dissection 571
2. External Anatomy of the Kidney 573
3. Internal Anatomy of the Kidney 573

The urinary system of the cat is similar to that of humans. In your dissection of the feline urinary system, trace the pathway of urine from its site of formation in the kidney through its passage via the urinary bladder and the urethra to the exterior of the body.

Safety Tip: Cat Dissection Basics

Before beginning this dissection, review the Safety Tip presented at the beginning of Dissection Exercise 1. ▲

LAB ACTIVITY 1 Preparing the Cat for Dissection

If the ventral body cavity has not been opened on your dissection specimen, complete the following procedures. Otherwise, skip to Lab Activity 2. Use Figure D8.1 as a reference as you dissect the ventral body cavity.

Materials

- ☐ Gloves
- ☐ Safety glasses
- ☐ Dissecting tools
- ☐ Dissecting tray
- ☐ String
- ☐ Preserved cat, skin removed

Procedures

1. Put on gloves and safety glasses and clear your workspace before obtaining your dissection specimen.
2. Secure the specimen ventral side up on the dissecting tray by spreading the limbs and tying them flat with lengths of string passing under the tray. Use one string for the two forelimbs and one string for the two hindlimbs.
3. If the ventral body cavity has not been opened, use scissors to cut a midsagittal section through the muscles of the abdomen to the sternum.
4. To avoid cutting through the bony sternum, angle your incision laterally approximately 0.5 inch and cut the costal cartilages. Continue the parasagittal section to the base of the neck.

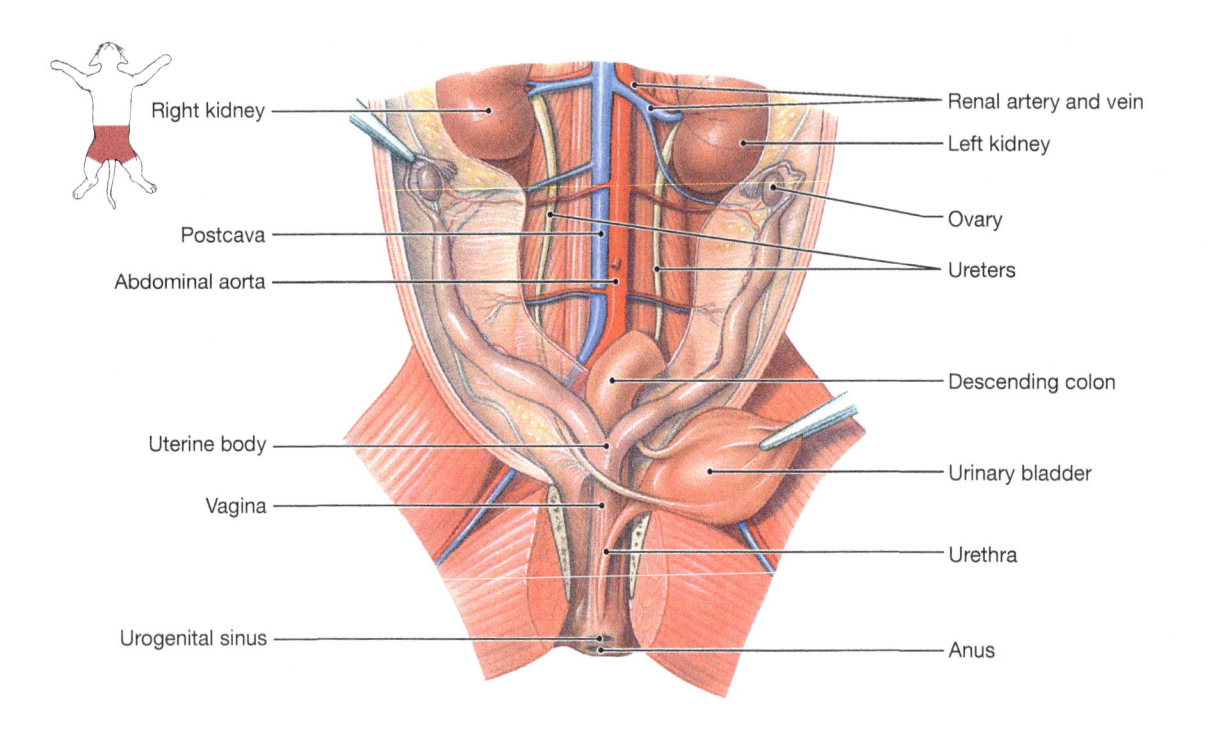

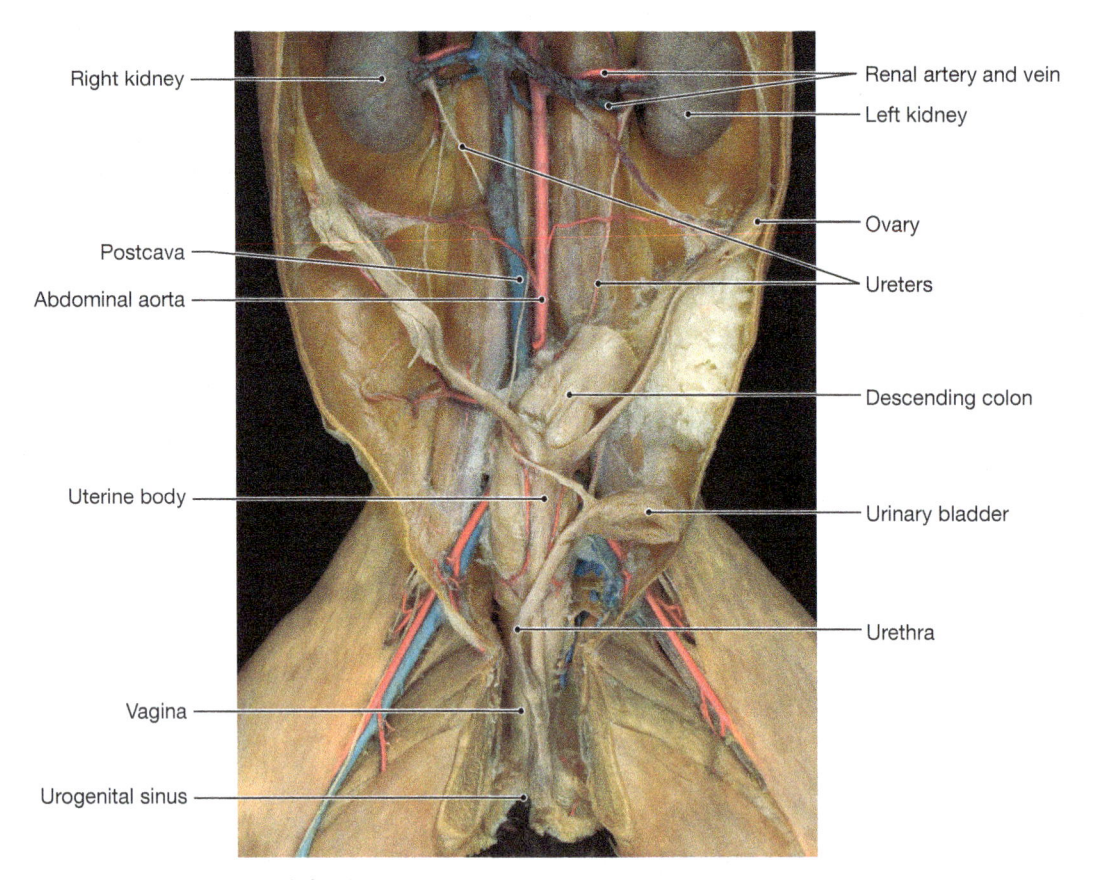

Figure D8.1 Urinary System of the Cat
Illustration and photo.

572

5. Make a lateral incision on each side of the diaphragm. Use care not to damage the diaphragm or the internal organs. Spread the thoracic walls to observe the internal organs.

6. Make a lateral section across the pubic region and angle toward the hips. Spread the abdominal walls to expose the abdominal organs. ∎

LAB ACTIVITY 2 — External Anatomy of the Kidney

Use Figure D8.1 as a guide during your dissection. Take care in handling and repositioning organs.

Materials

- Gloves
- Safety glasses
- Dissecting tools and tray
- Preserved cat, skin removed

Procedures

1. Reflect the abdominal viscera to one side of the abdominal cavity, and locate the large, bean-shaped kidneys. As in humans, the kidneys are **retroperitoneal** (outside the peritoneal cavity). Each kidney is padded by **perirenal fat** that constitutes the **adipose capsule**. Remove the fat to expose the kidney. Deep to the adipose capsule, the kidney is encased in a fibrous sac called the **renal capsule.**

2. Locate the **adrenal glands**, superior to the kidneys and close to the aorta. Identify the **suprarenal arteries** (in red if your specimen has been injected with latex) and the **suprarenal veins** (injected blue).

3. Finish exposing the kidney by removing the surrounding peritoneum and then carefully opening the renal capsule with scissors.

4. Identify the three structures that pass through the **hilus** which is the concave medial surface of the kidney. These three structures are the **renal artery** (injected red), the **renal vein** (injected blue), and the cream-colored tube known as the **ureter**.

5. Follow the ureter as it descends posteriorly along the dorsal body wall to drain urine into the **urinary bladder**. Examine the bladder and locate the **suspensory ligaments** that attach the bladder to the lateral and ventral walls of the abdominal cavity. The ligaments are not visible in Figure D8.1.

6. Distinguish the various regions of the bladder, starting with the **fundus**, which is the main egg-shaped region. Then pull the bladder anteriorly to observe the region where the fundus narrows into the **neck** and continues as the **urethra**, the tube through which urine passes to the exterior of the body.

7. Note where the urethra terminates. If your specimen is male, follow the urethra as it passes into the penis. If your specimen is female, notice how the urethra and the vagina empty into a common **urogenital sinus**. ∎

LAB ACTIVITY 3 — Internal Anatomy of the Kidney

Use Figure D8.2 as a guide during your dissection. Take care in handling and repositioning organs.

Materials

- Gloves
- Safety glasses
- Dissecting tools and tray
- Preserved cat, skin removed

Procedures

1. Cut any vessels or ligaments holding the kidney in the abdominal cavity. Remove the kidney from the cavity and place it in the dissecting tray. Make a frontal (coronal) section through the kidney so that the section passes through the middle of the hilus. Separate the two halves of the kidney.

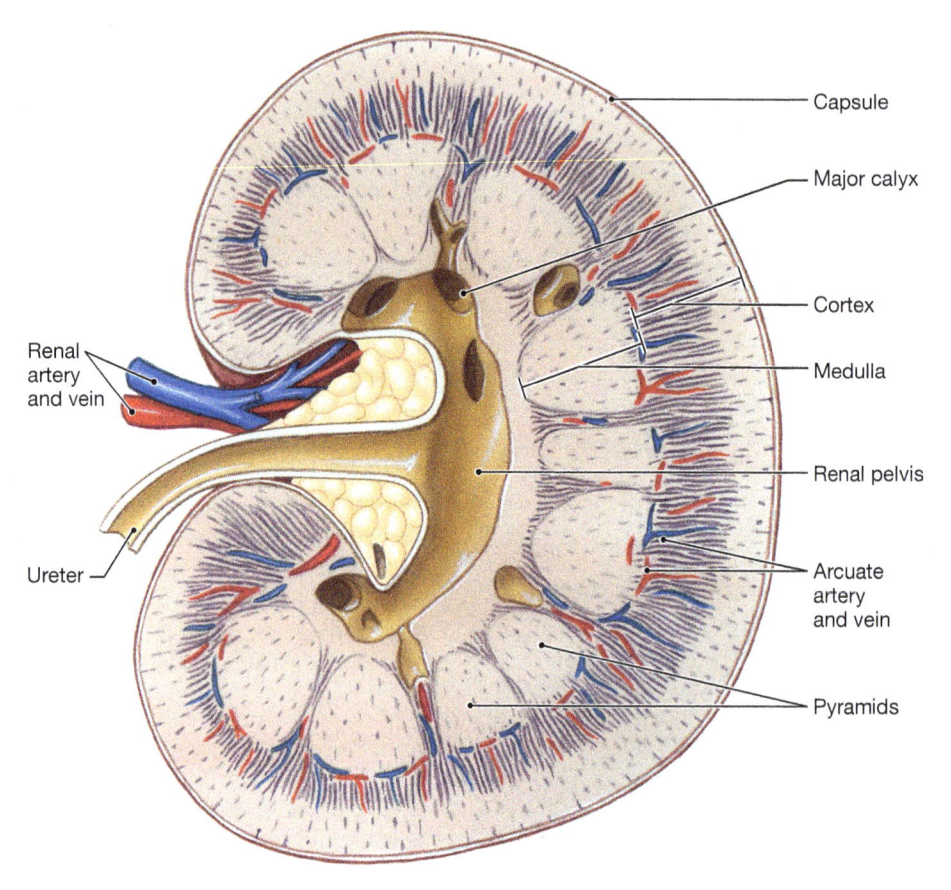

Figure D8.2 Cat Kidney

2. Locate the outer, lighter **cortex** and the inner, darker **medulla**. The medulla region contains numerous triangular **renal pyramids**, with each two adjacent pyramids separated by a **renal column**.

3. The hollow interior of the kidney—the part not occupied by the renal pyramids and columns—is called the **renal sinus.** Observe the expanded terminus of the ureter, the **renal pelvis**, that enters this region from the hilus side of the kidney.

4. The renal pelvis enters the kidney and branches into several **major calyces** (singular *calyx*) that, in turn, branch into many **minor calyces**. A minor calyx surrounds the tip of a pyramid where a wedge-shaped renal **papilla** projects into the calyx. Urine drips out of the papilla and into the minor calyx, the major calyx, and the renal pelvis. From there, it migrates out the kidney into the ureter. ∎

⚠️ *Safety Tip:* **Storing the Cat and Cleaning Up**

To store your specimen, wrap it in the skin and moisten it with fixative. Use paper towels if necessary to cover the entire specimen. Return it to the storage bag and seal the bag securely. Label the bag with your name, and place it in the storage area as indicated by your instructor.

Wash all dissection tools and the tray, and set them aside to dry.

Dispose of your gloves and any tissues from the dissection as indicated by your laboratory instructor. ▲

LAB REPORT

DISSECTION EXERCISE 8

Cat Urinary System

Name _____
Date _____
Section _____

A. Matching

Match each term in the left column with its correct description from the right column.

_____ 1. renal papilla
_____ 2. cortex
_____ 3. renal pelvis
_____ 4. renal sinus
_____ 5. renal pyramid
_____ 6. urogenital sinus
_____ 7. renal column
_____ 8. renal capsule
_____ 9. ureter
_____ 10. hilus
_____ 11. minor calyx
_____ 12. major calyx

A. fibrous covering of kidney
B. space within kidney that contains renal pelvis
C. concave region of kidney surface
D. extends into minor calyx
E. drains into renal pelvis
F. tissue between adjacent pyramids
G. transports urine to bladder
H. outer layer of kidney
I. drains into major calyx
J. major portion of the medulla
K. drains major calyces
L. site where female urethra empties

B. Short-Answer Questions

1. Describe the location of the kidneys.

2. The kidneys are retroperitoneal. How are they protected?

575

EXERCISE 8 — LAB REPORT

3. Trace a drop of urine from the renal papilla to its exit from the body.

4. Describe the blood supply to, and drainage of, the kidneys.

C. Analysis and Application

1. How is the urinary system of the female cat different from that of the female human?

2. Describe how the feline urinary bladder is supported.

DISSECTION EXERCISE 9

Cat Reproductive System

OBJECTIVES

On completion of this exercise, you should be able to:

- Identify the penis, prepuce, and scrotum of the male cat.
- Identify the testes, ducts, accessory glands, and urethra of the male cat.
- Identify the structures of the feline female reproductive tract.
- Identify the ovaries, ligaments, uterine tubes, and uterus of the female cat.
- Identify the vagina and the features of the vulva.

LAB ACTIVITIES

1. Preparing the Cat for Dissection 577
2. The Reproductive System of the Male Cat 578
3. The Reproductive System of the Female Cat 580

The function of the reproductive system is to produce the next generation of offspring. The reproductive systems of cats and humans are, in general, very similar, although there are differences in the uterus and in where the urethra empties. Because of the similarities, this exercise complements the study of the human reproductive system.

Be sure to observe the organs of both male and female cats. If your class does not have both a female and a male cat for each dissection team, your instructor will arrange for you to observe from time to time as some other team dissects a cat of the sex opposite that of your dissection specimen.

Safety Tip: Cat Dissection Basics

Before beginning this dissection, review the Safety Tip presented at the beginning of Dissection Exercise 1. ▲

LAB ACTIVITY 1 — Preparing the Cat for Dissection

If the ventral body cavity has not been opened on your dissection specimen, complete the following procedures. Otherwise, skip to Lab Activity 2 if your dissection specimen is male and to Lab Activity 3 if your specimen is female.

1 Materials

- ☐ Gloves
- ☐ Safety glasses
- ☐ Dissecting tools
- ☐ Dissecting tray
- ☐ String
- ☐ Preserved cat, skin removed

Procedures

1. Put on gloves and safety glasses, and clear your workspace before obtaining your dissection specimen.
2. Secure the specimen ventral side up on the dissecting tray by spreading the limbs and tying them flat with lengths of string passing under the tray. Use one string for the two forelimbs and one string for the two hindlimbs.

3. Make a longitudinal midline incision through the muscles of the neck, the bones of the sternum, and the muscles of the thorax and the abdominal wall.
4. Make a lateral incision on the cranial side of the diaphragm and another on the caudal side of the diaphragm. Use care not to damage the diaphragm or the internal organs. Spread the thoracic walls to observe the internal organs. ■

LAB ACTIVITY 2 — The Reproductive System of the Male Cat

The feline male reproductive tract is very similar to its counterpart in human males. As in humans and other mammals, the feline **testes** produce spermatozoa that are outside the body cavity and housed inside a covering called the **scrotum**. Ventral to the scrotum is the **penis**, which is the tubular shaft through which the urethra passes. Although in the human the penis contains no bone, the feline penis has a small bone called the **os penis** near one side of the urethra. Refer to Figure D9.1 during your dissection.

2 Materials

- Gloves
- Safety glasses
- Dissecting tools and tray
- Preserved cat, skin removed

Procedures

1. Identify the scrotum ventral to the anus and the penis ventral to the scrotum. Carefully make an incision through the scrotum and expose the testes. The testes are covered by a peritoneal capsule called the **tunica vaginalis**.
2. On the lateral surface of each testis, locate the comma-shaped **epididymis**, where spermatozoa are stored.
3. Locate the **spermatic cord**, which is covered with connective tissue and consists of the **spermatic artery**, **spermatic vein**, **spermatic nerve**, and **ductus deferens** (vas deferens). The ductus deferens carries spermatozoa from the epididymis to the urethra for transport out of the body.
4. Trace the spermatic cord through the **inguinal canal** and **inguinal ring** into the abdominal cavity.
5. Free the connective tissue around the components of the spermatic cord. Trace the ductus deferens into the pelvic cavity and note how this tube loops over the ureter and passes posterior to the bladder.
6. To observe the remaining reproductive structures, use bone cutters to cut through the pubic symphysis at the midline of the pelvic bone. Cut carefully, because the urethra is immediately dorsal to the bone. After cutting, split the pubic bone apart by spreading the thighs. This action exposes the structures within the pelvic cavity. Tease and remove any excess connective tissue.
7. Locate the **prostate gland**, a large, hard mass of tissue that surrounds the urethra. Trace the ductus deferens from the prostate to its merging with the spermatic cord structures. Here note one difference between the feline and human male systems: the seminal vesicles present in humans are absent in cats.
8. Trace the **urethra** to the proximal end of the penis. The urethra consists of three parts: the **prostatic urethra** that passes through the prostate, the **membranous urethra** that passes between the prostate gland and the penis, and the **penile urethra** (also called the spongy urethra) that passes through the penis.
9. Note the **bulbourethral glands** located on either side of the membranous urethra.
10. Identify the **prepuce**, a fold of skin that covers the expanded tip of the penis, the **glans**. Make a transverse section of the penis and locate the penile urethra. Identify the cylindrical erectile tissues of the penis: the **corpus spongiosum** around the urethra and the paired **corpora cavernosa** on the dorsal side.
11. Locate the os penis, also called the **baculum** (*bacul*, a rod, staff), the small bone in the glans penis. This bone stiffens the tip of the penis. ■

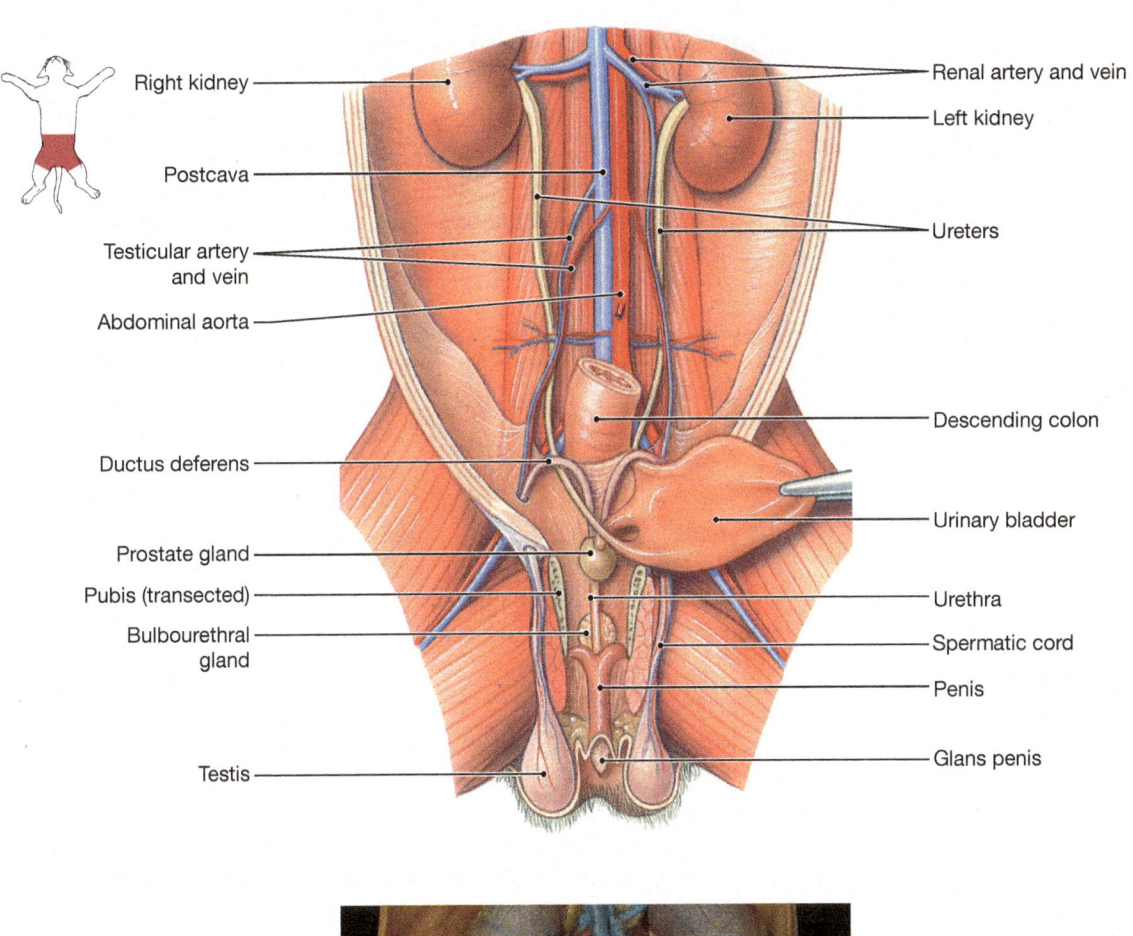

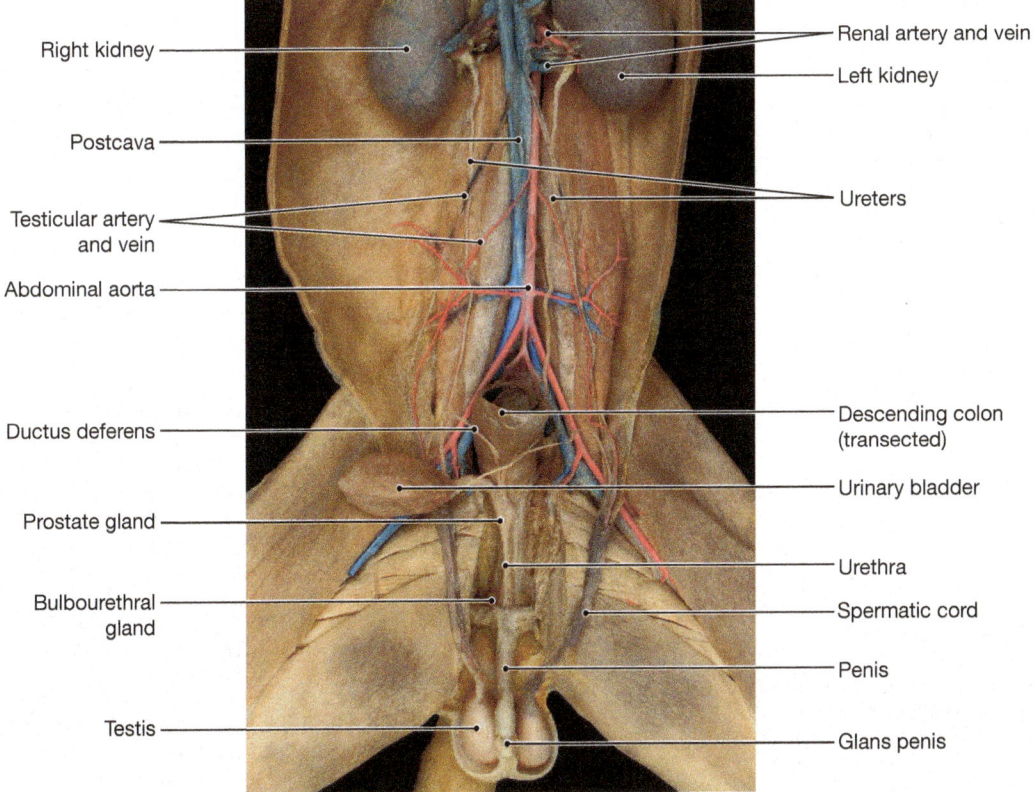

Figure D9.1 **Reproductive System of the Male Cat**
Illustration and photo.

580 DISSECTION EXERCISE 9 Cat Reproductive System

> ⚠ **Safety Tip: Disposal of the Cat**
>
> Because this is the last dissection exercise, you will probably be disposing of the cat.
>
> 1. To dispose of your specimen, first pour any excess fixative from the storage bag into a chemical collection container provided by your instructor.
> 2. Place the cat and the bag in the biohazard box as indicted by your instructor. ▲

LAB ACTIVITY 3 The Reproductive System of the Female Cat

The reproductive systems of female cats and female humans are similar, but there are several important differences. Because cats gestate litters of multiple twins (same mother but possibly different fathers), the feline uterus is branched into right and left horns. Humans typically gestate and give birth to a single offspring, and the uterus is not branched. Another difference between female cats and humans is that the feline urethra and vagina join as a common reproductive and urinary passageway. In humans, females have separate urethral and vaginal openings. Refer to Figure D9.2 during your dissection.

3 Materials

- ☐ Gloves
- ☐ Safety glasses
- ☐ Dissecting tools and tray
- ☐ Preserved cat, skin removed

Procedures

1. Reflect the abdominal viscera to one side and locate the paired, oval **ovaries**, lying on the dorsal body wall lateral to the kidneys.
2. On the surface of the ovaries, find the small, coiled **uterine tubes**, also called **oviducts**. Their funnel-like **infundibulum**, with finger-like tips called *fimbriae,* curves around the ovary and partially covers it to catch ova released during ovulation.
3. Note that, unlike the pear-shaped uterus of the human, the uterus of the cat is Y-shaped (bicornate) and consists of two large **uterine horns** joining a single **uterine body**. Each uterine tube leads into a uterine horn. The horns are where the fertilized ova are implanted for gestation of the offspring.
4. Identify the **broad ligament** that aids in anchoring the uterine horn to the body wall. This ligament is a peritoneal fold with three parts: the **mesovarian** suspends the ovary, the **mesosalpinx** is the peritoneum around the uterine horns, and the **mesometrium** supports the uterine body and horns.
5. To observe the remaining female reproductive organs, use bone cutters to section the midline of the pubic symphysis. Cut carefully, because the urethra and vagina are immediately dorsal to the bone. After cutting, split the pubic bone apart by spreading the thighs. This action exposes the structures within the pelvic cavity. Tease and remove any excess connective tissue.
6. Return to the uterine body and follow it caudally into the pelvic cavity, where it is continuous with the **vagina**.
7. Locate the **urethra** that emerges from the urinary bladder. The vagina is dorsal to the urethra. At the posterior end of the urethra, the vagina and urethra unite at the **urethral orifice** to form the **urogenital sinus** (vestibule), the common passage for the urinary and reproductive systems.
8. Lastly, note the urogenital sinus opening to the outside at the **urogenital aperture**. This opening is bordered by folds of skin called the **labia majora**. Together the urogenital aperture and the labia majora are considered external genitalia, and the collective name for them is the **vulva**. ■

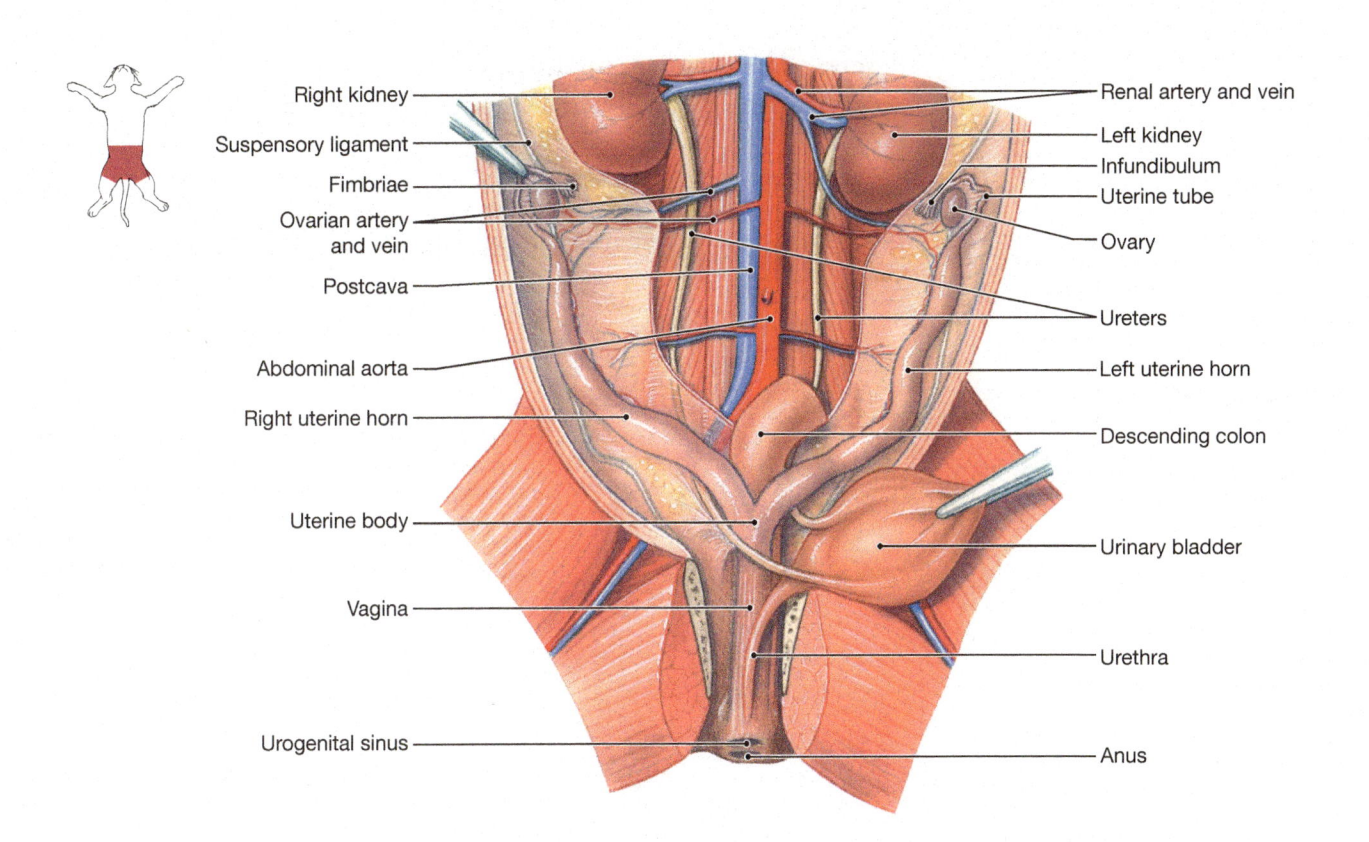

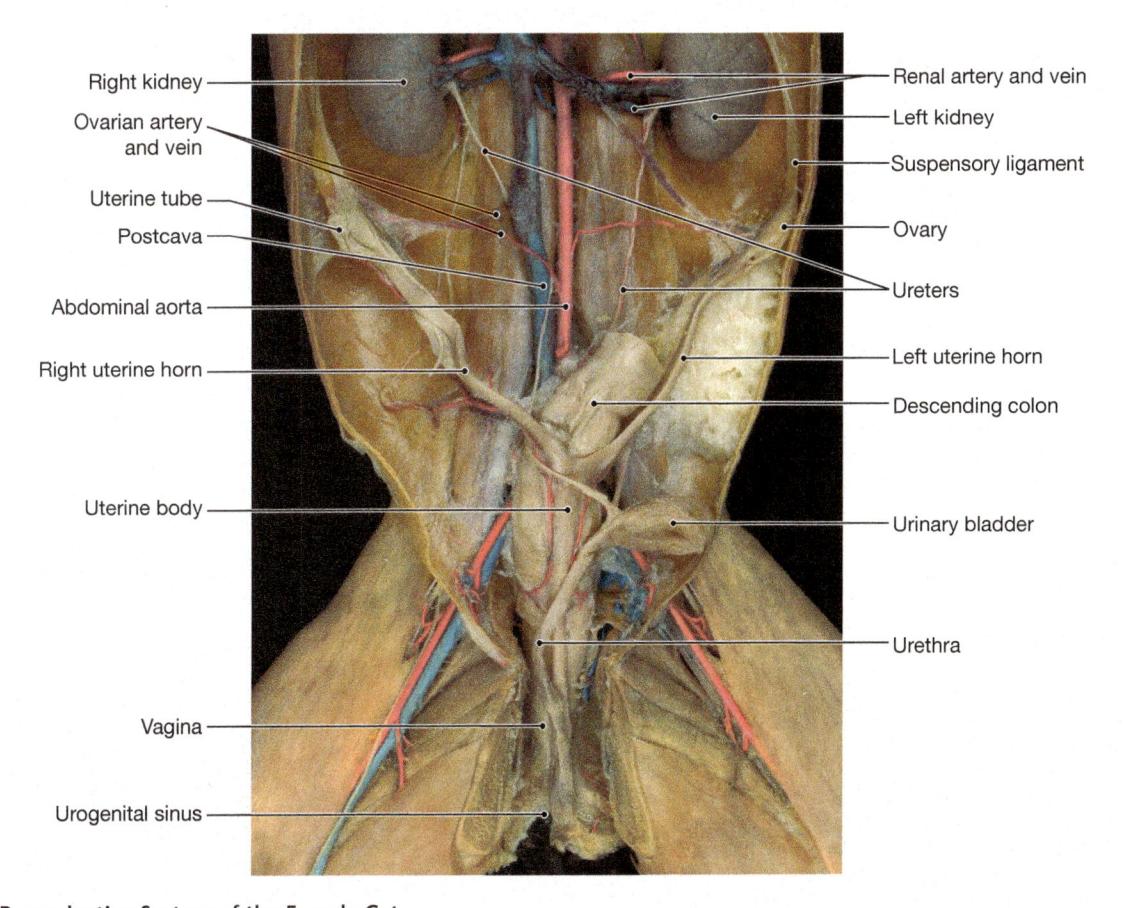

Figure D9.2 Reproductive System of the Female Cat
Illustration and photo.

581

> ⚠ **Safety Tip: Disposal of the Cat**
>
> Because this is the last dissection exercise, you will probably be disposing of the cat.
>
> 1. To dispose of your specimen, first pour any excess fixative from the storage bag into a chemical collection container provided by your instructor.
> 2. Place the cat and the bag in the biohazard box as indicated by your instructor. ▲

DISSECTION

LAB REPORT

EXERCISE 9

Name _____

Date _____

Section _____

Cat Reproductive System

A. Matching

Match each structure in the left column with its correct description from the right column.

_____ 1. broad ligament
_____ 2. bulbourethral glands
_____ 3. corpora cavernosa
_____ 4. ductus deferens
_____ 5. epididymis
_____ 6. infundibulum
_____ 7. labia majora
_____ 8. scrotum
_____ 9. urogenital sinus
_____ 10. uterine horn
_____ 11. testes
_____ 12. vulva

A. common urinary and reproductive passageway
B. folds around urogenital aperture
C. major branch of uterus
D. paired erectile cylinders
E. pouch that contains testes
F. site of spermatozoa production
G. site of spermatozoa storage
H. small glands in pelvic floor
I. supports uterine horns
J. female external genitalia
K. transports spermatozoa to urethra
L. receives ova released during ovulation

B. Short-Answer Questions

1. Trace the route of spermatozoa through the male reproductive system.

2. Describe the three parts of the broad ligament in the female cat.

3. Trace the route of an egg from the site of ovulation to where it will implant if it is fertilized.

583

EXERCISE 9 — LAB REPORT

C. **Analysis and Application**

1. How is the feline female reproductive tract different from that of the human female?

2. How is the cat uterus different from the human uterus?

3. Describe the ways in which the feline male reproductive system differs from that of the human male.

4. Compare the location of the gonads of males and females.

Appendix | WEIGHTS AND MEASURES

Table 1 The U.S. System of Measurement

Physical Property	Unit	Relationship to Other U.S. Units	Relationship to Household Units
Length	inch. (in.)	1 in. = 0.083 ft	
	foot (ft)	1 ft = 12 in. = 0.33 yd	
	yard (yd)	1 yd = 36 in. = 3 ft	
	mile (mi)	1 mi = 5,280 ft = 1,760 yd	
Volume	fluidram (fl dr)	1 fl dr = 0.125 fl oz	
	fluid ounce (fl oz)	1 fl oz = 8 fl dr = 0.0625 pt	= 6 teaspoons (tsp) = 2 tablespoons (tbsp)
	pint (pt)	1 pt = 128 fl dr = 16 fl oz = 0.5 qt	= 32 tbsp = 2 cups (c)
	quart (qt)	1 qt = 256 fl dr = 32 fl oz = 2 pt = 0.25 gal	= 4 c
	gallon (gal)	1 gal = 128 fl oz = 8 pt = 4 qt	
Mass	grain (gr)	1 gr = 0.002 oz	
	dram (dr)	1 dr = 27.3 gr = 0.063 oz	
	ounce (oz)	1 oz = 437.5 gr = 16 dr	
	pound (lb)	1 lb = 7000 gr = 256 dr = 16 oz	
	ton (t)	1 t = 2000 lb	

APPENDIX Weights and Measures

Table 2 The Metric System of Measurement

Physical Property	Unit	Relationship to Standard Metric Units	Conversion to U.S. Units	
Length	nanometer (nm)	1 nm = 0.000000001 m (10^{-9})	= 3.94×10^{-8} in.	25,400,000 nm = 1 in.
	micrometer (μm)	1 μm = 0.000001 m (10^{-6})	= 3.94×10^{-5} in.	25,400 mm = 1 in.
	millimeter (mm)	1 mm = 0.001 m (10^{-3})	= 0.0394 in.	25.4 mm = 1 in.
	centimeter (cm)	1 cm = 0.01 m (10^{-2})	= 0.394 in.	2.54 cm = 1 in.
	decimeter (dm)	1 dm = 0.1 m (10^{-1})	= 3.94 in.	0.25 dm = 1 in.
	meter (m)	standard unit of length	= 39.4 in.	0.0254 m = 1 in.
			= 3.28 ft	0.3048 m = 1 ft
			= 1.093 yd	0.914 m = 1 yd
	kilometer (km)	1 km = 1000 m	= 3280 ft	
			= 1093 yd	
			= 0.62 mi	1.609 km = 1 mi
Volume	microliter (μl)	1 μl = 0.000001 l (10^{-6}) = 1 cubic millimeter (mm^3)		
	milliliter (ml)	1 ml = 0.001 l (10^{-3}) = 1 cubic centimeter (cm^3 or cc)	= 0.0338 fl oz	5 ml = 1 tsp
				15 ml = 1 tbsp
				30 ml = 1 fl oz
	centiliter (cl)	1 cl = 0.01 l (10^{-2})	= 0.338 fl oz	2.95 cl = 1 fl oz
	deciliter (dl)	1 dl = 0.1 l (10^{-1})	= 3.38 fl oz	0.295 dl = 1 fl oz
	liter (l)	standard unit of volume	= 33.8 fl oz	0.0295 l = 1 fl oz
			= 2.11 pt	0.473 l = 1 pt
			= 1.06 qt	0.946 l = 1 qt
Mass	picogram (pg)	1 pg = 0.000000000001 g (10^{-12})		
	nanogram (ng)	1 ng = 0.000000001 g (10^{-9})	= 0.000000015 gr	66,666,666 mg = 1 gr
	microgram (μg)	1 μg = 0.000001 g (10^{-6})	= 0.000015 gr	66,666 mg = 1 gr
	milligram (mg)	1 mg = 0.001 g (10^{-3})	= 0.015 gr	66.7 mg = 1 gr
	centigram (cg)	1 cg = 0.01 g (10^{-2})	= 0.15 gr	6.67 cg = 1 gr
	decigram (dg)	1 dg = 0.1 g (10^{-1})	= 1.5 gr	0.667 dg = 1 gr
	gram (g)	standard unit of mass	= 0.035 oz	28.4 g = 1 oz
			= 0.0022 lb	454 g = 1 lb
	dekagram (dag)	1 dag = 10 g		
	hectogram (hg)	1 hg = 100 g		
	kilogram (kg)	1 kg = 1000 g	= 2.2 lb	0.454 kg = 1 lb
	metric ton (kt)	1 mt = 1000 kg	= 1.1 t	
			= 2205 lb	0.907 kt = 1 t

Temperature	Centigrade	Fahrenheit
Freezing point of pure water	0°	32°
Normal body temperature	36.8°	98.6°
Boiling point of pure water	100°	212°
Conversion	°C → °F: °F = (1.8 × °C) + 32	°F → °C: °C = (°F − 32) × 0.56

Photo Credits

Frontmatter
About the author photo, Beverly J. Poplin

Exercise 1
1.4a,b,c Custom Medical Stock Photo, Inc.

Exercise 2
2.1; 2.6 Olympus America Inc.

Exercise 3
3.3a,b,d,e,f Ed Reschke/Peter Arnold, Inc.
3.3c James Solliday/Biological Photo Service
3.3g Centers for Disease Control and Prevention (CDC)

Exercise 4
4.2a Ward's Natural Science Establishment, Inc.
4.2b Pearson Benjamin Cummings
4.2c; 4.3a Frederic H. Martini
4.3b,c Gregory N. Fuller, M. D. Anderson Cancer Center, Houston, TX
4.4a,b Frederic H. Martini
4.5a,b Project Masters, Inc./The Bergman Collection
4.6a Ward's Natural Science Establishment, Inc.
4.6b Frederic H. Martini
4.6c Ward's Natural Science Establishment, Inc.
4.7a Robert B. Tallitsch/Pearson Education
4.7b Robert B. Tallitsch/Pearson Education
4.7c; 4.9 Frederic H. Martini
4.10a Robert Brons/Biological Photo Service
4.10b Photo Researchers, Inc.
4.10c Ed Reschke/Peter Arnold, Inc.
4.11 Frederic H. Martini
4.12a Robert B. Tallitsch/Pearson Education
4.12b Robert B. Tallitsch/Pearson Education
4.12c; 4.13b Pearson Benjamin Cummings
4.14a,b,c,d,e,f; 4.15a,b,c,d,e,f Michael G. Wood

Exercise 5
5.2 Robert B. Tallitsch/Pearson Education
5.3; 5.4a,b Frederic H. Martini
5.5b Manfred Kage/Peter Arnold, Inc.
5.7; 6.1a Ralph T. Hutchings

Exercise 6
6.2b Robert B. Tallitsch/Pearson Education
6.3 Ralph T. Hutchings
6.4 Robert B. Tallitsch/Pearson Education

Exercise 7
7.1; 7.2; 7.3; 7.4; 7.5a,b,c; 7.6a,b,c; 7.7a,b; 7.8a,b; 7.9a,c; 7.10; 7.11; 7.12a,b; 7.13 Ralph T. Hutchings
7.14b Frederic H. Martini
7.16b; 7.18a,b,c,d; 7.19a,b; 7.20a,b; 7.21a,b,c; 7.22a,b,c; 7.23; 7.24 Ralph T. Hutchings

Exercise 8
8.1a,b,c; 8.2a,b,c; 8.3a,b; 8.4a,b,c,d; 8.5a,b; 8.7a,b; 8.8a,b; 8.9a,b; 8.10a,b; 8.11a,b,c; 8.12 Ralph T. Hutchings

Exercise 9
9.3a,b,c,d; 9.4a,b; 9.5a,b,c,d,e,f Ralph T. Hutchings
9.6b Patrick M. Timmons
9.7d Ralph T. Hutchings

Exercise 10
10.3b Don W. Fawcett/Photo Researchers, Inc.
10.4d Ed Reschke/Getty Images Inc.

Exercise 11
11.1b; 11.7; 11.10a,b Ralph T. Hutchings

Exercise 12
12.3b; 12.4b; 12.8b; 12.10b,c; 12.11c Ralph T. Hutchings

Exercise 13
13.3; 13.5; 13.6; 13.7b Michael G. Wood

Exercise 14
14.1b,c Ralph T. Hutchings
14.2b Michael J. Timmons
14.3a Ralph T. Hutchings
14.4 Patrick M. Timmons
14.7; 14.9a,b Ralph T. Hutchings
14.10 Shawn Miller & Mark Nielsen, Organ & Animal Dissector; Pearson Education/Benjamin Cummings Publishing Company

Exercise 15
15.3b; 15.4c Ralph T. Hutchings
15.6c Pat Lynch/Photo Researchers, Inc.
15.6e Michael J. Timmons
15.7a,b; 15.9a,bB Ralph T. Hutchings
15.9bT Ward's Natural Science Establishment, Inc.
15.10a Ralph T. Hutchings
15.11; 15.12; 15.13 Shawn Miller & Mark Nielsen, Organ & Animal Dissector; Pearson Education/Benjamin Cummings Publishing Company
15.14; 15.16 Ralph T. Hutchings

Exercise 16
16.1d, f Frederic H. Martini
16.2; 16.3 Michael G. Woo

Exercise 17
17.1c Michael G. Wood
17.2cT Pearson Benjamin Cummings
17.2cM Robert B. Tallitsch/Pearson Education
17.3 Michael G. Wood

Exercise 18
18.1a Ralph T. Hutchings
18.3d Michael J. Timmons
18.4a Ed Reschke/Peter Arnold, Inc.
18.4c Custom Medical Stock Photo, Inc.
18.5 William C. Ober, M.D.
18.6a,b,c Shawn Miller & Mark Nielsen, Organ & Animal Dissector; Pearson Education/Benjamin Cummings Publishing Company

Exercise 19
19.3d; 19.5a Michael G. Wood
19.6 William C. Ober, M.D.

Exercise 20
20.1b Manfred Kage/Peter Arnold, Inc.
20.2b,c; 20.3b,c; 20.4c,d Frederic H. Martini
20.5c Ward's Natural Science Establishment, Inc.
20.6b Michael G. Wood
20.6c Ward's Natural Science Establishment, Inc.
20.7 Ralph T. Hutchings

Exercise 21
21.2a David Scharf/Peter Arnold, Inc.
21.2b Susumu Nishinaga/Science Source
21.3b,c,d,e,f Ed Reschke/Peter Arnold, Inc.
21.5 Karen Peterson
21.6; 21.7 Michael G. Wood

Exercise 22
22.2c Ed Reschke/Peter Arnold, Inc.
22.4a,b; 22.5b Ralph T. Hutchings
22.7a,b; 22.8 Shawn Miller & Mark Nielsen, Organ & Animal Dissector; Pearson Education/Benjamin Cummings Publishing Company

Exercise 23
23.1 Biophoto Associates/Photo Researchers, Inc.
23.12; 23.13 Ralph T. Hutchings

Exercise 24
24.2b Frederic H. Martini
24.4 Ralph T. Hutchings

24.5 Biophoto Associates/Photo Researchers, Inc.
24.6c Frederic H. Martini

Exercise 25
25.4b CNRI/Science Source
25.5b Biophoto Associates/Science Source
25.6c Ward's Natural Science Establishment, Inc.
25.7a; 25.8 Ralph T. Hutchings

Exercise 26
26.3b Frederic H. Martini
26.5a Alfred Pasieka/Peter Arnold, Inc.
26.5b Astrid and Hanns-Frieder Michler/SPL/Photo Researchers, Inc.
26.6b Ralph T. Hutchings
26.7c Robert B. Tallitsch/Pearson Education
26.9d Michael G. Wood
26.11b; 26.13b Ward's Natural Science Establishment, Inc.
26.15c Frederic H. Martini
26.16 Ralph T. Hutchings

Exercise 27
27.1b Ralph T. Hutchings
27.2b,c,d Pearson Benjamin Cummings
27.5a Ward's Natural Science Establishment, Inc.
27.5b,c Frederic H. Martini
27.7 Shawn Miller & Mark Nielsen, Organ & Animal Dissector; Pearson Education/Benjamin Cummings Publishing Company
27.8 Ralph T. Hutchings

Exercise 28
28.2a Don W. Fawcett, Harvard Medical School
28.2c Ward's Natural Science Establishment, Inc.
28.4bL Ward's Natural Science Establishment, Inc.
28.4c,d,e Frederic H. Martini
28.8T Frederic H. Martini
28.8M C. Edelmann/La Villete/Photo Researchers, Inc.
28.8B Mike Peres/Custom Medical Stock Photo, Inc.
28.10b Ward's Natural Science Establishment, Inc.
28.12b Frederic H. Martini
28.13a,b Ralph T. Hutchings

Exercise 29
29.1a Francis Leroy/Photo Researchers, Inc.
29.5a Arnold Tamarin
29.5b,c,d Lennart Nilsson/Albert Bonniers Forlag AB

Dissection Exercise 1
D1.3, D1.4, D1.5, D1.6, D1.7, D1.8, D1.9, D1.10, D1.11 Shawn Miller/Mark Nielsen/Pearson Education/Benjamin Cummings Publishing Company

Dissection Exercise 2
D2.1, D2.2 Shawn Miller/Mark Nielsen/Pearson Education/Benjamin Cummings Publishing Company

Dissection Exercise 4
D4.1, D4.2, D4.3, D4.4, D4.5 Shawn Miller/Mark Nielsen/Pearson Education/Benjamin Cummings Publishing Company

Dissection Exercise 6
D6.1, D6.2 Shawn Miller/Mark Nielsen/Pearson Education/Benjamin Cummings Publishing Company

Dissection Exercise 7
D7.1, D7.3 Shawn Miller/Mark Nielsen/Pearson Education/Benjamin Cummings Publishing Company

Dissection Exercise 8
D8.1b Shawn Miller/Mark Nielsen/Pearson Education/Benjamin Cummings Publishing Company

Dissection Exercise 9
D9.1, D9.2 Shawn Miller/Mark Nielsen/Pearson Education/Benjamin Cummings Publishing Company

Index

Note: An italicized letter following a page number indicates that the item is referenced in a figure ("*f*") or a table ("*t*") on that page.

A

A band, 157, 158*f*
Abdomen, 7*f*, 432
Abdominal, 7*f*
Abdominal aorta, 330*f*, 334*f*, 367, 368*f*, 370, 371*f*, 372*f*, 431*f*
Abdominal cavity, 10, 11*f*
Abdominal region muscles, 180*t*
Abdominal wall, 487*f*
Abdominopelvic cavity, 10, 11*f*
Abdominopelvic quadrants, 8*f*
Abdominopelvic regions, 6, 8*f*, 483
Abducens nerve, 268, 268*f*, 269*t*, 271*t*, 274*f*
Abduction, 143*f*, 146
Abductor digiti minimi, 201, 203*t*, 204*f*, 214, 215*t*, 217*f*
Abductor hallucis, 213, 215*t*, 217*f*
Abductor muscles, 198
Abductor pollicis brevis, 201, 203*t*, 204*f*
Abductor pollicis longus, 200*f*, 201*t*, 202*f*
ABO blood group, 342–343
Accessory cells, 282*f*
Accessory glands, 454–455
Accessory hemiazygos vein, 377*f*
Accessory nerve, 246*f*, 268*f*, 269*t*, 270–271
Accessory pancreatic duct, 332*f*, 430, 431*f*
Accommodation curves, 102
Acetabular fossa, 124, 125*f*
Acetabular notch, 124, 125*f*
Acetabulum, 124, 124*t*, 125*f*
Acetylcholine (ACh), 159, 159*f*, 232, 233*f*
Acetylcholinesterase (AChE), 159
ACh, 159, 159*f*, 232, 233*f*
ACh receptor site, 159*f*
AChE, 159
Achilles tendon, 213
Acid reflux, 418
Acidophils, 323
Acinar cells, 431*f*
Acini, 430
Acini cells, 430
Acne, 71
Acoustic nerve, 269*t*
Acromial, 7*f*
Acromial end, 115, 116*f*
Acromioclavicular joint, 116, 116*f*
Acromion, 117, 117*f*, 484*f*, 485, 485*f*, 486*f*
Actin, 157
Actinin, 157
Action, 163, 196, 198, 299*f*

Action potential, 59, 155
Action potential path, 159*f*
Acute marginal branches, 359
Adam's apple, 402
Adaptation, 289
Adduction, 143*f*, 146
Adductor brevis, 205*t*, 206*f*, 207, 382*f*
Adductor group muscles, 205*t*, 206*f*, 207
Adductor hallucis, 215*t*
Adductor longus, 194*f*, 205*t*, 206*f*, 207, 210*f*, 211*f*, 219*f*, 382*f*, 490*f*
Adductor magnus, 205*t*, 206*f*, 207, 209*f*, 210*f*, 211*f*, 219*f*, 491*f*
Adductor muscles, 198, 200
Adductor pollicis, 201, 203*t*, 204*f*
Adductor tubercle, 126, 127*f*
Adenohypophysis, 322, 322*f*
Adenoid, 392
Adenosine triphosphate (ATP), 30, 325, 399
ADH, 322*f*, 323
Adipocytes, 47*t*, 48, 49*f*
Adipose capsule, 435
Adipose cushion, 304*f*
Adipose tissue, 47, 47*t*, 48, 49*f*, 55*f*, 72*f*, 244*f*, 436*f*
Adrenal glands, 3*f*, 235, 329–331, 334
Adrenal medulla, 235
Adrenalin, 334
Adrenocorticoids, 330
Adventitia, 366*f*, 414, 419*f*
Afferent arteriole, 439*f*, 440, 440*f*
Afferent division, 225, 226*f*
Afferent lymphatic vessels, 391*f*, 392
Afterbirth, 478
Agglutinates, 343
Agglutination, 343*f*
Agglutinins, 342
Agglutinogens, 342
Aggregate lymphoid nodules, 423, 427*f*
Agonist, 163
Agranular leukocytes, 338, 340*f*
Airways, 409
Ala, 107*f*
Alar cartilage, 401*f*
Albuterol, 404
Aldosterone, 330
Allantois, 476, 477*f*
Alpha cells, 332, 333
Alveolar ducts, 404, 405*f*
Alveolar nerve branches, 417*f*
Alveolar part, 98*f*
Alveolar process, 95, 96*f*, 98
Alveolar sacs, 405*f*, 408
Alveolar vessel branches, 417*f*

Alveoli, 4*f*, 405*f*, 408
   mammary, 464, 465*f*
Alveolus bone, 417*f*
Amacrine cells, 301*f*, 302
Amnion, 475*f*, 476, 477*f*
Amniotic cavity, 472, 473*f*, 474*f*, 475*f*, 477*f*
Amniotic fluid, 476, 477*f*, 479*f*
Amphiarthroses, 138, 138*t*
Amphimixis, 470, 471*f*
Ampulla (duodenum), 428, 429*f*
Ampulla (ductus deferens), 453, 455*f*
Ampulla (ear), 312*f*, 313
Ampulla (uterine), 460, 461*f*, 472*f*
Amygdaloid body, 262*f*, 264
Anal canal, 426*f*
Anal columns, 425, 426*f*
Anal triangle, 184*f*, 185*t*
Anaphase, 32, 32*f*
Anastomose, 370
Anastomoses, 360
Anatomical adjectives, 7*f*
Anatomical landmarks, 6, 8*f*, 483
Anatomical neck, 118, 119*f*
Anatomical position, 2, 5–6
Anatomical terms, 7*f*. *See also* Terminology
Anatomy, 1–2
Anconeus, 197, 199*t*, 200*f*, 202*f*, 488*f*
Androgens, 331
Angle, 97, 98*f*
Angular movements, 141, 143*f*
Ankle, 7*f*, 82, 128–130
   appendicular muscles, 213–216
Ankle extension, 146*f*
Ankle flexion, 146*f*
Annular ligament, 147, 148*f*, 404*f*
Annulus fibrosus, 103
ANS. *See* Autonomic nervous system
Ansa cervicalis, 246*f*
Ansa cervicalis superior, 246*t*
Antagonistic muscle, 163
Antebrachial, 7*f*
Antebrachial cutaneous nerve, 247*f*, 248*t*
Antebrachial interosseous membrane, 148*f*
Antebrachial vein, 375*f*, 376, 377*f*, 489*f*
Antebrachium, 7*f*
Antecubital, 7*f*
Antecubitis, 7*f*, 486
Anterior, 2, 5*f*, 11*f*
Anterior arch, 104, 105*f*
Anterior cardiac veins, 359*f*
Anterior cavity, 299*f*, 300
Anterior cerebral artery, 367, 369*f*
Anterior cervical triangle, 483, 484*f*, 485*f*
Anterior chamber, 300
Anterior clinoid process, 93, 94*f*

Anterior commissure, 261, 263f
Anterior communicating artery, 369f, 370
Anterior cranial fossa, 89
Anterior cruciate ligament, 149, 150f
Anterior fontanel, 101, 101f
Anterior gluteal line, 124, 125f
Anterior gray commissure, 241f, 242
Anterior gray horn, 241, 241f
Anterior horn of lateral ventricles, 258f, 262f
Anterior humeral circumflex artery, 371f
Anterior inferior cerebellar artery, 369f
Anterior inferior iliac spine, 125f
Anterior interosseous artery, 371f
Anterior interosseous nerve, 247f
Anterior interosseous vein, 377f
Anterior interventricular branch, 359, 359f
Anterior interventricular sulcus, 355, 356f, 361f
Anterior leg compartment, 218t, 219f
Anterior lobe (cerebellum), 265, 266f
Anterior lobe (pituitary), 322, 322f
Anterior median fissure, 240f, 241, 241f, 244f
Anterior mediastinum, 352f
Anterior neck, 173–175
Anterior scalene, 178f, 179, 180t
Anterior scalene slips, 178f
Anterior semicircular duct, 312f
Anterior spinal artery, 369f
Anterior superior iliac spine, 124, 125f, 210f, 487f
Anterior thigh compartment, 218t, 219f
Anterior tibial artery, 219f, 368f, 373f, 374
Anterior tibial border, 491f
Anterior tibial margin, 128, 129f
Anterior tibial vein, 219f, 375f, 378, 379f
Anterior tubercle, 105f
Anterior ulnar recurrent artery, 371f
Anterior white column, 242
Anterior white commissure, 241f, 242
Antibodies, 337, 343f, 387
Antidiuretic hormone (ADH), 322f, 323
Antigens, 337, 342, 387, 394
Antiserum, 343
Antrum, 84t, 460f
Anus, 184f, 425, 426f, 446f, 450f, 458f, 463f, 465f
Aorta, 352f, 354, 355f, 361f, 362f, 380f, 393f, 426f, 432f, 479f
   abdominal, 330f, 334f, 367, 368f, 370, 371f, 372f, 431f
   ascending, 356f, 357f, 367, 368f, 371f
   descending, 356f, 357f, 367, 368f
   thoracic, 181f, 371f, 372f
Aortic arch, 356f, 367, 368f, 371f, 409f
Aortic semilunar valve, 362f
Aortic valve, 357f, 358
Aperture (microscope), 19
Apex (heart), 351, 352f, 353f
Apex (lungs), 406, 407f
Apical foramen, 416, 417f
Apocrine gland cells, 71f
Apocrine sweat glands, 71f, 72
Appendicular division, 80

Appendicular muscles, 191–219
   ankle, 213–216
   arm, 195–196
   fingers, 197–204
   foot, 213–217
   forearm, 197–200
   hand, 197–203
   leg, 208–213
   pectoral girdle, 191–194
   thigh, 204–208
   toes, 213–217
   wrist, 197–202
Appendicular skeleton, 80, 115–132
   lower limb, 126–131
   pectoral girdle, 115–118
   pelvic girdle, 123–126
   shoulder, 132
   upper limb, 118–123
Appendix, 8f, 425, 426f
Aqueduct of midbrain, 258, 258f, 259f, 263f, 266f, 275f
Aqueous humor, 299f, 300
Arachnoid granulations, 259, 259f
Arachnoid mater, 241f, 243, 244f, 252f, 257f, 258, 259f, 276f
Arachnoid trabecula, 259f
Arbor vitae, 265, 266f, 275f, 409f
Arcuate arteries, 440, 440f, 462f
Arcuate line, 123, 123f, 125f
Arcuate veins, 440f, 441
Areola, 464, 465f, 486f
Areolar connective tissue, 47, 47t, 48, 49f, 55f, 353f
Arm, 7f, 80, 118, 489f
   appendicular muscles, 195–196
Arm bud, 475f
Arm (microscope), 18, 18f
Armpit, 7f
Arrector pili, 68f, 72, 72f
Arteries, 68f, 80f, 351, 365, 366f
   abdominopelvic cavity/lower limb, 370, 372–374
   branching, 370, 371f
   compared with other vessels, 365–367
   head/neck/upper limb, 367–370
   major, 367–374
   naming, 370
   unpaired, 370, 372
Arteriole, 80f, 389f, 405f, 423f
Arthritis, 139
Arthrology, 137
Articular capsule, 132f, 140, 148f, 150f
Articular cartilage, 78, 140, 140f, 148f
Articular facets, 108, 109f, 116f
   inferior, 103f, 104, 107f
   superior, 103f, 104, 105f, 106f, 107f
Articular processes, 103f, 104, 105f, 106f, 107f
Articular surface, 98
Articular tibial surface, 128
Articular tubercle, 90, 92f
Articulations, 137–150
   angular movements, 141, 143f
   axial skeleton, 144t

diarthroses, 141, 146–147
elbow, 147, 148f
joint classification, 137–139
knee, 147, 149, 150f
lower limb, 145t
movements, 141–143, 146
pectoral girdle, 145t
pelvic girdle, 145t
rotational movements, 141, 144f
special movements, 146
synovial joints, 140, 142f
upper limb, 145t
Aryepiglottic fold, 401f, 403f
Arytenoid cartilages, 402, 402f
Ascending aorta, 356f, 357f, 367, 368f, 371f
Ascending colon, 372f, 380f, 423f, 425, 426f
Ascending limb of nephron, 438
Associated bones, 87
Association areas, 261
Asthma, 404
Astral rays, 32f
Astrocytes, 227, 227f, 228f, 230f
Atlas, 104, 105f, 142f, 409f
   articulations, 144t
ATP, 30, 325, 399
Atrioventricular (AV) valves, 357f, 358
Audition, 281, 287
Auditory association area, 261f
Auditory cortex, 261f
Auditory nerve, 270
Auditory ossicles, 309–310, 310f
Auditory tubes, 309, 310f, 311f, 400, 401f, 415f
Auricle (ear), 309, 310f, 484f
Auricle (heart), 355, 356f, 361f
Auricular surface, 106, 107f, 124, 125f
Auris, 7f
Autolysis, 30
Autonomic ganglia, 233
Autonomic nervous system (ANS), 226, 226f, 232–235
   innervation distribution, 234f
   parasympathetic (craniosacral) division, 226, 226f, 233, 235
   pathways, 232, 233f
   sympathetic (thoracolumbar) division, 226, 226f, 233–235
Autonomic plexuses, 234f, 235
Autonomic (sympathetic) ganglion, 244f
AV valves, 357f, 358
Avascular, 39
Axial division, 80
Axial muscles, 163–185
   anterior neck, 173–175
   eye, 167–169
   facial expression, 164–167
   mastication, 169–170
   oblique/rectus, 179–183, 421f
   pelvic region, 183–185
   pharynx, 170–172
   tongue, 170–171
   vertebral column, 175–179

INDEX

Axial skeleton, 80, 87–109
  articulations, 144t
  cranial bones, 87–95. See also Cranial bones
  cranium, 87–90. See also Cranium
  facial bones, 87, 95–98. See also Facial bones
  fetal skull, 101
  hyoid bone, 99
  skull, 80, 87. See also Skull
  thoracic cage, 108–109
  vertebral column, 10, 102–108. See also Vertebral column
Axilla, 7f, 381f, 486f
Axillary, 7f
Axillary artery, 249f, 368f, 369f, 370, 371f, 381f
Axillary lymph nodes, 388f, 395f
Axillary nerve, 246, 247f, 248t, 249f, 488f
Axillary vein, 132f, 375f, 376, 376f, 377f, 395f
Axis, 104, 105f, 142f
Axolemma, 229
Axon, 60, 60f, 156f, 159, 159f, 228, 229f, 266f
  myelinated, 230f, 231f
Axon hillock, 228, 229f
Azygos vein, 377f, 390f

B

B cells, 340, 387, 391f
Baby teeth, 416
Back, 7f, 483
Back muscles, 175, 193f
Balance, 309
Ball-and-socket joints, 141, 142f
Baroreceptors, 281
Basal cell, 43f, 287, 288f, 290f
Basal compartment, 451f
Basal lamina, 40, 41, 42f, 43f, 44f, 69f, 71f
Basal nuclei, 262f, 264
Base (heart), 351, 352f, 353f
Base (lungs), 406, 407f
Base (microscope), 18f, 19
Basilar artery, 268f, 369f, 370
Basilar layer of endometrium, 462f
Basilar membrane, 313, 314f, 315f
Basilar zone, 461
Basilic vein, 375f, 376, 377f, 489f
Basophil, 52f, 323, 338, 339, 340f
Beta cells, 332
Biaxial joints, 138t, 139, 142f
Biceps, 197
Biceps brachii, 148f, 199t, 202f, 249f, 381f, 486, 486f, 488f, 489f
  long head, 194f, 197, 198f
  short head, 194f, 197, 198f
Biceps brachii tendon, 202f
Biceps femoris, 208, 210f, 212t, 219f, 250f
  long head, 206f, 209f, 211f, 491f
  short head, 206f, 209f, 211f, 216f, 491f
Biceps femoris tendon, 150f, 490f
Biceps tendon, 148f
Bicuspid valve, 357f, 358, 362f

Bicuspids, 416, 417f
Bifid, 104, 105f
Bile, 428
Bile canaliculi, 428
Bile duct, 429f
Bile ductules, 428, 429f
Binocular microscope, 19
Bipolar cells, 300, 301f
Birth, 478, 480f
Blastocoele, 470, 472f, 473f, 474f, 476, 477f
Blastocyst, 470, 472f, 473f
Blastocyst formation, 470–472, 472f
Blastodisc, 472, 473f, 474f
Blastodisc organization, 474f
Blastomere, 470, 471f, 472f
Blastula, 34
Blind spot, 301f, 302
Blood, 3f, 47t, 50, 52f, 334, 337–348, 351
  ABO blood group, 342–343
  coagulation, 50
  formed elements of, 50, 52f, 337, 338f, 387
  hematocrit, 346–348
  oxygen-carrying capacity, 346–348
  plasma, 47, 50, 337, 338f, 435
  RBCs, 337–338. See also Red blood cells
  Rh blood group, 344
  safety in handling, 344–346
  typing, 343f, 345–346
  WBCs 337, 338–340. See also White blood cells
  whole, 337–341
Blood capillaries, 389f
Blood clotting, 50
Blood collection, 345
Blood flow, 355f
Blood gases, 337–338
Blood group distribution, 342t
Blood plasma, 47. See also Plasma
Blood pressure, 366, 370
Blood type, 342
Blood typing, 343f, 345–346
Blood vessels, 3f, 48f, 54f, 156f, 231f, 327f, 351
  composition, 365–367
Body cavities, 10–11, 11f
Body position sense, 281
Body stalk, 475f, 476, 477f
Body tube (microscope), 18f, 19
Bolus, 415
Bone, 3f, 47, 52, 54f, 56f, 124t. See also Skeleton
Bone articulations, 84t
Bone histological organization, 79–80
Bone markings, 83–84
Bone marrow, 3f
Bone openings, 84t
Bone shaft, 78f
Bone structure, 77–79
Bone tissue, 47t
Bony cochlear wall, 315f
Bony fusion joints, 138t
Bony labyrinth, 310f, 311, 312f
Bowman's capsule, 437f, 438

Bowman's glands, 287
Brachial, 7f
Brachial artery, 202f, 249f, 368f, 370, 371f, 381f
Brachial plexus, 245f, 246–248, 381f, 484f, 485f
  spinal segments, 248t
Brachial vein, 375f, 376, 377f
Brachialis, 148f, 197, 198f, 199t, 202f, 486, 488f
Brachiocephalic trunk, 356f, 357f, 359f, 367, 368f, 369f, 371f
Brachiocephalic veins, 374, 375f, 376, 376f, 377f, 409f
Brachioradialis, 197, 198f, 199t, 200f, 202f, 488f, 489f
Brachium, 7f, 118
Brain, 3f, 10, 60f, 226f, 255–276
  blood flow to, 370
  cerebellum, 255, 265–267. See also Cerebellum
  cerebrum, 255, 260–264. See also Cerebrum
  cranial meninges, 256–258
  cranial nerves, 267–271. See also Cranial nerves
  diencephalon, 255, 256f, 264, 265f, 267
  dissection, 271–275
  medulla oblongata, 255, 264–265. See also Medulla oblongata
  mesencephalon, 255, 264. See also Mesencephalon
  pons, 255, 264. See also Pons
  protection/support, 256–258
  ventricles, 258–260
Brain freeze, 283
Brain stem, 265f, 267, 276
Breast, 7f, 391, 395
Brim, 123
Bristle, 466
Broad ligament, 461, 461f
Bronchi, 4f, 403, 404f, 405f
Bronchial artery, 405f
Bronchial nerve, 405f
Bronchial tree, 403–406
Bronchioles, 400f, 403, 404, 405f
Bronchodilator, 404
Bronchomediastinal trunk, 390f
Bronchopulmonary segment, 405f, 406, 408
Bronchus, 352f, 400f, 404f, 405f, 407f
Brown fat, 48
Brush border, 423
Bucca, 7f
Buccal, 7f
Buccal cavity, 415
Buccinator, 74f, 164f, 165, 165f, 166t
Bulbospongiosus, 184f, 185t, 466
Bulbo-urethral glands, 450f, 454, 455f, 457f
Burns, 70
Bursa, 140, 148f, 150f
Bursitis, 196
Buttock, 7f
Buttock fold, 490f

C

C cells, 323, 324f
C thyrocytes, 323, 324f
Calcaneal, 7f
Calcaneal tendon, 213, 216f, 219f, 489, 491f
Calcaneus, 7f, 81f, 128, 130f, 216f, 491f
Calcitonin (CT), 323, 325
Calf, 7f
Canal, 84f, 84t
Canal of Schlemm, 298
Canaliculi, 47t, 54, 54f, 80f
    bile, 428
    bone, 79
    lacrimal, 296f
Cancer, 32
Canine, 416
Canthus, 295, 296f
Capacitation, 470
Capillaries, 80f, 156f, 351, 354, 391f, 324f, 451f
    compared with other vessels, 365–367
    lymphatic, 389, 389f
    peritubular, 440f, 441
    renal/glomerular, 437f, 439f
Capillary beds, 355f, 405f
Capillary network, 423f
Capitate bone, 121, 122f
Capitulum, 108, 109f, 118, 119f, 120f, 148f
Capsular epithelium, 438
Capsular space, 437f, 438, 439f
Capsule, 282f
Carbon dioxide (CO_2), 399
Cardia, 419, 420f, 421f
Cardiac centers, 233f
Cardiac muscle cells, 352, 353f
Cardiac muscle tissue, 56, 57, 58f, 59f, 353f
Cardiac notch, 406
Cardiac plexus, 234f
Cardiac veins, 356f, 357f, 359, 359f
Cardiocyte, 57, 352
Cardiovascular system, 3f, 351
Carina, 403, 404f
Carotene, 69
Carotid canal, 89f, 91, 369f
Carotid pulse, 483, 485f
Carotid sinus, 367, 369f
Carpal, 7f
Carpal articular surface, 120f
Carpal bones, 80, 81f, 121, 145t
Carpal tunnel, 197
Carpal tunnel syndrome, 197
Carpus, 7f
Cartilage, 3f, 47, 52
Cartilaginous joints, 138t
Cat, 495–582
    anatomical position/location terminology, 495
    anatomy. *See* Cat—anatomy
    arteries, 532–538. *See also* Cat—arteries
    digestive system, 561–568. *See also* Cat—digestive system
    disposal, 580, 582
    dissection basics, 495–496, 517
    muscles. *See* Cat—muscles
    preparing for dissection, 496–498, 517, 525, 532, 547, 553–554, 561–562, 571, 573, 577–578
    reproductive system, 577–581
    respiratory system, 553–557
    skinning, 496–498
    storing, 498, 514, 521, 528, 542, 550, 557, 568, 574
    urinary system, 571–574
    veins, 539–542. *See also* Cat—veins
Cat—anatomy
    abdominal aorta, 536, 537f, 572f, 579f, 581f
    abdominal cavity, 564–568
    acromiodeltoid, 498, 499f, 501f
    acromiotrapezius, 498, 499f, 501f
    adductor femoris, 509, 510f, 511f
    adductor longus, 509, 510f
    adipose capsule, 573
    adrenal artery, 536
    adrenal cortex, 527
    adrenal glands, 526f, 527, 537f, 573
    adrenal medulla, 527
    adrenolumbar arteries, 536, 537f
    adrenolumbar vein, 537f, 540
    ampulla, 565, 567
    anterior tibial artery, 538f
    anterior tibial vein, 538f
    anus, 565, 572f, 581f
    aorta, 532
    aortic arch, 532, 533f
    arachnoid membrane, 521
    arcuate artery, 574f
    arcuate vein, 574f
    arytenoid cartilage, 555
    ascending colon, 565, 566f, 567f
    auditory tube, 554f
    axillary artery, 519f, 533f, 534
    axillary nerve, 519f
    axillary vein, 533f, 539
    azygous vein, 533f, 539
    baculum, 578
    biceps brachii, 505, 507f
    biceps femoris, 511f, 512, 513f
    brachial artery, 507f, 519f, 533f, 535, 535f
    brachial plexus, 518, 519f
    brachial vein, 533f, 535f, 540
    brachialis, 508f, 509
    brachiocephalic artery, 532, 533f
    brachiocephalic vein, 539
    brachioradialis, 507, 508f
    broad ligamentum, 580
    bronchial arteries, 534
    bulbo-urethral glands, 578, 579f
    calcaneus, 513f
    canine tooth, 563f
    capsule, 574f
    cardia, 564–568, 564f
    carotid artery, 502f
    cauda equina, 520f
    caudal artery, 537f, 538f
    caudal fold, 556f
    caudal vein, 537f, 540
    caudate lobe, 566
    caudofemoralis, 511f
    cecum, 565, 566f, 567f
    celiac trunk, 536, 537f
    cephalic vein, 535f, 540
    choanae, 554
    cisterna chyli, 548f, 549
    clavobrachialis, 498, 508f
    clavodeltoid, 498, 499f, 501f, 502f, 504f, 507f
    clavotrapezius, 498, 499f, 501f, 502f
    colic artery, 536, 537f
    collateral ulnar artery, 535f
    collateral ulnar vein, 535f
    colon, 565
    common bile duct, 566f, 567
    common carotid arteries, 533f, 534, 556f
    common fibular nerve, 520f, 521
    common hepatic duct, 566–567
    common iliac vein, 537f
    common peroneal nerve, 521
    coronary artery, 533f
    coronary vein, 533f
    corpora cavernosa, 578
    corpus spongiosum, 578
    cortex, 574
    costocervical artery, 534
    costocervical vein, 539
    cranial lobe, 556f
    cricoid cartilage, 555
    cystic artery, 536
    cystic duct, 566, 566f
    deep femoral artery, 537f, 538f
    deep femoral vein, 537f, 538f, 542
    deltoid group, 498
    descending colon, 537f, 565, 566f, 567f, 572f, 579f, 581f
    diaphragm, 537f, 555, 556f, 566f, 567f
    digastric, 502f, 503
    dorsal roots, 521
    dorsal thoracic nerve, 519f
    duct of Wirsung, 567
    ductus deferens, 578, 579f
    duodenum, 564f, 565, 566f, 567, 567f
    dura mater, 521
    endocrine glands, 527
    endocrine system, 526–528, 526f
    epididymis, 578
    epiglottis, 554f, 555
    epitrochlearis, 505, 507f, 508f
    esophageal arteries, 534
    esophagus, 533f, 537f, 555, 556f, 562, 564, 564f
    extensor carpi radialis, 507
    extensor carpi radialis brevis, 507, 508f
    extensor carpi radialis longus, 507, 508f
    extensor carpi ulnaris, 508f
    extensor digitorum communis, 508f
    extensor digitorum lateralis, 508f
    extensor digitorum longus, 513f, 514
    external carotid artery, 533f, 534
    external iliac arteries, 537f, 538f

INDEX

external iliac vein, 537f, 540
external intercostals, 505, 506f
external jugular vein, 502f, 504f, 519f, 533f, 539, 556f
external nares, 554, 554f
external oblique, 503, 504f, 506f, 510f, 511f
falciform ligament, 566, 566f, 567f
fascia, 511f
fascia lata, 511f
female reproductive system, 580–581
femoral artery, 504f, 510f, 537f, 538f
femoral nerve, 510f, 520f, 538f
femoral vein, 504f, 510f, 537f, 542
fibularis brevis, 512, 513f
fibularis group, 512
fibularis longus, 513f, 514
fibularis tertius, 514
fimbriae, 580, 581f
flexor carpi radialis, 507, 508f
flexor carpi ulnaris, 507, 508f
flexor digitorum longus, 512, 513f
flexor digitorum profundus, 508f
flexor hallicus longus, 513f
flexor retinaculum, 508f
frenulum, 563f
fundus, 564, 564f, 573
gallbladder, 541f, 566–567, 566f, 567f
gastric artery, 536
gastrocnemius, 512, 513f
gastrocnemius tendon, 513f
gastroduodenal artery, 536
gastrosplenic vein, 541f, 542
glans penis, 578, 579f
glottis, 554, 554f
gluteus maximus, 509, 511f
gluteus medius, 509, 511f
gonad, 526f
gonadal artery, 536, 537f
gonadal veins, 537f, 540
gonads, 528
gracilis, 509, 510f, 513f
gray horns, 521
greater curvature, 564, 564f
greater omentum, 564, 564f, 566f, 567f
hamstring group, 512
hard palate, 554f, 562, 563f
heart, 567f
hepatic artery, 536
hepatic portal vein, 540, 541f
hepatic vein, 540
hilus, 555, 573
humeral circumflex artery, 535
ileocecal sphincter, 565, 566f, 567f
ileum, 565, 566f, 567f
iliolumbar arteries, 536, 537f
iliolumbar vein, 537f
iliopsoas, 510f
inferior mesenteric artery, 536, 537f
inferior mesenteric vein, 542
inferior phrenic artery, 537f
inferior phrenic vein, 537f, 542
inferior thyroid artery, 533f
infraspinatus, 500, 501f

infundibulum, 580, 581f
inguinal canal, 578
inguinal ring, 578
intercostal arteries, 533f, 534
intercostal group, 505
intercostal vein, 533f, 539
internal carotid artery, 533f, 534
internal iliac artery, 537f, 538f
internal iliac vein, 537f, 540
internal intercostals, 505
internal jugular vein, 502f, 533f, 539, 548f, 556f
internal mammary artery, 533f, 534
internal mammary vein, 533f, 539
internal maxillary artery, 534
internal nares, 554
internal oblique, 503, 504f, 506f
internal spermatic artery, 537f
internal spermatic vein, 537f, 540
internal thoracic vein, 539
isthmus, 527
jejunum, 565, 566f, 567f
kidney, 526f, 527, 572f, 573–574, 574f, 579f, 581f
labia majora, 580
large intestine, 565
laryngopharynx, 554, 554f, 562
larynx, 554f, 555
lateral head, 508f, 509
lateral lobes, 527, 566, 566f, 567f
latissimus dorsi, 498, 499f, 501f, 504f, 506f
lesser curvature, 564, 564f
lesser omentum, 564, 564f, 566f, 567f
levator scapulae ventralis, 498, 499f, 501f
ligamentum arteriosum, 532
linea alba, 503, 504f
lingual frenulum, 562
liver, 541f, 566–568
long head, 508f, 509, 535f
long thoracic artery, 533f, 535
long thoracic nerve, 519f
long thoracic vein, 533f, 539
lumbar arteries, 536
lumbar vein, 540
lumbar nerve roots, 520f
lungs, 555–557, 567f
lymph, 547
lymph nodes, 547, 548f, 549, 562, 563f
lymphatic ducts, 548f, 549
lymphatic nodules, 547
lymphatic system, 547–550
lymphatic vessels, 547
major calyces, 574, 574f
male reproductive system, 578–579
mandible, 502f, 563f
masseter, 502f, 503, 563f
medial head, 509
medial lobe, 566, 566f, 567f
medial sacral artery, 538f
median cubital artery, 535f
median cubital vein, 535f
median nerve, 507f, 518, 519f, 535f

medulla, 574
membranous urethra, 578
mesentery, 565, 566f, 567f
mesocolon, 565
mesometrium, 580
mesosalpinx, 580
minor calyces, 574
musculocutaneous nerve, 518, 519f
mylohyoid, 502f, 503
nasopharynx, 554, 554f, 562
nerve roots, 519f
nervous system, 517–521
occipital artery, 533f
oral cavity, 554f, 562–564
oropharynx, 554, 554f, 562
os penis, 578
ovarian arteries, 536, 581f
ovarian veins, 540, 581f
ovaries, 526f, 528, 572f, 580, 581f
oviducts, 580
palate, 550
palatine tonsils, 550
palmaris longus, 507, 508f
pancreas, 526f, 527, 566f, 567, 567f, 568
pancreatic duct, 567
pancreaticoduodenal veins, 541f
papilla (renal), 574
papillae, 562
parathyroid gland, 526f, 527
parietal pleura, 555
parotid duct, 562, 563f
parotid gland, 502f, 562, 563f
pectineus, 510f, 511
pectoantebrachialis, 503, 504f, 506f, 507f, 508f
pectoralis group, 503
pectoralis, 519f
pectoralis major, 503, 504f, 506f, 507f
pectoralis minor, 504, 504f, 506f
penile urethra, 578
penis, 578, 579f
perirenal fat, 573
peroneal nerve, 511f
peroneus muscles, 512
pharynx, 554, 562, 564
phrenic artery, 536
phrenic nerve, 557
pia mater, 521
plantaris, 513f
pleura, 555
pleural cavity, 555
pleural fluid, 555
popliteal artery, 538f
popliteal vein, 538f, 542
postcava, 533f, 537f, 539, 540, 542, 548f, 572f, 579f, 581f
posterior cerebral artery, 534
posterior lobe, 566
posterior tibial artery, 538f
posterior tibial vein, 538f
precava, 533f, 539
premolar tooth, 563f
prepuce, 578

Cat—anatomy, Cont.
  primary bronchi, 555
  pronator teres, 507, 508f
  prostate gland, 578, 579f
  prostatic urethra, 578
  pubis, 579f
  pulmonary artery, 532
  pulmonary ligament, 556f
  pulmonary trunk, 532, 533f
  pulmonary vein, 539
  pyloric sphincter, 564, 564f, 566f, 567f
  pylorus, 564, 564f
  quadriceps femoris group, 511
  quadriceps group, 510f
  radial artery, 535, 535f
  radial nerve, 518, 519f, 535f
  radial vein, 535f, 540
  rectum, 565
  rectus abdominis, 503, 504f, 506f
  rectus femoris, 510f, 512
  renal artery, 536, 537f, 572f, 573, 574f, 579f, 581f
  renal capsule, 573
  renal column, 574
  renal pelvis, 574, 574f
  renal pyramids, 574, 574f
  renal sinus, 574
  renal vein, 537f, 540, 572f, 573, 574f, 579f, 581f
  reproductive system, 577–582
  retinaculum, 508f
  rhomboideus capitis, 500, 501f
  rhomboideus group, 500
  rhomboideus major, 500, 501f
  rhomboideus minor, 500, 501f
  rib cage, 556f
  rugae, 564f, 565
  sacral nerve roots, 520f
  sacral plexus, 518, 520–521
  salivary glands, 562–564
  saphenous artery, 538f
  saphenous nerve, 538f
  saphenous vein, 538f, 542
  sartorius, 509, 510f, 511f, 513f
  scalenus medius, 505, 506f
  scapular spine, 501f
  sciatic nerve, 511f, 520f, 521
  scrotum, 578
  semimembranosus, 510f, 511f, 512, 513f
  semitendinosus, 510f, 511f, 512, 513f
  serratus dorsalis, 505, 506f
  serratus ventralis, 505, 506f
  small intestine, 565
  soft palate, 554f, 562
  soleus, 512, 513f
  spermatic arteries, 536, 578
  spermatic cord, 537f, 578, 579f
  spermatic nerve, 578
  spermatic vein, 578
  spinal cord, 520f, 521
  spinal nerves, 517
  spinodeltoid, 498, 499f, 501f
  spinotrapezius, 498, 499f, 501f
  spleen, 541f, 547, 550, 565, 566f, 567f
  splenic artery, 536
  splenius, 500, 501f
  spongy urethra, 578
  sternohyoid, 502f, 503, 504f
  sternomastoid, 500, 502f, 504f
  sternothyroid, 502f
  stomach, 541f, 564, 564f, 566f, 567f
  subarachnoid space, 521
  subclavian artery, 532, 533f
  subclavian veins, 533f, 534, 548f
  sublingual gland, 562
  submandibular duct, 564
  submandibular gland, 562, 563f
  subscapular artery, 533f, 535
  subscapular vein, 533f, 539
  superficial inguinal nodes, 548f
  superior articular artery, 538f
  superior articular vein, 538f
  superior hemorrhoidal artery, 536, 537f
  superior mesenteric artery, 536, 537f
  superior mesenteric vein, 541f, 542
  superior thyroid arteries, 533f, 534
  suprarenal arteries, 573
  suprarenal veins, 573
  supraspinatus, 500, 501f
  sural artery, 538f
  sural vein, 538f
  suspensory ligaments, 573, 581f
  teeth, 562
  tendon, 508f
  tensor fasciae latae, 509, 510f, 511f
  teres major, 500, 501f
  testes, 528, 578, 579f
  testicular artery, 579f
  testicular vein, 579f
  thoracic aorta, 532, 533f, 548f
  thoracic duct, 548f, 549
  thoracodorsal artery, 535
  thymus gland, 526f, 527, 547, 548f, 550, 556f
  thyrocervical artery, 533f, 534
  thyrocervical vein, 533f
  thyrohyoid, 502f
  thyroid cartilage, 502f, 555
  thyroid gland, 526f, 527, 533f, 556f
  tibia, 510f, 513f
  tibial nerve, 511f, 520f, 521
  tibialis anterior, 512, 513f
  tongue, 554f, 562, 563f
  tonsils, 547, 550
  trachea, 502f, 526f, 533f, 555, 556f
  tracheal rings, 555
  transverse abdominis, 503, 504f, 506f
  transverse colon, 565, 566f
  transverse jugular vein, 533f, 539
  transverse scapular vein, 539
  trapezius group, 498
  triceps, 508f, 535f
  triceps brachii, 507f, 508f, 509
  tunica vaginalis, 578
  ulnar artery, 535, 535f
  ulnar nerve, 507f, 518, 519f, 535f
  ulnar vein, 535f, 540
  umbilical artery, 538f
  ureters, 537f, 572f, 573, 574f, 579f, 581f
  urethra, 572f, 573, 578, 579f, 580, 581f
  urethral orifice, 580
  urinary bladder, 537f, 566f, 567f, 572f, 573, 579f, 581f
  urinary system, 571–574
  urogenital aperture, 580
  urogenital sinus, 572f, 573, 580, 581f
  uterine horns, 580, 581f
  uterine tubes, 580, 581f
  uterus, 572f, 580, 581f
  vagina, 572f, 580, 581f
  vagus nerve, 502f
  vascular system, 531–542
  vastus intermedius, 512
  vastus lateralis, 510f, 511, 511f
  vastus medialis, 510f, 512
  ventral roots, 521
  ventral thoracic artery, 533f, 534
  ventral thoracic nerve, 519f
  ventral thoracic vein, 533f, 539
  vertebra, 520f
  vertebral artery, 534
  vertebral vein, 533f, 539
  vestibular ligaments, 555
  vestibule, 562
  villi, 565
  visceral pleura, 555
  vocal ligaments, 555
  vulva, 580
  white columns, 521
  xiphihumeralis, 505, 506f
Cat—arteries, 532–538
  abdominal cavity, 536–538
  head/neck/thorax, 532–535
  hindlimb, 538
  shoulder/forelimb, 535
Cat—digestive system, 561–568
  abdominal cavity, 564–565
  esophagus, 564
  gallbladder, 566–567
  large intestine, 565–566
  liver, 566–568
  oral cavity, 562–564
  pancreas, 566–567
  pharynx, 562–564
  salivary glands, 562–564
  small intestine, 565–566
  spleen, 565
  stomach, 564–565
Cat—muscles
  back/shoulder, 498–500
  forelimb, 505–509
  lower hindlimb, 512–514
  neck/abdomen/chest, 500–505
  thigh, 509–512
Cat—respiratory system, 553–557
  bronchi/lungs, 555–557
  larynx/trachea, 555
  nasal cavity/pharynx, 554
Cat—veins, 539–542
  abdominal cavity, 540–542
  forelimb, 540
  head/neck/thorax, 539
  hindlimb, 542
Cauda equina, 239, 240f
Caudal, 5, 495

Caudate lobe, 428, 428f
Caudate nucleus, 262f, 263f, 264
Cavernous sinus, 376f
Cecum, 8f, 423f, 425, 426f
Celiac ganglion, 234f, 235, 334f
Celiac trunk, 330f, 334f, 368f, 372, 372f, 431f
Cell, 27, 469
Cell anatomy, 27–31
Cell body, 60, 60f, 228, 229f
Cell division, 31–34
Cell level, 2
Cell life cycle, 32f
Cell membrane, 27, 28f
Cell theory, 27
Cellular trophoblast, 472, 473f, 474f, 476, 477f
Cement, 417f
Cementum, 416
Central adaptation, 289
Central arteries, 393f, 394
Central canal, 54, 54f, 79, 80f, 240f, 241f, 242, 258f, 259f
Central diaphragmatic tendon, 181f
Central incisor, 416, 417f
Central incisors, 417f
Central nervous system (CNS), 10, 59, 225, 226f, 227f. *See also* Brain; Spinal cord
 glial cells, 227–228
 preganglionic exit points, 233
Central sulcus, 256f, 259f, 260, 261f, 263f
Central vein, 428, 429f
Centrioles, 28f, 29, 32f
Centromere, 31
Centrosome, 28f, 29
Centrum, 103
Cephalic, 5, 7f, 495
Cephalic vein, 132f, 375f, 376, 377f, 489f
Cephalon, 7f
Cerebellar artery, 369f
Cerebellar cortex, 265, 266f
Cerebellar fossa, 92f
Cerebellar hemispheres, 265, 266f
Cerebellar nuclei, 265, 266f
Cerebellar peduncles, 265f
Cerebellum, 255, 256f, 257f, 259f, 262f, 263f, 265–267, 268f, 273f, 275f
Cerebral arterial circle, 369f, 370
Cerebral arteries, 367, 369f, 370
Cerebral cortex, 255, 257f, 276f
Cerebral crest, 92f
Cerebral hemispheres, 255, 256f, 258f, 259f, 261f, 273f, 275f
Cerebral peduncles, 263f, 264, 265f, 275f
Cerebral veins, 376f
Cerebrospinal fluid (CSF), 243, 244f, 258f, 259, 259f, 260, 276
Cerebrum, 255, 256f, 260–264, 266
Ceruminous glands, 309
Cervical, 7f
Cervical artery, 174f, 369f
Cervical canal, 461, 461f, 480f
Cervical curve, 102f
Cervical enlargement, 239, 240f
Cervical lymph nodes, 388f
Cervical nerves, 245, 246f, 246t

Cervical plexus, 174f, 245–246, 245f, 246f, 249f
 spinal segments, 246t
Cervical plug, 479f
Cervical region, 102f
Cervical region muscles, 180t
Cervical spinal cord, 240f
Cervical spinal nerves, 240f, 245f
Cervical vertebrae, 102, 105f
Coccygeal vertebrae, 102
Cervicis, 7f
Cervix, 446f, 458f, 461, 461f, 465f, 478, 479f, 480f
Cheek, 7f, 415, 415f
Cheekbones, 95, 483
Chemical detection sense, 281
Chemoreceptors, 281
Chest, 7f, 483
Chief cells, 326, 327f, 421f, 422
Chin, 7f
Chondroblasts, 47t, 52, 52f
Chondrocytes, 47t, 52, 52f, 53f
Chorda tympani nerve, 311f
Chordae tendineae, 357f, 358, 362f
Chorion, 476, 477f
Chorionic villi, 475f, 476, 477f
Choroid, 298, 299f, 301f
Choroid plexus, 259, 259f, 260, 262f, 265f, 409f
Chromatids, 31, 32f, 451
Chromatin, 28f, 29
Chromatophilic substance, 229f
Chromophils, 322–323
Chromophobes, 322, 323
Chromosomal microtubule, 32f
Chromosomes, 28, 32f
Chyme, 422
Cilia, 29, 44f
Ciliary body, 298, 299f, 304f
Ciliary ganglion, 234f, 235
Ciliary glands, 297
Ciliary muscle, 298
Ciliary process, 298
Cingulate gyrus, 263f
Circle of Willis, 370
Circular muscle, 421f
Circular muscle layer, 414f, 420f, 423f, 427f
Circulation, 354, 365–382. *See also* Blood; Cardiovascular system
Circumduction, 143f, 146
Circumferential lamellae, 79, 80f
Circumflex branch, 356f, 359, 359f
Circumvallate papillae, 290f, 291
Cisterna chyli, 388f, 389, 390f
Claustrum, 262f, 263f
Clavicle, 80, 81f, 115, 116f, 142f, 164f, 173f, 174f, 246f, 249f, 324f, 369f, 376f, 381f, 400f, 484f, 485, 485f, 486f
 acromial end, 488f
Clavicular articulations, 109f, 145t
Cleavage, 470–472, 472f
Cleavage furrow, 32, 32f
Clitoris, 4f, 458f, 463f, 464
CNS. *See* Central nervous system
CO₂, 399
Coagulation, 50

Coarse adjustment knob, 18f, 19
Coccygeal bones, 144t
Coccygeal cornua, 106
Coccygeal ganglia, 234f
Coccygeal nerve, 240f, 245, 245f
Coccygeal region, 102f
Coccygeal vertebrae, 102
Coccygeus, 183, 184f, 185t
Coccyx, 5, 81f, 102, 106, 107f
 articulations, 144t
 sex differences in, 124t
Cochlea, 310f, 311, 312f, 314f
Cochlear branch, 314f, 315f
Cochlear chambers, 315f
Cochlear duct, 312f, 313, 314f, 315f
Cochlear nerve, 315f
Cochlear wall, 315f
Coelom, 10
Colic artery, 372f, 426f
Colic vein, 380f, 426f
Collagen fibers, 47, 47t, 49f, 50, 51f, 53f, 282f, 283f, 391
Collagenous sheath, 457f
Collarbones, 115
Collateral branches (axon), 228, 229f
Collateral ganglia, 235
Collecting duct, 437f, 438, 439f
Colon wall, 427f
Columnar epithelium, 40, 40f, 41, 42f, 43f, 44–45, 44f, 46f, 423f, 427f, 462f
Common bile duct, 332f, 428, 428f, 429f, 431f
Common carotid artery, 174f, 324f, 369f, 372f, 404f
 left, 356f, 357f, 359f, 367, 368f, 371f
 right, 249f, 367, 368f, 371f, 409f
Common fibular nerve, 250f, 491f
Common hepatic artery, 334f, 372, 372f, 429f, 431f
Common hepatic duct, 428, 429f
Common iliac arteries, 368f, 372f, 373f, 374
Common iliac vein, 375f, 377f, 378, 379f, 479f
Compact bone, 77, 78f, 79, 80f, 140f
Compartment syndrome, 218
Compartments, 218
Compensation curves, 102
Compound lens system (microscope), 19
Compound microscope, 17. *See also* Microscope
Concentric lamellae, 79
Condenser, 18f, 19
Condenser adjustment knob, 19
Condylar head, 98
Condylar joints, 141, 142f
Condylar process, 97, 98f
Condyle, 84f, 84t, 118, 119f
Condyloid fossa, 89f, 92f
Cones, 300, 301f, 302
Conjunctivitis, 297
Connecting tubule, 437f, 438
Connective tissue, 39, 42f, 43f, 46–56, 442f, 455f
 coverings, 155–156
 dermis, 71f

Connective tissue, *Cont.*
    fluid, 47, 47*t*, 50, 52*f*
    loose, 42*f*, 43*f*, 44*f*, 47, 49*f*, 79, 276*f*, 389*f*
    mucous, 47, 48*f*
    proper, 47–50, 47*t*
    supportive, 47, 47*t*, 52–55
Connective tissue capsule, 327*f*, 451*f*
Connective tissue proper, 47–50, 47*t*
Conoid tubercle, 116, 116*f*
Contraction, 57, 157, 163, 414
Conus arteriosus, 357*f*
Conus medullaris, 239, 240*f*
Convoluted tubules, 437*f*, 438, 439*f*
Coracobrachialis, 194*f*, 195, 195*t*, 198*f*, 249*f*, 489*f*
Coracoid process, 117, 117*f*, 198*f*
Cornea, 298, 299*f*, 304*f*
Corneal limbus, 296*f*, 298, 299*f*
Corniculate cartilages, 402, 402*f*, 403*f*
Corona, 456, 457*f*
Corona radiata, 460*f*, 470, 471*f*
Coronal section, 9
Coronal suture, 87, 88*f*, 91*f*, 96*f*, 101*f*
Coronary arteries, 356*f*, 357*f*, 359, 359*f*
Coronary circulation, 359–360
Coronary ligament, 428, 428*f*
Coronary sinus, 356*f*, 357*f*, 359, 359*f*
Coronary sulcus, 355, 356*f*
Coronoid fossa, 118, 119*f*
Coronoid process, 97, 98*f*, 118, 120*f*
Corpora cavernosa, 456
Corpora quadrigemina, 263*f*, 264, 265*f*, 273*f*, 275*f*
Corpus albicans, 458, 460*f*
Corpus callosum, 261, 262*f*, 263*f*, 275*f*
Corpus cavernosum, 450*f*, 457*f*, 466
Corpus luteum, 458, 460*f*
Corpus spongiosum, 450*f*, 456, 457*f*, 466
Corrugator supercilii, 74*f*, 164*f*, 165*f*, 166, 166*t*
Cortex
    bone, 78*f*, 79
    cerebellar, 265, 266*f*
    cerebral, 255, 257*f*, 276*f*
    hair, 72, 72*f*
    kidney, 436, 436*f*, 437*f*, 445*f*
    lymphatic, 391, 391*f*
    motor, 261*f*
    prefrontal, 261*f*
    premotor, 261, 261*f*
    sensory, 261*f*
    suprarenal, 329, 330*f*
    thymus, 328, 328*f*
Cortical nephrons, 436, 437*f*, 440*f*
Cortical radiate arteries, 440*f*
Corticosterone, 331
Cortisol, 331
Costae, 108
Costal cartilage, 108, 109*f*, 181*f*
    articulations, 144*t*
Costal facet, 104–105, 106*f*, 109*f*
Costal margin, 486*f*, 487*f*
Costal process, 105*f*, 109*f*
Costal tuberosity, 116, 116*f*

Cowper's glands, 454
Coxal bone, 81*f*, 123
CPR, 483
Cranial, 5, 7*f*
Cranial bones, 87–95. *See also specific bones*
    articulations, 144*t*
    ethmoid, 93
    frontal, 90
    occipital, 90
    parietal, 90
    sphenoid, 91–93
    temporal, 90–91
Cranial cavity, 10, 11*f*
Cranial meninges, 256–258
Cranial nerve function tests, 271*t*
Cranial nerves, 226, 235, 243, 246*f*, 267–271
    branches, 269*t*
Cranial sutures, 101*f*
Craniosacral division, 233
Cranium, 7*f*, 10, 87–90, 101, 257*f*, 276
    sex differences in, 124*t*
Cremaster, 453, 453*f*
Cremasteric fascia, 453*f*
Crest, 84*f*, 84*t*
Cribriform plate, 89*f*, 93, 95*f*, 288*f*
Cricoid cartilage, 324*f*, 401*f*, 402, 402*f*, 409*f*, 484*f*
Cricothyroid, 173*f*
Cricothyroid ligament, 402*f*
Cricotracheal ligament, 402*f*
Crista galli, 89*f*, 93, 95*f*, 96*f*, 276
Cristae, 30, 312*f*, 313
Crossing over, 452
Cross-sections, 9
Crown, 416, 417*f*
Crural, 7*f*
Crus, 7*f*, 457*f*
Crying, 270
Crypts of Lieberkuhn, 423
CSF, 243, 244*f*, 258*f*, 259, 259*f*, 260, 276
CT, 323, 325
Cubital fossa, 486*f*, 489*f*
Cubital vein, 375*f*, 376, 377*f*, 486, 486*f*, 489*f*
Cuboid bone, 129, 130*f*
Cuboidal epithelial cells, 40, 40*f*
Cuboidal epithelium, 40, 40*f*, 41, 42*f*, 43*f*, 44, 46*f*, 324*f*
Cuneiform bones, 128–129, 130*f*
Cuneiform cartilages, 402, 403*f*
Cupula, 312*f*, 313
Cusp, 357*f*
Cuspids, 416, 417*f*
Cutaneous membrane, 69
Cutaneous plexus, 68*f*
Cystic artery, 372*f*
Cystic duct, 428, 429*f*
Cystic vein, 380*f*
Cytokines, 339
Cytokinesis, 31, 32*f*, 471*f*
Cytology, 2
Cytoplasm, 27, 42*f*, 43*f*, 44
Cytoskeleton, 28*f*, 29
Cytosol, 27, 28*f*

D

Dartos, 450, 453*f*
De Humani Corporis Faberica (Vesalius), 1
Decidua basalis, 476, 477*f*
Decidua capsularis, 476, 477*f*
Decidua parietalis, 476, 477*f*
Deciduous dentition, 416, 417*f*
Decussate, 265
Deep, 6
Deep artery of penis, 457*f*
Deep brachial artery, 371*f*, 381*f*
Deep cerebral vein, 376*f*
Deep cortex (lymphatic), 391, 391*f*
Deep dermis, 51*f*
Deep femoral artery, 368*f*, 373*f*, 374, 382*f*
Deep femoral vein, 375*f*, 378, 379*f*
Deep layer (embryo), 472
Deep layer (vertebral column muscles), 176*t*, 177
Deep leg compartment, 218*t*, 219*f*
Deep palmar arch, 370, 371*f*
Deep radial nerve, 247*f*
Deep transverse perineal, 184*f*, 185*t*
Deep veins, 377*f*
Deferential artery, 453*f*
Delta cells, 332
Deltoid, 132*f*, 193*f*, 194*f*, 195, 195*t*, 200*f*, 249*f*, 485, 486*f*, 488*f*, 489*f*
Deltoid tuberosity, 118, 119*f*
Dendrites, 60, 60*f*, 228, 229*f*, 266*f*, 282*f*
Dendritic cells, 391*f*
Dendritic process, 282*f*, 283*f*
Dendritic spines, 229*f*
Dens, 104, 105*f*
Dens of axis, 409*f*
Dense connective tissue, 47–48, 47*t*
Dense fibrous layer, 353*f*
Dense irregular connective tissue, 50, 51*f*, 56*f*
Dense regular connective tissue, 50, 51*f*, 56*f*
Denticulate ligament, 243, 244*f*
Dentin, 416, 417*f*
Dentition, 562
Depression, 84*t*, 146*f*, 147
Depressor anguli oris, 74*f*, 164*f*, 165, 165*f*, 166*t*
Depressor labii inferioris, 74*f*, 164*f*, 165, 165*f*, 166*t*
Depth of field, 22–23
Dermal papillae, 69
Dermis, 50, 51*f*, 67, 68*f*, 69–70, 69*f*, 71*f*, 72*f*, 73*f*, 282*f*, 283*f*
Descending aorta, 356*f*, 357*f*, 367, 368*f*
Descending colon, 380*f*, 423*f*, 425, 426*f*
Descending genicular artery, 368*f*, 373*f*
Descending limb of nephron, 438
Desmosomes, 68
Detrusor, 441, 442*f*, 443*f*
Development. *See* Human development
Diabetes mellitus, 333
Diabetic retinopathy, 300
Diagonal branches, 359, 359*f*
Diaphragm, 11*f*, 180*t*, 181*f*, 182, 328*f*, 352*f*, 368*f*, 390*f*, 393*f*, 400*f*, 407*f*, 420*f*, 421*f*, 432*f*

INDEX

Diaphragma sellae, 257f, 322f
Diaphysis, 77, 78, 78f
Diarthroses, 138t, 139, 141, 146–147
Diencephalon, 255, 256f, 264, 265f, 267
Differentiation, 470
Diffuse lymphoid tissues, 390, 391, 392
Digastric, 173, 173f, 174t
Digestion, 413
Digestive epithelium, 424f
Digestive organs, 12
Digestive system, 4f, 413–432. *See also specific organs*
    esophagus, 418–419
    gallbladder, 428–430
    histology, 413–415
    large intestine, 425–427
    liver, 428–430, 432
    mouth, 415–418
    pancreas, 430–431
    pharynx, 418–419
    small intestine, 422–424
    stomach, 419–422
    upper abdomen, 432
Digestive tract, 413
Digital, 7f
Digital arteries, 370, 371f
Digital nerves, 247f
Digital veins, 375f, 376, 377f, 379f
Digits, 7f
Dilation stage, 478, 480f
Diploë, 78f, 79
Diploid, 450, 451f, 456
Dissection, 1, 495–496. *See also* Cat
    brain, 271–275
    eye, 305–306
    heart, 360–362
    kidney, 444–445
    spinal cord, 251–252
Distal, 5, 5f
Distal convoluted tubule, 437f, 438, 439f
Distal extremity (radius), 120f
DNA replication, 451f, 459f
Dorsal, 2, 5f, 7f
Dorsal arch, 373f, 374
Dorsal artery, 457f
Dorsal body cavity, 10
Dorsal interosseous, 203t, 204f, 215t, 217f
Dorsal ramus, 244f, 245
Dorsal root, 240, 240f, 241f, 244f
Dorsal root ganglion, 240, 240f, 241f, 244f, 252f
Dorsal venous arch, 375f, 378, 379f, 491f
Dorsalis pedis artery, 368f, 373f, 491f
Dorsiflexion, 146f, 147, 214t
Dorsum, 7f
Dorsum of nose, 401f
Dorsum of tongue, 415f
Dorsum sellae, 94f
Double gland, 430
Dual innervation, 232
Ductless glands, 321
Ducts of Rivinus, 415
Ductus arteriosus, 356
Ductus deferens, 4f, 450f, 452–455, 455f, 457f, 466

Duodenal ampulla, 428, 429f
Duodenal artery, 372f
Duodenal papilla, 428, 429f, 431f
Duodenum, 332f, 420f, 422, 423f, 424f, 431f
Dura mater, 240f, 241f, 243, 244f, 252f, 257, 257f, 259f, 276f, 299f
Dural layer, outer, 257
Dural sinus, 257f, 258

E

E, 331, 334
Ear, 7f, 309–318, 475f
    bones, 87
    external, 309–311, 484f
    inner, 311–316
    middle, 309–311
    muscles, 167
    tympanic membrane, 317–318
Eardrum, 309
Eccrine glands, 72
Ectoderm, 474, 474f
Efface, 478
Effectors, 59, 225, 226f
Efferent arteriole, 439f, 440, 440f
Efferent division, 225–226, 226f
Efferent lymphatic vessels, 391f, 392
Egg nests, 456, 460f
Ejaculatory duct, 450f, 454, 455f, 457f, 466
Elastic cartilage, 47t, 52, 53f, 56f, 310f
Elastic fibers, 47–48, 47t, 49f, 51f, 53f, 366f, 405f
Elastic ligament, 51f
Elasticity, 155
Elbow, 7f, 119, 120f
    articulations, 147, 148f
    muscles, 199t
Elevation, 146f, 147
Elevations, 84t
Ellipsoidal joints, 141
Embryo, 469, 472f, 475f, 477f
Eminences, 201
Enamel, 416, 417f
Encapsulated lymph organs, 390–391
Endocardium, 352, 353f
Endocrine cells, 431f
Endocrine glands, 321
Endocrine system, 3f, 321–334. *See also specific glands*
    pancreas, 332–333
    parathyroid glands, 326–327
    pituitary gland, 322–323
    suprarenal glands, 329–331, 334
    thymus gland, 328–329
    thyroid gland, 323–325
Endocrine tissues, 3f
Endoderm, 474, 474f
Endolymph, 311, 312f, 314f
Endolymphatic duct, 313
Endolymphatic sac, 312f, 313
Endometrium, 458f, 461, 461f, 462f, 465f, 472, 474f, 476
    uterine, 446f
Endomysium, 155, 156f
Endoneurium, 231, 231f

Endoplasmic reticulum (ER), 28f, 29
Endosteal layer, 257, 257f, 259f
Endosteum, 78
Endothelial cells, 389f
Endothelium, 41, 353f, 366, 366f
Eosinophils, 52f, 338, 339, 340f
Ependymal cells, 227f, 228
Epiblast, 472, 473f, 474f
Epicardium, 352, 352f, 353f
Epicranial aponeurosis, 74f, 164f, 165f, 167, 276f
Epicranium, 166t
Epidermal derivatives, 70
Epidermis, 41, 67–69, 69f, 71f, 72f, 73f, 282f, 283f
Epididymis, 4f, 450f, 452, 453, 453f, 466
Epidural block, 243
Epidural space, 243, 244f
Epigastric region, 8f
Epiglottis, 401, 401f, 402f, 403f, 409, 409f, 415f
Epimysium, 155, 156f
Epinephrine (E), 331, 334
Epineurium, 231, 231f
Epiphyseal cartilage, 78
Epiphyseal line, 79
Epiphysis, 77–78, 78f
Episiotomy, 464
Epithalamus, 263f
Epithelia, 39
Epithelial tissue, 39–46
    pseudostratified, 44–45
    simple, 41–42
    stratified, 41, 43–44
    transitional, 44, 45
Epithelium, 39, 44f, 438
    columnar, 40, 40f, 41, 42f, 43f, 44–45, 44f, 46f, 423f, 427f, 462f
    cuboidal, 40, 40f, 41, 42f, 43f, 44, 46f, 324f
    digestive, 424f
    mucosal, 414f
    mucous, 421f
    olfactory, 287, 288f, 289f
    parietal, 437f, 438, 439f
    pharyngeal, 392
    pseudostratified, 40, 44–45, 44f, 46f
    respiratory, 404f, 405f
    simple, 40, 40f, 41, 42f, 46f, 427f, 462f
    squamous, 40, 40f, 41, 42f, 43f, 46f, 67, 419f, 442f
    stratified, 40, 40f, 41, 42f, 43–44, 43f, 46f, 67, 419f, 442f
    transitional, 40, 44, 45, 46f, 442f
    visceral, 437f, 438, 439f
Eponychium, 73, 73f
Equilibrium, 281, 287, 311
ER, 28f, 29
Erector spina, 178f
Erector spinae muscle group, 177, 178f, 181f, 193f, 486f
Erythroblastosis fetalis, 344
Erythrocytes, 50, 56f, 337
Esophageal hiatus, 418
Esophageal mucosa, 419f

Esophageal vein, 377f
Esophagus, 4f, 172f, 181f, 352f, 380f, 400f, 401f, 404f, 409, 409f, 418–419, 419f, 420f, 421f
Estrogen, 458
Ethmoid, 87, 88f, 89f, 93, 95f, 100f, 288f
Ethmoid notch, 91f
Ethmoidal air cells, 93, 99, 100f
Ethmoidal labyrinth, 93, 95f
Ethmoidal sinus, 93, 99, 100f
Eustachian tube, 309
Eversion, 146, 146f
Excitable, 59, 155
Exocrine cells, 332f, 431f
Exocrine glands, 321
Exocytosis, 30
Exophthalmos, 325
Expulsion stage, 478, 480f
Extensible, 155
Extension, 143f, 146
Extensor carpi radialis brevis, 199t, 200, 200f, 202f, 488f
Extensor carpi radialis brevis tendon, 204f
Extensor carpi radialis longus, 199t, 200, 200f, 202f, 488f
Extensor carpi radialis longus tendon, 204f
Extensor carpi ulnaris, 199t, 200, 200f, 202f, 488f
Extensor carpi ulnaris tendon, 200f, 204f
Extensor digiti minimi, 200f, 201t, 202f
Extensor digiti minimi tendon, 202f, 204f
Extensor digitorum, 200, 200f, 201t, 202f, 488f
Extensor digitorum brevis, 213, 215t
Extensor digitorum brevis tendon, 217f
Extensor digitorum longus, 213, 214t, 216f, 219f, 491f
Extensor digitorum longus tendon, 217f, 489, 491f
Extensor digitorum tendon, 202f
Extensor expansion, 217f
Extensor hallucis brevis, 217f
Extensor hallucis brevis tendon, 217f
Extensor hallucis longus, 213, 214t, 216f, 219f
Extensor hallucis longus tendon, 216f, 217f, 489, 491f
Extensor indicis, 201t, 202f
Extensor indicis tendon, 204f
Extensor muscles, 175, 200, 212t, 214t, 486
Extensor pollicis brevis, 200f, 201, 201t, 202f
Extensor pollicis brevis tendon, 204f
Extensor pollicis longus, 200f, 201t
Extensor pollicis longus tendon, 202f, 204f
Extensor retinaculum, 200, 200f, 202f, 204f, 216f, 217f
Extensor tendon, 150f
External acoustic meatus, 88f, 89f, 90, 92f, 309, 311f
External anal sphincter, 183, 184f, 185t, 425, 426f
External carotid artery, 174f, 367, 369f
External ear, 309–311, 484f

External elastic membrane, 366f
External genitalia (female), 463, 463f, 466
External genitalia (male), 466
External iliac artery, 368f, 372f, 373f, 374
External iliac vein, 375f, 377f, 378, 379f
External intercostal muscles, 179, 180t, 181f, 194f
External jugular vein, 174f, 374, 375f, 376f, 377f, 409f, 484f, 485f
External nares, 399, 401f
External oblique aponeurosis, 182f, 194f
External oblique, 180, 180t, 181f, 182f, 193f, 194f, 487f
External occipital crest, 90, 92f, 409f
External occipital protuberance, 89f, 90, 92f
External os, 446f, 461f, 465f, 479f
External root sheath, 72f
External table, 79
External urethral orifice, 443f, 446f, 450f, 456, 457f, 465f
External urethral sphincter, 184f, 185t, 441, 443f
Extracellular fluid, 27
Extraembryonic membranes, 476–477, 477f
Extraglomerular mesangial cells, 439f
Extraocular muscles, 166–169, 298f, 304f
Extrinsic eye muscles, 167
Extrinsic muscle, 167
Eye, 7f, 295–306, 475f
  accessory structures, 296f
  diseases/infections, 297, 300
  dissection, 305–306
  external anatomy, 295–298
  extraocular muscles, 166–169, 298f, 304f
  internal anatomy, 298–300
  muscles, 166t, 167–169
  retina, 295, 299f, 300–304, 304f
  retinal observation, 303–304
Eyeball, 295
Eyelashes, 296f, 297
Eyelid, 296f

F

F cells, 332
Face, 7f, 74, 95
Facet, 84f, 84t, 104
Facial, 7f
Facial artery, 74f, 165f, 367, 369f
Facial bones, 87, 95–98. *See also specific bones*
  articulations, 144t
  inferior nasal conchae, 95, 97
  lacrimal bone, 88f, 95, 96f, 97
  mandible, 95, 97–98
  maxillae, 95
  nasal bones, 88f, 95, 96f, 97, 101f
  palatine bones, 95, 97
  vomer, 95, 97
  zygomatic bones, 95, 97
Facial expression muscles, 164–167
Facial nerve, 165f, 268, 268f, 269f, 270, 271t, 274f, 310f

Facial vein, 74f, 165f, 174f, 376f
Facies, 7f
Falciform ligament, 407f, 428, 428f, 432f
Fallopian tubes, 460
False pelvis, 123
False ribs, 108, 109f
False vocal cords, 402
Falx cerebelli, 257f, 258
Falx cerebri, 257f, 258
Fascia, 50, 382f
Fascicles, 155, 156f, 231, 231f, 242
Fat, 68f, 296f, 356f
Fatty apron, 422
Fauces, 400, 415, 415f
Femoral, 7f
Femoral artery, 210f, 211f, 219f, 368f, 373f, 374, 382f, 490f
Femoral articular surface, 127f
Femoral circumflex artery, 373f
Femoral circumflex vein, 379f
Femoral condyle, 126, 127f
Femoral cutaneous nerve, 248, 250f, 251t
Femoral epicondyle, 126, 127f
Femoral nerve, 210f, 211f, 219f, 249, 250f, 251t, 382f
Femoral triangle, 490f
Femoral vein, 210f, 219f, 375f, 378, 379f, 382f, 490f
Femoral vessels, 211f
Femur, 7f, 81f, 82, 84f, 126–127, 127f, 150f
  articulations, 145t
Fertilization, 469–471, 471f, 472f
Fetal skull, 101
Fetus, 469, 479f
Fibrinogen, 454
Fibroblasts, 47t, 48, 50
Fibrocartilage, 47t, 52, 53f, 56f
Fibrocyte, 49f, 51f, 282f, 451f
Fibrous capsule, 436f
Fibrous joint capsule, 140f
Fibrous joints, 138, 138t
Fibrous layer, 407f
Fibrous pericardium, 356f, 407f
Fibrous tendon sheath, 217f
Fibrous tunic, 298, 299f
Fibula, 81f, 82, 128, 129f, 150f, 219f
  articulations, 145t
  head, 129f, 150f, 216f, 490f
Fibular artery, 368f, 373f, 374
Fibular collateral ligament, 149, 150f
Fibular nerve, 216f, 219f, 250f, 251t
Fibular vein, 375f, 378, 379f
Fibularis brevis, 213, 214t, 216f, 219f
Fibularis brevis tendon, 216f, 217f, 491f
Fibularis longus, 213, 214t, 216f, 219f, 491f
Fibularis longus tendon, 216f, 219f, 491f
Fibularis muscles, 489
Field of view, 17
Fight-or-flight response, 232, 233f, 331
Filaments, 57
Filiform papillae, 290f, 291
Filtrate, 436
Filtration slits, 438, 439f
Filum terminale, 239, 240f, 244f, 259f

Fimbriae, 460, 461f
Fine adjustment knob, 18f, 19, 22f
Fingers, 7f, 80, 121
    appendicular muscles, 197–204
First bicuspid, 416
First molar, 416, 417f
First polar body, 456, 459f
First premolar, 417f
First rib, 369f, 376f, 390f
First trimester, 469–478
Fissure, 84f, 84t
Fixed ribosomes, 28f, 29
Flagellum, 29
Flat bones, 77, 78f, 79
Flexion, 143f, 146
Flexor carpi radialis, 197, 198f, 199t, 200f, 202f, 489f
Flexor carpi radialis tendon, 200f, 204f, 489f
Flexor carpi ulnaris, 197, 198f, 199t, 200f, 202f, 488f
Flexor carpi ulnaris tendon, 204f, 489f
Flexor digiti minimi brevis, 201, 203t, 204f, 215t, 217f
Flexor digitorum brevis, 213, 215t, 217f
Flexor digitorum brevis tendon, 217f
Flexor digitorum longus, 213, 214t, 216f, 219f
Flexor digitorum longus tendon, 217f, 491f
Flexor digitorum profundus, 148f, 200f, 201t, 202f
Flexor digitorum profundus tendon, 204f
Flexor digitorum superficialis, 148f, 197, 198f, 200f, 201t, 202f
Flexor digitorum superficialis tendon, 202f, 204f, 489f
Flexor hallucis brevis, 213–214, 215t, 217f
Flexor hallucis longus, 213, 214t, 216f, 219f
Flexor muscles, 175, 200, 212t, 214t
Flexor pollicis brevis, 203t, 204f
Flexor pollicis longus, 200f, 201t, 202f
Flexor pollicis longus tendon, 204f
Flexor retinaculum, 197, 198f, 200f, 202f, 204f
Floating kidney, 442
Floating ribs, 108, 109f
Flocculonodular lobe, 265, 266f
Fluid connective tissue, 47, 47t, 50, 52f
Focal depth (microscope), 22
Focal plane (microscope), 22, 22f
Folia, 265, 266f
Follicle cavities, 324f
Follicle cells, 323, 460f
Follicles, 456, 460f
Follicles (thyroid), 323
Follicular development, 460f
Follicular fluid, 460f
Fontanels, 101
Foot, 7f, 82, 128–130, 489
    appendicular muscles, 213–217
Foramen, 84f, 84t
Foramen lacerum, 89f, 93
Foramen magnum, 89f, 90, 92f, 245f, 409f

Foramen of Monro, 258
Foramen ovale, 89f, 93, 94f
Foramen rotundum, 89f, 93, 94f
Foramen spinosum, 89f, 93, 94f
Foramina, 95f
Forearm, 7f, 80, 118, 489f
    appendicular muscles, 197–200
Forebrain, 475f
Forehead, 7f, 124t
Formed elements, 50, 52f, 337, 338f, 387
Fornix, 261, 262f, 263f, 275f, 299f, 458f, 463
Fossa, 84f, 84t
Fossa ovalis, 356, 357f
Fossae, 89
Fourth ventricle, 258, 259f, 263f, 265f, 266f, 275f
Fovea, 299f, 301f, 302
Fovea capitis, 126
Free edge, 73, 73f
Free macrophage, 49f
Free nerve endings, 282f, 283
Free ribosomes, 28f, 29
Free surface, 39
Frenulum, 415, 415f, 416f
Frons, 7f
Frontal, 7f
Frontal air cells, 91f
Frontal bone, 87, 88f, 89f, 90, 91f, 96f, 100f, 101f, 168f, 298f
Frontal crest, 91f
Frontal lobe, 256f, 260, 261f, 263f
Frontal plane, 9f
Frontal section, 9
Frontal sinus, 96f, 99, 100f, 400f, 401f
Frontal squama, 90
Frontal suture, 90, 91f, 101f
Frontalis, 74f
Frontonasal suture, 88f
Fructose, 454
Functional layer, 462f, 473f
Functional zone, 461
Fundus (gallbladder), 429f
Fundus (stomach), 419, 420f, 421f
Fungiform papillae, 290f, 291

G

G cell, 421f
Gallbladder, 4f, 8f, 428–430, 428f
Gametes, 449, 459f, 469
Ganglia, 228, 232–233, 233f, 234f, 235, 244f
Ganglion cells, 300, 301f
Ganglionic fibers, 232
Ganglionic neuron, 232, 233f, 234
Gas exchange, 355f, 399
Gastric area, 393f
Gastric artery, 334f, 372, 372f, 429f
Gastric glands, 421f, 422
Gastric pit, 421f, 422
Gastric vein, 380f
Gastrocnemius, 150f, 213, 214t, 216f, 219f, 490f, 491f
    lateral head, 209f, 216f, 250f, 491f
    medial head, 209f, 210f, 216f, 250f, 491f

Gastroduodenal artery, 372f, 431f
Gastroepiploic artery, 372f
Gastroepiploic vein, 380f
Gastroepiploic vessels, 420f
Gastrosplenic ligament, 393f
Gastrulation, 472, 474, 474f
Gelatinous material, 312f, 313
Gemellus, 205t, 206f, 207, 250f
Gene, 29
General senses, 261, 281–284
    general-sense receptors, 281–284
    two-point discrimination test, 284
Geniculate nucleus, 265f
Genioglossus, 170, 171f, 171t
Geniohyoid, 172t, 173, 173f, 174t, 246f
Genitofemoral nerve, 248, 250f, 251t, 453f
Germ layers, 474
Germinal center, 391, 391f, 392f
Germinative cells, 71f
Gestation, 469
Gingiva, 415f, 416, 417f
Gingival sulcus, 416, 417f
Glans, 456, 457f
Glans of clitoris, 463f, 464
Glassy membrane, 72f
Glaucoma, 300
Glenoid cavity, 117, 117f, 132f
Glenoid fossa, 117
Glenoid labrum, 132f
Glial cells, 59, 159f, 227–228
Gliding joints, 141
Globus pallidus, 262f, 263f, 264
Glomerular capillary, 439f
Glomerular capsule, 439f
Glomerular epithelium, 438
Glomerulus, 438, 440f
Glossopharyngeal nerve, 268f, 269t, 270, 271t, 274f
Glottis, 401, 401f, 403f
Glucagon, 333
Glucocorticoids, 330
Gluteal, 7f
Gluteal aponeurosis, 209f
Gluteal artery, 373f
Gluteal group muscles, 204, 205t, 206f, 207
Gluteal injection site, 490f
Gluteal nerves, 250f, 251t
Gluteal region, 490f
Gluteal tuberosity, 126, 127f
Gluteal vein, 379f
Gluteus, 7f
Gluteus maximus, 184f, 193f, 205t, 206f, 207, 209f, 210f, 219f, 250f, 489, 490f
Gluteus medius, 193f, 194f, 205t, 206f, 207, 209f, 210f, 250f, 490f
Gluteus minimus, 205t, 206f, 207, 250f
Glycogenolysis, 333
Goblet cells, 41, 423f, 427f
Goiter, 325
Golgi apparatus, 28f, 29, 30, 229f
Golgi tendon organs, 282
Gomphosis, 138, 138t
Gonadal arteries, 334f, 368f, 372
Gonadal veins, 334f, 375f, 377f, 378

Gonads, 3f, 449
Graafian follicle, 458
Gracilis, 194f, 205t, 206f, 207, 209f, 210f, 211f, 250f, 490f, 491f
Gracilis tendon, 209f
Granular leukocytes, 338
Granulocytes, 338, 340f
Granulosa cells, 460f
Graves' disease, 325
Gray horns, 241, 241f
Gray matter, 229–230, 241f, 242, 244f, 252f, 266f
Gray ramus, 245
Great auricular nerve, 246f, 246t
Great cardiac vein, 356f, 357f, 359, 359f
Great cerebral vein, 376f
Great pancreatic artery, 431f
Great saphenous vein, 211f, 375f, 378, 379f, 382f, 491f
Greater curvature (stomach), 419, 420f, 421f
Greater femoral trochanter, 206f, 250f, 490f
Greater horn, 99, 99f
Greater omentum, 420f, 421f, 422, 426f
Greater palatine foramen, 89f, 97
Greater sciatic notch, 124, 125f
Greater trochanter, 126, 127f
Greater tubercle, 118, 119f, 132f
Greater vestibular glands, 458f, 463
Greater wings, 91, 94f, 101f
Groin, 7f
Gross anatomy, 1
Ground substance, 46–47
Growth, 478–479
Gustation, 281, 287, 290–292
Gustatory cells, 290, 290f
Gustatory cortex, 261f
Gynecology, 456
Gyrus, 255, 256f, 261, 261f, 263f

H

H band, 158f
H zone, 157
Hair, 3f, 72–73, 72f, 282f
Hair bulb, 72f
Hair cells, 312f, 315f
Hair follicle, 68f, 71f
Hair papilla, 72, 72f
Hair root, 72, 72f
Hair shaft, 68f, 72, 72f
Hallux, 129
Hamate bone, 121, 122f
Hamstrings, 208, 489, 490f
Hand, 7f, 80, 118, 121–123
    appendicular muscles, 197–203
    extensor muscles, 199t
    flexor muscles, 199t
Haploid, 450, 451f, 456, 459f
Hard palate, 400, 401f, 409, 409f, 415f, 417f
Hassall's corpuscles, 328
Haustra, 425, 426f
Haustrum, 427f
Haversian systems, 79

Head, 7f, 84f, 84t, 475f, 483–485
Head (microscope), 18f
Head fold, 474, 477f
Hearing, 281, 287, 309
Heart, 3f, 10, 328f, 351–362, 371f, 475f. See also Cardiovascular system
    anatomy, 354–358
    coronary circulation, 359–360
    dissection, 360–362
    position of, 351, 352f
    serous membranes, 12
    wall, 351–353
Heartburn, 418
Heel, 7f
Hematocrit, 346–348
Hemiazygos vein, 377f, 390f
Hemoglobin, 338
Hemolysis, 343f
Hemolytic disease of the newborn, 344
Hepatic artery branch, 429f
Hepatic artery proper, 372f, 428f, 429f
Hepatic duct, 428, 429f
Hepatic portal vein, 378, 380f, 426f, 428f, 429f
Hepatic veins, 334f, 375f, 377f, 378, 380f, 428f
Hepatocytes, 428
Hepatopancreatic sphincter, 428, 429f
Hilum, 391f, 393f, 407f, 428f, 436f
Hilus, 392, 406, 435–436
Hinge joints, 141, 142f
Hips, 489
Histological stains, 323
Histology, 2, 39
HIV, 41
Homeostasis, 2, 27, 59, 225, 321
Homologous chromosomes, 451
Hooke, Robert, 27
Horizontal cells, 301f, 302
Horizontal fissure, 406, 407f
Hormones, 321
Human development, 469–480
    birth, 478, 480f
    blastocyst formation, 470–472
    cleavage, 470–472
    extraembryonic membranes, 476–477
    fertilization, 469–471
    first trimester, 469–478
    gastrulation, 472, 474
    implantation, 472, 473f
    labor, 478, 480f
    placenta, 476–478
    second trimester, 469, 478–479
    third trimester, 469, 478–479
Humeral circumflex artery, 371f
Humeral epicondyle, 198f, 488f
Humeroradial articulation, 147
Humeroulnar articulation, 147
Humerus, 80, 81f, 84f, 118, 119f, 120f, 142f, 148f, 198f
    articulations, 145t
    head, 118, 119f, 132f
Hyaline cartilage, 47t, 52, 53f, 56f, 78
Hyaline cartilage plate, 405f
Hyaluronidase, 470, 472

Hydrocephalus, 260
Hydroxyapatite, 52
Hymen, 463, 463f
Hyoglossus, 170, 171f, 171t, 172t
Hyoid bone, 80, 87, 99, 99f, 171f, 173f, 174f, 324f, 400f, 401f, 402f, 404f, 409f, 415f, 485f
Hyperextension, 146
Hyperthyroidism, 325
Hypoblast, 472, 473f, 474f
Hypochondriac region, 8f
Hypodermis, 68f, 69
Hypogastric region, 8f
Hypoglossal canal, 89f, 90, 92f, 96f
Hypoglossal nerve, 246f, 268f, 269t, 271, 271t, 274f
Hyponychium, 73
Hypophyseal fossa, 91, 96f
Hypophysis, 322
Hypothalamus, 255, 256f, 263f, 264, 275f, 322f
Hypothenar eminence, 201

I

I bands, 157, 158f
Ileocecal valve, 422, 425, 426f
Ileocolic artery, 372f
Ileocolic vein, 380f
Ileum, 422, 423f, 426f
Iliac crest, 124, 125f, 193f, 206f, 209f, 211f, 487f, 489, 490f
Iliac fossa, 123, 123f, 124t, 125f
Iliac notch, 125f
Iliac spine, 124, 125f, 490f
Iliac tuberosity, 125f
Iliacus, 205t, 206f, 208, 210f, 382f
Iliococcygeus, 184f, 185t
Iliocostalis cervicis, 176t, 177, 178f
Iliocostalis group, 176t
Iliocostalis lumborum, 176t, 177, 178f
Iliocostalis thoracis, 176t, 177, 178f
Iliofemoral ligament, 126
Iliohypogastric nerve, 250f, 251t
Ilioinguinal nerve, 250f, 251t
Iliolumbar artery, 373f
Iliopsoas, 194f, 211f
Iliopsoas group muscles, 205t, 206f, 208
Iliotibial tract, 206f, 207, 209f, 216f, 489, 490f
Iliotibial tract muscle, 206f
Ilium, 81f, 82, 123, 124t, 125f
Immune system, 338, 387
Implantation, 472, 472f, 473f
Incisive fossa, 89f, 95
Incisors, 417f
Incus, 309, 311f
Infarctions, 360
Infection, 41, 391
Inferior, 2, 5f, 495
Inferior angle, 116–117, 117f
Inferior articular surfaces, 129f
Inferior carotid triangle, 483, 485f
Inferior colliculi, 263f, 264, 265f, 266f, 273f
Inferior gluteal line, 124, 125f

INDEX

Inferior mesenteric ganglion, 235
Inferior mesenteric vein, 380f, 426f
Inferior nuchal line, 90, 92f
Inferior oblique muscles, 168, 168f, 168t, 296f, 297, 298f
Inferior orbital fissure, 88f, 93
Inferior pharyngeal constrictor, 170, 172f, 172t
Inferior rectus, 167, 168f, 168t, 296f, 297, 298f
Inferior sagittal sinus, 257f, 258, 376f
Inferior serratus posterior, 180
Inferior temporal line, 88f, 90, 92f
Inferior tibiofibular joint, 128, 129f
Inferior trunk, 247f
Inferior vena cava, 181f, 330f, 334, 334f, 354, 356f, 357f, 360, 372f, 375f, 377f, 378, 380f, 390f, 426f, 428f, 432f
Inferior vertebral notch, 104, 106f
Infraglenoid tubercle, 117, 200f
Infrahyoid muscles, 173
Infra-orbital foramen, 88f, 95, 96f
Infrapatellar bursa, 150f
Infrapatellar fat pad, 150f
Infraspinatus, 132, 132f, 193f, 195t, 196, 486f, 488f
Infraspinous fossa, 117, 117f
Infundibulum, 264, 268f, 274f, 275f, 322, 322f, 460, 461f
Inguen, 7f
Inguinal, 7f
Inguinal canal, 453, 453f, 487f
Inguinal hernia, 453
Inguinal ligament, 206f, 210f, 211f, 373f, 382f, 453f, 487f, 490f
Inguinal lymph nodes, 388f
Inguinal region, 8f
Inguinal ring, 194f, 453f
Initial segment (axon), 228, 229f
Inner cell mass, 470, 472f, 473f
Inner dural layer, 258
Inner ear, 309, 310f, 311–316
Inner hair cells, 313, 315f
Innervation, 232, 234f
Innominate artery, 367
Insertion, 163, 166t, 168t, 169t, 175t, 196
    appendicular muscles, 192t, 195t, 199t, 201t, 203t, 205t, 212t, 214t, 215t
    axial muscles, 166t, 168t, 169t, 171t, 172t, 174t, 175–176t, 180t, 185t
    lower limb compartments, 218t
Insula, 261, 261f, 262f, 263f
Insulin, 322, 333
Integument, 67
Integumentary system, 3f, 67–74
    accessory structures, 70–73
    dermis, 69–70. *See also* Dermis
    epidermis, 67–69. *See also* Epidermis
    hair, 72–73. *See also* Hair
    nails, 3f, 73, 73f
    sebaceous glands, 70–71. *See also* Sebaceous glands
    sweat glands, 3f, 71–72, 71f
Interatrial septum, 356, 357f
Intercalated discs, 57, 58f, 352, 353f

Intercondylar eminence, 128, 129f
Intercondylar fossa, 126, 127f
Intercostal artery, 371
Intercostal nerves, 248
Intercostal thoracis muscles, 179
Intercostal veins, 375f, 377f, 390f
Interlobar arteries, 440, 440f, 445f
Interlobar veins, 440f, 441, 445f
Interlobular arteries, 440
Interlobular septum, 405f, 429f
Interlobular veins, 440f, 441
Intermediate hairs, 72
Intermediate mass, 275f
Internal acoustic meatus, 89f, 91, 92f, 96f
Internal anal sphincter, 425, 426f
Internal capsule, 262f, 263f, 264
Internal carotid artery, 367, 369f
Internal elastic membrane, 366, 366f
Internal iliac artery, 368f, 372f, 373f, 374
Internal iliac vein, 375f, 377f, 378, 379f
Internal intercostals, 179, 180t, 181f, 194f
Internal jugular vein, 324f, 374, 375f, 376f, 377f, 390f
Internal nares, 400, 400f, 401f
Internal oblique, 180t, 181, 181f, 182f, 193f, 194f
Internal occipital crest, 92f
Internal occipital protuberance, 92f
Internal os, 446f, 461f, 465f
Internal table, 79
Internal urethral orifice, 466
Internal urethral sphincter, 441, 443f
International Federation of Associations of Anatomists, 1
Interosseous membrane, 120, 120f
Interphase, 31, 32f
Interspinales muscles, 176t, 177, 178f
Interstitial cells, 450, 451f
Interstitial fluid, 389f
Interstitial lamellae, 79, 80f
Interthalamic adhesion, 264
Intertransversarii, 176t, 177, 178f
Intertrochanteric crest, 126, 127f
Intertrochanteric line, 126, 127f
Intertubercular groove, 118, 132f
Intertubercular sulcus, 119f
Intertubercular synovial sheath, 197
Interventricular foramen, 258, 258f, 263f
Interventricular septum, 356, 357f, 362f
Intervertebral discs, 103, 103f
Intervertebral foramen, 103f, 104
Intervertebral muscles, 178f
Intestinal arteries, 372f, 426f
Intestinal crypt, 423f, 427f
Intestinal glands, 423, 424f
Intestinal lining, 42f
Intestinal lumen, 429f
Intestinal trunk, 390f
Intestinal veins, 380f, 426f
Intestines
    large, 4f, 8f, 425–427
    small, 4f, 8f, 332f, 372f, 380f, 422–424
Intramural ganglia, 233f, 235
Intrinsic eye muscles, 167
Intrinsic muscle, 167

Invagination, 474
Inversion, 146, 146f
Involuntary, 57
Iris, 298, 299f
Iris diaphragm, 19
Iris diaphragm lever, 18f, 19
Irregular bones, 77
Ischial ramus, 124, 125f, 457f
Ischial spine, 123f, 124, 124t, 125f
Ischial tuberosity, 124, 125f, 206f
Ischiocavernosus, 184f, 185t
Ischium, 81f, 82, 123, 125f
Islets of Langerhans, 332, 332f
Isthmus, 323, 324f

J

Jejunum, 422, 423f
Joint capsule, 150f
Joint cavity, 140, 140f, 148f
Joint classification, 137–139
Joint movements, 141–143, 146
    angular, 143f
    rotational, 144f
Joints, 137
Jugular foramen, 89f, 91
Jugular notch, 92f, 109f, 483, 485f, 486f
Jugular trunk, 390f
Junctional fold, 159f
Juxtaglomerular complex, 438, 439f
Juxtamedullary nephrons, 437f, 438, 439f, 440f

K

Keratin, 41
Keratinized, 41
Keratinocytes, 67
Keratohyalin, 68
Kidney tubule, 42f
Kidneys, 4f, 8f, 12, 330f, 334f, 432f, 435–441
    blood supply, 440–441
    nephrons, 436–439, 440f
Knee, 7f, 489
    articulations, 147, 149, 150f
    muscles, 150f, 212t
Knee extensors, 150f, 212t
Kneecap, 7f, 82
Knee-jerk reflex, 239
Knob, 288f

L

L_3, 181f
L_5, 144t
Labia, 4f, 415
Labia majora, 446f, 458f, 463, 463f, 465f
Labia minora, 446f, 458f, 463, 463f, 465f
Labor, 478, 480f
Labor stages, 478, 480f
Laboratory safety, 251–252, 305, 344, 360, 444, 495–496, 514, 521
Labyrinthine artery, 369f
Lacrimal apparatus, 297
Lacrimal bone, 88f, 95, 96f, 97
Lacrimal canaliculus, 296f

Lacrimal canals, 297
Lacrimal caruncle, 295, 296f
Lacrimal ducts, 297
Lacrimal fluid, 297
Lacrimal fossa, 90, 91f
Lacrimal gland ducts, 296f
Lacrimal glands, 296f, 297
Lacrimal groove, 88f
Lacrimal punctum, 296f, 297
Lacrimal sac, 296f, 297
Lacrimal sulcus, 96f
Lactation, 464
Lacteal, 423, 423f, 424f
Lactiferous ducts, 464, 465f
Lactiferous sinuses, 464, 465f
Lacunae, 47t, 52, 52f, 53f, 54f, 472, 473f, 474f
  bone, 79, 80f
Lambdoid suture, 87, 88f, 89f, 96f, 101f
Lamellae, 47t, 54, 282f
Lamellated corpuscles, 68f, 69, 282f, 283, 283f
Lamina, 103f, 104, 105f, 106f, 107f
Lamina propria, 287, 288f, 414, 414f, 419f, 421f, 423f, 424f, 442f
Lamp, 19
Landmarks, 93
Large intestine, 4f, 8f, 425–427
Laryngeal cartilage, 409f
Laryngeal ligaments, 401
Laryngeal prominence, 402f
Laryngopharynx, 400, 401f, 409f, 415f, 418
Larynx, 4f, 164f, 400f, 401–403, 404f
Lateral, 5, 5f
Lateral angle, 116–117, 117f
Lateral antebrachial cutaneous nerve, 247f
Lateral apertures, 258, 259f
Lateral border, 116, 117f
Lateral canthus, 295, 296f
Lateral condyle, 150f
Lateral epicondyle, 118, 119f
Lateral excursion, 170
Lateral facet, 128
Lateral femoral circumflex artery, 373f, 382f
Lateral femoral condyle, 126, 127f
Lateral femoral epicondyle, 126, 127f
Lateral fibular malleolus, 128, 129f, 217f, 489, 491f
Lateral flexion, 146f, 147
Lateral gray horns, 241f, 242
Lateral humeral epicondyle, 488f
Lateral incisor, 416, 417f
Lateral leg compartment, 218t, 219f
Lateral malleolus, 216f
Lateral masses, 93, 95f
Lateral meniscus, 147, 150f
Lateral nail fold, 73f
Lateral nasal cartilage, 401f
Lateral patellar facet, 128f
Lateral plantar artery, 373f
Lateral plantar nerve, 250f
Lateral plate, 93, 94f
Lateral rectus, 167, 168f, 168t, 297, 298f
Lateral rotation, 144f, 146

Lateral rotator group muscles, 205t, 206f, 207
Lateral sacral artery, 373f
Lateral sacral crest, 106, 107f
Lateral sacral vein, 379f
Lateral semicircular duct, 312f
Lateral sulcus, 256f, 260
Lateral supracondylar ridge, 126, 127f
Lateral sural cutaneous nerve, 250f
Lateral thoracic artery, 371f
Lateral tibial condyle, 128, 129f
Lateral tibial tubercle, 128
Lateral tubercle, 129f
Lateral umbilical ligament, 443f
Lateral ventricles, 256f, 258, 258f, 262f, 263f, 275f
Lateral white column, 241f, 242
Latissimus dorsi, 175, 181f, 193f, 194f, 195t, 196, 200f, 486f, 487f, 488f
LCA, 356f, 357f, 359, 359f
Left atrioventricular (AV) valve, 357f, 358
Left atrium, 352f, 354, 355f, 359f, 361f, 362f
Left bronchomediastinal trunk, 390f
Left colic flexure, 425, 426f
Left coronary artery (LCA), 356f, 357f, 359, 359f
Left internal jugular vein, 390f
Left lower quadrant (LLQ), 8f
Left rotation, 144f, 146
Left splenic flexure, 425, 426f
Left subclavian trunk, 390f
Left upper quadrant (LUQ), 8f
Left ventricle, 352f, 354, 355f, 356f, 357f, 359f, 361f, 362f
Leg, 7f, 82, 126, 489
  appendicular muscles, 208–213
Leg bud, 475f
Leg compartments, 218t, 219f
Lens, 304f
Lens (microscope), 17, 18, 18f, 298, 299f
Lentiform nucleus, 262f, 263f, 264
Lesser cornu, 402f
Lesser curvature (stomach), 419, 420f, 421f
Lesser horn, 99, 99f
Lesser occipital nerve, 246f, 246t
Lesser omentum, 421f, 422, 429f
Lesser palatine foramen, 89f
Lesser sciatic notch, 124, 125f
Lesser trochanter, 126, 127f
Lesser tubercle, 118, 119f, 132f
Lesser wings, 91, 94f
Leukocytes, 50, 56f, 338. See also White blood cells
Levator ani, 183, 184f, 185t
Levator labii superioris, 74f, 164f, 165, 165f, 166t
Levator palpebrae superioris, 166f, 167, 168f, 296f, 297, 298f
Levator scapulae, 192, 192t, 193f
Levator veli palatini, 170, 172f, 172t
Levels of organization, 2
Ligaments, 3f, 50, 51f
  attachment, 84t
  knee, 149
Ligamentum arteriosum, 355, 356f, 357f

Ligamentum capitis femoris, 126
Ligamentum nuchae, 191
Light control knob, 18f, 19
Limbs, 381–382
Line, 84f, 84t
Linea alba, 181, 181f, 182f, 486f, 487f
Linea aspera, 126, 127f
Lingual artery, 369f
Lingual frenulum, 415, 415f, 416f
Lingual tonsils, 392, 392f, 401f, 415f
Lip, 415f
Liposuction, 50
Liver, 4f, 8f, 372f, 420f, 428–430, 428f, 432, 432f
LLQ, 8f
Lobules, 328, 328f
  liver, 429f
  lungs, 405f, 408
  pancreas, 332f, 431f
  pulmonary, 405f
  testis, 450
Location, 11f
Locomotion, 155
Loin, 7f
Long bones, 77, 78f
Long preganglionic neuron, 233f
Long thoracic nerve, 248t
Longissimus capitis, 176t, 177, 178f
Longissimus cervicis, 176t, 177, 178f
Longissimus group, 176t
Longissimus thoracis, 176t, 177, 178f
Longitudinal arch (ankle), 130f
Longitudinal fissure, 255, 259f, 263f
Longitudinal layer, 423f
Longitudinal muscle, 421f
Longitudinal muscle layer, 414f, 420f, 427f
Longus capitis, 176t, 177, 178f
Longus colli, 176t, 177, 178f, 179
Loop of Henle, 438
Loose connective tissue, 42f, 43f, 44f, 47, 49f, 79, 276f, 389f
Lower esophageal sphincter, 418
Lower extremity, 126
Lower jaw, 417f
Lower limb, 7f, 80, 82, 126–131, 489, 490f, 491f
  ankle, 128–130. See also Ankle
  articulations, 145t
  femur, 126–127. See also Femur
  fibula, 128. See also Fibula
  foot, 128–130. See also Foot
  lymphatics, 388f, 389
  patella, 126, 128. See also Patella
  tibia, 128. See also Tibia
Lower respiratory system, 399, 400f
Lumbar, 7f
Lumbar arteries, 374
Lumbar curve, 102f
Lumbar enlargement, 239
Lumbar lymph nodes, 388f
Lumbar nerves, 245
Lumbar plexus, 245f, 248–251, 251t
Lumbar region, 8f, 102f
Lumbar spinal nerves, 240f, 245f

INDEX

Lumbar trunk, 390f
Lumbar veins, 375f, 377f, 378
Lumbar vertebrae, 102, 107f
Lumbosacral enlargement, 240f
Lumbosacral plexus, 248
Lumbrical muscle, 203t, 204f, 215t, 217f
Lumbus, 7f
Lumen, 71f
  artery, 366f
  bronchus, 405f
  duct, 43f
  ductus deferens, 455f
  hair, 71f
  intestinal, 429f
  seminal gland, 455f
  seminiferous tubule, 451f
  trachea, 404f
  urethra, 442f
  urinary bladder, 442f
  vein, 366f
Luminal compartment, 451f
Luminal surface, 421f
Lunate bone, 121, 122f
Lunate surface, 124, 125f
Lung tissue, 404f
Lungs, 4f, 12, 352f, 400f, 406–408
Lunula, 73, 73f
LUQ, 8f
Lymph, 47t, 50, 387
Lymph node artery, 391f
Lymph node vein, 391f
Lymph nodes, 4f, 174f, 387, 388f, 390–393, 390f, 395f, 485f
Lymphatic capillaries, 389, 389f
Lymphatic capsule, 391, 391f, 393f
Lymphatic ducts, 387, 388f, 389, 390f
Lymphatic medulla, 391, 391f
Lymphatic system, 4f, 387–395
  female breast, 395
  lymph nodes, 390–393. See also Lymph nodes
  lymphatic vessels, 389–390. See also Lymphatic vessels
  lymphoid tissues, 390–393
  spleen, 393–394. See also Spleen
Lymphatic trunks, 389
Lymphatic valves, 389, 389f
Lymphatic vessels, 4f, 387, 389–390, 389f, 391, 391f, 392, 405f, 414f, 421f, 423f
Lymphatics, 388f, 389
Lymphocytes, 47t, 49f, 50, 52f, 328, 328f, 339–340, 340f
Lymphoid nodules, 392, 392f, 423f
Lymphoid tissues, 390–393
Lysosome, 28f, 30
Lysozyme, 297

M

M line, 157, 158f
M phase, 31, 32f
Macrophages, 48, 49f, 387
Macula densa, 438, 439f
Macula lutea, 301f, 302
Maculae, 312f, 313

Major calyx, 436, 436f
Malleus, 309, 311f
MALT, 388f
Mamillary bodies, 263f, 264, 266f, 268f, 274f, 275f, 322f
Mamma, 7f
Mammary, 7f
Mammary gland lobes, 465f
Mammary glands, 4f, 395, 395f, 464–465, 465f
  lymphatics, 388f
Mandible, 74f, 88f, 95, 96f, 97–98, 98f, 101f, 165f, 169f, 171f, 173f, 174f, 401f, 409f
  articulations, 144t
  body, 97, 484f
  sex differences in, 124t
Mandibular angle, 484f, 485f
Mandibular branch, 268
Mandibular dental arcade, 417f
Mandibular foramen, 98f
Mandibular fossa, 89f, 90, 92f, 98
Mandibular notch, 97, 98f
Manual, 7f
Manubrium, 108, 109f, 116f, 142f, 409f, 484f, 486f
Manus, 7f
Marrow cavity, 78
Massa intermedia, 264
Masseter, 74f, 164f, 165f, 169, 169f, 169t, 174f
Mast cells, 47t, 48, 49f, 339
Mastectomy, 395
Master gland, 323
Mastication, 169–170, 416
Mastication muscles, 169–170
Mastoid air cells, 90
Mastoid fontanel, 101, 101f
Mastoid foramen, 89f
Mastoid process, 88f, 89f, 90, 92f, 96f, 484f, 485f
Mastoiditis, 310
Maternal mitochondria, 30
Matrix, 30, 46, 47, 47t, 52, 53f, 54f
  bone, 79
  hair, 72
Maxillae, 88f, 89f, 95, 96f, 100f, 101f, 168f, 298f
  articulations, 144t
Maxillary artery, 367, 369f
Maxillary branch, 268
Maxillary dental arcade, 417f
Maxillary sinus, 99, 100f
Maxillary vein, 376f
Meatus, 84f, 400
Mechanical stage, 18f, 19
Mechanical stage controls, 18f, 19
Mechanoreceptors, 281
Medial, 5, 5f
Medial border, 116, 117f
Medial condyle, 150f
Medial cord, 381f
Medial epicondyle, 118, 119f, 120f, 202f, 486f, 489f

Medial facet, 128
Medial meniscus, 147, 150f
Medial plate, 93, 94f
Medial rectus, 167, 168f, 168t, 297
Medial rotation, 144f, 146
Medial tubercle, 129f
Median aperture, 258, 259f
Median eminence, 322f
Median nerve, 202f, 246, 247f, 248t, 249f, 381f
Mediastinal vein, 377f
Mediastinum, 10, 11f, 351, 352f
Medulla, 475f
  adrenal, 235
  hair, 72, 72f
  kidney, 436, 436f, 437f
  lymphatic, 391, 391f
  suprarenal, 329, 330f
  thymus, 328, 328f
Medulla oblongata, 255, 256f, 257f, 258f, 263f, 264–265, 265f, 266f, 268f, 274f, 275f
Medullary cavity, 140f
Medullary cords, 391, 391f
Medullary sinuses, 391–392, 391f
Megakaryocytes, 340
Meiosis, 450, 451f, 470, 471f
Meiosis I, 450, 451, 451f, 456, 459f
Meiosis II, 450, 451f, 452, 456, 459f
Meissner corpuscles, 69, 283
Melanin, 68
Melanocytes, 68
Melanocyte-stimulating hormone (MSH), 322f, 323
Membranous labyrinth, 311, 312f
Membranous organelles, 28, 29–30
Membranous urethra, 443f, 454, 457f, 466
Meningeal layer, 257, 257f, 259f
Meninges, 10
Menstruation, 461
Mental, 7f
Mental foramen, 88f, 98, 98f
Mental protuberance, 88f, 98, 98f, 484f
Mentalis, 164f, 165, 166t
Mentis, 7f
Merkel cells, 282f, 283
Merocrine gland cells, 71f
Merocrine sweat glands, 71f, 72
Mesencephalon, 255, 256f, 263f, 264, 265f, 266f, 274f
Mesenchymal cell, 49f
Mesenchyme, 47, 48f
Mesenteric artery, 330f, 334f, 368f, 372, 372f, 414f, 426f, 431f
Mesenteric vein, 380f, 414f, 426f
Mesenteries proper, 422
Mesentery, 414f
Mesoderm, 474, 474f, 476, 477f
Mesothelium, 41, 353f
Mesovarium, 461, 461f
Metacarpal bones, 80, 81f, 121, 122f, 142f
  articulations, 145t
Metaphase, 32, 32f, 471f
Metaphase plate, 32, 32f

Metaphysis, 78, 78f
Metastasis, 32
Metastasize, 395
Metatarsal bones, 81f, 82, 129, 130f, 491f
	articulations, 145t
Metopic suture, 90, 91f
Microanatomy, 2
Microglia, 227f, 228
Microscope, 17–34
	care/handling, 17–18
	depth of field observation, 22–23
	field diameter, 23–24, 23f
	focusing, 21
	light intensity control, 21
	magnification, 19, 22f, 23–24, 23f
	magnification control, 21
	ocular lens adjustment, 21
	parts of, 18–19
	setup, 20
	using, 20–22
	wet-mount slides, 21–22
Microtubules, 29
Microvilli, 28f, 29, 42f, 290, 290f, 423
Midbrain, 264
Middle cardiac vein, 359, 359f
Middle cerebral artery, 367, 369f
Middle clinoid process, 94f
Middle colic artery, 372f, 426f
Middle colic vein, 380f, 426f
Middle cranial fossa, 90
Middle ear, 309–311, 311f
Middle lobe, 406
Middle phalanx, 121, 122f, 129, 130f
Middle pharyngeal constrictor, 170, 172f, 172t
Middle scalene, 174f, 178f, 179, 180t
Middle trunk, 247f
Midline, 5
Midsagittal section, 9
Milk, 465f
Milk teeth, 416
Mineralocorticoids, 330
Minor calyx, 436, 436f, 437f, 440f, 445f
Mitochondria, 28f, 30, 156f, 159f, 229f
Mitosis, 31–32, 32f, 71f, 450, 451f, 456, 459f
Mitral valve, 357f, 358
Mitral valve prolapse, 358
Mixed nerves, 226, 240
Moderator band, 357f, 358
Molars, 416, 417f
Monaxial joints, 138t, 139, 142f
Monocular microscope, 19
Monocytes, 52f, 339, 340, 340f
Mons pubis, 446f, 463, 463f, 465f
Morphogenesis, 469
Morula, 470, 472f
Motor cortex, 261f
Motor end plate, 159, 159f
Motor nerves, 240
Motor neuron, 159, 159f
Mouth, 7f, 415–418
	muscles, 165–166, 166t
MSH, 322f, 323

Mucin, 415
Mucosa, 413, 414f, 419f, 420f, 423f, 424f, 442f
Mucosa-associated lymphoid tissue (MALT), 388f
Mucosal epithelium, 414f
Mucosal glands, 414f
Mucous cells, 416f, 421f
Mucous connective tissue, 47, 48f
Mucous epithelial glands, 442f
Mucous epithelium, 421f
Mucous layer, 288f
Mucous neck cells, 421f
Multiaxial joints, 139, 142f
Multifidus, 176t, 177, 178f
Multinucleated, 57
Muscle belly, 155
Muscle cell, 57
Muscle compartments, 218–219, 219f
Muscle fascicle, 156f
Muscle fiber orientation, 182
Muscle fibers, 57, 58f, 155, 156f, 157f, 160f
Muscle group, 163
Muscle layers, 175
Muscle modeling, 163–164
Muscle spindles, 282
Muscle tissue, 39, 56–59, 155, 156f, 157f
Muscles. *See* Skeletal muscles
Muscular system, 3f
Muscularis externa, 413, 414, 414f, 419f, 423f, 424f, 427f
Muscularis mucosae, 413, 414, 414f, 419f, 421f, 423f, 424f, 427f
Musculocutaneous nerve, 246, 247f, 248t, 249f
Myelin sheath, 228, 229f
Myelin sheath gap, 229f
Myelinated axon, 230f, 231f
Myelinated internode, 229
Myelination, 228–230
Myenteric plexus, 414, 414f, 421f, 423f
Mylohyoid, 173, 173f, 174t
Mylohyoid line, 98, 98f
Myoblasts, 57
Myocardium, 352, 353f
Myoepithelial cell, 71f
Myofibril, 156, 156f, 157f, 159f
Myometrium, 446f, 458f, 461, 461f, 462f, 465f, 477f
Myoneural junction, 159
Myosatellite cell, 156f
Myosin, 157

N

Nail bed, 73
Nail body, 73, 73f
Nail root, 73, 73f
Nails, 3f, 73, 73f
Naming. *See* Terminology
Nasal, 7f
Nasal bones, 88f, 95, 96f, 97, 101f
Nasal cartilages, 399
Nasal cavity, 4f, 100f, 288f, 399, 400f, 401f, 415f

Nasal conchae, 95, 97, 400f, 401f
	inferior, 88f, 95, 96f, 97, 100f, 296f, 400, 401f, 409f
	middle, 88f, 93, 95f, 96f, 100f, 400, 401f
	superior, 93, 95f, 400, 401f
Nasal septum, 96f, 97, 100f, 400
Nasal vestibule, 399, 401f
Nasalis, 74f, 164f, 165f, 166f, 167
Nasolacrimal duct, 296f, 297
Nasopharynx, 400, 400f, 401f, 409f, 415f, 418
Nasus, 7f
Navel, 7f
Navicular bone, 128, 130f
NE, 232, 233f, 331, 334
Neck, 7f, 84f, 84t, 409, 483–485
	muscles, 166t, 167, 173–175, 193f
Neck (tooth), 416, 417f
Negative feedback, 321
Nephron, 436–439, 440f
Nephron loops, 437f, 439f
Nephroptosis, 442
Nerve, 156f, 159f, 225
	anatomy, 231–232
Nerve fibers, 68f, 312f, 315f
Nerve roots, 240, 244f, 246f, 268f
Nerve tissue, 59
Nervous system, 3f
	ANS, 232–235. *See also* Autonomic nervous system
	CNS, 225. *See also* Central nervous system
	histology, 227–230
	nerve anatomy, 231–232
	organization, 225–235
	PNS, 225. *See also* Peripheral nervous system
Neural folds, 475f
Neural part, 299f, 300, 301f
Neural plate, 475f
Neural tissue, 39, 59–60, 60f
Neural tunic, 299f, 300–304
Neurilemma, 229
Neurofilament, 229f
Neuroglia, 60f, 227
Neurohypophysis, 322, 322f
Neuromuscular junction, 159–160, 160f
Neuromuscular synapse, 159f
Neurons, 59, 60f, 227, 228–230, 230f, 232
	ganglionic, 232, 233f, 234
	preganglionic, 232, 233, 233f, 234
	motor, 159, 159f
	parasympathetic, 234
	postsynaptic, 228
	presynaptic, 228
	sympathetic, 233, 233f
Neurotransmitter, 228
Neutrophils, 52f, 338–339, 340f
Nipple, 464, 465f, 483, 486f
Nissl bodies, 228
Nociceptors, 281
Nodes, 229, 229f
Nodes of Ranvier, 229
Nonaxial joints, 139, 142f

Nonkeratinized, 41
Nonmembranous organelles, 27–28, 29
Norepinephrine (NE), 232, 233f, 331, 334
Nose, 7f, 288f, 399–401
    muscles, 166t, 167
Nosepiece (microscope), 18f, 19
Nostrils, 399
Nuchal region, 484f
Nuclear envelope, 28f, 29
Nuclear pores, 28f
Nucleolus, 28f, 29, 60f, 229f
Nucleoplasm, 28f
Nucleus, 27, 28, 29, 32f, 42f, 43f, 44f
    B cells, 391f
    basal cell, 290f
    bipolar cells, 301f
    cardiac muscle cell, 353f
    caudate, 262f, 263f, 264
    cones, 301f
    epithelial cells, 405f
    fibrocyte, 51f
    ganglion cells, 301f
    geniculate, 265f
    gustatory cell, 290f
    lentiform, 262f, 263f, 264
    muscle tissue, 58f, 156f, 157f
    neural tissue, 60f
    neuron, 229f
    primary oocyte, 460f
    rods, 301f
Nucleus pulposus, 103
Nurse cells, 451f

O

O_2, 399
Objective lens (microscope), 18, 18f, 19
Oblique fissure, 406, 407f
Oblique muscle layer, 420f
Oblique muscles, 179–183, 421f
Oblique tendon, 296f
Obturator artery, 373f
Obturator externus, 205t, 206f, 207
Obturator foramen, 124t, 125f, 126
Obturator groove, 125f
Obturator internus, 205t, 206f, 207
Obturator nerve, 250f, 251t
Obturator vein, 379f
Obtuse marginal branch, 359
Occipital artery, 369f
Occipital bone, 87, 89f, 90, 92f, 96f, 101f, 245f
    articulations, 144t
Occipital condyle, 89f, 90, 92f
Occipital fontanel, 101
Occipital lobe, 256f, 260, 261f
Occipital region, 484f
Occipital sinus, 376f
Occipital vein, 376f
Occipitofrontalis, 164f, 165f, 166f, 167
Occipitomastoid suture, 89f
Occipitoparietal suture, 87
Occlusal surface, 416
Ocular, 7f
Ocular conjunctiva, 297, 299f

Ocular lens (microscope), 18f, 19, 21
Oculomotor muscles, 167
Oculomotor nerve, 267, 268f, 269t, 271t, 274f
Oculus, 7f
Odontoid process, 104
Oil-immersion lens (microscope), 19
Olecranal, 7f
Olecranon, 7f, 120f, 148f, 488f
Olecranon fossa, 118, 119f
Olecranon process, 118
Olfaction, 281, 287–290
Olfactory adaptation, 289–290
Olfactory bulb, 267, 268f, 274f, 275f, 288f
Olfactory cilia, 288f
Olfactory cortex, 261f
Olfactory epithelium, 287, 288f, 289f
Olfactory foramina, 93
Olfactory glands, 287, 288f
Olfactory nerve, 267, 268f, 269t, 271t, 288, 288f
Olfactory nerve fibers, 288f
Olfactory receptor cells, 287, 288f
Olfactory tract, 267, 268f, 274f, 288, 288f
Oligodendrocytes, 227, 227f
Omental appendices, 426f, 427f
Omohyoid, 173, 173f, 174f, 174t, 246f, 485f
Oocyte, 456, 459f, 460f, 470, 471f
Oogenesis, 450, 456, 458–459, 459f
Oogonia, 456, 459f
Ophthalmic artery, 367, 369f
Ophthalmic branch, 268
Ophthalmoscope, 303, 303f
Opponens digiti minimi, 201, 203t, 204f
Opponens pollicis, 201, 203t, 204f
Opposition, 146f, 147
Optic canal, 88f, 93, 96f
Optic chiasm, 263f, 267, 268f, 274f, 275f, 322f
Optic disc, 299f, 301f, 302, 304f
Optic groove, 93, 94f
Optic nerve, 168f, 267, 268f, 269t, 271t, 274f, 298f, 299f, 301f, 304f
Optic tract, 265f, 274f
Ora serrata, 299f, 300
Oral, 7f
Oral cavity, 401f, 415
Orbicularis oculi, 74f, 164f, 165f, 166, 166f, 296f, 297
Orbicularis oris, 74f, 164f, 165, 165f, 166t
Orbit, 100f
Orbital, 7f
Orbital fat, 296f
Orbital fissure, 88f, 93, 94f, 96f
Orbital part, 91f
Orbital surface, 91, 94f
Organ of Corti, 312f, 313, 314f, 315f
Organ position, 479f
Organ system level, 2
Organ systems, 2–4, 67
Organelles, 27, 156f, 228
Organization, levels of, 2
Organogenesis, 474

Origin, 163
    appendicular muscles, 192t, 195t, 199t, 201t, 203t, 205t, 212t, 214t, 215t
    axial muscles, 166t, 168t, 169t, 171t, 172t, 174t, 175–176t, 180t, 185t
    lower limb compartments, 218t
Oris, 7f
Oropharynx, 400, 401f, 409f, 415f, 418
Os coxae, 82, 123, 124
    articulations, 144t, 145t
Ossification, 78
Osteoblasts, 54, 54f, 77
Osteoclasts, 78
Osteocytes, 47t, 54, 54f, 77, 79
Osteon, 47t, 79, 80f
Osteopenia, 82
Osteoporosis, 82
OT, 322f, 323
Otic, 7f
Otic ganglion, 234f, 235
Otitis media, 310
Otolith, 312f, 313
Otoscope, 317, 317f
Outer connective tissue layer, 442f
Outer cortex, 391, 391f
Outer ear, 309
Outer hair cells, 313, 315f
Ova, 449
Oval window, 310, 310f, 311f, 314f
Ovarian artery, 461f
Ovarian follicle, 458f
Ovarian ligaments, 461, 461f
Ovarian vein, 461f
Ovaries, 3f, 4f, 8f, 322f, 446f, 449, 456, 458, 458f, 465f
Ovulation, 460f, 470, 471f
Ovum, 449, 459f, 469
Oxygen (O_2), 399
Oxyphil cells, 326, 327f
Oxytocin (OT), 322f, 323

P

Pacinian corpuscles, 69, 283
Packed red cell volume, 346–348
Pain, 281, 283
Palatal process, 89f
Palate, 392f
Palatine bones, 88f, 89f, 95, 96f, 97
Palatine process, 95
Palatine tonsils, 392, 392f, 401f, 415f
Palatoglossal arch, 415, 415f
Palatoglossus, 170, 171f, 171t
Palatopharyngeal arch, 415, 415f
Palatopharyngeus, 170, 172f, 172t
Palm, 7f
Palma, 7f
Palmar, 7f
Palmar arch arteries, 368f
Palmar carpal ligament, 198f, 202f
Palmar digital nerves, 247f
Palmar interosseous, 203t, 204f
Palmar venous arches, 375f, 376, 377f
Palmaris brevis, 203t, 204f, 489f
Palmaris longus, 197, 198f, 199t, 200f, 202f

Palmaris longus tendon, 198f, 200f, 204f, 489f
Palpebrae, 295, 296f, 304f
Palpebral conjunctiva, 297, 299f
Palpebral fissure, 296f
Pampiniform plexus, 453f
Pancreas, 3f, 4f, 8f, 332–333, 372f, 393f, 430–431, 431f, 479f
Pancreatic acini, 332, 332f, 431f
Pancreatic artery, 372f, 431f
Pancreatic duct, 332f, 429f, 430, 431f
Pancreatic islets, 332, 332f, 430, 431f
Pancreatic lobules, 332f, 431f
Pancreatic vein, 380f
Pancreaticoduodenal artery, 372f, 431f
Pancreaticoduodenal vein, 380f
Papillae, 290f, 291, 291f, 292f
    dermal, 69
    duodenal, 428, 429f, 431f
    hair, 72, 72f
    renal, 436, 436f, 437f, 445f
Papillary duct, 437f, 438
Papillary layer, 68f, 69
Papillary muscles, 357f, 358, 362f
Paranasal sinuses, 4f, 99–100
Parasagittal section, 9
Parasternal lymph node, 395f
Parasympathetic division, 226, 226f, 233, 234f, 235
Parasympathetic ganglia, 233f, 233–234, 235
Parasympathetic ganglionic neurons, 234
Parasympathetic pathways, 233f
Parasympathetic preganglionic neurons, 234
Parathyroid cells, 327f
Parathyroid glands, 326–327, 327f
Parathyroid hormone (PTH), 326
Paraurethral glands, 458f, 464
Paravertebral ganglia, 234–235
Parfocal, 21
Parietal bones, 87, 88f, 89f, 91, 92f, 93, 94f, 96f, 101f
Parietal cells, 421f, 422
Parietal eminence, 90, 92f
Parietal epithelium, 437f, 438, 439f
Parietal layer, 11–12, 453f
Parietal lobe, 256f, 260, 261f
Parietal pericardium, 11f, 12, 352, 352f, 353f, 356f
Parietal peritoneum, 12, 393f, 432f
Parietal pleura, 12, 390f, 405f
Parietal region, 484f
Parieto-occipital sulcus, 259f, 263f
Parotid ducts, 415, 415f, 416f
Parotid glands, 165f, 415
Parotid salivary gland, 74f, 174f, 416f
Pars distalis, 322, 322f
Pars intermedia, 322, 322f
Pars nervosa, 322
Pars tuberalis, 322, 322f
Parturition, 478
Patella, 7f, 81f, 82, 126, 128, 128f, 150f, 206f, 210f, 211f, 216f, 490f, 491f

Patellar apex, 128, 128f
Patellar articular surface, 128f
Patellar base, 128, 128f
Patellar facet, 128f
Patellar ligament, 128f, 149, 150f, 206f, 210f, 211f, 216f, 490f, 491f
Patellar retinaculae, 149, 150f
Patellar surface, 126, 127f
Pectinate muscles, 356, 357f
Pectineal line, 125f, 126, 127f
Pectineus, 194f, 205t, 206f, 207, 210f, 211f, 382f
Pectoral fat pad, 465f
Pectoral girdle, 80, 115–118, 145t
    appendicular muscles, 191–194
Pectoral lymph nodes, 395f
Pectoral nerves, 248t
Pectoralis major, 132f, 174f, 182f, 194f, 195, 195t, 249f, 381f, 395f, 465f, 483, 486f, 487f, 489f
Pectoralis minor, 192, 192t, 194f
Pedal, 7f
Pedicels, 438, 439f
Pedicle, 103f, 104, 105f, 106f, 107f
Pelvic, 7f
Pelvic apex, 107f
Pelvic base, 107f
Pelvic brim, 123
Pelvic cavity, 10–11, 11f
Pelvic diaphragm, 183, 184f
Pelvic floor muscles, 184f
Pelvic girdle, 80, 82, 123–126, 145t
Pelvic inlet, 123, 124t
Pelvic lymph nodes, 388f
Pelvic nerves, 234f, 235
Pelvic outlet, 123, 123f
Pelvic region muscles, 183–185
Pelvic surface, 107f
Pelvis, 84f, 123, 123f, 443f, 466, 489, 490f
    sex differences in, 124t
Penile body, 456, 457f
Penile bulb, 457f
Penile root, 456, 457f
Penis, 4f, 450f, 453f, 456, 457f, 466
Perforating canal, 79, 80f
Perforating fibers, 80f
Pericardial cavity, 10, 11f, 351, 352f, 353f
Pericardial fluid, 352f
Pericarditis, 12
Pericardium, 11f, 12, 352, 352f, 353f, 356f, 407f
Perichondrium, 47t, 52, 52f
Perikaryon, 228, 229f
Perilymph, 311, 312f, 314f
Perimetrium, 458f, 461, 461f, 462f
Perimysium, 155, 156f
Perineal raphe, 453f
Perineum, 464
Perineurium, 231, 231f
Periodontal ligament, 416, 417f
Periosteum, 54, 54f, 77, 80f, 140f, 276f
Peripheral adaptation, 289
Peripheral nerves, 3f, 225, 226f

Peripheral nervous system (PNS), 225, 226f, 227f, 233, 239
    afferent division, 225
    efferent division, 225–226
    glial cells, 228
    nerves, 245
Peristalsis, 414
Peritoneal cavity, 11f, 12
Peritoneum, 12, 334f, 443f
Peritonitis, 12
Peritubular capillaries, 440f, 441
Permanent dentition, 416
Peroneal artery, 374
Peroneal vein, 378
Peroneus muscles, 213
Peroxisome, 28f, 30
Perpendicular plate, 88f, 93, 95f, 96f
Pes, 7f
Petrosal sinuses, 376f
Petrous part, 91, 92f, 310f, 311f
Peyer's patches, 423
Phagocytes, 338, 391, 394
Phalangeal, 7f
Phalanges, 7f, 73f, 80, 81f, 82, 121, 122f, 129, 130f
    articulations, 145t
Pharyngeal arches, 475f
Pharyngeal epithelium, 392f
Pharyngeal tonsil, 392, 392f, 400, 401f, 415f
Pharyngotympanic tube, 309
Pharynx, 4f, 170–172, 399–401, 418–419
Phasic receptors, 281
Phenylthiocarbamide (PTC), 292
Photoreceptors, 300, 301f
Phrenic arteries, 330f
Phrenic nerve, 246f, 246t, 352f
Phrenic veins, 377f, 378
Physiology, 2
Pia mater, 241f, 243, 244f, 252f, 257f, 258, 259f, 276f
Pigmented part, 299f, 300, 301f
Pineal gland, 262f, 263f, 264, 265f, 273f, 275f
Pinna, 309
Piriformis, 205t, 206f, 207
Pisiform bone, 121, 122f, 489f
Pituicytes, 322
Pituitary gland, 3f, 257f, 264, 275f, 322–323, 322f, 369f
Pivot joints, 141, 142f
Placenta, 476–478, 479f, 480f
Placental formation, 477f
Placental stage, 478, 480f
Plane joints, 141, 142f
Plane of section, 9, 9f
Planes, 6, 9
Planta, 7f
Plantar, 7f
Plantar aponeurosis, 217f
Plantar arch, 368f, 373f, 374
Plantar artery, 373f
Plantar flexion, 146f, 147
Plantar interosseous muscles, 215t

Plantar nerve, 250f
Plantar venous arch, 375f, 378, 379f
Plantaris, 150f, 206f, 213, 214t, 216f, 219f, 250f
Plasma, 47, 50, 337, 338f, 435
Plasma composition, 338f
Plasma membrane, 27
Plasmalemma, 28f
Plasmocytes, 391f
Platelets, 47t, 50, 52f, 337, 338f, 340
Platysma, 164f, 166f, 167, 194f, 485f
Pleura, 11f, 12, 390f, 405f
Pleural cavities, 10, 11f, 352f, 405f, 409f, 432f
Pleurisy, 12
Pleuritis, 12
Plexus, 245
Plica circulares, 423f
Plicae, 414f, 423
PNS nerves, 245
Podocytes, 438, 439f
Polar body, 456, 459f, 471f, 472f
Pollex, 7f, 121
Polymorphonuclear leukocytes, 338
Pons, 255, 256f, 258f, 263f, 264, 265f, 268f, 274f, 275f
Pontine artery, 369f
Popliteal, 7f
Popliteal artery, 209f, 250f, 368f, 373f, 374, 489f, 490f, 491f
Popliteal fossa, 250f, 489, 490f, 491f
Popliteal ligament, 149, 150f
Popliteal line, 128
Popliteal region, 250f
Popliteal surface, 126, 127f
Popliteal vein, 209f, 375f, 378, 379f
Popliteus, 7f, 150f, 212, 212t, 216f, 219f
Pore, 68f
Portal area, 429f
Postcentral gyrus, 256f, 261, 261f, 263f
Posterior, 2, 5f, 11f
Posterior arch, 104, 105f
Posterior cavity (eye), 299f, 300
Posterior cerebral arteries, 369f, 370
Posterior cervical triangle, 483, 484f, 485f
Posterior chamber, 300
Posterior clinoid process, 91, 94f
Posterior communicating arteries, 369f, 370
Posterior cord, 381f
Posterior cranial fossa, 90
Posterior cruciate ligament, 149, 150f
Posterior fontanel, 101f
Posterior gluteal line, 124, 125f
Posterior gray commissure, 241f, 242
Posterior gray horns, 241, 241f
Posterior horns (lateral ventricles), 258f, 262f
Posterior interventricular branch, 359, 359f
Posterior interventricular sulcus, 355, 356f, 361f
Posterior left ventricular branch, 359, 359f
Posterior lobe (cerebellum), 265, 266f
Posterior lobe (pituitary), 322, 322f
Posterior median sulcus, 240f, 241, 241f

Posterior scalene, 178f, 179, 180t
Posterior tubercle, 104, 105f
Posterior vein of left ventricle, 359, 359f
Posterior white column, 241f, 242
Postsynaptic cell, 228
Postsynaptic neuron, 228
Power switch, 18f
Precentral gyrus, 256f, 261, 261f, 263f
Prefrontal cortex, 261f
Preganglionic exit points, 233
Preganglionic fibers, 232
Preganglionic neuron, 232, 233, 233f, 234
Preganglionic sympathetic neurons, 233f
Pregnancy, 479f
Premolars, 416, 417f
Premotor cortex, 261, 261f
Prepatellar bursa, 150f
Prepuce, 456, 457f
Prepuce of clitoris, 463f
Pressure detection sense, 281
Presynaptic neuron, 228
Primary bronchi, 403, 404f
Primary curves, 102
Primary fissure, 265, 266f
Primary follicles, 456, 460f
Primary motor cortex, 261f
Primary oocytes, 456, 459f
Primary sensory cortex, 261f
Primary spermatocytes, 451
Primary teeth, 416
Primitive streak, 472
Primordial follicles, 456, 460f
Primordial oocyte, 460f
Principal cells, 326
Procerus, 74f, 164f, 165f, 166f, 167
Process, 84f, 84t
Progesterone, 458
Projections, 84t
Pronation, 144f, 146
Pronator quadratus, 197, 198f, 199t, 202f
Pronator teres, 148f, 197, 198f, 199t, 202f, 489f
Prone, 2
Pronucleus, 470, 471f
Pronucleus formation, 471f
Prophase, 31–32, 32f
Proprioception, 281
Proprioceptors, 281
Prostate gland, 4f, 443f, 450f, 454, 455f, 457f, 466
Prostatic glands, 455f
Prostatic urethra, 441, 443f, 450f, 454, 455f, 457f, 466
Protein fibers, 47
Protraction, 146f, 147
Proximal, 5, 5f
Proximal convoluted tubule, 437f, 438, 439f
Proximal limb muscles, 194f
Proximal nail fold, 73f
Proximal phalanx, 121, 122f, 129, 130f
Pseudostratified ciliated columnar epithelium, 44f, 45
Pseudostratified columnar epithelium, 44–45
Pseudostratified epithelium, 40, 46f

Psoas major, 181f, 205t, 206f, 208
Pterygoid canal, 93, 94f
Pterygoid muscles, 169f, 169t, 170
Pterygoid processes, 89f, 93, 94f
Pterygopalatine ganglia, 234f, 235
PTH, 326
Pubic, 7f
Pubic angle, 123, 123f, 124t
Pubic crest, 125f
Pubic ramus, 125f, 126
Pubic symphysis, 123, 125f, 210f, 443f, 446f, 450f, 457f, 458f, 465f, 466, 479f, 480f, 487f
Pubic tubercle, 125f, 126, 210f
Pubis, 7f, 81f, 82, 123, 125f
Pubococcygeus, 184f, 185t
Pudendal artery, 250f, 373f
Pudendal nerve, 249, 250f, 251t
Pudendal vein, 379f
Pulmonary arteries, 352f, 354, 355f, 356f, 357f, 405f, 407f
Pulmonary circuit, 354, 355f
Pulmonary lobule, 405f
Pulmonary plexus, 234f
Pulmonary semilunar valve, 362f
Pulmonary trunk, 352f, 354, 356f, 357f, 359f, 361f, 368f
Pulmonary valve, 357f, 358
Pulmonary veins, 352f, 354, 355f, 356f, 357f, 361f, 407f
Pulmonary vessels, 365
Pulp, 416
Pulp cavity, 416, 417f
Pupil, 296f, 298, 299f
Pupillary constrictor muscles, 299f
Pupillary dilator muscles, 298, 299f
Pupillary sphincter muscles, 298
Purkinje cell, 266f
Putamen, 262f, 263f, 264
Pyloric antrum, 420f
Pyloric canal, 420f
Pyloric sphincter, 419, 420f
Pyloric valve, 419
Pylorus, 419, 420f, 421f
Pyramids, 265

Q

Quadrants, 6, 8f
  mouth, 416
Quadrate lobe, 428, 428f
Quadratus femoris, 205t, 206f, 207
Quadratus lumborum, 176t, 177, 178f, 179, 181f
Quadratus plantae, 215t
Quadriceps femoris, 150f, 212
Quadriceps muscles, 212
Quadriceps tendon, 150f, 210f, 211f

R

Radial artery, 148f, 368f, 370, 371f, 381f, 462f
Radial collateral ligaments, 147
Radial fossa, 119f
Radial groove, 118, 119f

Radial nerve, 246, 247f, 248t, 249f
Radial notch, 118, 120f
Radial pulse, 489f
Radial styloid process, 120, 120f, 122f, 488f
Radial tuberosity, 120, 120f
Radial vein, 375f, 376, 377f
Radical mastectomy, 395
Radioulnar joint, 120, 120f, 147
Radius, 80, 81f, 118, 120, 120f, 122f, 142f, 148f, 198f, 200f, 202f
   articulations, 145t
Rami communicantes, 244f, 245
Ramus, 84f, 84t, 97, 98f
RBCs. *See* Red blood cells
RCA, 356f, 359, 359f
Receptive field, 284, 284f
Rectal artery, 372f, 426f
Rectal vein, 380f
Rectouterine pouch, 458f
Rectum, 372f, 423f, 425, 426f, 443f, 446f, 450f, 458f, 465f, 466, 479f
Rectus abdominis, 180t, 181f, 182, 182f, 194f, 466, 487f
Rectus femoris, 194f, 206f, 210f, 211f, 212, 212t, 219f, 382f, 489, 490f, 491f
Rectus muscles, 179–183
Rectus sheath, 181f, 182f, 194f
Red blood cells (RBCs), 47t, 50, 52f, 327, 337–338, 338f, 339f, 340f
   volume, 346–348
Red marrow, 79
Red pulp, 393, 393f
Reduction division, 452
Referred pain, 283
Reflect, 500
Regions, 6, 8f, 483
Regulatory hormones, 323
Regurgitation, 418
Remodeling, 78
Renal area, 393f
Renal arteries, 330f, 334f, 368f, 372, 440, 440f, 445f
Renal capsule, 435
Renal columns, 436, 436f, 445f
Renal corpuscle, 437f, 438, 439f
Renal fascia, 435
Renal lobe, 436f
Renal papilla, 436, 436f, 437f, 445f
Renal pelvis, 436, 436f, 445f
Renal pyramids, 436, 436f, 445f
Renal sinus, 436, 436f
Renal tubule, 438
Renal veins, 330f, 334f, 375f, 377f, 378, 440f, 441, 445f
Reproductive organs, 8f
Reproductive system, 4f, 449–466
   accessory glands (male), 454–455
   comparison, 465–466
   ductus deferens, 4f, 450f, 452–455, 455f, 457f, 466
   epididymis, 4f, 450f, 452, 453, 453f, 466
   female, 456, 458–465, 465f
   male, 449–456, 465f, 466
   mammary glands, 464–465. *See also* Mammary glands
   oogenesis, 450, 456, 458–459, 459f
   ovaries, 456, 458. *See also* Ovaries
   pelvis, 466. *See also* Pelvis
   penis, 4f, 450f, 453f, 456, 457f, 466
   spermatogenesis, 449–452
   testes, 3f, 4f, 322, 449–452
   uterine tubes, 4f, 446f, 458f, 460–462, 465f
   uterus, 460–462. *See also* Uterus
   vagina, 463–464. *See also* Vagina
   vulva, 463–464
Resolution, 17
Respiratory bronchioles, 404, 405f
Respiratory epithelium, 404f, 405f
Respiratory system, 4f, 399–409
   bronchial tree, 403–406
   head/neck, 409
   larynx, 401–403. *See also* Larynx
   lungs, 406–408. *See also* Lungs
   nose, 399–401. *See also* Nose
   pharynx, 399–401. *See also* Pharynx
   trachea, 403–406. *See also* Trachea
Rest-and-repose response, 232, 233f
Reticular cells, 328, 328f
Reticular fibers, 49f, 50
Reticular layer, 68f, 69
Reticular tissue, 47, 47t, 49f, 50, 55f
Reticulocytes, 50
Retina, 295, 299f, 300–304, 304f
Retinal artery, 299f, 301f
Retinal vein, 299f, 301f
Retraction, 146f, 147
Retractor, 261f
Retroperitoneal, 12, 435, 573
Rh blood group, 344
Rh factor, 344
Rh negative, 344
Rh positive, 344
RhoGam, 344
Rhomboid major, 192, 192t, 193f
Rhomboid minor, 192, 192t, 193f
Rib cage, 87, 109f
Ribosomes, 28f, 29
Ribs, 80, 81f, 108–109, 109f, 352f, 369f, 390f, 393f, 400f, 486f
   articulations, 144t
Right atrioventricular (AV) valve, 357f, 358
Right atrium, 352f, 354, 355f, 356f, 357f, 361f, 362f
Right bronchomediastinal trunk, 390f
Right colic flexure, 425, 426f
Right coronary artery (RCA), 356f, 359, 359f
Right hepatic flexure, 425, 426f
Right hypochondriac region, 8f
Right inguinal region, 8f
Right lower quadrant (RLQ), 8f
Right rotation, 144f, 146
Right upper quadrant (RUQ), 8f
Right ventricle, 352f, 354, 355f, 356f, 357f, 359f, 361f, 362f
Risorius, 164f, 165, 166t
RLQ, 8f
Rods, 300, 301f, 302
Root
   dorsal, 240, 240f, 241f, 244f
   hair, 72, 72f
   lung, 404f
   nail, 73, 73f
   nerve, 240f, 244f, 246f, 268f
   penis, 456, 457f
   tongue, 403f, 415f
   tooth, 416, 417f
   ventral, 240, 240f, 241f, 244f, 252f
Root canal, 416, 417f
Root-hair plexus, 72, 282f, 283
Rotation, 146
Rotational movements, 141, 144f
Rotator cuff, 132, 196
Rotatores cervicis, 176t, 177
Rotatores lumborum, 176t, 177
Rotatores muscles, 176t, 177, 178f
Rotatores thoracis, 176t, 177, 178f
Rough ER, 28f, 29
Round ligaments, 428, 428f, 429f, 461, 461f
Round window, 310f, 311f, 313, 314f
Ruffini corpuscles, 282f, 283
Rugae, 420f, 421f, 422, 441, 443f
RUQ, 8f

S

S phase, 31, 32f
Saccule, 312f, 313
Sacral canal, 106, 107f
Sacral cornu, 107f
Sacral crest, 106, 107f, 490f
Sacral curve, 102f, 107f
Sacral foramina, 106, 107f, 245f
Sacral hiatus, 106, 107f
Sacral nerves, 240f, 245, 245f
Sacral plexus, 245f, 248–251, 251t
Sacral promontory, 107f, 480f
Sacral region, 102f
Sacral spinal nerves, 240f, 245f
Sacral tuberosity, 106, 107f
Sacral vein, 377f
Sacral vertebrae, 102
Sacroiliac joint, 106, 124
Sacrotuberous ligament, 184f
Sacrum, 81f, 102, 106, 107f, 115, 206f, 210f, 240f
   articulations, 144t
   sex differences in, 124t
Saddle joint, 141, 142f
Safe sex, 41
Safety. *See* Laboratory safety
Sagittal plane, 9f
Sagittal section, 9
Sagittal suture, 87, 88f, 92f, 101f
Salivary gland duct, 43f
Salpingopharyngeus, 170, 172t
Saphenous nerve, 250f, 251t, 381
Saphenous vein, 211f, 375f, 378, 379f, 381, 382f, 491f
Sarcolemma, 57, 156, 156f, 157f, 159f
Sarcomeres, 157, 158f
Sarcoplasm, 156, 156f, 157f
Sarcoplasmic reticulum, 156, 157f
Sartorius, 194f, 206f, 208, 209f, 210f, 211f, 212t, 250f, 382f, 489, 490f, 491f

INDEX

Satellite cells, 227f, 228
Scala media, 313, 315f
Scala tympani, 313, 315f
Scala vestibule, 313
Scalene muscle group, 179, 180t
Scalp, 72f, 276f
    muscles, 166t, 167
Scaphoid bone, 121, 122f, 142f
Scapula, 80, 81f, 115, 116, 116f, 117, 117f, 142f, 485
    articulations, 145t
Scapular angle, 486f, 488f
Scapular notch, 117
Scapular nerve, 247f, 248t
Scapular spine, 117, 117f, 193f, 485, 486f, 488f
Schwann cells, 227f, 228, 229f, 231f
Sciatic nerve, 209f, 211f, 219f, 245f, 249, 250f, 251t, 490f
Sclera, 167, 296f, 298, 299f, 301f, 304f
Scleral venous sinus, 298
Scrotal cavity, 453f
Scrotal septum, 453f
Scrotal skin, 453f
Scrotum, 4f, 449, 450f, 457f, 466
Sebaceous follicles, 70, 71f
Sebaceous glands, 68f, 70–71, 71f, 72f
Sebum, 70, 71f
Second bicuspid, 416
Second molar, 416, 417f
Second polar body, 456, 459f
Second premolar, 417f
Second rib, 369f
Second trimester, 469, 478–479
Secondary bronchi, 403, 404f, 405f
Secondary curves, 102
Secondary dentition, 416
Secondary follicles, 456, 460f
Secondary oocyte, 456, 459f, 460f
Secondary spermatocytes, 452
Secondary teeth, 417f
Secretory alveoli, 465f
Secretory pockets, 455f
Secretory vesicles, 28f, 30
Sectioning, 9
Sections, 9
Segmental arteries, 440, 440f, 445f
Segmental bronchi, 403
Sella turcica, 89f, 91, 94f, 96f, 257f, 322f
Semen, 449, 454
Semicircular canals, 310f, 311, 312f, 314f
Semicircular ducts, 312f, 313
Semilunar notch, 118
Semilunar valves, 358
Semimembranosus, 206f, 208, 209f, 210f, 212t, 250f, 490f, 491f
Seminal fluid, 454
Seminal gland, 450f, 455f, 457f
Seminal gland duct, 455f
Seminal vesicles, 4f, 454
Seminiferous tubules, 450, 451f
Semispinalis capitis, 176t, 177, 178f, 193f
Semispinalis cervicis, 176t, 177, 178f
Semispinalis group, 176t
Semispinalis thoracis, 176t, 177, 178f

Semitendinosus, 206f, 208, 209f, 210f, 211f, 212t, 219f, 250f, 490f, 491f
Semitendinosus tendon, 490f
Sense organs, 3f
Sensory nerve fiber, 282f
Sensory nerves, 282f, 312f
Sensory receptors, 225, 226f
Septa, 328f
Septum pellucidum, 258, 262f, 263f
Serosa, 413, 414, 414f, 421f, 423f, 424f, 461
Serous cells, 416f
Serous fluid, 12, 352
Serous membrane, 10, 11–12
Serratus anterior, 181f, 182f, 192, 192t, 193f, 194f, 381f, 487f
Serratus posterior inferior, 180t, 181f, 193f
Serratus posterior muscles, 180
Serratus posterior superior, 180t, 193f
Sesamoid bones, 77
Sex cells, 449
Sex differences, 124t
Sheep dissection. See Dissection
Shin, 128
Short bones, 77
Short ganglionic neuron, 233f
Short preganglionic neuron, 233f
Shoulder, 7f, 132, 191, 381f, 485, 488f, 489f
Shoulder muscles, 193f
Sigmoid artery, 372f, 426f
Sigmoid colon, 372f, 423f, 425f, 426f, 458f, 466
Sigmoid flexure, 426f
Sigmoid sinus, 92f, 376f
Sigmoid vein, 380f, 426f
Simple columnar epithelium, 40f, 41, 42f, 46f, 427f, 462f
Simple cuboidal epithelium, 40f, 41, 42f, 46f
Simple epithelium, 40, 40f, 41, 42f
Simple squamous epithelium, 40f, 41, 42f, 46f
Sinus, 84f, 84t, 124t
Sinus congestion, 100
Sinuses, 374
Sinuses of skull, 4f, 99–100
Sinusoids, 428
SITS, 196
Skeletal muscle fibers, 159f
Skeletal muscle innervation, 159–160, 159f
    appendicular muscles, 192t, 195t, 199t, 201t, 203t, 205t, 212t, 214t, 215t
    axial muscles, 166t, 168t, 169t, 171t, 172t, 174t, 175–176t, 180t, 185t
Skeletal muscle tissue, 56, 57, 58f, 59f
Skeletal muscles, 3f, 155–160
    appendicular muscles, 191–219. See also Appendicular muscles
    arrangement, 155–158
    axial muscles, 163–185. See also Axial muscles
    neuromuscular junction, 159–160, 160f
    organization, 155–160

Skeletal system, 3f, 80–82
    appendicular, 80, 115–132. See also Appendicular skeleton
    axial, 80. See also Axial skeleton
    bone histological organization, 79–80
    bone markings, 83–84
    bone structure, 77–79
    organization, 77–84
Skeleton, 80–82, 124t
Skene's glands, 464
Skin, 3f, 453f. See also Integumentary system
Skull, 7f, 10, 80, 81f, 84f, 87, 88f, 89f
    articulations, 144t
    fetal, 101
    sex differences in, 124t
    sinuses of, 4f, 99–100
Slides, 17, 21–22
Small cardiac vein, 359, 359f
Small intestine, 4f, 8f, 332f, 372f, 380f, 422–424
Small saphenous vein, 375f, 378, 379f
Smell, 281, 287
Smooth ER, 28f, 29
Smooth muscle, 44f
Smooth muscle cell, 421f
Smooth muscle tissue, 56–57, 58f, 59f, 366f, 389f, 405f, 423f, 442f, 455f
Soft palate, 400, 401f, 409, 409f, 415f
Sole of foot, 7f
Soleal line, 129f
Soleus, 150f, 213, 214t, 216f, 219f, 490f, 491f
Soma, 60, 60f, 228, 229f
Somatic cells, 450
Somatic effectors, 225
Somatic motor, 241f
Somatic motor association area, 261f
Somatic nervous system, 226, 226f
Somatic sensory, 241f
Somatic sensory association area, 261f
Somatic sensory receptors, 226f
Somites, 474, 475f
Special movements, 146
Special senses, 281, 287–292
    gustation, 281, 287, 290–292
    olfaction, 281, 287–290
    olfactory adaptation, 289–290
Special sensory receptors, 226f
Speculum, 317
Sperm cell, 449
Spermatic cord, 8f, 453, 453f
Spermatids, 451f, 452
Spermatocytes, 451, 451f, 452
Spermatogenesis, 449–452
Spermatogonia, 450, 451f
Spermatozoa, 449, 451f, 453, 469, 471f
Spermiogenesis, 451f
Sphenoid bone, 87, 88f, 89f, 91, 93, 94f, 96f, 101f, 322f
Sphenoid sinus, 100f
Sphenoidal fontanel, 101, 101f
Sphenoidal sinus, 94f, 96f, 99, 400f
Sphenoidal spine, 94f
Sphenoparietal suture, 87
Sphincter of Oddi, 428

Spinal accessory nerve, 271t, 274f
Spinal blood vessel, 244f
Spinal cavity, 10, 11f
Spinal cord, 3f, 11f, 60f, 181f, 226f, 234f, 239–244, 244f, 245f, 257f, 258f, 259f, 268f, 273f, 274f, 275f, 409f
    dissection, 251–252, 252f
    gross anatomy, 239–242
    organization, 241f
    sectional anatomy, 241
    spinal meninges, 243
    spinal nerves, 243, 245–251
    spinal segments, 239–240, 246t
Spinal cord axis, 475f
Spinal curves, 102f
Spinal flexors, 176t, 177, 179
Spinal meninges, 243, 244f
Spinal nerve roots, 244f
Spinal nerves, 226, 232f, 240f, 243, 245–251, 245f
Spinal reflexes, 239
Spinal tap, 243
Spinalis cervicis, 175t, 177, 178f
Spinalis group, 175t
Spinalis thoracis, 175t, 177, 178f
Spindle fibers, 29, 32f
Spindle formation, 471f
Spine, 84f, 84t, 102
Spinous process, 103, 103f, 105f, 106f, 107f, 178f, 409f, 486f
Spiral artery, 462f
Spiral ganglia, 313, 314f, 315f
Spiral organ, 313, 315f
Splanchnic nerves, 235
Spleen, 4f, 8f, 372f, 380f, 388f, 391, 393–394, 393f, 420f, 432f
Splenic artery, 334f, 372, 372f, 393f, 431f
Splenic nodule, 393f
Splenic vein, 380f, 393f, 426f
Splenius capitis, 175t, 177, 178f, 193f
Splenius cervicis, 175t, 177
Spongy bone, 77, 78f, 79, 80f, 140f
Spongy urethra, 443f, 450f, 456, 457f, 466
Squamous epithelium, 40, 40f, 41, 42f, 43f, 46f, 67, 419f, 442f
Squamous part, 90, 91f
Squamous superficial cells, 43f
Squamous suture, 87, 88f, 92f, 96f, 101f
Stabilizing ligament, 311f
Stage (microscope), 18, 18f, 19
Stage clips, 20
Stapedius, 310, 311f
Stapes, 310, 311f, 314f
Statoconia, 312f, 313
Stem cell, 40, 43f, 79
Stensen's ducts, 415
Sternal body, 108
Sternal end, 115, 116f
Sternal facet, 116f
Sternoclavicular joint, 116, 116f
Sternocleidomastoid, 74f, 165f, 173, 173f, 174t, 193f, 194f, 484f, 485f, 486f
    clavicular head, 164f, 173f, 174f, 249f, 484f
    sternal head, 164f, 173f, 174f, 249f, 484f

Sternocleidomastoid region, 485f
Sternohyoid, 173, 173f, 174f, 174t, 246f
Sternothyroid, 173, 173f, 174f, 174t, 246f
Sternum, 11f, 80, 81f, 87, 108, 109f, 116f, 173f, 324f, 484f, 486f
    articulations, 144t, 145t
    body, 109f, 409f, 486f
Steroids, 330
Stomach, 4f, 8f, 372f, 393f, 419–422, 432f, 479f
    body, 419, 420f
    histology, 422
    lining, 421f
    neck, 421f
    regions, 419
    wall, 422
Straight artery, 462f
Straight sinus, 376f
Strata, 67
Stratified columnar epithelium, 40f, 43f, 44, 46f
Stratified cuboidal epithelium, 40f, 43f, 44, 46f
Stratified epithelium, 40, 40f, 41, 42f, 43–44
Stratified squamous epithelium, 40f, 41, 43f, 46f, 67, 419f, 442f
Stratum basale, 68
Stratum corneum, 69, 69f
Stratum germinativum, 68, 69f
Stratum granulosum, 68, 69f
Stratum lucidum, 69, 69f
Stratum spinosum, 68, 69f
Stress, 235
Striation, 58f
Sty, 297
Styloglossus, 170, 171f, 171t
Stylohyoid, 172t, 173, 173f, 174t
Styloid process, 88f, 89f, 90–91, 92f, 96f, 171f
    of ulna, 118
Stylomastoid foramen, 89f, 91
Stylopharyngeus, 170, 172f, 172t
Subarachnoid space, 243, 244f, 257f, 258, 259f, 276f
Subclavian arteries, 249f, 356f, 357f, 359f, 367, 368f, 369f, 371f, 381f
Subclavian lymph node, 395f
Subclavian trunk, 390f
Subclavian veins, 374, 375f, 376f, 377f, 390f
Subclavius, 192, 192t, 194f
Subclavius nerve, 248t
Subcostal nerve, 250f
Subcutaneous adipose tissue, 72f
Subcutaneous layer, 68f, 69, 71f
Subdural space, 243, 257f, 258, 259f
Sublingual ducts, 415, 416f
Sublingual salivary glands, 415, 416f
Submandibular ducts, 415, 415f, 416f
Submandibular fossa, 98, 98f
Submandibular ganglia, 234f, 235
Submandibular lymph nodes, 485f
Submandibular salivary glands, 174f, 415, 416f, 485f

Submandibular triangle, 483, 485f
Submucosa, 413, 414, 414f, 419f, 421f, 423f, 424f, 427f, 442f
Submucosal artery, 423f
Submucosal glands, 414f, 423, 424f
Submucosal plexus, 414, 414f, 423f
Submucosal vein, 423f
Subscapular artery, 371f, 381f
Subscapular fossa, 117, 117f
Subscapular lymph nodes, 395f
Subscapular nerves, 248t
Subscapular sinus, 391
Subscapular space, 391f
Subscapularis, 132, 132f, 194f, 195, 195t
Substage light, 18f
Substantia nigra, 263f
Sudoriferous glands, 71
Sulcus, 84f, 84t, 255
Superciliary arch, 91f
Superficial, 6
Superficial columnar cells, 43f
Superficial layer (embryo), 472
Superficial layer (vertebral column muscles), 175–176t, 177
Superficial palmar arch, 370, 371f
Superficial scrotal fascia, 453f
Superficial temporal artery, 367, 369f
Superficial transverse perineal, 184f, 185t
Superficial veins, 377f
Superior, 2, 5f, 495
Superior angle, 116–117, 117f
Superior border, 116, 117f
Superior carotid triangle, 483, 485f
Superior colliculi, 263f, 264, 265f, 266f, 273f
Superior lobes, 406, 407f
Superior mesenteric ganglion, 234f, 235
Superior nuchal line, 89f, 90, 92f
Superior oblique muscles, 168, 168f, 168t, 297, 298f
Superior pharyngeal constrictor, 170, 172f, 172t
Superior rectus, 167, 168f, 168t, 296f, 297, 298f
Superior sagittal sinus, 257f, 258, 259f, 374, 376f
Superior serratus posterior, 180
Superior temporal line, 88f, 90, 91f, 92f, 169f
Superior tibiofibular joint, 128, 129f
Superior trunk, 247f
Superior vena cava, 352f, 354, 356f, 357f, 375f, 376, 376f, 377f, 390f
Supination, 144f, 146
Supinator, 198, 198f, 199t, 202f
Supine, 2
Supporting cells, 287, 288f, 312f
Supportive connective tissue, 47, 47t, 52–55
Supraclavicular fossa, 485f
Supraclavicular nerve, 246f, 246t
Supracondylar ridge, 126, 127f
Supraglenoid tubercle, 117, 117f
Suprahyoid muscles, 173
Suprahyoid triangle, 483, 485f

Supra-orbital foramen, 88f, 90, 91f, 96f
Supra-orbital margin, 91f, 484f
Supra-orbital notch, 90, 91f
Suprapubic bursa, 150f
Suprarenal arteries, 330f, 368f, 372, 440f
Suprarenal cortex, 329, 330f
Suprarenal glands, 329–331, 334, 334f
Suprarenal medulla, 329, 330f
Suprarenal veins, 330f, 334, 334f, 375f, 377f, 378
Suprascapular artery, 174f, 369f, 371f
Suprascapular nerve, 174f, 247f, 248t
Suprascapular notch, 117f
Supraspinatus, 193f, 195t, 196
Supraspinous fossa, 117, 117f
Suprasternal notch, 484f
Sura, 7f
Sural, 7f
Sural cutaneous nerve, 250f
Sural nerve, 250f
Surface anatomy, 483–491
　head/neck/trunk, 483–485, 487f
　lower limb, 489, 490f, 491f
　pelvis, 489, 490f
　shoulder/upper limb, 485, 488f, 489f
　thorax, 486f
Surface antigens, 343f
Surface markings, 83
Surfactant cells, 408
Surgical neck, 118, 119f
Suspensory ligaments, 298, 299f, 446f, 461, 461f, 465f
Sustentacular cells, 287
Sutural bones, 77
Sutures, 77, 138, 138t
Swallowing, 418
Sweat, 72
Sweat gland duct, 68f
Sweat glands, 3f, 71–72, 71f
Sympathetic chain ganglia, 234, 234f
Sympathetic division, 226, 226f, 233–235
Sympathetic ganglia, 233, 233f
Sympathetic nerves, 234f
Sympathetic preganglionic neurons, 233
Symphyseal surface, 125f
Symphyses, 138–139, 138t
Synapse, 228
Synapsis, 451
Synaptic cleft, 159, 159f, 228
Synaptic knob, 228
Synaptic terminal, 159, 159f, 228
Synaptic vesicles, 159, 159f, 228
Synarthroses, 137, 138t
Synchondroses, 138, 138t
Syncytial trophoblast, 472, 473f, 474f, 476, 477f
Syndesmoses, 138, 138t
Synergists, 163
Synostoses, 138, 138t
Synovial fluid, 140, 140f
Synovial joints, 138t, 139, 140, 140f, 141–150
Synovial membrane, 140, 140f, 150f
Synovial sheath, 204f
Systemic arteries, 354

Systemic capillaries, 354
Systemic circuit, 354, 355f, 365–382
　arteries, 365, 367–374
　backflow, 366
　blood vessel comparison, 365–367
　limbs, 381–382
　veins, 365, 374–380
Systemic veins, 354

T

T cells, 339, 391f
T thyrocytes, 323, 324f
T tubules, 156, 157f
T_1, 109f
T_3, 323, 325
T_4, 323, 325
T_{10}, 181f
Tactile corpuscles, 68f, 69, 282f, 283, 283f
Tactile discs, 282f, 283
Tactile receptors, 282–283, 282f
Taenia coli, 425, 426f, 427f
Tail, 5, 475f
Tail fold, 474, 477f
Talar trochlea, 130f
Talus, 128, 130f, 145t
Tapetum lucidum, 304f, 306
Target cells, 321
Tarsal, 7f
Tarsal bones, 81f, 82, 128, 145t
Tarsal plates, 296f
Tarsus, 7f
Taste, 281, 287
Taste buds, 290, 290f, 291f, 292f
Taste hairs, 290f
Taste pore, 290f, 291
Tears, 297
Tectorial membrane, 313, 314f, 315f
Teeth, 4f, 98f, 416, 417f, 562
　articulations, 144t
　sex differences in, 124t
Telodendria, 228, 229f
Telophase, 32, 32f
Temperature, 281
Temporal artery, 367, 369f
Temporal bones, 87, 88f, 89f, 90–91, 92f, 94f, 96f, 101f, 310f, 311f
　articulations, 144t
Temporal lobe, 256f, 261, 261f, 263f
Temporal process, 88f, 96f, 97
Temporal region, 484f
Temporal squama, 89f
Temporal vein, 376f
Temporalis, 164f, 169f, 169t, 170
Temporomandibular joint, 98
Temporomandibular joint capsule, 169f
Temporoparietalis, 74f, 164f, 165f, 166f, 167
Tendinous inscriptions, 181f, 182, 182f, 487f
Tendon sheath, 217f
Tendons, 3f, 50, 51f, 155–156
　attachment, 84t
Tensor fasciae latae, 194f, 205t, 206f, 207, 209f, 210f, 211f, 382f, 489, 490f
Tensor tympani, 310, 311f
Tensor veli palatini, 170, 172f, 172t

Tentorium cerebelli, 257f, 258
Teres major, 193f, 194f, 195t, 196, 200f, 486f, 488f
Teres minor, 193f, 195t, 196
Terminal bouton, 228, 229f
Terminal bronchioles, 404, 405f
Terminal cisternae, 156, 157f
Terminal ganglia, 233f, 235
Terminal hairs, 72
Terminologica Anatomica, 1
Terminology, 1
　bone markings, 84
　directional, 2, 5–6, 495
　elbow, 119
　muscles, 57, 163–164, 177
　regional, 6–8
Tertiary bronchi, 403, 405f
Tertiary follicle, 458, 459f, 460f
Testes, 3f, 4f, 184f, 322, 449–452, 453f, 466
Testicle, 450
Testicular artery, 453f
Testicular vein, 453f
Testosterone, 450
Tetanus, 165
Tetrad, 451f, 452, 459f
Thalamus, 255, 256f, 262f, 263f, 265f
Thecal cells, 460f
Thenar eminence, 201
Thermoreceptors, 281
Thick ascending limbs, 437f
Thick filaments, 157, 157f, 158f
Thigh, 7f, 82, 126, 382f, 489, 490f
　appendicular muscles, 204–208
Thigh compartments, 218t, 219f
Thin descending limbs, 437f
Thin filaments, 157, 157f, 158f
Third molar, 417f
Third trimester, 469, 478–479
Third ventricle, 258, 258f, 259f, 262f, 263f, 265f, 275f, 322f
Thoracic, 7f
Thoracic aorta, 181f, 371f, 372f
Thoracic artery, 369f, 371f
Thoracic cage, 108–109
Thoracic cavity, 10, 11f
Thoracic curve, 102f
Thoracic duct, 388f, 389, 390f
Thoracic lymph nodes, 390f
Thoracic nerves, 240f, 245, 248t
Thoracic region, 102f
Thoracic region muscles, 180t
Thoracic spinal nerves, 240f, 245f
Thoracic vein, 374, 376f, 377f
Thoracic vertebrae, 102, 106f, 144t, 486f
Thoracis, 7f
Thoracoacromial artery, 371f
Thoracodorsal fascia, 178f
Thoracodorsal nerve, 248t
Thoracolumbar division, 233
Thoracolumbar fascia, 193f
Thorax, 7f, 483, 486f, 488f
Thumb, 7f, 121, 142f
Thymic corpuscles, 328, 328f
Thymus gland, 4f, 328–329, 328f, 387, 388f, 391

Thyrocervical trunk, 324f, 369f, 371f
Thyroglobulin, 323, 324f, 325
Thyrohyoid, 172t, 173, 173f, 174t, 246f
Thyrohyoid ligament, 402f
Thyrohyoid membrane, 402f
Thyroid artery, 324f, 369f
Thyroid cartilage, 164f, 173f, 324f, 401f, 402, 402f, 409f, 484f, 485f
Thyroid follicles, 324f, 327f
Thyroid gland, 3f, 323–325, 328f, 352f, 401f, 404f
    left lobe, 327f
    superior horn, 402f
Thyroid veins, 324f
Thyroxine (T_4), 323, 325
Tibia, 81f, 82, 128, 129f, 150f, 216f, 219f
    articulations, 145t
Tibial artery, 219f, 368f, 373f, 374, 491f
Tibial collateral ligament, 149, 150f
Tibial condyle, 128, 129f
Tibial interosseous border, 129f
Tibial malleolus, 128, 129f, 217f, 491f
Tibial nerve, 209f, 219f, 250f, 251t
Tibial tubercle, 128
Tibial tuberosity, 128, 129f, 150f, 210f, 216f, 490f, 491f
Tibial vein, 219f, 375f, 378, 379f
Tibialis anterior, 213, 214t, 216f, 219f, 489, 491f
Tibialis anterior tendon, 217f, 219f, 491f
Tibialis posterior, 150f, 213, 214t, 216f, 219f
Tibialis posterior tendon, 491f
Tissue, 2, 39
    connective, 46–56. See also Connective tissue
    endocrine, 3f
    epithelial, 39–46
    muscle, 56–59. See also Muscle tissue
    neural, 39, 59–60, 60f
Tissue level, 2
Titin, 158f
Toes, 7f, 213–217
Tongue, 4f, 43f, 170–171, 400f, 401f, 415, 415f
Tongue intrinsic muscles, 291f
Tonic receptors, 281
Tonsillectomy, 392
Tonsillitis, 392
Tonsils, 4f, 388f, 392, 392f, 400, 401f, 415f
Total magnification (microscope), 19
Touch, 281
Trabeculae, 79, 80f, 391, 391f
Trabeculae carneae, 357f, 358
Trabecular arteries, 393f, 394
Trabecular veins, 393f
Trachea, 4f, 44f, 324f, 328f, 352f, 400f, 401f, 402f, 403–406, 409f
Tracheal cartilages, 402f, 403, 404f, 409f
Trachealis, 403, 404f
Tracts, 242
Transitional cell, 290f
Transitional epithelium, 40, 44f, 45, 46f, 442f
Transverse arch (ankle), 130f

Transverse colon, 423f, 425, 426f, 479f
Transverse fibers, 263f
Transverse foramen, 104, 105f
Transverse ligament, 104
Transverse lines, 107f
Transverse perineal, 184f, 185t
Transverse plane, 9f
Transverse process, 103, 103f, 105f, 106f, 107f, 178f
Transverse section, 9
Transverse sinus, 257f, 258, 376f
Transverse tubules, 156
Transversospinalis muscles, 176t, 177
Transversus abdominis, 180t, 181, 181f, 182f, 194f
Transversus thoracis, 179, 180t, 181f
Trapezium, 121, 122f, 142f
Trapezius, 164f, 174f, 175, 181f, 191, 192t, 193f, 194f, 484f, 485f, 486f
Trapezoid bone, 121, 122f
Triad, 156, 157f
Triaxial joints, 138t, 139, 142f
Triceps brachii, 148f, 193f, 195t, 197, 486
    lateral head, 199t, 200f, 486f, 488f
    long head, 197, 198f, 199t, 200f, 486f, 488f, 489f
    medial head, 197, 198f, 199t, 202f, 488f
    short head, 197
Triceps brachii tendon, 148f, 202f, 488f
Tricuspid valve, 357f, 358
Trigeminal nerve, 268, 268f, 269t, 271t, 274f
Trigone, 441, 443f, 457f
Triiodothyronine (T_3), 323, 325
Triquetrum, 121, 122f
Trochanter, 84f, 84t
Trochlea, 84f, 84t
    eye, 168f, 297, 298f
    humerus, 118, 119f, 120f
Trochlear nerve, 267–268, 268f, 269t, 271t, 274f
Trochlear notch, 118, 120f
Trophoblast, 470, 472f, 473f
Tropic hormones, 323
True pelvis, 123
True ribs, 108, 109f
True vocal cords, 402
Trunk, 483–485
Trunk muscles, 194f
Tubal ligation, 461
Tubercle, 84f, 84t, 108
Tuberculum sellae, 91, 94f
Tuberosity, 84f, 84t
Tubular pole, 439f
Tubulin, 29
Tubuloalveolar glands, 455f
Tunica externa, 365
Tunica intima, 366, 366f
Tunica media, 366, 366f
Tunica vaginalis, 453f
Two-point discrimination test, 284
Tymosin, 329
Tympanic cavity, 310f, 311f
Tympanic duct, 312f, 313, 314f, 315f

Tympanic membrane, 309, 310f, 311f, 317–318
Tympanum, 309, 311f
Type I diabetes, 333
Type II diabetes, 333

U

Ulna, 80, 81f, 118, 120f, 122f, 142f, 148f, 198f, 200f, 202f
    articulations, 145t
    head, 118, 120f, 488f, 489f
    styloid process, 120f, 122f
Ulnar artery, 148f, 368f, 370, 371f, 381f
Ulnar collateral artery, 371f
Ulnar collateral ligaments, 147, 148f
Ulnar nerve, 246, 247f, 248t, 249f, 488f
Ulnar notch, 120, 120f
Ulnar olecranon, 200f, 202f, 486
Ulnar recurrent artery, 371f
Ulnar tuberosity, 118, 120f
Ulnar vein, 375f, 376, 377f
Umami, 290f, 291
Umbilical, 7f
Umbilical arteries, 476
Umbilical cord, 475f, 476, 477f, 479f, 480f
Umbilical ligament, 443f
Umbilical region, 8f
Umbilical stalk, 477f
Umbilical vein, 476
Umbilicus, 7f, 182f, 483, 486f, 487f
Uninucleated, 352
Universal acceptors, 343
Universal donors, 343
Upper abdomen, 432
Upper extremity, 118
Upper jaw, 417f
Upper limb, 7f, 80, 118–123, 485, 488f, 489f
    articulations, 145t
    hand, 121–123. See also Hand
    humerus, 118. See also Humerus
    lymphatics, 388f
    radius, 120. See also Radius
    ulna, 118. See also Ulna
    wrist, 121–123. See also Wrist
Upper respiratory system, 399, 400f
Urachus, 443f
Ureteral openings, 443f
Ureters, 4f, 8f, 334f, 436f, 441–443, 443f, 445f, 450f, 455f, 457f
Urethra, 4f, 184f, 441–443, 443f, 446f, 458f, 465f, 479f
Urethral opening, 463f
Urinary bladder, 4f, 8f, 44f, 441–443, 443f, 446, 446f, 450f, 453f, 455f, 458f, 465f, 479f
    pregnancy and, 446
Urinary system, 4f, 435–446
    dissection, 444–445
    histology, 442f
    kidney, 435–441. See also Kidneys
    ureters, 441–443. See also Ureters
    urethra, 441–443. See also Urethra
    urinary bladder, 441–443, 446. See also Urinary bladder

Urine, 435
Urogenital diaphragm, 184f, 443f, 455f, 457f
Urogenital triangle, 184f, 185t
Uterine artery, 462f
Uterine cavity, 461, 461f, 462f, 473f, 477f
Uterine endometrium, 446f
Uterine glands, 462f, 473f
Uterine tubes, 4f, 446f, 458f, 460–462, 465f
Uterus, 4f, 443f, 446, 458f, 460–462, 477f, 479f
   body, 446f, 461, 461f, 465f
   fundus, 446f, 461, 461f, 465f, 479f
   isthmus, 460, 461f
   lumen, 477f
   wall, 462f
Utricle, 312f, 313
Uvea, 298, 299f
Uvula, 400, 401f, 409f, 415, 415f

V

Vagina, 4f, 184f, 443f, 446f, 458f, 461f, 463–464, 465f, 479f, 480f
Vaginal artery, 461f
Vaginal entrance, 463f
Vaginal orifice, 463
Vaginal rugae, 461f
Vagus nerve, 233f, 268f, 269t, 270, 271t, 274f, 420f
Vas deferens, 453
Vasa recta, 437f, 441
Vascular pole, 439f
Vascular tunic, 298, 299f
Vasectomy, 453–454
Vastus intermedius, 211f, 212, 212t
Vastus lateralis, 206f, 210f, 211f, 212, 212t, 216f, 219f, 489, 490f, 491f
Vastus medialis, 210f, 211f, 212, 212t, 489, 490f, 491f
Veins, 68f, 80f, 351, 365, 366f, 375f
   compared with other vessels, 365–367
   head/neck/upper limb, 374–377
   lower limb/abdominopelvic cavity, 378–380
   major, 374–380
Vellus hairs, 72
Venae cavae, 355f
Ventral, 5, 5f
Ventral body cavity, 10, 11f
Ventral ramus, 244f, 245
Ventral root, 240, 240f, 241f, 244f, 252f
Ventricles, 258–260
Ventricular fold, 409f
Venule, 80f, 389f, 423f, 440f
Vermis, 265, 266f

Vertebra prominens, 104, 105f, 486f
Vertebrae, 80, 81f, 102, 104–106, 107f, 244f
Vertebral arch, 103f, 104, 105f
Vertebral artery, 268f, 367, 368f, 369f, 371f
Vertebral body, 103, 103f, 105f, 106f, 107f, 244f
Vertebral border, 486f, 488f
Vertebral canal, 103f
Vertebral column, 10, 102–108
   articulations, 144t
   cervical vertebrae, 104
   coccygeal vertebrae, 106
   lumbar vertebrae, 105–106
   muscles, 175–179
   sacral vertebrae, 106
   thoracic vertebrae, 104–105
   vertebral anatomy, 103–104
Vertebral foramen, 103f, 104, 105f, 106f, 107f
Vertebral regions, 102f
Vertebral vein, 375f, 376f, 377f
Vertebrochondral ribs, 108, 109f
Vertebrosternal ribs, 108
Vertical section, 9
Vesalius, Andreas, 1
Vesicle, 30
Vesicouterine pouch, 458f
Vestibular branch, 314f
Vestibular duct, 312f, 313, 314f, 315f
Vestibular ligaments, 402, 402f, 403f
Vestibular membrane, 313, 314f, 315f
Vestibule
   ear, 310f, 311, 312f
   mouth, 415, 415f
   pelvic, 443f, 446f
   vaginal, 463f, 464, 465f
Vestibulocochlear nerve, 268f, 269t, 270, 271t, 274f, 310f, 314f
Villi, 414f, 423, 423f
Visceral effectors, 225
Visceral epithelium, 437f, 438, 439f
Visceral layer, 11
Visceral motor, 241f
Visceral pericardium, 11f, 12, 352, 352f, 353f
Visceral peritoneum, 12, 393f, 414, 414f, 442f
Visceral pleura, 12, 405f
Visceral sensory, 241f
Visceral sensory receptors, 226f
Vision, 281, 287
Visual association area, 261f
Visual cortex, 261f
Vitreous body, 300
Vitreous chamber, 299f, 300
Vitreous humor, 304f

Vocal cords, 402
Vocal fold, 401f, 409f
Vocal ligaments, 402, 402f, 403f
Voluntary, 57
Vomer, 88f, 89f, 95, 96f, 97, 100f
Vulva, 463–464

W

Water receptors, 290f
WBCs. *See* White blood cells
Wet-mount slide, 21–22, 22f
Wharton's ducts, 415
Wharton's jelly, 47, 48f
White blood cells (WBCs), 47t, 50, 52f, 56f, 337, 338–340, 338f, 340f
White fat, 48
White matter, 229–230, 240f, 241f, 244f, 252f, 266f
White pulp, 393, 393f, 394
White ramus, 245
Whole blood, 337–341
Windpipe, 403
Wisdom tooth, 416
Working distance, 19
Wormian bones, 77
Wrist, 7f, 80, 118, 121–123, 489f
   appendicular muscles, 197–202

X

Xiphoid process, 108, 109f, 181f, 483, 486f, 487f

Y

Yellow marrow, 78
Yolk sac, 474f, 476, 477f
Yolk stalk, 476, 477f

Z

Z lines, 157, 158f
Zona fasciculata, 329, 330, 330f
Zona glomerulosa, 329, 330, 330f
Zona pellucida, 460f, 471f
Zona reticularis, 329, 330f, 331
Zone of overlap, 157, 158f
Zygomatic arch, 88f, 89f, 90, 169f, 484f
Zygomatic bones, 74, 88f, 89f, 95, 96f, 97, 100f, 483, 484f
Zygomatic process, 88f, 90, 92f
Zygomaticofacial foramen, 88f, 96f, 97
Zygomaticus major, 164f, 165, 165f, 166t
Zygomaticus minor, 164f, 165, 165f, 166t
Zygomaticus muscles, 74, 74f
Zygote, 450, 469